MW00761104

Multi-dimensional Optical Storage

Duanyi Xu

Multi-dimensional Optical Storage

Duanyi Xu
Tsinghua University
Beijing
China

ISBN 978-981-10-0930-3   ISBN 978-981-10-0932-7   (eBook)
DOI 10.1007/978-981-10-0932-7

Jointly published with Tsinghua University Press

Library of Congress Control Number: 2016939108

Printed on acid-free paper

This Springer imprint is published by Springer Nature
The registered company is Springer Science+Business Media Singapore Pte Ltd.

Foreword

Professor Duanyi Xu was director and founder of Optical Memory National Engineering Research Center (OMNERC) at Tsinghua University, chief scientist of National Basic Research Project and an excellent scientist on optical storage in China. He led and achieved Chinese National Basic Research Planning on "Super-density and Super-speed Optical Memory" etc. national key projects. Professor Xu began to study principles of multidimensional optical storage from 1990s early, and based on medium absorb photon of different frequencies and intensity to carry out the multi-wavelength and multi-level (MW/ML) optical storage, that is earliest result of multidimensional optical storage in this country. Since then, he proposed the theory of photo-induced electron transfer in medium, and employed this principles to study and develop super-resolution MW/ML photochromic optical storage and 3-dimensional MW/ML optical solid state memory cell with more space to present a variety of new principles and techniques for optical storage, to win China National Science and Technology Award that gave OMNERC, Tsinghua University a position of national prominence in multidimensional optical storage.

The multidimensional optical storage is a developing interdisciplinary research field of photonics, photophysics, photochemistry, materials and information science. So the book presents all accomplishments of author's research team with every field experts more 20 years in optical storage and up-to-date literatures in the world that introduces the principles systematically of multidimensional optical storage and engineering implementation, examines the impact to future technologies and challenges, summary and offers a thought-provoking and entertaining vision about "Multidimensional Optical Storage". Meanwhile this book devoted to description of the key building principles, experimental and testing technology for multidimensional optical storage, to perform the first detailed study on the photochemistry materials, binary optical ML lasers/Photodetectors hybrid integration, self-assembled ultrathin films, dual-mode molecular modulator, multidimensional codes and MW/ML CD-ROM duplication technology etc.

This book is the first provide a framework for thinking about the future Multidimensional Optical Storage, it create a structure for strategic planning and development for exploring potential development road map to the optical memory and should appeal to the readers in universities and industry to understand the fundamental principles of Multidimensional Optical Storage. I am confident it will be an important valuable resource for readers and future specialists who engaged upon information science and technology.

Bingkun Zhou
President of the Chinese Optical Society
Members of the Chinese Academy of Sciences
Professor of Department of Electronic Engineering
at Tsinghua University

Preface

The world is challenged by exploding amounts of data. Around the world, vast quantities of data are generated for video, audio, pictures, documents, newspapers, and other publication from ZB (10^{21} byte) to YB (10^{24} byte), even much more for Internet every day. So far, all of these data are stored with magnetic storage, optical storage, and IC solid state memory successfully by improving storage capacity with minifying recording elements. However, all of these two-dimensional storage methodologies are reaching their physical limit. Therefore, the expansion of storage capacity has to come from the advancement of multi-dimensional storage technology as new direction. Since there is a lot of physical and chemical interaction between light and matter that can be used for information recording, optical storage is more feasible than magnetic storage or IC solid state memory in achieving multi-dimensional storage. The Optical Memory National Engineering Research Center (OMNERC) at Tsinghua University started to research multi-dimensional optical storage in 1991. The research investigated several methods to increase the recording capacity: use multi-wavelength (different frequencies); multi-level (different optical density); different polarization of light (different spatial angle); and different recording element geometry (length, width, and depth). The research projects covering the recording principles, recording medium, and engineering applications are part of the National Key Basic Research Projects and have obtained a series of achievements and patents.

Interest in this field has been growing continuously in recent years. Dhawat E. Pansatiankul and Alexander A. Sawchuk proposed "Multidimensional Modulation Codes and Error Correction for Page-Oriented Optical Data Storage" in 2001; Erez Louidor, Tze Lei Poo, etc. proposed "Maximum Insertion Rate and Capacity of Multi-dimensional Constraints" in 2008; and Min Gu and Xiangping Li proposed "The Road to Multi-Dimensional Bit-by-Bit Optical Data Storage" in 2010. Their papers are undoubtedly important references for studying multi-dimensional optical storage. Since most of our past research works were published in Chinese, some foreign friends frequently request us to introduce our works (patents) in more detail, and if possible, publish them as a book in English.

A preliminary review of the literature shows that numerous related research projects on this field have been carried out, but that there was no systematic study on multi-wavelength and multi-level optical storage yet. My book will let people obtain a comprehensive and balanced picture of this field.

Multi-dimensional storage is based on photonics, photochemistry, photophysics, photochromism and materials science, with the reversible transfer between photon and collective atomic excitation in the medium. This important study will be introduced to general reader in detail with accurate expressions derived by using density matrix equations of motion, coupled wave equations for different frequency photon, photo-nonlinear reaction in medium, stereochemistry and isomerization, preservation of photon energy during storage, margin analysis based on rigorous modeling, conversion efficiency nano-crystalline file, photochromic dye in amorphous state, electron delocalization valence, error correction and application probabilities, the advantages like more performance with less energy consumption and high sensitivity (less noise, higher data rate and more flexible cases), as well as the disadvantages like more temporary nature, incompatibility, and manufacturing problems. In conclusion, photons and light seem to be better than electrons and electric current to carry and store information.

The book is organized as following: Chap. 1 presents an introduction to overview the history of optical digital data storage development and up to date achievements of optical storage in China, as well as the big data storage and application, the frontier science and technologies related to the optical storage and advanced efforts for multi-dimensional optical storage. Chapter 2 presents the mechanism of multi-dimensional optical storage, including the principles of photophysics, photochemistry, photo-induced electron transfer process, and reaction control, as well as other actions of photon with inorganic materials, organometallic polymers, and synthetic photosynthesis. Based on Chap. 2, Chap. 3 continues to discuss the recording process of multi-dimensional optical storage, including photochromic reaction mechanism for multi-dimensional storage, recording model, and quantitative evaluation for absorbable spectrum of medium. Chapter 4 introduces another key part: laser sources and super-resolution technology for multi-dimensional optical storage, such as micro-aperture semiconductor laser and optical injected quantum dot laser etc. Chapter 5 focuses on fundamentals and configurations of multi-wavelength and multi-level (MW/ML) with super-resolution mask storage system, including super-resolution MW/ML disc, optical channel characteristics, modulation and coding, MW/ML error code correction and multi-level run-length limited coding, rewritable multi-level storage, multi-level Blue-ray Disc (BD) drive, multi-level CD-ROM, and duplication technology. Chapter 6 deal with some new mechanism of three-dimensional (3D) multi-level storage: two-photon absorption 3D optical storage, vertical resolution and adaptive aberration correction, multi-channel and multi-layer parallel read/write, 3D MW/ML optical solid state memory, and multi-valued polarization-sensitive storage. Chapters 7 and 8 present principles of the volume holographic storage and multi-wavelength volume holographic storage and devices, including various dynamic static speckle multiplexing, polarized volume holographic storage, and orthogonal polarized dual channel system based on photochemistry and optically injected quantum dot laser, and advanced polarized volume holographic storage with dithienylethenes and nonintegrated photonic materials. The introduction part of every chapter provides a guide those who may have less knowledge of the field. Each chapter ends with highlights of the content covered in the chapter, as a review of the material. The multi-dimensional optical storage brings out more expandable development space, as a result of a lot of new research achievement on the fundamental theory of the interaction of light with medium to nano-photonics integration process technology and equipment for next-generation optical storage. Meanwhile, various appendixes of mathematical symbols and physical and chemical constants used in this book, and more than 500 related articles are listed in the end of the book as reference for readers.

The book is based on systematic research work, thesis papers, as well as collection of some related research from abroad. The book also discusses the nano-integrated photonic materials and nano-photonics integration process, including principles of optical physics and their affection for data storage, ensuring photons storage, generation, and measurement, broadband waveguide, and margin analysis based on rigorous modeling and preservation process. This book presents both principles and applications to expand the storage space from 2D to 3D or even multi-dimensional with light gray scale (intensity), color (different wavelengths), polarization, and coherence of light, which are used to improve the density, capacity, and data transfer rate of optical data storage. Moreover, the book also discusses the application implementation technologies to make mass data storage devices. Some new high-sensitivity mediums, which have linear absorption characteristics for different wavelength and intensity to light, are introduced too.

Primarily, this book is a textbook on optical engineering for graduate students, which is intended to provide information about the most advanced progress and future development in the field of optical storage. I have successfully used some preliminary versions of selections from this book, such as "Optical super resolution mask 3D storage," "Optical solid state memory based on photochemistry," "Big data multi-dimensional codes," "Theory of nanostructure fabrication for photonics integration devices," in courses at Tsinghua University, and other universities as academic exchange or short course, and also as some subjects of SPIE international conferences. It can be used for training future researchers at both undergraduate and graduate levels, as well as post-education and training program. Most of the materials covered in this book were developed during the teaching of the courses and refined with valuable feedback from these course participants. It is hoped that this book could be used both as a textbook and as a reference for researchers and engineers. Some developing researches, such as photon rate equations, multi-wavelength lasers, highly efficient broadband modulation, vertical resolution, three-dimensional qubit scattering, color free-electron laser, nanoscale nonlinear optical processes, interactions between electrons and photons, excitation dynamic devices, microcavity-based devices, combination of photonic crystals and plasmonics to various linear, nonlinear optical functions, and light-activated nanoparticles medium, have been mentioned as references. The book is also valuable for industries and business: It provides a critical evaluation of the current status of multi-dimensional optical storage technologies, materials, nano-photonic devices manufacturing, and testing equipment. The book also introduces diversified novel multi-dimensional optical storage principles, various laser sources, organic/inorganic materials, and particular experimental results.

Considering different levels of readers, the book introduces concepts with minimal mathematical details, and examples are provided to illustrate principles and applications. It can enable a newcomer to this field to acquire the necessary background knowledge to undertake research and development of advanced optical storage. In addition, I hope this book to be helpful to promote recent optical memory development and make them accessible to engineers in this field. At several places in the text, tables are provided to facilitate conversion to other units. The book attempts to present the topic of each chapter as self-contained unit and to provide up-to-date research evidence to the greatest extent possible. So this book highlights and illustrates the background and the current status of multi-dimensional optical storage and systematically describes many aspects of the MW/ML optical storage, from the basic principles to engineering applications. It can be references to researchers, engineers, and graduate and undergraduate students in related field.

Using conventional optical manufacturing equipment with precision injection duplication, OMNERC at Tsinghua University created multi-dimensional optical storage devices like the multi-wavelength and multi-level disc and photochemistry optical solid state memory cell which was built on the silicon process with optical interconnections. It simplified the process to make an all-optical memory and made it easier to commercialize and integrate it into application-specified devices, which had optical correlators for detecting signals. A new generation of hybrid nano-materials, which involved different levels of integration of organic and inorganic structures, holds considerable new fundamental science and novel technologies to prepare inorganic nanostructures for multi-dimensional optical storage. Those new materials were developed by our main partner Physics Chemistry Institute of CAS in China, which made it flexible to fabricate components with diverse functions and heterogeneous characteristics. Novel easy synthetic technique and processing of nano-materials, such as new types of molecular nanostructures and supramolecular assemblies with varied nano-architectures, and self-assembled periodicity to induce multi-functionality and cooperative effects were developed for applications. These accomplishments made favorable condition to research on multi-dimensional optical storage. So this book is the result of all researchers in OMNERC and in Physics Chemistry Institute of CAS, as well as other collaboration experts. I would like to express my sincere thanks to these experts, professors, graduate students, and my colleagues. In addition, I would express my sincere appreciation to Prof. Bingkun Zhou and the editors of Springer Press and Tsinghua University Press. This book could not be accomplished and published successfully without their strong support and help.

November 2015   Duanyi Xu

Contents

Chapter 1
Introduction

So far has been used in big data storage is the magnetic storage and optical storage or semiconductor memory successfully. The basic technique progress to improve capacity is minified the symbol size of the record information, i.e., increasing storage density, whether it is the semiconductor memory, magnetic storage, or optical storage. But those storage technologies would be closed to their limits. Therefore, for expansion of capacity, to increase the storage space, especially exceeded conventional 3-dimensional geometric coordinate system to extend multidimensional space is very important for further increasing storage density and capacity. Due to the interaction between light and materials with more physical and chemical action, that could be used for information storage that expanded a great approach to enhance capacity. The Optical Memory National Research Center (OMNERC) at Tsinghua University in China comes into research on multidimensional optical storage from 1991, including multilayer, multiplexing volume holography, multi-wavelength (different frequencies), multilevel (different light density or absorptivity), different polarization of light (different spatial angle), different geometry shapes of recording symbol (length, width, and deepness of recording mark), etc., that can increase the storage capacity and data rate at one time in particular. Meanwhile, the multidimensional codes could increase capacity exceedingly, it will be huge influence the mode and capacity of data storage in the future.

Dhawat E. Pansatiankul and Alexander A. Sawchuk proposed "Multidimensional modulation codes and error correction for page-oriented optical data storage" in 2001 [1], Erez Louidor, Tze Lei Poo, etc., proposed "Maximum Insertion Rate and Capacity of Multidimensional Constraints" in 2008 [2], Min Gu and Xiangping Li proposed a "5-D device that could hold up to 2000 times more data than a conventional DVD." A team of Boffins at the Swinburne University of Technology in the Queendom of Down under have tested a new type of "five-dimensional" optical storage medium that they estimated might hold up to 2000 times more data than a conventional DVD also [3–5]. They discuss this and other milestones on the road to multidimensional optical memory with petabyte capacity that was basilic reference points to development of multidimensional optical storage. Nanomaterials are

© Tsinghua University Press and Springer Science+Business Media Singapore 2016
D. Xu, *Multi-dimensional Optical Storage*,
DOI 10.1007/978-981-10-0932-7_1

photoreactive and adjust their shape according to different colors of the visible spectrum, which were illuminated by lasers in this case. The team of OMNERC, then followed up by applying multiple polarizations to the same physical disc space, effectively writing the data at different angles in the same place [6–8].

As the traditional technology road map to improve the optical storage density is to reduce the size of the recording spot by reducing the laser wavelength and increase the number of the value of numerical aperture of the objective lens and shorter wavelength of the recording laser to increase optical storage density and capacity. But continue to reduce the laser wavelength will face many flinty technical barriers that are restricted by materials and process to make shorter wavelength semiconductor laser in the first. Otherwise, if the wavelength of lasers are ultraviolet or deep ultraviolet (DUV), most of the current optical materials, including optical system of pickup, substrate of disc-based materials will have a series of technical problems, as strong absorption for example. Continue to increase the numerical aperture is also facing more problems. Now, the numerical aperture has been 0.85, so potential improvement is very small. If it continues to increase the numerical aperture that the disc must be removed the layer of protection, the disc will loss it is an important advantage, it cannot run in the general environment and long-term preservation.

I studied on multidimensional optical memory with different absorption rates of medium to various wavelengths to actualize multi-wavelength and multilevel optical storage from late nineteenth century and proposed adoption multidimensional optical storage to replace conventional 2-dimensional or 3-dimensional optical storage [10]. Of course, the multidimensional optical storage could be an important potential to continue improvement of optical storage capacity. Of course, there are a lot of new principles, materials and technology, such as sensitivity, reliability, crosstalk, and lifetime of recording medium, as well as laser sources and manufacturing technology, process, and equipment expect to research and develop. This book based on existing results to summarize a feasible technology road map for multi-dimensional optical storage as follows:

1. The space of storage from a flat two-dimensional extended to three-dimensional even multi-dimensional (as frequency, gray scale, polarization, photorefractive, and other photophysics and photochemistry effect) for increasing the reference frame of the storage space, to promote the growth of capacity from linear function turn to the exponential function;
2. Employ the optical nonlinear absorption, super-resolution, negative refractive effect, photo-refractive and coherence of light in medium to control or reduce the effective size of the recording mark, to increasing density of storage;
3. The photochromic material replaced the conventional dye and phase change medium, i.e., photochemistry effects replaced photo-thermal effect that can be super- speed recording with high data rate;
4. Based on the principles of multifunction matrix modulation/coding with multi-dimensional optical storage replaced traditional binary system, that the object function will include gray scale, refractive index, polarization angle, wavelength, intensity of light, etc., to establish new style modulation and coding system.

5. The recording process, which from a single photon effect turn to the two-photon effect, that can achieve the spatial three-dimensional storage or optical solid state storage devices easily;
6. Using a variety of optical parameters to storage data, it can write/read parallel with multichannel parallel, that the access speed (data rate) can be increased synchronous;
7. A new multilevel run-length limited code and error correction system replaced the traditional run-length limited code to improve the sector guidance and supporting code sharing and reduce redundancy;
8. Bring forward principles and methods of synthesis design, integration, and manufacturing with new medium, multi-wavelength laser resources and photodetector, new modulation and detection system to research and development next-generation optical storage devices with larger capacity and more function.

1.1 Look Back on Optical Digital Data Storage

The information (data) storage has great important status in information society. Optical digital data storage has been contributing to world nearly half a century.

1.1.1 Progress of Digital Data Storage

Optical storage has long-term history if to include photography which is over 1000 years, i.e., Alhazen (965–1040) studied the camera obscure and pinhole camera beginning to the first permanent photograph was an image produced in 1826 by the French inventor Joseph Nicéphore that experienced more 800 years. But the advanced optical storage especially optical digital data storage with computer application is about 50 years only. It covers a wide range from optics, photonics, materials science, information science, device physics/chemistry, precision engineering and nanotech applications, so that is a typical highly multidisciplinary field, which could be a reference for development of other disciplines also. This chapter would like to summarize the experience of the optical digital data storage technology development and discuss the future road map.

Also optical disc and drive are very common and may be declined now. But current trends in optical computing emphasize communications, for example, the use of free-space optical interconnects as a potential solution to alleviate bottlenecks experienced in electronic architectures, including loss of communication efficiency in multiprocessors and difficulty of scaling down the IC technology to nanometer levels. Light beams can travel very close to each other, and even intersect, without observable or measurable unwanted signals, therefore, dense arrays of interconnects can be built using optical systems. In addition, risk of noise is further reduced, as light is immune to electromagnetic interferences. Finally, as light travels fast and it has extremely large spatial bandwidth and physical channel density, photons are uncharged and do not interact with one another as readily as electrons, light beams may pass through one another in full duplex operation, for example, without distorting the information carried. It appears to be an excellent media for information transport and hence can be harnessed for data processing. This high bandwidth capability offers a great deal of architectural advantage and flexibility. Based on the technology now available, future systems could have each channel clocked at Tb per second, giving aggregate data capacity in the parallel optical highway of more that TB per second, even this could be further increased to Pb. Free-space optical techniques are also used in scalable systems, which allow arbitrary interconnections between inputs and outputs. Optical interconnects are used in asynchronous transfer modes or packet routing and in shared memory multiprocessor systems. In optical computing two types of memory are applications. One consists of arrays of one-bit-store elements and the other is mass storage. Future research may be expected to lead to compact, high-capacity, rapid/random-access, radiation-resistant for future intelligent spacecraft and massive capacity and fast-access terrestrial data archives. As multimedia applications and services become more and more prevalent, entertainment and data storage companies are looking at ways to increase the amount of stored data and reduce the access time. Recent results of Tsinghua University in China have shown the multi-wavelength and multilevel optical storage technology can increase 5–15 times of capacity. By multilevel and closely controlling the groove depth is expected to produce removable optical drives with capacities to 200 GB on a multi-level Blue disk. Peter Zijlstra, James W.M. Chon presented the multiplexed optical recording that can provides an unparalleled approach to increasing the information density beyond Tb per cm^3 by storing multiple, individually addressable patterns within the same recording volume [3, 9–11]. The multilayer holography system could be gotten PB level storage capacity with page or picture data frame. At the same time the new achievements of photonics technology will take an immensity development space for optical memory, especially for optical solid state memory with application of nanotechnology and photonic integration.

Optical Data Storage has come a long way in the past 45 years. World's First MO Optical Disc Recorder MnBi film coated optically flat disc on air bearing spindle with He–Ne laser, E–O modulator and deflector. By the early 1970s analog video disc systems were commercially available. These were closely followed by 12″ write-once (WORM) drives and media. In 1982 Sony and Philips announced the 120 mm diameter compact disc (CD-DA) followed by the CD-ROM in 1984. In 1995 the DVD-ROM was announced and in 2002 Blu-ray Disc (BD) [12, 13]. Each of these technologies increased capacity significantly and mainly supported important consumer electronics applications. Also in 1995, the EIDE/ATAPI standard was promulgated, which allowed these drives to become a standard part of a PC's storage suite. Other types of optical storage of various disc diameters and storage mechanisms were extant in the 1980–1995, as shown in Fig. 1.1 that the diameter of optical disc from 12 in. decreased to 3 in.

Fig. 1.1 Four generation optical disc that diameter from 12 in. down to 3 in. but capacity increasing 1000×

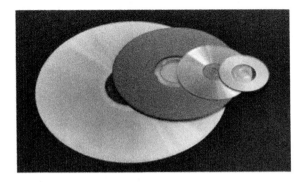

From 2012, nearly 30 years after the introduction of the CD, classical optical data storage has perhaps reached, or even passed. Solid state flash drives, portable hard drives and downloading of music and video from Internet have begun to erode significantly the optical data storage market share. Moreover, optical data storage technology appears to have reached some fundamental physical limits. The utility of optical data storage is derived from how small a diffraction-limited laser beam can be focused to writing and reading spot size. From basic optical theory to know that optical spot size is proportional to wavelength λ and inversely proportional to the effective numerical aperture (NA) of the optical system and storage densities is proportional to $(\lambda/NA)^2$. But optical data storage has several nonconventional means that may permit the technology to reach new capacity plateaus. These range from multilayer discs and near-field recording (NFR) to UV lasers, negative refraction, plasmaron lenses, and multidimensional storage. The future of optical storage will be analyzed in terms of advanced technologies and meet difficulty of implementation, cost, impact on manufacturing yield, and market need. Some related data storage technologies that promise multi-TB capacities are developed. The engineering challenges of these advanced optical read/write methods on lasers, media, optical pickups, servos, and read/write channels will be surmounted or achievable. They could be done if optical data storage is to survive that can confidently predict the future of optical storage capacity will be over many TB. As the Holographic Versatile Disc (HVD) capacity has been to 5 TB even more. High densities are possible by moving these closer on the tracks: 100 GB at 18 μm separation, 200 GB at 13 μm, 500 GB at 8 μm, and most demonstrated of 5 TB for 3 μm on a 10 cm disc. The system uses a green laser, with an output power of 1 W which is high power for a consumer device laser. Possible solutions include improving the sensitivity of the polymer used, or developing and commoditizing a laser capable of higher power output while being suitable for a consumer unit. The relative curves between storage density and average price per megabyte from 1985 to 2015 is shown in Fig. 1.2 that proves the fastest technological progress (the storage density will be increased synchronously) the average price per megabyte dropped faster, the sale revenue is growing fastest at the same time [14, 15].

From 1985 and especially since the early 1990, Japanese companies have heavily invested in optical data storage and enjoy a comfortable lead in this area.

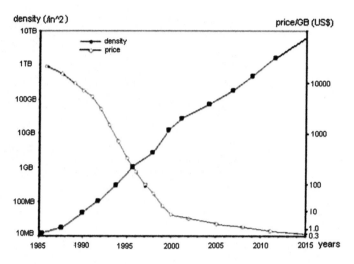

Fig. 1.2 Curves of storage density and average price/GB

Many big companies, such as Sony, Hitachi, Matsushita, Fujitsu, Toshiba, and Canon dominate the property in Japan. Similar advances have been made with optical disk systems as they have progressed through three generations of products (CD, DVD, Blu-Ray, and CBHD—China Blue High-definition Disc) in about 35 years. The success of the CD and DVD family of products has resulted in a 20 % annual revenue growth rate in 1990–2000 [16]. Some universities of U.S. take on an important role in conventional optical storage, since a significant amount of know-how on optical disk heads and metrology exists at these institutions still. The research, in the meantime, has been improved up by the storage giant Samsung in Korea also. Because little U.S. government funding is available in this area, some of this effort is sponsored by the smaller U.S. companies. However, in addition to IBM, significant know-how and intellectual property reside in U.S. universities and small U.S. R&D companies on several unconventional long-term optical data storage approaches that promise data densities approaching Tb/in^2. These include near-field and solid immersion lens approaches, volumetric (multilayer and holographic) storage, and probe storage techniques. In addition, in recent years, under government funding, the United States has gained an advantage on certain potentially enabling technologies such as vertical cavity lasers (VCLs), array optics, and MEMS. These powerful technologies may impact or become affected by optical data storage. Solid immersion lens-based approaches appear promising in the short term. Optical storage technology has been studying from 1970 early in universities and research institutes in China. From 1991, as VCD products put into market successfully that China become the largest manufacturing and consuming country of optical disc and player in the world that the growth rate could be over 50 % average year. Meanwhile, Chinese government provided more financing and set up Optical Memory National Research Center to focus on fundamental research and development of new products for optical disc industry in China. In 1999 Chinese government fund the National Key Basic Research Project (i.e., called "973" project in China) on optical storage to long-term research and should cooperate with Japan and United States in this area [17–19].

Scientists and researchers are working on first things, about how to do this before working out how to make people believe it is worth entrusting lots of valuable data to it. The technology should be ready within the next ten years, but it is also possible to see its application for long-term preservation information of the future. For the longer term and significant gain in market share, development of suitable hardware that exploits parallel readout to facilitate content-based data search may point to a potential opportunity. In addition, investing in microme-chanics for microactuators and creating a new infrastructure to support future data storage approaches certainly appears compelling at this time. Optical data storage research is devoted to researching solutions to the problem of storing and accessing information at densities of 100 atoms per bit or higher. The real power of optical storage is yet to be exploited. It is the only technology that can easily operate in the frequency domain using techniques of a spectroscopic nature. This gives it the power to precisely select very small elements within a larger volume. In principle it can extend down to the atomic level. Some interest is focused on multidimensional optical storage that has been demonstrated able to reversibly store information in the frequency domain, as multi-wavelength memory for example. Simultaneously studying new optoelectronic inertia-less access systems to complete the transfor-mation of optical storage systems are thrown up.

Hard drives store data in the magnetic polarization of small patches of the surface coating on a metal disk. The maximum areal density is defined by the size of the magnetic particles in the surface, as well as the size of the magnetic head used to read and write the data. The areal density of disk storage devices has increased dramat-ically since IBM introduced the RAMAC, the first hard disk in 1956. Commercial hard drives in 2013 typically offer densities between 400 and 500 Gbit/in^2 using perpendicular recording technology, an increase of about 200 million times over the RAMAC. Toshiba and Seagate expected that perpendicular recording technology can scale to about 1 Tbit/in^2 at its maximum. A number of technologies are attempting to surpass the densities of all of these media. IBM's Millipede memory is attempting to commercialize a system at 1 Tb/in^2 in 2007. This is about the same capacity that perpendicular hard drives are expected to be, and Millipede technology has so far been losing the density race with hard drives. The latest demonstrator with 2.7 Tb/in^2 is seemed promising that the IBM technology, racetrack memory uses an array of many small nanoscopic wires arranged in 3D, each holding numerous bits to improve density. Although exact numbers have not been mentioned, IBM news articles talk of "100 times" increases. Various holographic storage technologies are also attempting to leapfrog existing systems, but they too have been losing the race, and are estimated to offer 1 Tb/in^2 as well, molecular polymer storage has been shown to store 10 Tb/in^2 [20, 21]. By far the densest type of memory storage experimentally to date is electronic quantum holography. By superimposing images of different wavelengths into the same hologram, a Stanford research team was able to achieve a bit density of 35 bit/electron, that is, approximately 3 EB/in^2, which was demonstrated using electron microscopes and a copper medium.

Of the many potential applications of nanotechnology, one of the most promising ones is in data storage, particularly the hard disk drives. This is because the physical size of the recording bits of hard disk drives is already in the nanometer scope, and continues shrinking due to the ever-increasing demand for higher recording densities. If the pace of areal density increase is maintained at the current level for about ten years, the dimension of the recording bit will reach the sub ~ 10 nm scope. The rapid shrinkage of bit size poses formidable challenges to the read sensors. Its sensitivity must be improved continuously so as to compensate the loss in signal-to-noise ratio due to the decrease in the bit size. The former has to rely heavily on the advance of nanotechnology and spintronics. The combination of two fields has played an important role in advancing the areal density of magnetic recording from 500 Gb/in^2 to some Tb/in^2 and capacity will be 100 TB or more. In addition to hard disk drives, the technologies developed have also been applied to magnetic random access memories (MRAMs). In 2008, HP announced that it had actually built a memristor, that is basically a fourth class of electrical circuit, joining the resistor, the capacitor, and the inductor, and exhibit their unique properties primarily at the nanoscale. Theoretically, Memristors, a concatenation of "memory resistors," are a type of passive circuit elements that maintain a relationship between the time integrals of current and voltage across a two-terminal element. Thus, a memristors resistance varies according to a devices memristance function, allowing, via tiny read charges, access to a "memory" of applied voltage. The material implementation of memristive effects can be determined in part by the presence of hysteresis (an accelerating rate of change as an object moves from one state to another) which, like many other nonlinear "anomalies" in contemporary circuit theory, turns out to be less an anomaly than a fundamental property of passive circuitry.

Further advance in the two fields is the key to realizing Tb/in^2 hard disk drives and nonvolatile memories within this decade. Spintronics is positioned in the hierarchy of various different types of data storage technologies. The data storage devices can be categorized into three major groups, that is, magnetic and optical data storage and solid-state memory. Each of them can be divided further into several subgroups also. Among the magnetic storage devices, the hard disk drive (HDD) is the dominant secondary mass storage device for computers, and very likely also for consumer electronic products in the near future. The ever-increasing demand for higher areal densities has driven the read head evolving from a thin-film inductive head to an anisotropic magnetoresistive head, recently the giant magnetoresistive spin-valve (GMR SV) head. There are two different forms of GMR SVs, depending on whether the sense current flows in the film plane (CIP) or perpendicular to the film plane (CPP). Currently, CIP SVs are dominant, but CPP SVs are more promising for future extremely high-areal-density heads. Although the effect on performance is most obvious on rotating media, similar effects come into play even for solid-state media like Flash RAM or DRAM. In this case the performance is generally defined by the time it takes for the electrical signals to travel though the computer bus to the chips, and then through the chips to the individual "cells" used to store data. One defining electrical property is the resistance of the wires inside the chips. As the cell size decreases, through the improvements in semiconductor fabrication that lead to Moore's law, the resistance is reduced and less power is needed to operate the cells. That less electrical current is needed for operation, and thus less time is needed to send the required amount of electrical charge into the system. In DRAM in particular the amount of charge that needs to be stored in a cell's capacitor also directly affects this time. As fabrication has improved, solid-state memory has improved dramatically in terms of performance. Modern DRAM chips had operational speeds on the order of nanosecond and the minimal line width less 8 nm. An obvious effect is that as density improves, the number of DIMMs needed to supply any particular amount of memory decreases, which in turn means less DIMMs overall in any particular computer. This often leads to improved performance as well, but there is less bus traffic. However, this effect is generally not linear. In fact that overall price has remained fairly steady has led to the common measure of the price/performance ratio in terms of cost per bit. In these terms the increase in density of hard drives becomes much more obvious. IBM's RAMAC from 1956 supplied US$12,000 per megabyte. In 1989 a typical 40 MB hard drive from Western Digital retailed for US$36/MB. But in 2015, the 5 TB hard drive is selling for less than US$150, an improvement of 0.3 million times more since 1989, and 0.17 billion times since the RAMAC as shown in Fig. 2.2. So, solid-state storage has seen similar dramatic reductions in cost per bit also. In this case the primary determinant of cost is yield, the number of working chips produced in a unit time. Chips are produced in batches lithography on the surface of a single large silicon wafer, which is then cut up and nonworking examples are discarded. To improve yield, modern fabrication has moved to ever-larger wafers, and made great improvements in the quality of the production environment. Other factors include packaging the resulting wafer, which puts a lower limit on this process of about US$1 per completed chip. The relationship between information density and cost per bit can be illustrated as follows: a memory chip that is half the physical size means that twice as many units can be produced on the same wafer, thus halving the price of each one. As a comparison, DRAM was first introduced commercially in 1971, a 1 Kb part that cost about US$50 in mass produce, about 5 cents per bit. But the 8 GB Dual Channel Kit DDR3 1066 PC3 8500 204-Pin SO-DIMM memory parts price is US$50 only also in 2015, which cost was down to about 0.1 microcents per bit. The magnetization reversal Fe (100) disk can storage 10 Terabyte would be 1000–100,000 times the capacity of Blue-Ray, disk drives, or tape drive. That is 1000 times any state-of-the-art hard disk technology with 100 gigabytes on one disk. Hard drive technology has exceeded 20 terabytes on a disk in 2015. Atomic holographic optical image data storage bandwidth is 400,000 times faster than binary bit text processing bandwidths used in other storage technology. Colossal storage of spintronic, polymer molecular and Memory Molecular Image (MMI) technology has been shown to store also. The size of drive could be reduced with MEOMs constantly. The size of a postage stamp, Toshiba introduced a 0.75 in. hard drive for mobile devices and shipped 16 GB units, 32 and 64 GB (as shown in Fig. 1.3).

Fig. 1.3 A diameter 0.75 in. hard divec from Toshiba Corporation

However, solid state flash memory card have long surpassed 512 GB and size smaller than it [22, 23].

Review the current state of computer storage technology, included solid-state (mostly silicon) devices, such as RAM and DRAM chips, magnetic devices (ranging from magnetic tapes to disks to magneto-optic devices) and optical systems including optical disks and volume optical memories such as holographic and two-photon devices. The reason for the continuing competition among these disparate technologies is that each brings its own strengths and weaknesses to the field and none fulfills every need. Among the performance attributes considered for storage technology are capacity, access time, throughput or transfer rate, reliability, removability, erasable, long lifetime, robustness, and most importantly cost. In the past few years a number of comprehensive reviews of these technologies have been published.

1.1.2 From VCD to Blu-Ray Disc

Optical digital disc from first CD developed to the Blu-ray Disc step to step as Table 1.1. Blu-ray Disc (official abbreviation BD) is an optical disc storage medium designed to supersede the DVD format. The plastic disc is 120 mm in diameter and the same size as DVDs and CDs. Blu-ray Discs contain 25 GB per layer, with dual layer discs (50 GB) being the norm for feature length video discs. Triple layer discs (100 GB) and quadruple layers (128 GB) are available for BD-XL rewriter drives. The first Blu-ray Disc prototypes were unveiled in October 2000, and the first prototype player was released in April 2003 in Japan. Afterward, it continued to be developed until its official release in June 2006. The name Blu-ray Disc refers to the blue laser used to read the disc, which allows information to be stored at a greater density than is possible with the longer wavelength red laser for DVDs. The Blu-ray Disc specification requires the testing of resistance to scratches by mechanical abrasion. In contrast, DVD media are not required to be scratch resistant, but since development of the technology, some companies, such as Verbatim, implemented hard-coating for more expensive lineups of recordable DVDs. There are many variants for HD DVD, as CBHD later. Their recording speed is shown in Table 1.2 [17, 24].

Table 1.1 Development of digital optical disc

Video tape	Analog	Philips VCR (1972) V-Cord (1974) IVC (1975) VHS (1976) SVR (1979) Video 2000 and CVC (1980) VHS-C (1982) Video8 (1985) MII (1986) S-VHS (1987) Hi8 (1989)
	Digital	D1 (1986) D2 (1988) D3 (1991) DCT (1992) Digital Betacam (1993) D5 (1994) Digital-S (D9) (1995) Betacam SX (1996) Digital-8 (1999) MicroMV (2001)
	High definition	W-VHS (1994) HDCAM (1997) D-VHS (1998) D6 HDTV VTR (2000) HDCAM SR and HDV (2003)
Video disc	Analog	Ampex-HS (1967) TeD (1975) Laserdisc (1978) CED (1981) VHD (1983) CD Video (1987)
	Digital	VCD (1993) DVD and MiniDVD (1996) DVD-video (1997) CVD and SVCD (1998) EVD (2003) High-definition versatile disc (HVD) (2004) FVD, UMD and VMD (2006)
	High definition	MUSE Hi-Vision LD (1994) HD DVD/Blu-ray Disc (2001) Holographic versatile disc (HVD) and CBHD (2007)
Digital optical disc	Media agnostic	DV and DVCPRO (1995) DVCAM (1996) DVCPRO50 (1997) DVCPRO HD (2000)
	Tapeless	Editcam (1995) XDCAM (2003) MOD (2005) AVCHD and AVC-Intra (2006) TOD (2007) iFrame (2009)
	Solid state drive (SSD)	P2 solid state memory-4 GB(2004) Sony SxS solid state memory-32 GB (2007) SATA III solid state drive-120 GB(2010) Dell PCIe SSD hard drive-3.2 TB(2015)
Video recorded to film	Kinescope (1947) Electronicam kinescope (1950s) Electronic video recording (1967)	

(continued)

Table 1.1 (continued)

High-definition (HD) disc (2010)	
Media formats	Blu-ray disc, China blue high-definition disc (CBHD), D-VHS, HD DVD and holographic versatile disc
Promoter	Blu-ray disc association, China high-definition DVD industry association, HD DVD promotion group and HVD forum
Interactivity	Advanced content, BD-Java
Recordable formats	BD-R. BD-RE, HD DVD-R, HD DVD-RAM, HD DVD-RW
Comparison	Comparison of high-definition optical disc formats
Copy prevention	AACS (BD, FVD, and HD DVD), BD + (BD), HDCP (BD and HD DVD) and ROM Mark (BD)
Blu-ray disc players	Total media theater, cyberlink powerDVD, playstation 3 and sony BDP-S1
HD DVD players	Xbox 360 HD DVD drive
Concepts of high-definition video, high-definition television, ultra high-definition television and high-definition audio	
Analog broadcast	(All defunct): 819 line system, HD MAC, MUSE (Hi-Vision)
Digital broadcast	ATSC, DVB, ISDB, SBTVD, DMB-T/H.
Audio	Dolby Digital (5.1) and MPEG-1 Layer II, PCM, LPCM, DXD, DSD and AAC
Filming and storage	HDV and DCI
HD media and compression	Blu-ray Disc, CBHD, HD DVD, HD VMD, D-VHS, Super Audio CD, DVD-Audio, MPEG-2, H.264, VC-1and MVC

Blu-Disc recordable refers to two optical disc formats that can be recorded with an optical disc recorder. BD-Rs can be written to once, whereas BD-REs can be erased and re-recorded multiple times. The current practical maximum speed for Blu-Discs is about 12×. Higher speeds of rotation (10,000 + rpm) cause too much wobble for the discs to be written properly, as with the 20× and 52× maximum speeds, respectively, of standard DVDs and CDs [18].

High-definition video may be stored on BD-ROMs with up to 1920 × 1080 pixel resolution at up to 60 fields per second, if interlaced. Alternatively,

Table 1.2 The recording speed and data rate of Blu-Disc

Drive speed	Data rate		Theoretical write time for Blu-disc (minutes)	
	Mbit/s	MB/s	Single layer	Dual layer
1×	36	4.5	90	180
2×	72	9	45	90
4×	144	18	22.5	45
6×	216	27	15	30
8×	288	36	11.25	22.5
10×	360	45	9	18
12×	432	54	7.5	15

progressive scan can go up to 1920 × 1080 pixel resolution at 24 frames per second, or up to 1280 × 720 at up to 59.94 frames per second. For audio, BD-ROM players are required to support Dolby Digital (AC-3), DTS, and linear PCM. Players may optionally support Dolby Digital Plus and DTS-HD High Resolution Audio as well as lossless formats Dolby TrueHD and DTS-HD Master Audio. BD-ROM titles must use one of the mandatory schemes for the primary soundtrack. A secondary audio track, if present, may use any of the mandatory or optional codes. All frame rates are properly listed in frames per second. Some manufacturers will list field rate for interlaced material, but this is incorrect industry practice. To avoid confusion, only FRAME rates should ever be listed. MPEG-2 at 1440 × 1080 was previously not supported in a draft version of the specification from March 2005. For users recording digital television programming, the recordable Blu-ray Disc standard's initial data rate of 36Mbit/s is more than adequate to record high-definition broadcasts from any source (IPTV, cable/satellite, or terrestrial). BD Video movies have a maximum data transfer rate of 54 Mbit/s, a maximum AV bitrate of 48 Mbit/s (for both audio and video data), and a maximum video bit rate of 40 Mbit/s. This compares to HD DVD movies, which have a maximum data transfer rate of 36/s, a maximum AV bitrate of 30.24 Mbit/s, and a maximum video bitrate of 29.4 Mbit/s. The Blu-ray Disc region coding is shown in Table 1.3 and the diameter, layers, capacity, and recording speed, etc., specification of various BD is shown in Table 1.4.

In circumvention of region coding restrictions, stand-alone Blu-Disc players are sometimes modified by third parties to allow for playback of Blu-ray Discs (and DVDs) with any region code. Instructions describing how to reset the Blu-ray region counter of computer player applications to make them multiregion indefinitely are also regularly posted to video enthusiast websites and forums. Unlike DVD region codes, Blu-ray region codes are verified only by the player software, not by the optical drive's firmware. The BD-ROM specification of various BD is shown in Table 1.4.

The BDXL format supports 100 and 128 GB write-once discs and 100 GB rewritable discs for commercial applications. It was defined in June 2010. BD-R 3.0 Format Specification (BDXL) defined a multilayered disc recordable in BDAV format with the speed of 2× and 4×, capable of 100/128 GB and usage of UDF2.5/2.6. BD-RE 4.0 Format Specification (BDXL) defined a multilayered disc rewritable in BDAV with the speed of 2× and 4×, capable of 100 GB and usage of

Table 1.3 The blu-disc region coding

Region code	Area
A	Includes most North, Central and South American and Southeast Asian countries plus the Republic of China (Taiwan), Japan, Hong Kong, Macau and Korea
B	Includes most European countries, African and Southwest Asian countries plus Australia and New Zealand
C	Includes the remaining central and South Asian countries, as well as the People's Republic of China and Russia

Table 1.4 Specifications of BD series

Type	Diameter (cm)	Layers	Capacity GB
Standard disc/single layer	12	1	25
Standard disc/dual layer	12	2	50
Standard disc/XL 3 layer	12	3	100
Standard disc/XL 4 layer	12	4	128
Mini disc/single layer	8	1	7.8
Mini disc/dual layer	8	2	15.6

Recording speed				
Drive speed	Data rate		Write time (minute)	
	Mb/s	MB/s	Single layer	Dual layer
1×	36	4.5	90	180
2×	72	9	45	90
4×	144	18	22.5	45
6×	216	27	15	30
8×	288	36	11.25	22.5
10×	360	45	9	18
12×	432	54	7.5	15

UDF2.5 as file system. BDXL discs are not compatible with existing BD drives though a firmware update may be available for some newer drives. "Blu-ray Disc recordable" refers to two optical disc formats that can be recorded with an optical disc recorder. BD-Rs can be written to once, whereas BD-REs can be erased and re-recorded multiple times. The current practical maximum speed for Blu-ray Discs is about 12×. Higher speeds of rotation (10,000 + rpm) cause too much wobble for the discs to be written properly, as with the 20× and 52× maximum speeds, respectively, of standard DVDs and CDs. 100 GB BDXL Triple layer disc made by Sharp. The BDXL format supports 100 and 128 GB write-once discs and 100 GB rewritable discs for commercial applications. It was defined in June 2010. BD-R 3.0 Format Specification (BDXL) defined a multilayered disc recordable in BDAV format with the speed of 2× and 4×, capable of 100/128 GB and usage of UDF2.5/2.6. BD-RE 4.0 Format Specification (BDXL) defined a multilayered disc rewritable in BDAV with the speed of 2× and 4×, capable of 100 GB and usage of UDF2.5 as file system. BDXL discs are not compatible with existing BD drives though a firmware update may be available for some newer drives. The IH-BD (Intra-Hybrid Blu-ray) format includes a 25 GB write-once layer (BD-R) and a 25 GB read-only layer (BD-ROM), designed to work with existing Blu-ray Discs. Ritek had successfully developed a High-Definition optical disc process that capacity increases to 250 GB with ten layers. However, the major obstacle is that current read/write technology does not support the additional layers. Pioneer Corporation unveiled a 400 GB Blu-ray Disc (containing 16 layers, 25 GB each) that will be compatible with current players after a firmware update. Its planned

launch is frames for ROM and for rewritable Blu-ray discs with 1 TB in 2013,but cannot see it in market yet.

BD-Live requires the Blu-ray Disc player to have an Internet connection to access Internet-based content. BD-Live features have included Internet chats, games, downloadable featurettes, quizzes and movie trailers that while some players may have an Ethernet port, these are used for firmware updates and are not used for Internet-based content. In addition, Profile 2.0 also requires more local storage in order to handle this content, with the exception of the latest players and the Play Station 3. BD-Live players combine the features available in other Blu-ray Disc players—breathtaking video and audio playback, BD-J powered interactive features, and the picture-in-picture (PiP) and secondary audio mixing with an internet connection that serves as a gateway to additional content and web-enabled features. With a BD-Live player and a BD-Live compatible title, viewers can enjoy special online benefits and features that might include extra downloadable content, online bonus features, live transactions, release updates and social interaction centered on their favorite films. In addition, some BD-Live players can be updated or using the BD-Live connection. Studios are releasing movies in combo packs with Blu-ray Discs and DVDs as well as Digital copies that can be played on computers and iPods. Other strategies are to release movies with the special features only on Blu-ray Discs and none on DVDs. In 2009 the Blu-ray Disc Association (BDA) officially announced 3D specs for Blu-ray Disc, allowing backward compatibility with current 2D Blu-ray players. The Blu-ray 3D specification calls for encoding 3D video using the "Stereo High" profile defined by Multiview Video Coding (MVC), an extension to the ITU-T H.264 Advanced Video Coding (AVC) codec currently supported by all Blu-ray Disc players. MPEG4-MVC compresses both left and right eye views with a typical 50 % overhead compared to equivalent 2D content, and can provide full 1080p resolution backward compatibility with current 2D Blu-ray Disc players. This means the MVC (3D) stream is backward compatible with H.264/AVC (2D) stream, allowing older 2D devices and software to decode stereoscopic video streams, ignoring additional information for the second view. Sony has released a firmware upgrade for PlayStation 3 consoles that enables 3D Blu-ray Disc playback. They previously released support for 3D gaming on April 21, 2010 (followed by the availability of 3D movies). Since the version 3.70 software update in 2011, the PlayStation 3 can support DTS-HD Master Audio and DTS-HD High Resolution Audio while playing 3D Blu-ray. Dolby TrueHD is used on a small minority of Blu-ray 3D releases, and bitstreaming is supported by slim PlayStation 3 models only. It can play 3D Blu-ray content at full 1080p.

The best 3D Blu-ray players not only contain the latest and fastest HDMI ports, but they also feature older connection methods such as component, composite and coaxial. Additionally, should make sure that other options such as USB and S-Video are available so can connect with computer, flash drive or mobile device to the 3D Blu-ray player. A 3D Blu-ray player is the perfect complement to HDTV 3D TV. The TV set with 3D capabilities is a waste of money without a player with the capabilities listed above. The basic emphasis of the potential of ODS to increase significantly disc capacity. This is an exciting area for both research and product development to focus on the status and means for capacity increases for optical discs using the Blu-ray disc (BD) model, and providing background/historical information.

1.2   Research and Development of Optical Storage in China

There are more 30 universities and institutes to participate research and development of optical storage from 1970 early in China, which are distributed in Beijing, Tianjing, Shanghai, Guangzhou and Taipei, etc., cities in China. Tsinghua University in Beijing is the eldest one of research and development optical storage in China. The Optical Memory National Engineering Research Centre (OMNERC) has been set up at Tsinghua University by the Planning Committee of the State of China and constructed by China National Science and Technology Fund supporting in 1990. It is a biggest engineering research center for optical memory in China that braches are built in Shenzhen and Hefei also. Meanwhile the Physics Department, Chemistry Department, Materials Science Department And Information Engineering Institute are engaged this program in this University. The research achievements from OMENRC are transfered to Tong Fang Optical Disk Company which is under Tsinghua University for mass production [19, 25, 26].

Another main institute and university engaged this field study are following: Physics Chemistry Institute of Chinese Academy of Sciences (CAS) in Beijing for storage medium, Shanghai Institute of Optics and Fine Mechanics of CAS and Fudan University in Shanghai for M-O storage and application systems, Beijing University of Aeronautics and Astronautics in Beijing for medium, University of Electronic Science and Technology in Chengdu for atom storage, Taiwan Jiaotong University in Taipei for optical systems and medium, University of Science and Technology of China in Hefei and Nanjing University in Nanjing for photon quantum memory and application, Xian Institute of Optics and Fine Mechanics in Xian and Chnagchun Institute of Optics and Fine Mechanics in Changchun of CAS for SIL system and South China University of Technology in Guangzhou for superresolution and molecule storage, etc. They carried out a number of studies in the field as ultrahigh density magneto-optical storage, holographic storage, near-field storage, nanoprobe storage technology, and other new storage principles and various storage media that have made a considerable achievements including: the ultra-high density optical–magnetic deposit hybrid digital information storage in Huazhong University, optical storage and magnetic combining record in Fudan University. Prof. Guangcan Guo and his team of University of Science and Technology proposed a new type of platform which allows the simulation of 2D Bosonic system under arbitrary syntheticgauge field. Quantum simulation is one of the most important research fields in quantum information science, which not only allows the study of the existed physical systems, but also new physical modes with new phenomena for quantum memory simulation. Optics and Fine Mechanics of CAS and Shanghai Institute in Shanghai have made significant progress on research of new type superresolution M-O storage media, experiment system and architecture mechanism also.

A national project on "Advanced Super-density and High-speed Optical Data Storage and Processing" (the project No. G199903300) was arranged with Chinese National Fundamental Research Planning (called "973" program in China) and was responsible by OMENRC at Tsinghua University in 1999. Prof. Xu Duanyi of Tsinghua University is chief scientist of the project and 36 famous professors or experts who are from a lot of main research universities and institutes in all of China. The national fund of the project "973" and fund from Ministry of Education (i.e., "985" program) all together that make of most funds for optical storage in China. Tsinghua University takes on holography storage system, SIL design and manufacturing, multi-wavelength multi-level optical storage systems and PB level integration optical storage application systems. Other subprojects are arranged as Physics Chemistry Institute of CAS for organic medium, Beijing University of Technology and Beijing Polytechnic University for holography system, stability and reliability of holography storage, Nankai University in Tainjin for inorganic medium, Harbin Institute of Technology in Herbin for crystal medium to holography storage, Huanzhong Science and Technology University in Wuhan for encoding, decoding and data superspeed transfer, University of Science and Technology of China for quantum memory, multilayer and polarization state storage, Shanghai University for optical system design and manufacturing. This project has been finished in 2005 and gotten 153 patents in China or other countries, that some have used in China to develop next-generation optical storage products as Blu-disc CBHD and PB database application system for example. Other most important achievement is established the foundation of multidimensional optical storage as multi-wavelength, multi-level, multi-layer and volume holographic storage with dynamic static speckle multiplexing, etc., that will be presented in this book in detail [20, 21, 27, 28]. The main accomplishments and achievements are:

1. Established the theory of multidimensional optical storage based on multi-wavelength, multi-level, multi-layer storage with superresolution technology and volume holographic storage with superposed multiplexing technology.
2. The study on near-field optical storage principles to established a complete theoretical model to carry out record information and the interaction between light and medium, non-radiation field transfer and some physical problems of nanointerface control for the near-field optics.
3. In superresolution CD-ROM research, using nonlinear absorption media to be mask layer and achieve superresolution recording with resolution higher than 100 nm and application for multi-wavelength storage.
4. A new-type crystalline, polymer, organic-inorganic materials composite and photochromic compounds were designed, synthesized and optimized to make up series of the refractive index modulation 10^{-4}, the sensitivity of 10^{-5} dual-doped LN crystal materials for value holography storage.

5. Developed supermolecular dye-sensitized initiator medium with photosensitive wavelength of 400–780 nm, and two-photon absorption cross section of 26.4×10^{-46} cm^4 s/photon for multi-wavelength, multi-level and multi-layer optical storage.
6. Accomplished the multi-wavelength and multi-level recording experimental system with capacity of 200 GB, a volume holography storage experimental system with capacity of 6 TB, and a 1.8 PB big data storage application experimental system for China Knowledge Resource Integrated Database.

Meanwhile, on the micromechanical dynamics of superhigh speed serve mechanism and theory of nanoscale measurement in fast moving state, binary micro-optical multifunction devices, micro-optical mechanical and electronic hybrid integrated devices completed the miniaturization mass data storage system and high-speed data exchange interface of I/O. Those achievements have gotten China National Natural Science Award, National Invention Award and China National Science and Technology Progress Award. The typical results detail as follows:

1.2.1   Multi-wavelength and Multilevel Storage

From the early 90 s of last century, Tsinghua University has cooperated with Physics Chemistry Institute of CAS in research on multi-wavelength and multilevel optical storage, as we consider the photochromism has a reversible stabilized transformation between different absorption spectra by photoirradiation that may be great potential application for multidimensional optical storage. Numerous organic photochromic compounds could be used to optical data memory with different absorption spectrum, photon-mode erasable recording, fluorescence and nonlinear optical properties. Meanwhile the photochromic compounds has certain advantages over the thermal effects medium, as higher resolution, sensitivity and better multiplex recording capability.

If an optical storage system while using n wavelengths and m level to storage data at same position that capacity and data transfer rate can be increased by $n \times \ln m$ times at the same time. Based on a lot of experiments of interaction between photon and recording materials with different wavelength completed a three wavelengths (405 nm/532 nm/650 nm) and eight level write-once optical storage system and a tow wavelength multilayer storage system in Tsinghua University in 2003. The optical system of 3-wavelengths and 8-level is shown in Fig. 1.4, where 1-lasers$_1$, 2-collimating lens, 3-polarize beam splitter, 4-quarter waveplate, 5-photo detector, 6-lens, 7 and 8-prism group, 9- achromatic objective lens, 10-recording medium with sensitivity to 405 nm/532 nm/650 nm wavelength respectively. Meanwhile based on nonlinear absorption of medium to photon completed the 3-wavelengths/8-level write-once optical disc with superresolution mask layer, that

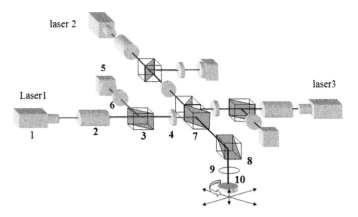

Fig. 1.4 The three wavelengths and multilevel modulation optical storage experimental system

the experimental results show its resolution can increase about 21 %, the project is composed of:

1. design of the new storage optical system, superresolution mask mechanism with inorganic and dielectric nonlinear absorption;
2. established pupil function model for multi-level optical recording and channel features characteristic model for multi-wavelength and multi-level coding;
3. multi-level and multi-wavelength medium process and experimental disc with superresolution mask layer making;
4. key correlative equipment: multi-wavelength multi-level mastering system, replication system and process technology.

The experimental results show that the capacity can be to 100 GB per layer and the data rate is over 40 MB/s. Consider application in the future, lasers in the system are commercial products in the optical disc industry. The hardware structure of drive is same with traditional optical disc drive in general. So this technology can be used to make red laser (650 nm) or blue laser (405 nm) multilevel CD-R disc and drive. The multilevel CD-ROM can be realized with changing of surface topography (depth and width of the "pit" on disc surface), that could be duplication with conventional optical disc replication manufacturing line. But there are two problems have to resolve that are mastering equipment and multilevel encoding. So OMNERC researched and developed the multi-wavelength and multilevel mastering system and established the multilevel code system, which is a multi-wavelength and multi-level run-length coding, details see "5.6 Multi-level run-length limited code" of Chap. 5 in this book.

In order to miniaturization of the multi-wavelength and multilevel optical storage system, the pick up have to be integration in the first. Based on the binary optics, subwavelength grating principles and coupled wave theory to design and develop a micro prototype of multi-wavelength and multilevel pick up, as shown in Fig. 1.5.

Fig. 1.5 The three
wavelength optical pick up of
hybrid integration and
miniaturization with
subwavelength grating, binary
optical devices, lasers and
photodetctors integration

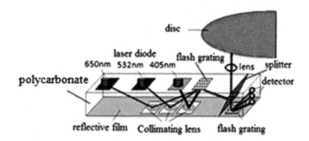

This is a typical multi-wavelength optical pick up with binary optics and MEOMs technology that is utilizing polycarbonate to be the substrate, which optical properties are similar very much each other. The miniaturization triple-wavelength pick up is composed of 405, 532, 650 nm lasers and corresponding photodetectores that are integrated on a polycarbonate substrate with precision injection process. The parallel laser beams are incident on the wavelength selector with different numerical apertures and focused on the disc by objective lens with focusing and tracking system. Meanwhile, Tsinghua University and Physics Chemistry Institute of CAS completed various high sensitivity and suitable for 405, 532 and 650 nm spectrum organic recording materials and for super resolution mask inorganic materials. The experiment shows that the coupling efficient and reliability for multi-wavelength is better than above (see Fig. 1.4) optical system with discrete optical components and carry out commercial product rather easily in the future.

The multi-level technology can be used to single-wavelength optical storage systems of course. A multi-level disc drive with 650 nm wavelength and VCR have been developed by OMNERC and Tong-Fang optical disc Inc. as show in Fig. 1.6. The disc was using write-once organic photochromic material with capacity of 12.5 GB.

In order to duplicate of multi-level CD-ROM, OMNERC research group developed a three-wavelength and multi-level photometric mastering system as shown in Fig. 1.7. The three wavelength are 405 nm/532 nm/650 nm that can write at the same time to make three wavelength multilevel mask and single wavelength

Fig. 1.6 Multi-level
write-once optical disc video
cassette recorder

Fig. 1.7 The three wavelengths and multilevel mastering system with NA1.5 liquid immersion apochromatic objective lens and piezoelectricity drive autofocusing system

multi-level photometric mask or gray scale mask for multi-level CD-ROM. It can be used to make conventional optical disc master also. Its object lens is an apochromatic immersion liquid lens with refractive coefficient 1.516, so that effective numerical aperture can be to over 1.5. As the lens is rather weight especially to develop a piezoelectricity drive autofocusing system.

As the laser power have to be suitable for three kinds of material different written spectrum absorption characteristics, a high frequency pulse modulation writing energy control system was introduce to the system. Meanwhile the system has testing function for mask and recording medium. So the system can be used to experiment for variety photochromic, gray scale, photo-refractive index materials and their optical parameters testing [29, 30].

The duplicate of multi-wavelength and multilevel disc could use similar IC manufacture technology as introduction in Chap. 5 of this book in detail.

1.2.2   Volume Holographic Storage

Based on study of the system performance, multiplexing principles, dynamic range, response time and recording materials characters and optimized to complete a miniaturization two-dimensional dynamic speckle multiplexing volume holographic storage system with capacity of 6 TB as shown in Fig. 1.8. The multitrack overlapping one-dimensional diffuser mobile dynamic static speckle multiplexing

Fig. 1.8 The miniaturization orthogonal polarization dual-channel volume holography system with dynamic speckle angle mixed multiplexing

(DSSM) technology and the crystal shift of the single point with the angle multiplexing was used to this system, that can record image of 2030 on a point, and the storage density can be to 650 GB/cm^3. The channel density and the recognition rate is greater than 4500 Frames/s, accuracy is higher 95 % on the cationic ring-opening polymer (CROP) medium.

Otherwise the group completed the holographic recording disc diameter of 120 mm and thickness of 0.8–1.2 mm for nonvolatile thermal fixing high-density holographic storage, including online and offline precision temperature control heating device. The thermal and light fixing the volatile holographic storage in photorefractive materials can improve the electronic grating dark storage performances obviously.

The system used wavelength of 532 nm, power 150 mW diode-pumped solid-state lasers (diode-pumped Nd: YAG frequency-doubled lasers) with dimensions of 150 mm × 50 mm × 50 mm and divergence angle of 2×10^{-4} rad, reflecting spatial light modulator (LSM) with pixels of 1280 × 1280 (pixel size 13.2 × 13.2 μm^2) and high-speed CMOS sensor with readout rates of 200 Mb/s. When the defocus between 8 and 15 mm, the spectral intensity is distributed uniformity still. The distortion of Fourier transform lens is within 0.1704×10^{-4} in full field, resolution is 235 lines/mm with MTF > 0.15 and the relative distortion < 0.2×10^{-4}. The system uses a 1.3 M pixel CMOS image sensor and high-speed Field-Programmable Gate Array (FPGA) chip addressing, readout rate is 200/s of holographic image and readout data rate of 100 Mb/s with delay less 15 ns. In order to SLM to CCD pixel 1:1 match alignment, the errors will be compensated automatically with control software that realized 25 data pages storage with data error rate is within 1.67×10^{-3}. For big image, pattern and Chinese manuscript storage and searched, based on correlation coefficient to establish recognition algorithm and complete 1000 images or 1020 Chinese character image recognition, and the recognition rate is greater than 98 % in this system. The group also developed a two-wavelength nonvolatile light fixed volume holography experimental system, that using ultraviolet light and space interference modulated red light irradiated medium to record. The technical details will be described in Chap. 7 of this book.

1.2.3 Near-Field Optical Storage

Near-field optical storage technology based on solid immersion lens (SIL), can greatly reduce the recording information symbol size to enhance the storage density. Near-field optical storage research group in OMNERC established a SIL-based near-field optical storage experimental system, carried out a series principal experiments for dynamic and static near-field optical data storage, completion of the larger numerical aperture objective lens and superhemisphere SIL lens design and production. Based on SIL theoretical models and the actual detection get the numerical simulation curve of the optical field distribution is shown in Fig. 1.9a and the curve of numerical simulation of immersion lens and the electromagnetic field distribution on the optical axis section as shown in Fig. 1.9b that approved the recording mark size can be down to under 50 nm with flying control system with positive and negative pressure.

The half peak width and peak intensity achieved the computer simulation of the microporous pupil function of incident beam parameters and got the relationship between interference field and near-field gap within 40–70 nm, main achievements as follows: [31–33].

1. Based on the vector diffraction theory to establish the total reflection near-field optical storage system model that was used to design optical system for near-field coupling on the SIL, aberrations simulation analysis, determined of the gap and disc structure and energy distribution of incident light.
2. Achieved near-field optical nanoscale flight gape dynamic load/unload process and control, including dynamics analyze model, start and stop parameters, stability of dynamic, accuracy and efficiency analysis with modified Reynolds equation to calculate nanoscale flight gape and system optimization.
3. Developed the slider dynamics numerical simulation program, equations of motion of the folding arm and lubrication equation simultaneously. Considering hot deformation, particulate pollution and other factors for posture of the flight

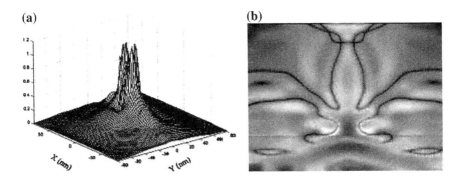

(a)   **(b)**

Fig. 1.9 Curves of numerical simulation of the optical field distribution of immersion lens (**a**) and the electromagnetic field distribution on the optical axis section

slider to amendment the transnational Reynolds equation to design and manufacture nanoflying head with CMOS process. It is composed of opening main
ventilation slot, ear-like track, airflow mediation track, rail protection column
and dust articles structure to ensure stability and stiffness of the system as shown
in Fig. 1.10. The micro-flying head by CMOS design is shown in Fig. 1.10a. On
the underside, there are tracks of 1–3, 5, 7–9, shallow ventilation slot track 4, the
center of the main ventilation slot 6, underside of the square orbit 9 which is the
underside of SIL. When the airflow speed is 18.85 m/s or more, the micro-flying
head of the largest near-field spacing can be to 50 nm and the undulate is within
3.1 nm, i.e., the stability of the CMOS flight system is very stabilization when
airflow velocity changing and corresponding to different load. The pressure
distribution on underside of CMOS micro-flying head is shown in Fig. 2.11b,
the boundary airflow first enter the guide rails 4, then on guide rail 2,7 and SIL
underside 9 form positive pressure to support micro-flying head, and on the
main ventilation slot 6 form the negative pressure to attract the micro-flying
head. The simulation experimental results of the micro-flying head at steady
state are shown in Table 1.5. Corresponding to a larger airflow velocity,

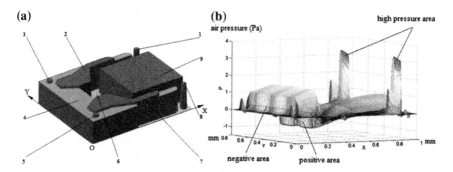

Fig. 1.10 a micro-flying head with CMOS process **b** pressure distribution on surface of the
vacuum micro-flying head of SIL with positive and negative pressure

Table 1.5 The simulation experimental results of the micro-flying head at steady state

Characteristics	Results	
Boundary air line speed	$V_1 = 9.425$ m/s	$V_2 = 18.85$ m/s
Minimum spacing gap h_{min}	45.75 nm	63.28 nm
Maximum spacing h_{max}	62.19 nm	78.87 nm
Spacing gap on SIL center	50.27 nm	64.17 nm
Positive pressure	119.68 mN	129.86 mN
Negative pressure	25.59 mN	35.84 mN
Bearing capacity	94.09 mN	94.02 mN
Pitch angle	40.43 μrad	30.15 μrad
Roll angle	0.61 μrad	0.03 μrad

dynamic pressure bearing stiffness of the gas film is much satisfaction. As the area of bottom surface of the SIL is greater than the area of enter side of the track, so the airflow changing caused the pressure increment on bottom surface of SIL, resulting smaller pitch angle on micro-flying head and increasing spacing gap.

4. Completed a dynamic SIL storage experiment system with miniature flying slider and microaperture laser at first in China that is shown in Fig. 1.11. The system adopted material of the refractive index 1.926 to make SIL and 405 nm blue ray laser for light source. The system used to static experimental exposure on thickness of 50 nm photoresist film to get minimum recording symbol of diameter small than 45 nm. The system load/unload (starts/stops) smooth in dynamic experiments show that the spin speed of the disc can be to 10,600 prm, the airflow of 46 m/s, the carrying capacity of 94 mN, airflow velocity the 9.3–48 m/s, and the change of spacing of the SIL flying is smaller than 5 nm.

Furthermore, the system has been employed to super resolution exposure experiment with microaperture laser. The wavelength of laser is 405 nm with aperture of 60 nm which structure is in Fig. 1.12b. The result of dynamic exposure experiment shows that symbol size is about 50–70 nm [34–36].

1.2.4 Big Data Storage System Integration

China is a big country that needs supercapacity and super data rate information storage systems as CCTV want capacity over PB system to storage video program for example. From the perspective of information society, the demand to information storage capacity will be more than EB (Exabyte = 10^{18} B), any unit storage devices cannot achieve this requirement. Therefore, big data storage system integration is a very great requirement with miniaturization and integrated. Of course, increasing density and capacity of each storage unit is base for further huge capacity

(a)   **(b)**

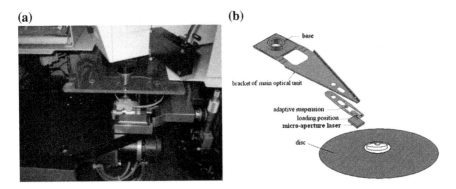

Fig. 1.11 **a** the near-field optical storage experimental system, **b** positive and negative flying slide head with microaperture laser

by various new storage mechanism and technology, as multidimensional optical storage. But some special technologies have to resolve, so OMNERC study out following research projects and results for big data storage and application for Chinese market.

1.2.4.1 Sub-integration Unit of Optical Disc

Based on static/dynamic micro drive, microstructure measurement devices, microsensor, no mechanical high speed drive parts and precision positioning technology to design and develop a lot of integration unit for different optical disc as in Fig. 1.12. Figure 1.12a is a optical array unit with 12 drives which one of them is to calibration, so its capacity and data access rate is increased 11 times to single drive, but the error rate can be down to 1×10^{-15}. Figure 1.12b is an optical disc cluster with 120 discs and matching to the optical array unit to be a sub-jukebox. It is integrated with disc cupboard, multi-manipulator, high speed magnetic bearing serving positioning system. The CD clusters and array are completed a sub-integration unit with data with capacity of 600 GB–2.4 TB each (for different disc) and data rate of 220 MB/s by parallel writing/reading. If using multi-level optical disc or volume holography optical disc in this system, the capacity could be great increasing. Figure 1.12c is an integration CD copy system with 12 drive that the productivity of 70–150 discs copy per hour average. Above three products have been mass produced in China for consumer to database backup.

(a)   (b)   (c)

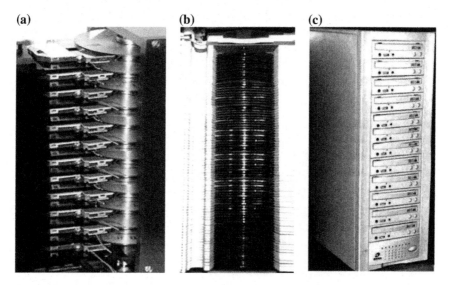

Fig. 1.12 Sub-integration units of optical disc **a** optical array, **b** optical disc cluster, **c** integrated optical disc copy tower

1.2.4.2 Integration System for Big Data Storage

Based on above sub-integration units of optical array and optical disc cluster (sub-jukebox) completed the big data storage system, as optical disc library for example. The system is buildup with 500–3000 ($N \times M$) arrays of sub-integration units that capacity can be achieved much PB as shown in Figs. 1.13 and 1.14. It is according to user actual needs to design to different applications. For the digital library, the access frequency higher data can be storage into a buffer, which was composed of many hard discs and servers for multiuser simultaneous access a same files, it is equivalent a virtual mirror of actual optical disc. So the system I/O is used ATA interface protocol that the handshaking and the transmission efficiency are superior to the SCSI device. But the ATA bus is not suitable for integrated applications so to establish an ATA parallel digital transmission models in it. The part-parallel transfer mode proposed a ATA handshaking mode that data is divided to many chunk to keep the ATA data transfer efficiency and improve the cache hit ratio. The system is an adaptive prefetching scheme for I/O also that is improved the continuous or semi-continuous transmission performances effectively. Cache data scheme is based on the full prefetch CPR RAID5 to balance the I/O between arrival rate and buffer space, so RAID5 has good I/O performances and response characteristics in heavy load with small cache still [37–39].

The system can be addition of volume holographic optical storage with fast image recognition and addressing characteristics that is suitable for image and Chinese manuscript file storage and retrieval. In this system, based on a optical fiber channel data hardware and transfer software developed a two-dimensional modulation codes, error correction encoding, decoding and holographic data transfer protocol with data rate of 360 MB/s. Meanwhile using the multiplex bus system and serial Ultra2 SCSI chip embedded microprocessor connect the common data of SCSI protocol for other

Fig. 1.13 The optical disc big data storage integration system: **a** mainframe and data management system **b** sub-jukebox, i.e., sub-integration units with optical disc array and optical disc cluster

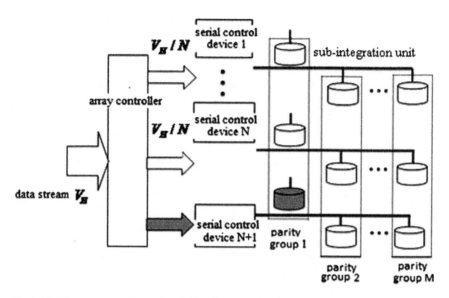

Fig. 1.14 The structure diagraph of big data storage integration system with 500–3000 sub-integration units to buildup $N \times M$ array with optical disc cluster and floating parity calibration group. Total capacity can be much more PB

optical disc storage devices. As an optical fiber network platform, it will split the high-speed data flow deceleration to format parallel high-speed data stream to connection with various capacity external storage devices and different rate of continuous data recording. The system adopted field-programmable logic devices (FPGAs), SCSI controller, embedded CPU to build an array of parallel RAID1 + 0 or RAID3 + 0 composite array data transfer model as in Fig. 1.15, that is greatly improving the system reliability. The array parallel data transmission system can be flexible parallel records of high-speed data transmission. There is a dual-page buffer in this system with pre-switching logic to ensure that a page to accept the recorded data stream and the other one to write data into system to high speed data stream transmission and recording continuously. Experiments show that the single channel continuous recording speeds is 50 MB/s, for 8 channels parallel data of continuous recording speed can be to 400 MB/s.

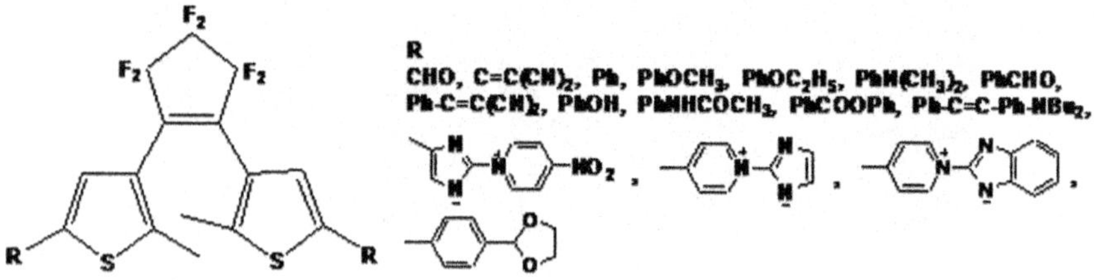

Fig. 1.15 The typical molecule structure of diarylethene and fulgide photochromic materials

For improving data security in this system, there are a lot of "floating parity group" in this $N \times M$ array when any storage device failure can be adjusted to another one soon automatically as shown in Fig. 1.14. So the system may fail only when the redundant resources of the all system are exhausted, that ensured the reliability, effective data transfer rate to external memory and total capacity substantially. The system has been applied for various high-speed big data storage as meteorology, population and resource management, knowledge resource database, aeronautical information services, aerospace remote sensing telemetry and audio/video/TV in TV station, etc., as application to China Knowledge Resource Integrated Database for example, its total capacity is 1.8 PB, data rate of 160 MB/s.

1.2.5 New Storage Materials and Medium

The storage materials and medium are very important for optical storage. There are research groups to study storage materials and medium in Tsinghua University, Physics Chemistry Institute of CAS, Shanghai Institute of Optics and Fine Mechanics of CAS, Fudan University, Beijing University of Aeronautics and Astronautics, University of Electronic Science and Technology in Chengdu, Taiwan Jiaotong University, University of Science and Technology of China, Nanjing University, Huazhong Science and Technology University and Harbin Industry University, etc., in China. Main achievements are including:

1.2.5.1 Organic Photochromic Materials Design and Synthesis

Because the versatility of photochromic polymeric materials can be used to multi-wavelength and multilevel optical storage excellently. Tsinghua University and Physics Chemistry Institute of CAS based on 13 kinds of diarylethene and fulgide photochromic materials designed and synthesized series of diaryl molecular photochromic materials, its absorption wavelength from UV to near-infrared region. The molecular structure of typical diarylethene and fulgide photochromic compounds is shown in Fig. 1.15. Experiments show that materials match with the current industrial production semiconductor lasers of 405, 532, 650, and 780 nm wavelength very well, their spectral overlap is less than 10 %, and the quantum yield is higher 92 %. Based on the engineering of synthetic superlattices and update organic/inorganic photorefractive composite thin film technology developed the typical medium that have a good sensitivity, chemical thermal stability, fatigue resistance and lower crosstalk. Two resorcinarene molecular structure and photo-chemic characteristics revealed the diarylethene DT-3 isomer phenomenon and the two-photon fluorescence effect, its halo reaction (π, π^*) is transition to zero-order reaction on constant light intensity and the open-loop reaction. The absorption experiment with wavelength of 400–780 nm shows that the diarylethene of photochromic fulgide is sutable for multi-wavelength and multilevel optical storage much well. After repeat read out of 10^5 times, the attenuation of the read out signal is less than 25 % in multi-wavelength multi-order storage experiments.

1.2.5.2 Two-Photon 3D Storage Materials

The 3D multi-layer optical storage materials are bifunctional photochromic characters as spirobenzopyran and spirooxazine. Research group of Physics Chemistry Institute of CAS utilizing biheterocycle substituted photochromic compound in fulgide category and synthetic developed one wavelength two-photon 3D and double wavelength two-photon storage materials, it can be dissolved in high-molecular polymer with cyclohexanone or toluene. Based on relationship between excitation fluorescence emission and photochromical response calculated Raman spectra molecular vibration by photochromic reaction. And based on the polarized Raman spectra and polarized UV absorption spectroscopy determined the relationship between the two resorcinarene crystal microscopic molecular orientation and electronic transition dipole moment of diarylethene crystals to the short-wavelength photochromic reaction. The typical molecular structure and synthesis reaction of perfluorinated resorcinarene crystal is shown in Fig. 1.16. Its closed-loop state absorption spectrum is 400–700 nm and two-photon absorption cross section is 17.3×10^{-49} cm^4 s/photon, decomposition temperature is higher than 200 °C higher than two mono-substituted compounds 2–3 times. It was made thickness of 100 μm PMMA film and used to multilayer recording experiment, as reference Sect. 5.6.3 of Chap. 5 in this book.

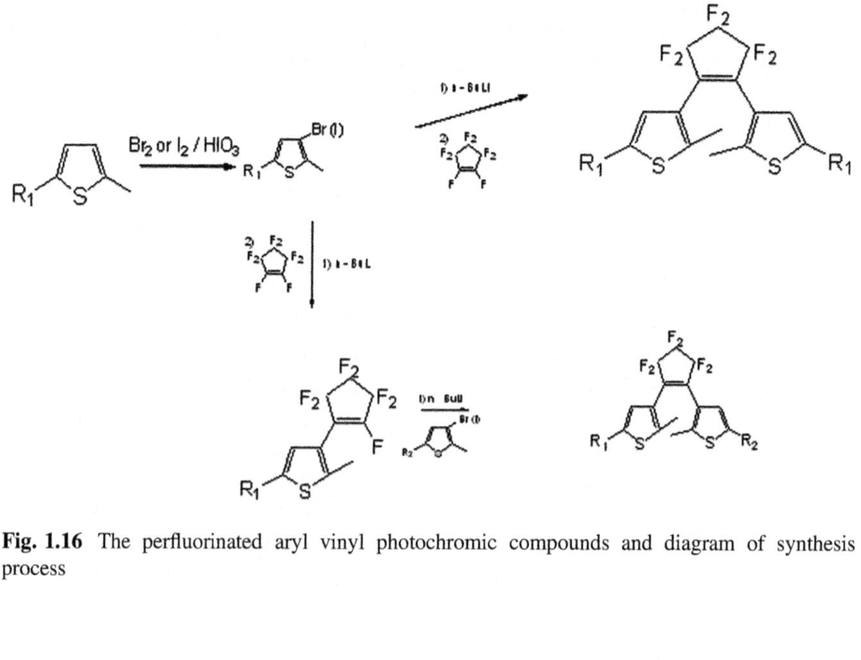

Fig. 1.16 The perfluorinated aryl vinyl photochromic compounds and diagram of synthesis process

1.2.6   Chinese Blue High Density Disc

OMNERC of Tsinghua University, Shinco In and TCL in China cooperated to develop the Chinese Blue High Density (CBHD) drive and disc independently, including front servo and back-end decoding, that the capacity is 15 GB (single layer) or 30 GB (two layer structure) in 2009. CBHD disc and player are based on Chinese independent intellectual property rights and the most advanced blue laser technology, that supports 1920 × 1080P resolution high-definition movies or TV programs, and compatible with DVD discs. China Record Corporation and China Film audiovisual Publishing House produced the CBHD high-definition programming disc. The CBHD can maximize to employ the existing CD industrial base in China. The international DVD Forum has approved the related patents and the full specification, as well as admitted the CBHD format standards.

1.3   The Advanced Efforts for Multi-dimension Optical Storage

1.3.1   3-Dimension (3D) Optical Storage

3D optical data storage can give to any form of optical data storage in which information recorded and read with three dimensional resolution. Recording and readback are achieved by focusing lasers within the medium. Mark M. Wang and Sadik C. Esener proposed a three-dimensional optical data storage in afluorescent dye-doped photopolymer, that the recoding layer can be to more 100 [40, 41]. However, because of the volumetric nature of the data structure, the laser light must travel through other data points before it reaches the point where reading or recording is desired. Therefore, some kind of nonlinearity is required to ensure that these other data points do not interfere with the addressing of the desired point. Traditional optical data storage media store data as a series of reflective marks on an internal surface of a disc. In order to increase storage capacity, it is possible for discs to hold two or even more of these data layers, but their number is severely limited since the addressing laser interacts with every layer that it passes through on the way to and from the addressed layer. These interactions cause noise that limits the technology to under 10 layers. The addressing methods of 3D optical data storage have to be specifically addressed volumetric pixel interacts substantially with the addressing light only. This necessarily involves nonlinear data reading and writing methods, in particular nonlinear optics.

3D optical data storage is related to holographic data storage also. Traditional examples of holographic storage do not address in the third dimension, and are therefore not strictly "3D", but more recently 3D holographic storage has been realized by the use of microholograms. Layer-selection multilayer technology (where a multilayer disc has layers that can be individually activated, e.g., electrically) was used 100 layer storage system in the first in 2003 by Call & Recall Inc in United State of America.

1.3.2 5-Dimensional Optical Storage

Peter Zijlstra, James W.M. Chon, and Min Gu proposed a 5-dimensional (5D) optical recording mediated by surface plasmons in gold nanorods in 2009 [3, 26]. The multiplexed optical recording provides an approach to increasing the information density beyond 10^{12} bits per cm^3(1 Tbit cm^{-3}) by storing multiple, individually addressable patterns within the same recording volume. The major hurdle is the lack of a suitable recording medium that is extremely selective in the domains of wavelength and polarization and in the three spatial domains, so as to provide orthogonality in all five dimensions. A 5-dimensional optical recording is exploited the unique properties of the longitudinal surface plasmon resonance (SPR) of gold nanorods. The longitudinal SPR exhibits an excellent wavelength and polarization sensitivity, whereas the distinct energy threshold required for the photothermal recording mechanism provides the axial selectivity. The recordings were detected using longitudinal SPR-mediated two-photon luminescence, which possess an enhanced wavelength and angular selectivity compared to conventional linear detection mechanisms. Combined with the high cross section of two-photon luminescence, this enabled nondestructive, crosstalk-free readout. This technique can be applied to optical patterning, encryption and data storage, where higher data densities are pursued. The concept of 5D optical storage principles is illustrated in Fig. 1.17, the sample consists of thin recording layers of spin-coated polyvinyl alcohol doped with gold nanorods, on a glass substrate as Fig. 1.17a. When the right polarization and wavelength is chosen, the patterns can be read out individually without crosstalk as in Fig. 1.17b. The sample consists of a multilayered stack in which thin recording layers (~ 1 µm) are separated by a transparent spacer

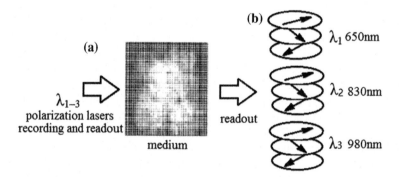

Fig. 1.17 5D optical storage principles with three wavelength polarization laser beams and surface plasmon resonance recording sample

(~ 10 μm). In both the wavelength and polarization domains, three-state multiplexing is illustrated to provide a total of nine multiplexed states in one recording layer. The key is a recording material that the recording material based on plasmonic gold nanorods meets all the above criteria. Gold nanorods have been extensively used in a wide range of applications because of their unique optical and photothermal properties. The narrow longitudinal SPR linewidth of a gold nanorod (100–150 meV, 45–65 nm for the near-infrared as Fig. 1.17), combined with the dipolar optical response, allows us to optically address only a small subpopulation of nanorods in the laser irradiated region. These recording layers were spaced by a transparent pressure-sensitive adhesive with a thickness of 10 mm. In the recording layers, patterned multiple images used different wavelengths (λ_{1-3}) and polarizations of the recording laser. When illuminated with unpolarized broadband illumination, a convolution of all patterns will be observed on the detector as in Fig. 1.17a (filters attenuate the reflected readout laser light). This selectivity achieved longitudinal SPR-mediated recording and readout governed by photothermal reshaping and two-photon luminescence (TPL) detection respectively.

1.3.3   Negative Refraction Index Materials Application to SIL 3-Dimensional Storage

The solid immersion lens (SILs) has high resolution and have been used in optical storage systems with very high density, but it cannot be applied for 3 dimensional (3D) storage, as its air-gap h_a (see Fig. 1.19) is very smaller (less than 50 nm generally).

All the rays pass perpendicularly through the upper spherical surface of the SIL without changing their propagation directions. The focal plane lies right on the lower (planar) surface of the SIL. The effective numerical aperture (NA_{eff}) of the whole system can be expressed as $NA_{eff} = n_s \cdot \sin \theta_m$, where n_s is the refractive index of the SIL. But a solid immersion lens (SIL) has higher magnification and higher numerical aperture than common lenses by filling the object space with a high-refractive-index solid material. There are two types of SIL: Hemispherical SIL–Theoretically capable of increasing the numerical aperture of an optical system by n, the index of refraction of the material of the lens. Weierstrass SIL (super-hemispherical SIL or super SIL) the height of the truncated sphere is $(1 + 1/N)r$, where r is the radius of the spherical surface of the lens. Theoretically capable of increasing the numerical aperture of an optical system is by N^2. So the size of the focused spot is proportional to λ_0/NA_{eff} (here λ_0 is the working wavelength in vacuum). Thus, to achieve a small focused spot at the lower surface of the SIL, a large NA_{eff} is preferred (usually $NA_{eff} > 1$). However, such a small focused spot can only exist near the SIL–air interface since the high frequency components in the angular spectrum of the focused beam become exponentially decaying (i.e., the evanescent waves) in the air-gap besides the divergence of the low-frequency components. Therefore, in a conventional SIL system see the Fig. 1.18a, the laser

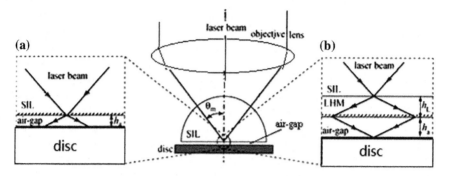

Fig. 1.18 Comparison of the near-field optical storage system: **a** conventional SIL system, **b** SIL system with the negative refraction index LHM to get large air-gap h_a

beam diverges very fast in the air-gap. Thus, in order to achieve a large signal contrast and a high storage density, the medium should always be placed very close to the lower surface of the SIL, as the air-gap h_a is smaller than 50 nm. This will bring some difficulty or inconvenience to 3D multilayer storage. Many efforts have been made to increase the air-gap while keeping the signal contrast and the storage density. Milster et al. showed that the signal contrast can be improved by placing a pupil filter in front of the SIL. Hirota et al. filled the air-gap with a liquid lubricant to increase the total internal refraction angel at the interface of the SIL. It increased the focal depth of an SIL with a numerical aperture in front of the SIL. In all these approaches, the smallest beam spot of the whole system is still located at the lower surface of the SIL.

Liu [20] proposed a new type of left-handed material (LHM) which has negative refraction index with negative effective permittivity and permeability over a certain frequency band simultaneously, such as the subwavelength focusing ability of an LHM slab. A homogeneous LHM (or atomic level) at optical frequencies has also been investigated. Theoretically, an ideal LHM slab can image an object with a perfect resolution. However, some intrinsic aspects such as the losses and the finite lateral size will degrade the image quality. Nevertheless, subwavelength imaging/focusing can still be achieved for a thin LHM slab. A new near-field optical storage system with an LHM slab, it is under the lower surface of the SIL which refer to this system as an L-SIL system, is shown in Fig. 1.18b. The new near-field optical storage system utilizing LHM is introduced by a LHM slab to the lower surface of a conventional solid immersion lens (SIL).

The performances of the present storage system are compared with a conventional SIL system through numerical simulation as Fig. 1.19a. The LHM slab in the new storage system image focused spot at the lower surface of the SIL to the surface of medium. But it allows a large air-gap for the 3D (multilayer) storage while keeping a large signal contrast and a high storage density. Meanwhile, the tolerance of the air-gap is improved also. The dependence of the signal contrast (normalized with its value of $f_s = 0$) on the spatial frequency f_s for different values

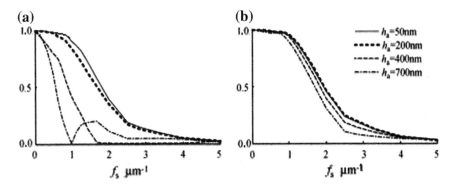

Fig. 1.19 The relationship curves between the normalized signal contrast and spatial frequency f_s for different values of air-gap h_a in (**a**) the conventional SIL system and (**b**) the present L-SIL system (when $h_L = h_a$)

of the air-gap in the conventional SIL system and the present L-SIL system. It can get the response of a low-frequency filter. The pass-band width of the curve that the dependence of the signal contrast (normalized with its value at $f_s = 0$) on the spatial frequency f_s for different values of the air-gap in the conventional SIL system and the present L-SIL system. Each curve in Fig. 1.19 gives the frequency response of a low-frequency filter. The pass-band width of the curve usually is determined the storage density. From Fig. 1.19a clearly see that the pass-band width of the conventional SIL system becomes very narrow when the air-gap h_a increases to 700 nm. Consideration of both the signal contrast and the storage density, the air-gap in the conventional SIL system without the LHM slab should be very small (less than 100 nm). However, in the new L-SIL system with the LHM slab, that the pass-band width h_a can be to 700 nm,but the signal contrast is decreased very small as comparison to that for $h_a = 50$ nm (see Fig. 1.19b). The tolerances of the air-gap for the conventional SIL system with $h_a = 50$ nm and the present L-SIL system with $h_L = h_a = 700$ nm so that can be used to multilayer storage expediently.

1.3.4   The Higher Order Radially Polarized Beams Application in SIL

Radially polarized incident light can generate a more confined longitudinal electric field on a focal plane in near-field (NF) optics than focusing circularly polarized light. It is feasible to reduce beam spot size on storage media to increase the areal density of optical data storage. A radially polarized beam generates a beam spot which is 20 % more confined on the 1st surface of medium than that of circularly polarized light. However, the peak intensity of total electric field sharply decreases and its transverse component is much more dominant inside the media stack. This confirms that radially polarized optics can be a candidate not for an NF recording system but for an NF read-only memory (ROM) system, the results could be useful to understand the effect of radial and circular polarizations inside and outside medium for various applications of NF optics potentially.

For improving the recording ability of a near-field optical storage system by higher order radially polarized beams, distributions of the optical field in a solid immersion lens recording system are calculated for higher order radially polarized modes of the incidence. Results show that two higher order radially polarized modes of R-TEM11 and R-TEM21 are useful to near-field optical recording, but further higher order modes such as R-TEM31, R-TEM41, and R-TEM51 are not useful due to the strong side-lobe intensity. Compared with R-TEM01 beam focusing, the full width at half-maximum of the recording spot is decreased markedly and the focal depth is increased substantially by using R-TEM11 beam focusing. The effect of the beam width of the R-TEM11 mode is also discussed.

Thorough analysis of the focal field distribution of a solid immersion lens system of arbitrary thickness can discover that the cases of linearly and radially polarized illumination are examined and accurate expressions derived for the electric field in the image space. The effect of the spherical interface on both transverse and axial intensity profiles is stronger. The performance and practicality of configurations deviating from the hemispherical and aplanatic cases shows that optimal resolution is obtained at focal positions between the hemispherical and aplanatic points when radially polarized illumination is applied. The behavior of the electric field in a focal plane consisting of a solid immersion lens (SIL), an air-gap and a measurement sample for radially polarized illumination in SIL-based near-field optics with an annular aperture. The analysis of SIL was based on the Debye diffraction integral and multiple beam interference. For SIL-based near-field optics whose NA is higher than unity, radially polarized light generates a smaller beam spot on the bottom surface of a SIL than circularly polarized light, however, the beam spot on the measurement sample is broadened with a more dominant transverse electric field. By introducing an annular aperture technique, it is possible to decrease the effects of the transverse electric field, and therefore the size of the beam spot on the measurement sample can be small. It could have various applications in near-field optical storage, near-field microscopy and lithography.

1.3.5 Other Schemes for Multi-value Optical Storage

1.3.5.1 Metal and Its Oxides Medium

On the film of metal and its oxides makes a physical hole or pit with semiconductor etching process to instead injection process to manufacture optical disc. This M-DISC looks just like conventional CD, but its size or figure could be diversification and may allows higher storage density and capacity as shown in Fig. 1.20. If the material can be decompose at a certain temperature that can be writeable and add this function to conventional CD drive. It can read back over the long term

Fig. 1.20 The construction
of M-DISC

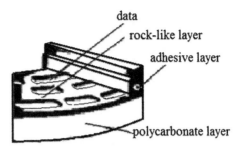

requires that compatible drives remain available, especially if the data or information want to preserve, even simple files like family pictures, but hard disk drives,
tape, and typical CDs and DVDs won't hold data for an infinite period of time.
They all degrade over time and most writable DVDs can lose data after just a couple
of years. M-DISC, designed to work just like a standard writeable DVD but without
decay that can't be overwritten, erased or otherwise corrupted. Most optical storage
uses either dye or phase-change media to record data, while hard drives and tape
flip bits on and off magnetically. M-DISC is composed of the rock-like layer,
adhesive layer and polycarbonate layer. M-DISC and M-READY disc storage
technology that permanently etches data onto the write layer of the disc for use
anytime and for generations to come without data lost. The M-DISC will not
degrade over time and is usable on a daily basis. M-DISC is a perfect storage
solution for music, photos, videos, genealogical records, business records, data loss
prevention, permanent file backup, medical imaging, government usage, and for
archival purposes [21, 27].

The M-DISC is backward compatible non-dye-based CD optical technology
constructed of inorganic materials that are known to last centuries. M-DISC
compatible drives are a high quality optical drive specifically designed to laser-etch
digital information onto the M-DISC. This combination allows information to be
written once and read over time and offers the best permanent data storage solution
in the industry.

1.3.5.2   Multilevel and Multilevel Optical Recording
with Different Reflectance

The multilevel optical recording method is shown in Fig. 1.21, it implemented four
level or more recording schemes in an inorganic material disc that used either two
or three recording layers. Each of the recording regions represent two-bits data, and
every bit has a different reflectance for easy recognition. The recording density can
be double or triple, depending of the schemes. However, the mark size difference
between two or three recording layers is also enlarged due to excessive heat diffusion. The difference in mark size in different layers affects the quality of the
readout signal which makes the readout process more difficult to be recognized.

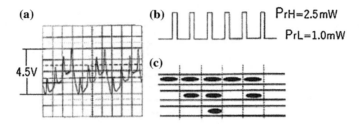

Fig. 1.21 a Readout signal of repetitive recording regions with three recording layers by **b** pulse-read detection **c** recording mark

1.4   Frontier Science and Technologies Related to Optical Storage

Many potential technology for optical storage may be application and competitive each other. Future optical storage will be based on some new principles and technologies. The holographic memories and Millipede, etc., are very difficult to be commercially viable. The plasmatic, negative refraction optical devices, nanotech, molecular self-assembling, integration optics, quantum optics, etc., principles will be applied in this area that just like semiconductor memory, magnetic disk, optical storage will keep increasing within a certain time. By this way, the three kinds of storage principles of optics, magnetic and semiconductor will be combined to create new perfects storage devices as optical solid state storage, optics–magnetic storage devices and magnetoresistive random access memory for example and may be becoming a true universal memory. Summary of various potential optical storage technologies are shown in Table 1.6 [21, 27].

The traditional optical disc technology development model to reduce mark size and increase areal density that has reached the classical technology end of life with BD discs and drives at $\lambda = 405$ nm and NA = 0.85. But extending the definition of classical optics, optical storage used very small part of electromagnetic spectrum in the 400–700 nm range. However, extending the meaning of "optical" to include UV, DUV (deep ultraviolet) and X-radiation, opens new frontiers for high-density data storage, the challenges faced to reach 1 TB or more capacity. Multi-layer solutions are feasible 16-layer $\sim$ 100 more layer discs are proven. NFR is also feasible, but needs to be proven outside the lab. The nanotech is determined qualification really. Electron Beam Mastering and Near Field Read 120 mm Capacity is 220 GB, storage density is 144.8 Gb/in^2. The track pitch is 108 nm, the minimum mark size is 58 nm by present nanotechnology threshold. Near term multidimensional optical storage, near-field recording $\rightarrow$ UV-X-ray -Atomic/Quantum Mechanisms application to memory [30, 31].

Table 1.6 Main technology roadmap to increase density and capacity optical storage

Types (feature)	Min. mark bit (nm)	Track (nm)	Density (Gb/in^2)	Capacity(GB 120 mm disc)	Principles and technologies
Blu-ray and 4–8 layers disc	112–149	320	18.1	100–200	Multi-layer, Multi-wavelength and Multi-level
Near-Field (NFR) read only	41.3–58	108	144.8	220	EB mastering and near-field read out
UV laser diodes and NFR	20–25	75	351	500	EB mastering and doubled 405–202.5 nm
Plasmonic optical storage	15–20	45	1×1000	1.7×10^3	Plasmonic optical pick-up with negative refraction lenses
Nanotech	10–15	30	1.8×10^3	3×10^3	DUV and EB Mastering nanoimprinting and nanodetector
Optical solid state storage	~20	~20	$1–2 \times 10^3$	$1–10 \times 10^3$/ unite	Nanotech, molecular self-assembling quantum optics, multi-wavelength and multi-level
Holograms recordable storage	–	–	$2.5–5.2 \times 10^3$	$5.4–10.7 \times 10^3$	NA 0.65/λ 407 nm, multi-wavelength and medium thickness > 1.5 mm
Molecular polymer storage	8–12	8–12	$10–12 \times 10^3$	$15–20 \times 10^3$	Self-assembling polymer arrays
X-ray digital holograms storage	–	–	160×10^3	50×10^3	X-ray laser (λ 0.5 nm), system NA 0.5
Nanoprobe storage	1	2–3	645×10^3	1.13×10^6	Dynamic contact electrostatic force microscopy (DC-EFM)
Electronic quantum holography	35 bit/electron	–	3×10^6	$5–10 \times 10^6$	Written in electron waves and read out with the scanning tunneling microscope

By continuation of the classical optical roadmap, potential future ODS technologies is UV disc that requires UV laser diodes, next X-ray disc as digital holography for example. Atomic/Molecular data storage means of configuration or quantum state or both. Some enabling means negative refraction (spot sizes less than the diffraction limit), variable focus lenses for multi-layer discs to correct for spherical aberration. The nanotech as superhigh storage densities, self assembly, patterned media, nanophotonics modulators, gratings implemented in Silicon—plasmonics (spot sizes less than the diffraction limit—photon) sieves for far UV and X-ray spot formation [28–30].

1.4.1 Ultraviolet Optical Storage

Ultraviolet optical storage was classical optical data storage end with $\lambda = 405$ nm when the technology uses near and mid-range ultraviolet (UV). Much of UV optical storage technology will be adapted from semiconductor UV and EUV lithography. UV optical storage will be far more challenging than near-IR and visible optical storage ever was. Front surface recording layer and reflection component OPU (optical pickup unit) required [31, 32]. Using frequency doubled 650 nm can get $\lambda = 325$ nm laser for example that will increases capacity to 39 GB/layer and reach 100 GB for 3 layers per disc. For $\lambda = 202.5$ nm in vacuum UV regime use frequency doubled 405 nm that the areal density increase $4\times$ to BD. This increases the capacity per layer to 100 GB. However, the technology will probably be abandoned before reaching $\lambda = 202.5$ nm, owing to cost and complexity. UV optical storage challenges UV laser diodes, including UV optical components for OPU, low noise UV media, mastering and replication processes that not be proportional of cost to complexity for capacity increase. UV laser diodes technology is still immature and only a few commercial products that engineering samples from Nichia is 200 mW CW output 375 nm, 340–360 nm is current research sweet spot of 240–260 nm demonstrated in the lab. The diode-pumped solid state (DPSS) lasers, which can be frequency tripled or quadrupled, but must be greatly reduced in size and cost to be candidates. Other options to UV laser diodes and DPSS (for example, KrF or F_2 fiber) have no possibility of meeting size and cost requirements for UV optical storage. Nanotech may hold the key to long-term prospects on structural enhancements and materials improvements. So UV laser diodes are in about the same position for solutions need a long term [33].

1.4.2 Plasmonic Optical Storage

Plasmonics is a branch of physics in which surface plasmon resonances of metals are used to manipulate light at the subwavelength scale. Surface plasmon polaritons (SPPs) are collective oscillations of electron density at an interface of a metal and dielectric. Because SPPs can be excited and strongly coupled with incident light, they have many potential applications in high-resolution optical imaging and storage and lithography. Some metals (gold and silver, for example) exhibit strong SPPs resonance in certain wavelength ranges, and therefore can be used to guide and concentrate light to nanoscale spots less than the classical diffraction limit. Some of the SPPs resonant structures can produce a field irradiance (W/m^2) at the near field that is greater by orders of magnitude than the incident light. Some resonant optical antennas can concentrate laser light into < 25 nm FWHM size spots (as defined by gap widths). Plasmonic optical pick-up utilized negative refraction variable focus lenses which needed to aid layer to layer focusing for ML discs. The above enabling technologies may provide the means to write/read significantly smaller marks. Plasmonic optical pick-up (POPU) needed to aid layer to layer focusing for ML discs and similar to NFR that flies are 20 nm or thin on the disc surface. The spot sizes can be to 25 nm or smaller. This corresponds to an areal density is about 1 Tb/in^2, or a capacity 17 TB on a 120 mm disc to increase about 85× for Blu-ray Disc. Plasmonic optical storage will permit multichannel read/write and integrated POPU per side optical disc drives. Major challenges will be servo control of POPU flying height and tracking with the air bearing surface (ABS) slide. Volume holography storage theoretical maximum volume storage density is about $1/\lambda^3$ for binary data, i.e., if λ = 500 nm calculates density about to be 8 bits/μm^3 or 131 Tb/in^3 when the NA of optical system is large enough. The corresponding theoretical areal density is about 8 N bits/μm^2 ($\sim$5.2 N Gb/in^2). N is the number of page that can be independently stacked in a common volume of the storage medium. For N = 1000, the areal storage density is about 5.2 Tb/in^2.

1.4.3 X-Ray Storage

X-ray storage concept is designed for x-radiation with $\lambda \leq$ 1 nm 1D or 2D computer generated Fourier transform holograms Select page size (N or $N \times N$ pixels) and offset angle compute and sample analog interference pattern apply data coding and EDAC modulate and scan write spot to form hologram parallel read by means of holographic reconstruction position read beam over hologram project N or $N \times N$ pixels onto photodetector array and process and format serial data stream. For X-ray storage challenges are safe and inexpensive compact X-ray laser, compact photodetector arrays, all optics must be reflective and new mastering (write) and replication methods. The advantages are needless to page composer (SLM), 3D media and incoherent superposition (stacking) of holograms and can apply method to all media formats that read out servo may be same as today's DVD. Meanwhile, X-ray storage performance potential of digital Fourier transform holograms, for a disc of 50 mm diameter and a recording area of 1600 mm^2, λ 0.5 nm and NA 0.5, the storage density 250 Gb/mm^2 (160 Tb/in^2), capacity 50 TB unformatted access time < 10 ms, read out data rate, which is function of pixels, read power, detector sensitivity and scan speed, could be 50 Gbps or higher. State-of-the-art X-ray laser from University of Hamburg, the free-electron laser (FEL) may be suitable for optical storage application. Next-generation light sources, based on Free-Electron Lasers (FEL) will be capable of producing X-rays of such extreme brilliance, that new possibilities of focusing the radiation emerge. An optical element, based on the simple concept of an array of pinholes (a photon sieve), exploits the monochromaticity and coherence of light from a free-electron laser to focus soft X-rays with unprecedented sharpness (high irradiance levels) [38].

1.4.4 Nanoprobe and Molecular Polymer Storage

The nanoprobe storage has higher storage density that bits written on ferroelectric film the marks are roughly on 25 nm centers corresponding to a storage density of about 1 TB/in^2. Writing and reading used a voltage nanoprobe with switching between two polarization states. Imaging is done with Dynamic Contact Electrostatic Force Microscopy (DC-EFM), density can be to 645 Tb/in^2, i.e., 1130 TB (unformatted) on a 120 mm disc with 1 nm bit and track pitches. The highest storage density could be atom storage as Hydrogen (H) atom and Fluorine (F) atom level storage that storage density can be to 51.6 Pb/in^2 and represent is about 0 s and F atoms represent 1 s only.

1.4.5 Molecular Polymer Storage

Molecular polymer storage builds on existing approaches by combining the lithography techniques traditionally used to pattern microelectronics with novel self-assembling materials called block copolymers. Molecular polymer storage has been shown to store 10 Tbit/in^2. When added to a lithographically patterned surface, the copolymers' long molecular chains spontaneously assemble into the designated arrangements. There are information encoded in the molecules that results in getting certain size and spacing of features with certain desirable properties. Thermodynamic driving forces make the structures more uniform in size and higher density than traditional materials. The block copolymers pattern the resulting array down to the molecular level, offering a precision unattainable by traditional lithography-based methods alone and even correcting irregularities in the underlying chemical pattern. Such nanoscale control also allows the researchers to create higher resolution arrays capable of holding more information than those produced. In addition, the self-assembling block copolymers only need one-fourth as much patterning information as traditional materials to form the desired molecular architecture, making the process more efficient. The large potential gains in density offered by patterned media make it one of the most promising new technologies on the horizon for future hard disk drives. In its current form, this method is very well-suited for designing hard drives and other data storage devices, which need uniformly patterned templates—exactly the types of arrangements the block copolymers form most readily. With additional advances, the approach may also be useful for designing more complex patterns such as microchips. These results have profound implications for advancing the performance and capabilities of lithographic materials and processes beyond current limits.

A number of technologies are attempting to surpass the densities of all of these media. IBM's Millipede memory is attempting to commercialize a system at 1 Tbit/in^2. This is about the same capacity that perpendicular hard drives are expected to "top out" at, and Millipede technology has so far been losing the density race with hard drives. Development appears to be moribund since 2007, although the latest demonstrator with 2.7 Tbit/in^2. A newer IBM technology, racetrack memory, uses an array of many small nanoscopic wires arranged in 3D, each holding numerous bits to improve density. Although exact numbers have not been mentioned, IBM news articles report that is increased 100 times. Various holographic storage technologies are also attempting to leapfrog existing systems, but they too have been losing the race, and are estimated to offer 1 Tbit/in^2 as well, with about 250 GB/in^2 being the best demonstrated to date- for non-quantum holography systems.

1.4.6   Persistent Spectral Holeburning

Persistent spectral holeburning allows information to be stored in a material by changing the characteristics of a broad absorption band. Data is encoded as a wavelength-dependent modulation of the absorption band and recalled by appropriate illumination of the band with laser light. For successful storage and readout, several conditions are required. The absorption linewidth of the individual absorbers, the homogeneous width must be very narrow. The inhomogeneous absorption linewidth must be much broader. This linewidth results from random frequency shifts of the individual absorbers, resonances in local crystal fields and strains in the material.

Materials fulfilling these two conditions, known as in homogeneously broadened materials, come in a number of forms, one of which is produced by doping absorber ions into a crystalline or glass host. Upon illumination by single-frequency narrow-band laser light, absorbers in resonance with the excitation transition to an excited electronic state. In most materials, when the excitation is removed, the excited absorbers decay back to their ground state. However, in some materials there exists an alternate state into which the some of the absorbers may decay. If this alternative state has a different energy than the original ground state, and is sufficiently long lived, then these absorbers fall out of in resonance with the original excitation. In some materials, decay into this reservoir state may even be assisted by an additional gating radiation field at a different wavelength than the excitation wavelength. This process is termed "gated storage" and it has the advantage that the reading process is nondestructive and the writing process is typically more permanent.

Reading of the stored data is achieved by frequency scanning across the absorption band with a laser. A spectral "hole" is created at the wavelengths where absorbers have relaxed into reservoir states causing decreased absorption. The number of these spectral holes allowable in a single physical location is roughly given by the ratio of the inhomogeneous linewidth to the homogeneous linewidth. This number can be as high as 10^7 in some materials at liquid helium temperatures. In practical materials this ratio may be only 10^3, but this still represents an enormous increase in information storage density compared to the storage densities possible with conventional two-dimensional spatial storage techniques. Frequency tuning the laser can give gigabyte storage and access times of submicroseconds with the access times determined by how fast the laser tunes. Electro-optic (or acousto-optic) scanning could give of up to terabytes of data with increased access time (tens of microseconds). Transfer rates would be as high as tens of MHz per page.

1.4.7   Coherent Transient Memory

Coherent transient memories rely on these same selective absorption properties of inhomogeneously broadened materials. They work using the output signal emitted from an inhomogeneously broadened two-level absorber after excitation by three temporally encoded light pulses. The first two waveforms, temporally separated and angled with respect to each other, form the reference and data pulses. As long as these waveforms satisfy three conditions, the medium will respond to their combined power spectrum. These conditions require the waveforms to be within the inhomogeneous bandwidth of the absorbing transition, be shorter than the transition's homogeneous lifetime, and to not saturate the two-level transition. The resulting power spectrum contains an interference term proportional to the product of the Fourier transforms of the waveforms. This interference is stored in the spectral population distribution of ground state absorbers. It forms a spectral hologram analogous to the spatial hologram produced by the interference of two spatially modulated light beams with angular separation. These interference patterns can be stored permanently, enabling long-term memory or continuous data processing. The resultant grating spectrally filters the Fourier components of subsequent optical inputs and can, under appropriate conditions, give an output mimicking the original data pulse, hence the common term "photon echo". The requirements for the output pulse to mimic the input pulse include that the first and third pulses be either temporally brief with respect to the modulation of the data pulses, or identically frequency chirped pulses with chirp bandwidth greater than the data bandwidth, or identical pseudo-random broadband pulses. A coherent transient memory system can store frequency, phase or amplitude modulated encoded data but the bandwidth is ultimately limited by the inhomogeneous bandwidth of the absorbing transition.

These types of memory are projected to have capacities exceeding hundreds of Tb per cubic centimeter. Data bandwidths ranging from hundreds of Mb to Tb per second are possible. As with other memory types there is a tradeoff between capacity and bandwidth. A major stumbling block in the inhomogeneously broadened memory has been materials development. A new concept in holeburning memory, swept carrier, is a hybrid of the time-domain and frequency-domain approaches that adapts to maximize a particular material's attributes thus greatly increasing the number of potential practical materials. Even so, the requirement of a large inhomogeneous to homogeneous bandwidth requires cryogenic temperatures (typically less than 20 K). Practical systems will only be built once the temperature limitations are overcome or cryogenic operations become more widespread and usable. This trend may be on the immediate horizon [39, 40].

1.4.8 Molecular Storage

The concept of molecular computing and storage is really in its infancy and as such will not be discussed much here other than to mention it is currently being looked at by researchers. The concept relies on the observation that molecular reactions occur very fast and in parallel. The field might be considered to have originated with a talk by R. Feynman on the possibility of building "submicroscopic" computers, but it gained renewed interest in 1995 when L. Adelman built a "computer" in a test tube of DNA. In this "computer" the chemical reactions that a particular synthe-sized DNA sequence underwent produced a molecule that when decoded, answered a specific well-posed problem. Storage is viewed as the simplest task for a DNA computer. Data would be encoded into DNA sequences and the DNA could be stored. To retrieve data a DNA strand designed so that its sequence attaches to the key word wherever it appears in the stored DNA would need to be synthesized. This type of memory would be able to hold thousands of terabytes in a test tube, (1 bit per cubic nanometer or 10^{21} bits/cm^3 according to Adelman), but its access time would be very slow (minutes to hours). Over the time frames of interest to this study, this technology is not expected to be sufficiently developed to be practice.

Using the projected improvement numbers for the competing technologies have plotted the projected performances in 2005. The solid state memory will probably still be the fastest, and although its densities will have improved substantially, they will not approach the total capacities available on disks. Magnetic and optical disk performances will be quite similar. Magnetic tape stays the same. The new entries for optical memories, both volume bit-wise or holographic and coherent transient types fill a new niche—that for faster access, with much higher capacities.

Certain ideas and techniques are being developed outside the field of optical/magnetic/electronic recording, but the storage community could benefit from these developments once become sufficiently familiar with the new concepts and methodologies. Developments in the areas of nano- and biophotonics, fluorescence microscopy, quantum-dots, optical tweezers, micro- and nanofluidics, femtosecond lasers, etc., have the potential to influence future generations of data storage systems.

1.4.9 All-Optical Magnetic Recording

Magnetization reversal in thin films of GdFeCo has been induced by ultrashort, circularly polarized laser pulses ($\tau \sim 40$ fs, $\lambda = 800$ nm, $f = 1$ kHz). No external magnetic field is required for switching, and the stable final state of the

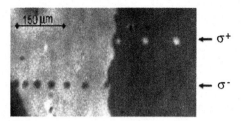

Fig. 1.22 Effect of single 40 fs circularly polarized laser pulses on the magnetic domains of $Gd22Fe_{74.6}Co_{3.4}$. The 20 μm domains were obtained by sweeping at 5 cm/s a *right* (σ^+) or *left* (σ^-) circularly polarized beam ($\sim 29p$ J/μm^2) over the surface

magnetization is determined by the helicity of the laser pulse, as shown in Fig. 1.22. This finding reveals an ultrafast and efficient pathway for writing magnetic bits at record-breaking speeds, paving the way for a new generation of ultrafast magnetic recording devices [41].

1.4.10 Electronic Quantum Holography

The electronic quantum holography is the densest type of memory storage experimentally so far. By superimposing images of different wavelengths into the same hologram, a Stanford research team was able to achieve a bit density of 35 bit/electron, i.e., approximately 3 EB/in^2. This was demonstrated using electron microscopes and a copper medium as reported in the Stanford Report in 2009. The initials for Stanford University are written in electron waves on a piece of copper and projected into a tiny hologram. On the two-dimensional surface of the copper, electrons zip around, behaving as both particles and waves, bouncing off the carbon monoxide molecules the way ripples in a shallow pond might interact with stones placed in the water. The ever-moving waves interact with the molecules and with each other to form standing "interference patterns" that vary with the placement of the molecules. In a traditional hologram, laser light is shined on a 2-dimensional image and a ghostly 3D object appears. In the new holography, the 2-dimensional "molecular holograms" are illuminated not by laser light but by the electrons that are already in the copper in great abundance. The resulting "electronic object" can be read with the scanning tunneling microscope. Several images can be stored in the same hologram, each created at a different electron wavelength. The researchers read them separately, like stacked pages of a book. The experience is roughly analogous to an optical hologram that shows one object when illuminated with red light and a different object in green light [42].

1.4.11 Scanning Electron Beam Record

An ultra-high density record bits as amorphous region using scanned electron beam which is reported by Gibson and the concept is illustrated in Fig. 1.23a, an electron beam is scanned over the surface of the diode and generates electron–hole pairs. Writing process to form an amorphousbit is done by using a high power electron optical beam to heat and quench a small region. Retrieving data is carried out using a readout beam to scan over the surface at a low-power density that will not cause accidental erasure and generates electron–hole pairs within the recording material. As reported which is shown in Fig. 1.23b, it is possible to achieve a written mark size as small as 200 nm in diameter using this method. However, the disadvantage of using this method is the power of laser beam used for writing is extremely high, and the number of recrystallization cycles that is able to perform is less than 1000 times. Another disadvantage of this design is that the amorphous bits are closely spaced that will diminish the signal from neighboring crystalline regions, thus significantly reduce signal-to-noise ratio for sub-150 nm bits [43, 44].

1.4.12 3D Integration Solid State Memory

Researchers of California Institute of Technology announced that they have constructed a memory circuit from molecules and nanometer-sized wires that is as dense as what manufacturers expect to be building in 2020. The circuit, which stores 0 s and 1 s by switching clusters of molecules between two states, contains 1.6×10^4 bits jammed together at a density of 10^{11} bits per square centimeter. Conventional microchips are at least 10 times less dense. The prototype is not yet as stable or reliable as commercial computer memory, and building it would require manufacturers to learn to harness materials other than silicon, the workhorse of computing technology. But the scale of the device dwarfs any electronic circuit previously constructed using nanotechnology.

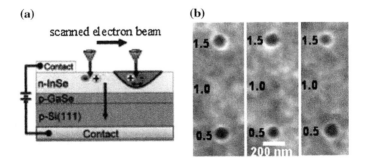

Fig. 1.23 **a** Schematic drawing of phase-change diode storage medium and **b** recording marks

The group of California Institute of Technology built the device that major goal here was never to just make a memory circuit to develop a manufacturing technique that could work at molecular dimensions as shown in Fig. 1.25a. The device has pushed far beyond previous limits of integration density and bit numbers realized previously in the field of molecular electronics. Researchers are exploring nanosize electronics systems because silicon circuits cannot be packed with wires at increasing densities—yielding higher numbered Pentium processors—forever. Eventually, electrons will start seeping between wires and lithography techniques for stamping out silicon circuits may reach their physical limit. A 3-D Skybridge fabric representation IC memory is shown in Fig. 1.24b that reveal a 30–60× density, 3.5× performance/watt benefits, and 10X reduction in interconnect lengths scaled 16-nm CMOS. Fabric-level heat extraction features are shown to successfully manage IC thermal profiles in 3-D. Skybridge can provide continuous scaling of integrated circuits beyond CMOS. This 3-D circuit mapping is made compatible with heat extraction and manufacturing requirements. The figure shows layout of Skybridge fabric incorporating all fabric components. Logic and signal nanowires are separated, and are interleaved with each other. Logic nanowires contain transistor stacks, and have power rail contacts at top, middle and bottom. Signal nanowires carry signals themselves and also facilitate routing through coaxial routing structures and bridges. Coaxial routing structures have dedicated GND signal layer for noise shielding. Heat extraction features ensure thermal management. Heat extraction junctions are placed on selective places on logic nanowires. This idea is very important consult for 3D optical solid state memory [45, 46].

The Caltech group combined two approaches: molecular electronics (transistors made of molecules) and nanowire crossbars, which are perpendicular junctions of ultrathin wires. To make their device, the team laid down a tightly packed series of 400 parallel silicon wires (separated by just 30 nm) and coated them with a layer of barbell-shaped rotaxane molecules. They created a grid of wires by covering the molecule layer with 400 more platinum wires, resulting in groups of molecules sandwiched between each node formed by the crisscrossed wires.

(a)   **(b)**

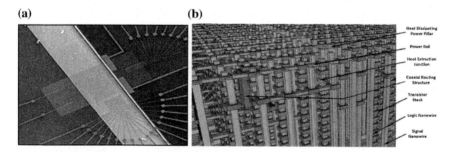

Fig. 1.24 a Ultradense integrated memory circuit with nanowires **b** 3-D Skybridge fabric representation IC memory

To switch between 0 and 1 the researchers applied a voltage across a group of molecules at a node, which toggled the molecules between two states. The rotaxane molecules each contain a ring around the "handle" of the barbell. A voltage applied across the molecule caused the ring to slide up or down, changing the electrical conductivity of the molecule. The wires were so crowded that the team could not build conventional electrodes capable of electrifying only two wires at a time (those that define a node), that instead they switched the junctions on and off in groups of nine. The junctions routinely broke down after being switched more than about 10 times. The molecules spontaneously flipped back to their previous state after nearly an hour, which is another limitation for a memory device. Commercial flash memory is stable for up to years. The molecules were also slow to switch between states that although this time can probably be improved, the speed of such a memory circuit would not come from switching one junction at a time. Instead it would result from switching many junctions at once.

1.4.13 Spintronics Data Storage

Spin is a fundamental quantum attribute of electrons and other subatomic particles. It is often regarded as a bizarre form of nanoworld angular momentum, and it underlies permanent magnetism. It can assume one of two states relative to a magnetic field, typically referred to as up and down, and can use these two states to represent the two values of binary logic to store a bit, as in Fig. 1.25a. The development of spin-based electronics, or spintronics, promises to open up remarkable possibilities. In principle, manipulating spin is faster and requires far less energy than pushing charge around, and it can take place at smaller scales. The key technology in the field is transfers magnetic information from ferromagnet into silicon through oxide layer to be spin transistor as in Fig. 1.25b. Chips built out of spin transistors would be faster and more powerful than traditional ones and may feature such new and remarkable properties as the ability to change their logic functions. The transconductance of a flash memory transistor is presented. The devices were programmed and erased with ± 10 V 100 ms pulses, whereby a large memory window = 1 V is observed. Memory (charge) retention characteristics at

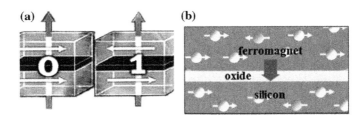

Fig. 1.25 The spintronics data storage device: **a** two states of spin can represent the two values of binary logic, **b** transfers magnetic information from ferromagnet into silicon through oxide layer

room temperature are stable and wide memory window, it is seen to be maintained. Control devices that are fabricated without the nanocrystals showed a much smaller 0.1 V hysteresis, implying that charge storage is indeed in the nanocrystals. The memory window movement and closure is most likely caused by gradual oxide degradation from generation and filling of traps with the tunneling program/erase stress. Optimization of control oxide quality and the pulsing scheme of endurance testing can possibly yield superior endurance characteristics. The chips exploit spin in a more modest way are already available as random access memory or MRAM of spintronic memory. Some companies and university research groups—are investigating MRAM technology now. Today's computers often use four kinds of storage. Dynamic random access memory, or DRAM, has high density but needs to be constantly refreshed and consumes lots of power. Static random access memory, or SRAM, used in caches, is fast to read and write but takes up considerable space on a chip. Flash, unlike SRAM and DRAM, is nonvolatile but is quite slow to write. And then there are hard disk drives, these have high density but rely on moving parts, which impose size and speed limitations. MRAM is attractive because it could, in principle, replace all other kinds of memory.

Beyond the investigation of ferromagnetic surfaces, thin films, and epitaxial nanostructures with unforeseen precision, it also allows the achievement of a long standing real space of atomic spins in antiferromagnetic surfaces. The phenomena in surface magnetism in most cases could not be imaged directly. After starting with a brief introduction to basics of the contrast mechanism, recent major achievements will be presented, like the direct observation of the atomic spin structure of domain walls in antiferromagnets and the visualization of thermally driven switching events in superparamagnetic particles consisting of a few hundred atoms only. Recently observed complex spin structures containing 15 or more atoms will be presented. But the transfers magnetic information into silicon through oxide layer at room temperature is the vital step forward for future spintronics storage [47].

1.4.14   Two-Photon 3D Optical Memory

Two-photon optical memory systems represent another important optical memory technology. In the system the memory material consists of a photochromic dye embedded in a polymer that changes its molecular state upon illumination with the appropriate light. Writing to the material requires the simultaneous excitation by two photons from two separate light beams. Reading is accomplished by either illumination with one or two photons which causes the written molecule to fluoresce. The requirement of two simultaneous photons allows addressing individual volume elements in the medium by suitably directed optical beams. One beam can be configured to represent a bit, a vector or an entire plane of data such as impressed. The second beam addresses the recording area by the location of its intersection with the data beam. The beams can intersect orthogonally or can counter propagate. The address can be selected by appropriate timing of the intersection. In this case, sufficiently short pulses are required to delineate the plane of memory. High repetition rates allow fast recording times, but there is a tradeoff between high pulse powers and repetition rates. These three (pulse power, repetition rate and recording times) may need to be traded off in the ultimate system. In addition, to the previously mentioned requirements of good wavefront quality, pulse-to-pulse uniformity and lateral mode intensity uniformity are necessary. Due to the volume nature of two-photon memory materials, the theoretical capacities are projected to be very high, on the order of TB/cm^3. The access times are only limited by the technology required to read out the detector array since reading and writing times are so fast. For counter propagating beams, the access time could be as fast as a microsecond, the time required for an optical pulse delay device. In the orthogonal architecture, the access time is determined by the device used to focus on the desired image plane. For a dynamic focusing lens, this could be around a millisecond but faster accessing technologies would speed this up. There is a tradeoff between data transfer rate and capacity, however. The fastest transfer rates rely on the parallelism that arises with reading full vectors or images at a time. However, primarily due to the limitations of diffraction, as one goes from single pixel images to large two dimensional arrays, the volume density decreases. This is illustrated in 6.1 Two-photon absorption multi-layer storage of Chap. 6 in this book.

1.4.15   Macromolecular Data Storage

The typical macromolecular data storage is the massively parallel writing of data into a DNA backbone. As shown in Fig. 1.26, using the four natural bases of DNA (identified by the letters G, C, A, T) one can create, for example, 1024 different segments using 5-base sequences. Each data bit (0 or 1) is associated with a short segment of DNA that complements the address of a specific location on the template. The "1" bits have an attachment (e.g., a protein), while the "0" bits have no attachments at all. A 5120-base-long sequence then forms a standard 1 kilo-bit template, where the location of each data bit (0 or 1) in a 1 Kb block will have a unique address. The data bits are short DNA segments that complement the address

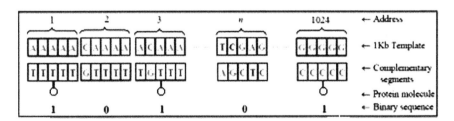

Fig. 1.26 Massively parallel writing of binary data onto a single-stranded DNA sample. The G, C, A and T represent the four natural bases of DNA. The 1 Kb sample is a string of 1024 distinct 5-base sequences

of each location in the template (C and A complement G and T, respectively). For instance, if a given location's designated address is TCGAG, the data bit associated with that location will be AGCTC. The "1" bits have an attachment (e.g., a protein molecule), while the "0" bits have no attachments at all.

To create a specific sequence of 0 s and 1 s, the short segments having protein attachments must be released into a solution that contains a single copy of the template. The released segments automatically find their complements on the template and get attached at the intended locations. It is this automatic binding of the data bits onto the template that forms the basis for massively parallel writing in the proposed scheme of macromolecular data storage. An enzyme (DNA polymerase) then fills the gaps that are left open between the "1" segments by "writing" onto the template the complements of the remaining segments, i.e., 0 s of the binary sequence. In the final step, the double-stranded DNA molecule with its complete sequence of 1 s and 0 s, with and without attached proteins is transferred to an assigned location within the macromolecular storage system. That increasing the size of a data block from 1 Kb to 2 MB, would require the assignment of only 12 DNA bases to each segment corresponding to a single bit, because $4^{12} = 16,777,216$ bits = 2 MB. Creating a practical macromolecular data storage system is a challenge that will require advances in microfluidics, nanoscale integration, biochemistry on a chip, and advanced opto-electronic methods of single-molecule detection and manipulation.

1.4.16   3D Micron Ships Phase-Change Memory

Phase change random access memory (PRAM) is one of the strongest candidates for next-generation nonvolatile memory for flexible and wearable electronics, its construction of section is shown in Fig. 1.27a. In order to be used as a core memory

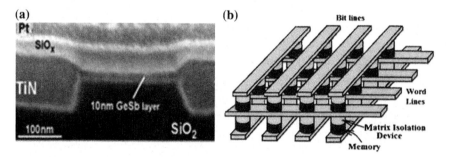

Fig. 1.27 a The cross-sectional micrograph of the phase-change random access memory bridge, **b** stacked array of 3D micron ships phase-change memory

for flexible devices, the most important issue is reducing high operating current. The effective solution is to decrease cell size in submicron region as in commercialized conventional PRAM. However, the scaling to nanodimension on flexible substrates is extremely difficult due to soft nature and photolithographic limits on plastics, thus practical flexible PRAM has not been realized yet. Recently, the flexible PRAM was enabled by self-assembled block copolymer (BCP) silica nanostructures with an ultralow current operation on plastic substrates. BCP is the mixture of two different polymer materials, which can easily create self-ordered arrays of sub-20 nm features through simple spin-coating and plasma treatments. BCP silica nanostructures successfully lowered the contact area by localizing the volume change of phase-change materials and thus resulted in significant power reduction. Furthermore, the ultrathin silicon-based diodes were integrated with phase-change memories (PCM) to suppress the intercell interference, which demonstrated random access capability for flexible and wearable electronics. Another way to achieve ultralow-powered PRAM is to utilize self-structured conductive filaments (CF) instead of the resistor-type conventional heater. The self-structured CF nanoheater originated from unipolar memristor can generate strong heat toward phase-change materials due to high current density through the nanofilament. This ground-breaking methodology shows that sub-10 nm filament heater, without using expensive and noncompatible nanolithography, achieved nanoscale switching volume of phase change materials, resulted in the PCM writing current of below 20 µA, the lowest value among top-down PCM devices. This self-structured conductive filament nanoheater for chalcogenide phase transition. In addition, due to self-structured low-power technology compatible to plastics, recently succeeded in fabricating a flexible PRAM on wearable substrates. The demonstration of low power PRAM on plastics is one of the most important issues for next-generation wearable and flexible nonvolatile memory. The methodology represents the strong potential for commercializing flexible PRAM [48–50].

In addition, PRAM is a promising candidate to achieve scalable in the upcoming 3D-stacking technology as in Fig. 1.27b. The higher temperature of 3D chips is beneficial to PRAM power savings due to its unique, heat-driven programming mechanisms, that the Through Silicon Vias (TSVs) ubiquitously used in 3D implementations contribute further PRAM power savings due to their substantially lower resistance to the high PRAM programming current. To effectively integrate PRAM into a conventional memory hierarchy, that the architecture and OS support to address its write latency and reliability disadvantages. A hybrid PRAM/DRAM memory architecture and exploit an OS-level paging scheme can improve PRAM write performance and lifetime. Moreover, the error-correcting capability of strong ECC codes expand PRAM lifespan and use wear-out aware OS page allocation to minimize ECC performance overhead. Experimental results show that the design reduces the overall power consumption of the memory system by 54 % with 6 % performance degradation, consequently alleviating the thermal constraint of 3D chips by up to 4.25 °C and achieving a speedup of up to 1.1×. The lifetime can be improved by a factor of 114× using endurance optimization schemes.

1.4.17   Compositive Application of the Different Principles

The integration all of the different principles and technologies enable to improve the density and capacity of optical storage and make a road map in the future is shown in Table 1.7. The basic idea in seeking outside the traditional engineering optical technology to improve the resolution of the system, give full play to the advantages of optical storage that can be expanded to full electromagnetic spectrum, multidimensional space and more interaction mechanism of light with medium. Meantime optical storage have to combine with other up to date technologies as nanotech, semiconductor memory and magnetic memory tech to explore new development roadmap. The optical storage from current plane two-dimensional extended to three-dimensional, even multidimensional storage. Various nanotechnology will be application in optical storage: as near-field optics, super-RENS, nonlinear optics, UV, EUV, DUV and electron beam technology can reduce the effective size of the storage symbol incessantly. New optical storage materials replace the traditional dye and phase change medium, that the recording mechanism will be changed from present photothermal effect convert to photon quantum effects, great reducing the recording power and increasing data rate. A variety principles and methods of modulation and coding, especially multidimensional coding with various optical parameters to constitutes multidimensional coding mode with multilevel matrix will bring forth excellence advantage of optical storage [51–54].

Table 1.7 Comparison between traditional storage principles/methods and new principles/technology roadmap for improving optical storage density and capacity

Traditional methods and technology	Further principles and the implementation
Two dimensional record	Three-dimensional space and multidimensional storage as frequency dimension for example
Visible light sources (405–650 nm)	UV, EUV, DUV, FEL (free-electron lasers)
Traditional far field optics	Near-field optics and superresolution technology
Single parameter changes	Variety of parameter changes: optical interference, coherence, polarized, reflectivity, refractive index, absorptivity or magnetic declination, while taking advantage of chroma, gray scale, photochemistry and quantum efficient, etc.
Thermal effects to write	Effects of photon quantum and electron transfer
Single disc form	Multidisc or solid state storage form
Single channel to write/read	Multichannel parallel reading and writing
Linear encoder	Serial multidimensional and multi-level matrix code
Traditional precision manufacturing (sub-micro)	Nanotech and MEOMs process(a few nanometer)

1.5 Big Data Storage and Application

The term big data refers to the massive amounts of digital information companies and governments collect all society and surroundings that create 2.5×10^{18} bytes of data every day. So much that 90 % of the data in the world today has been created in the last two years alone. Security and privacy issues are magnified by velocity, volume and variety of big data, such as large-scale cloud infrastructures, diversity of data sources and formats, streaming nature of data acquisition and high volume intercloud migration. The use of large-scale cloud infrastructures, with a diversity of software platforms, spread across large networks of computers, also increases the attack surface of the entire system. Traditional security mechanisms, which are tailored to securing small-scale static (as opposed to streaming) data, are inadequate. For example, analytics for anomaly detection would generate too many outliers. Similarly, it is not clear how to retrofit provenance in existing cloud infrastructures. Streaming data demands ultrafast response times from security and privacy solutions. Big data brought following high-priority opportunities and challenges:

1. Web Databases/Business Intelligence (BI): Businesses are turning to cloud services for access to data and analytics that help them make smarter business decisions. Because users do not want to wait for anything, solid state storage technology can improve their experience by accelerating web application response times. It allows BI SaaS companies to deliver complex, real-time analytics and data visualizations without slowing page load time. Best of all, it allows Software as a Service (SaaS) providers to scale to meet demand quickly and cost-effectively without sacrificing performance.
2. Data Mining and Analytics: Like SaaS providers, online data mining services need to be responsive. Solid state storage technology helps increase application performance so providers can process more jobs and more complex queries in less time. I/O intensive log files and frequently accessed tables can bog down an analytics engine. But PCIe-based flash storage puts that data closer to the processor, ensuring a fast response time. It also allows for complex queries at massive scale, while remaining cost-effective and easy to install.
3. Social Media: Social media creates a growing data challenge because of the vast amount of unstructured and structured data sets. Social media sites have to provide access to videos, photos, audio files, status updates, tweets and other online transactions that people want, quickly and accurately. Solid state storage technology provides the low latency, scalable, high-performance storage platform that allows social media companies to deliver the experience their users demand. At the same time, it reduces infrastructure capital expenditures (Capex) —including hardware, data center footprint, and power and cooling—while shrinking operational expenses (Opex) year after year.
4. Server Virtualization: Organizations have turned to virtualization to maximize server utilization, reduce hardware expenses, and improve the responsiveness and resiliency of application delivery. Solid state storage technology can improve the performance of virtualized environments by combining hardware

with software caching to reduce latency. This allows more data to quickly get to the CPU, increasing overall throughput and utilization. The result? IT can run more virtual machines (VMs), while improving the performance across all VMs.

5. The Cloud: The next logical step for virtualization is the cloud, which extends the virtual environment online to provide organizations and consumers with self-service access to managed solutions. But in order to do so effectively, IT needs to ensure persistent performance, high availability, high capacity and low latency. Solid state storage technology can help by delivering predictable, sustained response times, even on data-intensive workloads. And because SSDs can work with existing hot-swappable storage system designs, no "forklift" upgrades are required. With solid state storage technology, IT can deliver higher performance with less SAN infrastructure—reducing power consumption and cooling costs, as well as Capex and Opex.

6. Software as a Service (SaaS): For many businesses, the ultimate expression of the "cloudification" of service delivery is to provide SaaS. Delivering this model effectively can be difficult, as organizations must find ways to deploy hardware that ensures an optimal user experience without sacrificing profitability. Not easy if their system is bogged down by I/O constraints driven by simultaneous queries by multiple users. Solid state storage technology can help by lowering latency so that requests can be quickly fulfilled, providing an improved customer experience. It also allows SaaS providers to support a greater number of databases while allowing them to run more VMs on each server.

1.5.1   Cloud Storing

Cloud infrastructures can be classified private cloud, community cloud, public cloud and hybrid cloud. Private cloud infrastructure is operated solely for an organization or personal. It may be managed by the organization or a third party and may exist on premise or off premise. Community cloud infrastructure is shared by several organizations and supports a specific community that has shared concerns (e.g., mission, security requirements, policy, and compliance considerations). It may be managed by the organizations or a third party and may exist on premise or off premise also. Public cloud infrastructure is made available to the general public or a large industry group and is owned by an organization selling cloud services. Hybrid cloud infrastructure is a composition of two or more clouds (private, community, or public) that remain unique entities but are bound together by standardized or proprietary technology that enables data and application portability (e.g., cloud bursting for load balancing between clouds). Below take the private cloud for example of a typical.

Private Cloud provides business intelligence on data from several business-critical data sources. To keep up with customer growth and technology infrastructure, that wanted to consolidate the rapidly-growing volumes of data for reporting, trending, and analytical purposes. This Cloud Software to power a cloud-based big data solution while reducing costs and improving operational efficiency. Data points included customer account data, usage and billing information, with business intelligence toolset interoperability from informatics. From an operational level, the overall data became unmanageable once important information like monitoring, response, and support metrics came in from dedicated, virtual, and cloud devices. Daily reporting became a time consuming and resource-intensive process, only occurring nightly and with a 24-h data point lag time. Commercial database licensing and hardware costs were rising in a disproportionate manner as the EBI team worked with database administrators to quickly increase capacity during peak hours. Finally, the legacy set up did not handle unstructured data very well, and the team wanted to be able to apply different best-of-breed technologies alone or in combination depending upon the type and size of data they wanted to store and analyze. To continue serving the business efficiently and effectively, EBI put together requirements for a new solution. Named the Analytic Compute Grid (ACG), the solution would act as the backbone and needed to be able to house an ever-growing set of data collected in different formats, structured and unstructured, from multiple business units, rapidly dynamically scale resources up and down to efficiently meet business demands, add new resources on the fly without waiting for new hardware provisioning during peak hours, big data technologies for storing, managing, analyzing and distributing data on one technology platform, embrace open cloud and open source technologies. These requirements led to design and build a stack based on open source technologies from infrastructure to big data software to allow for rapid growth and scale. The underlying infrastructure platform was Rackspace Private Cloud. The solution was dubbed as Analytic Compute Grid. The big data management software platform built on Rackspace Private Cloud software. As a key benefit, it provides a consolidated and flexible solution to store, analyze, distribute and present the data based on the type of the data (structured or unstructured), operation (storing or analyzing the data) and the consumer's skill set (data scientist accessing via APIs or a marketing analyst using BI tools to run reports.) By creating a single holistic platform utilizing open source technologies, the enterprise business intelligence storing analytic compute grid can handle the storage, analysis and distribution of data at scale in a timely manner. The big data tools available today helped solve the problem but required new ways of thinking about the underlying infrastructure, processes and data structures to make it a reality. Built using Rackspace Private Cloud powered tools were resulted in improvement in data processing speeds and a significant reduction in overall capex and opex. Multiple business units that can now make near real-time decisions that can directly benefit Private Cloud, that allows users to run a Rackspace Cloud in data center. The fastest way for enterprise to leverage open cloud technologies at scale is to choose a knowledgeable cloud provider that understands and uses it every day—and is standing ready to help match business needs with the appropriate open cloud solution [55, 56].

Cloud storing is a most important part for cloud computing. Its definition, use cases, underlying technologies, issues, risks, and benefits will be refined and better understood with a spirited debate by the public and private sectors. This definition, its attributes, characteristics, and underlying rationale will evolve over time that the cloud computing model or providing cloud services. Cloud computing is a model for enabling ubiquitous, convenient, on-demand network access to a shared pool of configurable computing resources (e.g., networks, servers, storage, applications, and services) that can be rapidly provisioned and released with minimal management effort or service provider interaction. This cloud model promotes availability and is composed of five essential characteristics, three service models and four deployment models.

At the same time, network-based cloud computing is rapidly expanding as an alternative to conventional office-based computing. As cloud computing becomes more widespread, the energy consumption of the network and computing resources that underpin the cloud will grow. This is happening at a time when there is increasing attention being paid to the need to manage energy consumption across the entire information and communications technology (ICT) sector. While data center energy use has received much attention recently, there has been less attention paid to the energy consumption of the transmission and switching networks that are keys to connecting users to the cloud. The analysis of energy consumption in cloud computing that analysis considers both public and private clouds, and includes energy consumption in switching and transmission as well as data processing and data storage. That energy consumption in transport and switching can be a significant percentage of total energy consumption in cloud computing. Cloud computing can enable more energy-efficient use of computing power, especially when the computing tasks are of low intensity or infrequent. However, under some circumstances cloud computing can consume more energy than conventional computing where each user performs all computing on own personal computer (PC). A consumer can unilaterally provision computing capabilities, such as server time and network storage, as needed automatically without requiring human interaction with each service's provider. Capabilities are available over the network and accessed through standard mechanisms that promote use by heterogeneous thin or thick client platforms (e.g., mobile phones, laptops, and PDAs). The provider's computing resources are pooled to serve multiple consumers using a multi-tenant model, with different physical and virtual resources dynamically assigned and reassigned according to consumer demand. There is a sense of location independence in that the customer generally has no control or knowledge over the exact location of the provided resources but may be able to specify location at a higher level of abstraction (e.g., country, state, or datacenter). Examples of resources include storage, processing, memory, network bandwidth, and virtual machines. Capabilities can be rapidly and elastically provisioned, in some cases automatically, to quickly scale out, and rapidly released to quickly scale in. To the consumer, the capabilities available for provisioning often appear to be unlimited and can be purchased in any quantity at any time.

Cloud systems automatically control and optimize resource use by leveraging a metering capability1 at some level of abstraction appropriate to the type of service (e.g., storage, processing, bandwidth, and active user accounts). Resource usage can be monitored, controlled, and reported, providing transparency for both the provider and consumer of the utilized service. Cloud Software as a Service (SaaS). The capability provided to the consumer is to use the provider's applications running on a cloud infrastructure. The applications are accessible from various client devices through a thin client interface such as a web browser (e.g., web-based email). The consumer does not manage or control the underlying cloud infrastructure including network, servers, operating systems, storage, or even individual application capabilities, with the possible exception of limited user-specific application configuration settings. Cloud Platform as a Service (SaaS). The capability provided to the consumer is to deploy onto the cloud infrastructure consumer-created or acquired applications created using programming languages and tools supported by the provider. The consumer does not manage or control the underlying cloud infrastructure including network, servers, operating systems, or storage, but has control over the deployed applications and possibly application hosting environment configurations.

Cloud systems achieve economies of scale by serving multiple customers from a shared pool of resources; each customer (which could be a company or enterprise) is a tenant of the cloud infrastructure. Physical resource pooling enables load balancing, homogeneity for management and much higher utilization rates. Sharing of physical resources is important for compute clouds as well as storage-centric clouds. In storage-centric clouds, the pooled resources include the physical media and the servers controlling the media. In a cloud where all physical resources are pooled, any given device may have data from multiple, unrelated tenants. A major concern expressed by many businesses over moving to a public cloud delivery model is security. This concern stems from the commingling of the data of different tenants on shared physical resources. It is common for cloud storage systems to provide application-level security, in which components that authenticate and process user requests run with sufficient privileges to access any tenant's data; the code of each component is responsible for authorizing requests based on the requester's credentials. This architecture is used by OpenStack Swift and other publicly available cloud storage systems. Application-level security only provides a single level of defense; a vulnerability that allows bypassing the security check, such as a confused deputy attack, can compromise all data stored in the cloud. This is very weak isolation. If each tenant had its own segregated physical resources, it would be much less likely that a single vulnerability could jeopardize all tenants' data. To provide security similar to physical isolation while allowing complete pooling of resources, propose Secure Logical Isolation for Multi-tenancy (SLIM). SLIM adds an orthogonal tenant isolation mechanism over existing application-level security by leveraging the Linux process isolation mechanisms that have been thoroughly tested for over 20 years by the Linux community and enhanced with mechanisms such as SELinux. SLIM therefore enables resource pooling while decreasing the likelihood that a single vulnerability could jeopardize all tenants data. In particular, SLIM provides this additional isolation across tenants by following the principle of least privilege: each system component runs with the least set of privileges required to complete its task. Moreover, such privileges are designed to be tenant-specific: for example, that define separate privilege classes to access authentication material of tenant A and tenant B. As a consequence, whenever possible, SLIM contains breaches within a tenant by leveraging process isolation. For the remaining components, which need to be trusted to operate on data of multiple tenants, we minimize their attack surface. Employing these principles in an efficient manner is difficult. First, each request must be correctly associated with a tenant and properly split into appropriate sub-pieces each of which is processed under the principle of least privilege. For instance, we use different privileges for authenticating a user and for accessing a disk. Second, need to address security as we move between components in the cloud implementation, either between different processes implementing different parts of the function on a single server or communicating with a different server, e.g., for data replication. In both cases that need to correctly track the tenant identity and use the appropriate privilege. In real world systems may need to use existing resources like a distributed cache or a shared data store but cannot implement all of principles. SLIM addresses these issues using the following mechanisms:

1. A security gateway ensures a request is handled with the right tenant-specific privilege.
2. A proxy/guard mechanism prevents escalation of privileges as move between components.
3. A gatekeeper provides a secure wrapper for existing resources.

These SLIM components are relevant for almost any cloud-based object store system, and can be easily integrated to provide end-to-end isolation between tenants. For implementation of SLIM for OpenStack Swift, first complete security model and set of principles for safe logical isolation is between tenant resources in a cloud storage system. Define a set of mechanisms for implementing secure, logically isolated cloud storage systems. Implement SLIM for swift and present initial performance results, determining that process recycling is the most critical factor influencing performance, various approaches to security isolation for multi-tenancy and provides background on swift.

1.5.2 EB-Scale Distributed Storage

To provide high storage capacity, large-scale distributed storage systems have been widely deployed in enterprises. In such systems, data is striped across multiple nodes (or servers) that are interconnected over a networked environment. Ensuring data availability in distributed storage systems is critical, since node failures are prevalent. Data availability can be achieved via erasure codes, which encode original data and stripe encoded data across multiple nodes. Erasure codes can tolerate multiple failures and allow the original data to remain accessible by decoding the encoded data stored in other surviving nodes. Compared to replication, erasure codes have less storage overhead at the same fault tolerance.

In addition to tolerating failures, another crucial availability requirement is to recover any lost or unavailable data of failed nodes. To achieve high-performance recovery, one approach is to minimize the recovery bandwidth (i.e., the amount of data transfer over a network during recovery) based on regenerating codes, in which each surviving node encodes its stored data and sends encoded data for recovery. In the scenario where network capacity is limited, minimizing the recovery bandwidth can improve the overall recovery performance. In this work, they explore the feasibility of deploying regenerating codes in practical distributed storage systems. However, most existing recovery approaches are designed to optimize single failure recovery. Although single failures are common, node failures are often correlated and co-occurring in practice, as reported in both clustered storage and wide-area storage. In addition, concurrent recovery is beneficial to delaying immediate recovery. That can perform recovery only when the number of failures exceeds a tolerable limit. This avoids unnecessary recovery should a failure be transient and the data be available shortly (e.g., after rebooting a failed node). Given the importance of concurrent recovery, thus pose the following question: It can achieve bandwidth saving, based on regenerating codes, in recovering a general number of failures including single and concurrent failures that complete system, which supports both single and concurrent failure recovery and aims to minimize the bandwidth of recovering a general number of failures. It augments existing optimal regenerating code constructions, which are designed for single failure recovery, to also support concurrent failure recovery. A key feature is that it retains existing optimal regenerating code constructions and the underlying regenerating-coded data. It adds a new recovery scheme atop existing regenerating codes. That achieves the minimum recovery bandwidth for a majority of concurrent failure patterns, and can achieve suboptimal bandwidth saving even for the remaining concurrent failure patterns. Implement and experiment the prototype on a Hadoop Distributed File System (HDFS) tested with up to 20 storage nodes, that compared to erasure codes, that achieves recovery throughput gains with up to $3.4\times$ for single failures and up to $2.3\times$ for concurrent failures. In modern EB-scale storage systems, data is stored on a storage cluster including many servers that are directly accessed by clients via the network, while metadata is managed separately by a metadata server (MDS) cluster consisting of a number of dedicated metadata servers. By separating file data access and metadata transactions, object-based storage architecture is a prevalent system architecture for EB-scale storage systems. The dedicated metadata server cluster manages the global namespace and the directory hierarchy of file system, the mapping from files to objects, and the permissions of files and directories. The MDS cluster just allows for concurrent data transfers between large numbers of clients and storage servers instead of being responsible for the storage and retrieval of data. Meanwhile, it provides efficient metadata service performance with specific workloads, such as renaming a large directory near the root of the hierarchy and thousands of clients updating to the same directory or accessing the same file. The main problem of designing metadata server cluster is how to partition metadata efficiently among MDS cluster to provide high-performance metadata services. Metadata server cluster is involved in moving metadata to keep metadata servers storage load balancing when MDSs are added or removed dynamically, in which the overhead of metadata migration should be minimized. In order to keep good namespace locality, some MDSs are heavily overloaded in storage load, while other MDSs are lightly overloaded. A well-designed metadata server cluster should be able to achieve satisfactory storage load balancing. In addition, we have to efficiently organize and maintain very large directories, each of which may contain billions of files. Internet applications such as Facebook already have to manage hundreds of billions of photos. As there are millions of new files uploaded by users every day, the total number of files increases very rapidly and will soon be more than one trillion. Meanwhile, to provide high-performance metadata services for a large-scale storage system with hundreds of billions or trillions of files. For example, Facebook serves over one million images per second at peak, and one billion new photos per week. Compared to the overall data space, the size of metadata is relatively small, and it is typically 0.1–1 % of data space, but it is still large in EB-scale storage systems, e.g., 1–10 PB for 1 EB data. Besides, 50–80 % of all file system accesses are to metadata. Therefore, in order to achieve high performance and scalability, a careful metadata server cluster architecture must be designed and implemented to avoid potential bottlenecks caused by metadata requests. To efficiently handle the workload generated by a large number of clients, metadata should be properly partitioned so as to evenly distribute metadata traffic by leveraging the MDS cluster efficiently. At the same time, to deal with the changing workload, a scalable metadata management mechanism is necessary to provide highly efficient metadata performance for mixed workloads generated by tens of thousands of concurrent clients. The concurrent accesses from a large number of clients to large-scale distributed storage will cause request load imbalance among metadata servers and inefficient use of metadata cache. Caching is a popular technique to handle request load imbalance, and it is both orthogonal and complementary to the load balancing technique proposed in this paper. In this paper, A novel metadata server cluster architecture named Dynamic Ring Online Partitioning (DROP). It is a highly scalable and available key-value store using chain replication, and it provides a simple interface: lookup under put and get operations. In DROP, use locality preserving hashing (LPH) to improve namespace locality, thus increasing put/get success rate depending on fewer MDSs and upgrading put/get performance involving fewer lookups. Maintaining metadata locality improves availability and performance of metadata substantially, but it causes storage load imbalances of MDSs. An efficient Histogram-based Dynamic Load Balancing (HDLB) mechanism in DROP was proposed in current research. Finally, to evaluate DROP and its competing metadata management strategies by simulations from multiple perspectives including namespace locality and load distribution, and demonstrate that DROP converges to load balancing quickly with different MDS cluster sizes. The results demonstrate that DROP is more effective than traditional state-of-the-art metadata management approaches, and it is also highly scalable. The rest of the paper is organized as follows describes related work. The proposed mechanism of preserving namespace locality and the histogram-based dynamic load balancing mechanism and present performance evaluation results of DROP.

To provide high storage capacity, large-scale distributed storage systems have been widely deployed in enterprises. In such systems, data is striped across multiple nodes (or servers) that are interconnected over a networked environment. Ensuring data availability in distributed storage systems is critical, since node failures are prevalent. Data availability can be achieved via erasure codes, which encode original data and stripe encoded data across multiple nodes. Erasure codes can tolerate multiple failures and allow the original data to remain accessible by decoding the encoded data stored in other surviving nodes. Compared to replication, erasure codes have less storage overhead at the same fault tolerance. In addition to tolerating failures, another crucial availability requirement is to recover any lost or unavailable data of failed nodes. To achieve high-performance recovery, one approach is to minimize the recovery bandwidth (i.e., the amount of data transfer over a network during recovery) based on regenerating codes, in which each surviving node encodes its stored data and sends encoded data for recovery. In the scenario where network capacity is limited, minimizing the recovery bandwidth can improve the overall recovery performance. The feasibility of deploying regenerating codes in practical distributed storage systems. However, most existing recovery approaches are designed to optimize single failure recovery. Although single failures are common, node failures are often correlated and co-occurring in practice, as reported in both clustered storage and wide-area storage [57, 58].

In addition, concurrent recovery is beneficial to delaying immediate recovery. That can perform recovery only when the number of failures exceeds a tolerable limit. This avoids unnecessary recovery should a failure be transient and the data be available shortly (e.g., after rebooting a failed node).

1.5.3 Air Traffic Management

Air traffic one of is most busyness transportation in the world today. An important area with both commercial and military interest is in the field of Air Traffic Management (ATM). ATM primarily consists of three distinct activities:

1.5.3.1 Air Traffic Control

The process of Air traffic control by aircraft are safely separated in the sky as they fly and at the airports where they land and take off again. Tower control at airports is a familiar concept but aircraft are also separated as they fly en route; Europe has many large Air Traffic Control Centers (ATCC) for example which guide aircraft to and from terminal areas around airports.

1.5.3.2 Air Traffic Flow Management

The Air traffic flow management is an activity that is done before flights take place. Any aircraft using air traffic control, from a business aeroplane to an airliner, files a flight plan and sends it to a central repository. All flight plans for flight into, out of and around Europe are analyzed and computed. For safety reasons, air traffic controllers cannot handle too many flights at once so the number of flights they control at any one time is limited. Sophisticated computers used by Air Traffic Flow Management calculate exactly where an aircraft will be at any given moment and check that the controllers in that airspace can safely cope with the flight. If they cannot, the aircraft has to wait on the ground until it is safe to take off.

1.5.3.3 Aeronautical Information Services

These services are responsible for the compilation and distribution of all aero-nautical information necessary to airspace users. These include information on: safety, navigation, technical, administrative or legal matters and their updates. The information can take the form of maps showing the air routes and air traffic control centers and the areas that they are responsible for, or it can be notices, information circulars or publications. Some of these publications contain orders which must be carried out. Some are simply to give useful information—on prevailing weather, for instance: all of them are aimed at promoting the safety, regularity and efficiency of air navigation.

Every activities of ATM is need of betimes and credible information with big data. Because of storage limitations, not everything can be recorded, so they are limited to recording all the Windows commands that go to the screen. The desire is to have enough capacity to record an entire 8 h shift so the controller could report for his or her shift, plug in the recording device and remove and take it along when the shift is over. With the current terminal sizes and refresh rates (2000×2000 pixels operating at 60 Hz) this could amount to recording up to 600 GB during a shift, assuming no compression is used.

Another more ATM-related application for mass storage is also related to accident investigations. At the current time, the ground stations providing infor-mation to the cockpits are not required to be certified for flight critical components. The cockpit has been separated from the information that comes in from the ground. There has been limited display space in the cockpit as well as limited communi-cation bandwidth. However, a current interest lies in giving flight crew more information to allow them more autonomy in determining their flight path. This vision includes using phased array antennas to provide a high bandwidth data link to the ground to bring large amounts of situational awareness information to the cockpit. This could include map and terrain data, information on all other aircraft in the area and weather information. As in some current cockpits, the flight crew will not be able to monitor all the available information, but will need to have a lot of it stored and automatically analyzed to provide only the most critical information to the crew. In such a scenario, the ground station (where the information originates) and the cockpit become more tightly coupled, and some believe that the whole system (ground and cockpit) will need to undergo joint certification. In such a case, there would be a requirement for recording the information transmitted for accident analysis. Such a system could easily require TB or more of capacity over transcontinental or transoceanic trips.

The upgrades for increased functional capability listed above are currently being studied. Each of these imposes storage requirements on the avionics system. The sum could easily exceed many TB. On-board storage capacity will influence the effectiveness safety system in terms of probability of survival and probability of died. The quantification of effectiveness sensitivity to storage capacity is currently underway. They cite the need to not move much TB around the aircraft, but rather have the storage available where needed, particularly at the offensive and defensive system operator stations of air force. They are enthusiastic about the capacity, and low cost of the two-photon media and its inherent ruggedness and survivability. In particular its apparent insensitivity to electromagnetic pulses (EMP), as compared to solid state memories, is considered by them as key advantage. Large, fast data storage is a critical technology for ATM. As part of the mission planning function, while on the ground, the pilot uses a helmet mounted display and a data storage device to plan and rehearse the mission. That same data storage device is then taken on the plane to provide mission management, situation display aids, in-flight replanning and even embedded training. During the mission, the data storage device also needs to record all the signals and information coming to the pilot which is then replayed during mission debriefing. Thus the data storage device needs to be removable and rugged. It is critical in the pre-flight stage, during the flight and post flight.

1.5.4   Data Exploitation and Communications (DEC)

The Data Exploitation and Communications (DEC) ground system currently uses three separate systems (one for data collection and two for mission planning) that contain 8 TB disks. The data collection involves storing "video snapshots" about 20 GB per second. This represents a fairly minimal requirement of about 400 MB. Map data for planning consists of a few maps such as Digital Terrain Elevation Data (DTED) of about 320 MB. More map data storage would be desirable, and the capability of full-motion video, not just snapshots would be useful. This could raise the capacity requirements substantially. The main environmental concern for this deployment is dust. Used in a field station, where personnel are not paying particular attention to cleanliness, good sealing against dust is their prime problem now. The program personnel are currently thinking about architectures for future systems in the future that the project a need for up to 1000 times more storage. Among the desired functions are more map data and more image processing.

However, a big question in the whole UAV scenario remains as to how much processing is to be deployed on the vehicles themselves and how much is left at the ground stations. The answer to this question depends critically on how fast data can be moved around. Faster communication links might minimize requirements for on-board processing and put more demands on the ground station. This may also be a more cost effective solution, depending on how many UAV's are controlled by a given ground station.

In addition ground, ocean airborne and space applications were reviewed and system requirements identified. Current and future society security and military programs in the areas of surveillance, reconnaissance and intelligence have increasing need to gather, assimilate and analyze large amounts of data. These requirements stem from both ground-based applications such as battlefield situational awareness, mission logistics and intelligence data and airborne requirements such as avionics, requirements for in-flight diagnostics and on-board maintenance, as well as mission-specific requirements. Among specific types of memory intensive applications are those that require image analysis, or use sensors with very wide bandwidths and high speed data buses. It is generally recognized that currently available memory systems will not fulfill all the needs of future systems. Development must begin now for insertion into new systems over the next decade. Promising new technologies include optical multidimensional storage, photon and quantum memory. Considers commercial and military programs in the future that included both commercial electronic and industry systems will be stringent. Current and future information programs in the areas of surveillance and intelligence have increasing need to gather, assimilate and analyze large amounts of data for any city or country.

Optical big data storage is driving much research effort in the areas of materials, system architectures and optical interconnection schemes. The development of photon memory technology could develop field able systems include electronic, magnetic and optical based. For each include a discussion of their mission objectives and current state as well as projected requirements over the next decade. After reviewing all the programs surveyed, and the requirements drawn from them, present a roadmap proposing the path needed to take the technology from today's state to the required state of the next decade. Then include a list of additional programs worthy of consideration for memory requirements or as potential customers to serve as field test programs. The first task was a technology requirements assessment. This consisted of a survey of the current field in storage technology. This was accomplished mainly by standard traditional methods using published articles in technical and trade journals. A second part of task to survey promising, appropriate with needs in this area and then determine their current status and future needs. A final nontraditional source of data is the World Wide Web of the Internet. This vast collection of computer files contains a wealth of information resource that has been useful. To develop a technology roadmap for the development of mass storage systems into a field able technology, that the plan is based on the study of the technology.

1.5.5 *China Knowledge Resource Integrated Database*

China Knowledge Resource Integrated Database (CKRID) is an experimental platform for big data storage and application of OMNERC at Tsinghua University. It has
special bonding current international strategic plan that creates exceptional research,
teaching and learning environment for global citizenship. Tsinghua-Tongfang
Knowledge Network Technology Co., Ltd set up a digital content and academic
library of China Academic Journals in 1999. It has been making the transition to the
wealth of digital resources and providing access to them on and off the Tsinghua
University campus. To support its studies, the library along with local and international partners, has been preserving and creating impressive resources and tools in
digital format. Up to early 2015, CKRID digital repository has been more 70 million
Chinese information and resource from all of China and overseas, that is including
China Academic Journals Full-text Database from 1915 onwards 15.7 million
records, China Doctoral Dissertations Full-text Database from 1984 onwards 1.2
million, China Masters' Theses Full-text Database from 1984 onwards 2.8 million,
China Proceedings of Conference Full-text Database from 1953 onwards 2.6 million,
International Proceedings of Conference Full-text Database from 1981 onwards 2.3
million, China Core Newspapers Full-text Database from 2000 onwards 17.1 million,
China Yearbooks Full-text Database from 1912 onwards 23.5 million and abroad
literature 7.9 million in total. OMNERC designed and developed a hybrid (optical
disk, hard disk and tape) big data storage system was used for China knowledge
resource integrated database from 2005. The latest development of this system is
comprised digital special collections, digital learning projects and tools, digital
community outreach programs and collaborative digital research projects, as well as
corresponding software [59–61].

A file data storage requirements of a typical University is shown in Table 1.8,
including respective 19 schools, 55 departments and 292 research institutes or
centers that compilation of forms completed by researchers. The charts that follow
provide a breakdown of the various types of data storage across the different
schools and research centers. It is clear from those charts that there are some rather
significant gaps, and that the results presented here only provide an incomplete
overview of the University's research data storage requirement. IT services central
file store: history—long-term and active storage and active science and technology
file 763 TB, management 120 TB. IT services web service active data 139 TB,
network file store active 938 TB, long-term store 1060 TB, staff computer
long-term 260 TB and active 200 TB [62, 63]. The survey identified a storage need
of about 1.4 PB for active data with an annual increase of approximately 11 %, and
a long-term storage need of approximately 600 Tb with an annual increase of 17 %.
The 1.4 PB of active data storage include a requirement of 176 TB for working
copies of data produced by a High-Performance Computing Cluster in the School of
Mathematics and 350 TB of scratch space for users. It provides a breakdown of
current data storage split into the use of different options, that only a very small
proportion of the University's research data is stored on centrally provided systems.

Table 1.8 The typical file data storage requirements of a University

Category	Active data (GB)		Long-term storage (GB)		Continue education (GB)	
	Current need	Annual increase	Current need	Annual increase	Current need	Annual increase
ITS central file store	369	89	232	67	355	151
ITS web servers	2066	1220	2200	800	2770	1560
Networked file store (non-ITS)	938,721	37,100	153,000	27,600	3350	1330
Non-ITS web servers	39,510	19,600	50,104	16,500	–	–
Staff computers	208,253	28,900	164,030	732	–	–
External storage media	114,635	38,290	181,361	47,812	6700	1231
External data centre	41,200	11,310	12,500	4700	8912	4520
External cloud services	54,593	11,270	–	–	1850	921
Total	1,399,347	147,779	563,427	98,211	23,937	9713

An approximate 20 % of research data is currently stored on staff computers, and a further 15 % is on external storage media. Only 9 % of research data is stored in the public cloud. A few used an external data centre to look after their data that some of their research data is stored on a privately owned computer at home.

In addition combination of multimedia and textbook lessons maintain student interest and increase need of the long-term memory of the lessons. So the resource for continue education could grow and apply on internet rapidly in the future.

A review of the current state of storage technologies presents the well known gaps in storage types. These gaps are increasing as technologies develop at differing rates. For a large number of military and commercial systems, as well as standard consumer products, the computing limitation continues to be the memory technology. New promising types of memory under development have the potential to fill these gaps. Among the most promising are volume optical systems, including both holographic and two-photon optical. Later on, coherent optical techniques may become useful. Among the many systems with a need for more and faster memory over the next decade are both ground-based and airborne. The CBT applications have some of the largest requirements, yet at the same time have some of the least restrictive operating environments. These are applications where initial fielded systems could be tried. They could have improved performance from the nearest to delivery product.

This could help in the material development. Ultimately, the RWE devices are exceedingly important for future military missions. Once some packaging issues have been addressed, these systems can and should be flight tested on military jets. Both the ATIMS and B-IB program personnel have expressed interest in being involved in these sorts of tests.

Bandwidth considerations remain a concern, but further architecture studies should help in this area. At the same time, further developments in auxiliary components such as writing lasers and readout devices are required. These components are being developed for other markets such as medical technologies and consumer electronics. As these markets drive the development, performance should improve and prices should continue to fall. The performance is important for military and commercial applications, but cost remains the overall driver. SAMCOM's systems as discussed in the main text have much more challenging requirements.

References

1. D.E. Pansatiankul, A.A. Sawchuk, Multidimensional modulation codes and error correction for page-oriented optical data storage, SPIE 4342. Opt. Data Storage **2001**, 393 (2002)
2. E. Louidor, T. L. Poo, P. Chaichanavong, B.H. Marcus, Maximum insertion rate and capacity of multidimensional constraints (2008)
3. P. Zijlstra, J.W.M. Chon, M. Gu, Five-dimensional optical recording mediated by surface plasmons in gold nanorods. Nat. Lett. **459**(21), 410–413 (2009)
4. L.H. Tingab, X.S. Miaoa, M.L. Leea et al., Optical and magneto-optical characterization for multi-dimensional multi-level optical recording material. Synth. React. Inorg. Met.-Org., Nano-Met. Chem. **38**(3), 284–287 (2008)
5. Gu Min, Xiangping Li, The road to multi-dimensional bit-by-bit optical data storage. Opt. Photonics News **21**(7), 28–33 (2010)
6. X. Duan-yi, Q. Guosheng, Z. Hui, Three wavelength multi-level optical disc drive, Tsinghua University, CN01134740.6 (2002)
7. X. Duan-yi, L. He-xiong, P. Jiang, Q. Guosheng, Multilevel recording optical disc, Tsinghua University, CN03100602.7 (2004)
8. G. Yuan, W.L. Tan, L.T. Ng, Multi-dimensional multi-level optical pickup head. J. Appl. Phys. **47**, 5933 (2008). doi:10.1143/JJAP.47.5933
9. E.P. Walker, W. Feng, Y. Zhang, H. Zhang, F. McCormick, S. Esener, 3-D parallel readout in a 3-D multilayer optical data storage system, in *ISOM/ODS meeting* (2002)
10. D. Xu, High density optical data storage, ISBN-730206282X, 97873,02062820. Tsinghua University Press, in Beijing (2003), p. 17
11. M.M. Wang, S.C. Esener, Three-dimensional optical data storage in afluorescent dye-doped photopolymer. Appl. Opt. **39**(11), 1826–1834 (2000)
12. White paper blu-ray disc rewritable format audio visual application format specifications for BD-RE version 3.0 (2010)
13. Blu-ray Disc Association. White paper, Blu-ray Disc Read-Only Format, 2.B Audio Visual Application Format Specifications for BD-ROM Version 2.4. BD-ROM (2011)
14. W. Martyn, Blu-ray disc to support MPEG-4, VC-1, Pcworld.com (2011)
15. X.P. Li, J.W.M. Chon, S.H. Wu, R.A. Evans, M. Gu, Rewritable polarization encoded multilayer data storage in 2,5-dimethyl-4-(p-nitrophenylazo) anisole, doped polymer. Opt. Lett. **32**, 277–279 (2007)
16. D. Xu, D. Yi, L. Zhijun, P. She, L. Rong, Z. Qichen, Multi-wavelength and multi-level optical disc with photochromic materials, Tsinghua University, CN03102676.1 (2004)
17. P. Joseph, 1st HD DVD players to decode all mandatory, optional audio codecs. TWICE.com. (2011)
18. X. Duan-yi, Y. Ni, P.L. Fa, C. Ken, X. Jianping, L. Da, W. Heng, M. Jianshe, P. Jing, Z. Qichen, Replication method of multi-level CD-ROM, Tsinghua University, CN200510053509.3 (2005)

19. X. Duan-yi, L. Rong, S.P. Lei Zhijun, A sandwich structure with nonlinear mask super-resolution high-density optical disc, Tsinghua University, CN01134864.X (2002)
20. L. Liu, S. He, Near-field optical storage system using a solid immersion lens with a left-handed material slab. Opt. Express **12**(20), 4835–4840 (2004)
21. D. Walter, Industrial processing of various materials using ultrashort pulsed laser sources. Appl. Technol. Page AM2 K.1 (2015)
22. X. Zhang, K. Davis, S. Jiang, Transformer: Using SSD to improve disk scheduling for high-performance I/O, in *Proceedings of the 26th IEEE International Parallel and Distributed Processing Symposium (IPDPS'2012)*, 2012
23. M. Grupp, J.D. Davis, S. Swanson, The bleak future of NAND flash memory, in *Proceedings of the 10th USENIX Conference on File and Storage Technologies (FAST)*, 2012
24. R. Fontana, S. Hetzler, G. Decad, Technology roadmap comparisons for TAPE, HDD, and NAND flash: implications for data storage applications. IEEE Trans. Magn. **48**(5), 1692–1696 (2012)
25. A. Darafsheh, C. Guardiola, Optical super-resolution imaging by high-index microspheres embedded in elastomers. Opt. Lett. **40**(1), 5–8 (2015)
26. G. Min, X. Li, Y. Cao, Optical storage arrays: a perspective for future big data storage. Sci. Appl. **3**, e177 (2014). doi:10.1038/lsa.2014.58
27. X. Duan-yi, Y. Ni, P.L. Fa, C. Ken, X. Jianping, L. Da, S. Jie, Run length limited multi-level recordable discs, CN200510011474.7 (2005)
28. B. Lebeau, C. Marichal, A. Mirjol, G.J. de A.A. Soler-Illia, R. Buestrich, M. Popall, L. Mazerolles, C. Sanchez, Synthesisof highly ordered mesoporous hybrid silica from aromatic fluorinated organosilane precursors. New J. Chem. **27**(166) (2003)
29. D. Loke et al., Ultrafast switching in nanoscale phase-change random access memory with superlattice-like structures. Nanotechnology. **22**, 254019-1-6 (2011)
30. L. Liu, Z. Shi, S. He, Analysis of the polarization-dependent diffraction from a metallic grating by use of a three-dimensional combined vectorial method. J. Opt. Soc. Am. A **21**, 1545–1552 (2004)
31. M. Jianshe, Y. Ji-gang, P.L. Fa, X. Duanyi, Testing methods of dynamic parameters for multi-level optical pickup actuator, Tsinghua University and Jiangsu Yinhe Electronics Co., Ltd.: CN200510066028.6 (2005)
32. M. Jianshe, Y. Ji-gang, P. L. Fa, W. Jianming, X. DuanYi, J. Jian-Dong, Testing system of optical pickup actuator dynamic parameters, Tsinghua University and Jiangsu Yinhe Electronics Co., Ltd., CN200510066029.0 (2005)
33. C.D. Stanciu et al., All-optical magnetic recording with circularly polarized light. Phys. Rev. Lett. **99**, 047601 (2007)
34. X. Duan-Yi, Y. Haibo, Z. Qichen, Auto-focusing two-photon three-dimensional disk storage and tracking devices, Tsinghua University, CN200510011853.6 (2006)
35. E.N. Glezer, M. Milosavljevic, L. Huang, R.J. Finlay, T.H. Her et al., Three-dimensional optical storage inside transparent materials. Opt. Lett. **21**, 2023–2025 (1996)
36. M. Jianshe, Y. Ji-gang, P.L. Fa, W. Jianming, X. Duan-Yi, J. Jian-Dong, Z. Standard, Z. Jianyong, S. Hongwei, Detection axial rotation defect of the optical pickup actuator, Tsinghua University and Jiangsu Yinhe Electronics Co., Ltd.,: CN200510066030.3 (2005)
37. P.L. Fa, H. Heng, N. Yi, P. Jing, X. Haizheng, L. Da, X. Duan-yi, H. Hua, Multi-level run-length data conversion method and apparatus, and blue multilevel optical storage device, Tsinghua University,: CN200610169823.2 (2007)
38. X. Duan-yi, L. Zhijun, S. Peng, L. Rong, A rewritable phase-change optical disc with multi-layer and multi-level recording medium, CN01139830.2 (2002)
39. X. Duanyi, *Principles and Design Optical Storage Systems* (National Defence Press, Beijing, 2000)
40. H. Mamoru, Hitachi demonstrates 4 layer BD playback Using 'Standard Drive' (2011)
41. J. Jacob, I.I. Smolyaninov, E.E. Narimanov, Broadband Purcell effect: radiative decay engineering with metamaterials. Appl. Phys. Lett. **100**, 181105 (2012)

42. R. Wilson, Samsung's already awesome HD disc hybrid BD-UP5000 Upgraded to Profile 1.1 (Bye Bye Format Bitching). Gizmodo.com. June 15, 2011
43. S. Patil, G.A. Gibson, Scale and concurrency of gigabyte: File system directories with millions of files, in FAST, pp. 177–190 (2011)
44. H. Lei, Understanding nanoscale magnetization reversal and spin dynamics by using advanced transmission electron microscopy, State University of New York at Stony Brook (2010)
45. R. Harris, Quantum holographic storage, IBM Newsletters, 3 Feb 2009
46. K. Rashmi, N. Shah, P. Kumar, K. Ramchandran, Explicit construction of optimal exact regenerating codes for distributed storage, in *Proceedings of Allerton Conference on Communication, Control and Computing*, 2009
47. C. Carraher, *Introduction to Polymer Chemistry* (Taylor & Francis, New York, 2007)
48. Y. Hai-Bo, X. Duan-Yi, M. Jian-She, Design and implementation of a PRML detection system for HD-DVD. Acta Physica Sin. **57**(2), 867–872 (2008)
49. X. Duan-yi, L. Rong, S.P. Lei Zhijun, Super-resolution high-density optical disc with nonlinear mask sandwich structure, Tsinghua University, CN01270982.4 (2002)
50. I. Raicu, I.T. Foster, P. Beckman, Making a case for distributed file systems at exascale, in *LSAP* (2011), pp. 11–18
51. Y. Chen, R. Sion, To cloud or not to cloud musings on costs and viability, in *ACM Symposium on Cloud Computing (SOCC)*, 2011
52. Y.Hu, H. Chen, P. Lee, Y. Tang, NCCloud: Applying Network coding for the storage repair in a cloud-of-clouds, in *Proceedings of USENIX FAST* (2012)
53. C. Huang, H. Simitci, Y. Xu, A. Ogus, B. Calder, P. Gopalan, J. Li, S. Yekhanin, Erasure coding in windows azure storage, in *Proceedings of USENIX ATC* (2012)
54. H. Adam, T.J. Hsu, Pioneer showcases 16-layer 400 GB optical disc. Digitimes.com. (2011)
55. Holographic data storage. IBM journal of research and development. Retrieved 28 Apr 2008
56. J. Zhang, J. Ma, D. Xu, Influence of optical pick-up assembly error on tracking signal. J. Optoelectron. Laser **17**(8), 953–957 (2006)
57. Z. Jian-Yong, M. Jian-She, X. Duan-Yi, Analysis of the influence of photo detector integrated circuit position error on differential phase detection signal. Opt. Tech. **32**(6), 909–911
58. X. Duanyi, *Super-density and Super-speed Optical Data Storage, ISBN 978-7-5381-6248-6* (Liaoning Science and Technology Publishing House, Liaoning, China, 2009)
59. O. Khan, R. Burns, J. Plank, W. Pierce, C. Huang, Rethinking erasure codes for cloud file systems: minimizing I/O for recovery and degraded reads, in *Proceedings of USENIX FAST* (2012)
60. R. Li, J. Lin, P. Lee, CORE: Augmenting regenerating coding based recovery for single and concurrent failures in distributed storage systems. Ar Xiv, preprint arXiv:1302.3344 (2013)
61. H. Hua, X. Duan-Yi, H. Heng, Run-length-limited codes for optical storage system. Opt. Tech. **32**(3), 323–326+329 (2006)
62. H. Heng, H. Pan Long-Fa, L.Z.J. Hua, X. DY, Research progress of multilevel optical storage. Opt. Tech. **32**(2), 270–273 (2006)
63. Z. Qi-Cheng, X. Duan-Yi, S. Jie, Readout signal of optical disc with multi-level run-length limited coding. Opt. Tech. **32**(6), 842–847 (2006)

Chapter 2
Mechanism of Multidimensional Optical Storage

The principles of multidimensional storage involve a lot of research area. The chapter focuses on the discussion of physical and chemical mechanisms of interaction of light with materials in the optical storage.

2.1 Photophysics and Photochemistry

Photophysics and photochemistry are basic principles for multidimension storage as both deals with the impact of energy in the form of photons on materials. Photochemistry focuses on the chemistry involved as a material is impacted by photons, where as photophysics deals with physical changes that result from the impact of photons. The essential theory of photophysics and photochemistry will be applied in multidimension optical storage. There are lot of books introduced as N.S. Allen, Photochemistry in 2010, Vincenzo Balzani and Paola Ceroni etc., Photochemistry and Photophysics in 2014, C. Carraher, Polymer Chemistry in 2008, R. Dessauer, Photochemistry in 2006, D. Neckers and Advances in Photochemistry in 2007, D. Phillips and Polymer Photophysics in 2007, V. Ramamurthy Semiconductor Photochemistry and Photophysics in 2003, N. Turro, V. Ramamurthy and J. Scaiano, Principles of Molecular Photochemistry in 2009 which are important significance for this research [1–4].

This section will focus on some of the basic principles related to photophysics and photochemistry followed by general examples. Finally, these principles will be related to photosynthesis. In many ways, there is a great similarity between a material's behavior when struck by photons, whether the material is small or macromolecular. Differences are related to size and the ability of polymers to transfer the effects of radiation from one site to another within the chain or macromolecular complex. The importance of the interaction with photons in the natural world can hardly be overstated. It forms the basis for photosynthesis converting carbon dioxide and water into more complex plant-associated structures.

© Tsinghua University Press and Springer Science+Business Media Singapore 2016
D. Xu, *Multi-dimensional Optical Storage*,
DOI 10.1007/978-981-10-0932-7_2

Polymer photochemistry and physics have been recently reviewed, and readers are encouraged to investigate this further in the suggested readings given at the end of the chapter. Here, introduce some of the basic concepts of photophysics and photochemistry, and illustrate the use of photochemistry and photophysics in the important area of solar energy conversion.

Photophysics involves the absorption, transfer, movement, emission of electromagnetic and light energy without chemical reactions. By comparison, photochemistry involves the interaction of electromagnetic energy that results in chemical reactions. Briefly review the two major types of spectroscopy with respect to light. In absorption, the detector is placed along the direction of the incoming light and the transmitted light is measured. In emission studies, the detector is placed at some angle, generally $90°$, away from the incoming light. When absorption of light occurs, the resulting polymer P^*, contains excess energy and is excited as

$$P + hv \rightarrow P^* \tag{2.1}$$

The light can be simply reemitted:

$$P^* \rightarrow hv + P \tag{2.2}$$

Of much greater interest is light migration, either along the polymer backbone or to another chain. This migration allows the energy to move to a site of interest. Thus, for plants, the site of interest is chlorophyll. These 'light-gathering' sites are referred to as antennas. Natural antennas include chlorophyll, carotenoids, and special pigment-containing proteins. These antenna sites harvest the light by absorbing the light photon and storing it in the form of an electron, which is promoted to an excited singlet energy state (or other energy state) by the absorbed light.

Bimolecular occurrences can occur, leading to an electronic relaxation called quenching. In the approach P^* find another molecule or part of the same chain A, and transferring the energy to A.

$$P^* + A \rightarrow P + A^* \tag{2.3}$$

Generally, the quenching molecule or site is initially in its ground state. Eliminating chemical rearrangements, quenching is most likely ends with electronic energy transfer, complex formation, or increased nonradioactive decay. Electronic transfer involves an exothermic process, in which part of the energy is absorbed as heat and part is emitted as fluorescence or phosphorescence radiation. Polarized light is taken on in fluorescence depolarization, also known as luminescence anisotropy. Thus, if the chain segments are moving at about the same rate as the reemission, part of the light is depolarized. The extent of depolarization is then a measure of the segmental chain motions.

Complex formation is important in photophysics. Two terms need to be described here. First, an exciplex is an excited state complex formed between photophysics and photochemistry two different kinds of molecules, that one is excited and other one is in ground state. The second term, excimer is similar, except the complex is formed between like molecules. Here focus on excimer complexes that form between two like polymer chains or within the same polymer chain. Such complexes can be formed between two aromatic structures. Resonance interactions between aromatic structures, such as two phenyl rings in polystyrene, give a weak intermolecular force formed from attractions between the π-electrons of the two aromatic entities. Excimers involving such aromatic structures give strong fluorescence. Excimer formation can be described as follows where $[PP]^*$ is the excimer:

$$P^* + P \rightarrow [PP]^* \tag{2.4}$$

The excimer decays, giving two ground-state aromatic sites and emission of fluorescence by:

$$[PP]^* \rightarrow h\nu + 2P \tag{2.5}$$

As always, the energy of the light emitted is less than that originally taken on. Through studying the amount and energy of the fluorescence, radiation decay rates, depolarization effects, excimer stability, and structure can be determined.

2.1.1 Light Absorption

Light is composed of particles known as photons, each of which has the energy of Planck's quantum, hc/λ; where h is Plank's constant, c is velocity of light, and λ is the wavelength of the radiation. Light has dualistic properties of both waves and particles; ejection of electrons from an atom as a result of light bombardment is due to the particle behavior, whereas the observed light diffraction at gratings is attributed to the wave properties. The different processes related to light interactions with molecules can be represented as in Fig. 2.1. The absorption of light by materials produces physical and chemical changes. On the negative side, such absorption can lead to discoloration generally as a response to unwanted changes in the material's structure. Absorption also can lead to a loss in physical properties, such as strength. In the biological world, it is responsible for a multitude of problems, including skin cancer. It is one of the chief modes of weathering by materials. Here focus on the positive changes effected by the absorption of light. Absorption of light has intentionally resulted in polymer cross-linking and associated insolubilization. This forms the basis for coatings and negative-lithographic resists. Light-induced chain breakage is the basis for positive-lithographic resists. Photoconductivity forms the basis for photocopying, and photovoltaic effects form the basis for solar cells being developed to harvest light energy. It is important to remember that the basic laws governing small and large molecules are the same.

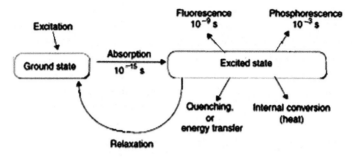

Fig. 2.1 Different processes associated with light interaction with a molecule

The Grotthus–Draper law states that photophysical/photochemical reactions occur only when a photon of light is absorbed. This forms the basis for the First Law of Photochemistry—that is, only light that is absorbed can have a photophysical/photochemical effect and can write as following:

$$M + \text{light} \rightarrow M^* \qquad (2.6)$$

where M^* is M after it has taken on some light energy acquired during a photochemical reaction. The asterisk is used to show that M is now in an excited state. Optical transmittance T is a measure of how much light that enters a sample is absorbed as:

$$T = I/I_\infty \qquad (2.7)$$

If no light is absorbed then $I = I_0$. Low transmittance values indicate that lots of the light has been absorbed.

Most spectrophotometers give their results in optical absorbency A, or optical density which is defined by:

$$A = \log(I/I_0) \qquad (2.8)$$

so that

$$A = \log(1/T) = -\log T \qquad (2.9)$$

Beer's law states that A, the absorbance of chromophores, increases in proportion to the concentration of the chromophores, where k is a constant.

$$A = kc \qquad (2.10)$$

Beer's law predicts a straight-line relationship between absorbance and concentration and is often used to determine the concentration of an unknown after construction of the known absorbance verses concentration line.

The optical path l, is the distance the light travels through the sample. This is seen in looking at the color in a swimming pool, where the water is deeper colored at the deep end because the optical path is greater. This is expressed by Lambert's law, where k' is another empirical constant as:

$$A = k'l \tag{2.11}$$

To the eye some colors appear similar but may differ in intensity, when c and l are the same. These solutions have a larger molar absorption coefficient, ε, meaning they adsorb more. The larger the adsorption coefficient the more the material adsorbs. The Beer–Lambert law combines the two laws and giving by:

$$A = \varepsilon l c \tag{2.12}$$

The proportionality constant in the Lambert's law is ε. The extinction coefficients of chromophores vary widely from $100 < \text{M cm}^{-1}$, for a so-called forbidden transition, to greater than 10^5 M cm^{-1} for fully allowed transitions, can redefine the elements of the Beer–Lambert law, where l is the sample thickness and c is the molar concentration of chromophores. This can be rearranged to determine the penetration depth of light into a polymer material. Here l is defined as the path length, where 90 % of the light of a particular wavelength is absorbed so A approaches l is giving:

$$l(\text{in } \mu\text{m}) = 10^4 \varepsilon c \tag{2.13}$$

This relationship holds when the polymer chromophore (or any chromophore) is uniformly distributed in a solution or bulk. In polymers with a high chromophore concentration, l is small and the photochemical/photophysical phenomenon occurs largely in a thin surface area. As examine the color of a red wine that the wine contains color sites or chromophores. The photons that are not captured pass through and give us the red coloration. Can see color became a chromophore interacts with light. Molecules that absorb photons of energy corresponding to wavelengths in the range 190 to ~ 1000 nm absorb in the UV-VIS region of the spectrum.

The molecule that absorbs a photon of light becomes excited. The energy that is absorbed can be translated into rotational, vibrational, or electronic modes. The quantized internal energy E_{int} of a molecule in its electronic ground or excited state can be approximated, with sufficient accuracy for analytical purposes by:

$$E_{\text{int}} = E_{\text{el}} + E_{\text{vib}} + E_{\text{rot}} \tag{2.14}$$

where E_{el}, E_{vib}, and E_{rot} are the electronic, vibrational, and rotational energies, respectively. According to the Born–Oppenheimer approximation, electronic transitions are much faster than atomic motion. Upon excitation, electronic transitions occur in about 10^{-15} s, which is very fast compared to the characteristic time scale

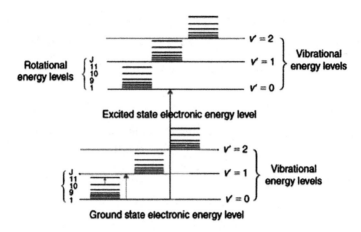

Fig. 2.2 The relative ordering of electronic, vibrational and rotational energy levels

for molecular vibrations of 10^{-10} to 10^{-12} s. Hence, the influence of vibrational and rotational motions on electronic states should be almost negligible. Franck–Condon stated that electronic transition is most likely to occur without changes in the position of the nuclei in the molecular entity and its environment. It is then possible to describe the molecular energy by a potential energy diagram in which the vibrational energies are superimposed upon the electronic curves as in Fig. 2.2.

For most molecules, only one or two lower energy electronic transitions are normally postulated. Thus one would expect that the UV-VIS spectrum would be relatively simple. This is often not the case. The question shows that why are many bands often exhibiting additional features. The answer lies in the Franck–Condon principle, by which vibronic couplings are possible for polyatomic molecules. Indeed, both vibronic and electronic transitions will be observed in the spectrum, generating vibrationally structured bands, and sometimes even leading to broad unresolved bands. Each resolved absorption peak corresponds to a vibronic transition, which is a particular electronic transition coupled with a vibrational mode belonging to the chromophore. For solids (when possible) and liquids, the rotational lines are broad and overlapping, so that no rotational structure is distinguishable [5].

To apply this concept for a simple diatomic molecule, the example is given in Fig. 2.3. At room temperature, according to the Boltzman distribution, most of the molecules are in the lowest vibrational level v of the ground state, i.e., $v = 0$. The absorption spectrum presented in Fig. 2.3b exhibits, in addition to the pure electronic transition that the so-called 0–0 transition, several vibronic peaks whose intensities depend on the relative position and shape of the potential curve.

The transition from the ground to the excited state, where the excitation goes from $v = 0$ (in the ground state) to $v = 2$ (in the excited state), is the most probable for vertical transitions because it falls on the highest point in the vibrational probability curve in the excited state. Yet many additional transitions occur, so that

(a)   **(b)**

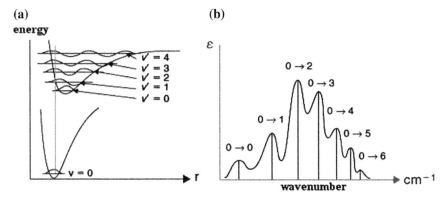

Fig. 2.3 a Potential energy diagram for a diatomic molecule, illustrating the Franck–Condon excitation. **b** Intensity distribution among vibronic bands as determined by the Franck–Condon principle

the fine structure of the vibronic broad band is a result of the probabilities for the different transitions between the vibronic levels.

There are two kinds of spectra, i.e., excitation and absorption in this action. The absorption and excitation spectra are distinct but usually overlap, energy levels sometimes to the extent that they are nearly indistinguishable.

The excitation spectrum is the spectrum of light emitted by the material as a function of the excitation wavelength. The absorption spectrums the spectrum of light absorbed by the material as a function of wavelength. The origin of the occasional discrepancies between the excitation and absorption spectra are due to the differences in structures between the ground and the excited states or the presence of photo reactions or the presence of nonradiative processes that relax the molecule to the ground state without passing through the luminescent states, i.e., S_1 and T_1.

Visible color is normally a result of changes in the electron states. Molecules that reside in the lowest energy level are said to be in the ground state or unexcited state. Restricted the attention to the electrons that are in the highest occupied molecular orbital (HOMO) and the lowest unoccupied molecular orbital (LUMO). These orbitals are often referred to as the frontier orbitals. Excitation of photons results in the movement of electrons from the HOMO to the LUMO as shown in Fig. 2.4. Photon energies can vary. Only one photon can be accepted at a time by an orbital.

Fig. 2.4 A photon being absorbed by a single molecule of chromophore

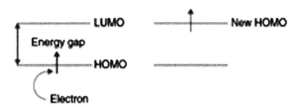

This is stated in the Stark–Einstein law also known as the Second Law of Photochemistry—if a species absorbs radiation, then one particle (molecule, ion, atom, etc.) is excited for each quantum of radiation (photon) that is absorbed. [6]

Remember that a powerful lamp will have a greater photon flux than a weaker lamp. Further, photons enter a system one photon at a time. Thus, every photon absorbed does not result in bond breakage or other possible measurable effect. The quantum yield φ is a measure of the effectiveness for effecting the desired outcome, possibly bond breakage, and formation of free radicals:

$$\varphi = N_m/N_p \tag{2.15}$$

where N_m is number of molecules of reactant consumed, N_p is number of photo consumed.

Quantum yields can provide information about the electronic excited state relaxation processes, such as the rates of radiative and nonradiative energy gap.

Moreover, they can also find applications in the determination of chemical structures and sample purity. The emission quantum yield can be defined as the fraction of molecules that emits a photon after direct excitation by a light source. So emission quantum yield is also a measure of the relative probability for radiative relaxation of the electronically excited molecules. Quantum yields vary greatly, the photons range from very ineffective (10^{-6}) to very effective (10^6). If values >1 indicate that some chain reaction, such as in a polymerization. To differentiate between the primary quantum yield, which focuses on only the first event, here the quantum yield cannot be >1 and secondary quantum yield, which focuses on the total number of molecules formed via secondary reactions that the quantum yield can be high. The common emission quantum yield measurement involves the comparison of a very dilute solution of the studied sample with a solution of approximately equal optical density of a compound of known quantum yield, i.e., standard reference. The quantum yield of an unknown sample is related to that of a standard by Eq. (2.16) as:

$$\Phi_u = \left[\frac{(A_s F_u n^2)}{(A_u F_s n_0^2)}\right]\Phi_s \tag{2.16}$$

where, the subscript u refers to 'unknown,' and s to the comparative standard, Φ is the quantum yield, A is the absorbance at a given excitation wavelength, F is the integrated emission area across the band, and n and n_0 are the refractive indices of the solvent containing the unknown and the standard, respectively. For the most accurate measurements, both the sample and standard solutions should have low absorptions (≤ 0.05) and have the similar absorptions at the same excitation wavelength.

2.1.2 Luminescence

Luminescence is a form of cold body radiation. Older TV screens operated on the principle of luminescence, by which the emission of light occurs when they are relatively cool. Luminescence includes phosphorescence and fluorescence. In a TV, electrons are accelerated by a large electron gun sitting behind the screen. In the black-and-white sets, the electrons slam into the screen surface, which is coated with a phosphor that emits light when hit with an electron. Only the phosphor that is hit with these electrons gives off light. The same principle operates in the old-generation color TVs, except the inside of the screen is coated with thousands of groups of dots, each group consisting of three dots (red, green, and blue). The kinetic energy of the electrons is absorbed by the phosphor and reemitted as visible light to be seen. Fluorescence involves the molecular absorption of a photon that triggers the emission of a photon of longer wavelength, i.e., less energy as in Fig. 2.2. The energy difference ends up as rotational, vibrational, or heat energy loses. Here excitation is described as

$$S_0 + h\nu_{ex} \rightarrow S_1 \tag{2.17}$$

and emission as

$$S_1 \rightarrow h\nu_{em} + S_0 \tag{2.18}$$

where S_0 is the ground state and S_1 is the first excited state. The excited state molecule can relax by a number of different, generally competing pathways. One of these pathways is conversion to a triplet state that can subsequently relax through phosphorescence or some secondary nonradiative step. Relaxation of the excited state can also occur through fluorescence quenching. Molecular oxygen is a particularly efficient quenching molecule because of its unusual triplet ground state.

Watch hands that can be seen in the dark allow to read the time without turning on a light. These watch hands typically are painted with phosphorescent paint. Like fluorescence, phosphorescence is the emission of light by a material previously hit by electromagnetic radiation. Unlike fluorescence, phosphorescence emission persists as an afterglow for some time after the radiation has stopped. The shorter end of the duration for continued light emission is 10^{-3} s but the process can persist for hours or days. An energy level diagram representing the different states and transitions is called a Jablonski diagram or a state diagram. The Jablonski diagram was first introduced in 1935, a slightly modified version is presented in Fig. 2.5. The different energy levels are given in this figure, where S_0 represents the electronic ground state and S_1 and S_2 represent the first and second singlet excited states, respectively. The first and second triplet states are denoted T_1 and T_2, respectively. In the singlet states, all electron spins are paired and the multiplicity of this state is 1. The subscript indicates the relative energetic position (electronic level) compared to other states of the same multiplicity. On the other hand, in the triplet states, two electrons are no longer antiparallel and the multiplicity is 3. The triplet state is more

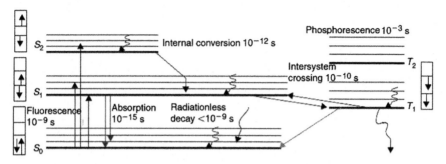

Fig. 2.5 Jablonski diagram showing the various processes associated with light absorption and their time scale. *Arrows* in boxes, the relative spin states of the paired electrons

stable than the singlet counterpart (S) and the source for this energy difference is created by the difference in the Coulomb repulsion energies between the two electrons in the singlet versus triplet states and the increase in degree of freedom of the magnetic spins. Because the electrons in the singlet excited state are confined within the same orbital, the Coulomb repulsive energy between them is higher than in the triplet excited state where these electrons are now in separate orbitals. The splitting between these two states (S–T) also depends on the nature of the orbital.

Consider a case, where the two orbitals involved in a transition are similar, i.e., two p-orbitals of an atom, or two π-orbitals of an aromatic hydrocarbon. For this situation, the overlap between them may be high, and the two electrons will be forced to be close to each other resulting in the S–T splitting being large. The other situation is the case where the two orbitals are different (i.e., $n-\pi^*$ or $d-\pi$ transitions), resulting in a small overlap. Because the overlap is small that the two electrons will have their own region of space in which to spread, resulting in a minimization of the repulsive interactions between them, and hence the S–T splitting will be small. Absorption occurs on a time scale of about 10^{-15} s. When inducing the promotion of an electron from the HOMO to the LUMO, the molecule passes from an electronic ground singlet state S_0 (for diamagnetic molecules) to a vibrational level of an upper singlet or triplet excited state S_n or T_n, respectively. The energy of the absorbed photon determines which excited state is accessible. After a while, the excited molecule relaxes to the ground state via either radiative with emission of light or nonradiative without emission of light processes. The radiative processes for diamagnetic molecules include either the spin-allowed fluorescence or spin-forbidden phosphorescence. Nonradiative processes include intersystem crossings (ISCs), a process allowing a molecule to relax from the S_n to the T_n manifolds, and internal conversions (IC and IP), a stepwise (vibrational) energy loss process relaxing molecules from upper excited states to any other state without or with a change in state multiplicity, respectively. An internal conversion (IC) is observed when a molecule lying in the excited state relaxes to a lower excited state. This is a radiationless transition between two different electronic states of the same multiplicity and is possible when there is a good overlap of the vibrational wave functions or probabilities that are involved between the two states to beginning and final.

2.2   Photoinduced Electron Transfer Process

Internal conversion occurs on a time scale of 10^{-12} s, which is a time scale associated with molecular vibrations. A similar process occurs for an internal conversion, (IP) when it is accompanied by a change in multiplicity such as triplet T_1 down to S_0. Upon nonradiative relaxation, heat is released. This heat is transferred to the media by collision with neighboring molecules. Fluorescence is a radiative process in a diamagnetic molecule involving two states (excited and ground states) of the same multiplicity, for example S_1–S_0 and S_2–S_0 is shown in Fig. 2.6. Fluorescence spectra show the intensity of the emitted light versus the wavelength. A fluorescence spectrum is obtained by initial irradiation of the sample, normally at a single wavelength, where the molecule absorbs light. The lifetime of fluorescence is typically on the order of 10^{-8} to 10^{-9} s, i.e., an ns time scale for organic molecules and faster for metal-containing compounds (10^{-10} s or shorter). In general, the fluorescence band, typically S_1–S_0, is a mirror image of the absorption band (S_0–S_1), as illustrated in Figs. 2.6 and 2.7. This is particularly true for rigid

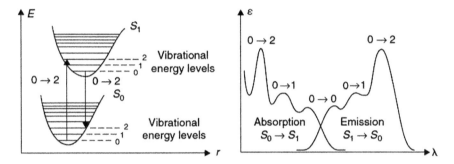

Fig. 2.6 Potential energy curves and vibronic structure in fluorescence spectra

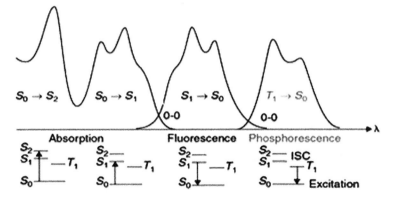

Fig. 2.7 Relative positions of absorption, fluorescence and phosphorescence The 0–0 is common to both absorption and fluorescence spectra (see Fig. 2.6). ISC, Intersystem crossing

molecules, such as aromatics. Once again, the Franck–Condon principle is applicable, and hence the presence of vibronic bands is expected in the fluorescence band. However, there are numerous exceptions to this rule, particularly, when the molecule changes geometry in its excited state. Another observation is that the emission is usually red shifted in comparison with absorption. This is because the vibronic energy levels involved are lower for fluorescence and higher for absorption, as illustrated in Fig. 2.6. The difference in wavelength between the 0–0 absorption and the emission band is usually known as the Stokes shift. The magnitude of the Stokes shift gives an indication of the extent of geometry difference between the ground and excited states of a molecule as well as the solvent–solute reorganization. Another nonradiative process that can take place is known as intersystem crossing from a single to a triplet or triplet to a single state. This process is very rapid for metal-containing compounds. This process can take place on a time scale of $\sim 10^{-6}$ to 10^{-8} s for an organic molecule, while for organometallics it is $\sim 10^{-11}$ s [7, 8].

This rate enhancement is due to spin-orbit coupling present in the metal-containing systems—that is, an interaction between the spin angular momentum and the orbital angular momentum, which allows mixing of the spin angular momentum with the orbital angular momentum of S_n and T_n states. Thus, these singlet and triplet states are no longer "pure" singles and triplets, and the transition from one state to the other is less forbidden by multiplicity rules. A rate increase in intersystem crossing can also be achieved by the heavy atom effect, arising from an increased mixing of spin and orbital quantum number with increased atomic number. This is accomplished either through the introduction of heavy atoms into the molecule via chemical bonding (internal heavy atom effect) or with the solvent (external heavy atom effect). The spin-orbit interaction energy of atoms grows with the fourth power of the atomic number Z. In addition to the increase in the intersystem crossing rate, heavy atoms exert more effects, which can be summarized as follows. Their presence acts (1) to decrease the phosphorescence lifetime due to an increase in the nonradiative rates, (2) to decrease the fluorescence lifetime, and (3) to increase the phosphorescence quantum yield. The presence of a heavy atom affects not only the rate for intersystem crossing but also the energy gap between the singlet and the triplet states, where the rate for the intersystem crossing increases as the energy gap between S_1 and T_1 decreases. Moreover, the nature of the excited state exerts an important effect on the intersystem crossing. For example, the $S_1(n, \pi^*) \rightarrow T_2(\pi, \pi^*)$ (e.g., as in benzophenone) transition occurs almost three orders of magnitude faster than the $S_1(\pi, \pi^*) \rightarrow T_2 (\pi, \pi^*)$ transition for example in anthracene.

Relaxation of triplet state molecules to the ground state can be achieved by either internal conversion (nonradiative IP) or phosphorescence (radiative). Emissions from triplet states, i.e., phosphorescence that exhibit longer lifetimes than fluorescence. These long-lived emissions occur on time scale of 10^{-3} s for organic samples and 10^{-5} to 10^{-7} s for metal-containing species. This difference between the fluorescence and the phosphorescence is associated with the fact that it involves a spin-forbidden electronic transition. Moreover, as already noted, the phosphorescence bands are always red shifted in comparison with their fluorescence counterpart because of the relative stability of the triplet state compared to the singlet manifold as shown in Fig. 2.7. Nonradiative processes in the triplet states increase exponentially with a decrease in triplet energies (energy gap law). Hence phosphorescence is more difficult to observe when the triplet states are present in very low energy levels. It is also often easier to observe phosphorescence at lower temperatures, at which the thermal decay is further inhibited.

2.2.1   Emission Lifetime

The luminescence lifetime is the average time the molecule remains in its excited state before the photon is emitted. From a kinetic viewpoint, the lifetime can be defined by the rate of depopulation of the excited (singlet or triplet) states following an optical excitation from the ground state. Luminescence generally follows first-order kinetics and can be described as follows.

$$[S_1] = [S_1]_0 e^{-\Gamma t} \tag{2.19}$$

where $[S_1]$ is the concentration of the excited state molecules at time t, $[S_1]_0$ is the initial concentration and Γ is the decay rate or inverse of the luminescence lifetime. Various radiative and nonradiative processes can decrease the excited state population. Here, the overall or total decay rate is the sum of these rates:

$$\Gamma_{total} = \Gamma_{radiative} + \Gamma_{nonradiative} \tag{2.20}$$

For a complete photophysical study, it is essential to study not only the emission spectrum but also the time domain because it can reveal a great deal of information about the rates and hence the kinetics of intramolecular and intermolecular processes. The fundamental techniques used to characterize emission lifetimes of the fluorescence and the phosphorescence are briefly described next.

When a molecule is excited (Eq. 2.21), it is promoted from the ground to the excited state. This excited molecule can then relax to the ground state after emission lifetime loosing its extra energy gained from the exciting source via a radiative (Eq. 2.22) and nonradiative (Eq. 2.23) processes:

$$A_0 + hv \rightarrow A^* \tag{2.21}$$

$$A^* \rightarrow A_0 + hv' \tag{2.22}$$

$$A^* \rightarrow A_0 + heat \tag{2.23}$$

Therefore, can write as:

$$-\frac{d[A^*]}{dt} = (k_r + k_n)[A^*]t = -\frac{t}{\tau} \tag{2.24}$$

where $[A^*]$ is the concentration of the species A in its excited state at a given time t and k_r and k_n are the rate constants for the radiative and nonradiative processes, respectively. The relative concentration of A^* is given by

$$\ln\frac{[A^*]_t}{[A^*]_{t=0}} = -(k_r + k_n)t = -\frac{t}{\tau} \tag{2.25}$$

Hence, the mean lifetime (τ) of $[A^*]$ is

$$\tau = 1/(k_r + k_n) \tag{2.26}$$

where k_r and k_n are the rate constants for the radiative and nonradiative processes, respectively, represented by Eqs. (2.22) and (2.23). Thus, the measured unimolecular radiative lifetime is the reciprocal of the sum of the unimolecular rate constants for all the deactivation processes. The general form of the equation is given by

$$\tau = \frac{1}{\sum_i k_i} \tag{2.27}$$

where τ is observed radiative lifetime and the rate constant k_i represents the unimolecular or pseudo-unimolecular processes that deactivate A^*. The lifetime can be measured from a time-resolved experiment in which a very short pulse excitation is made, followed by measurement of the time dependent intensity, as illustrated in Fig. 2.8. The intensity decays are often measured through a polarizer oriented at some angle such as about from the vertical z-axis to avoid the effects of Introduction to Photophysics and Photochemistry anisotropy on the intensity decay. Then, the log of the recorded intensity is plotted against time to obtain a straight line predictable from the integration of the Eq. (2.24). The slope of this line is the negative reciprocal of the lifetime. When more than one lifetime is present in the decay traces, then there is more than one radiative pathway to relaxation. This often signifies that more than one species is emitting light at the excitation wavelength. The analysis of such multicomponent decays involves the deconvolution of an equation of the same form of Eq. (2.24) where a weighing factor for each component is added to each component. One possible explanation for the polyexponential curves can be an exciton process. The exciton phenomenon is a delocalization of excitation energy through a material. A description of this is given in Fig. 2.9. It shows a one-dimensional coordination or organometallic polymer denoted by -|M_n|-|M_n|-|M_n|-|M|-, where M_n represent a mononuclear ($n > 1$) or polynuclear center ($n = 1$). The incident irradiation is absorbed by a single

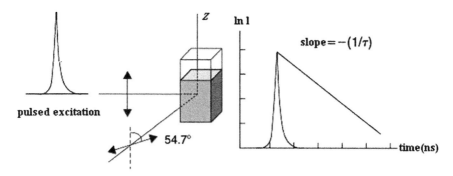

Fig. 2.8 Time-domain lifetime measurement

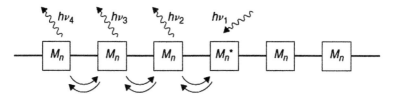

Fig. 2.9 The excitation phenomenon and process

chromophore, $|M_n|$, along the backbone, and then this stored energy is reversibly transmitted via an energy transfer process to the neighboring chromophore (with no thermodynamic gain or loss; i.e., $\Delta G^0 = 0$). This newly created chromophore can reemit, or not, the light (hv_2, hv_3, hv_4, ...) at a given moment. The interactions between the different units in the excited states are called excimers. These excimers can be excited dimers, trimers, tetramers, etc. These excited oligomers have different wavelengths and emission lifetimes. The extent of the interactions in the excited state (dimers, trimers, tetramers) is hard to predict because it depends on the amplitude of the interactions and the relaxation rates. Hence, the lifetime decay curve will have a polyexponential nature.

2.2.2 Ground and Excited State Molecular Interactions

Ground-state intermolecular interactions are present in some systems and require measurements of the binding constants. These interactions are manifested by the spectral changes experienced in the absorption spectra. Therefore, these changes can be monitored as a function of the concentration of the substrates leading to the extraction of the binding constants. On the other hand, intermolecular and intramolecular excited state interactions refer to the energy and electron transfer operating in the excited states of different dyad orpolyad systems. These can also be excimers, dimers, or oligomers that are formed only in the excited states. Studies of photoinduced energy and electron transfers involve the measurement of their corresponding rates. The theory and methods used to characterize the different types of interactions are described next. Binding constant considerations are described elsewhere.

2.2.3 Energy and Electron Transfer

The energy and electron transfer is excited state interactions and reactions indeed. The possible deactivation pathways of the excited state are summarized in Fig. 2.10. The fluorescence and phosphorescence relaxation pathways and the thermal deactivation processes. A transfer of the excitation energy from the donor to the acceptor will occur when an energy acceptor molecule is placed at the proximity of an excited energy donor molecule. After energy transfer, the donor relaxes to its ground state and the acceptor is promoted to one of its excited states. A photoinduced electron transfer can be initiated after photoexcitation when an excited single electron in the LUMO of the electron donor is transferred to a vacant molecular orbital (LUMO) of the acceptor. The mechanisms for the energy and electron transfers are outlined below [9, 10].

1. Energy Transfer

In presence of a molecule of a lower energy excited state (acceptor), the excited donor (D^*) can be deactivated by a process known as energy transfer which can be represented by the following sequence of equations.

$$D + hv \rightarrow D^* \tag{2.28}$$

$$D^* + A \rightarrow D + A^* \tag{2.29}$$

For energy transfer to occur, the energy level of the excited state of D^* has to be higher than that for A^* and the time scale of the energy transfer process must be faster than the lifetime of D^*. Two possible types of energy transfers are known—namely, radiative and nonradiative (radiationless) energy transfer. Radiative

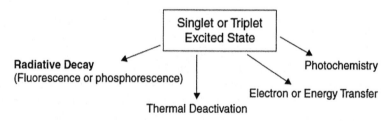

Fig. 2.10 Different pathways for the reactivation of the excited state

transfer occurs when the extra energy of D^* is emitted in form of luminescence and this radiation is absorbed by the acceptor (A).

$$D^* \rightarrow hv + D \tag{2.30}$$

$$hv + A \rightarrow A^* \tag{2.31}$$

For this to be effective, the wavelengths where the D^* emits need to overlap with those where A absorbs. This type of interaction operates even when the distance between the donor and acceptor is large (100 Å). However, this radiative process is inefficient because luminescence is a three-dimensional process in which only a small fraction of the emitted light can be captured by the acceptor. The second type, radiationless energy transfer, is more efficient. There are two different mechanisms used to describe this type of energy transfer: the Förster and Dexter mechanisms.

The energy transfer action is according to Förster mechanism. The Förster mechanism is also known as the coulombic mechanism or dipole-induced dipole interaction. It was first observed by Förster. Here, the emission band of one molecule (donor) overlaps with the absorption band of another molecule (acceptor). In this case, a rapid energy transfer may occur without a photon emission. This mechanism involves the migration of energy by the resonant coupling of electrical dipoles from an excited molecule (donor) to an acceptor molecule. Based on the nature of interactions present between the donor and the acceptor, this process can occur over a long distances (30–100 Å). The mechanism of the energy transfer by this mechanism is illustrated in Fig. 2.11. In Fig. 2.11, an electron of the excited donor placed in the LUMO relaxes to the HOMO, and the released energy is transferred to the acceptor via coulombic interactions. As a result, an electron initially in the HOMO of the acceptor is promoted to the LUMO. This mechanism operates only in singlet states of the donor and the acceptor. This can be explained on the basis of the nature of the interactions (dipole–induced dipole) because only multiplicity conserving transitions possess large dipole moments. This can be understood considering the nature of the excited state in both the singlet and the triplet states. The triplet state has a diradical structure, so it is less polar, making it difficult to interact over long distances (i.e., Förster mechanism). The rate of energy transfer (k_{ET}) according to this mechanism can be evaluated by the Eq. (2.32):

$$k_{ET} = k_D R_F^6 \left(\frac{1}{R}\right)^6 \tag{2.32}$$

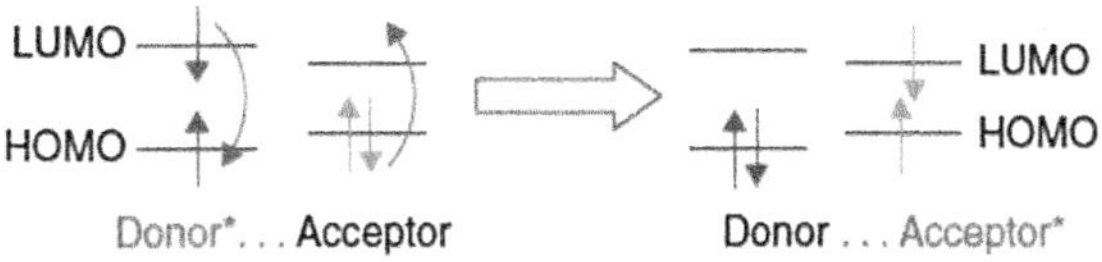

Fig. 2.11 Mechanism of energy transfer action according to Förster

where k_D is the emission rate constant for the donor, R is the interchromophore separation, and R_F is the Förster radius, which can be defined as the distance between the donor and the acceptor at which 50 % of the excited state decays by energy transfer—that is, the distance at which the energy transfer has the same rate constant as the excited state decay by the radiative and nonradiative channels ($k_{ET} = k_r + k_{nr}$). R_F is calculated by the overlap of the emission spectrum of the donor excited state (D^*) and the absorption spectrum of the acceptor (A).

The energy transfer action is according to the Dexter mechanism also. The Dexter mechanism is a nonradiative energy transfer process that involves a double electron exchange between the donor and the acceptor (Fig. 2.12). Although the double electron exchange is involved in this mechanism, no charge-separated state is formed.

The Dexter mechanism can be thought of as electron tunneling, by which one electron from the donor's LUMO moves to the acceptor's LUMO at the same time as an electron from the acceptor's HOMO moves to the donor's HOMO. In this mechanism, both singlet–singlet and triplet–triplet energy transfers are possible. This contrasts with the Förster mechanism, which operates in only singlet states. For this double electron exchange process to operate, there should be a molecular orbital overlap between the excited donor and the acceptor molecular orbital. For a bimolecular process, intermolecular collisions are required as well. This mechanism involves short-range interactions that is about 6–20 Å or shorter. Because it relies on tunneling, it is attenuated exponentially with the intermolecular distance between the donor and the acceptor. The rate constant can be expressed by the following:

$$k_{ET} = \frac{2\pi}{\hbar} V_0^2 J_D \exp\left(-\frac{2R_{DA}}{L}\right) \tag{2.33}$$

where R_{DA} is distance between the donor and the acceptor, J_D is the integral spectral overlap between the donor and the acceptor, L is the effective Bohr radius of the orbitals between which the electron is transferred, $\hbar$ is Plank's constant, and V_0 is the electronic coupling matrix element between the donor and acceptor at the contact distance. Comparing the two energy transfer mechanisms, the Förster mechanism involves only dipole–dipole interactions, and the Dexter mechanism operates through electron tunneling. Another difference is their range of interactions. The Förster mechanism involves longer range interactions (up to ~ 30–100 Å), but the Dexter mechanism focuses on shorter range interactions that is from 6 Å up to 20 Å because orbital overlap is necessary. Furthermore, the Förster

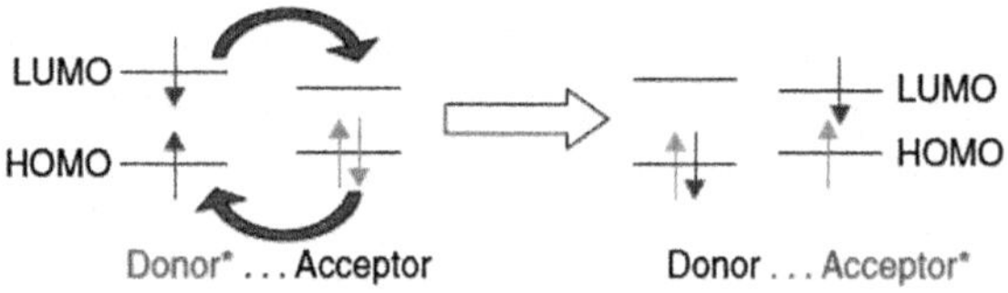

Fig. 2.12 Mechanism of energy transfer action according to the Dexter mechanism

Donor* ... Acceptor   Donor ... Acceptor*

mechanism is used to describe interactions between singlet states, but the Dexter mechanism can be used for both singlet–singlet and triplet–triplet interactions. Hence for the singlet–singlet energy transfer, both mechanisms are possible.

Simulated graphs using reasonable values for the parameters for the two mechanisms have been constructed for the purpose of distinguishing between the zones where Förster and Dexter mechanisms are dominant. The experimental values of the energy transfer rates in cofacial bisporphyrin systems were found to agree with the theoretically constructed graphs (see Fig. 2.13). In these graphs a Bohr radius value (L) of 4.8 Å (the value for porphyrin) is used in the Dexter equation. Also, the solid lines correspond to hypothetical situations in which only the Förster mechanism operates; the dotted lines are hypothetical situations for when the Dexter mechanism is the only process. The curved lines are simulated lines obtained with Eq. (2.32) (Förster) or Eq. (2.33) (Dexter) but transposed onto the other graph, i.e., Förster equation plotted against Dexter formulation and vice versa. These plots clearly suggest the presence of a crossing point between the two mechanisms. There is a zone in which one mechanism is dominant and vice versa. All in all, the relaxation of an excited molecule via energy transfer processes will use all the pathways available to it so the total rate for energy transfer can be better described as k_{ET} (total) 5 k_{ET} (Förster) 1 k_{ET} (Dexter). According to Fig. 2.13, the distance at which there is a change in dominant mechanism is about 5 Å.

2. Electron Transfer

Photoinduced electron transfer (PET) involves an electron transfer within an electron donor–acceptor pair. The situation is represented in Fig. 2.14. PET represents one of the most basic photochemical reactions and at the same time it is the

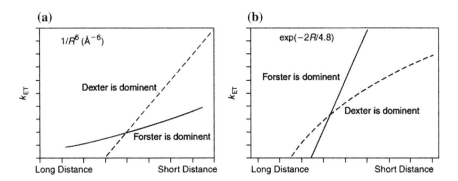

Fig. 2.13 Qualitative theoretical plots for **a** k_{ET} versus $1/R^6$ (Förster) and **b** k_{ET} versus exp $(-2R/4.8)$

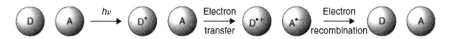

Fig. 2.14 Photoinduced electron transfer process

Fig. 2.15 Potential energy surfaces for the ground state (*DA*), the excited state (*DA**—reactant state), and the charge-separated state (D$^+$-A$^-$—product state), by Marcus's theory. λ—total reorganization energy, *TS*—transition state

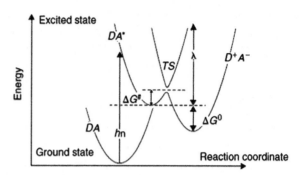

most attractive way to convert light energy or to store it for further applications. In Fig. 2.14, one can see a process taking place between a donor and acceptor after excitation, resulting in the formation of a charge-separated state, which relaxes to the ground state via an electron-hole recombination (back electron transfer).

A theory used to study and interpret the PET in solution was described by Marcus. In this theory, the electron transfer reaction can be treated by transition state theory where the reactant state is the excited donor and acceptor and the product state is the charge-separated state of the donor and acceptor (D$^+$-A$^-$), shown in Fig. 2.15. According to the Franck–Condon principle, the photoexcitation triggers a vertical transition to the excited state, which is followed by a rapid nuclear equilibration. Without donor excitation, the electron transfer process would be highly endothermic. However, after exciting the donor, electron transfer occurs at the crossing of the equilibrated excited state surface and the product state. The change in Gibbs free energy associated with the electron transfer event is given by the following relation as:

$$\Delta G^{\#} = \frac{(\lambda + \Delta G^0)^2}{4\lambda} \tag{2.34}$$

The total reorganization energy (λ), which is required to distort the reactant structure to the product structure without electron transfer, is composed of solvent (λ_S) and internal (λ_i) components ($\lambda = \lambda_i + \lambda_S$). The reaction free energy (ΔG_0), is the difference in free energy between the equilibrium configuration of the reactant (*DA**) and of the product states (D$^+$A$^-$). The internal reorganization energy represents the energy change that occurs in bond length and bond angle distortions during the electron transfer step and is usually represented by a sum of harmonic potential energies. In the classical Marcus theory, the electron transfer rate is given by equation

$$k_{ET} = \kappa_{ET} \nu_n \exp\left(\frac{-\Delta G^{\#}}{k_B T}\right) \tag{2.35}$$

where v_n is the effective frequency of motion along the reaction coordinate and κ_{ET} is the electronic transmission factor. The transmission factor is related to the transition probability (P_0) at the intersection of two potential energy surfaces, as given by the Landau–Zener theory.

$$\kappa_{ET} = \frac{2P_0}{1 + P_0} \tag{2.36}$$

A graph showing the change of the driving force for the electron transfer rate, calculated from Marcus theory, versus the rate constant is given in Fig. 2.16 (bottom). Using Eq. (2.35) to estimate the electron transfer rate, we can assign the Marcus normal region as that where the free reaction energy (ΔG^0) is decreased, leading to an increase of the electron transfer rate (k_{ET}). The second region that can be identified in Fig. 2.16 is the optimal or activationless region, where the driving force for electron transfer equals the reorganization energy—that is, $-\Delta G^0 = \lambda$. If ΔG^0 becomes even more negative, the activation barrier $\Delta G^{\#}$ reappears, resulting in a decrease in the values of k_{ET}. This last situation is observed over the region known as the inverted Marcus region and was first experimentally demonstrated by Closs and Miller. The potential energy illustrating the different Marcus regimes can be seen in Fig. 2.16 (top).

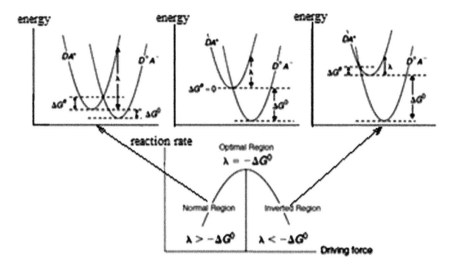

Fig. 2.16 The free energy regimes for electron transfer (*top*) and the corresponding reaction rate dependence on the free energy (*bottom*, driving force is $\Delta G^0 - \lambda$)

2.3   Other Actions of Photon to Materials

2.3.1   Nonlinear Optical Action

Nonlinear optics (NLO) involves the interaction of light with materials resulting in a change in the frequency, phase, or other characteristics of the light. There are a variety of frequency-mixing processes. Second-order NLO behavior includes second harmonic generation of light that involves the frequency doubling of the incident light. Frequency mixing where the frequency of two light beams are either added or subtracted. Electrooptic effects can occur where both frequency and amplitude changes and where rotation of polarization occurs. NOL behavior has been found in inorganic and organic compounds and in polymers. The structural requirement is the absence of an inversion center requiring the presence of asymmetric centers and/or poling. Poling is the application of a high voltage field to a material that orients some or all of the molecule dipoles generally in the direction of the field. The most effective poling in polymers is found when they are poled above the T_g (which allows a better movement of chain segments) and then cooled to lock in the poled structure. Similar results are found for polymers that contain side chains that are easily poled. Again, cooling helps lock in the poled structure. At times, cross-linking is also employed to help lock in the poled structure. Third-order NLO behavior generally involves three photons, resulting in effects similar to those obtained for second-order NLO behavior. Third-order NLO behavior does not require the presence of asymmetric structures. Polymers that have been already been found to offer NLO behavior include polydiacetylenes and a number of polymers with liquid crystal side chains. Polymers are also employed as carriers of materials that themselves are NLO materials. Applications include communication devices, routing components, and optical switches [11–13].

2.3.2   Photoconductive and Photonic Polymers

Some polymeric materials become electrically conductive when illuminated with light. For instance, poly (N-vinylcarbazole) is an insulator in the dark, but when exposed to UV radiation it becomes conductive as shown in Fig. 2.17a. The addition of electron acceptors and sensitizing dyes allows the photoconductive response to be extended into the visible and NIR regions. In general, such photoconductivity depends on the materials ability to create free-charge carriers, electron holes, through absorption of light, and to move these carriers when a current is applied.

Related to this are materials whose response to applied light varies according to the intensity of the applied light. This is nonlinear behavior. In general, polymers with whole-chain delocalization or large-area delocalization in which electrons are optically excited may exhibit such nonlinear optical behavior.

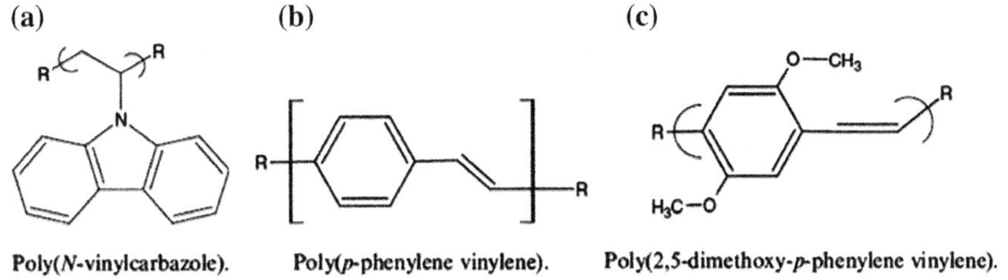

(a)   (b)   (c)

Poly(N-vinylcarbazole).   Poly(p-phenylene vinylene).   Poly(2,5-dimethoxy-p-phenylene vinylene).

Fig. 2.17 Molecular configuration of poly (N-vinylcarbazole) photonic polymers

A photoresponsive sunglass whose color or tint varies with the intensity of the sunlight is an example of nonlinear optical material. Some of the so-called smart windows are also composed of polymeric materials whose tint varies according to the incident light. Currently, information is stored using electronic means but optical storage is becoming common place with the use of CD-ROM and WORM devices. Such storage has the advantages of rapid retrieval and increased knowledge density (i.e., more information stored in a smaller space).

Since the discovery of doped polyacetylene, a range of polymeric semiconductor devices has been studied, including normal transistors, field-effect transistors (FETs) photodiodes, and light-emitting diodes (LEDs). Like conductive polymers, these materials obtain their properties from their electronic nature, specifically the presence of conjugated π-bonding systems. In electrochemical light-emitting cells, the semiconductive polymer can be surrounded asymmetrically with a hole-injecting material on one side and a low work function electron injecting metal (such as magnesium, calcium, or aluminum) on the other side. The emission of light may occur when a charge carrier recombines in the polymer as electrons from one side and holes from the other meet.

Poly(p-phenylene vinylene) (PPV) was the first reported in 1990 that polymer to exhibit electroluminescence. PPV is employed as a semiconductor layer. The layer was sandwiched between a hole-injecting electrode and electron injecting metal on the other. PPV has an energy gap of about 2.5 eV and thus produces a yellow-green luminescence when the holes and electron recombine. Today, many other materials are available that give a variety of colors as in Fig. 2.17b.

A number of poly (arylene vinylene) (PAV) derivatives have been prepared. The electron-donating substituents, such as two methoxy groups, act to stabilize the doped cationic form and thus lower the ionization potential. These polymers exhibit both solvatochromism (color change as solvent is changed) and thermochromism (color is temperature dependent) as in Fig. 2.17c. The introduction of metals into polymers that can exhibit entire chain electron delocalization is at the basis of much that is presented in this volume. These metal-containing sites are referred to as chromophores, and the combination of metal chromophores exhibiting metal to ligand charge transfer (MLCT) excited states opens new possibilities for variation of electronic and optical properties needed for the continual advancement in electronics and electronic applications. Application areas include light-emitting polymeric diodes, solar energy conversion and nonlinear optical materials (NLOs) and materials exhibiting photorefraction, electrochromism and electrocatalysis as shown in Fig. 2.17.

One of the major reasons for interest in this area is the ease with which the new hybrid materials' properties can be varied by changing the metal, metal oxidation state, metal matrix, and polymer. Multiple metal sites are readily available. This allows the metal-containing system to have a high degree of tunability. This is due to the often strong electronic interaction between the metal and the delocalized electron systems. The already noted variety of available metal sites is further leveraged by the increasingly capability of modern synthetic methodologies to achieve the desired structures. But the presence of metal atoms is at the heart of this.

2.3.3 Photosynthesis

The recent environmental issues related to the greenhouse effect and atmospheric contamination heightens the importance of obtaining energy from clean sources, such as photosynthesis. Photosynthesis also acts as a model for the creation of synthetic light-harvesting systems that might mimic chlorophyll in its ability to convert sunlight into usable energy. The basis of natural photosynthesis was discovered by Melvin Calvin. Using carbon-14 as a tracer, Calvin and his team found the pathway that carbon follows in a plant during photosynthesis. They showed that sunlight supplies the energy through the chlorophyll site, allowing the synthesis of carbon-containing units, mainly saccharides or carbohydrates. Chlorophyll is a metal embedded in a protein polymer matrix and illustrates the importance of metals in the field of photochemistry and photophysics. A brief description of the activity of chlorophyll in creating energy from the sun follows. The maximum solar power density reaching Earth is approximately 1350 W/m^2. When this energy enters the Earth's atmosphere, the magnitude reaching the surface drops approximately to 1000 W/m^2 owing to atmospheric absorption. The amount that is used by plants in photosynthesis is about seven times the total energy used by all humans at any given time, thus it is a huge energy source. Solar energy is clean and economical energy, but it must be converted into useful forms of energy. For example, solar energy can be used as a source of excitation to induce a variety of chemical reactions. Natural examples for conversion of light energy are plants, algae, and photosynthetic bacteria that used light to synthesize organic sugar-type compounds through photosynthesis. In photosynthesis, green plants and some bacteria harvest the light coming from the sun by means of their photosynthetic antenna systems. The light harvesting starts with light gathering by antenna systems are consist of pigment molecules, including chlorophylls, carotenoids, and their derivatives. The absorbed photons are used to generate excitons, which travel via Förster energy transfers towards the reaction centers (RCs). This overall series of processes is represented in Fig. 2.18.

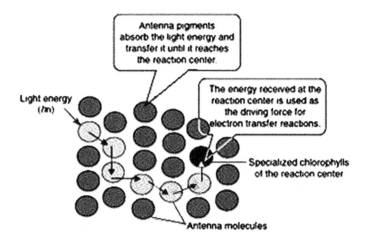

Fig. 2.18 Light is absorbed by the antenna, and the energy is transferred to the RC, where charge separation takes place to generate chemical energy

In RCs, this energy drives an electron transfer reaction, which in turn initiates a series of slower chemical reactions. Energy is saved as redox energy, inducing a charge separation in a chlorophyll dimer called the special pair (chlorophyll). Charge separation, which forms the basis for photosynthetic energy transfer, is achieved inside these RCs Eqs. (2.37), (2.38).

$$(Chlorophyll)_2 + Energy \rightarrow (Chlorophyll)_2^+ + e^- \qquad (2.37)$$

Specialized reaction center (RC) proteins are the final destination for the transferred energy. Here, it is converted into chemical energy through electron transfer reactions. These proteins consist of a mixture of polypeptides, chlorophylls (plus the special pair), and other redox-active cofactors. In the RCs, a series of downhill electron transfers occur, resulting in the formation of a charge-separated state. Based on the nature of the electron acceptors, two types of RCs can be described. The first type (photosystem I) contains iron-sulfur clusters (Fe_4S_4) as their electron acceptors and relays, whereas the second type (photosystem II) features quinones as their electron acceptors.

Both types of RCs are present in plants, algae, and cyanobacteria, whereas the purple photosynthetic bacteria contain only photosystem II and the green sulfur bacteria contain a photosystem I. To gain a better understanding of these two types of RCs each will be further discussed.

2.3.4   Purple Photosynthetic

1980s later, Deisenhofer reported his model for the structure of photosystem II for two species of purple photosynthetic bacteria (Rhodopseudomonas viridis and Rhodobacter) based on X-ray crystallography of the light-harvesting device II (LH II). Photosynthetic centers in purple bacteria are similar but not identical models for green plants. Because they are simpler and better understood, they will be described here. The photosynthetic membrane of purple photosynthetic bacteria is composed of many phospholipid-filled ring systems (LH II) and several larger dissymmetric rings (LH I) stacked almost like a honeycomb. Inside the LH I is a protein called the RC. The LH II complex antenna is composed of two bacteriochlorophyll a (BCHl) molecules, which can be classified into two categories. The first one is a set of 18 molecules arranged in a slipped face-to-face arrangement and is located close to the membrane surface perpendicularly to these molecules. The second ring is composed of BCHl in the middle of the bilayer. The first BCHl have an absorption maximum at 850 nm and are collectively called B850, while the second (9 BCHl) have an absorption maximum at 800 nm and are called B800. These structures are contained within the walls of protein cylinders with radii of 1.8 and 3.4 nm. Once the LH II complex antenna absorbs light, a series of very complex nonradiative photophysical processes are triggered.

First, the excitation energy migrates via energy transfers involving the hopping of excitation energy within almost isoenergetic subunits of a single complex. This is followed by a fast energy transfer to a lower energy complex with minimal losses (see Fig. 2.19). These ultrafast events occur in the singlet state (S_1) of the BCHl pigments and are believed to occur by a Förster mechanism. The energy collected by the LH II antenna is transferred to another antenna complex known as LH I, which surrounds the RC. The photosynthetic RCs of bacteria consist mainly of a protein that is embedded in and spans a lipid bilayer membrane. In the RC, a series of electron transfer reactions are driven by the captured solar energy. These electron transfer reactions convert the captured solar energy to chemical energy in Fig. 2.19. Two light-harvesting II (LH II) units is next to one light-harvesting I (LH I) unit. Gray circles, polypeptides, bars, rings of interacting bacteriochlorophylls a (called B850). In the middle of LH I, there is the RC, where the primary PET takes place

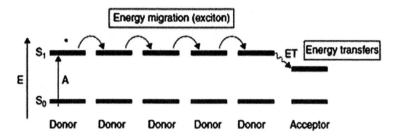

Fig. 2.19 The exciton and energy transfer processes

from the special pair of bacteriochlorophylls. Introduction to Photophysics and Photochemistry formed a charge separation process across the bilayer. The mechanism of this process is illustrated in Fig. 2.20. A special BCHl (P870) pair is excited either by the absorption of a photon or by acquiring this excitation energy from an energy transfer from the peripheral antenna BCHl (not shown in the figure for simplicity), triggering a photoinduced electron transfer inside the RC. Two photoinduced electrons are transferred to a plastoquinone located inside the photosynthesis membrane. This plastoquinone acts as an electron acceptor and is consequently reduced to a semiquinone and finally to a hydroquinone. This reduction involves the uptake of two protons from water on the internal cytoplasmic side of the membrane. This hydroquinone then diffuses to the next component of the apparatus, a proton pump called the cytochrome bc1 complex as in Fig. 2.20.

The oxidation of the hydroquinone back to a quinone and the energy released is used for the translocation of the protons across the membrane. This establishes a proton concentration and charge imbalance (proton motive force, pmf). Thus, the oxidation process takes place via a series of redox reactions triggered by the oxidized special pair BCHl, which at the end is reduced to its initial state. The oxidation process is ultimately driven, via various cytochrome redox relays, by the oxidized P870. Oxidized P870 becomes reduced to its initial state in this sequence. Finally, the enzyme ATP synthase allows protons to flow back down across the membrane driven by the thermodynamic gradient, leading to the release of ATP formed from adenosine diphosphate and inorganic phosphate (P_i).

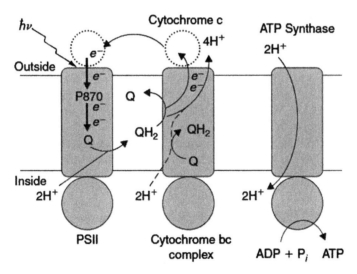

Fig. 2.20 A cross-section of the photosynthetic membrane in the purple photosynthetic bacteria. PS II-photosystem II; P870-special pair; Q-plastoquinone; QH2-dihydroplastoquinone; ADP-adenosine diphosphate; ATP-adenosine triphosphate

2.4   Photophysical Properties of Organometallic Polymers

Organic and organometallic polymers exhibit potential applications in photonics. Organometallic polymers have received a lot of interest because they could combine the advantages of the high luminescence of the organic moiety with the high carrier density, mobility, steady chemical properties, and physical strength of inorganic materials. Research on such materials is expanding because of their potential use as electric components, such as FETs, LEDs, and solar cells.

Much effort involving solar energy conversion is based on the natural chlorophyll system as a model. Here, a metal atom is embedded within a polymer matrix that exhibits high electron mobility (delocalization). Ruthenium, platinum, and palladium are the most employed metals. The use of materials containing the bis (2,2-bipyridine)ruthenium II moiety is common with ruthenium because this moiety absorbs energy in the UV region and emits it at energies approximating those needed to cleave water molecule bonds. The use of solar energy to create hydrogen that is harvested and later converted to useful energy has been a major objective. Here, we focus on a more direct conversion of solar energy into energy to charge batteries. For this purpose, metal-containing polymers can be classified into three types (types I, II, and III), as illustrated in Fig. 2.21. In type I, the metal centers are connected to the conjugated polymer backbone through saturated linkers, such as alkyl chains. Polymers of type I act as a conducting support. The electronic, optical, and chemical properties of the metal ions in this type of polymer remain the same as they would be if they were alone (i.e., unattached to the polymer backbone).

In the second type, the metal centers are electronically coupled to the conjugated polymer backbone. This affects both the polymer and the metal group properties. The metal centers for type III are located directly within the conjugated backbone. In this last type, there are strong interactions between the metal center and the organic bridge. For this arrangement, the electronic interactions between the organic bridge and the metal group are possible, and new properties can be obtained because of the combination of the characteristics of the organic polymers with the common properties of the transition metals. Heavy metal atoms in the polymer backbone increase the intersystem crossing rate of the organic lumophores due to enhanced spin-orbit coupling. This populates more of the triplet states facilitating the study of interactions on both singlet and triplet states. The study of energy

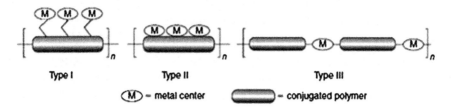

Fig. 2.21 Types of metal-containing polymers

transfer in organic and organometallic polymers is important. In fact, various types of organic and organometallic systems (oligomers and polymers) have been specifically designed for intramolecular energy transfer studies. Molecular architecture was found to play an important role in the efficiency of the energy transfer. The bridge between the donor and the acceptor chromophores exerts an important effect on the rates as well as the mechanism through which the energy transfer occurs. A through-bond mechanism operates very efficiently for the cases of rigid saturated hydrocarbon bridges, while through-space mechanism is efficient for flexible bridges.

The photophysical properties of macromolecules built on M-P and M-CN (isocyanide) bonds, including the metal in the backbone (The presence of the metal atom associated with the porphyrin moiety is examined here). Photosynthesis is a source of inspiration for scientists interested in nonnatural systems that convert light into chemical potential or electrical energy. Molecular wires, optoelectronic gates, switches, and rectifiers are typical examples of molecular electronic devices envisioned for use in energy or electron transfer processes. A basic device structure, stimulating the natural systems, needs a scaffold on which the energy or charge transfer can be induced. Such a scaffold is represented in Fig. 2.22. The approach for this system is the mimicry of the highly efficient photosynthesis process in biological systems, by which an antenna device collects the light energy that the energy and electron transfers lead to the synthesis of the plant's fuel. Porphyrins are an interesting class of compounds used for the study of energy- and electron transfer functions of the natural photosynthetic machinery.

The interest in porphyrins is motivated in part by their photocatalytic activity and electronic properties. Porphyrins are also structurally related to chlorophyll. Cofacial bisporphyrin systems use rigid spacers to provide a unique placement of two chromophores (donor and acceptor) at a given distance, inducing a through-space energy transfer as the shortest pathway for intermolecular interactions and communications.

Recently, the effect of the donor–acceptor separation has been studied. Both the fluorescence lifetime and quantum yield were found to decrease as the distance between the two porphyrins C_{meso}–C_{meso} (cd) and CC_{meso}–CC_{meso} (ab) decreases (Fig. 2.23). As the two rings get closer to each other, they interact more strongly,

Fig. 2.22 A scaffold for photoinduced intramolecular energy or electron transfer

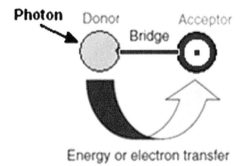

and hence nonradiative deactivation becomes more pronounced. The rate dependence for the S_1 energy transfer (S_1 ET) for such systems exhibits a dependence of the energy transfer (k_{ET}) rate on the C_{meso}–C_{meso} distance. The rate increases as the distance decreases. Face-to-face donor–acceptor separations are on the order of ~3.5 Å verses the corresponding various donor–acceptor separations in the living supramolecular structures (found in plants, algae, and cyanobacteria) (Figs. 2.23 and 2.24) which are found to have separation distances to ~20 Å. Despite this observation, the S_1 energy transfer data are strikingly slower (two orders of magnitude). Both through-space and through-bond energy transfer mechanisms are known, by which singlet–singlet energy transfer occurs through both Coulombic or dipole–dipole interaction (Förster) and double electron exchange (Dexter) mechanisms.

Different donor-bridge-acceptor based dyads based on metallated and free base porphyrins, by which singlet–singlet energy transfer occurs through both Förster and Dexter mechanisms are given in Fig. 2.25. The S_1 energy transfer in these systems occurs via a contribution from both coulombic and double electron exchange that have almost the same magnitude and are not affected by the donor and acceptor distances. The electronic interactions depend on the donor-bridge energy gap and the bridge conformation (planar or nonplanar). Studies of the

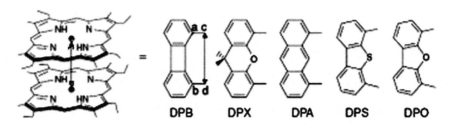

Fig. 2.23 Examples of cofacial face-to-face porphyrin systems with different spacers

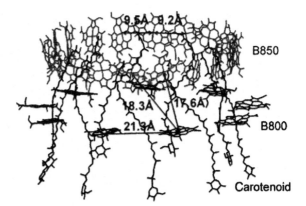

Fig. 2.24 A LH II ring showing only the chlorophyll for the B850 network, the noninteracting B800 bacteriochlorophylls and the rhodopin glucosides. Two of the B850 units are marked with arrows, representing the transition moments

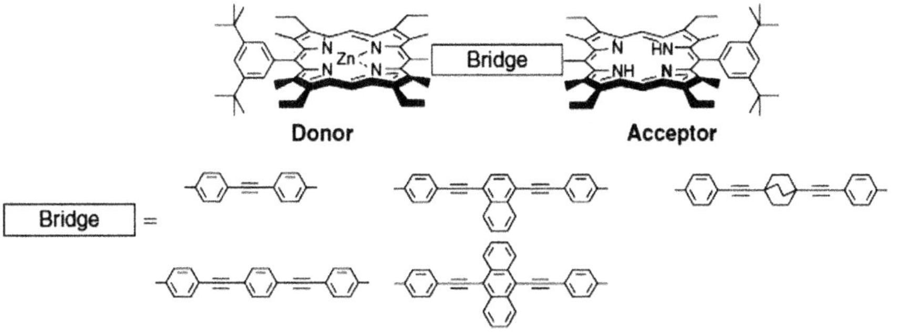

Fig. 2.25 Some donor-bridge-acceptor systems by which energy transfer occurs through both Förster and Dexter mechanisms

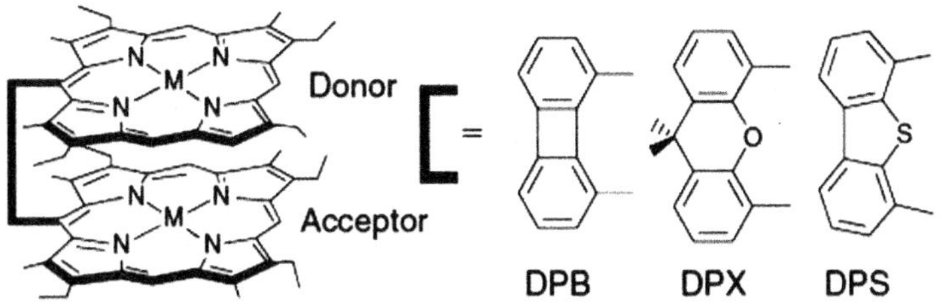

Spacer	DPB	DPB	DPB	DPX	DPX	DPX	DPX	DPS	DPS
M	Pt	2H	2H	Pt	2H	2H	Zn	2H	Zn
M'	Pt	Pt	Pd	Pt	Pt	Pd	Pd	Pd	Pd

Fig. 2.26 Examples of cofacial bisporphyrin systems containing heavy atoms

energy transfer rate as a function of the energy gap between the donor and the bridge have facilitated the separation of the two mechanisms. The rates observed for systems with the biggest energy gap were found to be almost equal to the Förster energy transfer rates [14, 15].

Harvey's group studied energy transfers arising from the longer-lived triplet states as well as from the singlet states. These studies involved porphyrins containing a heavy metal (e.g., Pt and Pd), as shown in Fig. 7.27. Spin-orbit coupling of the heavy atom increased the intersystem crossing rates, thus increasing the population of the triplet excited state. Triplet energy transfers can be analyzed only according to the Dexter mechanism because the Förster mechanism does not operate in the triplet excited states due to their diradicalnature and the multiplicity change during the process. Energy transfer for the Dexter mechanism occurs via a double electron exchange—HOMO (acceptor)-HOMO (donor) and LUMO (donor)-LUMO (acceptor)—between triplet states of the donor and acceptor. In these systems (as in Fig. 2.26), the Pd- and Pt-metallated chromophores act as triplet donors, whereas the free base and Zn-containing complexes are the energy acceptors.

Analyses of energy transfer rates revealed that no sensitive transfer was detected for systems in which the spacer was DPS. In contrast, for dyads with the DPB and DPX spacers containing dyads, energy transfer occurred. This result was explained on the basis that singlet states energy transfer occurs via both Förster and Dexter mechanisms in the DPB- and DPX-containing dyads: C_{meso}–C_{meso} = 3.80 and 4.32 Å, respectively. The singlet energy transfer mechanism proceeded predominantly via a Dexter mechanism. Conversely, singlet energy transfer in the DPS-containing dyad, C_{meso}–C_{meso} = 6.33 Å, operated predominantly according to the Förster mechanism. This latter mechanism is inactive in the triplet states. Thus, at such long distances, orbital overlap is poor and energy transfer is either weak or nil. This concept is of importance for designing molecular switches based on the distance separating the donor from the acceptor. Through-bond energy transfer was also observed for porphyrin systems (regardless whether it occurs via a Förster or a Dexter mechanism). Through-bond energy transfer was reported for the rhodium meso-tetraphenylporphyrin-tin, which exhibits a Rh-Sn bond length of 2.5069 Å and a 3.4 Å separation between the average macrocycle planes, as in Fig. 2.27.

A photophysical study of these porphyrin systems showed the presence of significant intramolecular triplet energy transfer with an estimated k_{ET} ranging between 10^6 and 10^8 s^{-1}. Rates for the through-bond process were found to be three to five orders of magnitude larger than the through-space energy transfer. Other examples for through-bond energy transfer are shown in Fig. 2.28. The intramolecular energy transfer rates within these systems were found to be slower than those estimated for cofacial systems by two or three orders of magnitude. These results can be helpful in predicting the rates for energy transfer (k_{ET}) for unknown systems.

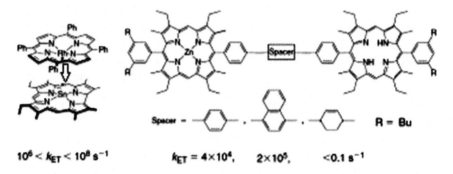

$10^6 < k_{ET} < 10^8$ s^{-1} $k_{ET} = 4{\times}10^4$, $2{\times}10^5$, <0.1 s^{-1}

Fig. 2.27 Porphyrin systems with through-bond energy transfer

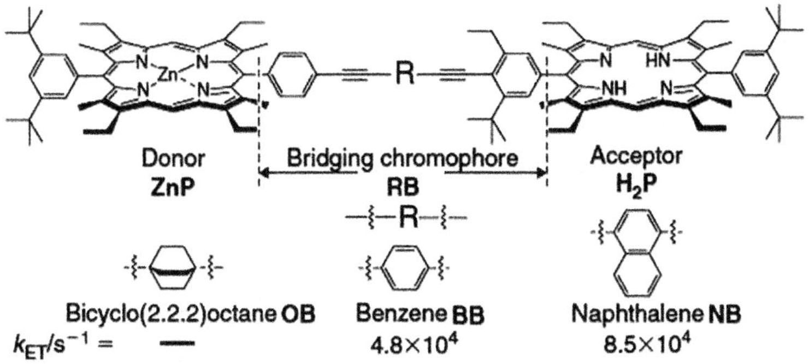

Fig. 2.28 Porphyrin systems with through-bond energy transfer

A similar observation was made by Albinson, using Zn(II) porphyrin as the donor and free base porphyrin as the acceptor. The solvent viscosity and temperature were investigated as factors affecting the donor–acceptor interactions (Fig. 2.28). In this example, and in agreement with Fig. 2.27, the rate increased with an increase of conjugation. Conversely, energy transfer is completely turned off when the conjugation is broken by the presence of the saturated system. This indicates that the through-bond energy transfer process occurs from the higher energy triplet state of Zn(II) porphyrin to the lower lying triplet state of the free base porphyrin. The triplet energy transfer rates were measured over temperatures from 295 to 280 K. The free energies of activation were found to be in the range of 1.0–1.7 kcal/mol (about 4–8 kJ/mol) in low-viscosity solvents, whereas in high-viscosity solvents, the temperature dependence is less pronounced. The triplet energy transfer was dependent on the solvent viscosity. Dramatically slower rates are observed in high-viscosity solvents due to smaller electronic coupling. The triplet excited donor porphyrin was suggested to adopt conformation in less viscous solutions, which have a much larger electronic coupling than is possible in highly viscous media. The porphyrins considered in the study are prone to conformational change in the triplet manifold. This was explained on conformational grounds.

In a donor-spacer-acceptor system (Fig. 2.28), with the ground state exhibiting a dihedral angle near 90°, the electronic coupling is changed in the triplet state to a situation in which the phenyl group should rotate towards the plane of the porphyrin macrocycle, leading to a considerable increase in the electronic coupling. This conformational freedom is lost when the solvent rigidifies, leading to a decrease in the coupling between the donor and the bridge. In solvents of low viscosity, another observation was made. Indeed, the change in temperature led to a triplet state distortion, inducing slower rates for triplet energy transfers.

All in all, the nature of the donor–acceptor linker is undoubtedly a controlling factor for the energy transfer, especially in the case of the triplet state interactions in which the mechanism of the interaction proceeds according to the Dexter mechanism (i.e., double electron exchange). This analysis illustrates the importance of studying different donor–acceptor spacers and their geometries during photoinduced energy transfer. Interest in the electronic properties of π-conjugated oligomers and polymers and polymers containing metal atoms continues to increase greatly. The metal site can offer chromophores that exhibit MLCT excited states in the π-conjugated polymers systems. This allows a variety of electronic and optical properties that are finding application in numerous areas, including solar energy conversion devices, NLOs, and polymer light-emitting diodes (PLEDs), with applications in physical and chemical sensing, electrochromism, and a wide scope of electrocatalysis.

The presence of the metal allows the synthesis of a wide variety of materials, with a variety of optical, electronic, chemical, and physical characteristics. The particular properties are changed and tuned by varying the metal, metal oxidation state, and metal environment. This volume describes some of these materials and applications for metal-containing sites embedded within polymer matrices and it suggests others.

2.5   Inorganic Photochemistry and Photocatalysis

Inorganic photochemistry and photocatalysis is one of important elements for multidimensional optical storage that offers some conceptions to investigate and develop new medium based on photoelectrochemistry and photocatalysis [16, 17].

2.5.1   Electron Storage on ZnO Particles

Zinc oxide is a semiconductor which has often been applied in photoelectrochemistry and photocatalysis. Small ZnO particles show typical size quantization effect, the onset of light absorption and position of the fluorescence band being shifted to shorter wavelengths with decreasing particle size. This effect has now been investigated in more detail to obtain the relationship between particle size and the wavelength of the absorption threshold.

The research in field of colloidal semiconductor nanocrystals has clarified and revealed new concepts and mechanisms in photophysics, electronic structure, magnetism, interfacial chemistry, redox chemistry, charge transport, crystal nucleation and growth, and self-assembly, and has introduced new approaches to energy harvesting and conversion, electronics, catalysis, chemical sensing, and imaging. The quantum dot field is growing rapidly and in many different directions. This new GRC to push forward research in this complex interdisciplinary field and will bring together making exciting contributions in both fundamental and applied research within the field of colloidal nanocrystals.

A improved methods for the preparation of colloidal ZnO solutions of different particle size are described, and the relation between absorption threshold and particle size is reported. CH_2OH radicals radiolytically generated transfer electrons to ZnO particles. The electrons are long-lived and cause a substantial blue shift of the absorption spectrum of ZnO in a wavelength range of 60 nm below the threshold. The wavelength of maximum bleaching is shifted to shorter wavelengths with decreasing particle size (size quantization effect). Maximum bleaching occurs with a negative absorption coefficient of $1.1 \times 10^5 \ M^{-1} cm^{-1}$. Electrons are also stored upon UV illumination of colloidal ZnO. The stored electrons react rather slowly with oxygen, the rate constant becoming lower with increasing particle size, and more rapidly with peroxy radicals.

After the preparation of ZnO as transparent colloidal solution became possible the photochemical studies on this material could be extended by the application of the fast kinetic methods of flash photolysis and pulse radiolysis. The previous studies have shown that electrons deposited on small particles of ZnO influence their optical absorption and fluorescence. Excess electrons can be generated either by UV light absorption in the colloidal particles or by electron transfer from reducing free radicals produced radio lytically in the bulk solution. Both methods are described in the present paper. Experiments of this type are of fundamental importance for the understanding of the mechanism of interfacial electron transfer in heterogeneous photocatalysis and of the mechanism of electron storage on semiconductor microelectrodes.

The preparation of ZnO sols has previously been described. In principle, Zn^{2+} ions are reacted with NaOH in alcohol solution, making use of the dehydrating properties of this solvent to prevent the formation of zinc hydroxide. However, it is crucial to use an alcohol having a certain water content that controls the rate of growth of the colloidal particles. The mechanism of ZnO formation is rather complex and poorly understood. Some improved methods of preparation are also described in the present paragraph. Small ZnO particles show typical size quantization effect, the onset of light absorption and position of the fluorescence band being shifted to shorter wavelengths with decreasing particle size. This effect has now been investigated in more detail to obtain the relationship between particle size and the wavelength of the absorption threshold.

ZnO forms stable colloids in methanol if either Zn^{2+} or OH^- ions are present in excess. For the preparations described below, two stock solutions were made: one is 0.2 M NaOH in CH_3OH by dissolving 2.0 g of NaOH in 250 mL of methanol, other one is 0.2 M $Zn(ClO_4)_2.6H_2O$ by dissolving 7.448 g of this salt in 100 mL of methanol p.a. (water content <0.05 %). ZnO Sols is with excess Zn^{2+} Ions. The opalescing solution was then stirred overnight at 20 °C. The opalescence disappeared, and a transparent 2×10^{-3} M ZnO solution with 4×10^{-4} M excess Zn^{2+} was obtained. This solution was stable for about 1 week and contained relatively small particles (mean diameter ~ 20 Å). Using 1:1 dilution with methanol led to a substantial increase in stability (stable for weeks). Figure 2.29 shows how the spectrum of the colloid developed after the addition of the stock solution.

The absorption threshold moved towards longer wavelengths as the particles grew, which is explained by the quantization of the electronic energy levels in the small particles. The mean size of the ZnO particles in the diluted methanol solution was about 20 Å. The solution had an absorption of only a few percent at 347 nm,

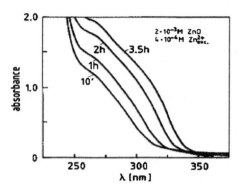

Fig. 2.29 Development of the absorption spectrum of ZnO in methanol at 20 °C and in the presence of 20 % excess ZnZ$^+$ ions. The long wavelength tail is due to scattering which becomes smaller with increasing reaction time

the wavelength of the frequency-doubled ruby laser with which the flash photolysis experiments were performed. A solution with a stronger absorption at 347 nm was obtained by mixing. The diluted methanol solution is with water (60 % water, 40 % solution) under argon. The particles grew in the mixture to about 40 Å.

For ZnO Sols with Excess OH$^-$ Ions, two procedures were applied where first a tetrahydroxozincate solution was made in which ZnO developed in the presence of a small amount of water. In the first procedure, 75 mL of stock solution A was diluted with 600 mL of methanol. A mixture of 15 mL of solution B and 70 mL of methanol was added under vigorous stirring. The zincate solution formed was transparent and stable. ZnO formation was started by adding a mixture of 5 mL of water and 95 mL of methanol under strong shaking. Methanol was added to bring the solution to a volume of 1 L and stirred for 24 h at 20 °C. The transparent ZnO sol, which was stable at 20 °C for weeks, contained 3×10^{-3} M ZnO and 9×10^{-3} M excess OH$^-$. Lower OH$^-$ concentration sled to less stable sols. Figure 2.30 shows the absorption spectrum during the development of the colloid. Again, one observes a shift of the absorption threshold to longer wavelengths.

Note that the threshold after long ripening of the colloid is still far below 372 nm, the threshold of macro crystalline ZnO. In the second procedure, a very concentrated ZnO sols was obtained which then was diluted with methanol. It could also be finished from American Chemical Society Colloidal Semiconductor is

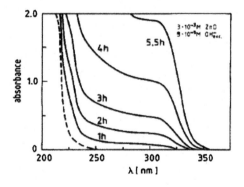

Fig. 2.30 Development of the absorption spectrum of ZnO in methanol at 20 °C and in the presence of 300 % excess OH$^-$ ions. The dashed line is the spectrum of the zincate solution

diluted with 2-propanol or 2-methyl-2-propanol to obtain sols where these diluents were the main solvents (>95 %).

A 10 mL sample of stock solution B was added dropwise to 50 mL of stock solution A. The transparent zincate solution obtained was stored in a closed vessel at 20 °C. ZnO formed over 2 days, the reaction of zincate being initiated by the water content of solution B (due to the water in $Zn(ClO_4)_2 6H_2O$). The sol finally obtained contained 3.3×10^{-2} MZnO, 0.2 M water, and 0.1 M excess OH^-. As it was stable for only a few days, it was diluted with 940 mL of methanol, giving a solution containing 2×10^{-3} MZnO and 6×10^{-3} M excess OH^- which was stable at 20 °C for weeks. When the concentrated sol was diluted with 2-methyl-2-propanolor 2-propanol, the particle sizes were determined by transmission electron microscopy. To prepare the samples, a drop of the colloid solution was applied to a copper mesh covered with a carbon film for 30 s and subsequently removed with a paper tip. Adhesion of the particles was promoted by exposing the carbon film to a glow discharge prior to this procedure. The granulation contrast originating from the amorphous carbon film was suppressed by an apodization technique.

Irradiations were carried out with a pulsed laser or a 4-MeV Graff generator. The radicals were produced in a low concentration of less than 10^{-6} M to avoid radical–radical deactivation. In some experiments a train of pulses with long intervals between the pulses was applied. The signals for eight pulses (or trains of pulses) were averaged. The base line was recorded every other pulse and finally subtracted from the recorded signals. The γ-irradiations were carried out with a source. The flash photolysis experiments were performed with a frequency-doubled ruby laser ($\lambda = 347.1$ nm, 15-ns pulse width). The data from several flashes were digitized and transferred to the results of a quantum mechanical calculation in Fig. 2.32 (-·-).

Fig. 2.31 Electron microscopic histograms of ZnO sample. The arrows point to the extrapolated values of the particle size

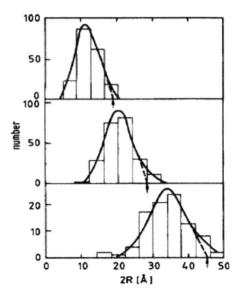

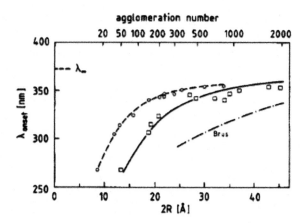

Fig. 2.32 Wavelength of absorption onset as a function of particle size: (*squared line*) experimental points obtained by extrapolation (see Fig. 2.32); (*circled line*) mean size of the samples (*doted line*) used quantum mechanical calculation

Colloidal ZnO is very sensitive towards UV light. It was therefore necessary to carry out the laser flash and electron pulse experiments with a low-intensity analyzing light beam.

2.5.2 Particle Size and Absorption

Figure 2.31 shows histograms of various ZnO colloids as determined by electron microscopy. The solution was investigated at different times after precipitation, and at different phases of particle growth. As has already been described for cadmium sulfide the particle size, which can be related to the onset of absorption, was obtained by extrapolating the steep part of the size distribution curve. The onset of light absorption was obtained by extrapolating the steep part of the rising absorption curve. In Fig. 2.32, the wavelength of the absorption threshold is plotted versus the particle diameter. Above 40 Å, the particles absorb close to 372 nm, where macro crystalline ZnO starts to absorb. With decreasing size, the onset is more and more rapidly shifted towards shorter wavelengths. Figure 2.32 also contains a curve which relates λ_{onset} with the mean particle size of the samples as determined by electron microscopy. This curve is of practical interest, when one wishes to obtain the mean agglomeration number or the concentration of the particles from the absorption spectrum.

2.5.3 Laser Flash Photolysis

As Fig. 2.33a, b shows the time profiles of the absorbance changes which were recorded after application of a single 347-nm flash of the ruby laser. Immediately after the flash, a negative signal was observed at 320 and 340 nm, respectively.

The higher concentration of oxygen in the solution, the more rapidly it decayed after the flash. In the case of the small particles as Fig. 2.33a the decay was much faster than for the larger ones as Fig. 2.33b. The spectra of the absorbance change are shown in the upper part of Fig. 2.34a, b. The lower part of the figures contains the absorption spectra of the solutions. It is seen that the signal is negative in a wavelength range of about 60 nm below the onset of absorption. In the previous studies on the continuous illumination of deaerated ZnO solutions long-lived bleaching had been observed in this wavelength range. The dashed curve in the lower part of Fig. 2.34b is the absorption spectrum of the solution after the laser flash as calculated from the original absorption spectrum and the changes in

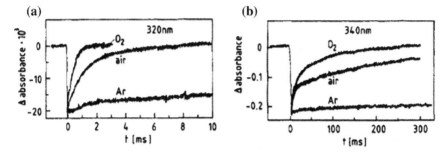

Fig. 2.33 **a** Change in absorption as a function of time after a laser flash. 1×10^{-3}M ZnO and 2×10^{-4} M excess Zn^{2+} in methanol. Concentration of absorbed photons: $4 \times$ M. Mean ZnO particle size: 17 Å. **b** Change in absorption as a function of time after a laser flash. 8×10^{-4} M ZnO and 1.6×10^{-4} M Zn^{2+} in methanol-water (40:60 vol.%). Concentration of absorbed photons: 3×10^{-4} M. Mean ZnO particle size: 40 Å

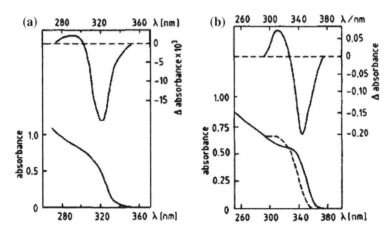

Fig. 2.34 **a** *Up* absorbance immediately after the laser flash as a function of wavelength. *Down* absorption spectrum of the ZnO solution which as in Fig. 2.33a. **b** *Up* absorbance after the laser flash immediately as a function of wavelength. *Down* absorption spectrum (*full line*) of the ZnO sol and spectrum (*dashed line*) of the solution after the laser flash. Solution is in Fig. 2.33**b**

Fig. 2.35 γ-irradiation of a deaerated 1×10^{-3} M ZnO sol in methanol containing 0.1 M formaldehyde and 3×10^{-3} M NaOH. Absorption spectrum is at different times of irradiation and dose rate is 2.45×10^4 rad/h

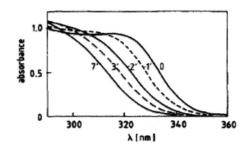

absorption. Note that the absorbance at the laser wavelength of 347 nm was much lower in the experiments of Fig. 2.34 than in those of Figs. 2.33b and 2.34b.

A deaerated solution of 1×10^{-3} M ZnO containing 3×10^{-3} M excess NaOH and 0.1 M formaldehyde as electron scavenger was γ-irradiated at a dose rate of 2.45×10^4 rad/h. Figure 2.35 shows the absorption spectrum at different times of irradiation. It is seen that the onset of absorption was shifted to shorter wavelengths, this effect becoming less and less pronounced with increasing irradiation time until a final shift was reached after about 7 min. γ-Ray absorption produces reducing organic radicals, CH_2OH, with a radiation chemical yield of 6.6 radicals per 100 eV of absorbed radiation energy, which are dissociated in the presence of NaOH: $CH_2OH + OH^- + CH_2O^- + H_2O$. The reaction of these radicals with the colloidal particles produced the same shift of absorption as the direct illumination with UV light.

The shifts persisted after the irradiation for about 20 min and then very slowly disappeared. Exposure of the irradiated solution to air led to an immediate recovery of the absorption spectrum. In Fig. 2.36, the change in absorbance at various wavelengths is plotted as a function of the radiation dose. The latter is given in terms of the concentration of CH_2OH radicals produced. It is recognized that a decrease in absorbance was observed at wavelengths above about 315 nm. At shorter wavelengths an increase occurred in the early stages of irradiation, followed by a decrease at higher doses.

Fig. 2.36 Change in absorbance at various wavelengths as a function of irradiation dose (dose expressed as concentration of radicals generated). Solution is shown in Fig. 2.35

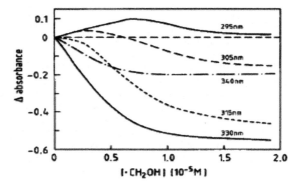

Two sols of different particle size were used in the pulse radiolysis experiments. Sol A contained particles of mean size 30 Å in water-methanol (60:40 vol.%) sol B contained 20 Å particles in methanol. The absorption spectra of both solutions are shown in the lower part of Fig. 2.37.

Two kinds of experiments were performed with these sols. In the first experiment, the solution was saturated with nitrous oxide and the intensity of the analyzing light beam kept very low. Under these conditions, CH_2OH radicals were generated which attacked the colloidal particles after the pulse. In most of the experiments, a train of eight pulses were used, the duration of a pulse being 1.5 ps and the interval between the pulses 50 ms. Typical kinetic traces are shown in Fig. 2.38 for sol A. It can be seen that the absorption at 340 nm decreased within a few milliseconds after each pulse, the decrease after the first pulse being noticeably smaller than after the following ones. At 320 nm, however, an increase after each pulse was observed. Similar observations were made with sol B. The absorbance

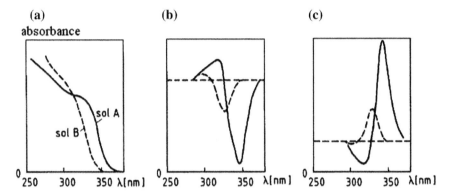

(a) **(b)** **(c)**

absorbance

sol A
sol B

Fig. 2.37 a Absorption spectrum of the two colloids (sol A and B), **b** difference spectra in the irradiation of the solutions under N_2O and air (**c**), with N_2O saturated solutions, the intensity of the analyzing light beam was very low and high intensity

Fig. 2.38 Changes in absorbance of sol A (30 Å, particles) upon irradiation with a train of pulses (irradiation under N_2O, low intensity of analyzing light beam)

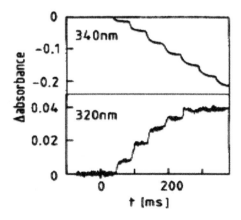

340nm

320nm

Δabsorbance

t [ms]

changes after the third pulse were used to construct the difference spectra shown in part b of Fig. 2.37. They are very similar to the spectrum obtained with laser illumination (see Fig. 2.35), i.e., bleaching was occurred in a certain wavelength range below the absorption to increase absorption at shorter wavelengths.

The difference spectrum was more intense for sol A ($\Delta \mathcal{E}_{cmax} = -1.1 \times 10^5 \text{ M}^{-1} \text{ cm}^{-1}$). Note that the observed changes in absorbance were independent of the analyzing light intensity provided that low intensities were applied. In the second experiment, the solutions contained air and the intensity of the analyzing beam was high. Under these circumstances oxidizing free radicals, such as O_2^- and O_2CH_2OH, were formed which reacted with the colloidal particles during the interval between the pulses. Figure 2.39 shows kinetic traces obtained with sol A in single-pulse experiments. Depending on the wavelength, bleaching or absorption signals were observed. Note that the half-life time of the buildup of these changes was independent of the wavelength of absorption. It amounted to 0.84 ms. The difference spectra for the two sols are shown in part of Fig. 2.37. It is seen that these spectra are mirror images of the spectra shown in part b. In other words, where bleaching occurred in the experiments of part b absorption signals were observed in the experiment of part a, and vice versa. Note that in these experiments the change in absorption became stronger with increasing intensity of the analyzing light. Experiments under N_2O were also carried out with solutions in which 2-propanol or 2-methyl-2-propanol were the solvents (besides a small amount of methanol; see experimental section). It was found that the $(CH_3)_2COH$ radicals, which were generated in 2-propanol, reacted as fast with the colloidal particles as the CH, OH radicals produced in methanol solution. The $CH_2(CH_3)_2COH$ radicals which were generated in 2-methyl-2-propanol did not react with the ZnO particles.

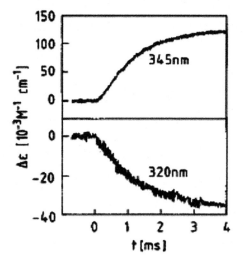

Fig. 2.39 Changes in absorption of sol A in a single-pulse experiment (irradiation under air, high intensity of analyzing light beam)

ZnO cannot be made as a colloid by reaction of Zn^{2+} ions with NaOH as the hydroxide which is formed does not dehydrate: $Zn(OH), -ZnO + H_2O$. In fact, it can be calculated from the thermodynamic data of the substances involved that the dehydration reaction is practically thermoneutral at room temperature. That ZnO is formed in alcoholic solution must be due to a substantially lower free enthalpy of the products of hydration. Kinetically, the process is very complex. It was observed that ZnO is formed in alcoholic solutions only in the presence of small amounts of water. Further in the preparations, first a stable solution of tetrahydroxozincate was made, and the formation of ZnO was initiated by the addition of water to this solution. It thus seems that water exerts a catalyzing effect on the transformation of zincate into ZnO in methanol solution. No detailed mechanism of this catalysis can be given at the present time.

2.5.4   Particle Size and Absorption Threshold

Quantum mechanical calculations of the shift of the band gap in small semiconductor particles were carried out. The lowest eigenstate of an exciton was calculated by solving Schrodinger's equation at the same level of approximation as is generally used in the analysis of bulk crystalline electron-hole states. Both the electron and the hole were considered as particles in a spherical box, and the usual values of the effective masses of the charge carriers were used. More recently, it was shown that the experimental observations on small CdS particles could be well described by wave mechanical one body calculations using a wave function of the form $\exp(-\gamma r)\,\psi_l(r)$, where the hydrogen-like factor takes account of the Coulomb attraction and $\psi_1(r) = 1/r\sin(\pi(r/R))$ is the lowest article in-a-spherical-box orbital (r = variational parameter, R = radius of the particle, r = distance from center of particle). For $r \geq R$, a potential increase of 3.8 eV was used. The curve in Fig. 2.32 was calculated by using the above wave function, expressing r and R in units of $\epsilon_\infty \hbar^2/\mu^* e^2$, where the reduced effective mass, μ^* was 0.1775 and the dielectric constant of ZnO (ϵ_∞) was 3.82, and again using a potential jump of 3.8 eV. Taking into consideration that certain unknowns exist in these calculations such as the exact shape of the particles and the applicability of the bulk values of μ^* and ϵ_∞, one may conclude that there is good agreement between the experimental observations and the theoretical calculations on small ZnO particles. The results of calculations are also included in Fig. 2.32. They over estimate the size quantization effect.

2.5.5   Blue Shift of Absorption upon Illumination

In the laser flash photolysis experiments (Figs. 2.33 and 2.34) similar observations were made as in the previous studies on the continuous illumination of ZnO sol. The onset of absorption was shifted to shorter wavelengths; i.e., an effect was observed as if the particles had become smaller under illumination. In fact, in our first explanation of the phenomenon, a mechanism of dissolution of illuminated ZnO was discussed and the absorption shift attributed to the decrease in size of the colloidal particles. However, after it was found later in the case of small CdS particles that storage of an electron affected the absorption spectrum in the same way, it was proposed that illumination of ZnO particles also produces excess electrons. The shift is explained by an increase in the energy of the exciton formed by light absorption due to the strong electric field caused by the excess electron. Perhaps, it could also be explained as the first step of "band filling," where by the excess electron fills the lowest state so that subsequent absorption requires higher photon energies in order to access empty states. The conclusion that the shift is caused in ZnO by excess electrons is corroborated by the fact that the shift in aqueous ZnO sols was much stronger when alcohol was present. The alcohol acts as a scavenger of the positive holes, simultaneously formed upon light absorption, and thus prevents the electrons from recombining with holes.

The excess electrons are long-lived in the absence of oxygen. In the presence of O, they are removed from the colloidal particles which explain the decay of the bleaching signal after the laser flash (Fig. 2.33). However, this reaction is relatively slow. From the half-life time of 0.8 ms in air-saturated solution as Fig. 2.33a, one calculates a rate constant of 3.2×10^6 M^{-1} s^{-1} for the reaction of O_2 with electrons stored on 17 Å ZnO particles. In the case of 40 Å particles to see Fig. 2.33b the specific rate is about 50 times lower. It thus seems that the electrons are present in traps of less negative potential in the case of large particles as is expected from the theory of size quantization. In the laser experiments, where the wavelength of the exciting light was not far from the onset of absorption, smaller particles in the solution were less excited, or even not at all, than the larger particles which possibly absorbed more than one photon. Under these conditions, a more detailed analysis of the data, such as a correlation between the amount of bleaching and the number of excess electrons, did not seem promising. Clearer conditions prevailed in the γ-radiolysis experiments where the number of stored electrons was known and no loss of electrons due to recombination with holes had to be feared. The γ-Radiolysis: Electron Injection from Free Radicals. When reducing radicals, which are formed in the γ-irradiation of methanol or 2-propanol solutions, react with ZnO particles, similar changes in the absorption spectrum were observed as in the UV illumination of such solution. The effects are explained in terms of electron injection from the organic radicals onto ZnO particles. In fact, CH_2OH and $(CH_3)_2COH$ radicals have reduction potentials (-1.0 and -1.5 V, respectively) which are much more negative than the lower edge of the conduction band in ZnO (-0.2 V). On the other hand, the $CH_2(CCH_3),OH$ radicals, formed in 2-methyl-2-propanol solution, did not react with ZnO which is understood in terms of the lower reducing power of such radicals.

From the initial slope of the curve for 330 nm, where maximum bleaching occurred in the experiments of Fig. 2.36, a negative absorption coefficient of 9×10^4 M cm^{-1} is calculated. The mean agglomeration number of the colloidal particles was 230. At an overall concentration of $8 \times$ M ZnO, the particle concentration was 3.5×10^{-6} M. The solution had an absorbance of 0.6 at 330 nm (curve 0 in Fig. 2.36), the absorption coefficient of the colloidal particles being $0.6/3.5 \times 10^{-6} = 1.7 \times 10^5$ M cm^{-1}. Thus, find that deposition of an electron on a small colloidal ZnO particle is accompanied by a decrease in its 330-nm absorption by almost a factor of 2. This indicates that the excess electron influences not just one ZnO molecule in the colloidal particle but an optical transition in which practically all the ZnO molecules in the particle are involved. It has made similar arguments in previous study on electron deposition on small CdS particles. When more than one electron is deposited per ZnO particle, i.e., when more than about $3.5 \times$ M free radicals were generated in the experiments of Fig. 2.36, the changes in absorbance no longer increased in a linear manner. At 330 and 340 nm, the absorbance decreased less and less strongly. At 315 nm, little absorption changes occurred during deposition of just one electron, but significant bleaching took place upon the storage of additional electrons. The shorter wavelengths occurred upon deposition of the first electron, which decrease further electron injection. The nature of the stored electrons is not known yet. They could be present in traps or in the form of monovalent zinc ions, Zn$^+$. The latter possibility does not seem very plausible because of the very negative redox potential of Zn$^+$ (<-2 V). On the other hand, Zn$^+$ is known to absorb close to 300 nm, i.e., the wavelength where the absorption increased upon deposition of one electron. When more than one electron is present, isolated Zn atoms may be formed.

2.5.6 *Reducing Reactions and Oxidizing Radicals*

As in the continuous irradiation experiments (Fig. 2.35), the reaction of pulse radiolytically generated CH$_2$OH radicals with ZnO particles caused bleaching at longer and absorption at shorter wavelengths (Figs. 2.37 and 2.38). Upon application of the first pulse (Fig. 2.38) the changes in absorbance were smaller than in the subsequent ones. This is attributed to traces of oxygen adsorbed on the colloidal particles which cannot be removed by bubbling the solution with another gas. The small radiation dose applied in the first pulse was sufficient to remove this residual oxygen. The rate constant of reaction of the radicals with ZnO particles was determined in a single-pulse experiment using a solution which had been preirradiated with two pulses. At a concentration of ZnO particles of 3.5×10^{-6} M the half-lifetime of the 340-nm bleaching, which obeyed pseudo-first-order kinetics was about 2 ms. A rate constant of 1×10^8 M^{-1} s^{-1} is calculated. This value is more than 10 times smaller than expected for a diffusion-controlled reaction.

It has already been mentioned that the intensity of the analyzing light beam had to be kept low in the laser flash and electron pulse experiments using N_2O-saturated solutions in order to avoid storage of electrons on the ZnO particles before the laser flash or electron pulse arrived. In experiments with aerated solutions as Figs. 2.37a and 2.39, the intensity of the analyzing light beam was as strong as in ordinary pulse radiolysis experiments with chemical systems that do not photolyze. The reaction of oxygen with stored electrons being rather slow as described above, illumination with the analyzing light beam led to a certain stationary concentration of stored electrons on the colloidal particles. The oxidizing radicals which were formed in the electron pulse via the processes that reacted with ZnO particles carrying electrons as following:

$$-CH_2OH + O_2 \rightarrow O_2CH_2OH \tag{2.38}$$

$$-O_2CH_2OH \rightarrow HO_2 + CH_2O \tag{2.39}$$

$$HO_2 \leftrightarrow H^+ + O_2^- \tag{2.40}$$

Removal of an electron by an oxidizing radical was accompanied by the recovery of the bleaching (at 345 nm) and of the absorption (at 320 nm) which were originally caused by this electron. The result being that an absorption signal at 345 nm and a bleaching signal at 320 nm were now observed (see Fig. 2.39). The mirror images of the difference spectra obtained after the attack of reducing and oxidizing radicals (as in Fig. 2.37) are thus readily understood. From the data in Fig. 2.39 it was calculated that the oxidizing radicals reacted with a specific rate of 3.2×10^6 M^{-1} s^{-1} with the colloidal particles.

The changes in absorbance were smaller when ZnO particles of smaller size were used in the experiments of Fig. 2.37. When an excess electron is deposited on a smaller ZnO particle, the number of ZnO molecules affected is smaller, this effect leading to a smaller change in absorbance. On the other hand, one would expect that the optical changes should also become smaller in the case of very large particles (larger than used in the present work) as the electric field produced by the excess electron would be rather weak in parts of the particle. These two opposing effects make believe that there might exist a definite particle size where the absorbance changes that accompany the deposition of one electron are most pronounced. The large bleaching coefficient of 1.1×10^5 M^{-1} cm^{-1}, observed for 40–50 Å particles, may be taken as an indication that this particle size is close to 40 Å. However, further experiments with particles of different size are required to check the validity of this supposition. It has recently been proposed that the fluorescence of colloidal ZnO in aerated solution is brought about by a "shuttle" mechanism. An electron generated by light absorption transfers to adsorbed O_2 to form O_2^-. The latter transfers the electron into a deeper trap on the surface. Recombination with a preexisting hole is accompanied by light emission. In the course of the present studies on solutions in which O_2^- is formed a few experiments were carried out to check this mechanism. In an experiment with the analyzing light beam shut off, fluorescence light should have been emitted, if O_2^- transferred an electron to ZnO and if recombination with a preexisting hole took place. No fluorescence was observed. Have to conclude that the more direct kinetic method of pulse radiolysis does not confirm the above mechanism. A shuttle mechanism in solutions containing methyl viologen, MV^{2+}, as additive has also been proposed, MV^+ acting as the shuttle agent, which is formed by electron transfer to MV^{2+} and then transfers its electron into a lower surface state of ZnO. More recent pulse radiolysis experiments in our laboratory showed that MV^+ does not react with ZnO particles. Again, it must be concluded that a shuttle mechanism is not operative.

2.5.7   Essential Criteria of Photochemical Reactions

Essential criteria for all photochemical reactions is including molecule must absorb light and radiation energy must match energy difference of ground and excited state. Typical absorption range of some important classes of organic compounds is including kinds of simple alkene (190–200 nm), acylic diene (220–250 nm), cyclic diene (250–270 nm), styrene (270–300 nm), saturated ketones (270–280 nm), α, β-unsaturated ketones (310–330 nm), aromatic ketones/aldehydes (280–300 nm), aromatic compounds (250–280 nm), etc. But the light is absorbed by a molecule also. The Franck–Condon principle says that the heavy atom nuclei do not change their positions. This leads to an initial geometry of the excited state which is usually not the energy minimum. During excitation the electron spin remains unchanged. Spin inversion during excitation is forbidden by quantum mechanics and therefore unlikely. Right after the excitation several things may happen. (1) Vibronic relaxation brings the molecule quickly into the new energy minimum structure for the excited state. Energy is released into the solvent. (2) Intersystem crossing leads to triplet states by spin inversion. The new energy minimum is reached by vibrational relaxation. (3) Emission of light and return to the ground state (luminescence, fluorescence, phosphorescence). (4) Quenching of the exited state: Energy is transferred to another molecule. Usually, we observe diffusion-controlled dynamic quenching by collision. Investigation of this is possible by the Stern–Vollmer plot (1/quantum yield *vs* concentration of quencher). Gives a strait line for diffusion-controlled quenching; large excess of quencher usually needed (1000 times excess). Molecule goes back to ground state by vibrational (thermal) deactivation (no light emission). The energy goes to the solvent/environment of molecule and pathways of photochemical processes as shown Fig. 2.40 [18, 19].

2.5.8 Main Types of Photochemical Reactions

The excited states are rich in energy. Therefore, reactions may occur that are highly endothermic in the ground state. Using the equation $E = \hbar v$ can correlate light of a wavelength of 350 nm with an energy of 343 kJ/mol. In the excited state anti-bonding orbitals are occupied. This may allow reactions which are not possible for electronic reasons in the ground state. Photochemical reaction can include singlet and triplet states. Thermal reactions usually show singlet states only. In photochemical reaction intermediates may be formed which are not accessible at thermal conditions as shown in Fig. 2.40.

The main influence complications to photochemical reaction and practical considerations for experiment as following:

1. Purity of starting materials is importance.

This prerequisite holds for many techniques in synthesis. While working with organometallic intermediates requires exclusion of air and moisture, photochemistry is very sensitive to colored or light absorbing impurities—either in the starting materials or formed during reaction. Both may interfere with the photo processes and may kill the reaction.

2. UV spectra of substrates.

Before starting a photochemical reaction a UV/vis spectrum of the "photoactive" compound is recorded. The "photoactive" compound is the molecule which should be electronically excited and undergo or initiate a reaction from its excited singlet or triplet state. From UV spectra recorded with different compound concentrations the extinction coefficients of all bands can be obtained, even for the weak absorptions which may be of importance. UV spectra of all reagents should be recorded to make sure that there is no or little interference in absorption with the "photoactive" compound. If available, a UV spectrum of the product should be recorded. UV spectra from the reaction mixture may help to identify ground-state interactions of compounds or CT complexes, and guide the way to the best reaction conditions. The UV spectrum of a photochemical reaction is shown in Fig. 2.41.

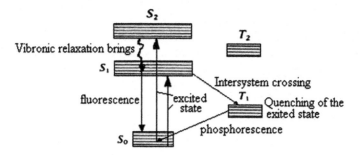

Fig. 2.40 The photochemical reaction may occur

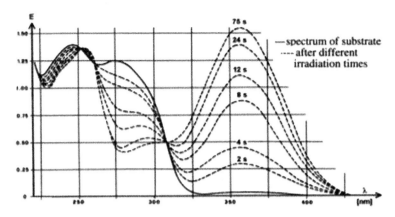

Fig. 2.41 UV spectrum of a photochemical reaction

2.5.9 Solvents

Photochemical reactions can be performed—in principle—in the gas phase, in solid state or in solution. For practical reasons most photochemical reactions are done in solution, therefore the choice of the right solvent is critical. At normal concentrations of a photochemical reaction the concentration of the "photoactive" compound is only 100–1000 times larger than the concentration of the solvent. If the extinction coefficient of the "photoactive" compound is only 10 times higher than that of the solvent at the irradiation wavelength, we will observe a significant filter effect of the solvent. The reaction is much slower than it could be.

Key selection criteria are including: solvent must dissolve reactants (try), solvent should be transparent at the irradiation wavelength and solvent must be free of impurities (analysis; add edta to complex trace metal ion content if necessary). If a reactive intermediate needs to be stabilized by the solvent, this has to be tried out until a solvent is found which matches all criteria. The solvents used for photoreaction is shown in Table 2.1.

2.5.10 Direct Sensitized Photolysis

The absorption characteristics do not tell anything about the behavior of a molecule in the excited state. It may rapidly deactivate via fluorescence or radiationless, it may undergo intersystem crossing into the triplet manifold. If available, data on fluorescence, phosphorescence, lifetimes, and quantum yield are very helpful to understand the processes.

From these data we can learn if it is necessary to sensitize the formation of the triplet state for a reaction or if the triplet is rapidly formed without our

Table 2.1 Solvents used for photoreaction

Solvent	Cut-off wavelength[a]	ε_r^b	ET (30)[c]
Water	185	78.30	63.1
Acetonitrile	190	35.94	45.6
N-hexane	195	1.88	31.0
Ethanol	204	24.5	51.9
Methanol	205	32.66	55.4
Cyclohexane	215	2.02	30.9
Diethyl ether	215	4.20	34.5
1,4-dioxane	230	2.21	36.0
Methylene chloride	230	8.93	40.7
Chloroform	245	4.81	39.1
Tetrahydrofuran	245	7.58	37.5
Ethyl acetate	255	6.02	38.1
Acetic acid	250	6.17	51.7
Carbon tetrachloride	265	2.23	32.4
Dimethylsulfoxide	277	46.45	45.1
Benzene	280	2.27	34.3
Toluene	285	2.38	33.9
Pyridine	305	12.91	40.5
Acetone	330	20.56	42.2

[a]Wavelength (nm) at which E is approximately 1.0 in a 10 mm cell
[b]Dielectric constant
[c]Dimroth Reichardt values (kcal/mol) for the longest-wavelength solvatochromic absorption based on a pyridinium-N-phenoxide betaine dye 30

help. Sometimes the solvent can be used as a sensitizer. Then it should adsorb at the irradiation wavelength and transfer the energy to the reactant. Acetone is a typical example of such a solvent (adsorbs up to 330 nm). Sensitizers and quenchers can help to investigate a photochemical reaction:

1. Pure singlet reactivity: No reaction in the presence of appropriate triplet sensitizers.
2. Pure triplet reactivity: enhanced product formation in the presence of appropriate sensitizers; no reaction in the presence of triplet quenchers.

Triplet as well as singlet reactivity: combination of methods (1) and (2) gives a product pattern corresponding to the specifically activated states. The sensitizers and quencher in nonpolar solvents were shown in Table 2.2, where (1) benzoic acid, (2) triplet energies in kJ/mol, (3) first excited singlet state energies in kJ/mol, (4) quantum yields for singlet-triplet intersystem crossing, (5) in polar solvents.

Table 2.2 The sensitizers
and quencher in nonpolar
solvents

Compound	E_T	E_S	Φ_{ISC}
Benzene	353	459	0.25
Toluene	346	445	0.53
Methyl benzoate	326	428	
Acetone	332	372	0.90/1.00
Acetophenone	310	330	1.00
Xanthone	310	324	
Benzaldehyde	301	323	1.00
Triphenylamine	291	362	0.88
Benzophenone	287	316	1.00
Fluorene	282	397	0.22
Triphenylene	280	349	0.86
Biphenyl	274	418	0.84
Phenanthrene	260	346	0.73
Styrene	258	415	0.40
Naphthalene	253	385	0.75
2-acetylnaphthalene	249	325	0.84
Biacetyl	236	267	1.00
Benzil	223	247	0.92
Anthracene	178	318	0.71
Cosine	177	209	0.33
Rose bengale	164	213	0.61
Methylene blue	138	180	0.52

2.6   Reaction Control

In the course of the reaction more and more product is formed which competes with
the starting material for light. If the product is available its UV spectrum gives
information about possible competition. The increasing absorption of irradiation by
the product may stop the reaction before complete conversion will be reached.
Therefore, it is important to follow the reaction by UV spectroscopy (see figure
above of UV spectra during reaction). The formation of colored byproducts may
stop a reaction before complete conversion, too. Chromatographic methods, such as
TLC, GC, or HPLC should be used to gain information about the course of the
reaction.

2.6.1   Side Reactions Can Easily Become the Major Track

Side reactions of photochemical reactions can in some cases become the major
reaction pathway. Examples are photosensitizers, which are use in catalytic or

stoichiometric amounts to mediate the wanted photochemical process, but they may also act as photoinitiator of a radical chain reaction. If solvents or starting materials are present that are susceptible to a radical chain process, this reaction will become dominant. Another origin of severe side reaction may be the presence of oxygen. Photooxygenation may be the desired photoprocess, but if not, it may be a side reaction. Oxygen should be excluded to avoid this. Even if only small amounts of peroxides are formed during the reaction they may become hazardous upon work up. The use of inert gas is not necessary, if the essential excited state is not efficiently quenched by triplet oxygen (which is often the case for short living singlet states). Free radicals produced during the reaction may cause side reactions. Radical scavengers, such as phenols, are added to trap them.

2.6.2   Quantum Yield and Chemical Yield

While reactions of "normal ground state" chemistry are described by the chemical yield of the reaction as one major indicator (there are others, in particular when it comes to describe the technical efficiency of a chemical transformation), for the chemistry of excited molecules another parameter has to be considered: The quantum yield of the reaction. The quantum yield is the number of events (e.g., photochemical induced transformation) divided by the number of absorbed photons in a specific system. Quantum yields can range from 0 to 100 or higher; if smaller than 0.01 conversion is very slow (chemical yield may still be high); for photoinitiated chain reactions the quantum yield can be as high as 100.000.

How much light goes into the reaction (determination of the amount of product formed) which has to calculate a quantum yield. A standard chemical procedure used since the 1950s is chemical actinometry. A compound that undergoes a defined photochemical transformation with known quantum yield is used to determine the light intensity. The concentration of the actinometric compound and the pathway of the exposure cell must be sufficiently high to make kinetics of the reaction approx. zero order. In this way, the rate of the reaction is not concentration dependent. The reaction of 2-nitrobenzaldehyde to 2-nitrosobenzaldehyd is a typical example in Fig. 2.42 for example. The quantum yield of the process is 0.5 for irradiation from 300–410 nm, and the $d_{Act}/dt = I_0\eta f$, where I_0 is light intensity, η is quantum yield and f is fraction of absorbed light.

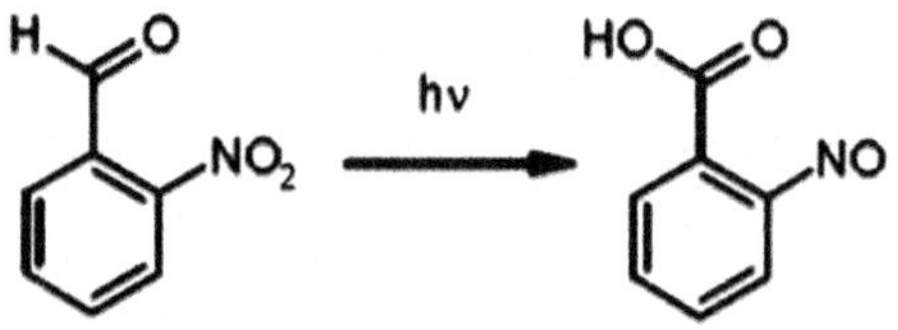

Fig. 2.42 The reaction of 2-nitrobenzaldehyde to 2-nitrosobenzaldehyd

Typical light sources for photochemistry are including:

1. Wavelength from 300 to1300 nm lasers, the sun is about also.
2. Low-pressure mercury (Hg approx. 10^{-5} atm) lamp: 185 nm (5 %); 254 (95 %) (see Fig. 2.43).
3. Rayonet lamps (specific emission wavelength from secondary fluorescence emission with coated layer as shown in Fig. 2.43c).
4. Medium pressure Hg lamps (Hg vapor pressure 5 atm) lamps (distinct lines between 250 and 600 nm),
5. High-pressure Hg lamps (Hg vapor pressure approx. 100 atm; expensive, easily damaged) (emission 360–600 nm, broad).
6. Low and high-pressure sodium lamps (emission around 600 nm).
7. High power light-emitting diodes (available at low cost for 650–400 nm; very narrow and intensive emission, long lifetime; UV-LED are currently still expensive. The emission properties of them are shown in Fig. 2.43).

Many types of reactors for photochemical reactions are known, the most typical are:

1. Apparatus for external irradiation (simplest case is an irradiated flask).
2. Immersion-well reactor in which the lamp is surrounded by the reaction solution.
3. Falling film apparatus.
4. Photo microreactor (similar to falling film, but easier to handle).

In all cases the light sources usually needs cooling to avoid its overheating and heating of the reaction solution. The material of the reactor depends on the irradiation wavelength necessary.

For irradiation at 254 nm quartz glass (expensive apparatus) is needed. For irradiation at 300 nm Pyrex glass is needed, and for irradiation >350 nm normal lab glass (window glass) is sufficient. The glass acts as a solid filter. Additional solid or liquid optical filters may be used to restrict the irradiation wavelength.

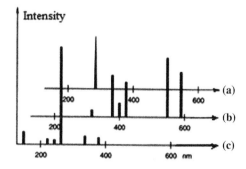

Fig. 2.43 The emission properties of *a* RPR-3000 Å for Rayoned photoreaction. *b* mercury low-pressure lamp and *c* mercury low-pressure lamp (strong 254 nm line)

Low-pressure mercury lamps have their main output at 254 nm. This light severely damages cells, eyes, and skin. Shield reactors; turn lamps off before checking the reaction. Never look into the beam of a high power LED; the lights very high intensity damage your eyes. Ozone generation: Short wavelength light may generate ozone from oxygen. Perform reactions always in a well ventilated fume hood. Lamps: Most lamps operate at high temperature and at high vapor pressure. Never move or touch lamps during operation. Never switch of the cooling right after switching of the lamp.

2.6.3   Photochemical Reactions

The absorption properties of ketones and aldehydes are convenient for irradiation around 300 nm ($n\pi^*$ 330–280 nm). Triplet–singlet energy gap is small (20–70 kJ/mol); intersystem crossing rates are high. Lifetime of first excited singlet state is in the nanosecond region for aliphatic aldehydes and ketones; in the sub-nanosecond region for aromatic aldehydes and ketones. Singlet photochemistry can be detected with aliphatic aldehydes or ketones, while aromatic substrates, such as benzophenone or acetophenone, react exclusively from their corresponding triplet states and are excellent triplet sensitizers.

1. Norrish Type I cleavage reaction (α-cleavage reaction)

This reaction type dominates gas phase photochemistry of many aldehydes and ketones. Less common in solution chemistry is only if no suitable C–H bonds, that are present to allow hydrogen atom abstraction. The examples of norrish type I cleavage reaction (α-cleavage reaction) as shown in Fig. 2.45 Special topic: 1 C–C Bond formation in organic crystals. Photochemical irradiation of crystalline (2R,4S)-2-carbomethoxy-4-cyano-2,4- diphenyl-3-butanone 1 led to highly efficient decarbonylation reactions. Experiments with optically pure and racemic crystals show that the intermediate radical pairs undergo a highly diastereomers and enantiospecific radical-radical combination that leads to the formation of two adjacent stereogenic centers in good chemical yield and with high chemical control.

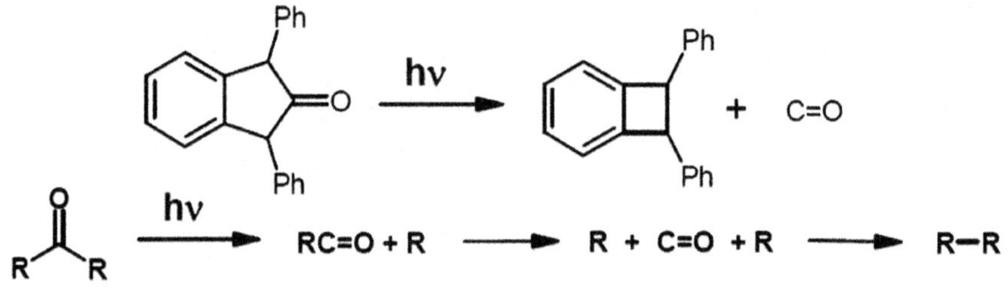

Fig. 2.44   The examples of Norrish type I cleavage reaction

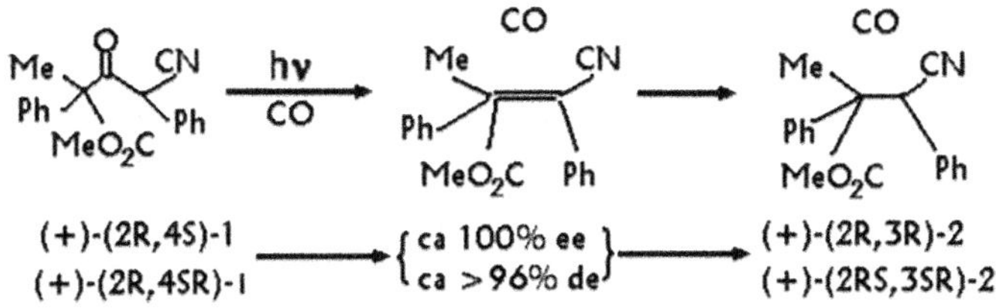

Fig. 2.45 Compound 1 was obtained in enantiomerically pure and racemic forms from 2-methyl-2-phenyl malonic acid monomethyl ester 3b. Samples of (+)-(R)-3b were prepared from the meso-diester 3a by enzymatic desymmetrization with pig liver esterase (90 %) or by resolution of acid (±)-3b with (-)-1-(1-naphthyl) ethylamine. Acid (+)-3b was converted into the corresponding acyl chloride, which was reacted with the anion of benzyl cyanide to give ketone (2R,4S)-1 and its diastereomer (2R,4R)-1a (not shown) in 85 % isolated yield in a 4:1 ratio

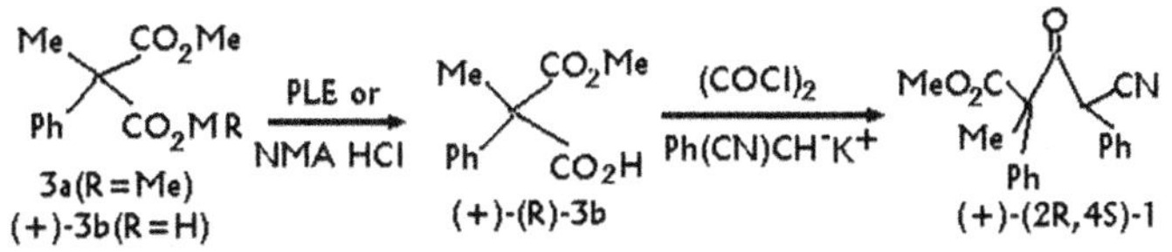

Fig. 2.46 Photochemical experiments with oxygen-free 0.1 M benzene solutions of (+)-(2R,4S)-1 and (±)-(2RS,4SR)-1 using a Pyrex filter (>300 nm) at 298 K led to complex product mixtures (as Fig. 2.47). The crystals of (+)-(2R,4S)-1 and (±)-(2RS,4SR)-1 (50 mg) irradiated under similar conditions resulted in a reaction with formation of a photoproduct in 40–60 % conversion and >95 % selectivity (in Fig. 2.46)

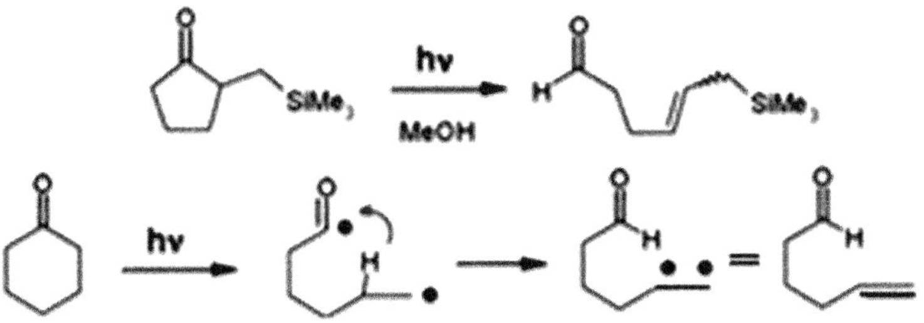

Fig. 2.47 The subsequent hydrogen migration

Reactions with chiral crystals occurred with quantitative enantiomeric yields and >95 % diastereomeric yields as shown in Figs. 2.44, 2.45 and 2.46.

2. Norrish type II photo elimination reaction

Beside carbon monoxide extrusion acyl radicals formed in a α-cleavage can be stabilized by subsequent hydrogen migration as an example is shown in Fig. 2.47.

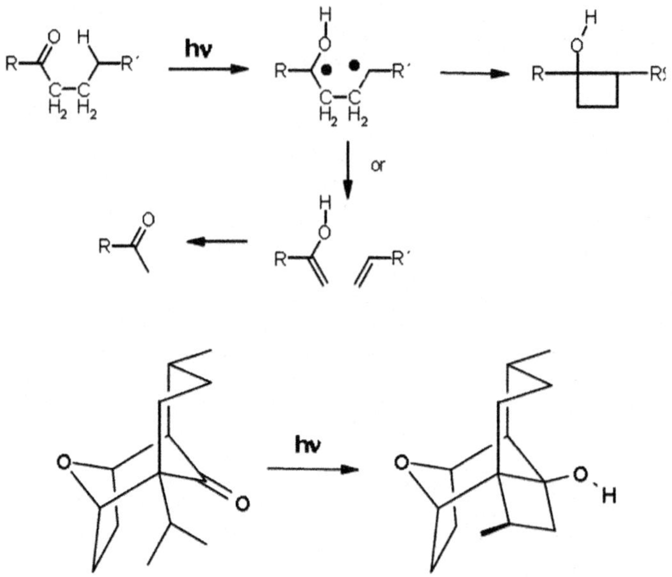

Fig. 2.48 The example the 1,4-diradical leads to cyclobutanols for C=O and C–H groups

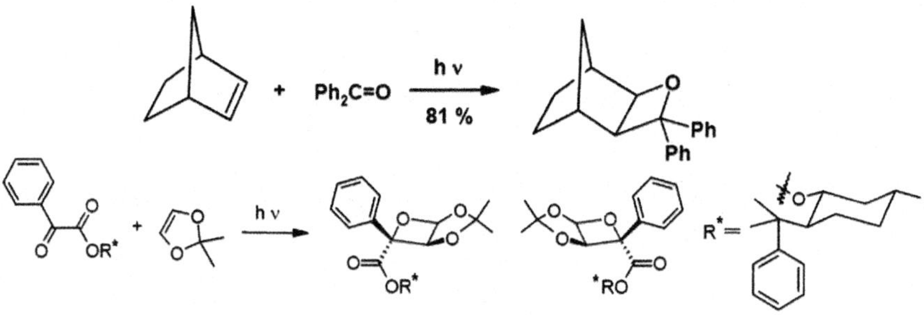

Fig. 2.49 The examples of high diastereoselectivities

3. Cyclobutanol formation

With an appropriate alignment of C=O and C–H groups and no secondary transformation prevents cyclization of the 1,4-diradical leads to cyclobutanols is shown in Fig. 2.48.

4. Photochemical [2 + 2] cycloaddition

The Photochemical [2 + 2] cycloaddition is photochemical [2 + 2] cycloaddition of an alkene and a carbonyl group reaction. Inter- and intra molecular examples are known. High diastereoselectivities can be observed in many examples in Fig. 2.49.

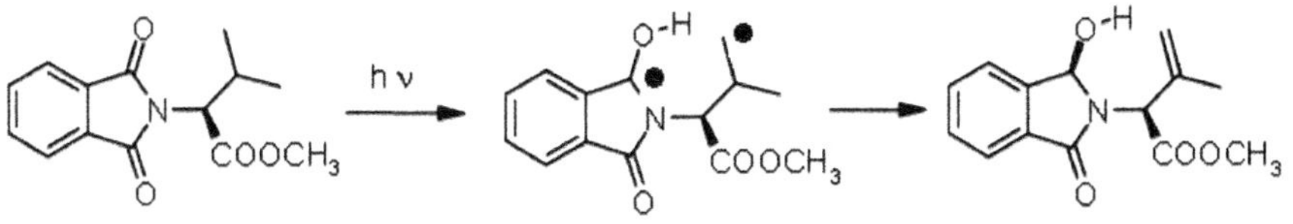

Fig. 2.50 An abstraction of hydrogen from the γ position

Fig. 2.51 The examples of photoreduction of an α, β-epoxy ketone

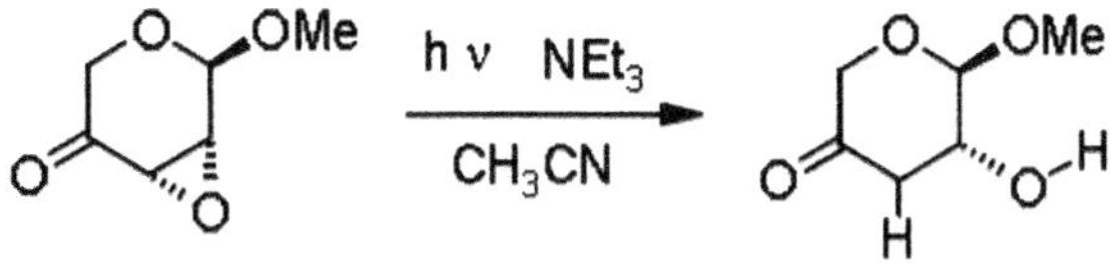

5. Photoisomerizations and photoreductions

Photoisomerizations and photoreductions are observed. Some examples are shown in Figs. 2.50 and 2.51:

In this reaction the first event is an abstraction of hydrogen from the γ position. Neither Norrish type II nor Yang reaction follow. Instead another hydrogen is stereoselectively transferred, now from the δ position.

The second example is a photoreduction of an α, β-epoxy ketone derived from carbohydrates. Triethylamine acts as the sacrificial electron and hydrogen donor. The configuration of one of the epoxy stereocenters remains intact as Fig. 2.51.

2.6.4 Alkenes

The formation of the lowest excited singlet state of simple alkenes arises from the allowed $\pi-\pi^*$ transition. This generally requires short wavelength irradiation extending to about 200–210 nm. Absorption of solvents and lack of suitable light sources make the use of simple alkenes in preparative photochemistry difficult. Substituted or conjugated derivatives are mainly used. Another possibility to circumvent the problem of absorption at short wavelengths is the use of sensitizers.

1. E,Z-Isomerizations

E,Z-Isomerizations of 1,2-disubstituted alkenes are well documented. A typical example is the photoisomerization of stilbene, where because of different absorption spectra the Z-isomere can be enriched in the photostationary state. A more complex example is the sensitized preparation of enantiomerically enriched trans-cyclooctene as Fig. 2.52.

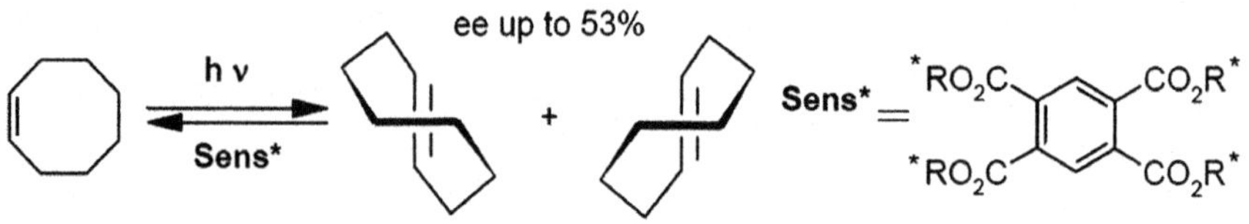

Fig. 2.52 The photoisomerization of stilbene

2. Sigmatropic shifts

These rearrangements involve a migration of a σ-bond across an adjacent π-system. The type of activation (thermal or photochemical) and the stereochemistry can often be predicted by the Woodward-Hoffmann rules. Processes involving stepwise biradical intermediates may be present in photochemical reactions as Fig. 2.53.

3. Di-π-methane rearrangement

This unique rearrangement was discovered later and belongs to the sigmatropic rearrangements of type. The rearrangement is stereospecific and has been used in organic synthesis as Fig. 2.54.

4. [2 + 2] Cycloaddition reactions

These transformations belong to the classic reactions that generally can be rationalized by the Woodward–Hoffmann rules. The [2 + 2] cycloaddition is photochemically allowed and a practical way to cyclobutane derivatives. Cyclobutane ring opening and hexatriene ring closure belong to the most frequently investigated

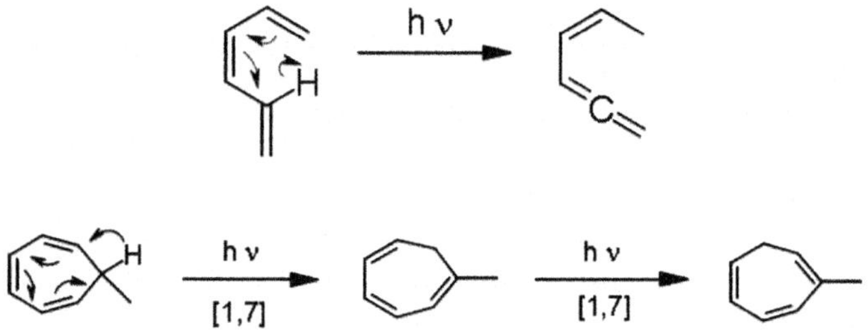

Fig. 2.53 Examples of processes involving stepwise biradical intermediates

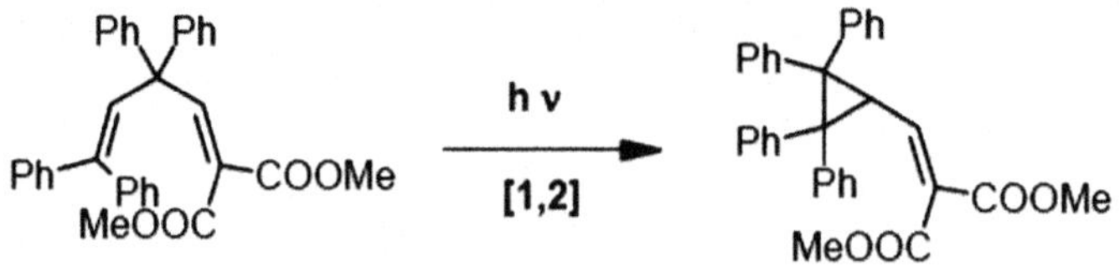

Fig. 2.54 Examples of the Di-π-methane rearrangement

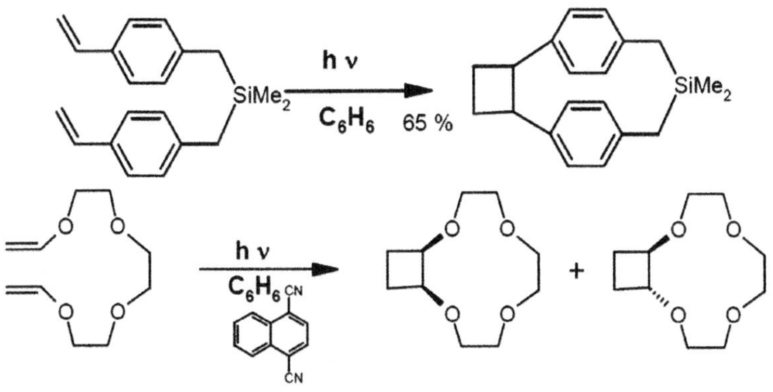

Fig. 2.55 Examples of intramolecular [2 + 2] photocycloaddition

electrocyclic reactions. The intramolecular [2 + 2] photocycloaddition is shown in Fig. 2.55. The length of the side chain leads to a complete control of the regioselectivity of the [2 + 2] photocycloaddition is shown in Fig. 2.56 and in Fig. 2.57.

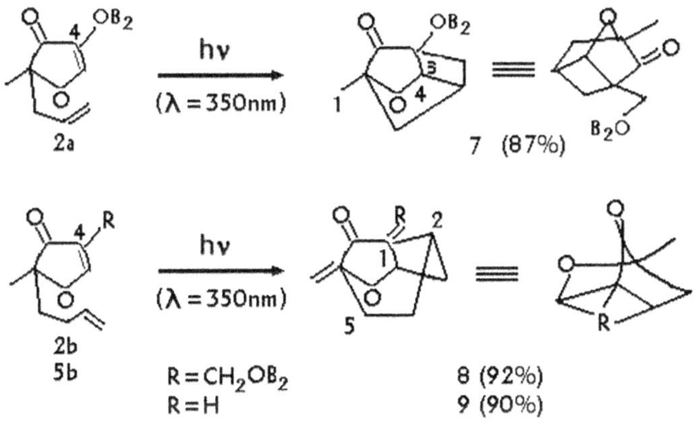

Fig. 2.56 The second example of intermolecular [2 + 2] photocycloaddition

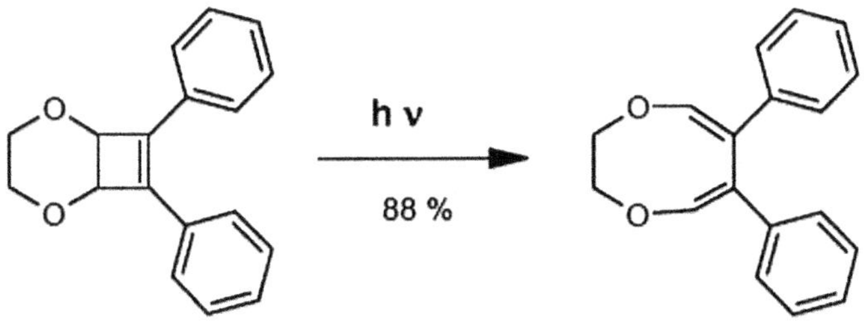

Fig. 2.57 The example of photoinduced electron transfer reactions with irradiation of 405 nm

2.7 Photoinitiation of Radical Reactions

In recent years, photoinitiated polymerization has received revitalized interest as it congregates a wide range of economic and ecological anticipations. For more than 30 years, photopolymerization has been the basis of numerous conventional applications in coatings, adhesives, inks, printing plates, optical wave-guides, and microelectronics. Some other less traditional but interesting applications, including production of optical disc and fabrication of 3D objects are also available. Many studies involving various photopolymerization processes have been continuously conducted in biomaterials for bones and tissue engineering, microchips, optical resins, and recoding media, surface relief gratings, anisotropic materials, polymeric photo-optical control materials, clay and metal nanocomposites, photoresponsive polymers, liquid crystalline materials, interpenetrated networks, microlens, multi-layers, surface modification, block and graft copolymerization, two-photon poly-merization, spatially controlled polymerizations, topochemical polymerization, solid-state polymerization, living/controlled polymerization, interfacial polymer-ization, mechanistically different concurrent polymerizations, pulsed laser poly-merization, polymerizations in microheterogenous media, and so forth. Interest has also grown in identifying the reactive species involved in the polymerization pro-cess by laser flash photolysis, time-resolved fluorescence and phosphorescence, and electron spin resonance spectroscopy as well as monitoring the polymerization itself by different methods including real-time IR spectroscopy, in-line NIR reflection spectroscopy, differential scanning calorimetry, in situ dielectric analysis, and recently developed optical pyrometry.

2.7.1 Photoinduced Electron Transfer Reactions

Photoinduced electron transfer reactions can be used to initiate radical reactions of alkenes. Both pathways, oxidative leading to a radical cation, and reductive, leading to a radical anion, are possible. The majority of reported examples involve oxidative pathway as Fig. 2.58, another example in Fig. 2.59.

1. Photoinduced cyclobutane ring opening

This reaction occurs in light dependent enzymatic DNA repair in bacteria. The enzymatic redox cofactor flavin is excited in its reduced form, transfers an electron onto the thymine cyclobutane dimer, which undergoes stepwise ring opening as radical anion. The electron is transferred back to the flavin and a new cycle can begin as shown in Fig. 2.60.

2. Alkene photocycloadditions in organic synthesis

Photochemical synthesis of a complex skeleton found in natural products. In contrast to simple homoallyl vinyl ethers, which have been shown to undergo

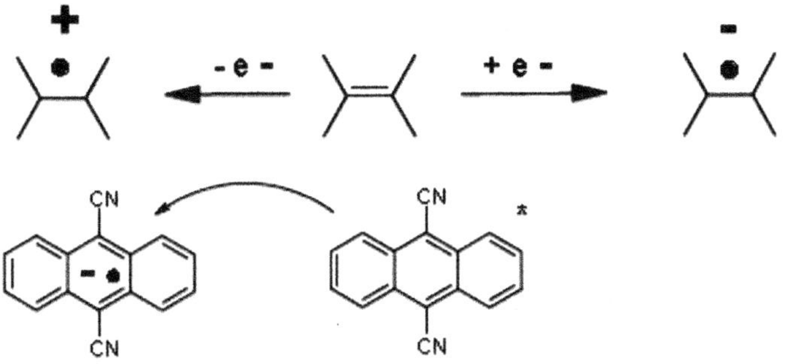

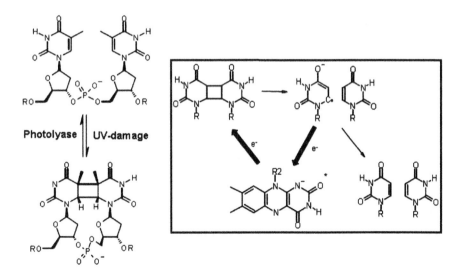

Fig. 2.58 Thee examples of oxidative pathway

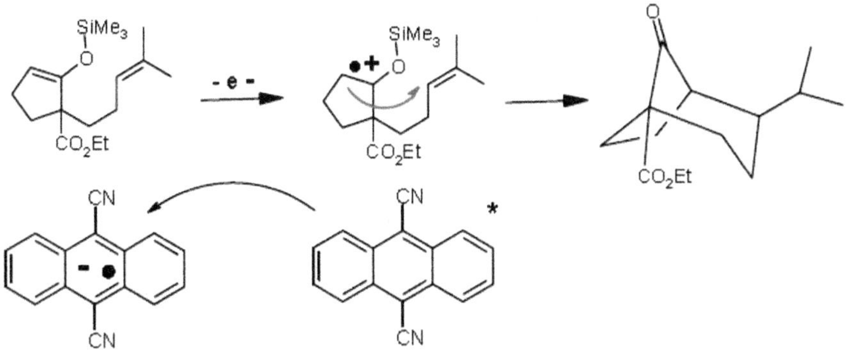

Fig. 2.59 Photooxidation of an enol silyl ether

Fig. 2.60 The electron is transferred back to the flavin

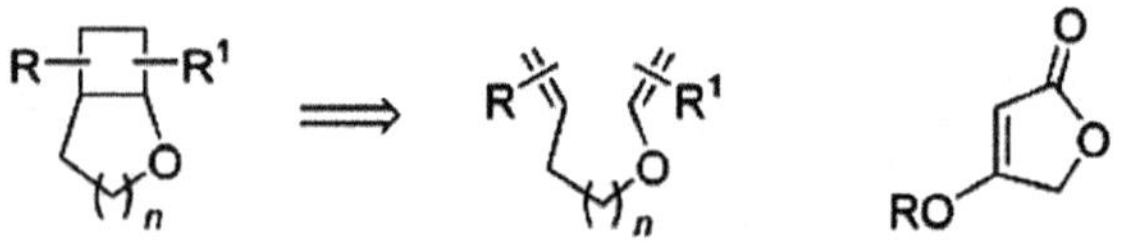

Fig. 2.61 The photo-
cycloaddition reactions

Cu-catalyzed [2 + 2] photocycloadditions, the more highly substituted substrates of type B do not react as desired, but decompose in the presence of the copper salts. As an alternative to enol ethers derivatives of tetronic acid (*1*, R=H) were tested in direct [2 + 2] photocycloaddition reactions as shown in Fig. 2.61. These building blocks deliver excellent results under irradiation conditions.

Synthesis of 2-oxabicyclo[3.2.0]heptanes by intramolecular [2 + 2] photocycload, tetronates of the general formula **2** are available from tetronic acid or its derivatives and alkenols by nucleophilic substitution. The intramolecular [2 + 2] photocycloaddition of the O-bridged dienes **2** proceeds smoothly yielding tetrahydrofuran **3a** and tetrahydropyran **3b** with excellent diastereoselectivity as photocycloaddition reactions as shown in Fig. 2.62. In each case, a single product is observed in diastereomerically pure form. The relative configuration was assigned by NOESY experiments and X-ray crystallographic analysis. The light source in the irradiation experiments was either a TNN 15/32 (Original Hanau, Heraeus Noblelight) low-pressure mercury arc or lamps of the type RPR-2437 Å (Rayonet).

Intramolecular [2 + 2] photocycloaddition of -alkenyl tetronates, analogous experiments in acetone as the solvent or with a triplet sensitizer (benzophenone, acetophenone) in diethyl ether or acetonitrile were less successful and led to the formation of side products. The intramolecular photocycloaddition is not restricted by the position of the alkenyl side chain. In compound 4 the -alkenyl chain is attached at C3, in compound 6 at C5. The cyclization of these substrates gives tricyclic products 5 (see Fig. 2.63) and 7 with high chemo-, regio-, and stereoselectivity. In all cases (Fig. 2.63 3, 5, 7), a single product was formed in which the cyclobutane ring and the annelated five- or six-membered rings are connected in a *cis* fashion. The facial diastereoselectivity in the reaction 6 to 7 can be understood as the attack of the terminal alkene to the face of the tetronate to which the alkyl chain is directed by the stereogenic center at C5 as shown in Figs. 2.63, 2.64 and 2.65.

Fig. 2.62 Tetronates of the
general formula

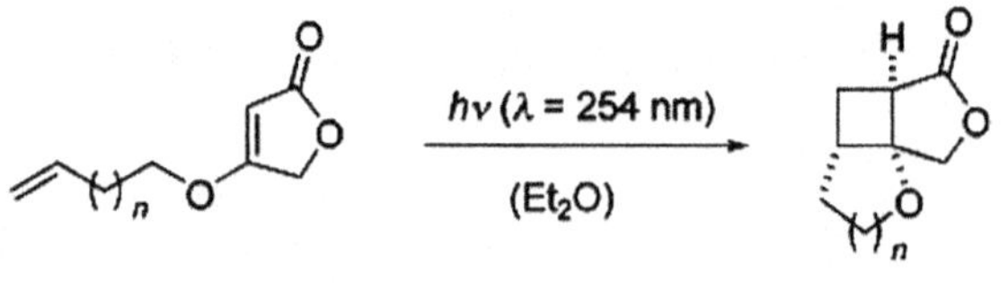

2a	*n* = 1	74%	**3a**
2b	*n* = 2	71%	**3b**

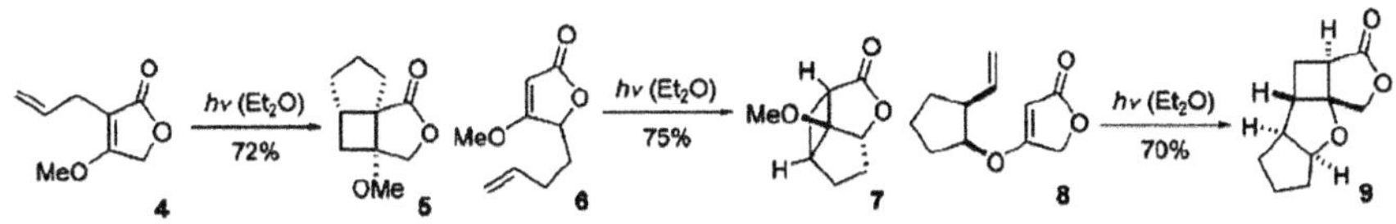

Fig. 2.63 Intramolecular [2 + 2] photocycloaddition of further substituted tetronates

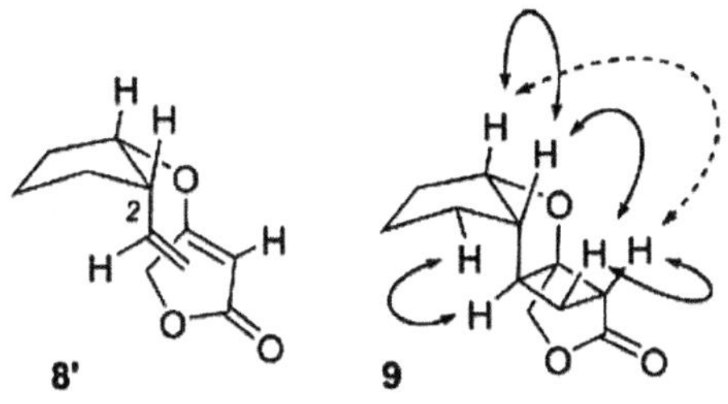

Fig. 2.64 Preferred conformation of compound 8 in the reaction to form 9 and NOESY contacts in the tetracyclic compound 9 (*dashed line* medium, *solid line* strong)

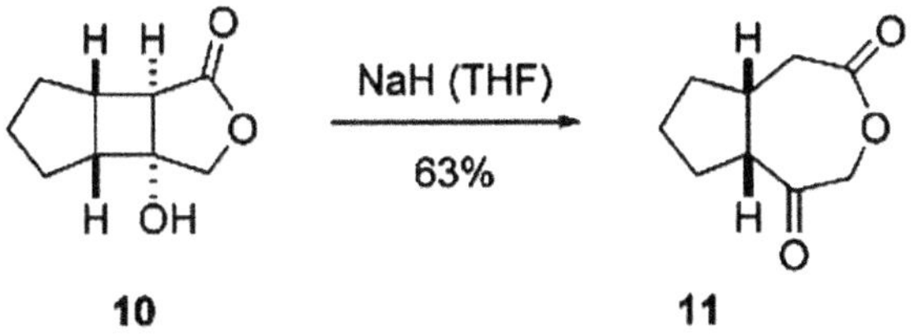

Fig. 2.65 Retroaldol reaction of the photocycloaddition product 10

The methodology was expanded to use the cyclobutane ring in the products for subsequent fragmentation reactions. A retroaldol reaction converts 10 into the ketolactone 11 upon treatment with a base.

2.7.2 Nitrogen-Containing Compounds

The presence of a nitrogen atom often alters the photophysical and photochemical properties of a given chromophore. Many nitrogen-containing compounds are much easier to oxidize than the corresponding nitrogen-free analogous. Aza-di-π-methane rearrangements as Fig. 2.66.

1. Decomposition of azo-compounds

As an alternative to heat or transition metal ions light can be used to generate carbenes from azo-compounds as Fig. 2.67.

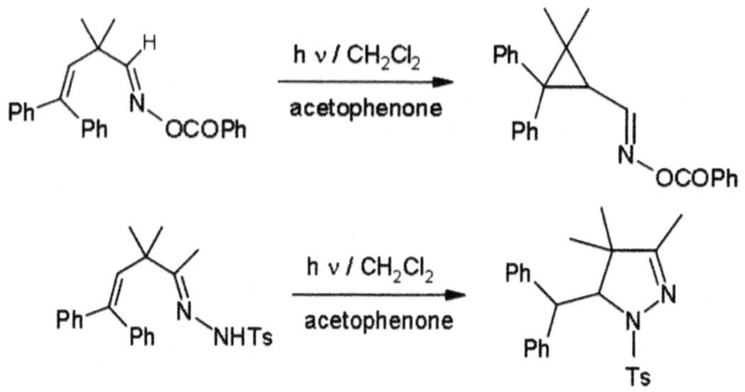

Fig. 2.66 Acetophenone is used as triplet sensitizer

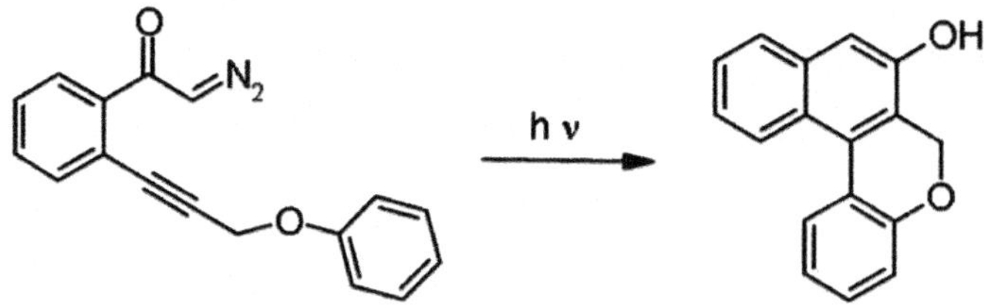

Fig. 2.67 Decomposition of azo-compounds

2. Aromatic compounds

The origin of reactivity of arenes (which are usually not very reactive) in the electronically excited state results from changes in the electron distribution. If benzene is irradiated with light of 254 nm small amounts for benzvalene and fulvene are formed. Upon irradiation with light of 203 nm formation of Dewar benzene is observed. Such isomerizations may be the initial steps of arene photoreactions as Figs. 2.68 and 2.69.

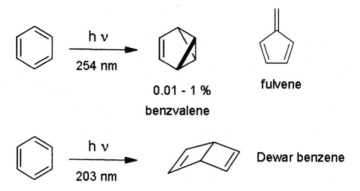

Fig. 2.68 The origin of reactivity of arenes

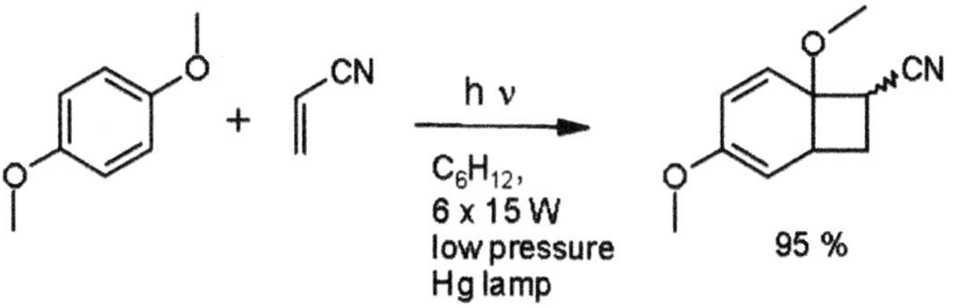

Fig. 2.69 Ortho-cycloaddition to benzene

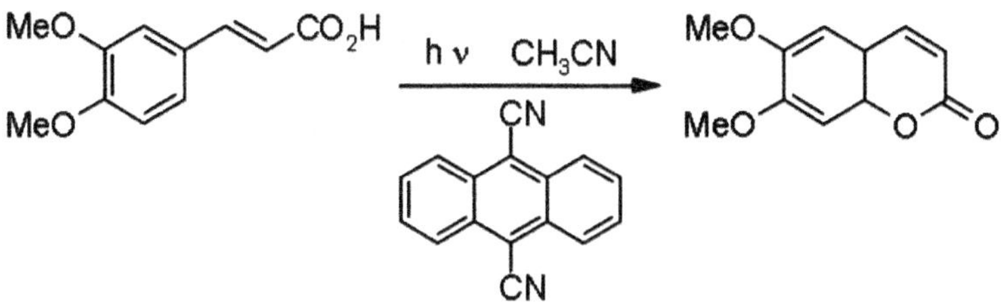

Fig. 2.70 Photoaddition by photoxidation: Synthesis of 6,7-dimethoxycoumarin Synthesis of 3-hydroperoxy-4-methyl-3-penten-2-ol

This is the photochemical key step of the synthesis of Pagodan. The photore-action gives a mixture with about 30 % of the cyclization product. Laborious purification is necessary as Fig. 2.70.

2.7.3 Photooxigenation and Photoreduction

Singlet oxygen is an easily available reagent. It can be generated from triplet oxygen in many solvents by a broad variety of sensitizers. The reaction of organic compounds with singlet oxygen can lead to reactive molecules, such as hydroperoxides, 1, 2-dioxetanes, and endoperoxides. These compounds are useful for subsequent transformations. Examples as shown in Fig. 2.71.

TPP means tetraphenylporphyrin. The mechanism can be described as an ene-type reaction. In general the reactivity of an alkene is in this reaction increases with alkyl substitution, because an electrophilic reagent is attacking. Terminal

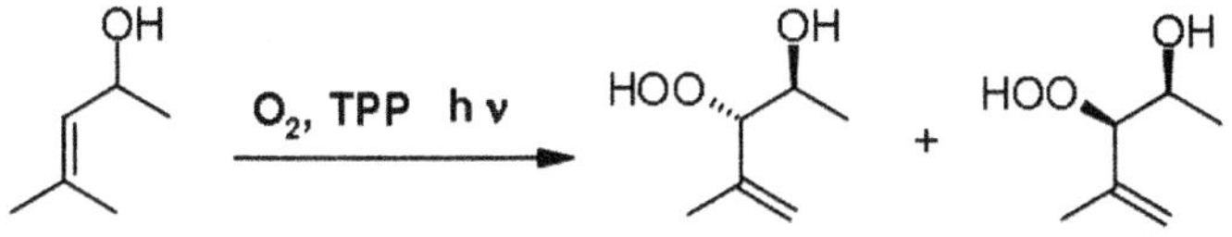

Fig. 2.71 Compounds are useful for subsequent transformations

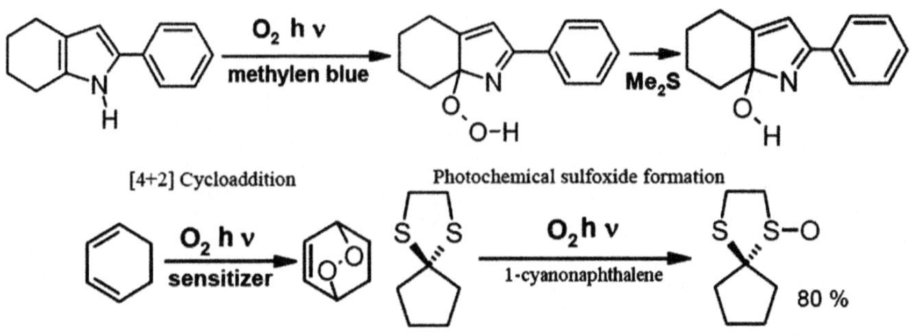

[4+2] Cycloaddition   Photochemical sulfoxide formation

80 %

Fig. 2.72 The examples of two alkyl substituents *vs* one hydrogen and one alkyl substituent

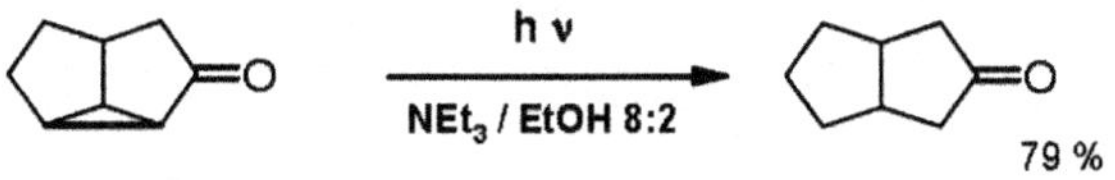

79 %

Fig. 2.73 The reduction of a strained bond in a tricyclic molecule. The irradiation is in a Rayonet photoreactor at 300 nm

alkenes usually do not react. If there is competition of several allyl positions for hydrogen abstraction, a general rule says that hydrogen abstraction occurs from the side of the double bond that is more substituted as Fig. 2.72.

Photoinduced electron transfer can lead to reduction processes, e.g. irradiation of the compound to be reduced in the presence of a sacrificial amine as electron donor. An example is the reduction of a strained bond in a tricyclic molecule as Fig. 2.73.

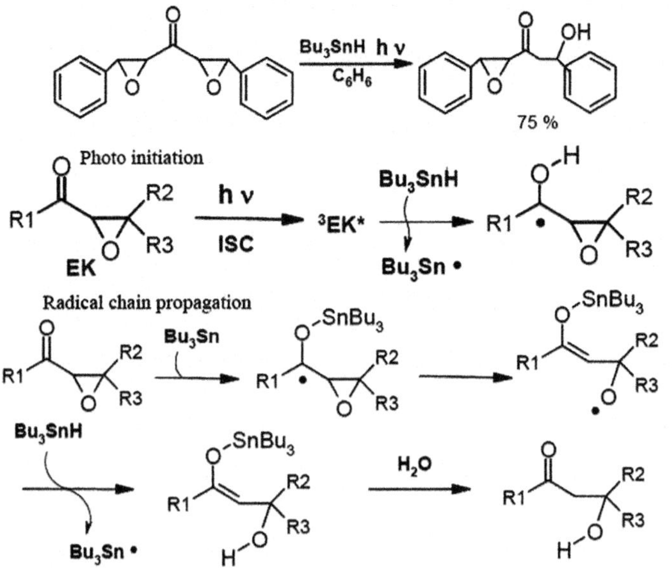

75 %

Photo initiation

Radical chain propagation

Fig. 2.74 An alternative reducing agent

An alternative reducing agent is tributyl tin hydride. The reductive expoide opening is photochemically initiated and proceeds by a radical chain mechanism as Fig. 2.74.

2.7.4 Photochemistry in Organized Systems

The selectivity of photochemical reactions can increase if the environment in which the reaction occurs has a specific geometry. This can either be the case in the solid state. The best known example is the [2 + 2] photocycloaddition of cinamic acid, which gives many isomers in solution, while irradiation of a crystal leads to fewer products. However, photochemistry in the solid state is difficult to predict in many cases. Therefore, modern approaches focus on topological reaction control in solution using templates to which the reactants are covalently or non-covalently bound as Fig. 2.75.

Photochemical reaction within a hydrogen-bonded aggregate is shown in Figs. 2.76 and 2.77.

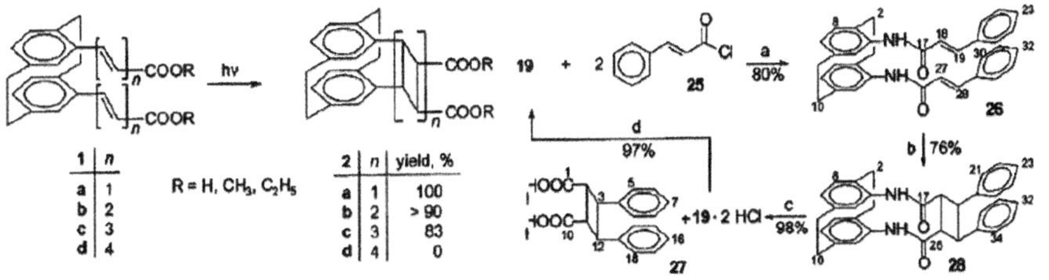

Fig. 2.75 Examples of Topochemical reaction control of [2 + 2] cycloadditions in solution. Compound 19 is a [2.2] paracyclophane-diamine

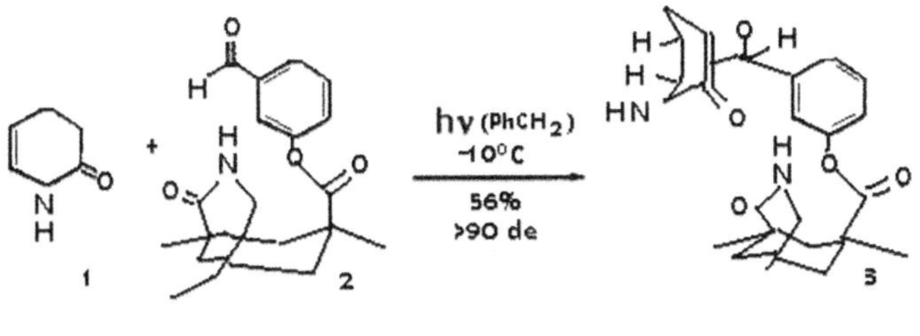

Fig. 2.76 The intra-assembly

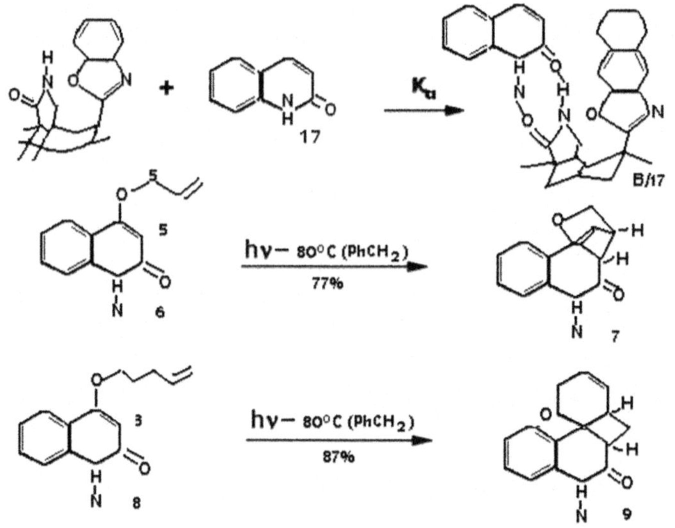

Fig. 2.77 Catalytic with template molecule: Enantioselective Norrish–Yang cyclization on a chiral template

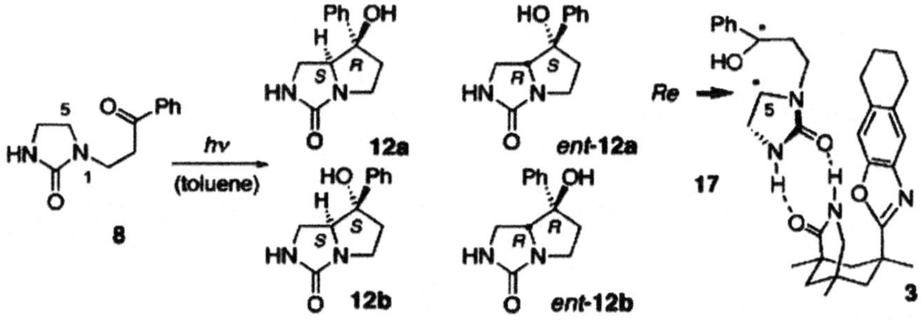

Fig. 2.78 Thestereoisomers of the reaction compound 12a

From all possible stereoisomers of the reaction compound 12a was formed preferentially if the reaction was performed in the presence of the chiral template as Figs. 2.78 and 2.79.

2.7.5 Photochromic Compounds

Photochromism characterizes reversible reactions in which one or both directions of the process can be triggered photochemically. The absorption spectra of starting material and product differ with respect to absorption wavelength and extinction;

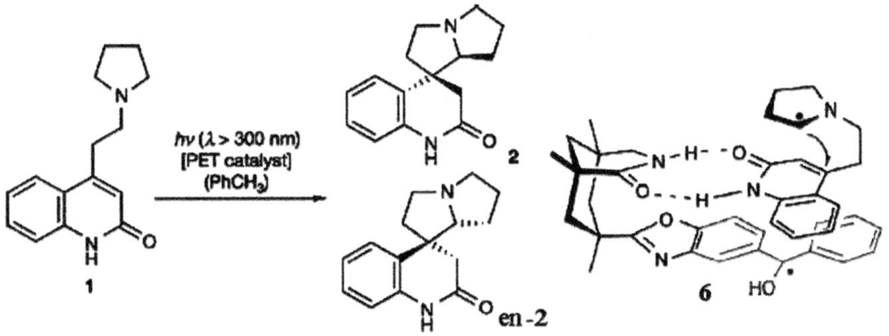

Fig. 2.79 Enantioselective catalytic PET-induced cyclization reaction

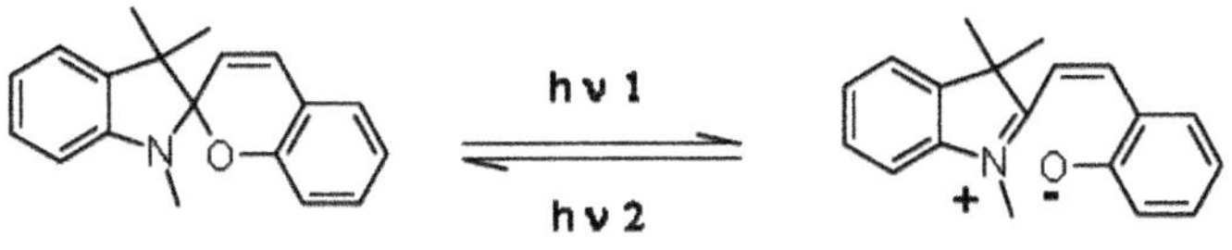

Fig. 2.80 Photochromic switches

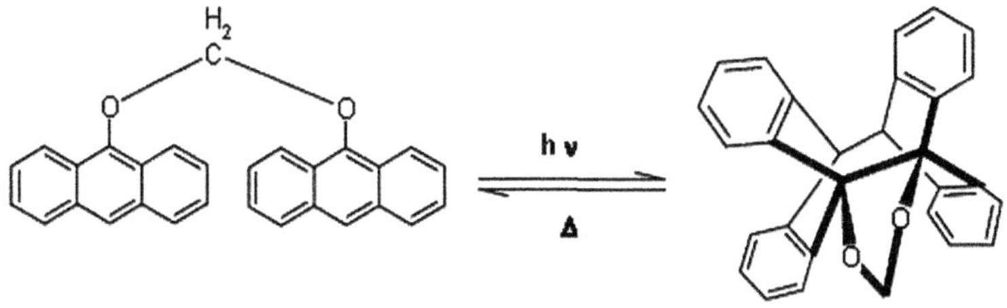

Fig. 2.81 Close: high-pressure Hg lamp in Pyrex. Open: heat diethyl ether solution with hairdryer for 15 min

therefore each process can be addressed more or less selectively. Photochromic switches can be regarded as a simple way of information storage. Many examples involve charge separation that leads to a different electronic structure with different absorption properties. The spiropyran—merocyanine dye system is one of the classics as Fig. 2.80.

The photodimerization of tethered anthracene has found to be thermally reversible by simple heating in diethyl ether is shown in Fig. 2.81.

The spiro compound—betaine switch is shown in Fig. 2.82.

The dithienylethylene allows reversible switching with UV and visible light is shown in Fig. 2.83.

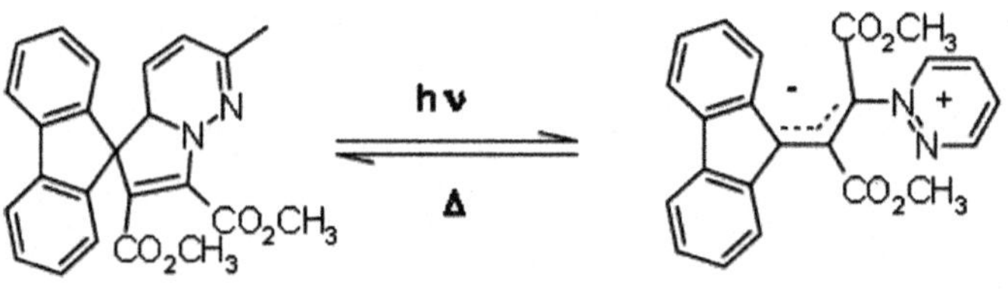

Fig. 2.82 Spiro compound—betaine switch Irradiation with visible light; half-life of the betaine at 20 °C: 14.4 min

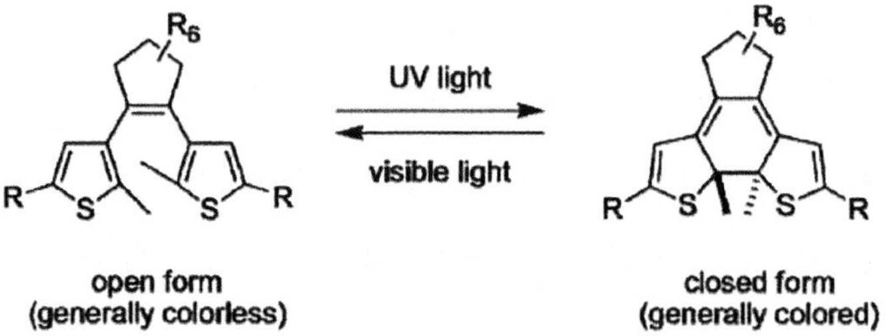

Fig. 2.83 The dithienylethylene allows reversible switching with UV and visible light. These switches can show (depending on R)

2.7.6 Photocleavable Protecting Groups and Linkers

The protecting groups as the norrish type II: ortho-nitrobenzyl alcohols is shown in Fig. 2.84.

For amino acids and heterocycles of protecting groups is shown in Fig. 2.85. For nucleic acids and sugars is shown in Fig. 2.86.

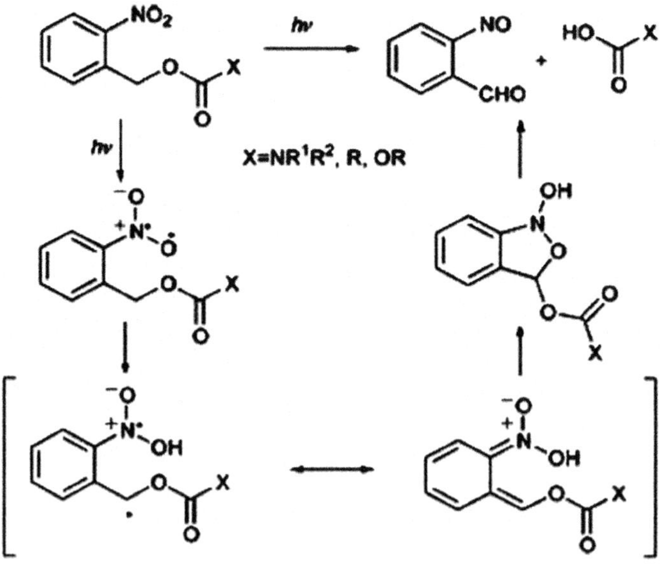

Fig. 2.84 The norrish type II: ortho-nitrobenzyl alcohols

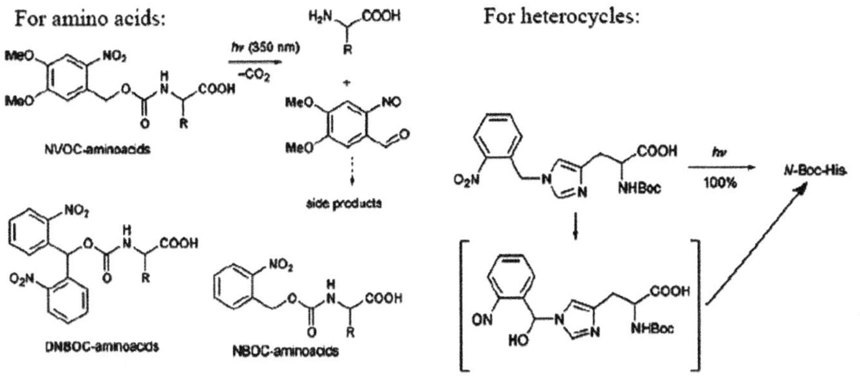

For amino acids: For heterocycles:

Fig. 2.85 For amino acids and heterocycles

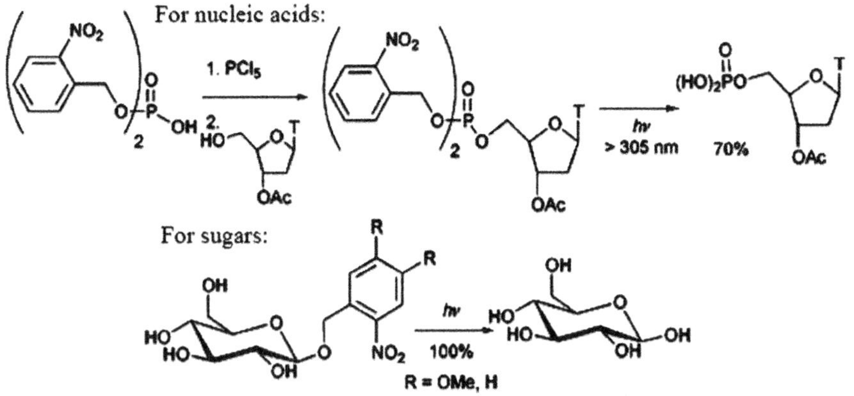

Fig. 2.86 For nucleic acids and sugars of protecting groups

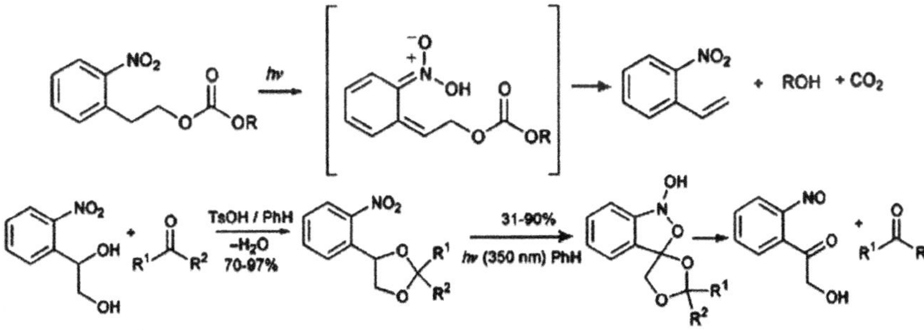

Fig. 2.87 The functional group to be protected is linked in β-position and protecting group for ketones

Different reaction pathway if functional group to be protected is linked in β-position and protecting group for ketones is shown in Fig. 2.87.

The PET example: Benzophenone as oxidant is shown in Fig. 2.88.

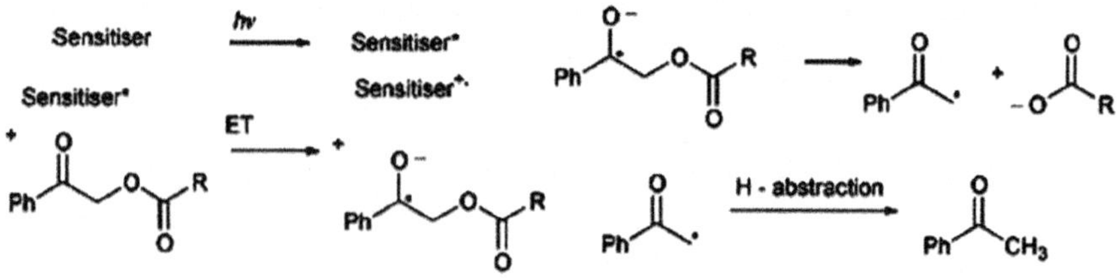

Fig. 2.88 Photoinduced electron transfer—Benzophenone as oxidant

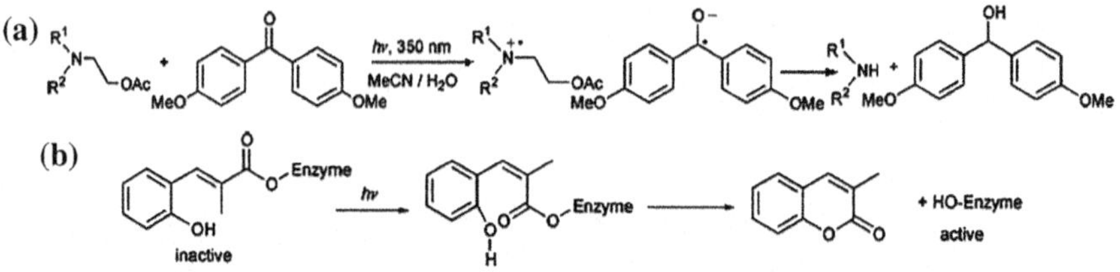

Fig. 2.89 **a** photodeprotection of an amine by wavelength of 350 nm light, **b** photoisomerization of *cis-trans*

The Photodeprotection of an amine (a) and the photoisomerization: *cis-trans* (b) is shown in Fig. 2.89.

2.8 Insertion Rate and Effective Capacity of Multidimensional Storage

Whether magnetic and optical data storage systems required constrained modulation codes also. These codes transform in a lossless manner, streams of arbitrary binary data into binary sequences that satisfy certain prescribed constraints. The set of words from which the code sequences may be drawn is referred to as a constrained system or simply a constraint. A constraint is characterized by a finite directed labeled graph, the paths of which generate the words in the set. These systems was initiated by Shannon who defined the capacity of a constrained system S as

$$\text{cap}(S) = \limsup_{n \to \infty} \frac{\log_2 |S(n)|}{n}, \tag{2.41}$$

where $S(n)$ denotes the number of sequences of S of length exactly n. Shannon showed that whenever there is a rate $p{:}q$ encoder (that is, an encoder which generates on average q output symbols for every p input bits) of a constrained system, then must have $p/q \leq \text{cap}(S)$. Furthermore, this bound is tight. The definition of constraint and capacity generalizes to higher-dimensional constraints. If S denotes a 2D constraint and $S(m, n)$ denotes the set of $m \times n$ arrays in S, and then

$$\text{cap}(S) = \limsup_{m,\, n \to \infty} \frac{\log_2 |S(m, n)|}{mn}. \tag{2.42}$$

So the insertion rate and effective capacity is an important research goal for multidimensional optical storage. The corresponding coding, error-correcting code, maximum insertion rate and capacity have been proposed. Typically in digital storage systems source data is encoded twice before written to the medium. First an error-correcting code (or ECC) encodes the source data into a codeword, and then a constrained (or modulation) code is used to transform the codeword into a sequence of bits that can be written reliably to the device. When reading back the data the process is reversed. As the decoder for the constrained code is typically a hard decoder, this scheme has the disadvantage of not making soft information available to the ECC decoder, thereby, limiting its error correction capability. A reverse concatenation scheme is considered where the user bits are first transformed into a constrained sequence in which certain entries are left unconstrained—i.e., filling them with any combination of bits would result in a sequence satisfying the constraint. Then, a systematic ECC of suitable rate is applied, placing the redundancy (parity-check) bits in these unconstrained positions. Since the error correction capability of the ECC depends on the number of redundancy bits, it is desirable that the number of unconstrained positions be as large as possible. On the other hand, increasing the number of unconstrained positions naturally reduces the rate of the constrained encoder. The tradeoff function defines for a given "density" of unconstrained positions the maximum rate of the constrained encoder. They also define the maximum insertion rate as essentially the largest asymptotic density of unconstrained positions possible for the constrained code. Extend some of these ideas to multidimensional constrained systems. Defined the maximum insertion rate of a multidimensional constraint and show that, for isotropic constraints, it is equal to the maximum insertion rate of the underlying one-dimensional constraint. The maximum insertion rate of an isotropic constraint is a lower bound on the limiting value of the capacity as the number of dimensions goes to infinity. Finally, that for isotropic constraints, whose underlying one-dimensional constraint has finite memory, when the maximum insertion rate is zero, the capacity decreases to zero exponentially fast as the number of dimensions goes to infinity [20–22].

2.8.1 Multidimensional Constraints with Unconstrained Positions

Multidimensional constraints given a finite directed graph whose edges are labeled with symbols from some finite alphabet, each path in the graph corresponds to a finite word attained by reading the labels of the edges of the path in sequence. The path is said to generate the word, and the set of all such generated words is a one-dimensional constrained system or a one-dimensional constraint. The graph is a presentation of the constraint, and its vertices are usually called states and edges, transitions. An example of a one-dimensional constrained system, commonly found in magnetic and optical storage systems, is the (d, k)-run length limited constraint, denoted by RLL (d, k), for nonnegative integers $d \leq k$. This is the set of all binary words that contain at least d zeros between pairs of adjacent '1's and do not contain more than k consecutive '0's anywhere. The k parameter is allowed to be ∞ which signifies no upper bound on the length of a run of zeros.

Consider multidimensional constraints of dimension D, constructed from one-dimensional constraints, by requiring that all the "rows," in a given direction, of a D-dimensional array satisfy a one-dimensional constraint. More precisely, fix a finite alphabet Σ and positive integer D. For a D-tuple $m = (m_1, \ldots, m_D)$ of positive integers, let Γ be a D-dimensional $m_1 \times m_2 \times \cdots \times m_D$ array of symbols of Σ whose entries are indexed an array a D-dimensional array of size m, as:

$$\mathbf{j} \in \{0, \ldots, m_1 - 1\} \times \{0, \ldots, m_2 - 1\} \times \ldots \times \{0, \ldots, m_D - 1\}, \qquad (2.43)$$

where

$$D - \text{tuples } (m_i^{(1)}, \ldots, m_i^{(D)})_{i=1}^{\infty}, \ (m_i^{(j)})_{i=1}^{\infty}, \quad j = 1, 2, \ldots, D. \qquad (2.44)$$

Given an integer $1 \leq i \leq D$, a row in direction i of Γ is a sequence of entries of Γ of the form

$$(\Gamma_{k_1, k_2, \ldots, k_{i-1}, j, k_{i+1}, \ldots, k_D})_{i=0}^{m_i - 1} \ 0 \leq k_l \leq m_l - 1; \quad 1 \leq l \leq D, \ l \neq i. \qquad (2.45)$$

where Γ satisfies a (one-dimensional) constraint in direction i, if every row in direction i of the array belongs to the constraint. Let Σ^m denote all D-dimensional arrays of size m over Σ, and define $\Sigma^{**\cdots*} = \Sigma^{*D}$, where the number of $*$'s in the superscript is D, as the union of Σ^m over all such D_m; that is, Σ^{*D} is the set of all finite-size D-dimensional arrays. Deal with subsets of Σ^{*D} defined in the following manner. Let $S_1, \ldots, S_D$ be D (one-dimensional) constraints over Σ. Denote by $S_1 \otimes S_2 \otimes \ldots \otimes S_D$ the set of finite D-dimensional arrays satisfying the constraint S_i in direction i; $i = 1, 2, \ldots, D$. A D-dimensional constrained system or D-dimensional constraint, and by saying that a D-dimensional array satisfies the constraint we mean that it belongs to the constraint. Often $S_1 = S_2 = \cdots = S_D = S$, in which case the constraint $S_1 \otimes \cdots \otimes S_D$ is called isotropic (or symmetric in some papers) and abbreviate it by $S^{\otimes D}$. Let S be a D-dimensional binary constrained system. For a D_m $m = (m_1, \ldots, m_D)$, denote by $N(S, m)$ the number of D-dimensional $m_1 \times m_2 \times \cdots \times m_D$ arrays satisfying S, and define

$$\mathcal{M}(S, \mathbf{m}) = (\log \mathcal{N}(S, \mathbf{m})) / \prod_{i=1}^{D} m_i,$$

where the base of the logarithm is 2 and take $\log(0)$ to be $-\infty$. The capacity of S, denoted cap (S), is defined by

$$\text{cap}(S) = \lim_{m \to \infty} \mathcal{M}(S, \mathbf{m}_i), \qquad (2.46)$$

i.e.

$$\text{cap}(S) = \inf_{\mathbf{m}} \mathcal{M}(S, \mathbf{m}). \qquad (2.47)$$

where $(m_i)_{i=1}^{\infty}$ is a sequence of D-tuples diverging to infinity. Since $\log N(S)$ is sub-additive, the above limit exists, is independent of the choice of $(m_i)_{i=1}^{\infty}$ and satisfies [23, 24].

2.8.2 Unconstrained Positions and the Maximum Insertion Rate

Restrict to the binary alphabet and set $\Sigma = \{0, 1\}$. Let $\widehat{\Sigma}$ denote the extended alphabet, $\{0, 1, \Psi\}$, and let m be a D-tuple of positive integers. For a D-dimensional array $\Gamma \in \widehat{\Sigma}^m$ with '0's, '1's, and 'Ψ's, define the set of all fillings of Γ by, $\Phi(\Gamma) = \{\Delta \in \Sigma^m : \text{for every index } j, \text{ if } \Gamma_j \neq \Psi \text{ then } \Delta j = \Gamma_j\}$; thus $\Phi(\Gamma)$ contains all the (binary) arrays attained by filling every 'Ψ' of Γ with '0' or '1' independently. Next we define, the set $\widehat{S}$ of D-dimensional arrays with entries in $\widehat{\Sigma}$ by,

$$\widehat{S} = \{\Gamma \in \widehat{\Sigma}^{*^D} : \Phi(\Gamma) \subseteq S\}, \qquad (2.48)$$

thus $\widehat{S}$ contains all the arrays over $\widehat{\Sigma}$ such that every filling of the 'Ψ's by '0's and '1's results in an array that satisfies S. If S is a binary one-dimensional constraint, shows how to construct a presentation for $\widehat{S}$. Thus, $\widehat{S}$ is a one-dimensional constrained system. If $S = S \otimes \ldots \otimes S_D$ for some D one-dimensional constraints $S_1, \ldots, S_D$, then it is easily verified that $\widehat{S} = \widehat{S}_1 \otimes \ldots \otimes \widehat{S}_D$, and hence $\widehat{S}$ is a D-dimensional constraint as well. For a binary constrained system S, its maximum insertion rate is the maximal asymptotic density of 'Ψ's in arrays of $\widehat{S}$ when the array size approaches infinity. More precisely, given an array $\Gamma \in \Delta\Sigma\Psi D$, let $\rho(\Gamma)$ denote the density of 'Ψ's in Γ, that is the ratio of the number of 'Ψ's to the total number of symbols in Γ. Define the maximum insertion rate of S, denoted $\mu(S)$, by

$$\mu(S) = \sup_{(\Gamma_i)_{i=1}^{\infty} \subseteq \widehat{S}} \limsup_{i \to \infty} \rho(\Gamma_i), \tag{2.49}$$

where the sup above is taken over all sequences of D-dimensional arrays in $\widehat{S}$ whose corresponding sequence of sizes diverges to infinity. For a binary one-dimensional constraint, S, let $\widehat{G}$ be a presentation for $\widehat{S}$. The density of squares in a path of G is the ratio of the number of edges in the path labeled with a 'Ψ', to the total number of edges in the path. It is shown that $\mu(S)$ is equal to the maximum density of 'Ψ's in a simple cycle of $\widehat{G}$, and is rational therefore. Additionally, it is shown that the maximum insertion rate of the RLL constraints is given by

$$\mu(\mathbf{RLL}(d,k)) = \frac{\left\lfloor \frac{k-d}{d+1} \right\rfloor}{\left\lfloor \frac{k+1}{d+1} \right\rfloor (d+1)} \tag{2.50}$$

$$\text{i.e. } \mu(\mathbf{RLL}(d,\infty)) = \frac{1}{d+1}, \tag{2.51}$$

for nonnegative integers $d \leq k$.

For multidimensional isotropic constraints constructed from a one-dimensional constraint S, it is somewhat surprising that the maximum insertion rate remains the same. This is stated in the following theorem:

$$\mu(S) \leq \min\{\mu(S_1), \ldots, \mu(S_D)\}$$
$$S_1 = \ldots = S_D = T \text{ then } \mu(S) = \mu(T) \tag{2.52}$$

Since the maximum insertion rate of a one-dimensional constraint can be computed from a presentation of $\widehat{S}$. Let S be a one-dimensional constraint as:

$$\text{cap}_{\infty}(S) \geq \mu(S). \tag{2.53}$$

That for $S = \text{RLL}(d, \infty)$, Eq. (2.53) holds with equality. This is also true for one-dimensional constraints S with maximum insertion rate zero and finite memory. This follows from the next theorem. A labeled graph G has finite memory m, if all paths of length m, generating the same word, terminate at the same vertex of G. A one-dimensional constraint has finite memory if it has a presentation with finite memory. The memory of such a constraint is the smallest memory of its finite memory presentations.

Let S be a one-dimensional constraint with finite memory m. If $\mu(S) = 0$ for all positive integers D,

$$\mathrm{cap}(S^{\otimes(D+1)}) \leq \frac{m}{m+1}\mathrm{cap}(S^{\otimes D}). \tag{2.54}$$

In particular $\mathrm{cap}_\infty(S) = 0$. Consider the constraint RLL(d, k) has memory k, and by Eqs. (2.50) and (2.51), $\mu(\mathrm{RLL}(d, k)) = 0$ if and only if $k \leq 2d$. Thus we can obtain the following results:

1. When d, k be nonnegative integers such that $d \leq k$. Then $\mathrm{cap}_\infty(\mathrm{RLL}(d, k)) = 0$ if and only if $k \leq 2d$. Another constraints found in magnetic and optical recording, is the multiple-spaced run length constraints denoted RLL(d, k, s). These are characterized by nonnegative integers d, k, s with $d \leq k$. A binary word satisfies RLL(d, k, s) if it satisfies RLL(d, k) and the length of each run of zeros is a multiple of s. A storage system is employing the RLL(2, 18, 2).
2. When d, k, s be nonnegative integers such that $d \leq k$ and $s > 1$. Then

$$\mathrm{cap}(\mathbf{RLL}(d, k, s)^{\otimes D}) \leq \left(\frac{k}{k+1}\right)^{D-1}\mathrm{cap}(\mathbf{RLL}(d, k, s)). \tag{2.55}$$

If $1 \leq j \leq D$ be an integer such that $\mu(Sj) = \min\{\mu(S1), \ldots, \mu(SD)\}$. Fix a positive real number ϵ, and let $(\Gamma_i)_{i=1}^\infty \subseteq \Psi\widehat{S}$ be a sequence of D-dimensional arrays with sizes diverging to infinity, such that

$$\limsup_{i\to\infty} \rho(\Gamma_i) \geq \mu(S) - \epsilon. \tag{2.56}$$

For a positive integer i, consider the set of rows in direction j of Γ_i. Since $\rho(\Gamma_i)$ is the average of the densities of 'Ψ's in these rows, there exist a row with density of squares at least $\rho(\Gamma_i)$. Let z_i be such a row. Clearly $z_i \in \widehat{S}_j$, and therefore

$$\mu(S_j) \geq \limsup_{i\to\infty} \rho(z_i) \geq \limsup_{i\to\infty} \rho(\Gamma_i) \geq \mu(S) - \epsilon. \tag{2.57}$$

Since ϵ is arbitrary, that have $\mu(Sj) \geq \mu(S)$. Fix a presentation $\widehat{G}$ for $\widehat{T}$, and let $w = w_0 \cdots w_{n-1}$ be a word in $\widehat{T}$ generated by a cycle of $\widehat{G}$ with density of 'Ψ's equal to $\mu(T)$. For each positive integer i, let Γ_i be the D-dimensional array of size with entries given by

$$(\Gamma_i)_{(m_1, m_2, \ldots, m_D)} = \mathcal{W}_{(m_1 + m_2 + \cdots + m_D)\bmod n}. \tag{2.58}$$

Then every row (in any direction) of Γ_i is a cyclic shift of i concatenations of ω, and in $\widehat{T}$. Thus, $(\Gamma_i)_{i=1}^\infty \subseteq \widehat{S}$ and $\rho(\Gamma_i) = \mu(T)$, for $i = 1, 2, \ldots$. It follows that $\mu(S) \geq \mu(T)$.

For any D-dimensional constraint S, $\mathrm{cap}(S) \geq \mu(S)$, that is $\mathrm{cap}(S^{\otimes D}) \geq \mu(S^{\otimes D}) = \mu(S)$ for any positive integer $\mathbf{D}$. Taking the limit as $D \to \infty$,

let $\mathcal{C}$ be a positive real number and let $(\Gamma_i)_{i=1}^{\infty} \subseteq \widehat{S}$ be a sequence of D-dimensional arrays, with sizes diverging to infinity, satisfying $\limsup_{i \to \infty} \rho(\Gamma i) \geq \mu(S) - \mathcal{C}$. Denote by m_i the size of Γ_i, and by n_i the product of entries in m_i (i.e., n_i is the number of symbols in Γ_i). Since each of the $2^{\rho(\Gamma_i)n_i}$ ways of replacing the 'Ψ's in Γ_i by '0's and '1's results in a distinct array of S, that $M(S, m_i) \geq \rho(\Gamma_i)$, and consequently as $\mathcal{C}$ is arbitrary

$$\text{cap}(S) = \lim_{i \to \infty} \mathcal{M}(S, \mathbf{m}_i) \geq \limsup_{i \to \infty} \rho(\Gamma_i) \geq \mu(S) - \epsilon. \tag{2.59}$$

If S is irreducible, i.e., has an irreducible (strongly connected) presentation. Need the following definition. Let $\Sigma = \{0, 1\}$. For a word $\omega \in \Sigma^*$ denote by $|\omega|$ the number of symbols (length) in ω. To one-dimensional constrained system S, (m, a)-determined, for nonnegative integers m and a, if there do not exist two words $x0y$, $x1y \in S$, such that $|x| = m$ and $|y| = a$. Thus, if ω is a word of an (m, a)-determined constraint, then each symbol placed sufficiently far from the beginning and end of ω is uniquely determined by its m preceding symbols and a succeeding symbols. Now, let S be a general one-dimensional constraint. If $\mu(S) > 0$, then, since there exist arbitrarily long words $x, y \in \Sigma^*$ such that $x, y \in \widehat{S}$, it follows that S is not (m, a)-determined for any nonnegative integers m. Let S be an irreducible one-dimensional constrained system with finite memory m. If $\mu(S) = 0$ then S is (m, m)-determined. Assume to the contrary that there exist two words $x0y$, $x1y \in S$, where x, y are binary words of length m. Let $\alpha_0 \to \alpha_1 \to \cdots \to \alpha_{2m+1}$, $\beta_0 \to \beta_1 \cdots \to \beta_{2m+1}$ be the paths in G generating $x0y$, $x1y$, respectively, where $\alpha_0, \ldots, \alpha_{2m+1}, \beta_0, \ldots, \beta_{2m+1}$ are states in G. Since G has memory m, it follows that $\alpha_m = \beta_m$ and $\alpha_{2m+1} = \beta_{2m+1}$. Since G is irreducible there exists a path $\alpha_{2m+1} \to \gamma_1 \to \gamma_2 \to \cdots \to \gamma_{l \to \alpha m}$ connecting α_{2m+1} to α_m, and generating some word z. Therefore, G contains the cycles $\alpha_m \to \alpha_{m+1} \to \cdots \to \alpha_{2m+1} \to \gamma_1 \to \cdots \to \gamma_{1 \to \alpha m}$ and $\alpha_m = \beta_m \to \beta_{m+1} \to \cdots \to \beta_{2m+1} \to \gamma_1 \to \cdots \to \gamma_l \to \beta_{m=\alpha m}$, that generate the words $0yz$ and $1yz$, respectively. Consequently, for any finite sequence of bits $b_1, b_2, \ldots, b_i \in \Sigma$, the word $b_{1yz}b_{2yz} \cdots b_{iyz} \in S$. Hence, $(\Psi_{yz})_i \in \widehat{S}$, for all positive integers i, contradicting the assumption that $\mu(S) = 0$. If S is an (m, a)-determined constraint, the capacity of $S^{\otimes D}$ decreases at least exponentially fast with D. This is a generalization where $S = \text{RLL}(d, k)$ with $k \leq 2d$ and for general (m, a)-determined constraints as well. If S be a one-dimensional (m, a)-determined constraint, for all positive integers D, then

$$\text{cap}(S^{\otimes(D+1)}) \leq \frac{b}{b+1}\text{cap}(S^{\otimes D}), \tag{2.60}$$

where $b = \min\{m, a\}$.

For the case $b = a$. The case $b = m$ is handled similarly. Let l be a positive integer. Denote by l the D-tuple with every entry equal to l, and let m_l be the $(D + 1) - e$ given by $m_l = (m + l(a + 1), l, l, \ldots, l)$. That $N(S^{\otimes D}, m_l)$ is the number of arrays in $S^{\otimes D+1}$ of size ml bound $N(S^{\otimes(D+10)}, m_l)$ from above. For an array Γ of size ml and integer $0 \leq i < m + l(a + 1)$, denote by $\Gamma^{(i)}$ the D-dimensional sub-array of Γ of size 1, consisting of the entries $\Gamma_{i,j}; j \in \{0, 1, \ldots, l - 1\}D$. Let $i_r = m + r(a + 1); r = 0, 1, \ldots, l - 1$, can set

$$A = \{0, 1, \ldots, m + l(a + 1) - 1\} \backslash \{i_r : r = 0, \ldots, l - 1\}. \qquad (2.61)$$

For each array Γ of size ml in $S^{\otimes(D+1)}$, each of the sub-arrays $\Gamma^{(i)}$ satisfies $S^{\otimes D}$. construct all such arrays Γ by first selecting the sub-arrays $\Gamma(k); k \in A$—each chosen from the set of arrays of size l in $S^{\otimes D}$ and then completing the entries of the sub-arrays $\Gamma^{(i_r)}; r = 0, 1, \ldots, l_{-1}$. that for each possible choice of the sub-arrays $\Gamma^{(k)}; k \in A$ there is at most one possibility to complete the sub-arrays $\Gamma^{(i_r)}; r = 0, 1, \ldots, l_{-1}$ such that the resulting array Γ satisfies $S^{\otimes(D+1)}$. For $j \in \{0, 1, \ldots, l - 1\}D$, consider the sequence of entries $\Gamma_j^{(0)}, \Gamma_j^{(1)}, \ldots, \Gamma_j^{(m+l(a+1)-1)}$. It is a row in direction 1 of Γ, and contains the yet unselected entries: $\Gamma_j^{(i0)}, \ldots, \Gamma_j^{(il-1)}$.

Complete these entries in sequence. The entry $\Gamma_j^{(i0)}$ is preceded by m, already selected, entries in the row and succeeded by a, already selected, entries in the row. Thus, since S is (m, a)-determined, there is at most one possibility to complete it. For $r \geq 1$, after completing the entries $\Gamma_j^{(ik)}$, for $0 \leq k < r$ the entry $\Gamma_j^{(ir)}$ is preceded by (at least) m completed entries in the row, and succeeded by a completed entries in the row. There is at most one possibility to complete it. Therefore, for each selection of the sub-arrays $\Gamma^{(k)}; k \in A$ there is at most one possibility to complete the rest of Γ so that the resulting array satisfies $S^{\otimes(D+1)}$. Since there are $N(S^{\otimes D}, l)^{m+la}$ ways to select the sub-arrays $\Gamma^{(k)}; k \in A$, it follows that $N(S^{\otimes(D+1)}, ml) \leq N(S^{\otimes D}, l)m + la$. Hence,

$$
\begin{aligned}
\mathcal{M}(S^{\otimes(D+1)}, \mathbf{m}_l) &= \frac{\log \mathcal{N}(S^{\otimes(D+1)}, \mathbf{m}_l)}{l^D(m + l(a + 1))} \\
&\leq \frac{(m + la) \log \mathcal{N}(S^{\otimes D}, 1)}{l^D(m + l(a + 1))} \qquad (2.62) \\
&= \frac{m + la}{m + l(a + 1)} \mathcal{M}(S^{\otimes D}, 1)
\end{aligned}
$$

Taking the limit is as $l \to \infty$, can obtain the result.

Let S be an irreducible one-dimensional constraint with finite memory m. S is (m, m)-determined. A reducible constraint with finite memory and $\mu(S) = 0$ need not be $(m.a)$-determined for any m and a. For example, consider the constraint S consisting of all sequences of the form $0^u 1^v$ where u, v are arbitrary nonnegative integers. This constraint is presented by a graph with two states a, b, a self loop labeled 0 at a, an edge from a to b labeled 1, and a self loop at b labeled 1. This presentation has finite memory, but S is reducible since a 0 can never follow a 1. And $\mu(S) = 0$ since any string in $\widehat{S}$ can contain at most one (at a transition from 0 to 1).

But S is not (m, a) determined for any m, a since $0^m \times 1^a$ can be filled in by $x = 0$ or $x = 1$. There can be much more complicated constraints with these features.

If S be any one-dimensional constraint with finite memory m and $\mu(S) = 0$. For each sequence x in S and positive integer N, let $V_N(x)$ be the set of sequences y in S such that $y_i = x_i$ whenever i is not divisible by N. By using the general structure of reducible graphs one can show that $|V_{m+1}(x)|$ is uniformly bounded over all x in S.

The maximum insertion rate of a constraint is a lower bound on its capacity, a simple upper bound on the maximum insertion rate for a constrained system. For isotropic D-dimensional constraints, it turns out that the maximum insertion rate is equal to the maximum insertion rate of the underlying one-dimensional constraint. For such constraints, the maximum insertion rate provides a lower bound on the limiting value of the capacity as the number of dimensions grows to infinity. That for isotropic constraints with maximum insertion rate 0, in which the underlying one-dimensional constraint has finite memory, the limiting value of the capacity is in fact 0 and the rate of convergence is exponential [25–27].

References

1. J. Song, D.-Y. Xu, G.-S. Qi, Multilevel read-only optical recording methods. Chin. Phys. **15** (8), 1788–1792 (2006)
2. N.S. Allen, J.F. Rabek (eds.), *New Trends in the Photochemistry* (New York, 2007)
3. V. Balzani, P. Ceroni, A. Juris, *Photochemistry and Photophysics: Concepts, Research, Applications* (Wiley, New York, 2014)
4. N.J. Turro, J.C. Scaiano, *Modern Molecular Photochemistry of Organic Molecules* (New York, 2010)
5. A. Albini, *Photochemistry* (Research Press, Champaign, 2011)
6. T. Tumkur, et al., Control of Förster energy transfer in vicinity of metallic surfaces and hyperbolic metamaterials. Faraday Discuss. F.D. **178** (2014)
7. B. Lebeau, C. Marichal, A. Mirjol, G.J.A.A. de Soler-Illia, R. Buestrich, M. Popall, L. Mazerolles, C. Sanchez, Synthesisof highly ordered mesoporous hybrid silica from aromatic fluorinatedorganosilane precursors. New J. Chem. **27**, 166 (2003)
8. J. Feldhaus, K. Dardis, H. Kavanagh, J. Luna, Pedregosa Gutierrez, Single-shot characterization of independent femtosecond extreme ultraviolet free electron and infrared laser pulses. App. Phys. Lett. **90**, 131108 (2007)
9. J.Y. Bae, O.H. Park, J.I. Jung, K.T. Ranjit, B.S. Bae, Photoionization of methylphenothiazine and photoluminescence of erbium 8-hydroxyquinolinate in transparent mesoporous silica films by spin-coating on silicon. Microporous Mesoporous Mater. **67**, 265 (2004)
10. N.S. Makarov, *Ultrafast Two-Photon Absorption in Organic Molecules: Quantitative Spectroscopy and Applications* (Montana State University, Bozeman, 2010)
11. H. Zhang, A.S. Dvornikov et al., Single-beam two-photon-recorded monolithic multilayer optical disks. Proc. SPIE **4090**, 174–178 (2000)
12. D. Xu, L. Ma, 3-Dimension digital storage technology. SPIE **4085**, 5–10 (2000)
13. X. Cheng, H. Jia, D. Xu, Vector diffraction analysis of optical disk readout. Appl. Opt. **39**(34), 6436–6440 (2000)
14. Y. Wang, B. Yao, N. Menke, Y. Chen, M. Fan, Optical image operation based on holographic polarization multiplexing of fulgide film. Chin. Opt. Lett. **9**(s1), s10302 (2011)

15. L.-P. Chang, Hybrid solid-state disks: combining heterogeneous NAND flash in large SSDs. in *Proceedings of the 13th Asia South Pacific Design Automation Conference (ASP-DAC)*, pp. 428–433 (2008)
16. Z. Zhang, D. Xu, Diffraction analysis of optical disk read-out signal deterioration caused by edge decline of groove. in *ISOS-5th International Symposium on Optical Storage*, pp. 19–25 (2000)
17. E.P. Walker, X. Zheng, F.B. McCormick, H. Zhang, N.H. Kim, J. Costa, A.S. Dvornikov, Servo error signal generation for 2-photon recorded monolithic muiltilayer optical data storage. ODS 2000 Proc. SPIE **4090**, 179–184 (2000)
18. H. Zhang, A.S. Dvornikov, E.P. Walker, N.H. Kim, F.B. McCormick, Single-beam two-photon-recorded monolithic multi-layer optical disks. ODS 2000 Proc. SPIE **4090**, 174–178 (2000)
19. H. Hua, D. Xu, L, Pan, Modulation code and PRML detection for multi-level run-length-limited DVD channels. in *2006 Optical Data Storage Topical Meeting*, pp. 112–114 (2006)
20. D. Kitayama, M. Yaita, Laminated metamaterial flat lens at millimeter-wave frequencies. Opt. Expr. **23**(18), 23348–23356 (2015)
21. J. Song, K. Chen, D.-Y. Xu, Simulation and experiment study on recording mask in mastering. Optoelectron. Laser **17**(3), 265–268 (2006)
22. J. Nunn, I.A. Walmsley, M.G. Raymer, K. Surmacz, F.C. Waldermann, Z. Wang, D. Jaksch, Phys. Rev. A **75**, 011401(R) (2007)
23. D.-Y. Xu, Q. Zhang, J. Jun, Z. Lei, Replication method for gray-degree recording optical disc, Tsinghua University, CN03119279.3, 2013
24. D.-Y. Xu, X. Fan, G. Qi, Q. Kun, Control method of laser drive for multi-level mastering and burning recording, Tsinghua University, CN03121143.7, 2003
25. D.-Y. Xu, Y. Ni, L. Pan, K. Chen, J. Xiong, D. Lu, H. Wu, J. Ma, J. Pei, Replication method of multi-level CD-ROM, Tsinghua University, CN200510053509.3, 2005
26. J.Y. Bae, J.I. Jung, O.H. Park, B.S. Bae, K.T. Ranjit, L. Kevan, Synthesis and characterization of mesoporous silicafilms by spin-coating on silicon: photoionization of methylphenothiazine and photoluminescence of erbium 8-hydroxyquinolinatein mesoporous silica films. Stud. Surf. Sci. Catal. **146**, 65 (2003)
27. D.Y. Xu, S. Jie, G. Qi, P. She, Digital color disc recording and reading method, Tsinghua University, CN200410037430.7, 2005

Chapter 3
Recording Materials and Process

The multi-dimensional optical storage is based on a lot of especial materials, devices, apparatuses, equipments, and process simultaneously. This chapter will introduce and discuss various inorganic or organic composite recording materials, photochromic reaction in medium, performances of recording process, model of writing, quantitative evaluation of the absorption spectrum for medium, influence of reflectivity to writing speed, crosstalk and non-destructive readout.

3.1 Inorganic–Organic Composite Materials

Since the book adopts mainly the principles and characteristics of spectral absorption and photosensitivity of medium for multi-wavelength and multilevel optical storage, that mean of these materials can record a lot of different color (wavelength) light and different intensity (gray scale), as very similar to color photos. In order to facilitate the promotion of future applications, the laser sources of the multi-dimension CD used the commercial semiconductor lasers with various wavelengths in CD industry in the first, and the hardware structure of drive is modular with traditional CD technology and products. For example, the multilevel CD-ROM can produce with traditional CD-ROM production line, the drive is compatible with traditional CD, DVD, and B-CD, the multilevel WORM drive can use traditional phase-change disc [1, 2].

The inorganic–organic composite materials were used in experiment in early 1900 in OMNERC. It was developed by Physics Chemistry Institute of CAS in China, including $C_{16}H_{33}NH_{34}Mo_8O_{26}\cdot(C_{16}$–Mo$)$ and $C18H_{37}NH_35HMo_7O_{24}$ $(C_{18}$–Mo$)$ inorganic–organic composite films, supramolecular into phosphomolybdic acid/dimethyl pairs of octadecyl amine superlattice thin films, etc. Experiments show that are better photochromic performance, as C_{16}–Mo for example, the composite film of methylene asymmetric stretching vibration appears at 2918 Å, so the hydrocarbon chain has been gathered into orderly crystalline structure. Amino, head of the organic

© Tsinghua University Press and Springer Science+Business Media Singapore 2016   155
D. Xu, *Multi-dimensional Optical Storage*,
DOI 10.1007/978-981-10-0932-7_3

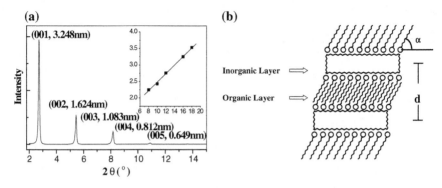

Fig. 3.1 Patterns of C_{16}–Mo composite film XRD: **a** The relationship between layer spacing and number of carbon atoms in organic amine. **b** The Cn–Mo composite film structure model

molecules exists in the quaternary ammonium-NH_3^+ form a composite film, composite film formation process is the inorganic components of the supramolecular template, the role of electrostatic attraction, hydrogen bonds, and other inorganic constituents composite. No units were mainly in the form of Mo_8O_{26} 4-accompanied trace amounts Mo_7O_{24}. Figure 3.1a C_{16}–Mo composite film XRD patterns clearly appear the five peaks, the spacing of 3.25, 1.62, 1.08, 0.812, and 0.649 nm, respectively, is a layered superlattice structure (001), (002), (003) (004) (005) diffraction peaks, layered superlattice interlayer spacing is 3.25 nm. In a series of different carbon chain length Cn–Mo composite system, n is the 18, 12, 10, and 8, there is a similar lamellar diffraction peak, the interlayer spacing of 3.53 nm, ($2\theta = 2.50°$), 2.76 ($2\theta = 3.20°$) nm, 2.43 nm, ($2\theta = 3.64°$), and 2.25 nm ($2\theta = 3.92°$). The interlayer spacing of the number of carbon atoms of the organic amine mapping is in good linear relationship (see Fig. 3.1a). After fitting, the slope of the line is 0.129 nm/C. The inclination of organic amine carbon chain α composite membrane is about 90° (Fig. 3.1b), the organic molecules between the inorganic layer are perpendicular to the two-dimensional inorganic network plane. The above fitting a straight line extrapolation to n equal to zero, the income intercept of 1.2 nm. Taking into account the ammonium group of 0.4 nm, the thickness of inorganic composition is 0.8 nm, about same as Mo_8O_{26}4-poly-anion size. Supramolecular templating prepared 16 ammonium 18 ammonium molybdate films have good photochromic properties, as shown in Fig. 3.2. The new system C_{16}–Mo composite film is colorless and translucent that becomes purple (λ max = 480 nm) after UV irradiation and very stable at room temperature in the dark, can save more than 1 year with good reversibility. If take -N-of $(CH_3)^{3+}$-based group to replace the amino-of NH^{3+}, organic/inorganic composite film did not the performance of the photochromic property, the amino-of NH^{3+} and multi-molybdate root ion composition of the inorganic layer formation of multiple hydrogen bonds,-N-(of CH_3 depends mainly on the electrostatic interaction)$^{3+}$ and the inorganic layer combination cannot provide a proton [3].

Using the atomic force microscope observation it was found that the interlayer spacing of the color before and after the film decreases from 3.25 to 3.11 nm. Is due

Fig. 3.2 The absorption
spectra of C$_{16}$–Mo composite
film. Initial film of A—initial
film, B—colored film, C—
coloring/achromatic
reversible cycle

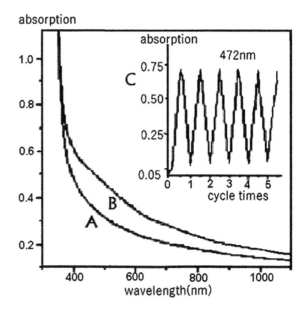

to the formation of the photochromic reaction by hydrogen bonds between the
charge-transfer complexes, the organic components, and inorganic components of
matter (about 0.28–0.30 nm) becomes the key (about 0.25 nm), about
$0.05 \times 2 = 0.10$ nm change in value; the other hand is due to proton transfer from
the organic components of the inorganic layer, the organic layer due to electrostatic
charge to reduce the organic molecules between the Coulomb repulsion weakened,
so organic molecules can be arranged more closely in order to achieve a new
exclusion to attract balance, or organic carbon chain has a certain degree of tilt, so
the interlayer spacing decreased.

3.1.1   Self-assembled Photochromic Materials

Physics Chemistry Institute of CAS self-assembly fabricated WO$_3$/4,4′-BAMBp (4,4′-
2(aminomethyl) biphenyl) WO$_3$/4,4′-BPPOBp (4,4′-2-(5-pyridyl pentyloxy) biphenyl)
WO$_3$/1,10-DAD   (1,10-Kuei-diamine),   PM$_{12}$/1,10-DAD   (PM$_{12}$ = H$_3$PM$_{12}$O$_{40}$,
M = Mo$_{42}$, W) and SiM$_{12}$/1,10-DAD (SiM$_{12}$ = H$_4$SiM$_{12}$O$_{40}$, M = Mo, W) have
photochromic properties as in Table 3.1. Self-assembled monolayer acid film mor-
phology using atomic force microscopy observations is shown in Fig. 3.3. It can be seen
that the adsorption of acid monolayer formed many island regions, each region ranged
in size from several nanometers to 10 nm. Monolayer thickness is about 4 nm. That
molybdate monolayer membrane based on poly phosphomolybdate anion composition,
rather than the form of a single molecular state adsorption from. The relationship
between layers and adsorption is shown in Fig. 3.4a. Figure 3.4b is made of

Table 3.1 Self-assembly medium and parameters

Composition of materials	$WO_3/$ 4,4-BAMBp	$WO_3/4,4$ BPPOBp	$WO_3/$ 1,10-DAD	$PMo_{12}/$ 1,10-DAD	$SiMo_{12}/$ 1,10-DAD	$PW_{12}/$ 1,10-DAD	$SiW_{12}/$ 1,10-DAD
Number of layer	100	100	50	40	40	40	40
Thickness of film (nm)	67	71	156	70	68	65	65
Change of absorption rate (%) after writing	1.5	2.1	8.5	16.8	15.1	3.0	0.5

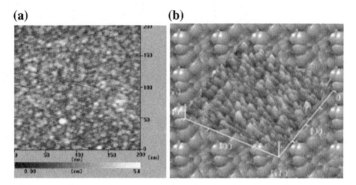

Fig. 3.3 AFM images of the phosphomolybdic acid monolayer: **a** plane scanning, **b** 3-dimensional picture

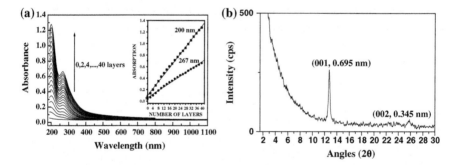

Fig. 3.4 a Relationship between layers and adsorption **b** X-ray diffraction pattern of self-assembled 40 layers of $WO_3/4,4'$- BAMBp multilayer

monocrystalline silicon 40 layers of $WO_3/4,4'$-BAMBp self-assembled multilayer X-ray diffraction pattern, which can be calculated from the diffraction peak position of the layer spacing of the multilayer films. In these self-assembled monolayers, organic molecules or tilt or vertical arrangement of the inorganic components, and the inorganic

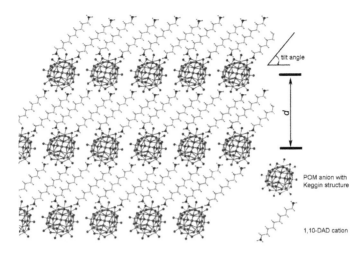

tilt angle

d

POM anion with
Keggin structure

1,10-DAD cation

Fig. 3.5 The multilayer structure model of Heteropoly acid/capric diamine

component in the composite film is undoubtedly a structural change, these polyanionic are adsorbed on the organic layer into the organic layer, the formation of inorganic/organic ion complexes. Heteropoly acid/capric diamine multilayer structure for the model is shown in Fig. 3.5 [4, 5].

Nanosized TiO_2 sol has excellent light response that can be used to improve the photochromic performance of WO_3, more photogenerated electrons (from WO_3 and TiO_2) is located in the capture of WO_3 within the band gap energy levels involved in the discoloration process. WO_3 and WO_3/TiO_2 in colloid inorganic–organic self-assembled multilayer films show good photochromic reversibility and high photochromic response. But the self-assembled composite system ($PMo_{12}/$ 1,10-DAD and $SiMo_{12}/$1,10-DAD) are higher photochromic response as shown in Table 3.1. 70-nm thick self-assembled films of photochromic response, equivalent to 1 μm thick MoO_3 Films (several tens of nanometers thick MoO_3 films have only a weak light-induced discoloration of response). Can be observed even in the amino protons of the substrate surface only when the deposition of a layer of phospho-molybdate photochromic phenomenon. UV radiation transfers the transparent film into blue light. All samples are colorless and transparent in front of the UV radiation, a strong absorption peak corresponding to the semiconductor band–band transitions in the short wavelength region as shown in Fig. 3.4. The WO_3/TiO_2 colloid becomes dark blue after UV radiation for 3 min, a strong absorption peak at 900 nm. After 3 min of UV radiation, although the absorbance has increased, WO_3 colloid has experienced a long period of radiation (such as more than 30 min), the naked eye can be directly observed to blue. It can be seen from the figure of $WO_3/$ TiO_2 and WO_3 colloid of ΔOD (optical density change) at 900 nm of 0.622 and 0.049, respectively, that the former is about 12.7 times, only TiO_2 and WO_3, the molar ratio of about 1:40. Regardless of whether the hole scavenger (oxalic acid),

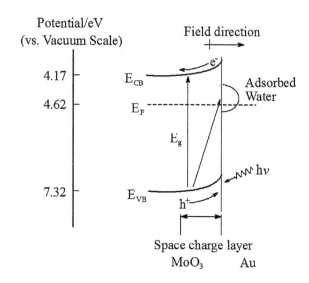

Fig. 3.6 Electronic energy band model and transfer diagram of assembled gold nanoparticle on MoO₃ surface that have two types of transitions: 1—from valence band to conduction band, 2—from valence band to surface states (adsorbed water)

ΔOD with the increase of the TiO_2 concentration increased, the maximum increase of 100 times [6].

MoO_3 and WO_3 is a n-type semiconductor, the Fermi level is located 4.3–4.9 eV (relative to vacuum level), As Au nanoparticles remained the metal properties, its work function is about 5.10 eV. So film of MO_3 (M = Mo, W) contacts with Au nanoparticles to form a space charge layer (i.e., ottky barrier) due to their different Fermi level at their interface. The binding electrons energy of composite film MO_3 valence band (O_{2p}) is higher than the valence MO_3 films that indicates the composite film MO_3. Fermi level moves down and the gold moves up, as MO_3/Au composite film for example as shown in Fig. 3.6. Since MoO_3/Au composite films formed at the interface with the upward bend after light excitation. It can suppress the photo-carriers process and conducive to photo-electron reduction of MoO_3 and light water hole oxidation process that great improve the photochromic reaction [7, 8].

3.1.2   Multi-wavelength Holography Materials

Polarization holographic storage with indolylfulgimide has a wide range of applications for storages and optical molecular switches. Nankai University and Harbin Industry University in China developed a lot of large dynamic range, high sensitivity, contraction of the small scattering, non-volatile storage materials including double-doped crystals of magnesium and iron lithium niobate crystal, zinc-doped iron lithium niobate crystal, indium-doped iron niobium acid lithium crystal, etc. The doped niobium lithium niobate, indium, iron concentration, and redox state can inhibition of photovoltaic electric field through the electrolyte solution, enlarge dynamic range and increasing, anti-light-induced scattering, and high

Fig. 3.7 Doped iron and manganese lithium niobate crystal non-volatile holographic memory write and read out the curves

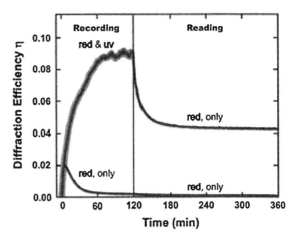

Fig. 3.8 Near-stoichiometric lithium niobate crystals photorefractive grating erasing curve (**a**) and writing curve (**b**)

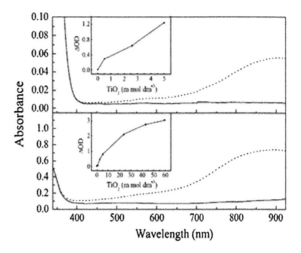

signal-to-noise ratio as shown in Fig. 3.7. Established for excellent photorefractive properties of double-doped lithium niobate crystals doped with ion species based on the band transport model to optimize the concentration ratio (low-doped indium-doped iron) as well as process parameters to growth of doped with 0.5 mol% of In, 0.06 wt% Fe lithium niobate. Utilizing the hydrogenation oxidation of indium-doped iron lithium niobate made the iron-doped lithium niobate crystal with large dynamic range and faster response. The pure near-stoichiometric LiNbO$_3$ crystal was developed by Nankai University that is strongly reduced in vacuum which photorefractive properties is shown in Fig. 3.8.

A short photorefractive response time of the order of 100 ms is measured at a wavelength of 514.5 nm, with incident light intensity of 1.6 W/cm^2, and possible corresponding mechanisms are discussed. The diffraction efficiency of a holographic grating written in this reduced crystal is low but can be enhanced by an externally applied electric field with wavelength of 488 nm to write. Photorefractive performance of a single iron-doped lithium niobate-doped lithium niobate, lithium niobate crystal has a strong photovoltaic effect, the performance of the crystal by the light, will form the open-circuit voltage. The photovoltaic effect is caused by the spatial asymmetry of the crystal microstructure. Photovoltaic electric field strength is generated by the iron-doped lithium niobate crystals, usually 10 kV/cm order of magnitude. The formation of photovoltaic electric field through the electro-optic effect will change the refractive index of the illuminated area of the crystal. Local hologram corresponds to different parts of the body surface of Prague offset angle, and the object beam in the crystal, which is approximated as a plane wave holographic method used in this experiment the object surface, the crystal, the divergence angle is very small, from the theoretically can be concluded that the uneven crystal photovoltaic electric field. Crystal placed in NaCl solution into the holographic storage experiments confirmed that the solution immersion method can effectively reduce the photovoltaic effect on the degradation of image quality, while significantly improving the storage capacity. Therefore, the electrolyte solution immersion method, can effectively inhibit the photovoltaic effect, to overcome the photovoltaic effect caused by the image quality is degenerate, the constraint decreasing exposure timing balanced diffraction efficiency and great increased density. Meanwhile, the copper and Ce doubly doped lithium niobate crystals achieved high-resolution non-volatile image storage and close to the diffraction efficiency of non-volatile raster storage. In the participating units with self-preparation of nearly one hundred kinds of doped lithium niobate crystal, including seven kinds of near-stoichiometric crystal that the near-stoichiometric lithium niobate crystal has good prospects of application in multi-wavelength holography storage as shown in Fig. 3.9. Doped zinc iron lithium niobate crystal has the advantages of high sensitivity and low noise, the main drawback is the poor dynamic body range. Single iron-doped 0.03 wt% lithium niobate crystal with excellent optical quality can be used to double wavelength holography [9].

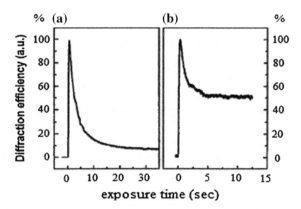

Fig. 3.9 Near-stoichiometric lithium niobate crystals photorefractive grating erasing curve (**a**) and writing curve (**b**)

The influence of composition on the photorefractive effect in pure LiNbO$_3$ crystals at low light intensity was investigated. The experimental results indicate that different defects dominate the photorefractive centers of pure LiNbO$_3$ with various compositions. Bipolarons are considered to be responsible for the enhanced photovoltaic field in reduced near-stoichiometric LiNbO$_3$, and their bulk photovoltaic constant κ is estimated to be $\sim 6.95 \times 10^{-32}$ m 3/V. Q polarons (composed of two bipolarons) are introduced to explain the photorefractive effect of congruent LiNbO$_3$ at both low and high light intensities.

3.1.3   Polymer Material for Volume Holographic Storage

Polymer material have larger dynamic range, high sensitivity, inexpensive, dry fixing, etc., advantages that become the most potential in the current volume holographic storage recording material. Physics Chemistry Institute of CAS developed a typical high photorefractive index and low shrinkage medium for volume holography. In terms of percentage by weight, 15–45 % of vinyl monomers, from 0.01 to 0.1 % of the photosensitizer, 0.1–0.9 % of photoinitiator, 0.1–1.0 % of the chain transfer agent, 0.1–1.0 % defoamers, 0.1–1.0 % of the leveling agent, 3–15 % of the epoxy reactive diluent, and 30–70 % of epoxy resin mixed with uniform transparent sol solution, then add 5–15 % of the epoxy curing agent mixed curing of light-induced polymer volume holographic storage material. It can be prepared into the storage media of any size and thickness, experiments show that it has high stability, low shrinkage, and large capacity as shown Fig. 3.10. In accordance with this structure was prepared by the contraction of the polymer material with high photorefractive index, vinyl monomers of the epoxy resin of high refractive index and low refractive index polymer light-induced volume holographic storage material.

3.2   Photochromic Reaction in Recording Materials

Photochromic reaction or photochromism is not a rigorous definition, but is usually used to be described a compounds that has reversible photochemical reaction when it absorbs certain wavelength light on the electromagnetic spectrum can change its characteristics of absorbtion dramatically in strength. The photochromic phenomenon

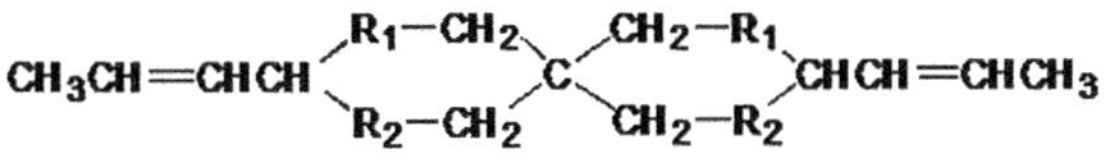

$$CH_3CH{=}CHCH \overset{R_1{-}CH_2}{\underset{R_2{-}CH_2}{<}} C \overset{CH_2{-}R_1}{\underset{CH_2{-}R_2}{>}} CHCH{=}CHCH_3$$

Fig. 3.10 Run out caused by the epoxy polymer of the open-loop and the light-induced acrylate monomer synthesis curable photopolymer with high photorefractive index and low shrinkage

first was discovered in the late 1880s, including work by Markwald, who studied the reversible change of color of 4-tetrachloro naphthalen-1(4H)-1 in solid state. He labeled this phenomenon as "phototropy", and this name was used until the 1950s when Yehuda Hirshberg of the Weizmann Institute of Science in Israel proposed the term "photochromism". Photochromism can take place in both organic and inorganic compounds, and also in biological systems of tetrahydronaphthalene. A phenomenon was known as the photoinduced thermodynamic reversible color change each other (i.e., phototropy). Yehuda Hirshberg researched this phenomenon indepth and clearly put forward the concept of photochromic reaction (photochromism) in 1956, and the establishment of the fineness and photobleaching cycle to constitute chemical memory model for the photochemical information storage. In particular, organic light-induced reaction of photochromic materials for laser signals to constitute a new generation of optical information storage materials that put forward the research to a new developed area. Since nearly half a century, the photochromic materials to carry out a lot of research projects that the photochromic materials and compounds were brought to broad and important applications. When the photochromic reaction of a compound or complex A with certain wavelength of light irradiation hv_1, the formation of the structure was became to another compound B. The compound B went back to the original structure A, when it was irradiation by wavelength light hv_2 simultaneity, the phenomenon as shown in Fig. 3.11 [10, 11].

As shown in Fig. 3.11, a photochromic materials has two completely different absorption spectrum of A and B when it absorbs different laser light. The two kinds of steady state of photochromic materials that can be represented the number "0" and "1" and rewritable, that is the basis using this photochemical phenomoena to record digital data.

Photochromism is a reversible transformation of a chemical species induced in one or both directions by absorption of electromagnetic radiation between two forms of A and B, which are different absorption spectra. A common example is the sunglasses that turn dark when exposed to sunlight.

Photochromism is derived from the Greek words: phos (light) and chroma (color). Photochromism may be used to molecular switches or optical data storage. A typical photochromic material and its absorption spectrum is shown in Fig. 3.12.

Fig. 3.11 The photochromic reaction, $\hbar$ is the Planck constant v_1, v_2 are different light frequency (wavelength)

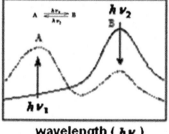

absorptivity

wavelength ($\hbar v$)

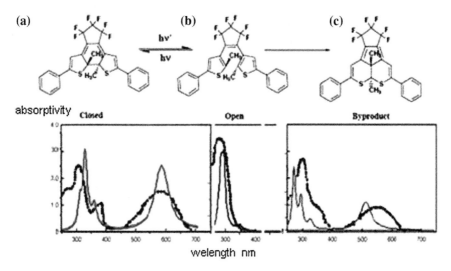

Fig. 3.12 Photochromic materials absorption spectrum of an organic material. When photochromic materials have different steady absorption spectrum of A and B when it absorbs different laser light

Fig. 3.13 Experiment readout signals of sensitivity wavelength for **a** 650 nm, **b** 532 nm and **c** 405 nm photochromic materials

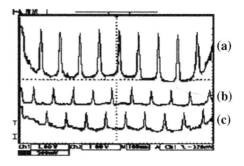

Actually, the first wavelength λ_1 light (known as the erase light) irradiated the recording media, that is transition from state A to B and the encoding information is written on the media. When λ_2 light on the area of the material, it was changed by the state B to state A. It is a binary-coded signal to representative "1" and "0". Not been λ_2 light exposure within the region material is still state B corresponds to the binary-coded "0". Testing absorptivity or reflectivity of the recording medium can get the readout signal by changing of the material absorptivity (or transmittance) to wavelength λ_2 light. As the absorptivity of coded "0" is very small to λ_2 light, but absorptivity of the coded "1" (state A) is much strong. Otherwise, using readout λ_2 light is very weak and innocuity to medium. Using absorption peaks were 650 nm/532 nm/405 nm recording materials, with same written conditions to experiment that results is shown in Fig. 3.13a–c. Obviously the sensitivity of (a) for 650 nm laser is highest and (c) lowest for 405 nm [12, 13].

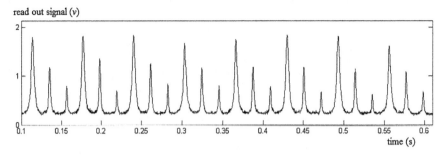

read out signal (v)

time (s)

Fig. 3.14 The read out single of multilevel recording experiment on the sensitivity wavelength for 650 nm material

The more sensitive photochromic material is diarylethylene which absorption peak is 650 nm. The multilevel storage experiment is shown in Fig. 3.14 using diarylethylene easily. It can be written by controlling the light intensity to get four-level storage. The medium contains three kinds of materials for different wavelength absorption peaks and were mixed into a recording layer that can achieve the three wavelength and four-level multilevel storage.

3.2.1 Classification of Photochromic Materials

Photochromic medium includes both inorganic and organic materials that molecular structure and photochromic principles are different absolutely. Inorganic photochromic medium are added with the compounds of metals (mainly transition period of heavy metals) ion valence, as well as compounds decompose and recombined to achieve photochromic reaction that can be divided into two kinds of the decomposition of the metal ions of variable valency and halide decompounds usually. Typical inorganic material is tellurium, that the material has a strong absorption and reflection characteristics of a variety of wavelength laser. But it is easy to produce the scaly cracking by prolonged exposure of light irradiation, and the control technology of thickness and uniformity are more complex in production, so that not be a comprehensive application. But latest achievements in investigations of photochromic effect in organic materials made an important breakthrough, that some chemical species can undergo reversible photochemical reactions to creates a possibility of application in optical data storage systems (photon-mode erasable optical recording media), optical memories and switches. For this purpose, organic photochromic compounds are often incorporated in polymers, liquid crystalline materials or other matrices. Developed organic photochromic compounds materials for digital optical storage have the following categories:

3.2.2   Spiropyran/Looxazine Compounds

This kind of photochromic reaction compounds are CS bond (or CO) of spiro compounds (closed-loop body), that can be cleaved and generated open-loop body at the short wavelength (UV) irradiation. It has a strong absorption and photochromic reaction in the visible region. Under visible light irradiation or in the case of dark, open-loop body to re-restore to a closed loop body that has good reversible reaction. Spiropyran has better photochromic properties also, the open-loop maximum absorption wavelength is less 600 nm, its main drawback is that the photochromic body of the open-loop is unstable. Looxazine thermal stability and fatigue resistance are better in the all, and its open-loop maximum absorption wavelength is great than the spiropyran with potential applications [14].

3.2.3   Fulgide Compounds

Fulgide compounds are used valence bond tautomerism intramolecular pericyclic reactions to be photochromic reaction. Its stability and fatigue resistance are better, but the absorption peak wavelength does not match with the commercial semiconductor lasers, that need to be adjusted. The representative structure of the fulgide molecule compounds is shown in Fig. 3.15.

3.2.4   Diarylethene Compounds

Diarylethene compounds is belonging to the *cis-trans* isomerization of photochromic compounds, due to the absorption spectra before and after the photochromism is smaller, the absorption spectrum will have a greater overlap that is not suitable for multi-wavelength storage. Irie et al. designed a synthesis with special structure of *cis*-diaryl heterocyclic ethylene in 1988. Experiments show that these compounds not only have good photochromic properties and the overlap of the absorption spectrum is also smaller, may be used for optical storage. Study also showed that these compounds thermal stability is good, and the open-loop body at 300 °C does not present the thermochromic phenomenon. During destructive testing, the closed-loop body may be stable at 80 °C in more than 3 months, and

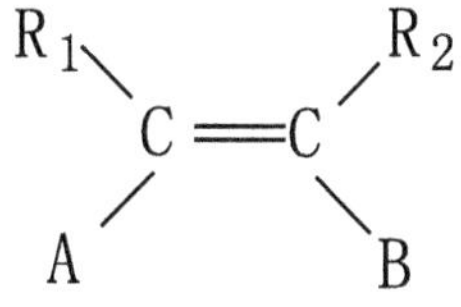

Fig. 3.15 Typical structure of the fulgide: R_1, R_2=H,CN; C(O),OC(O),$(CF_2)_n$, n (integer) = alkyl; A, B = aryl

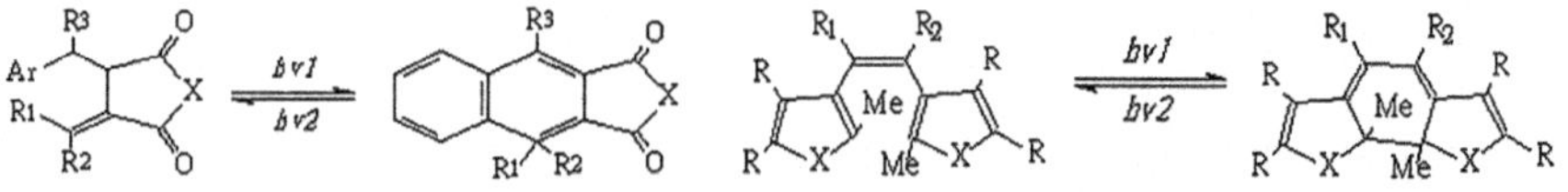

R = H, alkyl, King Kong-ene, norbornene, and heterocyclic: Ar = aryl; X = O, NR

Fig. 3.16 Typical molecular structures of diarylethene fulgide and photochromic reaction

fatigue resistance is also very good. Such compounds have a general form that used the valence tautomerism occurrence of intramolecular pericyclic reactions. Typical diarylethene molecular structure of photochromic reaction is shown in Fig. 3.16.

In addition, the performances of compounds with different substituents will be changed greatly, that the feature can be used to adjust design of the absorption peak as shown in Fig. 3.13, which will be the great development for application [15].

3.2.5   The Azo Compound

This kind of compounds is utilized the *cis-trans* isomerization of key. Its peak of the absorption can be adjust when the electronic grant or electronic receiver body are introduced to different parts of the molecule. The maximum absorption wavelength of some azo dyes can cover 700 nm. A typical molecular structure of the azo compound and its photochromic reaction are shown in Fig. 3.17.

The wavelength of the absorption spectra of azo compounds is shorter and that is changed very small before and after photochromic reaction, and thermal stability is poor. In other hand the sensitive peak of azo dyes cannot match with the laser of current commercial production, so it is applied in the experimental study of optical storage only.

3.2.6   Other Organic Photochromic Compounds

For light-induced photochromic storage materials are still much more indeed, that as long as the following conditions can meet photochromic optical storage. Where high sensitivity and good anti-fatigue performance, can repeatedly write, erase, and stable performance of the material is suitable for write-once optical storage. Best sensitive wavelength match with the industrial production of semiconductor lasers,

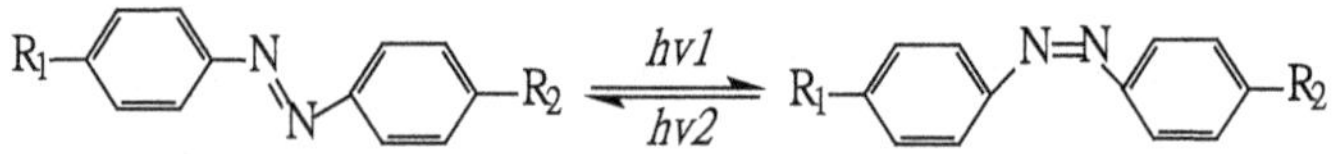

Fig. 3.17 Typical molecular structures of the azo compound and its photochromic reaction

good solubility, and can manufacture with spin coating method. It is one of most development prospects photochromic materials for optical storage.

For commercialize application products require materials with good thermal stability and long-term preservation also. Study of photochromic storage materials, including dedicated research of organic photochromic compounds, molecular assembly (such as MTCNQ and MTNAP norbornene, sulfur, indigo, etc.) and inorganic materials with good performance of the photochromic compounds have great potential development.

3.3 Structure and Testing of Medium

3.3.1 Structure of Medium

New achievements and progress in investigations of photochromic effect in organic materials. The absorption spectrum of the photochromic materials can change, choose and control within a narrow range that greatly reduces the absorption of the material except for its peak of absorption. Therefore, different wavelength lasers can be used to write and read in corresponding photochromic materials at one time to carry out multi-wavelength and multilayer parallel data storage, that structure the medium is shown in Fig. 3.18.

This is a multilayer and multi-wavelength storage program, the absorption peak of recording wavelength, respectively, to three kinds of photochromic materials, which can be coating on the CD with three-layer or one layer (three photochromic materials are mixed to dissolve in solvent). The absorption spectrum of the materials is required narrow, their spectrum has to be less than or close to the spacing between the neighboring recording laser wavelength, and the absorption spectrum of the material should not be overlap. Since each recording layer contains only a photochromic material, each recording layer is only sensitive to a recording wavelength laser, while the other wavelength lasers are transparent. For example,

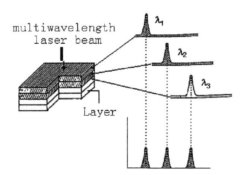

Fig. 3.18 Multi-wavelength and multilayer parallel information storage

recording laser λ_1 in Fig. 3.18, when the laser λ_1 writes the multi-wavelength photochromic plate, only absorption by this photochromic material that is sensitive to the wavelength λ_1, but other recording layers cannot be absorbed. When three wavelength synthesis coaxial lasers write the multi-wavelength photochromic plate at the same point of the disc, each wavelength laser will be absorbed by corresponding recording layer, that all recording layer material in the point are photochromic at the same times and response to multiple information. Experiment of this system can select recording lasers wavelength are 780, 650, 532 and 405 nm. These wavelengths of laser are typical lasers application for CD, DVD or BD and HD-DVD. Therefore, these lasers have good compatibility with existing CD produces and industrial scale production except the wavelengths 532 nm semiconductor laser. But it was researched and developed at Tsinghua University. Interval of these wavelengths of lasers is much balanced, easy to control of writing/reading and crosstalk is easement [16–18].

These recording photochromic materials are more mature, that they is not chemical reaction each other, so the disc can be used traditional spin coating process to manufacturing. In addition, to simplify the process, threes materials can be mixing dissolution in an efficient solvent, and spin coating a single-layer mixed recording media to obtain the same effect with the three-layer multi-wavelength CD, as shown in Fig. 3.19b.

The structure of this multi-wavelength photochromic disc is similar to the current CD-R/DVD-R, the CD/DVD-RW disk, it is composed of substrate, recording layer, reflective layer and protective layer. It has premolded track channel also, it is divided into multiple sectors. Each sector has the sector format code that used to identify the sector address to data storage. The difference is that the recording layer with variety of photochromic materials and sensitive to different wavelength to recording.

An obvious problem of the multi-wavelength photochromic storage system is the size of focus is different, if it does not used complex achromatic objective lens. Therefore, utilization super-RENS technology to control and reduce the size of focus, that the structure of this disc with super-RENS mask layer is shown in Fig. 3.20 (reference 4.1 Optical super resolution and micro-aperture laser in Chap. 4 of this book in detail). The material of mask layer is not transparent to sensitive wavelength laser of recording layer, but it can be made a hole using

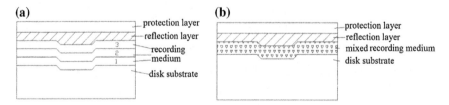

Fig. 3.19 The structure of multi-wavelength photochromic disk. **a** Multi-wavelength and multilayer. **b** Photochromic material mixed spin coating to a single recording layer

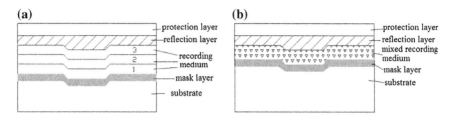

Fig. 3.20 The typical multi-wavelength multilayer storage disc with super resolution photochromic mask layer: **a** for multilayer recording medium **b** for mixed recording medium

another wavelength laser with nonlinear effect on mask layer, then recording wavelength lasers are enter recording layer through the openings hole on the mask layer. Therefore, it solved the problem that the focus spot size is difference by chromatic aberration of multiple wavelengths lasers. At the same time such that the uniform spot size on the super-RENS mask layer is reduced by 20–30 %, can further increase the total storage capacity.

The manufacturing process of the multi-wavelength disc with super resolution mask is: substrate is poly carbon acid resin material and molded with injection, then coating mask layer, mixed recording layer, reflection layer, and protection layer with spin coating method successively [19].

3.3.2   Testing of Medium

The experimental testing system of medium for multi-wavelength and multilevel is shown in Fig. 3.21. The figure shows $L_1 \ldots L_n$ are different wavelength lasers and beams collimating/shaping lens system, $D_1 \ldots D_n$ are photodetectors to $L_1 \ldots L_n$ laser

Fig. 3.21 Schematic of testing system for multi-wavelength and multilevel medium

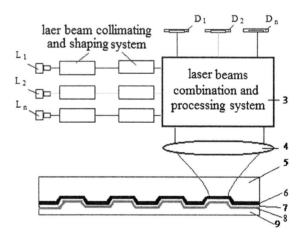

beam, 3—laser beams integrating and separating system, 4—complex achromatic objective lens, 5—substrate with pre-groove, 6—super-RENS mask layer, 7—multi-wavelength photochromic medium layer, 8—reflective layer and 9—protective layer. If employ n laser beams and m level modulation and coding to record experiment the storage capacity will be increased by N times ($N = n \times \ln m$) theoretically. Because the experimental testing system adopted complex achromatic objective lens, so it can be used to experiment for multi-wavelength photochromic medium disc without mask layer, but this objective lens is too intricacy and weight that cannot apply to commercial manufacturing.

The key component of the optical system is the laser beam coupling/splitting prism (LCSP) in the testing system. The Fig. 3.22 is a three beam LCSP for example that is composed of four cube corner prisms which are adhered each other. Two interfaces are coated with film series by which lights are coupling or splitting. The transmissivities of the two interfaces are shown in Figs. 3.23 and 3.24,

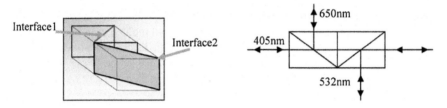

Fig. 3.22 The configuration of optical coupling/splitting prism (LCSP): Interface-1 is transparent to the light of 405 nm but reflecting the light of 650 nm. Interface-2 is transparent to the lights of 405 and 650 nm but reflecting the light of 532 nm

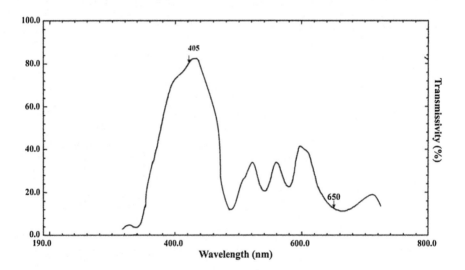

Fig. 3.23 Transmissivity of the interface-1 with the entrance angle of 45°

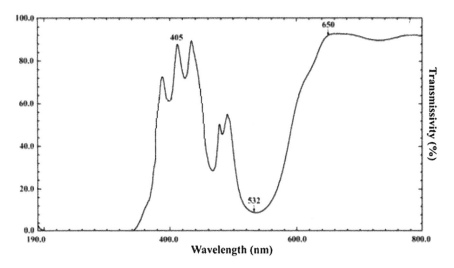

Fig. 3.24 Transmissivity of 405, 532 and 650 nm lasers on the interface-2 with the entrance angle of 45°

respectively. The transmissivity of the interface-1 is higher 82.06 % for 405 nm laser and lower 10.13 % (reflectivity higher 85.05 %) for 650 nm also, respectively, with the entrance angle of 45°. The transmissivity of the interface-2 is 89.35 % for 405 nm laser and 95.11 % for 650 nm laser, as well as lower 8.79 % (reflectivity higher 89.13 %) for 532 nm laser with the entrance angle of 45°.

In fact, the disc of photochromic materials can be manufactured with vacuum sputtering, that the quality is better than solvent spin coating. Thickness and uniformity of film are relatively easy to control, but it is high manufacturing costs and lower production efficiency, usually only used for the experimental study. The spin coating is a very mature method to preparation of uniform organic thin film, that the photochromic materials are directly dissolved in an organic solvent, then spin coating to film. The advantage is low cost, smooth thin film, uniform, easy to control the thickness of the medium, and widely used in optical disc manufacturing industry. Spin coating process is generally the first film-forming material soluble in a moderately volatile solvent, the formation of a certain viscosity liquid (or colloidal) spin coating liquid, so the selection of solvent is very important. Main requirements for organic solvents are shown in Table 3.2, that selection of solvent should meet the following conditions:

1. good solubility of the dyes and polymers;
2. volatile moderate boiling point of about 90–120 °C;
3. surface tension as small as possible, so that the preparation of the surface tension is less than the surface tension of the substrate;
4. insoluble substrate (the substrate is polycarbonate), nontoxic.

Table 3.2 Solvent and performances for the photochromic organic materials recording layer manufacturing

Organic solvents name	Boiling point (°C)	Surface tension (10^{-5} N/cm)	Viscosity (mPa s)	Solubility to PC
Diacetone alcohol	168.1	31.0	2.9	No
Ethyl-melting fiber agent	156.3	31.8	1.03	Very small
Cyclooctane	125.7	21.76	0.547	No
Toluene	110.6	30.92	0.773	No
Chloroform	61.1	27.14	0.563	Very small
Cyclohexanone	155.6	3450	2.2	Very small

Light-induced photochromic materials: diarylethene compounds, fulgide compounds have good photochromic properties is more suitable for optical storage. Light-induced photochromic multi-wavelength storage experiment is used diarylethene compounds mainly. Its samples are composed of the recording layer, reflective layer and the substrate special high flatness glass substrate which is a 1.1 mm special plane glass with size of 2.5 cm × 2.5 cm. After rigorous cleaning using vacuum sputtering to prepare complete reflective layer, which thickness is about 100 nm. The recording layer prepared with spin coating process as follows.

The polymethyl methacrylate (PMMA) was dissolved in chloroform with ultrasonic mixer, for photochromic compounds mixed into the solution to become homogeneous glue, and then using spin coating to formed film on the reflector layer. The layer thickness depends on the concentration of the solution and spin coating speed. In experiments is using the KW-4 spin coating machine with maximum speed of 6000 rpm. The spin coating process is that the photochromic compounds glue was put on center of substrate with speed of 50–80 rpm in beginning, then the speed rapidly increased to 2000 rpm. The photochromic compounds glue forms a symmetrical film on the substrate, that the excess glue from the edge of the substrate thrown. The spin coating machine is spinning sostenuto to about 40 s, that the solvent is volatilized all to obtain the film with thickness of about 500 nm. In addition, using the binary mixed solvent with a certain proportion of mixed-butyl ether solvent and 2,6-dimethyl heptanone auxiliary solvent can get better volatility and effectively prevent the aggregation of the film to the crystallization. In particular, it can help to balance the viscosity to fill into the pre-groove on the surface of the substrate. In production of light-induced photochromic medium for multi-wavelength optical storage experiment, in order to ensure a smooth and uniform of the recording films layer to prevent the aggregation and crystallization of the photochromic compounds, that have to study out a intact process indeed [20].

3.4 Model of Writing

In data writing process with photochromic materials to achieve multi-wavelength and multilevel optical storage, the photochromic recording medium appears photochromic reaction that is relation closely with a lot of factors, as exposure energy (time, speed and intensity of writing light), material sensitivity, absorption rate, absorption spectra and crosstalk etc. So it is very important to establish a model of photochromic recording process to quantify analysis for various affect factors. Actually the light-induced photochromic reaction is the discoloration of storage material in the exposure with certain wavelength that is from the closed-loop state transformation to the open-loop state. The transformation process caused to decrease number of the closed-loop state molecules in photochromic materials, it change its absorptivity of the recording medium to corresponding wavelength, i.e., the reflectivity of the disc manifested change of in experimental measurement. For discussion quantitative relationship between the reflectivity changes and exposure prerequisites of different materials to sensitive wavelength, have to analyze the absorption process of the photochemical reactions of medium in the first.

According to Beer's law, the absorptivity of medium is proportional to the number of photochemical molecules in the medium, i.e., only proportional to concentration C of the photochemical material in the medium. Meanwhile by Lambert's law, the absorptivity of incident light in the transparent medium is not any relationship to the incident light intensity and luminance. Based on about conception, a basic model can be established by the Beer–Lambert principles, as follows:

$$\frac{dI}{I} = -\alpha_v C dl \tag{3.1}$$

where I is the incident light intensity, C is the solute concentration (in mol/L), l is the thickness of medium (in cm), α_v is the constant of proportionality (for 10^3 cm^2/mol). The $C dl$ is the solution quality per unit area within dl absorption layer. The light intensity is influenced by the absorption of the dl layer. For a sample of certain length, the boundary conditions are: on the incident surface, as $l = 0$ so incident light intensity is I_0, but on the exit surface, i.e., $l = 1$, the light intensity is I therefore

$$\int_{I_0}^{I} \frac{dI}{I} = -\int_{0}^{l} \alpha_v C dl \tag{3.2}$$

Thus

$$\ln \frac{I}{I_0} = -\alpha_v C l \tag{3.3}$$

where α_v is the absorption coefficient, it is a function of radiation frequency and wavelength also.

Order $\varepsilon = \alpha_v/2.3$, known as the Molar extinction coefficient, its unit is 10^3 cm^2/mol. It is the absorption of 1 mol medium to certain wavelength light under fixed parameters conditions (temperature, concentration and optical path length). Thus available to the integral expression of the Beer–Lambert law:

$$\frac{I}{I_0} = e^{-2.3\varepsilon Cl} \tag{3.4}$$

where I_0 is the initial incident light, I is the light into the photochromic materials. The equation describes the remaining intensity of a parallel light beam of certain wavelength through the media which concentration is C and the optical path length is l. $D = \varepsilon Cl$ is defined extinction coefficient of the medium, also known as optical density.

In order to simplify the process of the multi-wavelength medium manufacturing, various photochromic materials are mixed uniformly, and dissolved in PMMA solvent, then spin coating on the substrate with aluminum reflective layer of reflectivity of R_f, so it is a single-layer disc without cover layer. Because of the role of the reflective layer, the writing light will be twice through recording layer, so the writing process can be simplified a model as shown in Fig. 3.25.

In the model, the layer 1 and layer 3 are recording medium and same thickness L. Layer 2 is an absorption layer which transmittance is its reflectivity R_f indeed but thickness is zero. For layer 1, I_0 is the incident light intensity, I_{t1} is the light intensity through the recording layer 1 and out to layer 2. I_{in1} is incident light intensity through the absorption layer 2 to recording layer 3. I_t is exit light intensity through the recording layer 3, i.e., reflected light or read out light intensity of the medium sample. In writing process experiments, if the sample has been written fully that the final reflectivity is defined "maximum reflectivity" R_{max}. The maximum reflectivity R_{max} is a very important concept to evaluate the performances of recording medium. The R_{max} of the final reflectivity is a constant for certain wavelength light, which has to be confirmed by experiment only. Clearly, the most important influence factor to R_{max} is the transmittance of the photochromic materials recording layer to a certain wavelength light and was defined T.

Fig. 3.25 The simplifying structure of photochromic medium samples for multi-wavelength storage experimental analysis

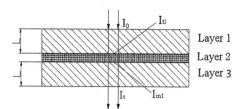

Introduction transmittance T, according to Eq. (3.4), the out light intensity through Layer 1 can express as

$$I_{t1} = I_0 T e^{-2.3D} \tag{3.5}$$

The out light intensity through absorbance layer 2 is

$$I_{in1} = I_{t1} R_f \tag{3.6}$$

Finally the out light intensity I_t through the absorbing layer Layer 3 is

$$I_t = I_{in1} T e^{-2.3D} \tag{3.7}$$

From (3.5) to (3.7) can get the light intensity through the simplifying structure sample (in Fig. 3.14), i.e., finally reflected light from the experimental sample is

$$I_t = I_0 T^2 R_f e^{-4.6D} \tag{3.8}$$

The proportion R between transmitted light intensity I_t and the initial light intensity I_0 is

$$R = \frac{I_t}{I_0} = T^2 R_f e^{-4.6D} \tag{3.9}$$

Taking into the R_{max} definition, when full writing $R_f = 1$ and molecular concentration of photochromic materials is 0, i.e., $D = 0$, thus

$$R_{max} = T^2 \tag{3.10}$$

Expression of the reflectivity can be obtained:

$$R = \frac{I_t}{I_0} = R_{max} R_f e^{-4.6D} \tag{3.11}$$

For a mixed various photochromic materials for multi-wavelength storage medium, consider the most complex cases, the absorption spectra of various materials of multi-wavelength are absorbed completely. Total of light transmittance the sample should be equal to the product of the light transmittances of each component separately in the mixed variety of photochromic materials, the total optical density is equal to the sum of all optical density of components and:

$$D_i = \sum_{j=1}^{n} \varepsilon_{j(\lambda_i)} C_j l \tag{3.12}$$

where, D_i is the total optical density for this system at the wavelength λ_i. The $\varepsilon_j (\lambda_i)$ is Molar extinction coefficient with j-th component in the wavelength λ_i light radiation. C_j is the concentration of material j. l is optical path, i.e., film thickness of photochromic materials and n is the number of wavelengths. So the reflectivity R_i to wavelength λ_i is

$$R_i = \frac{I_{t(\lambda_i)}}{I_{0(\lambda_i)}} = R_{\max(\lambda_i)} R_f e^{D_i} = R_{\max(\lambda_i)} R_f e^{-4.6 \sum_{j=1}^{n} \varepsilon_{j(\lambda_i)} C_j l} \qquad (3.13)$$

where $I_0 (\lambda_i)$ is the incident light intensity to wavelength λ_i. $I_t (\lambda_i)$ is reflective light intensity to wavelength λ_i. $R_{\max}(\lambda_i)$ is maximum reflectivity of the samples to wavelength of λ_i. R_f is reflectivity of the reflective layer. According to the model (as shown in Fig. 4.11), the light intensity $I_{ab(\lambda_j)1}$ after through the layer 1 of wavelength λ_j material is [21, 22]

$$I_{ab(\lambda_j)1} = I_{0(\lambda_j)} \left[1 - \left(\frac{R_j}{R_{\max(\lambda_j)} R_f} \right)^{1/2} \sqrt{R_{\max(\lambda_j)}} - \left(1 - \sqrt{R_{\max(\lambda_j)}} \right) \right] \qquad (3.14)$$

The incident light intensity $I_{in1(\lambda_j)}$ after through Layer 1 and Layer 2 to reach the Layer 3 surface of wavelength λ_j material is

$$I_{in1(\lambda_j)} = I_{0(\lambda_j)} \left(\frac{R_j}{R_{\max(\lambda_j)} R_f} \right)^{1/2} \sqrt{R_{\max(\lambda_j)}} R_f \qquad (3.15)$$

The light intensity $I_{ab(\lambda_j)}$ through Layer 3 with absorption wavelength λ_j is

$$I_{ab(\lambda_j)} = I_{0(\lambda_j)} \left(\frac{R_j}{R_{\max(\lambda_j)} R_f} \right)^{1/2} \sqrt{R_{\max(\lambda_j)}} R_f \left[\left(1 - \left(\frac{R_j}{R_{\max(\lambda_j)} R_f} \right)^{1/2} \sqrt{R_{\max(\lambda_j)}} \right) \right.$$
$$\left. - \left(1 - \sqrt{R_{\max(\lambda_j)}} \right) \right]$$

$$\qquad (3.16)$$

Thus, the light intensity $I_{ab(\lambda_j)}$ after absorption of photochromic materials to wavelength λ_j can be written:

$$I_{ab(\lambda_j)} = I_{ab(\lambda_j)1} + I_{ab(\lambda_j)3} = I_{0(\lambda_j)} \left(\sqrt{R_{\max(\lambda_j)}} - \sqrt{\frac{R_j}{R_f}} \right) \left(1 + \sqrt{R_f R_j} \right) \qquad (3.17)$$

Most of the light-induced discoloration (material achromatic response) can be regarded as a primary photochemical reaction. The multi-wavelength recording medium absorbs different wavelengths of light and react completely for each material to correspondence wavelength. The sum of the reaction rate of all materials is

$$-\frac{dC_i}{dt} = \sum_{j=1}^{n} \left(-\frac{dC_i}{dt}\right)_{\lambda_j} \tag{3.18}$$

The reaction rate of the material i to wavelength of λ_j is

$$\left(-\frac{dC_i}{dt}\right)_{\lambda_j} = \Phi_{i(\lambda_j)} \frac{\lambda_j}{Nhc} K_{i(\lambda_j)} \tag{3.19}$$

where $\Phi_i(\lambda_j)$ is the quantum yield of material i in the light irradiation of wavelength λ_j. $K_i(\lambda_j)$ is the absorption rate of material i to wavelength λ_j. N is Avogadro constant which equal 6.02214×10^{23} mol^{-1}. The $\hbar$ is the Planck constant which equal 6.62607×10^{-34} Js. The c is velocity of light in vacuum which equal 3×10^8 m/s. The reaction efficiency of the photochemical reaction materials can be used the quantum yield to characterize. Quantum yield Φ is photochemical reaction efficiency directly which is defined as $\Phi = N_r/N_m$. Here N_r is quantity of molecular for reactant, N_m is quantity of absorption photon of the photochemical material.

The quantum yield of photochemical reaction is relation with the photochemical material and the radiation wavelength, that is a constant to the certain photochemical material and certain wavelength of radiation.

Take (3.19) into (3.18) can obtain the total reaction rate:

$$-\frac{dC_i}{dt} = \sum_{j=1}^{n} \Phi_{i(\lambda_j)} \frac{\lambda_j}{Nhc} K_{i(\lambda_j)} \tag{3.20}$$

Absorption per unit volume absorbed light energy, i.e.,

$$K = \frac{E}{Stl} = \frac{I}{l} \tag{3.21}$$

where E is the radiant energy of certain wavelengths, S is the radiation area with unit of m^2, l is the optical path. Material i absorptivity to the wavelength λ_j is

$$K_{i(\lambda_j)} = I_{ab(\lambda_j)} \frac{\varepsilon_{i(\lambda_j)} C_i}{\sum_{k=1}^{n} \varepsilon_{k(\lambda_j)} C_k} \cdot \frac{1}{l} \tag{3.22}$$

Combined with Eqs. (3.17), (3.20) and (3.22) can get the function of the concentration C_i change of multi-wavelength light-induced discoloration material i to time t:

$$-\frac{dC_i}{dt} = \sum_{j=1}^{n} \Phi_{i(\lambda_j)} \frac{\lambda_j}{Nhc} I_{0(\lambda_j)} \left(\sqrt{R_{\max(\lambda_j)}} - \sqrt{\frac{R_j}{R_f}} \right) (1 + \sqrt{R_f R_j}) \cdot \frac{\varepsilon_{i(\lambda_j)} C_i}{\sum_{k=1}^{n} \varepsilon_{k(\lambda_j)} C_k} \cdot \frac{1}{l}$$

(3.23)

In multi-wavelength recording process, the reflection rate is decided by optical density of medium of various wavelengths. So the reflectivity is decided by concentration of all photochemical materials when the Molar extinction coefficient and the thickness of the recording medium are given. Therefore, concentration changes of the recording material can determine reflectivity in writing process and to aggregate to Eqs. (3.13) and (3.23), that can described the writing process of the multi-wavelength storage faultlessly.

Above model has taken into most complex situation in which the absorption spectrum of each material to cover all recorded wavelength. In the actual study on the material absorption spectrum as narrow as possible, the best absorption spectrum of each material is only covered with a record wavelength. In this case, the storage of multi-wavelength response model can be further simplified. Each component absorbed only with its corresponding wavelength of light, but a little or no absorption to other light, i.e., $\varepsilon_{i(\lambda_j)} \approx 0$ $(j \neq i)$. By Eqs. (3.13) and (3.21) are available:

$$R_i = R_{\max(\lambda_i)} R_f e^{-4.6\varepsilon_{i(\lambda_i)} C_i l}$$

(3.24)

$$-\frac{dC_i}{dt} = \Phi_{i(\lambda_i)} \frac{\lambda_i}{Nhc} I_{0(\lambda_i)} \left(\sqrt{R_{\max(\lambda_i)}} - \sqrt{\frac{R_i}{R_f}} \right) (1 + \sqrt{R_i R_f}) \cdot \frac{1}{l}$$

(3.25)

And from Eq. (3.22) can get the concentration of material i:

$$C_i = -\frac{1}{4.6\varepsilon_i l} \ln \frac{R_i}{R_{\max(\lambda_i)} R_f}$$

(3.26)

Derivative of both sides of the above equation to have:

$$\frac{dC_i}{dt} = -\frac{1}{4.6\varepsilon_{i(\lambda_i)} l} \cdot \frac{1}{R_i} \cdot \frac{dR_i}{dt}$$

(3.27)

The recording laser beams are uniform of intensity and parallel beams, namely

$$I = \frac{P}{S}$$

(3.28)

where P is the laser power with unit of W, S is light irradiation area, in unit of m^2.

Incorporated Eqs. (3.25), (3.27) and (3.28), can get the relationship equation between reflectivity and the irradiation time:

$$\frac{dR_i}{dt} = k_i \left(\sqrt{R_{\max(\lambda_i)}} - \sqrt{\frac{R_i}{R_f}} \right) (1 + \sqrt{R_i R_f}) R_i \tag{3.29}$$

where k_i is called to constant of writing time, it is a constant in writing process with unit s^{-1}. The constant characterizes the writing speed of wavelength λ_i laser on the sample, its value is

$$k_i = 4.6\varepsilon_{i(\lambda_i)} \Phi_{i(\lambda_i)} \frac{\lambda_i}{Nhc} \frac{P_i}{S_i} \tag{3.30}$$

Thus available:

$$t = \int_{R_{0(i)}}^{R_{f(i)}} \frac{1}{k_i} \cdot \frac{dR_i}{\left(\sqrt{R_{\max(\lambda_j)}} - \sqrt{\frac{R_i}{R_f}} \right) (1 + \sqrt{R_i R_f}) R_i} \tag{3.31}$$

where $R_{0\ (i)}$ and the $R_{f\ (i)}$ are initial reflectivity and the final reflectivity. The k_i is a constant to a recording medium and radiation conditions. Clearly, the reflectivity is changing for every wavelength independently over time, it do not affect each other in the simplified model. Of course, the simplified model can be used to description of the writing process to a single wavelength also. Above mathematical models based on the photochemical reaction principles of multi-wavelength photochromic storage process can be used to analyze crosstalk mechanism of light-induced, abatement crosstalk, optimization of write strategy to restrain crosstalk in the storage process in photochromic multi-wavelength recording, and then control crosstalk (have to be within less than 5 %) in multi-wavelength and multilevel readout signal.

3.5 Quantitative Evaluation for Absorbable Spectrum of Medium

The crosstalk between the different wavelengths is an important problem for multi-wavelength and multilevel storage. Crosstalk is emerged due to cross of absorption spectrum of the materials in the writing process mainly. The quantitative evaluation to absorbable spectrum of photochromic materials is groundwork account for the crosstalk. Although there are many special technical requirements control every material stringently. But the characteristics of the recording medium achieved perfect entirely without crosstalk, it is impossible almost. The absorption spectrum of the material and its cross extent is an important factor to determine the material can be practical or no. So it has to be established the analysis system to quantitative evaluation the cross-interference of photochromic materials. In the first, analysis of two kinds of materials absorption spectra without cross or very small for example, that are: 1,2 twofold (2-methyl-5-*n*-butyl-3-thienyl) perfluorinated cyclopentene and 1,2-twofold (2-methyl-5-(4-*N*,*N*-dimethyl-phenyl)-thiophene thiophene-3-yl) perfluorinated cyclopentene. Sensitive spectra of these two materials are 532 and 650 nm, uniform mixing of the two kinds of materials in PMMA solution to make the recording layer with spin coating, that the thickness is within 200 nm. Their absorption spectra and molecular structure are shown in Figs. 3.26 and 3.27. Before experiment, all used to experimental lasers have to be calibration with a reflectivity of 99.9 % sample.

In addition, the photochromic reaction of the photochromic material is depended on both of writing laser power and time. Such as 1,2-bis (2-methyl-5-(4-*N*,*N*-methyl-phenyl)-thiophene-3-yl) perfluorinated cyclopentene media writing experiment can obtain reflectivity of 0.8 different power 650 nm laser and exposure time, that the result of power—time curve is shown in Fig. 3.28. It can be seen using

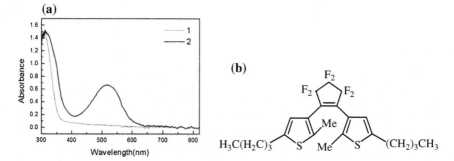

Fig. 3.26 Absorption spectra **a** and molecular structure **b** of the 1,2-bis (2- methyl-5-*n*-butyl-3-thienyl) perfluorinated cyclopentene

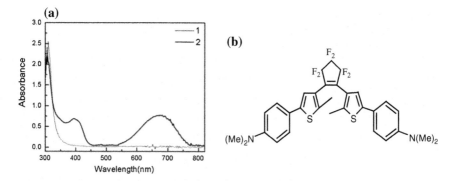

Fig. 3.27 Absorption spectra **a** and molecular structure **b** of 1,2-bis (2-methyl-5-(4-*N*,*N*-dimethyl-phenyl)–thiophenethiophene-3-yl) perfluorinated cyclopentene

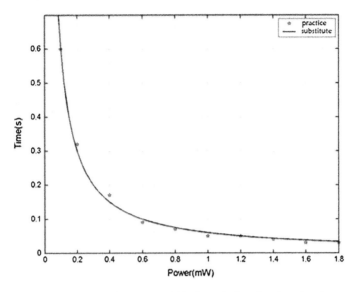

Fig. 3.28 Experimental power—time curve of thiophene-3-yl perfluorinated cyclopentene media with different power and exposure time. It obtains the reflection rate of 0.8 by 650 nm laser

different laser power and exposure time combination to get the same reaction. This feature of photochromic materials was used to improve or inhibit cross-interference in multi-wavelength and multilevel optical storage.

The experiments demonstrated also that the photochromic material reached a certain reflectivity that needs of exposure energy is determined. As above different writing power were reached reflectivity of 0.8 with a photochromic material for example. The product of writing power by exposure time (equivalent to the exposure energy) is changeless that is about 0.06 mJ, i.e., $P_t = 0.06$. It can be seen, when a photochromic material was written to reach a certain reflectivity, if the writing power is double, writing time will be reduced by half. Increasing the writing power can shorten the writing time with proportion. If materials without crossover of absorption spectrum, can choose the shortest exposure time to write. Above two materials without crossover of absorption spectrum, their reflectivity changes are independent also, so the writing model can be simplified. Using 532, 650 nm wavelength lasers write on a same sample, the experiment results reflectivity curve are shown in Figs. 3.29 and 3.30. The reflectivity of reflective layer of the sample is total reflection, eye $R_f = 1$, the initial reflectivity of 532 nm laser R_{ini1} is 0.589, initial reflectivity of 650 nm laser R_{ini2} is 0.38. The maximum reflectivity to 532 nm laser is $R_{max1} = 0.788$, to 650 nm laser is $R_{max2} = 0.86$ after full writing experiment. Based on the above parameters and the experimental data, using Eq. (3.31) can calculate writing time constant of perfluorinated cyclopentene material 1,2-bis (2-methyl-5-n-butyl-3-thiophene thiophene base) to be $k_1 = 5.61$ s^{-1}, and writing time constant of perfluorinated cyclopentene 1,2-bis (2-methyl-5-(4-N,N-dimethyl-phenyl)-thiophene-3-yl) material to be $k_2 = 5.34$ s^{-1}.

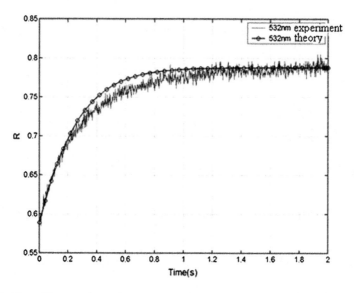

Fig. 3.29 The writing experimental and theoretical calculation results of relation between reflectivity and writing time to 1,2-bis (2-methyl-5-*n*-butyl-3-thiophene thiophene base) perfluorinated cyclopentene

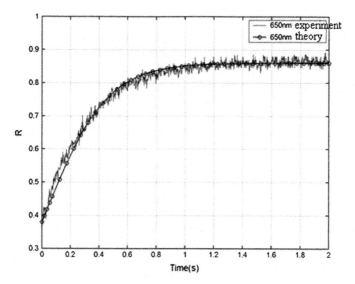

Fig. 3.30 The writing experimental and theoretical calculation results of relation between reflectivity and writing time to 1,2-bis (2-methyl-5-(4-*N*,*N*-dimethyl-phenyl)-thiophene-3-yl) perfluorinated cyclopentene

The comparison of experimental and theoretical calculation results with 532 and 650 nm lasers to writing are shown in Figs. 3.29 and 3.30, respectively [23–25].

The experimental results and theoretical calculations show that the simplified model can accurately describe the writing process in the multi-wavelength absorption spectra of various materials without cross. The k is an important parameter to characterize the write speed of light-induced discoloration of storage medium. The initial reflectivity R_{ini} and maximum reflectivity R_{max} are determined by the sample preparation process and measured by the experiment.

There are a lot of materials can be used to multi-wavelength and multilevel optical storage with very good characteristic of photochromic reaction, but most of them has some spectral cross often. For prevention of crosstalk in multi-wavelength and multilevel storage, first of all that should select the material without cross of absorption spectra, of course, but sometimes in order to take into account other characteristics of the material, such as sensitivity, stability and process and so on, have to choose some of the existence of certain cross-interference materials. After a long experimental demonstrated that their integrated parameters of the existence of certain cross-interference materials may have the desired results with control recording process also. So it is necessary to analysis the integrated speciality with certain cross-interference materials and impact each other. Following introduction an absorption spectra covering the whole fluorine cyclopentene materials for example that the molecular structure is shown in Fig. 3.31. Figure 3.31 shows the absorption spectrum of the material, its peak is 650 nm wavelength, but has some absorption at wavelength of 532 nm also. Therefore, this experiment is most representative for whole fluorine cyclopentene materials to research crosstalk of absorption with the common 650 and 532 nm wavelength.

As the material absorption spectrum cross it cannot description using simplify model, that must employ the Eqs. (3.13) and (3.23), more exact model for

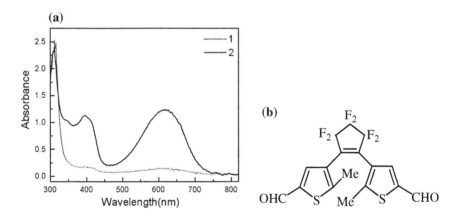

Fig. 3.31 **a** absorption spectra covering the whole fluorine cyclopentene materials (1 is open-loop absorption spectra, 2 is closed loop); **b** the molecular structure of whole fluorine cyclopentene materials

light-induced photochromic multi-wavelength storage. If the absorption wavelength of the materials-1 is $\lambda_1 = 532$ nm only. The absorption wavelength of materials-2 is both of $\lambda_1 = 532$ nm and $\lambda_2 = 650$ nm.

The reflectivity of the sample of the materials-1 to wavelength λ_1 is available present Eq. (3.13) as

$$R_1 = R_f R_{\max(\lambda_1)} e^{-4.6 \sum_{j=1}^{2} \varepsilon_{j(\lambda_1)} C_j l} = R_f R_{\max(\lambda_1)} e^{-4.6 l (\varepsilon_{1(\lambda_1)} C_1 + \varepsilon_{2(\lambda_1)} C_2)} \quad (3.32)$$

As material-1 absorbs wavelength λ_1 light only, its Molar extinction coefficient approximate to 0 to wavelength λ_2, the reflectivity of the sample to wavelength λ_2 is

$$R_2 = R_{\max(\lambda_2)} R_f e^{-4.6 \sum_{j=1}^{2} \varepsilon_{j(\lambda_2)} C_j l} = R_{\max(\lambda_2)} R_f e^{-4.6 \varepsilon_{2(\lambda_2)} C_2 l} \quad (3.33)$$

According to Eq. (3.32) can get:

$$\varepsilon_{1(\lambda_1)} C_1 + \varepsilon_{2(\lambda_1)} C_2 = -\frac{1}{4.6l} \ln \frac{R_1}{R_{\max(\lambda_1)} R_f} \quad (3.34)$$

$$\varepsilon_{1(\lambda_1)} \frac{dC_1}{dt} + \varepsilon_{2(\lambda_1)} \frac{dC_2}{dt} = -\frac{1}{4.6l} \cdot \frac{1}{R_1} \cdot \frac{dR_1}{dt} \quad (3.35)$$

From Eq. (3.33) can get:

$$C_2 = -\frac{1}{4.6 \varepsilon_{2(\lambda_2)} l} \ln \frac{R_2}{R_{\max(\lambda_2)} R_f} \quad (3.36)$$

$$\frac{dC_2}{dt} = -\frac{1}{4.6 \varepsilon_{2(\lambda_2)} l} \cdot \frac{1}{R_2} \cdot \frac{dR_2}{dt} \quad (3.37)$$

Based on the above four equations are available:

$$C_1 = -\frac{1}{4.6 \varepsilon_{1(\lambda_1)} l} \ln \frac{R_1}{R_f R_{\max(\lambda_1)}} + \frac{\varepsilon_{2(\lambda_1)}}{\varepsilon_{1(\lambda_1)}} \frac{1}{4.6 \varepsilon_{2(\lambda_2)} l} \ln \frac{R_2}{R_f R_{\max(\lambda_2)}} \quad (3.38)$$

Similarly, according to Eq. (3.23), the available material concentration

$$-\frac{dC_1}{dt} = \Phi_{1(\lambda_1)} I_{0(\lambda_1)} \frac{\lambda_1}{Nhc} \left(\sqrt{R_{\max(\lambda_1)}} - \sqrt{\frac{R_1}{R_f}} \right) (1 + \sqrt{R_1 R_f}) \cdot \frac{\varepsilon_{1(\lambda_1)} C_1}{\varepsilon_{1(\lambda_1)} C_1 + \varepsilon_{2(\lambda_1)} C_2} \cdot \frac{1}{l} \quad (3.39)$$

To the reflectivity of the sample of the material-2 with absorption wavelength of $\lambda_1 = 532$ nm and $\lambda_2 = 650$ nm is

$$-\frac{dC_2}{dt} = \Phi_{2(\lambda_1)} I_{0(\lambda_1)} \frac{\lambda_1}{Nhc} \left(\sqrt{R_{\max(\lambda_1)}} - \sqrt{\frac{R_1}{R_f}} \right)(1 + \sqrt{R_1 R_f}) \cdot \frac{\varepsilon_{2(\lambda_1)} C_2}{\varepsilon_{1(\lambda_1)} C_1 + \varepsilon_{2(\lambda_1)} C_2} \cdot \frac{1}{l}$$

$$+ \Phi_{2(\lambda_2)} I_{0(\lambda_2)} \frac{\lambda_2}{Nhc} \left(\sqrt{R_{\max(\lambda_2)}} - \sqrt{\frac{R_2}{R_f}} \right)(1 + \sqrt{R_2 R_f}) \cdot \frac{1}{l}$$

$$(3.40)$$

Sum Eqs. (3.34) to (3.40) and (3.28), can get the derivative dR_1/dt of reflectance R_1 of the sample to wavelength λ_1:

$$\frac{dR_1}{dt} = k_1 \left(\sqrt{R_{\max(\lambda_1)}} - \sqrt{\frac{R_1}{R_f}} \right)(1 + \sqrt{R_1 R_f}) \cdot \left(1 - \frac{\varepsilon_{2(\lambda_1)}}{\varepsilon_{2(\lambda_2)}} \frac{\ln \frac{R_2}{R_f R_{\max(\lambda_2)}}}{\ln \frac{R_1}{R_f R_{\max(\lambda_1)}}} \right) R_1$$

$$+ \frac{\varepsilon_{2(\lambda_1)}}{\varepsilon_{2(\lambda_2)}} \left[k_2 \left(\sqrt{R_{\max(\lambda_1)}} - \sqrt{\frac{R_1}{R_f}} \right)(1 + \sqrt{R_1 R_f}) \cdot \frac{\ln \frac{R_2}{R_f R_{\max(\lambda_2)}}}{\ln \frac{R_1}{R_f R_{\max(\lambda_1)}}} \right] \quad (3.41)$$

$$+ k_3 \left(\sqrt{R_{\max(\lambda_2)}} - \sqrt{\frac{R_2}{R_f}} \right)(1 + \sqrt{R_2 R_f}) \right] R_1$$

And the derivative of reflectivity R_2 to writing time t of this sample to wavelength λ_2:

$$\frac{dR_2}{dt} = \left[k_2 \left(\sqrt{R_{\max(\lambda_1)}} - \sqrt{\frac{R_1}{R_f}} \right)(1 + \sqrt{R_1 R_f}) \cdot \frac{\ln \frac{R_2}{R_f R_{\max(\lambda_2)}}}{\ln \frac{R_1}{R_f R_{\max(\lambda_1)}}} \right.$$

$$\left. + k_3 \left(\sqrt{R_{\max(\lambda_2)}} - \sqrt{\frac{R_2}{R_f}} \right)(1 + \sqrt{R_2 R_f}) \right] R_2$$

$$(3.42)$$

where

$$k_1 = 4.6\varepsilon_{1(\lambda_1)} \Phi_{1(\lambda_1)} \frac{\lambda_1}{Nhc} \cdot \frac{P_1}{S_1}$$

$$k_2 = 4.6\varepsilon_{2(\lambda_1)} \Phi_{2(\lambda_1)} \frac{\lambda_1}{Nhc} \cdot \frac{P_1}{S_1} \qquad (3.43)$$

$$k_3 = 4.6\varepsilon_{2(\lambda_2)} \Phi_{2(\lambda_2)} \frac{\lambda_2}{Nhc} \cdot \frac{P_2}{S_2}$$

Here k_n is writing time constant, k_2 is the writing time constant with cross of absorption spectrum of material. It is relation with the writing power only when the writing wavelength and materials have been determinate.

In order to analysis of above calculation correctly, make of two experimental samples with absorption spectrum of the 532 and 650 nm materials, respectively, to write experiment alone. The writing power to reach surface of sample: 532 nm wavelength is 0.07 mW, and 650 nm wavelength is 0.1 mW. Using simplified model to calculate the time constant are $k_1 = 5.61$ s^{-1}, $k_2 = 1.5$ s^{-1}, $k_3 = 4.51$ s^{-1}, respectively. Meanwhile, measured initial reflectance $R_{ini1} = 0.2$ for 532 nm light, and the initial reflectivity $R_{ini2} = 0.03$ for 650 nm light at same experimental conditions. The reflectivity of reflective layer of the experimental samples $R_f = 1$. The experimental results are shown in Fig. 3.32 with 532 and 650 nm light to write this sample.

According to the experimental curve with two wavelength while fully writing, maximum reflectivity of 532 nm $R_{max1} = 0.8$, the maximum reflectivity of 650 nm light $R_{max2} = 0.71$. In addition, the ratio of molar extinction coefficient of material-2 to wavelength of 532 and 650 nm is $\varepsilon_2(\lambda_1)/\varepsilon_2(\lambda_2) = 0.5$. Take above parameters into Eqs. (3.41) and (3.42), is available to get theoretical writing curve of mixed-material samples with spectrum cross-coefficient of two materials. The results of theoretical calculations compare with the experimental curve in Fig. 3.32, that shows the theoretical curve and experimental curve is consistent basically, and can accurately describe the absorption of various materials with spectrum cross to infection to the multi-wavelength photochromic storage. As can be seen, the time

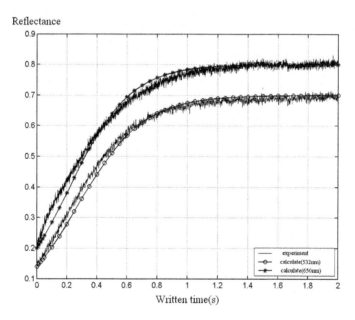

Fig. 3.32 Experimental and theoretical calculation curves of whole fluorine cyclopentene materials are written with 532 and 650 nm lasers, respectively

constant of the writing will influence to recording speed directly only. But the initial reflectivity, maximum reflectivity and reflectivity of reflective layer will affect whole recording process and quality. These differ relationships have great significance to research of reading and writing strategies in the future, following will be discussed separately.

3.6 The Reflectivity and Sensitivity

The reflectivity and sensitivity are very important parameters for medium, it is much different for single wavelength and multi-wavelength. So the section will introduce it, respectively, [26].

3.6.1 The Reflectivity and Sensitivity of Single Wavelength

1. The initial reflectivity and sensitivity

The initial reflectance is determined by the initial optical density or concentration of materials. It is lie on the manufacturing processes directly in the disc production. Of course, it can be controlled by adjusting the preparation process parameters also. For disc of multi-materials without cross of absorption spectra, its initial reflection is determined only by the concentration of the material. So change of initial reflectance of a certain material does not affect other reflectivity of materials. Here absorption spectra of the 650 nm material is analyzed for example. Changing the initial reflectivity of material of 650 nm light only, but other parameters and conditions is unchanged to calculate that the theoretical curves between reflectivity and writing time with different initial reflectance as shown in Fig. 3.33. It shows that the initial reflectance influenced to the writing speed or sensitivity of medium directly.

Figure 3.33 also shows that with the continuous reduction of the initial reflectance, writing time to achieve the maximum reflectivity is increased, i.e., the writing speed or sensitivity of medium is descend. In contrast, when the initial reflectance is increased, the reflectivity curve going to be gradient, and the nonlinear effect is more obvious.

The theoretical relational curves between sensitivity (derivative of reflectivity to writing time) and writing time of the medium of 650 nm with different initial reflectance are shown in Fig. 3.34. In beginning, sensitivity of the medium is increased with writing time and reached its maximum, then decreased gradually and slowed down to zero to reach full writing. Otherwise, the theoretical curves show also that the need of writing time to be peak of sensitivity (dR/dt) of the medium is increased with initial reflectance decreasing. Therefore, reducing the initial reflectance can delayed the sensitivity peak appearing time, resulting in lower writing sensitivity in the beginning, and the writing time to reach a certain reflectivity is

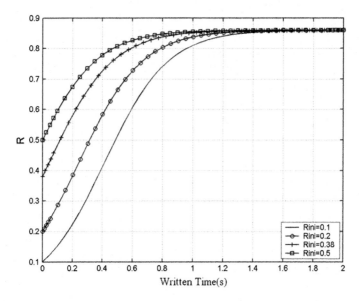

Fig. 3.33 The theoretical curves between reflectivity and writing time with different initial reflectance of medium of 650 nm

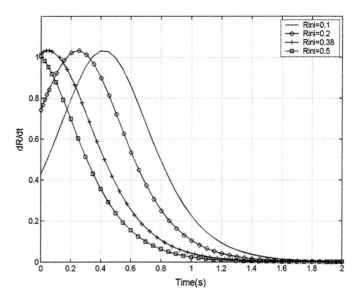

Fig. 3.34 The theoretical curves between sensitivity (derivative dR/dt of reflectivity to writing time) writing time with different initial reflectance of medium of 650 nm

increased, that can be used to improve other parameters of recording process and working life of medium. The medium has higher initial reflectance of higher and sensitivity, can reach the peak of fully written soon, and advanced writing speed.

Due to exist of impurities and partial loss of the photochromic molecules of medium in production process, even after full written, the practice reflectivity of the medium samples cannot reach the original reflectivity [27, 28].

2. Maximum reflectivity and sensitivity

The theoretical curves between reflectivity and writing time with different maximum reflectivity of medium are shown in Fig. 3.35. Figure 3.35 shows the maximum reflectivity changing do not affect the time to reach full written on the whole, but to influence the final value of the reflectivity after full writing.

The theoretical relational curves between sensitivity (derivative of reflectance to writing time) and writing time of the medium with different maximum reflectance are shown in Fig. 3.36. Figure 3.36 sows that the sensitivity of the medium is increased with writing time and reached its maximum soon with about same time. The theoretical curves shows that the sensitivity (dR/dt) of the medium is increased with maximum reflectance decreasing. Therefore, choosing the maximum reflectance suitability can adjust the sensitivity medium and recording time to improve and optimize other parameters of recording process. The higher maximum reflectance of medium has higher sensitivity and advance the writing speed in recording.

Meanwhile, the maximum sensitivity of the medium is not emergence at the same time in the theoretical curve of Fig. 3.36. It can be seen from this figure, that the

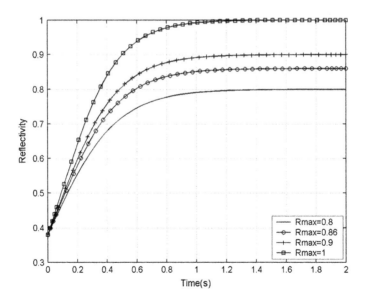

Fig. 3.35 The theoretical curves between reflectivity and writing time with different maximum reflectivity of medium and the maximum reflectivity influence to writing speed

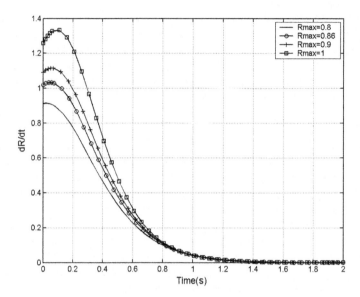

Fig. 3.36 The theoretical curves of sensitivity (dR/dt) and writing time of medium with different maximum reflectivity

maximum reflectivity influences the sensitivity of medium. The greater maximum reflectivity, the higher of the sensitivity and curtailed writing time to reach a certain reflectivity. But, if the maximum reflectivity was increased, its reflectivity peak will be postponed.

3. The reflectivity of reflective layer and sensitivity

The theoretical curves of the writing process are shown in Fig. 3.37 when the reflectivity of reflective layer is changed. The figure shows that if change of reflectivity of reflective layer will affect the final reflectivity of the full write only, but does not effect on the time to fully write. Of course, the higher reflectivity of reflective layer of the medium, the higher sensitivity is shown in Fig. 3.38. But the writing time, which the sensitivity obtain most, is same ultimately. Those specialties are very useful to optimize writing model of multiwavavlength photochromic medium.

3.6.2 The Reflectivity and Sensitivity of Multi-wavelength

1. The initial reflectance R_{ini}

The parameters of the molecular concentration of photochromic materials are influenced each other that is very complicated. Based on the theoretical model Eqs. (3.41) and (3.42) to analyze and calculate every parameters of the material

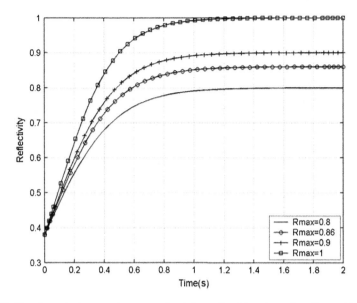

Fig. 3.37 The theoretical curves between reflectivity and writing time of medium with different reflectivity of reflective layer

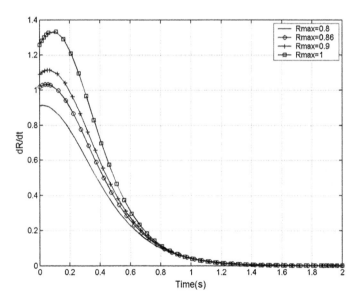

Fig. 3.38 The theoretical curves between sensitivity (dR/dt) and writing time of medium with different reflectivity of reflective layer

with cross of absorption spectra of 532 and 650 nm for example. The reflectance of the mixed material is affected by the 532 or 650 nm light together. As the reflectance of 650 nm material is affected by 532 and 650 nm light together also. Therefore, the initial reflectance of mixed materials with absorption spectra mutual cross has to be decided by both Molar extinction constant and concentration of two materials. If changing an initial reflectance of a wavelength but hope another initial reflectance of wavelength to unchange that have to regulate the ratio of the initial concentration of two materials. The theoretical curves of reflectivity in the writing process to keep the other parameters unchanged. The theoretical curves relationship between reflectance and writing time with different initial reflectance R_{ini} of 650 nm light only and other parameters unchanged is shown in Fig. 3.39. From Fig. 3.39 can see when heighten initial reflectivity of the 650 nm materials, that its writing time is shorten, but influence to writing process of the 532 nm light is not obvious. The other way round to reduce the initial reflectance of the 650 nm light, it influence to writing process of the 532 nm light very much [29–31].

Now analysis the relationship of writing sensitivity (dR/dt) and initial reflectance with cross of absorption spectra of 532 and 650 nm as shown in Fig. 3.40. In the theoretical curves of the writing sensitivity (dR/dt) to writing time to change the initial reflectivity R_{ini} of 650 nm laser only and keep other parameters unchanged. The peak of writing sensitivity (dR/dt) of 650 nm moves to left when its initial reflectivity increases to 0.3, i.e., the changing of sensitivity was faster (0.2 s) within beginning region of the 0–0.18 s, after then the change of sensitivity is slow gradually, and closed to 0 earlier to full written for 650 nm light. But the sensitivity

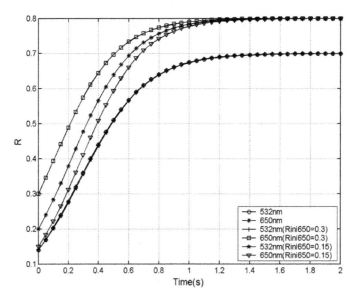

Fig. 3.39 The theoretical curves of reflectance and writing time when changing of the initial reflectivity R_{ini} of 650 nm light only, but to keep the other parameters unchanged

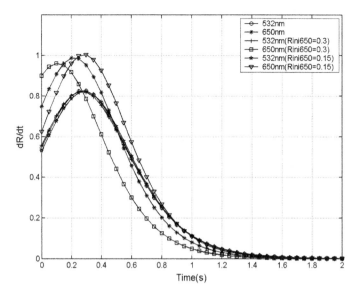

Fig. 3.40 The theoretical curves of the writing sensitivity (dR/dt) to writing time when keep other parameters unchanged but to change the initial reflectivity R_{ini} of 650 nm laser only

of 532 nm light is increased overall in writing process as a whole. When the initial reflectance of 650 nm is reduced to 0.15, writing sensitivity is change to the opposite, that the peak of sensitivity (dR/dt) of 650 nm moves to right and higher. The writing sensitivity is lower when writing time is within 0.2 s. The lower writing sensitivity of the medium induces exposure time of reaching a certain sensitivity to increase that is propitious to non-destructive read out and to improve the lifetime of medium.

The theoretical curves of reflectance to writing time with keeping of the other parameters unchanged but to change the initial reflectance R_{ini} of 532 nm light only is shown in Fig. 3.41. Such as the diagram, the initial reflectance of the 532 nm light is changed, that influenced the full writing time of 532 nm light, but it dos not effect to 650 nm light almost.

The theoretical curves of the writing sensitivity (dR/dt) of dual-wavelength medium keep the other parameters unchanged but to change the initial reflectance R_{ini} of 532 nm laser only is shown in Fig. 3.42. As Fig. 3.42, the writing sensitivity is changing with the initial reflectance of 532 nm over time. The sensitivity is declined when the initial reflectance increasing, but its sensitivity peak moves up. On the contrary, the sensitivity of peak is backward extension. The sensitivity of 650 nm is declined overall and the sensitivity peak is premised slightly, when the reflectivity of 532 nm increases. Therefore, change of the initial reflectance of the medium, will affect to writing process of both two lasers together, but the effective magnitude is depended on the sensitivity of the materials.

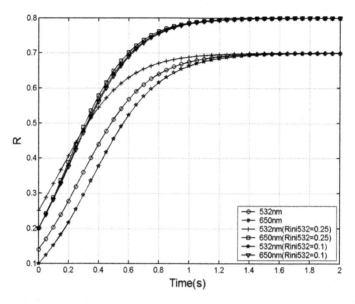

Fig. 3.41 The theoretical curves between reflectance and writing time when to change the initial reflectance R_{ini} of 532 nm laser only, but keep the other parameters unchanged in the writing process

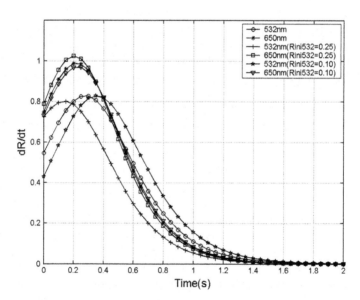

Fig. 3.42 The theoretical curves of the writing sensitivity (dR/dt) to writing time when keep the other parameters unchanged but to change the initial reflectance R_{ini} of 532 nm light only

2. Maximum reflectance $R_{\max}$

The theoretical curves of the writing process when keep the other parameters unchanged but to change the maximum reflectance $R_{\max}$ of 650 and 532 nm lasers is shown in Fig. 3.43. It can be seen from Fig. 3.43, change of maximum reflectance of 532 nm light is influence of the final reflectance of 532 nm for full writing, but does not affect the final reflectance of 650 nm light. Maximum reflectance of recording layer of tow photochromic materials decided the absorption of incident light energy in the all. So the reflectivity and energy of absorption of the two recording materials are effected also, and changing of maximum reflectance of 532 nm is influence the writing process of the 650 nm light slightly, but does not affect the time of fully written to two materials.

The theoretical curves between sensitivity (dR/dt) and writing time when keep the other parameters unchanged but to change the maximum reflectivity $R_{\max}$ of 650 and 532 nm lasers is shown in Fig. 3.44. From Fig. 3.44 we can see that increasing the maximum reflectance of 532 nm will affect the writing sensitivity of two wavelengths also.

Obviously, the larger of the maximum reflectance of 532 nm light, the higher writing sensitivity, that the time to achieve certain reflectivity is shorter, but influence to writing sensitivity of 650 nm light slightly. Of course, changing the maximum reflectance of 650 nm of this sample affects to the reflectivity of 532 nm light similarly. For example, when maximum reflectance of two materials are 0.6 and 0.9, respectively, the writing sensitivity (dR/dt) of the dual-wavelength sample are increased synchronous together, as in the theoretical curves in Fig. 3.44.

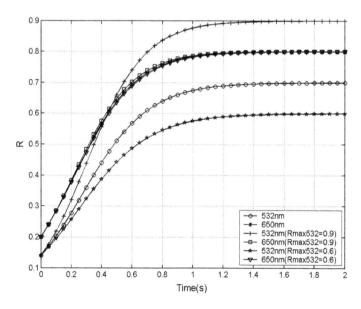

Fig. 3.43 The theoretical curves of the writing process when keep the other parameters unchanged but to change the maximum reflectance $R_{\max}$ of 650 nm and 532 nm lasers

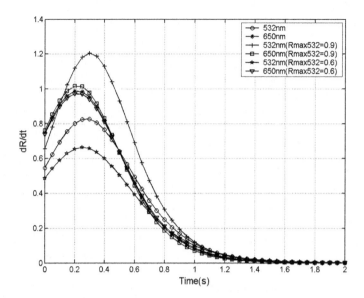

Fig. 3.44 The theoretical curves of the writing sensitivity to writing time when keep the other parameters unchanged but to change the maximum reflectivity R_{max} of 650 and 532 nm lasers

3. Reflectivity R_f of the reflective layer

The theoretical curves of the fully writing process when keep the other parameters unchanged but to change the reflectance of reflective layer is shown in Fig. 3.45. Clean seen, changing of reflectance of the reflective layer with materials absorption spectrum cross does not change the fully writing time, but affects the final reflectivity for two wavelengths also. Figure 3.46 is theoretical curves between sensitivity (dR/dt) and writing time with different reflectance of the reflective layer. Can be seen, that the sensitivity of writing process is increased clearly when reflectivity of the reflective layer of two wavelength are $R_f = 0.9$ and $R_f = 0.8$, respectively, and using the same laser write power to write also. The reflectivity of the reflective layer has a great influence to sensitivity of the dual-wavelength medium, and the higher of the reflectivity of the reflective layer of medium is higher sensitivity.

3.7 Written Time Constant and Crosstalk

3.7.1 Writing Time Constant k and Sensitivity

Writing time constant k is an important characterization for the photochromic material reactive speed, it is a constant when parameters of medium and writing conditions are certain. The time constant k to a certain writing wavelength is

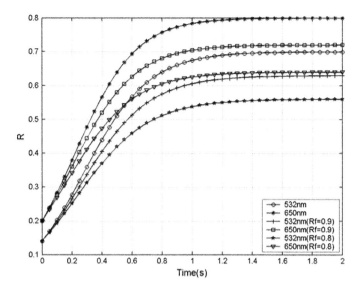

Fig. 3.45 The theoretical curves between reflectivity and writing time when changing reflectance of the reflective layer and to keep other parameters unchanged

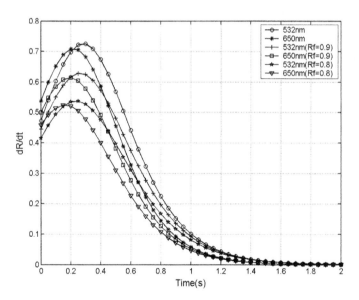

Fig. 3.46 Theoretical curves between sensitivity (dR/dt) and write time for different reflectance of reflective layer of the medium

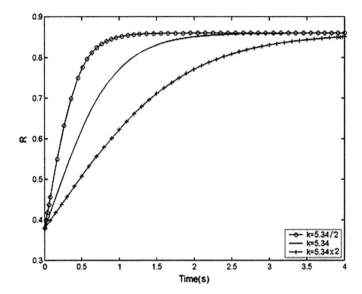

Fig. 3.47 The theoretical curves of relation between reflectivity and written time on same material with a different time constant k

determined by the properties of materials and the writing power. The theoretical curves between reflectivity and writing time with different constant k are shown in Fig. 3.47. It shows that changing of time constant k for a certain material can change the reflectance of the material. Figure 3.47 shows that the time constant k decreases that will delay completion of the process for full writing. The theoretical analysis shows also that the time to achieve a certain reflectivity is inversely proportional to time constant k, i.e., when k increases double, the write time is reduced by half. The theoretical curves of relationship between sensitivity (the derivative of reflectivity dR/dt) and written time of a material with different time constant k is shown in Fig. 3.48. If writing time constant k is increased that the sensitivity of the recording process could be improved simultaneity. When the write time k is very small, the sensitivity will be lower and the writing speed is reduced comparably. This speciality will be used to control for crosstalk and improve non-destructive readout.

3.7.2   Time Constants k

Materials of absorption spectrum, according to Eqs. (3.42) and (3.43) shows that the time constant k_1, k_2 are performances of proportion, when the materials and operating wavelength have been confirmed. For example, increase or reduce the writing power of the 532 nm laser, will change the time constants k_1 and k_2 together the as shown in Fig. 3.49.

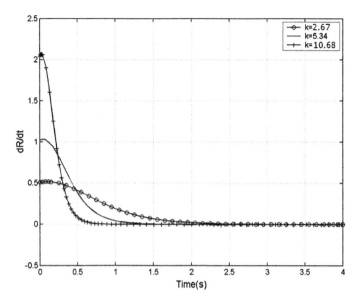

Fig. 3.48 Theoretical curves between sensitivity (dR/dt) and write time for different time constant k

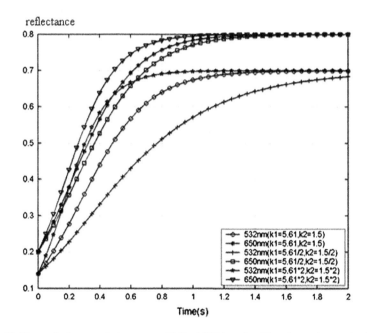

Fig. 3.49 Theoretical curves between sensitivity (dR/dt) and writing time when maintain other parameters but to change the time constants k_1 and k_2 (half or double)

Figure 3.49 is the theoretical curves of the writing process with other parameters are maintained but to change the time constants k_1 and k_2 (half or double). It shows that the k_1 and k_2 is increased or decreased, the writing time to a certain reflectivity is increase or decrease by inverse ratio. Theoretical calculations shows that also if write the time constants k_1 and k_2 are changed that will directly affect the relationship between reflectivity of the two wavelengths of light to writing time. As reflectance of any one wavelength of the two materials with absorption spectrum cross is influenced by writing energy at the same time. For example, increasing of 532 nm writing power, can increase writing time constants k_1 and k_2 and accelerate completion of the two wavelengths of light writing process, i.e., increased the writing speed. But changing of one of time constant k_1 or k_2 solely, writing speed is not inverse proportion to time constant k_n strictly.

Theoretical curves between sensitivity (dR/dt) and write time for different time constants k_1 and k_2 (half or double) is shown in Fig. 3.50. It is theoretical basis to actualize for low power and non-volatile readout in future. Changing of time constants of k_1 and k_2 will affect writing sensitivity (dR/dt). As the write power can change the time constant, so changing of the k_1, k_2 will affect writing sensitivity of 532 nm light also. For example, when k_1, k_2 is decreased, the writing sensitivity of 532 nm light is greatly reduced and its sensitivity peak was delayed, and the sensitivity of 650 nm light is down also as Fig. 3.50. When the time constant increasing, the sensitivity of 532 nm light will be increased, and its peak of sensitivity is in advance, but writing sensitivity of 650 nm light increases smaller.

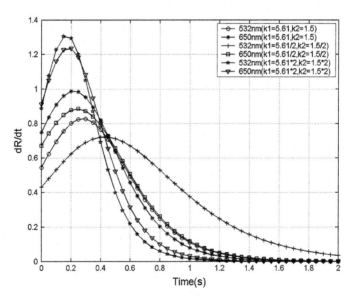

Fig. 3.50 Theoretical curves of relationship between sensitivity and write time to different time constants k_1 and k_2 (half or double)

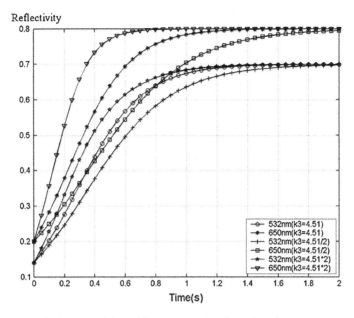

Fig. 3.51 Theoretical curves of the writing process when keep the other parameters unchanged but to change the time constants k_3 (half or double)

Figure 3.51 is the theoretical calculations curves of writing process when changing of writing time constant k_3 to dual-wavelength of 532 and 650 nm. It shows that increasing of the writing time constant k_3 will affect the writing process and speed to two wavelengths writing together, but it is greater to 650 nm. Changing of k_3 of 532 or 650 nm solely, the writing speed is not change inversely proportional to time constant strictly.

Writing time constant k_3 of dual-wavelength change to affect the writing sensitivity is shown in Fig. 3.52. It can affect sensitivity of dual-wavelength, but is larger for sensitivity of 650 nm light and smaller for write sensitivity of 532 nm light than. When the k_3 increases, that will accelerate the writing process to complete, the writing sensitivity of 650 nm light increase overall and its peak is in advance. The k_3 is changed solely and the writing speed is up quickly to achieve full writing. The writing sensitivity of 532 nm light increases also, and the sensitivity peak slightly ahead, but achieving full write time is reduced accordingly.

Theoretical curves of the reflectance to writing time when keeping other parameters unchanged, but proportion changing the time constants k_1, k_2 and k_3 at the same time is shown in Fig. 3.53. That the writing process in this figure shows that the three time constants are increased proportional, their reflectance is increased simultaneous greatly and the need of writing power is decreased. The relationship of distribution of the absorption of light energy to two materials is not proportionately. However, if a separate changing of any one of time constants, i.e., k_1, k_2

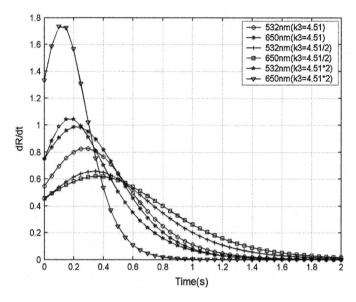

Fig. 3.52 Theoretical curves of the writing sensitivity to dual-wavelength when keep the other parameters unchanged but to change the time constants k_3 (half or double)

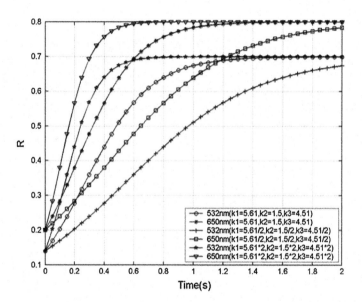

Fig. 3.53 Theoretical curves of the write process when keep the other parameters unchanged but to change the time constants k_1, k_2 and k_3 (half or double)

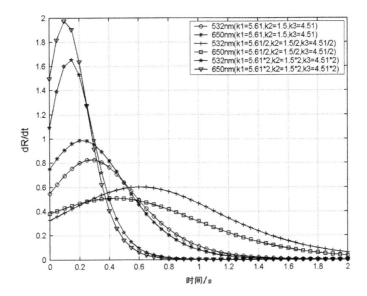

Fig. 3.54 Theoretical curves of the writing sensitivity when keep the other parameters unchanged but to change the time constants k_1, k_2 and k_3 (half or double)

and k_3 is changed incoordinately, the writing time to be same reflectance are not longer proportional changing.

The writing sensitivity for different time constants k_1, k_2, and k_3 is shown in Fig. 3.54. When the three time constants increase at same time, the writing sensitivity of the two wavelengths of light also much be increased greatly. Whether changing anyone of the writing time constants, writing sensitivity of the two wavelengths of light are affected for dual-wavelengths materials with absorption spectrum cross. Reducing the time constants of the dual-wavelength, the sensitivity will be decreased, that the fully written time required for growth and writing process is slowing. By contraries, increasing the time constants of the dual-wavelength writing process faster, writing sensitivity increases and fully written time is reduced. Therefore, in order to improve the writing speed, have to increase the time constants and the writing power. But in order to reduce the damage of read out the recorded data, time constants have to be minimized and using small power to read out.

3.7.3 Form Factors of Crosstalk

Light-induced discoloration of multi-wavelength storage uses photochromic materials for the memory medium which are different structures of molecular and have different absorptivity of light. The cross of absorption spectrum of the recording medium will bring crosstalk for readout signal. Through a systematic experimental study on the multi-wavelength photochromic storage process, analysis material absorption spectrum cross-writing process and quantitative evaluation for absorbable spectrum with mathematical model of the writing process in the first. Based on it to describe the photochromic multi-wavelength storage process quantificationally that could bring the theoretical basis to eliminate light-induced crosstalk in photochromic multi-wavelength storage.

The crosstalk of light-induced photochromic reaction in multi-wavelength storage is raised by cross of variety of absorption spectrum of recording materials indeed. Although to request of the absorption spectrum for each wavelength corresponding to the recording material that has to be as narrow as possible, i.e., to hope each recording material absorbed a wavelength only. But, in fact, a lot of organic photochromic materials with extremely advantageous characteristics are wider absorption spectrum sometimes, so these materials are adopted still. The more multi-wavelength photochromic recording materials are working together, elimination the crosstalk are more difficult. Under normal circumstances, there are part of absorption spectrum of the materials to cover each other always, so the crosstalk problem cannot avoid. The crosstalk affects the readout signal of multi-wavelength seriously, and to cause the recorded data to bring great error by crosstalk.

A typical recording experiment of photochromic multi-wavelength storage is shown in Fig. 3.55, and the experimental parameters are shown in Table 3.3. In this experiment, take six locations to write, where the point 1 and 4 was recorded by 650 nm laser and 532 laser at the same time, point 2 and 5 was recorded by 532 nm wavelength laser alone, but point 3 and 6 was recorded the by 650 nm laser alone. The read out power are 0.07 mW for 532 nm laser and 0.1 mW for 650 nm light simultaneously, readout moving speed is 10 mm/s. From the experimental results can be seen that the points 2, 5 are written with 532 nm laser only, but the read out reflectivity of 650 nm are close to the 650 nm laser alone to write at the same points, i.e., there are serious crosstalk at the two points.

Note point 1 and 4 are full written by 532 nm laser and 650 nm laser, the molecule of material 650 nm absorb 532 nm light cannot be discovery. But the point 3 and 6 are written by 650 nm laser only that is almost no absorption by 532 nm laser material, i.e., the crosstalk is small. Read out signal from above dual-wavelength recording experiment can also see, that the read out reflectivity with 650 nm laser of six point are about 0.8, i.e., six points are written by 650 nm laser. But in fact point 2, 5 are not written, that is a intolerable errors. Obviously the performances of photochromic materials affect the crosstalk in multi-wavelength storage fundamentally. In addition to solve this problem have to identify the optical absorption cross of materials, and the profiles of storage material absorption of certain cross. According to the trend of crosstalk select the appropriate write strategy, that can reduce or inhibition its impact to readout signal to the extent.

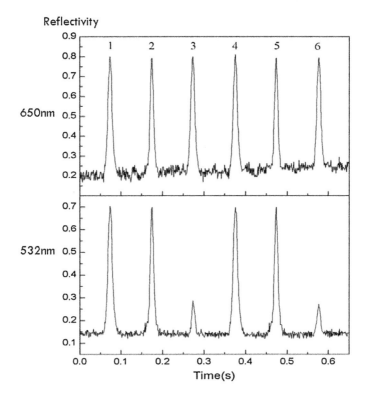

Fig. 3.55 A typical recording experiments with dual-wavelength of 532 and 650 nm

Table 3.3 Experimental writing parameters	Written position	Writing power (mW)		Writing time (ms)	
		532 nm	650 nm	532 nm	650 nm
	1, 4	0.07	0.1	2	2
	2, 5	0.07	0	5	0
	3, 6	0	0.1	0	4

3.7.4 Quantitative Evaluation for Crosstalk

Still based on above storage experimental results of two wavelengths recording for example, but the readout signal was adopted the data of [00], [01], and [11] to replace each point recording signal, that the readout signal can be show by following four possibilities:

1. $R[00]$, are not written by the two-wavelength laser, reflectivity close to zero, corresponding recorded data is [00];

2. $R[10]$, written after only one wavelengths of λ_1, corresponding to the recorded data is [10];
3. $R[01]$, only one wavelength laser of wavelength λ_2 written, corresponding to the recorded data is [01];
4. $R[11]$, written by two beams of λ_1 and λ_2 of two-wavelength laser, corresponding to the recorded data for [11];

Otherwise the readout signal in either point is the reflectivity R of the two wavelengths indeed and can be expressed as

$$R_1 = R_{1(00)} + [R_{1(10)} - R_{1(00)}]S_1(t) + [R_{1(01)} - R_{1(00)}](S_2(t) - S_1(t))S_2(t)$$
$$+ [R_{1(11)} - R_{1(10)}]S_1(t)S_2(t) \tag{3.44}$$

$$R_2 = R_{2(00)} + [R_{2(10)} - R_{2(00)}](S_1(t) - S_2(t))S_1(t) + [R_{2(01)} - R_{2(00)}]S_2(t)$$
$$+ [R_{2(11)} - R_{2(01)}]S_1(t)S_2(t) \tag{3.45}$$

Among them, $S_1(t)$ is the write pulse function of wavelength λ_1 laser, namely λ_1 written information; $S_2(t)$ is the write pulse function of laser wavelength of λ_2 writing information. Definition of reflectivity of a recorded point is full written alone by a wavelength and read out reflectivity by the same wavelength of light, i.e., recorded data. From Eqs. (3.44) and (3.45) can see that the read out reflectivity R_1 to wavelength λ_1 is determined by Eq. (3.46), where R_1 contains all crosstalk in $S_2(t)$. Recorded different data, the crosstalk can be divided into the crosstalk "0" and crosstalk "1". The so-called "1" crosstalk refers the crosstalk of read out signal of the point, that is brought by all wavelengths to participate to write. So the reflectivity of read out by the λ_1 light at this time will be greater than or equal to the effective reflectivity. Meanwhile, definition the crosstalk is the difference of read out reflectivity and effective reflectivity. The so-called [0] crosstalk is an additive signal by other wavelengths recording without λ_1 wavelength, i.e., all other wavelength light except λ_1 wavelength introduced crosstalk. It is defined the difference that is different between read out reflectivity and the initial reflectivity. The readout signals of dual-wavelength recording at the same time is [1] crosstalk for λ_1 and λ_2 also. So the crosstalk [1] of λ_1 read out signal is $R_1[11] - R_1[01]$. If the λ_2 wavelength laser alone write on the sample, the wavelength of λ_1 will bring "0" crosstalk to the readout signal, and it is $R_1[01] - R_1[00]$. Obviously, the property and impact of the two kinds of crosstalk are different, that affects to read out signal [1] is smaller, but to read out signal [0] even more. So selection of the writing strategy is difference for the two types of crosstalk.

Based on the above analysis can analysis more for wavelength λ_1 and wavelength λ_2, respectively.

1. The readout signal crosstalk of wavelength λ_1:
   when λ_2 laser writing only, the crosstalk [0] is $R_{1(01)} - R_{1(00)}$, and when two-wavelength lasers writing at the same time, the crosstalk [1] is $R_{1(11)} - R_{1(10)}$.

2. The readout signal crosstalk of wavelength λ_2:
   When only λ_1 laser to write, brought crosstalk [0] is $R_{2(10)} - R_{2(00)}$, and when two-wavelength laser write at the same time, brought of the crosstalk [1] is $R_{2(11)} - R_{2(01)}$.

The situation is more complicated in the three materials for multi-wavelength storage. The requirements of performances of materials are higher, especially for the absorption spectra is need of as narrow as possible. Here take the synthesis medium of (1) 1,2-bis (2-methyl-5-n-butyl-3-thienyl), perfluorinated cyclopentene; (2) 1,2-bis (2-methyl-5-(4-N,N-dimethyl) phenyl-thiophene-3-yl), perfluorinated cyclopentene and {1-[2-methyl-3-2-(1,3-dithiolane benzo thienyl)]} and (3) 2 − {2-methyl-5-[4-(2,2-two cyano-vinyl phenyl)]}, thiophene thiophene 3 perfluorinated cyclopentene, three materials for example. The absorption peaks of three kinds of materials are 532, 650 and 780 nm, respectively. The absorption spectra and molecular formula of the materials of 532 and 650 nm have been introduced in Figs. 3.26 and 3.27 in Sect. 3.5. The molecular formula and absorption spectra of material 780 nm are shown in Fig. 3.56.

Performances of these materials are better, that the absorption spectra of three materials are narrow correspondingly, and does not absorb each other and crosstalk is smaller generally. The three materials have been used to the three-wavelength storage experiment and achieved the better results.

The experimental parameters of three-wavelength mixed recording media is written and read out is shown in Table 3.4 (powers are actual power on the focal plane). The three-wavelength storage experiment results of readout signals (reflectivity) which is

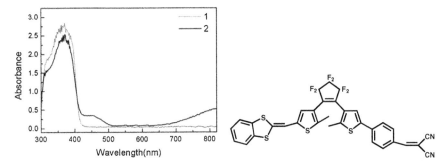

Fig. 3.56   The absorption spectra and molecular formula of 2-{2-methyl-5-[4-(2,2-two cyano-vinyl phenyl)]}, thiophene thiophene-3 perfluorinated cyclopentene for 780 nm laser

Table 3.4 Experimental parameters of three-wavelength mixed recording media is written and readout

Parameters	780 nm	650 nm	532 nm
Writing power (mW)	2	2.5	2.2
Writing time (s)	0.05	0.05	0.05
Readout power (mW)	0.1	0.1	0.07
Scan speed (m/s)	0.1		

Fig. 3.57 Readout signals (reflectivity) which is fully written in the same channel at the same time and same position synchronization with three wavelength of 532, 650, and 780 nm lasers

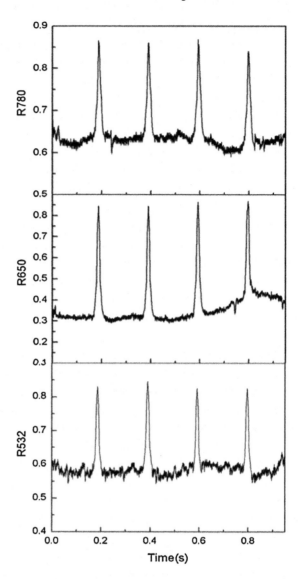

fully written in the same channel at the same time and same position synchronization with three wavelength of 532, 650 and 780 nm lasers are shown in Fig. 3.57.

It can be seen from Fig. 3.57 that in a three-wavelength readout on a point at the same time has approving reflectivity peaks, indicating that the molecules of three photochromic recording materials are transformed completely, and does not absorb other various wavelengths. The reflectivity of the three wavelengths on the experimental samples have been to maximum, and corresponded to "1" in the digital storage, i.e., achievement to record 3 bit at the same point. In Fig. 3.57, other parts except pulses is corresponded digital data "0", that has some fluctuations, which is caused by the recording layer unevenly.

Based on the Lambert–Beer photochemical law established mathematical model of read out reflectivity R_i of each wavelength is

$$R_i = R_f R_{\max(\lambda_i)} e^{-4.6 \sum_{j=1}^{3} \varepsilon_{j(\lambda_i)} C_j l} \qquad (3.46)$$

where $R_{\max(\lambda i)}$ is maximum reflectivity for wavelength λ_i of the sample, R_f is reflectivity of reflective layer, $\varepsilon_{j(\lambda i)}$ is the molar extinction coefficient of material j for the wavelength λ_i, C_j is concentration of material j, l is the optical path. Three materials mixed in a layer, so the optical path l is same also.

From Eq. (3.46) can see, the initial reflectivity to certain wavelength on the samples is determined by the concentration of material molecule and its molar extinction coefficient. As absorption spectra of the materials is not cross each other almost, thus the initial reflectivity of a wavelength only determined by the concentration of material and molar extinction coefficient.

To further analyze the crosstalk of various materials and the impact on signal quality, take another experiment to write and read under the same conditions. Modulation of three laser power, along the same channel parallel recorded on the experimental sample. The writing parameters are same with Table 3.4, but the position of every writing point of the three-wavelength laser is staggered. The read out signal is shown in Fig. 3.58, that each record of the three materials can be accurately read out on recording location of 532 nm laser, 650 and 780 nm laser cleanly. Confirmed 532 nm laser writing did not make the 650 nm materials and 780 nm materials achromatic response to introduction of crosstalk, and the 650 nm laser and 780 nm laser writing do not introduce crosstalk to other wavelengths also. So long as the absorption spectrum of the recording materials are not cross, that the mixture medium of three materials or any more materials at the same time recording and read out cannot present crosstalk. A few fluctuations of the readout signal are due to uneven of concentration of recording materials. So farther improving physical parameters of the recording layer materials and uniformity is very important for multi-wavelength photochromic optical storage. But the uniformity of concentration of the mixed materials are relationship with uniformity of solvent, spin coating process and the stability of the material itself and many other factors, that will be discussed more in the future.

3.8 Non-destructive Readout

Reaction of photochromic compounds is reversible usually. In photochromic optical storage, the colored state of photochromic materials to be initialized state, i.e., original state, after writing it was decolored, i.e., the recording state. The colored state molecule was decolored after writing and dos not absorb wring light again, that

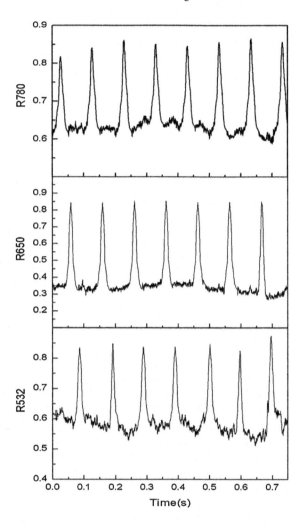

is transparence to writing light on this point. As the writing light was employed to read out, although it is very weak, that the some molecules of medium could be decolored in the non-writing area after much more readout. So the contrast of the reflectivity between writing point and not written point is decolored gradually. In the first, have to choose the photochromic materials which have a certain threshold for writing light. But experiments show that most photochromic materials does not exist obvious threshold to writing laser, even if relatively weak laser irradiation, a certain percentage of photochromic molecules will be decolored, resulting to destruction of the original written information still. In order to search non-destructive readout method, there are two way can come true non-destructive readout mainly as following: First is using other wavelength laser to read out, which cannot be absorbed by all materials in recording layer, so can achieve non-destructive readout in grain.

Another way is utilization of the some photochromic compounds which can be control by external trigger, such as light, electricity, magnetism, heat and chemical factors stimulate or zero quantum yield wavelength readout, etc. to restrain the photochromic reaction. This section will introduce an electric locking non-destructive readout method that belongs to a specific external excitation (electric field) method later.

Fumio MATSUI proposed zero quantum yield wavelength readout method, it is using change of the quantum yield of readout light to control achromatic reaction in the indole fulgide photochromic. Then employ 780 nm laser to readout, the indole fulgide is not sensitive to 780 nm wavelength, so the achromatic quantum yield is almost to zero, but there are still much enough to absorb for the detection. The experiment was succeed and achieved non-destructive read out. The disadvantage of this method is that have to use another wavelength laser, increasing the complexity of the read–write device especially for multi-wavelength storage systems. In addition, a temperature threshold method non-destructive readout proposed earliest by Fumio Tatazono et al. In this method, recording medium has a characteristics above room temperature threshold T_c, that using λ_1 laser of higher power (to get higher threshold T_c temperature) to write, read out using low power λ_2 (lower temperature), and using λ_1 and λ_2 two wavelength lasers erases at the same time. Because written location has stronger absorption to λ_2 only, but does not absorb in without written location, so can read out information not any damaging, only when the temperature is higher than the writing threshold T_c (using higher power λ_1 exposure), the recording data can be erased by λ_2. The research team used series of diarylethene photochromic compounds to experiment, its threshold temperature is about 85 °C that used 458 nm Ar$^+$ laser with power 1.6 mW and the pulse width 10 μs. In the experiments medium is recorded by Ar$^+$ laser (form of open-loop into a closed loop), and using 633 nm He–Ne laser (0.5 mW) read out and erase. But erase have to use both 633 nm laser and 458 nm laser at the same time. According to reports, this method was achieved non-destructive readout more than 10^6. The disadvantage of this method is the thermal effect that the responding speed is lower and loss the characteristics of high sensitivity and high-speed of the photon effect. In addition, this temperature threshold is not enough high over the environmental temperature, storage time cannot be too long.

The ultra-low power non-destructive read out the experiments was carried out earlier in Tsinghua University and has accumulated more data that is the main method currently. Tsuyoshi TSUJIOKA also carry out experimental research in Japan. Although this method every time readout can make part of the recording medium molecular restore to before writing state, but it is very limited due to the readout power is very low, so can readout a certain number and maintain the required signal-to-noise ratio still. Some theoretical calculations and experimental results demonstrate that effectively read out number can be to 10^6 when the laser power of read out under order level of 10 nW. The greatest advantages of the approach are that the structure of the system is relatively simple and ease to implementation. Main difficulty is that the readout power is very low, weak signal, signal-to-noise ratio is very small and need of special signal processing systems. In addition to above non-destructive readout eme, using of fluorescent luminescence properties to readout, the nature of intramolecular locking, mid-infrared laser to read out, etc., non-destructive readout method are during the study and exploration still.

Tsinghua University starts to study of the electric lock non-destructive read out early, that the program in the implementation is relatively easy to achieve. A team focuses on the study of electric lock non-destructive readout of photochromic multi-wavelength storage system wholly. This section will introduce the principles and experimental results of the non-destructive read of electric lock. This method of non-destructive readout refers electrochromic properties of photochromic molecules, when plus the positive voltage to the medium of written data that will be locked, as its absorption spectrum was changed that cannot be damage by writing light (i.e., readout light). The locked medium achieves security repeatable read out and no any restriction. The other way round, if plus negative voltage that the medium would be rewritable again. The method is capable application to three-state optical storage theoretically and the medium can be unlimited writing or erasing. Such as solution of diarylethene molecules has the property of the electric lock that can be conversion reversibly between three states (state A, state B and state C) as in Fig. 3.59. The three states are stable states also, and the absorption spectrum is not same that the conversion relationship between the three states is shown Fig. 3.59 in detail. When the medium is irradiated by UV (313–365 nm), that the colorless open-loop solution of the compound state A is becoming dark blue rapidly, and

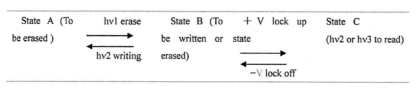

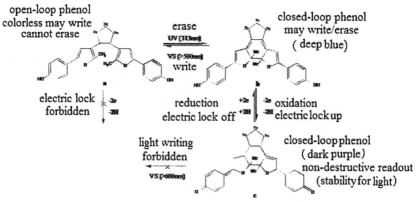

Fig. 3.59 Changing process of molecular structure in solution of diarylethene at three states with photochromic action and electric lock

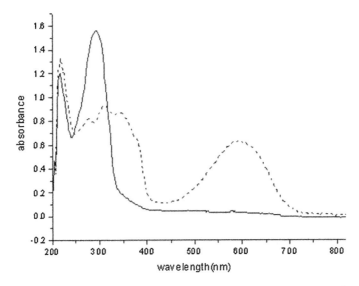

Fig. 3.60 Absorption spectra of two-state (state A: *solid line*, state B: *dotted line*) of acetonitrile

generated closed-loop phenol as state B as shown in Fig. 3.60. The state B is corresponded to absorption of 592 and 342 nm that can be used for writing and erasing. Closed-loop dark state B of dark blue solution can return to open-loop state A and the color was disappeared under visible light irradiation of wavelength larger than 510 nm. The A state and B state of the medium and their absorption spectrums are shown in Fig. 3.60. In order to add in the electric field, the experimental solution was mixed in the acetonitrile electrolyte. The mixing solution (medium) is put in between two transparent plate of electrode with electrode voltage of 1.5 V as shown in Fig. 3.62.

When solution of state A was UV radiation into the state B, the maximum absorption peak moved to 588 and 362 nm, and become dark blue. If voltage of 1.5 V adds to between two transparent plate of electrode, the solution was changed from dark blue state B to purple state C with maximum absorption of 550 and 380 nm, and the absorption intensity was increased as shown in Fig. 3.61. If the purple state C solution was strong irradiation by wavelength is greater than 540 nm, experimental result indicates that the state C is very stable still to keep the absorption spectrum does not change. Illustration the photochromic compounds generated very high stability state C with a voltage lock, and can use greater than 510 nm to non-destructive read out.

The experiment with solution is verified the principles and materials of the electric lock for non-destructive read out only, that confirmation of possibility to be used to photochromic recording really, and could be indepth study of the materials in thin film form of electric locking performances future.

The experimental sample with reflective layer was made of transparent plates of electrode and photochromic compounds solution, the structure is shown in

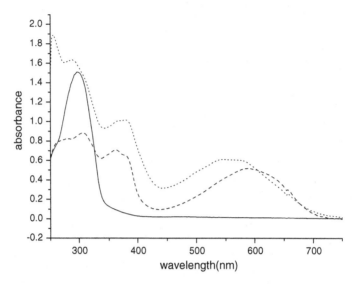

Fig. 3.61 Absorption spectra of three-state (state A: *solid line*, state B: *dotted line* and state C: *broken painted line*) of acetonitrile NBu4Br in electrolyte

Fig. 3.62 The experimental sample of photochromic compounds recording medium with electric locks layers

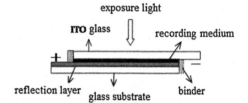

Fig. 3.62 in detail. The reflective layer of aluminum film on the glass substrate can instead one of electrodes. It is need of a special glass substrate of ITO conductive glass to be protective layer. The diarylethene compounds solution was spin-coated on the aluminum reflective layer and ITO conductive glass was sealed with binder pressed on the recording layer as a protective layer.

The reflective layer is a conductive film, when need of electric lock that the electrode voltage is applied on the reflective layer and the ITO electric glass. The samples have to be color with ultraviolet light in the first, and it was used to writing with 532 nm laser and effective power 0.1 mW.

Oscilloscope is applied to record and display the writing process, the reflectivity changes is shown in Fig. 3.63. It can be seen from the figure, the experimental sample has same reaction to other light-induced discoloration recording samples without electric lock. The sample absorbs the 532 nm laser that the reflectivity is creasing to maximum and becomes saturated eventually. Its reflectivity of the point changes from the initial 0.4 to the final reflectivity of about 0.9. As a comparative experiment, the sample was added lockout voltage of 3.2 V for 3 min, that the

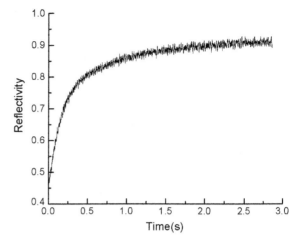

Fig. 3.63 Reflectivity of the recording photochromic compounds medium in writing process when electric lock is off

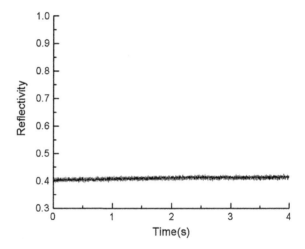

Fig. 3.64 Reflectivity of the recording photochromic compounds medium with electric locks when electric lock is on

sample from dark blue change to purple to appear the phenomenon of electric locks. Then using the same power wavelength of 532 nm laser write on the sample in another area that the changing of reflectivity is shown in Fig. 3.64. Figure 3.64 shows that the samples is not any variety of reflectivity after writing experiment with electric lock. As electric lock has taken the molecules into forbidden state C, that cannot happen photochromic reaction again, i.e., it is achieved the non-destructive readout completely. Reflectivity of not written point was 0.41, it is indicating the state C molecules, is not sensitive to light-induced reaction although, but still absorbs 532 nm light. The reflectivity of no written point with electric lock is much lower than the final reflectivity electric lock before. But in terms of the transmission sample or reflective samples experiment, the samples cannot transform from C state to B state again whether that it was imposed positive or negative voltage to electric lock. So implication this material cannot unlock and be erased to rewritable storage, that to be a write-once memory only.

These experiments show that the medium with electric locking can be used to non-destructive readout completely. Before recording, medium has to be colored by ultraviolet light, and into the recording state B for the absorption of 532 and 650 nm laser state with very good stability. Therefore, using 532 nm laser to write that the medium from state B conversion to state A at the written point. After writing, the medium was added electric lock and then can readout with 532 nm laser still. The written point of state C shows signals as "0", "1" that is achieved accurate non-destructive readout. State C molecules cannot get the H atom in the film, so it cannot return to the state B, and cannot write again. The absorption spectrum of photochromic compounds have wider absorption spectrum that easy to achieve matching of the semiconductor laser wavelength and industrialization, and use to multi-wavelength storage more easy relatively.

References

1. P. Wang, B. Huang, Y. Dai, M.-H. Whangbo, Plasmonic photocatalysts: harvesting visible light with noble metal nanoparticles. Phys. Chem. Chem. Phys. **14**, 9813–9825 (2012)
2. V.N. Peters, T.U. Tumkur, G. Zhu, M.A. Noginov, Control of a chemical reaction (photodegradation of the p3ht polymer) with nonlocal dielectric environments. Sci. Rep. **5**, Article number:14620 (2015)
3. D. Xu, N.F. Liao, M. Gong, G. Qi, P. Yan, Digital color multi-layer and multi-level signal read out method and apparatus, Tsinghua University, CN00103230.5, 2000
4. D. Xu, G. Qi, M. Gong, P. Yan, N.F. Liao, Multi-wavelength synthesized apparatus for color multi-layer and multi-level optical disc drive, Tsinghua University, CN00103231.3, 2000
5. A. Zeytunyan, Kevin T. Crampton, R. Zadoyan, Supercontinuum-based three-color three-pulse time-resolved coherent anti-Stokes Raman scattering. Opt. Express **23**(18), 24019–24028 (2015)
6. D. Xu, X. Chen, A digital color multi-wavelength and multi-order CD writing and reading method, Tsinghua University, CN00103501.0, 2000
7. D. Xu, L. Zhang, Photochromic disc with super resolution mask layer, Tsinghua University, CN00107368.0, 2000
8. J. Ma, L. Li, C. Yan, X. Cheng, L. Hair, R.C. Cui, L.F. Pan, H. Li, J. Lu, L. Man, M. Yang, A laser read head collimated optical structure, Tsinghua University, Dongguan. Digital Machinery Co., CN200610061767.0, 2007
9. D. Xu, H.X. Li, L. Ma, Three-dimensional multi-color storage optical head, Tsinghua University, CN00109128.X, 2000
10. M.K. Smit et al., A generic foundry model for InP-based Photonic ICs, OFC 2012, March 4–8, Los Angeles, Paper OM3E.3, 2012
11. J. Ma, J. Yu, L.F. Pan, J. Wu, D. Xu, J.-D. Ji, J. Zhang, H. Shi, L. Li, Y. Tang, Testing methods of multi-level optical pickup actuator dynamic parameters, Tsinghua University, Jiangsu Yinhe Electronics Co., Ltd., CN200510066028.6, 2005
12. J. Ma, L. Li, C. Yan, X. Cheng, L. Hair, L.L. Pan, H. Li, J. Lu, The blue laser beam structure shaping method for optical storage, Tsinghua University and Dongguan. Anwell Digital Machinery Co., Ltd., CN200610021437.9, 2007

13. W. Lee, X. Du, L. Li, W. Wang, A multi-function lens of compatible multi-dimensional optical drive pickup, Hong Kong Polytechnic University, Guo Weigang, Shenzhen Suncheon, CN201110251728.8, 2013
14. Q.-H. Shen, D. Xu, G.S. Qi, Blue-laser optical recording and its extended technology. Opt. Tech. **31**(6), 921–924 + 927 (2005)
15. A. Gurizzan, P. Villoresia, Ablation model for semiconductors and dielectrics under ultrafast laser pulses for solar cells micromachining. Eur. Phys. J. Plus **130**, 16 (2015)
16. K. Seger, *Compact Solid-State Lasers in the Near-Infrared and Visible Spectral Range* (Stockholm, Sweden 2013). ISBN 978-91-7501-764-8
17. D. Pinotsi, G.S. Kaminski ierle, C.F. Kaminski, Optical super-resolution imaging of β-amyloid aggregation in vitro and in vivo: method and techniques. Methods Mol. Biol. **1303**, 125–141 (2015)
18. Q. Zhang, Y. Ni, D. Xu, Multilevel run-length limited recording on read-only disc. Japan. J. Appl. Phys, Part 1 Regul. Pap. Short Notes Rev. Pap. **45**(5A), 4097–4101 (2006)
19. K. Ludge, E. Schöll, E.A. Viktorov, T. Erneux, Analytic approach to modulation properties of quantum dot lasers. J. Appl. Phys. **109**, 103112 (2011)
20. X. Chen et al., Polarization-independent grating couplers for silicon-on-insulator nanophotonic waveguides. Opt. Lett. **36**, 796 (2011)
21. J. Song, J. Pei, D. Xu, Microstructure measurement method of novel multi-level run-limited-length read only discs using the atomic force microscope. Japan. J. Appl. Phys. Part 1 Regul. Pap. Short Notes Rev. Pap.**45**(9A), 6958–6960 (2006)
22. H. Hu, D. Xu, Modulation code and PRML detection for multi-level run-length-limited DVD channels, in *Proceedings of SPIE*. Optical Data Storage, vol. 6282 (2006), p. 628228
23. A.V. Gorshkov et al., e-print quant-ph/0604037 2006. Phys. Rev. Lett. **98**, 123601 (2007)
24. H. Hieslmair, J. Stinebaugh, T. Wong, M. O'Neill, M. Kuijper, G. Langereis, 34 GB multilevel-enabled rewritable system using blue laser and high-NA optics, in *Joint International Symposium on Optical Memory and Optical Data Storage* (2002)
25. Q. Zhang et al., Multilevel run-length limited recording on read-only disc. Jpn. J. Appl. Phys. **45**, 4097–4101 (2006)
26. Y. Tang et al., Multi-level read-only recording using signal waveform modulation. Opt. Express **16**, 6156–6162 (2008)
27. H. Hu, L. Pan, J. Xiong, Y. Ni, New efficient run-length limited code for multilevel read-only optical disc. Jpn. J. Appl. Phys. **46**(6B), 3782–3786 (2007)
28. H. Hu, L. Pan, J. Xiong, 3-ary (2, 10) run-length limited code for optical storage channels. Electron. Lett. **41**(17), 972–973 (2005)
29. Q. Shen, D. Xu, Analysis of the differential phase detection signal in multi-level run-length limited read-only disk driver. Japan. J. Appl. Phys. Part 1 Regul. Pap. Short Notes Rev. Pap. **45**(7), 5764–5768 (2006)
30. Y. Zhang, A new three-zone amplitude-only filter for increasing the focal depth of near-field solid immersion lens systems. J. Mod. Opt. **53**, 1919–1925 (2006)
31. C. Liu, S.-H. Park, Numerical analysis of an annular-aperture solid immersion lens. Opt. Lett. **29**, 1742–1744 (2004)

Chapter 4
Super Resolution and Laser Sources

Super-RENS technology and Laser sources are very important for multiwavelength optical storage indeed. The electromagnetic spectrum has been used to various data storage technology, including hard disc, optical disc, UV holography, etc. is shown in Fig. 4.1. At present, conventional optical disc storage was restricted by resolution that employs electromagnetic spectrum is very limited in range. The multidimensional optical storage is need of larger bandwidth of electromagnetic spectrum, as in Fig. 4.1, from 350 to 1500 nm, i.e., from near ultraviolet to infrared radiation was. So it offers great development space for multidimensional optical storage. However, application of more bandwidth on electromagnetic spectrum will bring some problems, such as recording materials, optical diffraction limit resolution, laser sources, detectors, etc. in which the technology of photodetectors is succeed comparatively, other problems are under ravel out yet. The recording materials and process for multidimensional optical storage has been introduced in Chap. 3 of this book. So this Chapter will discuss the optical diffraction limit resolution and various wavelength laser sources in detail. In order to control the size of multiwavelength storage system with optical diffraction, a super-resolution near-field structure (super-RENS) mask has been developed to get smaller spots on the disc to keep the uniformity of different wavelength spots size and to increase the storage intensity and capacity. In addition consider to crosstalk of multiwavelength storage, that the absorption spectrum of the medium is great. So it is need of more different wavelength laser sources and format various wavelength spot size with super-RENS technology. Of course, the *micro-aperture laser* can solve the problem of different wavelengths storage density fundamentally, so it is an important supporting technology for multidimensional optical storage weightily [1].

© Tsinghua University Press and Springer Science+Business Media Singapore 2016   221
D. Xu, *Multi-dimensional Optical Storage*,
DOI 10.1007/978-981-10-0932-7_4

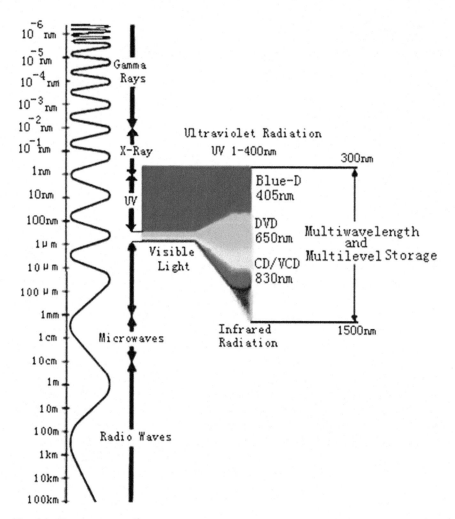

Fig. 4.1 The electromagnetic spectrum

4.1   Optical Super Resolution and Micro-aperture Laser

Optical super resolution deals with illumination and optical interaction by light emerging from a subwavelength aperture by a subwavelength metallic tip or nanoparticle, of an object in the immediate vicinity (or within a fraction of the wavelength of light) of the aperture or scattering source. The light in the near-field contains a large fraction of nonpropagating, evanescent field, which decays exponentially in the far field. The light is passed through a subwavelength aperture to form super resolution effect. Tominaga etc. proposed a super-RENS in 1998. They adopted a silver oxide (AgO_x) layer which has nonlinear optical characteristic that the resolution limit of super-RENS disc was greatly improved to less than 100 nm

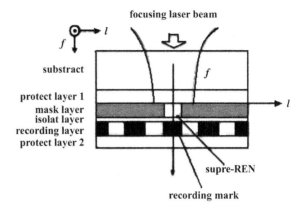

Fig. 4.2 An experimental sample structure of disc with near-field super-resolution mask

Table 4.1 The physical parameters of the disc structure with near-field super-resolution mask

Materials	Refractive index	Thickness (nm)
PC	1.52	infinite
SiN	2.1	170
Sb	3.11 + i5.66 (solid state)	15
	4.51 + i3.66 (melting state)	
SiN	2.1	20
Ge$_2$Sb$_2$Te$_2$	4.8 + i3.5 (crystalline)	15
	4.4 + i1.65 (amorphous)	
SiN	2.1	20

with near-field interaction. It suggested the nanosize Ag particles were generated in a small area, when AgO$_x$ layer was heated over the threshold temperature by the laser during the readout. The localized surface occured plasmon coupling effect, and made a closely recording layer under precipitated Ag particles, that is yielding strong near-field effect to increase intensity of light and reduce the recording mark size. After the laser beam is removed, the Ag and oxygen was formed the AgO$_x$ compound again to carry out the super resolution recording.

A new medium system with super-resolution mask was developed in OMNERC in 1999 that the structure of experimental disc sample with super-resolution mask is shown in Fig. 4.2. The film system actually can be divided into two parts: one is the aperture mask layer with SiN/Sb/Sin film to be super-RENS, another part is GeSbTe phase-change layer for recording. As SiN/Sb/Sin film will be formed a smaller transparence area by laser beam thermo-effect that realized super-RENS recording on the phase-change layer. The typical materials and the corresponding physical parameters of the disc are shown in Table 4.1. Meanwhile OMNERC developed a near-field super-resolution mask experimental system that is composed of the automatic focusing system, lasers power/exposure time control system and three-dimension precision worktable, its optical system is shown in Fig. 4.3.

The numerical aperture of objective lens of pickup is NA 0.65, and suited with wavelength of 1070, 830, 650, 532, and 405 nm semiconductor lasers.

Fig. 4.3 Optical system of
the experimental apparatus for
super-RENS recording

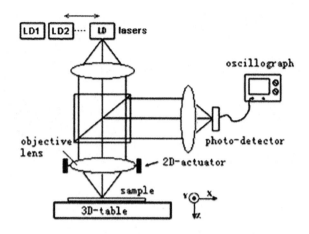

Based on the theory of diffraction by scattering mode, analysis, and calculation of the electromagnetic fields with super-resolution mask can established a mathematic model to description the electromagnetic field of super-resolution mask optical light scattering as:

$$\mu_0 \frac{\partial \vec{H}^{\text{scat}}}{\partial t} = -\nabla \times \vec{E}^{\text{scat}}$$

$$\varepsilon \frac{\partial \vec{E}^{\text{scat}}}{\partial t} = \nabla \times \vec{H}^{\text{scat}} - (\varepsilon - \varepsilon_0) \frac{\partial \vec{E}^{\text{inc}}}{\partial t} - \vec{J} \tag{4.1}$$

The substrate, protection layer, and recording layer of the Super-Lens disc are not scattering medium so that ε and σ unrelated with time, and then the polarization current is:

$$J = \sigma \left(\vec{E}^{\text{scat}} + \vec{E}^{\text{inc}} \right) \tag{4.2}$$

therefore, Eq. (4.1) can be described as follows:

$$\varepsilon_R(\omega) = n_L^2(\omega) = n_r^2 - n_i^2 - 2jn_r n_i$$

$$\frac{\partial \vec{J}}{\partial t} = -\Gamma \vec{J} + \omega_0^2 \left[\varepsilon_0 \chi_0 \left(\vec{E}^{\text{scat}} + \vec{E}^{\text{inc}} \right) - \vec{P}^L \right] \tag{4.3}$$

$$\frac{\partial \vec{P}^L}{\partial t} = \vec{J}$$

Utilized Eq. (4.3) simulation calculate the distribution of electromagnetic field intensity $E(x, y)$, the results of simulation parallel on cross section of the y_z coordinate is shown in Fig. 4.4, without mask layer as Fig. 4.4a and with mask layer as Fig. 4.4b. The effect of the micro aperture on mask layer cause the energy of the laser beam to centralization, so that the resolution recording size is reduced.

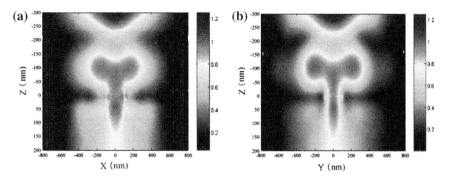

Fig. 4.4 The spatial distribution of the electric field strength $E(x, y)$ parallel to the y–z coordinate cross section: **a** without super-RENS mask layer **b** with super-RENS mask layer

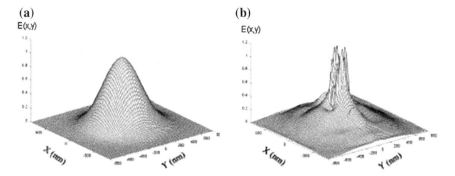

Fig. 4.5 The electromagnetic field distribution $E(x, y)$ of without mask layer (**a**) and with mask layer (**b**)

The simulation results of the mask layer effect to electromagnetic field are shown in Fig. 4.5 (a) electromagnetic field distribution $E(x, y)$ of the original beam spot without mask layer, (b) electromagnetic field distribution of laser beam through the mask layer. Comparison of Fig. 4.5a, b it can be seen that the incident field with Gaussian distribution has great changes, which diameter and area distribution was reduced evidently, and energy was more concentrated with the mask layer.

Through the above analysis can see that the super-resolution effect of the Super-RENS disc and the energy distribution of laser beam. In order to further analysis of the beam energy changes by the Super-RENS disc with super-resolution mask layer, establishment of the optical transmit model of focused laser beam through the films. The beam through k layers film optical transmitted process and related parameters are shown in Fig. 4.6. Where I_o is the incident energy, R is the reflectivity, T is transmissivity, $\vec{n}_1 \sim \vec{n}_k$ are the complex refractive index of incident surface of the films, $\vec{n}_0$ and $\vec{n}_{k+1}$ are complex refractive index of exit surface of the film, respectively. The light conduction model of optical properties for multilayer film can create the optical admittance matrix as follows:

Fig. 4.6 Physical parameters
of the film system

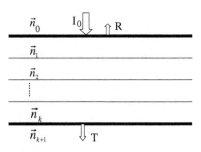

If the light normal incidence and the admittance of the j layer $\eta_j = n_j$, the admittance matrix of optical properties is:

$$M_j = \begin{pmatrix} \cos \vec{g}_j & (i/\vec{n}_j)\sin \vec{g}_j \\ i\vec{n}_j & \cos\vec{g}_j \end{pmatrix} \tag{4.4}$$

where the $\vec{g}_j$ is optical phase shift of j layer: $\vec{g}_j = 2\pi d_j \vec{n}_j/\lambda$, d_j is the thickness of j layer film, λ is the optical wavelength. According to the product of optical properties of every layer can obtain the optical parameters of k layer film as:

$$\begin{pmatrix} B \\ C \end{pmatrix} = \prod_{j=1}^{k} M_j \begin{pmatrix} 1 \\ \vec{n}_{k+1} \end{pmatrix} \tag{4.5}$$

$$R = \frac{\|\vec{n}_0 B - C\|^2}{\|\vec{n}_0 B + C\|^2} \tag{4.6}$$

$$T = 4\vec{n}_0\vec{n}_{k+1}/\|\vec{n}_0 B + C\|^2 \tag{4.7}$$

May have the absorption rate of energy on k layer film:

$$A = 1 - R - T \tag{4.8}$$

For further calculation of energy distribution of a focused laser beam on a certain layer, assume the linear speed of movement of disc is constant, the Gaussian distribution of incident laser beam intensity on the focal point is:

$$I(x,y,t) = \frac{2P}{\pi r_0^2} e^{-2\left((x-vt)^2 + y^2\right)/r_0^2} \tag{4.9}$$

where P is the laser power, r_0 is radius of the center cover energy e^{-2}, and $r_0 = 0.43 \frac{\lambda}{NA}$, x is movement direction of optical spot, y is the cross-track direction and t is the time. In accordance above analysis can create the photo-thermal effect model. In the first is the model of linear heat conduction equation of the Sb mask layer:

$$\nabla^2 T(r,t) + \frac{1}{k}g(r,t) = \frac{1}{\alpha}\frac{\partial T(r,t)}{\partial t} \tag{4.10}$$

where T is temperature, k is thermal conductivity, $g(r,t)$ is the thermal conductivity of the layer within per unit volume and time, $\alpha = \frac{k}{\rho C_p}$ is thermal diffusivity, ρ is density and C_p is specific heat. As only mask layer and the recording layer to absorb energy, so $g(r,t)$ can be written:

$$g(x,y,t) = \frac{I_1 + I_2}{d_{Sb}} \tag{4.11}$$

where I_1, I_2 are light intensity distribution of incident light and reflective light in the Sb mask layer:

$$I_1(x,y,t) = A_1 I(x,y,t)s(t)\exp(-\beta_1 z) \tag{4.12}$$

$$I_2(x,y,t) = A_2 I(x,y,t)s(t)\exp(-\beta_1(d_{Sb} - z)) \cdot \exp(-\beta_1 d_{Sb} - 2\beta_2 d_{Record}) \tag{4.13}$$

where A_1 and A_2 are the absorptivity of incident and reflected light in Sb layer, respectively, that can be calculated through Eqs. (4.3)–(4.8). Where xyz is Cartesian coordinate system, x is direction of spot movement, y is cross-track direction, d_{Record} is film thickness of mask layers Sb, d_{Record} is the recording layer thickness of $Ge_2Sb_2Te_5$, β_1 is attenuation coefficient in mask layers Sb, β_2 is attenuation coefficient of recording layer and $s(t)$ is recording layer power, respectively.

Based on the above heat conduction model can be to 3D numerical simulation analysis with finite element analysis software FEMLAB. Simulation with the boundary conditions and physics/chemic parameters is shown in Table 4.2, and the thermal parameters are shown in Table 4.3.

The simulation results of relationship between maximum temperature on mask layer Sb, laser power (~ 12 mW) and pulse width (40, 100, 140, 200 ns) is shown in Fig. 4.7. The simulation results of the temperature distribution with the laser power 6 mW, pulse width of 100 ns is shown in Fig. 4.8. When the energy of laser

Table 4.2 The substrate of super-RENS disc, film thickness and physics/chemic parameters

Materials		Optical constant	Thickness (mm)
Polycarbonate		1.52	1.2
SiN		2.1	170
Sb	Solid state	$3.11 - i5.66$	15
	Molten	$4.51 - i3.66$	
SiN		2.1	20
$Ge_2Sb_2Te_5$	Crystalline	$4.430 - i2.892$	15
	Crystalline	$4.071 - i1.462$	
SiN		2.1	20
Al		$2.701 - i8.588$	100

Table 4.3 Thermal parameters of super-RENS disc

ρ (kg/m^3)	C_p (J/kg K)	k (W/m K)	λ (nm)	NA	α (m^2/s)	T_m (K)	T_0 (K)
6380	190	24.3	780	0.45	2e−5	870	300

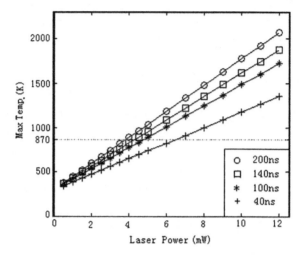

Fig. 4.7 The relationship between the maximal temperature on Sb film, laser power and pulse width

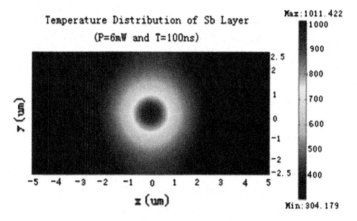

Fig. 4.8 The temperature distribution simulation on Sb film with laser power 6 mW and pulse width 100 ns

(product of pulse width and the power) exceeds the threshold of the mask layer will be formation of super-resolution aperture and its size will be increased with increasing power [2, 3].

Fig. 4.9 The EBM picture of recorded signal mark on GeSbTe media with super-RENS mask

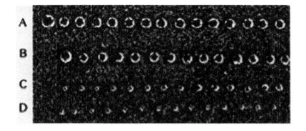

Fig. 4.10 The relationship between the signal modulation degree and readout laser power on the disc with super-resolution mask

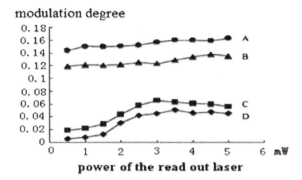

For further analysis the influence of laser power to Super-RENS mask, use 650 nm laser with same pulse width of 300 ns, but different power of 12, 10, 8, and 6 mW to write on a experimental sample as A, B, C, and D four groups. The experimental result was magnified by electron microscope, such as shown in Fig. 4.9, that are precise measurement, the diameters of recorded symbols are: A = 410 nm, B = 310 nm, C = 120 nm, and D = 90 nm averagely.

The read out experimental results of the above four groups (A–D) with difference laser power are shown in Fig. 4.10 that shows the degree of modulation is diversity extremely.

The contradistinction experiment of diameters of recording signal mark to super-resolution mask disc (ds) and without super-resolution mask disc (D) for different laser power is shown in Fig. 4.11. Where D is recording diameter of disc without mask layer, ds is recording diameter of disc with mask layer, using 650 nm laser power is fixed 7, 10, and 12 mW and with different pulse width modulation. From the results can see the proportion of D to ds is same whether any power and exposure time,but when the pulse width below 200 ns is rather poor.

With the extension of exposure time, D and ds dimensions were increase linear, but the proportion of D to ds does not change. The super-resolution effect does not affect to absolute value of the diameter of recording signal mark, but with exposure time shorter it is reduced monotonically. In this experiment within the parameters for different laser power, the super-resolution effect is not same completely. The relationship curves between proportion d/Ds (D is diameters of recording symbol

Fig. 4.11 The relationship curves between diameters (*ds*, *D*) of recording symbol and writing pulse width/power of laser

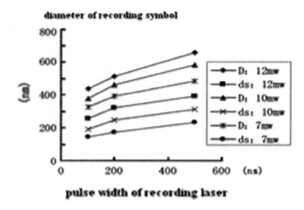

Fig. 4.12 The relationship curves between proportion *d/Ds* and writing pulse width

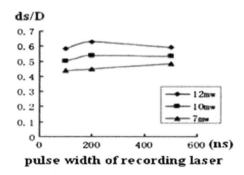

without super-resolution layer. The *ds* is diameters of recording symbol with super-resolution layer) and writing pulse width is shown Fig. 4.12.

With the extension of exposure time, *D* and *ds* dimensions were increase linear, but the proportion of *D* to *ds* does not change. The super-resolution effect does not affect to absolute value of the diameter of recording signal mark, but with exposure time shorter it is reduced monotonically. In this experiment within the parameters for different laser power, the super-resolution effect is not unanimous. The relationship curves between proportion *d/Ds* and writing pulse width is shown Fig. 1. 12, that the value of *d/Ds* is within 0.4–0.6, less than 1 significantly [4, 5].

Above experiment used 650 nm laser only. But a lot of wavelength lasers could be adopted in multiwavelength optical storage indeed. So another wavelength as 405, 532, and 830 nm laser evermore are applied in multiwavelength storage that have to be experiments necessarily. The amplification picture of the experiment results with above wavelength lasers is shown in Fig. 4.13. Four rows are record on a sample with mask layer for 405, 532, 650, and 830 laser. The power is 10 mW with writing pulse width of 100 ns all. The quality of the recorded symbol is different which minimum diameter of symbol for precise measurement is about 70 nm in top group as in Fig. 4.13 and its quality is better.

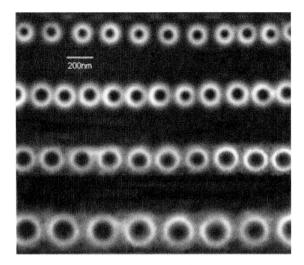

Fig. 4.13 The experimental results to record on super-resolution mask optical disc with 405 nm (*top*), 532, 650 and 830 nm, (*below*) laser, but laser power and pulse width are fixedness

(a)   **(b)**

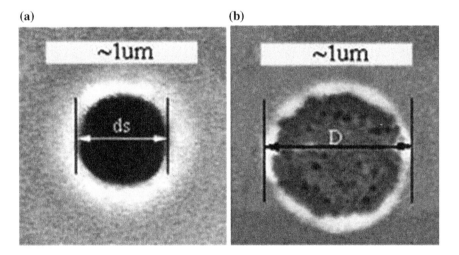

Fig. 4.14 The near infrared laser beam recording symbol on GeSbTe phase-change layer, **a** with super-resolution layer, **b** without super-resolution layer, with experiment system as in Fig. 4.3

In multiwavelength optical storage, the wavelength for super-resolution effect layer must be beyond recording wavelength in many cases. Especially, the sensitive spectrum of most medium of multiwavelength optical storage is visible area. So using a near infrared 1070 nm laser, power 10 mW, and pulse width of 100 ns write on a phase change disc with the mask layer and another disc without mask layer to experiment. The experimental result is shown in Fig. 4.14, which is magnification structure by scanning electron microscopy. The proportion of both recorded symbol sizes is $ds/D \approx 0.5$, it proves the effect of the super-resolution mask evidently. The focused spot diameter was reduced by 50 % approximately, and the boundary of the ablation zone and the amorphous regions is very clear. According to previously analysis, the super-RENS aperture size may be controlled by the exposure energy.

4.2   Micro-aperture Semiconductor Laser

4.2.1   Micro-aperture Laser for NFO Data Storage

The principles of near-field optics (NFO) can be used to laser beam writing directly on medium. But the very small aperture laser (VSAL) is an important light source used in the near-field optical storage system also. So it is necessary to study the near-field property of VSAL's output light. The optical characters and the intensity distribution in the near-field of the output region of the VSAL have been numerical simulated. The near-field optics is thus concerned with the scattering of electro-magnetic waves by nanoscopic systems, where atomic size structure is involved, and affected by the nanososcopic system embedding the atomic size structure. Evanescent waves are important in NFO because the typical size of the objects is comparable to λ and the decay of evanescent waves occurs within a range given by λ. The theoretical and experimental research will apply the VSAL to write and readout subwavelength data. Near-field distribution of a VSAL is essential for the application of such near-field devices. In recording the subwavelength data, the real resolution depends on the near-field spot size, the divergent angle of the beam and the distance from the aperture to the medium. Experimental results, including the near-field writing spot and detection of the subwavelength data by the VSAL. The two-dimensional scanning about the subwavelength data used the width 200 nm by a VSAL with a 100 nm $\times$ 100 nm aperture laser. The VSAL with different aperture shapes are fabricated indeed. Their far-field and near-field performances are ana-lyzed and tested by experiments commonly. The far-field performances, including the threshold current, the slope efficiency, the lasing ability, and the linear fre-quency modulation property are influenced by the front facet reflectivity. A factor method is used to analyze the lasing abilities of VSALs with different aperture shapes that the factor can diminish the discrepancies among the same type laser diodes. The near-field performance focuses on the confinement effect of the VSAL aperture. The near-field scanning optical microscopy can be used to measure the near-field intensity distribution from a VSAL and the near-field performances which was affected by the aperture shape. The analyzing results are possible application in the near-field optical recording, and some curves indicating the near-field optical characters of the output light will be presented in the section in detail.

More attention was paid on the ultra-high optical memory using near field optics (NFO) for future data storage. A common important issue of near-field optical storage is to realize high-output power source. The VSAL can be the high-output power light source for near field recording, that the VSAL demonstrated more than 10^4 times of increase of output power over coated tapered fibers with comparable aperture diameters. In addition, the near field characters of the micro-aperture vertical-cavity surface-emitting laser (VCSEL), that the calculated spot size of output is nearly as small as the aperture width of 80 nm when the wavelength is 650 nm. It indicated the potentiality of the VSAL as a near-field optical recording tool, so the validate analysis of the optical characters of the output in the VSAL's near field is necessary.

4.2.2   Vertical-Cavity Surface-Emitting Lasers (VCSEL)s

The Micro-aperture laser is based on vertical-cavity surface-emitting lasers (VCSEL)s technology. The paragraph introduced the nano-aperture VCSEL and epitaxial structure, the fabrication of the nano-aperture VCSELs. During the processing flow of the devices, electroluminescence spectrum of the VCSELs is measured before and after coating the VCSEL emission facet with metal coating to verify that the devices work as designed. The development of VCSELs has experienced very rapid progress. Starting as only laboratory novelties at the beginning of the 1990s, these devices have attracted much interest in both academia and industry due to their many advantages, such as low-cost wafer-scale fabrication and testing, easy application in arrays, circular beam shape, easy coupling to other optical elements, etc. There are already a number of products using VCSELs in the application of fiber-optical data communication, optical interconnects, optical storage, etc.

Conventional VCSELs typically consist of a top mirror, an active quantum well region and a bottom mirror. The top and bottom mirrors are composed of many pairs of quarter-wavelength thick semiconductor layers with alternating composition, namely distributed Bragg reflectors (DBR). For example, for mirrors using AlGaAs layers, each pair of alternating layers consist of one layer with low Al concentration and the other layer with high Al concentration so that the contrast of refractive index between these two layers is as large as possible. Figure 4.15 shows a typical structure of VCSELs. Due to the short cavity length, and hence, short gain length, VCSEL mirrors must have a very high reflection coefficient. For top emitting VCSELs, the reflectivity is typically over 99.9 % for the bottom mirror and about 99.5 % for the top mirror.

To provide current and optical mode confinement, there is usually an aluminum oxide layer buried inside VCSELs. This aluminum oxide layer is formed through the wet oxidation of an AlGaAs layer with very high Al concentration, typically up to 98 %. Exposing AlGaAs alloys to temperatures from 350 to 500 °C in a steam environment converts the semiconductor into a mechanically robust, chemically

Fig. 4.15 Schematic
structure of a conventional
VCSEL with oxide aperture

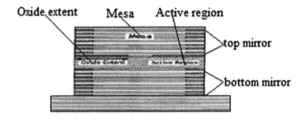

inert, insulating, and low refractive index oxide. The oxidation rate is highly dependent on the Al concentration. For example, at 420 °C, the oxidation rate for AlGaAs with 98 % Al is about an order of magnitude higher than that of AlGaAs with 92 % Al. Thus, this oxidation can be highly selective. A single layer of AlGaAs layer with high Al concentration (e.g., 98 %) can be used to form the oxide, while the other AlGaAs layers with lower Al concentration in the structure would not experience much oxidation. Precise control of this oxidation process can leave a small active region surrounded by an oxide aperture, as shown in Fig. 4.15. This insulating oxide aperture confines the current injected into the active region and its low refractive index helps confine the optical mode around the aperture region. Without the oxide aperture, the optical mode directly experiences the roughness of the etched mesa, which results in large optical loss. Also, the oxide aperture can be much smaller than the top mesa diameter. This makes the fabrication of a device with small active region easier, which is important for the fabrication of single-mode devices.

4.2.3 Modeling of Nano-aperture VCSELs

To develop a nano-aperture VCSEL, the easiest way one can think of is to build that on the basis of a conventional VCSEL. For example, one can deposit a SiO_2 film and then Au film on top of conventional VCSEL. By opening a nano-aperture in the Au film, one can obtain a nano-aperture VCSEL. The necessity of using the SiO_2 film will be discussed in detail later. Figure 4.16 shows the main part of the structure of such a nano-aperture VCSEL based on conventional VCSEL coated with SiO_2 and Au film.

Fig. 4.16 The structure of
nano-aperture VCSEL based
on a conventional VCSEL
(DBR stands for distributed
Bragg reflector)

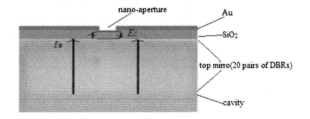

The problem with the above nano-aperture VCSEL structure is that the reflectivity of the top mirror in a conventional VCSEL is very high (typically about 99.5 %). The intensity transmitted through the top mirror, namely the intensity incident onto the nano-aperture is very low compared to the intensity inside the laser cavity. So the power coming out of the nano-aperture VCSEL will be very low. Simply using the conventional VCSEL structure for the nano-aperture VCSEL is not a good choice. Before starting to design a unique structure for the nano-aperture VCSEL, can built a model to evaluate the quantum efficiency η of the nano-aperture VCSEL. The η is defined as ratio of the number of photons coming out of the laser over the number of electrons injected into the laser per unit time. For the nano-aperture VCSEL the quantum efficiency η can be determined by

$$\eta = \eta_i T_{\text{aperture}} / \left(T_{\text{aperture}} + \alpha \right) \tag{4.14}$$

where η_i is the injection current efficiency which represents the fraction of injected carriers contributing to the emission process (some of the carriers can recombine nonradiatively). α is the total loss per round trip, which includes transmission through the top and bottom mirror, absorption and scattering loss, etc. T_{aperture} is the fraction of power transmitted through the nano-aperture per round-trip. The T_{aperture} is given by

$$T_{\text{aperture}} = T_{\text{SiO}_2} \cdot \frac{\text{PT} \times A_{\text{aperture}}}{A_{\text{mode}}} \tag{4.15}$$

where T_{SiO_2} is the intensity incident onto nano-aperture from the SiO_2 layer, normalized to the intensity incident from the laser cavity. PT is the power throughput of the nano-aperture, which is defined as the ratio of the power transmitted through the nano-aperture over the power incident onto then nano-aperture. A_{aperture} is the area of the nano-aperture. A_{mode} is the effective area of the optical mode.

From Eq. (4.14), it can be seen that there are three approaches to improve the quantum efficiency of the nano-aperture VCSEL. First, increase the intensity incident onto the nano-aperture, namely T_{SiO2}. Second, decrease the optical mode area A_{mode}. Third, reduce the total loss α. The first approach can be realized by reducing Eqs. (4.14) and (4.15) number of DBR (as in Fig. 4.16) pairs in the top mirror, whose reflectivity can be enhanced by the Au coating. This approach is first proposed by Robert Thornton et al. to improve the output power of nano-aperture VCSELs. The second approach can be realized by using a smaller oxide aperture to confine the optical mode. The third approach can also be realized by using the oxide aperture to reduce the scattering loss.

The output power from the nano-aperture VCSEL is then given by

$$P_{\text{out}} = \hbar \omega \eta (I - I_{\text{th}}) / q \tag{4.16}$$

where ω is the angular lasing frequency, q is the electron charge, η is the quantum efficiency, I is the injection current, and I_{th} is the lasing threshold current.

4.2.4 Design of the Nano-aperture VCSEL

Based on the principles discussed above, our top-emitting VCSELs are designed to operate around 972 nm and consist of 38.5 pairs of n-type $Al_{0.08}Ga_{0.92}As/Al_{0.92}Ga_{0.08}As$ DBR, three InGaAs/GaAsP quantum wells and 9.5 pairs of p-type $Al_{0.08}Ga_{0.92}As/Al_{0.92}Ga_{0.08}As$ DBRs. The number of p-type DBR pairs is only about half of that in conventional VCSELs, which is designed to increase the intensity incident onto the nano-aperture. The reflectivity of the top mirror is enhanced with a 150 nm thick Au coating. A half-wavelength thick SiO_2 film is inserted between the Au coating and the top DBR mirror to enhance the transmission through the nano-aperture. Here, $\lambda/(2 \times nSiO_2) = 972$ nm/ $(2 \times 1.5) = 324$ nm, where $nSiO_2$ is the refractive index of SiO_2 at a wavelength of 972 nm. So the thickness of half-wavelength thick SiO_2 film is 324 nm. The nano-apertures are etched through the Au coating using a Ga^+ focused ion beam (FIB). The nano-aperture needs to be placed in the center of the top mesa, which can be easily located in the lithographically defined circular mesa. This is far easier for the VCSEL compared to the edge-emitting laser where the nano-aperture requires a precise alignment with the quantum well region which is buried under the facet coatings and hard to locate. Figure 4.17 shows a schematic structure of the nano-aperture VCSEL.

Wet oxidation of $Al_{0.98}Ga_{0.02}As$ is used to obtain a 2.8 μm-diameter oxide aperture for current and mode confinement. This particular size of oxide aperture was chosen as a tradeoff between the optical mode area and the roll-over current. As shown before, a smaller oxide aperture leads to a smaller optical mode area, and hence can increase the quantum efficiency. However, if the oxide aperture is too small, the rollover current, which is the current beyond which the power output from the laser starts to decrease with increasing injection current due to excessive heating, decreases significantly. This limits the maximum output power that can be achieved with the nano-aperture VCSEL. Following Eq. (4.16), the maximum power is given by

$$P^{max}_{out} = \hbar\omega\eta(I_{rollover} - I_{th})/q \qquad (4.17)$$

where η is the quantum efficiency, $I_{rollover}$ is the rollover current, I_{th} is the threshold current. So although reducing the oxide aperture size can increase the quantum

Fig. 4.17 Nano-aperture VCSEL structure

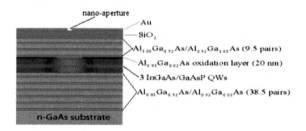

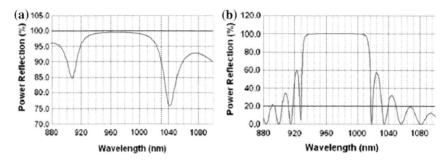

Fig. 4.18 Simulated power reflectivity of **a** the top mirror; **b** the bottom mirror

efficiency η, it also reduces the $I_{rollover}$. The oxide aperture size of 2.8 μm diameter is chosen as a balance between these two factors, as Eq. (4.17). Although the number of DBR pairs in the top mirror of the nano-aperture VCSEL structure is only half of that in conventional VCSELs, the reflectivity of the top mirror enhanced with the Au coating is comparable to that in conventional VCSELs. The bottom mirror consists of 38.5 pairs of DBRs and is similar to that in conventional VCSEL. For the designed structure shown in Fig. 4.17, the reflectivity of the top and bottom mirror is simulated using transfer matrix method. Figure 4.18 shows the simulated power reflectivity of the top and bottom mirror versus wavelength. At the designed lasing wavelength of 972 nm, the simulated power reflectivity is 99.49 % for the top mirror and 99.89 % for the bottom mirror.

One of the most important issues in designing this VCSEL structure is the phase matching condition. For the VCSEL to laze at the designed wavelength, the round trip phase for a photon has to be precisely equal to an integer number of 2π. In a conventional VCSEL, this condition is satisfied by designing the optical thickness of the laser cavity to be an integer number of half wavelengths. In our VCSEL structure, an additional Au coating is used to enhance the reflectivity of the top mirror. The reflection from the Au layer causes some additional phase shift. A special AlGaAs layer is used as part of the last DBR layers to provide phase matching to compensate for this phase shift. Under this phase matched condition, the peak of the standing wave pattern lies on the three quantum wells as shown in Fig. 4.19, and thus satisfies the lasing condition.

Fig. 4.19 E^2 distribution of the standing wave inside the laser cavity. Real part of refractive index of each layer is shown by the *black line*. The distance in *x*-axis starts from the topmost layer of the VCSEL epitaxial structure

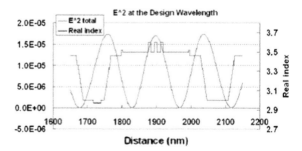

4.2.5   *Numerical Model*

The method to calculating on the typical structure of the vertical-cavity surface-emitting lasers (VCSEL)s resonant cavity of the VSAL is shown Fig. 4.20a. In order to be conveniently analyzed, the VSAL has been simplified to be the physical model showed in Fig. 4.20b. It involves two parallel plane mirrors, one of which is a noble metallic film with a micro-aperture.

This method is widely used to numerically obtain the eigenmodes of the side-opened resonators consisting of a pair of plane mirrors. For calculating, first set a field distribution of the pane wave on the metallic film as an initial condition then calculate the distribution on the other mirror by using the angular spectrum method. Moreover to calculate the field distribution on the metallic film in the same manner, can get the distribution of the base mode of the resonator by repeating the round-trip calculation of the field propagation. In the next, using the two dimension nonlinear finite-difference time-domain (FDTD) method simulate the light of the base mode propagating through the metallic film. The noble metals (gold, silver, or copper) in the optical regime process complex refractive indices in which the imaginary component is greater than that the requires permittivity to have a negative real component. The boundary condition on the tangential electric field cannot be satisfied with the standard FDTD method. So through combining the full-wave, vector, linear Maxwell equations solver with a Lorentz linear dispersion model, the nonlinear FDTD method has been carried out by Judkins and Ziolkowski. Solving the system of equations yields the following iterative expressions for YE mode. The expressions for TM mode are similar to the ones for TE mode:

$$Hx(i,j) = Hx(i,j) + \text{CD} \times (Ez(i,j) - Ez(i,j+1))$$
$$Hy(i,j) = Hy(i,j) + \text{CD} \times (Ez(i+1,j) - Ez(i,j)) \qquad (4.18)$$

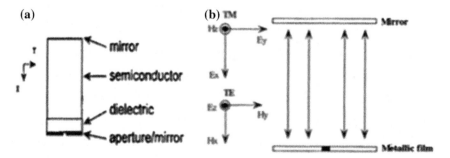

Fig. 4.20 a The structure of the resonant vertical-cavity of the VSAL. **b** Theoretic model and coordinates of the VSAL

where:

$$z(i,j) = Ez(i,j) + \frac{(c \times \varepsilon_0 \times \Delta t)}{\varepsilon}((Hy(i,j) - Hy(i,j)$$
$$+ Hx(i,j-1) - Hx(i,j))/\Delta s - J(i,j))Pz(i,j) \qquad (4.19)$$
$$= Pz(i,j) + \Delta t \times Jz(i,j)$$

And

$$Jz(i,j) = \frac{\frac{1}{\Delta t} - \frac{\Gamma}{2}}{\frac{1}{\Delta t} + \frac{\Gamma}{2}} \times Jz(i,j) + \frac{\omega_0^2}{\frac{1}{\Delta t} + \frac{\Gamma}{2}} \times \left(\frac{\chi_0 \times (Ez(i,j)}{c} - Pz(i,j) \right) \qquad (4.20)$$

$$\begin{cases} CA = \frac{1-0.5\sigma \times \Delta t}{1+0.5\sigma \times \Delta t} \\ CB = \frac{\varepsilon_0}{2\varepsilon + \sigma \times \Delta t} \\ CD = \frac{1}{2} \end{cases} \qquad (4.21)$$

where P is the polarization generated by the model, J is the polarization current ω_0, and Γ resonant frequency and the damping coefficient, respectively. The 2d-NL-FDTD algorithm employs a uniform mesh grid in both directions Δs is the unit length of the mesh grid. $\Delta t = \Delta s/2C$ is the unit time in the calculation, where C is the light velocity in the vacuum. In the code, the second-order absorbing boundary conditions are used at all boundary surfaces. Table 4.4 lists the parameters was used in the simulations calculation.

4.2.6 Calculation Results

In this calculation, the electric field perpendicular and parallel to the sheet have been defined as TE and TM modes, respectively. The coordinate systems are shown in Fig. 4.21a, b shows the intensity profile of base mode of the resonator with a micro-aperture l00 nm wide. Because the width of the mirror is nearly 100 nm, which is more wider than the aperture of 80 nm, and the length of the resonator is nearly 750 nm, the influence of the micro-aperture on the mode is negligible. But in the calculating region (l60 nm × 160 nm), the slim difference between the modes with different wide apertures can be shown in Fig. 4.21b. Because the 400 nm is

Table 4.4 Parameters for the simulation calculation

Phys. quantity	Symbol	Value
Wavelength	λ	980 nm
Velocity (vacuum)	C	3×10^8 m
Unit grid	Δs	9.8 nm
Unit time	Δt	1.6×10^{-17} s
Refraction index	n_{Au}	$0.175 - j4.91$

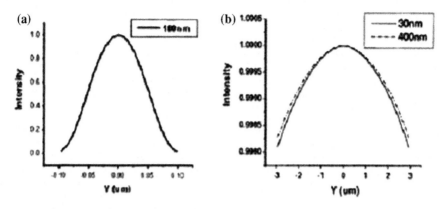

Fig. 4.21 a Intensity profile of the resonator with aperture of 100 nm wide. **b** With aperture width of 30 and 400 nm

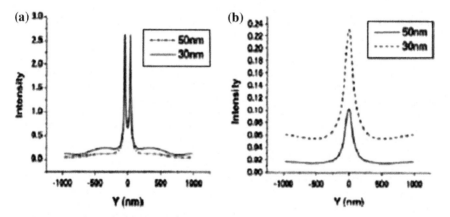

Fig. 4.22 a Intensity profile along Y of the output through 100 nm wide aperture with different thickness for TM mode. **b** For TE mode along Y

close to the wavelength (650 nm), the change of the mode is obvious. The modes with different aperture have been used in the FDTD simulation as the source.

Figure 4.22a shows the intensity profile of output through a 100 nm aperture with different thickness for TM mode at the boundary between Au and air. Two sharp peaks appear at both edges of the aperture. The reason for this result is the surface plasma enhancement of the metallic film for TM mode. The profiles of different thickness are similar each other, the spot size is as small as 108 nm, approximately equal to the width of aperture, which is much smaller than the wavelength. At the area away from the center of the curves, the intensity of the thinner film (30 nm) is larger than the thicker (50 nm) one, because the light can partially penetrate into metal especially when the thickness is much smaller than the wavelength. So the thinner thickness, the background noise is stronger. It is possible that this phenomenon can reduce the resolution of the spot. But the affect is negligible for TM mode because of the surface plasma enhancement.

The curves of intensity by the same calculation for TE mode is shown in Fig. 4.22b. One can see that the maximal point of the curves locate in the center of the aperture because of no surface plasma enhancement for TE mode, and the intensity of the thinner (30 nm) film is higher than the thicker (50 nm) one because of the partially penetrated light. The intensity of TM mode is over 10 times the one of the TE mode, so the intensity of the partially penetrated light is significant compared with the intensity of the spot. It means that the thickness of the metallic film is an important factor to defining the resolution of the outgoing spot for TE mode. Although the output intensity of the thinner film is high, the quality of the spot is not always good. From Fig. 4.22b, can see that the spot size (FWHM) of the film 50 nm thick is 136 nm, which is smaller than the one 30 nm and thick of 156 nm.

Figure 4.23a shows the calculated intensity distribution across the Au film 50 nm thick at the center of the aperture with different widths (30, 50, 100, 200, and 400 nm) for TM modes. The enhancement of the intensity of the electric field at the region of the Au film increases with decreasing the aperture width. The smaller the

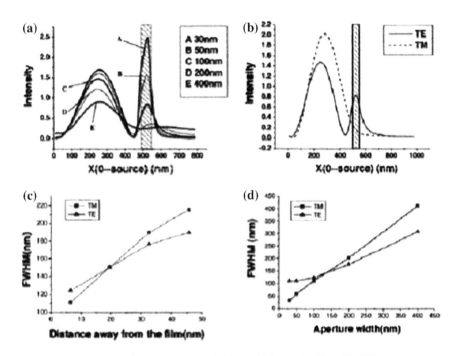

Fig. 4.23 **a** *Intensity* profile along *X* through 50 nm thickness Au film with different aperture at center for TM mode. **b** At 100 nm wide aperture for TM and TE mode. **c** The spot size (*FWHM*) as a function of the distance away from the film for TM and TE mode with thickness 50 nm and aperture 100 nm. **d** With thickness 50 nm and distance 6.53 nm

aperture width, the stronger is the attenuation of the near-field intensity of the output, and the nodes of the intensity profiles shift slightly to the left side of the Au film. It is obvious that the higher intensity of output decays more rapidly in the near-field region. It means that if to get high powerful output of VSAL that should realize the precision control of the distance between source and the recording medium.

The trend of the decay of the output in the near-field for TE mode is similar to the TM. But the important difference has been indicated through comparing the decay profiles for TM and TE modes. The curves have been shown in Fig. 4.23b. One can see that the intensity of the TM mode has an obvious enhancement in the Au film, none has the TE mode. In the near-field of the right side of the Au film, the attenuation of the output for TM mode is sharper than the TE. The intensity of TM mode is over the TE mode in the near-field because of the surface plasma enhancement. But these two modes have an approximately equal intensity in the far-field, because the enhancement is only significant in the near-field. Figure 4.23c shows the calculated spot size of the Au film of 50 nm thick with a 100 nm wide aperture as a function of the distance away from the metallic film for TM and TE modes. It can see that the spot size increases and the increase become slower with increasing the distance.

The spot size for the TM mode is smaller than the TE one when the distance less than or equal to 20 nm. The calculated spot size of the Au film 50 nm thick at the boundary between the film's right side and the air as a function of the aperture for two modes are also analyzed in Fig. 4.23d. That the spot size increases and the increase for the TM mode keeps steady while on the contrary the increase for TE mode becomes sharper with increasing the distance. The spot size of the TM mode is smaller than the TE one when the aperture width is less than 130 nm. The minimum spot size is about 32 nm at the 30 nm wide aperture for TM mode. Therefore, the output of the VSAL can produce the spot beyond the optical diffraction limit. And the TM mode source is more appropriate for near-field optical recording in compare with the TE mode because of its smaller size and higher power.

In conclusion, the method has been used to calculate the base mode of the simply resonator of the VSAL, and using the two dimensions NL-FDTD simulation, we have analyzed the near-field optical characters of the output of the VSAL. The intensity profile of the base mode has been shown. The calculated intensity distributions along the transversal and longitudinal directions for TM and TE modes have been presented, respectively. They show that the spot sizes all increase with increasing the distance away from the metallic film, and decrease with decreasing the aperture width. But the power of the output for TM mode is higher one order than the YE, and the decay of the intensity for former also sharper than the latter, because the surface plasma enhancement of the metallic film only appears in the TM mode. The spot size of the TE mode is smaller than the TM either the aperture width is greater than 110 nm or the distance is greater than 20 nm with 100 nm wide aperture at the right side of the film. But as a near-field recording source, the TM mode source is more suitable because of its higher power of the output.

4.2.7   Speciality of Spectrum

For multiwavelength optical storage, spectrum of laser source has to match with absorption spectrum of various medium that are required that can directly manip-ulate recording channels and their wavelength over the bandwidth of all materials to optical sensitivity. The bandwidth of wavelength is from near-ultraviolet to near-infrared, i.e., form 300 to 1800 nm as Fig. 4.1 in application of multiwave-length optical storage currently. In addition, there are a lot of special requirements for laser sources except wavelength, as linear, frequency response, threshold current density, stability, efficiency, structure size, lifetime, etc. Solutions based on non-linear processes have been proposed, but these suffer from having only low effi-ciencies as a result of low nonlinear susceptibilities. Here, demonstrate all-optical wavelength conversion of beam using a resonant nonlinear process within a tera-hertz quantum cascade laser. The process is based on injecting a low-power continuous-wave near-infrared beam in resonance with the interband transitions of the quantum cascade laser. This results in an enhanced nonlinearity that allows efficient generation of the difference and sum frequency, shifting the frequency of the near-infrared beam by the frequency of the quantum cascade laser. Efficiencies of 0.13 % are demonstrated, which are equivalent to those obtained using free electron lasers. As well as having important implications in its application in ultrafast wavelength shifting, this work also opens up the possibility of efficiently upconverting terahertz radiation to the near-infrared and enables the study of high terahertz–optical field interactions with quantum structures using quantum cascade lasers. The QD lasers now surpass the established planar quantum well laser technology in several respects. These include their minimum threshold current density, the threshold dependence on temperature, and range of wavelengths obtainable in given strained layer material systems. Self-organized QDs are formed from strained-layer epitaxy. Upon reaching such conditions, the growth front can spontaneously reorganize to form three-dimensional islands. The greater strain relief provided by the three-dimensionally structured crystal surface prevents the formation of dislocations. When covered with additional epitaxy, the coherently strained islands form the QDs that trap and isolate individual electron–hole pairs to create efficient light emitters. The materials and important characteristics for quantum dot lasers are shown in Table 4.5.

Semiconductor lasers are used to many technological products except optical storage including optical communication, laser printers, and other high technology schemes. The basis of laser operation depends on the creation of nonequilibrium populations of electrons and holes, and coupling of electrons and holes to an optical field, which will stimulate radiative emission. Calculations carried out in the early 1970s by C. Henry predicted the advantages of using quantum wells as the active layer in such lasers: the carrier confinement and nature of the electronic density of states should result in more efficient devices operating at lower threshold currents than lasers with "bulk" active layers. In addition, the use of a quantum well, with discrete transition energy levels dependent on the quantum well dimensions

Table 4.5 Materials, band gap, and output wavelength of quantum dot semiconductor lasers

Material	Energy gap at 300 K (eV)	$\lambda = ch/E_g$ (µm)
Diamond[a]	5.4	0.23
$In_{0.15}Ga_{0.85}N$	2.2	0.40
GaP	2.25	0.55
$In_{0.5}Ga_{0.5}P$	2.0	0.62
GaSb	0.68	0.73
GaAs	1.42	0.87
$Al_xGa_{1-x}As$ ($0 \leq x < 0.37$)	1.42–1.92	0.65–0.87
InP	1.35	0.92
$In_{0.8}Ga_{0.2}As_{0.34}P_{0.65}$	1.1	1.13
AISb	1.6	1.58
$In_{0.53}Ga_{0.47}As$	0.74	1.67
Germanium	0.66	1.88
Silicon	0.66	1.88
InAs	0.36	3.5
InSb	0.17	7.3

[a]Although diamond is an insulator it can be made conductive by irradiation with ultraviolet light

(thickness), provides a means of "tuning" the resulting wavelength of the material. The critical feature size-in this case, the thickness of the quantum well-depends on the desired spacing between energy levels. For energy levels of greater than a few tens of millielectron volts (meV, to be compared with room temperature thermal energy of 25 meV), the critical dimension is approximately a few hundred angstroms. Although the first quantum well laser demonstrated in 1975, was many times less efficient than a conventional laser, the situation was reversed by 1981 through the use of new materials growth capabilities (molecular beam epitaxy), and optimization of the heterostructure laser design. Even greater benefits have been predicted for lasers with quantum dot active layers. Arakawa and Sakaki predicted in the early that quantum dot lasers should exhibit performance that is less temperature-dependent than existing semiconductor lasers, and that will in particular not degrade at elevated temperatures. Other benefits of quantum dot active layers include further reduction in threshold currents and an increase in differential gain-that is, more efficient laser operation. Figures 4.19 and 4.20 illustrate some of the key concepts in the laser operation. Stimulated recombination of electron–hole pairs takes place in the GaAs quantum well region, that the confinement of carriers and of the optical mode enhanced the interaction between carriers and radiation as shown Fig. 4.19. In particular, note the change in the electronic density of states, as a function of the "dimensionality" of the active layer, shown in Fig. 4.20. The population inversion (creation of electrons and holes) necessary for lasing occurs more efficiently as the active layer material is scaled down from bulk (3-dimensional) to quantum dots (0-dimensional). However, the advantages in operation depend not only on the absolute size of the nanostructures in the active

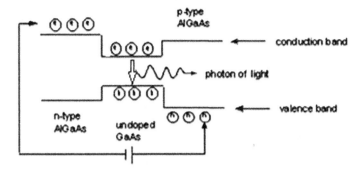

Fig. 4.24 The interaction between carriers and radiation of the semiconductor laser operation

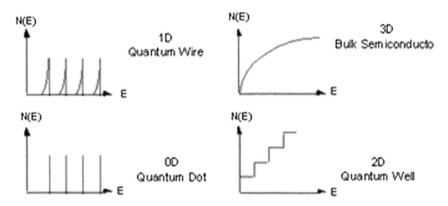

Fig. 4.25 Density of electronic states is the function of structure size and the change in the electronic density of states is a function of the dimensionality of the active layer

region, but also on the uniformity of size. A broad distribution of sizes "smears" the density of states, producing behavior similar to that of bulk material.

Thus, the challenge in realizing quantum dot lasers with operation superior to that shown by quantum well lasers is that of forming high quality, uniform quantum dots in the active layer. Initially, the most widely followed approach to forming quantum dots was through electron beam lithography of suitably small featured patterns (~ 200 Å) and subsequent dry-etch transfer of dots into the substrate material. The problem that plagued these quantum dot arrays was their exceedingly low optical efficiency: high surface-to-volume ratios of these nanostructures and associated high surface recombination rates, together with damage introduced during the fabrication itself, precluded the successful formation of a quantum dot laser (Figs. 4.24 and 4.25).

With the demonstration of the high optical efficiency self-assembled formation of quantum dots, formed without need of external processing and having the natural overgrowth of cladding material (which addressed issues of surface recombination),

Table 4.6 Commercial lasers and wavelengths

Wavelength (nm)	Laser diode and application
375	Excitation of hoechst stain, calcium blue
405	InGaN blue for violet laser, Blu-ray disc and HD DVD
445	InGaN deep blue laser multimode diode
473	Skyblue laser pointers, still very expensive
485	Excitation of GFP and other fluorescent dyes
510	(to ~ 525 nm) Green diodes
635	AlGaInP better red laser pointers, bright as 670 nm
640	AlGaInP high brightness red DPSS for laser pointers
657	AlGaInP for DVD drives, laser pointers
670	AlGaInP for cheaper red laser pointers
760	AlGaInP gas sensing for O_2
785	GaAlAs for compact disc drives
808	GaAlAs pumps YAG lasers
848	Laser mice
930	Single mode laser diode
980	InGaAs pump for optical amplifiers
1064	AlGaAs fiber-optic communication
1310	InGaAsP, InGaAsN fiber-optic communication
1480	InGaAsP pump for optical amplifiers
1512	InGaAsP Ga sensing for NH_3
1550	InGaAsP, InGaAsNSb fiber-optic communication
1625	InGaAsP fiber-optic communication, service channel

there ensued a marked increase in quantum dot laser research. The first demonstration of a quantum dot laser has high threshold density. Bimberg et al. achieved improved operation by increasing the density of the quantum dot structures, stacking successive, strain-aligned rows of quantum dots, and therefore, achieving vertical as well as lateral coupling of the quantum dots. In addition to utilizing their quantum size effects in edge-emitting lasers, self-assembled quantum dots have also been incorporated within vertical cavity surface-emitting lasers. Table 4.6 gives a summary of main achievements in quantum dot lasers. Commercial laser diodes have been applied diffusely and its wavelength are shown Table 4.6.

As with the demonstration of the advantages of the quantum well laser that preceded it, the full promise of the quantum dot laser must await advances in the understanding of the materials growth and optimization of the laser structure. Although the self-assembled dots have provided an enormous stimulus to work in this field, there remain a number of critical issues involving their growth and formation: greater uniformity of size, controllable achievement of higher quantum dot density, and closer dot-to-dot interaction range will further improve laser performance. Better understanding of carrier confinement dynamics and capture times, and better evaluation of loss mechanisms, will further improve device characteristics. It should be noted that the spatial localization of carriers brought about by the quantum dot confinement may play a role in the "anomalous" optical efficiency of the GaN-based materials, which is exceptional in light of the high concentration of threading dislocations (10^8–10^{10} cm^{-2}) that currently plague this material system. The localization imposed by the perhaps natural nanostructure of the GaN materials may make the dislocation largely irrelevant to the purely optical (but not to the electrical) behavior of the material.

4.3 Optically Injected Quantum Dot Lasers

Quantum dots (QD) can be made using a variety of methods but for real applications mainly three methods are used:

4.3.1 Epitaxy Growth (MBE, MOVPE)

Stranski–Krastanov (SK) growth is important when dealing with quantum size structures. The formation of the three-dimensional (3D) quantum structure (Volmer Weber equivalence) is driven by the strain that occur during growth when the deposited two-dimensional layers (2D) (Frank–van der Merwe equivalence) exceed a critical thickness. So SK growth is a three-dimensional growth of nano island (quantum dots). SK growth is usually established by strain formation between the substrate and the epilayer due to lattice mismatch between the two ($a_{sub} > a_{epi}$). The epilayer reacts to this strain by forming three-dimensional islands instead of two-dimensional flat surface. These dots are self-assembled and can have very small dimensions (<10 nm^3) as Fig. 4.26. They either form pretty randomly, on atomically flat substrates, or rather ordered at step edges on substrates with step edges. QD made using this method can be difficult to control both regarding size (volume), and density.

Microcrystallites in glass or polymers are useful for QD laser. So another way of making QD is by poring nano-particles into a molten glass or polymer and cooling it down so that the particles freeze in place (see Fig. 4.27). Artificially, patterned dots by electron-beam lithography can be used to QD also. In this way, the size and placement of the QD can be controlled in a very precise manner but the huge number of dots needed in a QD laser causes problems for this technique. Also, surface states can cause problems in etched dots.

The QD needs to be small in order for the carriers to be as three-dimensionally confined as possible in space (so get energy delta functions). The three-dimensional confinement potential needs to be significantly high in order for the carriers (electrons/holes) not to be thermally excited out of the QD. QDs need to be operational at and above room, meaning that this potential needs to be substantially larger than $k_{Troom} = 25$ meV that need many QDs which need to be as

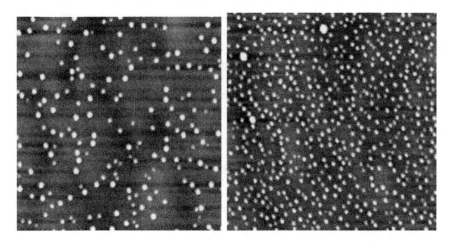

Fig. 4.26 InAs nano-islands. The two AFM figures are from the same sample, taken 3 cm apart. Reason for different QD concentrations is the different growth flux at the two positions

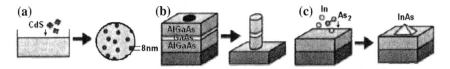

Fig. 4.27 Schematic representation of different approaches to fabrication of nanostructures: **a** microcrystallites in glass, **b** artificial patterning of thin film structures, **c** self-organized growth of nanostructures

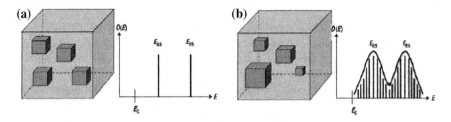

Fig. 4.28 The QD laser needs of smaller carriers (electrons/holes)

homogeneous as possible; otherwise will get a spread in discrete energy levels (as see Figs. 4.28 and 4.29).

Electrons/holes can be confined in all three dimensions in a dot or a quantum box. The situation is analogs to that of a hydrogen atom meaning that only discrete energy levels are possible. The density of states is given by

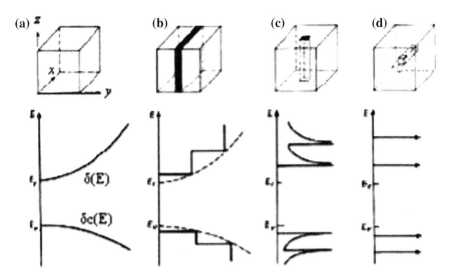

Fig. 4.29 Quantization of density of states: **a** original bulk, **b** quantum well, **c** quantum wire, **d** quantum dot

$$\rho_{0D} = 2\delta(E) \tag{4.22}$$

So end up with discrete energy levels:

$$E - E_C = E_{n,m,p} = \frac{\pi^2 h^2}{2m^*} \left(\frac{n^2}{L_x^2} + \frac{m^2}{L_y^2} + \frac{p^2}{L_z^2} \right) \tag{4.23}$$

As shown in Fig. 4.24.

4.3.2 QD-Lasers

In principles, QD lasers can be treated with the similar way as quantum well (QW) lasers and the laser structure is fabricated in a similar way, the only difference being that the optically active medium consists of QDs instead of QWs. The figure below shows a simple laser structure, consisting of an active layer embedded in a waveguide, surrounded by layers of lower refractive index to ensure light confinement. The active material consists of quantum wells or quantum dots where the band-gap is lower than that of the waveguide material as Fig. 4.30.

The active layer (QD or QW) is embedded in an optical waveguide (material with refractive index smaller than that of the active layer). Wavelength of the emitted light is determined by the energy levels of the QD rather than the band-gap energy of the dot material. Therefore, the emission wavelength can be tuned by

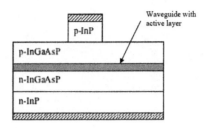

Fig. 4.30   Construction of QD lasers

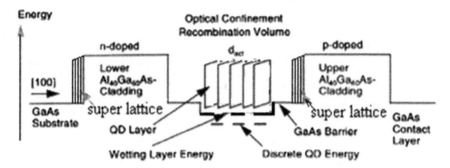

Fig. 4.31   The ideal QD laser consists of a 3D-array of dots with equal size and shape, surrounded by a higher band-gap material (confines the injected carriers). The barrier material forms an optical waveguide with lower and upper cladding layers (n-doped and p-doped $Al_{40}Ga_{60}As$)

changing the average size of the dots. Because the band-gap of the QD material is lower than the band gap of the surrounding medium can ensure carrier confinement. A structure like this, were carrier confinement is realized separately from the confinement of the optical wave, is called a separate-confinement heterostructure (SCH) as Fig. 4.31.

4.3.3   Optical Confinement Factor

For a QD array, the optical confinement factor G is on the order of total dot volume to total waveguide volume. It can be split into in-plane and vertical components. For consistency with a similar treatment for QW lasers, it should be pointed out that this notation is most appropriate for vertical-cavity quantum dot lasers where the light propagation is perpendicular to the active layer:

$$\Gamma = \Gamma_{xy}\Gamma_z \tag{4.24}$$

where $\Gamma_{xy} = N_D A_D A = \xi$ where N_D is the number of QD, A_D is the average in-plane size of the QD and A is the xy-area of the waveguide. The factor ξ is called the area coverage of QD.

The vertical component of the confinement factor is given by the ratio of the light intensity in the active layer (QD), averaged over area A, to the total light intensity in the whole heterostructure. This ratio characterizes the overlap between the QD and the optical mode.

$$\Gamma_z = \frac{1}{A} \int_{QD} |E(z)|^2 dz / \int_{whole} |E(z)|^2 dz \tag{4.25}$$

Example: For $N_D/A = 4 \times 10^{10}$ QD/cm^2 and volume of QD equal to $7 \times 7 \times 2$ nm^3, can get $\xi = 0.02$. For a 150 nm thick cavity a typical vertical optical confinement factor is 0.007. Hence, the total optical confinement factor is $1.4e^{-4}$.

It is worth noticing that since: $\Gamma_{xy} \propto N_D$ by increasing the number of dots (or more precisely by increasing the area coverage of QD, ξ), an increase in the total optical confinement factor. In a similar way since G_z is proportional to the thickness of the active layer, by increasing the number of QD layers in the active layer (increasing the number of active layers) an increase in the vertical optical confinement factor is achieved. Both these cases are illustrated in the figures below and both contribute to the total optical confinement factor.

4.3.4 Gain and Threshold

In order to achieve lasing need of population inversion and stimulated emission together with some sort of feedback provided usually by reflection by mirrors. The population inversion is obtained by electrically pumping the system with carriers (holes and electrons). In a simple twofold degenerate energy level system, this is achieved when enough current is pumped into the system in order to invert the ground-state population level. That is on average there is more than one electron–hole residing in the QD conduction-band state and more than one hole residing in the QD valence-band state. At a certain threshold current density (j_{th}), the lasing starts and by increasing the current above that threshold we increase the output power linearly with increasing current. The condition for lasing can be written as

$$g_{mod}(j_{th}) = \Gamma g_{mod}(j_{th}) = \alpha_{tot} \tag{4.26}$$

where g_{mod} is the modal gain of the system and α_{tot} is the total loss in the system consisting of internal losses in the active layer (α_i), losses in the waveguide (α_c) and losses at the reflectors ($a_{mirrors}$).

$$\alpha_{\text{tot}} = \Gamma\alpha_i + (1+\Gamma)\alpha_c + \frac{1}{2L}\ln\left(\frac{1}{R_1 R_2}\right) \tag{4.27}$$

So modal gain is given by:

$$g_{\text{mod}}(j) = \Gamma g_{\text{material}}(j) - \alpha_{\text{tot}} \tag{4.28}$$

Assuming that the we have a Gaussian distribution in QD volume size (P_g), and that spectral width of this distribution is much bigger than the width of the Lorentzian describing the intra-band relaxation times, the material gain g_{mat} of a QD ensemble can be calculated using the following formula:

$$g_{\text{mat}}(E) = C_g P(E)[f_c(E, E_{Fc}) - f_v(E, E_{Fv})] \tag{4.29}$$

where C is a constant and P_g is the Gaussian distribution and f is the Fermi distribution of the carriers. If one further assumes that the all the injected carriers are captured by QD and that overall charge neutrality exists (i.e., $f_v = 1 - f_c$ and $f_c - f_v = 2f_c - 1$), the total number (N) of carriers (electrons and holes) is:

$$N = 2f_c N_D \tag{4.30}$$

and these carriers are assumed to be equally distributed amongst all the QD. So end up with a material gain:

$$g_{\text{mat}}(E) = \frac{N - N_D}{N_D} C_g P(E) \tag{4.31}$$

If take a closer look at this expression for the material gain that it increases linearly with increasing number of carriers, N, from $-C_g P(E)$ (when the QD excited state has no carriers) to $C_g P(E)$ (when all QD excited states are filled with two electrons, $N^{\text{max}} = 2N_D$) as Fig. 4.32.

Fig. 4.32 QD excited states are filled with 2 electrons

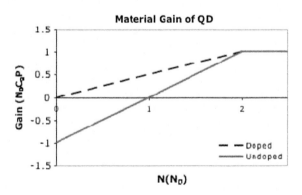

Fig. 4.33 Calculated material gain spectra (height of the QD peak and its width) for $In_{0.53}Ga_{0.47}As/InP$ quantum box, wire, well and bulk at $T = 300$ K. Electron density at 3×10^{18} cm^{-3}

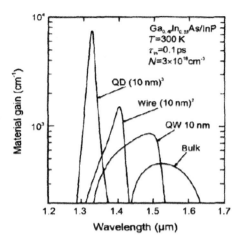

For typical values of C and $P(E)$ we obtain a maximum saturation value in the material gain in the order of $1e^5$ cm^{-1}, which is huge. However, due to the optical confinement factor one ends up with a much smaller modal gain ($g_{mod} = \Gamma g_{mat}$).

The shape of the gain function depends on the Gaussian distribution function P (E). This means that the shape of the gain function depends strongly on the uniformity of the QD volume size and shape. The more homogenous the system is, the sharper and higher the gain function is shown Fig. 4.33.

For $7 \times 7 \times 2$ nm^3 QD with $\zeta = 0.02$ (corresponding to 4×10^{-10} QD/cm^2), $\Gamma_z = 7 \times 10^{-3}$ for 150 nm waveguide and saturation material gain $g_{mat} = 1e^5$, can only get:

$$g_{mod}^{sat} = \Gamma g_{mat}^{sat} = \Gamma_{xy}\Gamma_z g_{mat}^{sat} = \zeta\Gamma_z g_{mat}^{sat} = 14 \text{ cm}^{-1}. \qquad (4.32)$$

In order to increase the gain saturation limit of given QD ensemble, the following steps could be used:

(1) Increase the modal gain by stacking layers of QD within the active layer.
(2) Increase the number of QD (N_D) in each sheet of QD in the active layer.
(3) Decrease the mirror loss by high reflection coatings or many mirrors. In VCSEL usually have small cavity length L. According to the expression for losses in mirrors this would mean larger loss the smaller the cavity. This could be compensated by having R_1 and R_2 high, either by using high reflective coatings or by using many sets of mirrors (20 mirrors in VCSEL give about 99 % reflection).

In order to have lasing, the carrier density at threshold, N_{th}, has to be at least:

$$N_{th} = N_D\left(1 + \frac{1}{\zeta}\frac{\sqrt{2\pi}\sigma_E\alpha_{tot}}{\Gamma_z C_g}\right) \qquad (4.33)$$

The maximum threshold current density is:

$$N_{\text{th}}^{\max} = 2N_D \tag{4.34}$$

So the minimum area dot coverage has to be:

$$\xi_{\min} = \frac{\sqrt{2\pi}\sigma_E \alpha_{\text{tot}}}{\Gamma_z C_g} \tag{4.35}$$

The relationship between carrier density N, and injection current j, is nontrivial but a simplified version, obtained using conventional rate equation models, gives:

$$j = \frac{5}{8}\frac{e}{A_D \tau_D}\xi\left(\frac{N}{N_D}\right)^2 \tag{4.36}$$

Using the above relation for threshold density of states, N_{th}, can obtain the following relation for the threshold current j_{th}:

$$j_{\text{th}} = \frac{5}{8}\frac{e}{A_D \tau_D}\xi\left(1 + \frac{\xi_{\min}}{\xi}\right)^2 \tag{4.37}$$

The curve of threshold current is shown in Fig. 4.29a. For $\xi < \xi_{\min}$ the current density goes to infinity and cannot get any lasing. For a_{tot} equal to 10 cm^{-1} and $\xi_{\min}$ equal to 0.013 can get a threshold current of 10 A/cm^2.

By increasing the number of active layers can obtain a decrease in threshold current as:

1. $J_{\text{th}} = 90$ A/cm^2 for 10 layers of In$_{0.5}$Ga$_{0.5}$As/GaAs.
2. $J_{\text{th}} = 62$ A/cm^2 for three layers of In$_{0.5}$Ga$_{0.5}$As/Al$_{0.15}$Ga$_{0.75}$As.
3. $J_{\text{th}} = 40$ A/cm^2 for three layers of InAs/GaAs.

The curve of threshold current j_{th} and number of layers is shown in Fig. 4.34b.

The relation between temperature and light-current characteristics is shown in Fig. 4.35. The modulation waveform at 10 Bbps at 20 and 70 °C with no current characteristics is shown Fig. 4.36.

The strained quantum well laser is a type of quantum-well laser, which was invented by Professor Alf Adams at the University of Surrey in 1986. The laser is distinctive for producing a more concentrated beam than other quantum well lasers, making it considerably more efficient. But the lasers are notable for usage in CD, DVD, and Blu-Ray drives. The modulation waveform of quantum dot (QD) laser and strained quantum well laser is shown Fig. 4.36, which is 10 Bbps at 20 and 70 °C with no current characteristics.

The characteristics of QD and QW laser are shown in Fig. 4.37: (a) relationship between output power and pump current, (b) relationship between output wavelength and temperature.

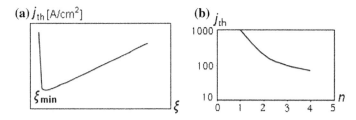

Fig. 4.34 **a** Threshold current density as a function of dot area coverage. **b** The curve of threshold current j_{th} and number of layers n

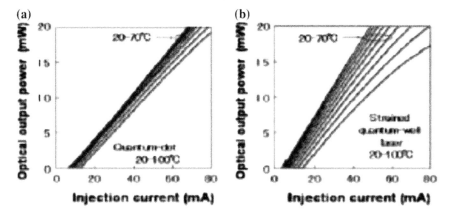

Fig. 4.35 Temperature dependence of light-current characteristics of **a** quantum dot (QD) laser and **b** strained quantum well (QW) laser

Fig. 4.36 Modulation waveform at 10 Bbps at 20 and 70 °C with no current characteristics

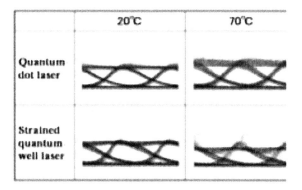

Optically injected quantum dot lasers: impact of nonlinear carrier lifetimes on frequency-locking dynamics carrier scattering is known to crucially affect the dynamics of quantum dot (QD) laser devices. The dynamic properties of a QD laser under optical injection are also affected by Coulomb scattering processes and can be

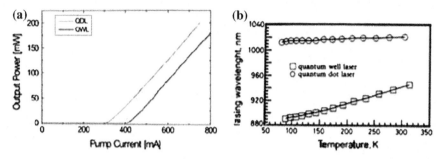

Fig. 4.37 Comparison between QD and QW laser: **a** curves of output power and pump current, **b** curves of output wavelength and temperature

optimized by band structure engineering. The nonlinear dynamics of optically injected QD lasers is numerically analyzed as a function of microscopically calculated scattering lifetimes. These lifetimes alter the turnon damping of the solitary QD laser as well as the complex bifurcation scenarios of the laser under optical injection. Furthermore, a pump current sensitivity of the frequency-locking range is directly related to the nonlinearity of the carrier lifetimes.

Self-organized semiconductor quantum dot (QD) structures have turned out to be very promising for applications in semiconductor lasers. An important feature of QD structures is the discrete energy levels that form the laser transition leading to unique properties such as very low threshold current, low linewidth enhancement factor and large temperature stability.

Further, these lasers show highly damped turn-on dynamics and, related to that, a low sensitivity to optical feedback. For the case of optical injection, where a laser is unidirectionally coupled to an injecting master laser, QD lasers show smaller chaotic regions and less complicated trajectories compared to other laser devices. The general features of optically injected semiconductor lasers have already been studied intensively. Based on experimental studies on QD lasers, simple rate equation approaches have been used to model their optical response under optical injection and to investigate the stability regions and bifurcation scenarios. Further models based on the standard rate equation approach that take into account the gain material and excited state filling typical of QD-based devices by an effective gain compression parameter have been proposed. They have been proven to successfully model the modulation response of quantum-dash lasers subject to optical injection. In these models, the gain compression parameter is obtained by fitting the nonlinear dependence of the relaxation oscillation frequency on the output intensity, which permits the modeling of nonconstant α-factors that depend nonlinearly on the pumpcurrent. The aim of this paragraph is to focus on the distinguishing properties of QD lasers and to understand the interplay between epitaxial structure and optical properties by linking the dynamics of the QD laser under injection directly to the carrier exchange dynamics between the QD and QW states without the need for fit parameters.

Extend microscopically based QD laser rate equation model, which treats electrons and holes separately, to the investigation of QD laser dynamics under optical injection and include the dynamics of the phase of the laser light. When subject to an injected field, the output frequency of the QD laser can be affected, and simulations shows pulsating and chaotic lasing behavior, connected by a variety of bifurcations. Simulations agree well with experimental data for QD and quantum-dash lasers. Furthermore, compare the dynamics of different QD lasers under optical injection by implementing different energetic band structures into the calculation of the microscopic carrier–carrier scattering rates. The influence of the scattering rates on the turn-on behavior and the relaxation oscillations of a solitary QD laser have been studied in. In this section, focus on their impact on the locking behavior and show how band structure engineering can lead to optimized operation conditions of optically injected QD lasers by shifting critical bifurcation points as a function of the carrier lifetimes. Our results obtained by direct numerical integration are supported by path continuation techniques. Further, prediction the current dependence of the locking region can be explained by the nonconstant, carrier density-dependent carrier lifetimes.

4.4 Quantum Dot Laser Model

The basic structure of the quantum dot (QD)s embedded in a quantum well (QW) laser structure is shown in Fig. 4.38, where $\hbar\upsilon$ labels the ground-state (GS) lasing energy. ΔE_e and ΔE_h mark the distance of the GS from the QW band edge for electrons and holes, respectively. Δ_e and Δ_h denote the distance to the bottom of the QD.

The crucial parameters are the confinement energies ΔE_e and ΔE_h that mark the energy differences between the QD ground-state (GS) and the band edge of the surrounding QW for electrons and holes, respectively. Our QD laser model is based on the model described in, which has shown good quantitative agreement with

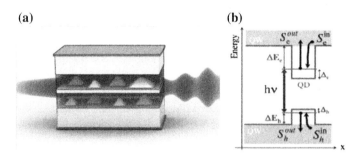

Fig. 4.38 **a** Scheme of the optically injected QD laser. QDs (*light pyramids*) are surrounded by a QW. **b** Energy diagram of the band structure across a QD

experiments regarding the turn-on behavior and the modulation response of QD lasers. To extend this model very similarly to the one for optical feedback in, but include optical injection instead of optical feedback. For the optical injection is consistent with approaches. Eliminating the laser's steady-state frequency v_L (which refers to an optical wavelength of $\lambda = 1.3$ μm), the field equation for the normalized slowly varying amplitude ε of the electric field E is:

$$E = \frac{1}{2}\left(\varepsilon e^{i2\pi v_L t} + \text{c.c.}\right) \qquad (4.38)$$

where ε is given by:

$$\dot{\varepsilon}(t) = \frac{(1+i\alpha)}{2}[g(\rho_e + \rho_h - 1) - 2\kappa]\varepsilon(t) + |\varepsilon_{\text{inj}}(t)|e^{i2\pi\Delta v_{\text{inj}}t} + F\varepsilon(t). \qquad (4.39)$$

Here $2k$ are the optical intensity losses and $g = 2\,g = 2\bar{W}Z_a^{QD}\,a$ is the linear gain coefficient for the processes of induced emission and absorption. The linewidth enhancement factor α is chosen to be constant although it is noted that it may vary slightly with the pump current. The gain coefficient is proportional firstly to the Einstein coefficient of induced emission $\overline{W}$ that measures the coherent interaction between the two-level system and the laser mode, and secondly to the number Z_a^{QD} of lasing QDs inside the waveguide (the factor 2 is due to spin degeneracy). The number of lasing QDs Z_a^{QD} is given by $Z_a^{QD} = a_L A N_a^{QD}$ where a_L is the number of self-organized QD layers, A is the in-plane area of the QW and N_a^{QD} is the density per unit area of the active QDs. As a result of the size distribution and material composition fluctuations of the QDs, the gain spectrum is in homogeneously broadened, and only a subgroup (density N_a^{QD}) of all QDs (N^{QD}) matches the mode energies for lasing. The ρ_e and ρ_h denote the electron and hole occupation probabilities in the confined QD levels, respectively, and E_{inj} is the normalized effective injected field and is given by:

$$\varepsilon_{\text{inj}}(t) = \sqrt{T_{\text{inj}}n_{\text{inj}}(t)A/(\tau_{\text{in}})}e^{i2\pi v_{\text{inj}}t} \qquad (4.40)$$

Using the approximation that the actual laser frequency v and the injected light frequency v_{inj} are very close, i.e., $v_{\text{inj}}/v \approx 1$. T_{inj} is the transmission coefficient of the cavity mirror, $T_{\text{inj}}n_{\text{inj}}$ is the injected photon density in the active region of the QD laser per cavity round trip time and $\Delta v_{\text{inj}} = v_{\text{inj}} - v_L$ is the input detuning. The complex Gaussian white noise term $F\varepsilon(t)$ models spontaneous emission. The time τ_{in} for one round trip of the light in the cavity of length is given by $\tau_{\text{in}} = 2L\sqrt{\varepsilon_{\text{bg}}}/c$ with background permittivity ε_{bg}. In order to obtain rate equations the complex stochastic differential equation (see Eq. 4.39) for the electric field, i.e., given by $\varepsilon = \sqrt{n_{\text{ph}}A}\exp(i\Phi)$, is transformed by an Ito transformation into two real stochastic differential equations for the photon density n_{ph} and the phase Φ. Since stochastic effects are beyond the scope of this paper, the noise terms are neglected in the following and only the deterministic spontaneous emission rate R_{sp} remains in the rate equations (see Eqs. 4.41–4.44). The final nonlinear, coupled and six-variable rate equation system including also the electron and hole occupation probabilities in the QDs, ρ_e and ρ_h, and the electron and hole densities in the QW respectively, is given by the following equations:

$$\dot{n}_{ph} = n_{ph}\left[2\overline{W}Z_a^{QD}(\rho_c + \rho_h - 1) - 2\kappa\right] + \frac{\beta}{A}2Z_a^{QD}R_{sp}(\rho_c, \rho_h) + \frac{2K}{\tau_{in}}\sqrt{n_{ph}n_{ph}^0}\cos\left(\Phi - 2\pi\Delta\nu_{inj}t\right),$$
(4.41)

$$\dot{\Phi} = \frac{\alpha}{2}\left[2\overline{W}Z_a^{QD}(\rho_c + \rho_h - 1) - 2\kappa\right] + \frac{K}{\tau_{in}}\sqrt{\frac{n_{ph}^0}{n_{ph}}}\sin\left(\Phi - 2\pi\Delta\nu_{inj}t\right),$$
(4.42)

$$\dot{\rho}_c = -\overline{W}A(\rho_c + \rho_h - 1)n_{ph} - R_{sp}(\rho_c, \rho_h) + S_c^{in}(w_c, w_h)(1 - \rho_c) - S_c^{out}(w_c, w_h)\rho_c,$$
(4.43)

$$\dot{\rho}_h = -\overline{W}A(\rho_c + \rho_h - 1)n_{ph} - R_{sp}(\rho_c, \rho_h) + S_h^{in}(w_c, w_h)(1 - \rho_h) - S_h^{out}(w_c, w_h)\rho_h,$$
(4.44)

$$\dot{w}_e = \frac{j}{e_0} - 2N^{QD}\left[S_e^{in}(w_e, w_h)(1 - \rho_e) - S_e^{out}(w_e, w_h)\rho_c\right] - \tilde{R}_{sp},$$
(4.45)

$$\dot{w}_h = \frac{j}{e_0} - 2N^{QD}\left[S_h^{in}(w_e, w_h)(1 - \rho_h) - S_h^{out}(w_e, w_h)\rho_h\right] - \tilde{R}_{sp}.$$
(4.46)

Here n_{ph}^0 denotes the steady-state photon density without injection ($K = 0$). It is introduced to normalize the injected photon density. This normalization is explained by the definition of the injection strength K:

$$K = \sqrt{\frac{T_{inj}n_{inj}}{n_{ph}^0}}.$$
(4.47)

As in Eq. (4.39) $\overline{W}$ is the Einstein coefficient for the coherent interaction and Z_a^{QD} is the number of active QDs inside the waveguide. The spontaneous emission from one QD is taken into account by $R_{sp}(\rho_e, \rho_h) = W\rho_e\rho_h$, where W is the Einstein coefficient for spontaneous emission resulting from the incoherent interaction of the QD with all resonator modes. Please note that the coefficients $\overline{W}$ and W differ by three orders of magnitude. β is the spontaneous emission factor, measuring the probability that a spontaneously emitted photon is emitted into the lasing mode. The in- and out-scattering rates for electrons and holes between QD and QW, S_e^{in}, S_e^{out} and S_h^{in}, S_h^{out} as depicted in Fig. 4.38, are calculated microscopically from Coulomb interaction and are connected by the detailed balance relation derived in:

$$S_e^{\text{out}}(w_e, w_h) = S_e^{\text{in}}(w_e, w_h) e^{-\frac{\Delta E_e}{kT}} \left[e^{\frac{w_e}{D_e kT}} - 1 \right]^{-1}, \tag{4.48}$$

$$S_h^{\text{out}}(w_e, w_h) = S_h^{\text{in}}(w_e, w_h) e^{-\frac{\Delta E_h}{kT}} \left[e^{\frac{w_h}{D_h kT}} - 1 \right]^{-1}. \tag{4.49}$$

Here, $\Delta E_e = E_e^{\text{QW}} - E_e^{\text{QD}}$ and $\Delta E_h = E_h^{\text{QD}} - E_h^{\text{QW}}$ are the energy differences between the QD levels E_e^{QD} and E_h^{QD} and the band edges of the QW E_e^{QW} and E_h^{QW} for electrons and holes, respectively. The carrier degeneracy concentrations are given by $D_{e/h} kT$, where $D_{e/h} = m_{e/h}/(\pi \hbar^2)$ are the 2D densities of state in the QW with the effective masses $m_{e/h}$ (see also Fig. 4.38). T is the temperature and k is the Boltzmann constant. The current density j is injected into the QW and is normalized by the elementary charge e_0. N^{QD} is the total QD density given by surface imaging techniques, which includes all nonlasing QDs that are present in the layer due to inhomogeneous broadening. The factor two in Eqs. (4.45) and (4.46) accounts for spin degeneracy of the QD levels. The spontaneous emission in the QW is incorporated by $\tilde{R}_{\text{sp}}(w_e, w_h) = B^S w_e w_h$, where B^S is the band–band recombination coefficient.

The carrier lifetimes τ_e and τ_h that result from Coulomb scattering between QDs and QW are defined by the nonlinear scattering rates as $\tau_e = 1/(S_e^{\text{in}} + S_e^{\text{out}})$ and $\tau_h = 1/(S_h^{\text{in}} + S_h^{\text{out}})$. It is crucial to note that these lifetimes are not constant but depend on the carrier densities in the surrounding QW and thus on the injected pump current. Their pump-dependent steady state values are shown in Fig. 4.39 for three different QD structures. Nonlinear steady-state carrier lifetimes τ_e and τ_h for electrons and holes, respectively, resulting from the scattering rates calculated microscopically for three different band structures of the QD–QW system (see Table 4.7). These simulations are carried out without optical injection ($K = 0$).

The different structures are modeled by using three different sets of confinement energies between QD and QW (as in Table 4.7 for details of the parameters). By

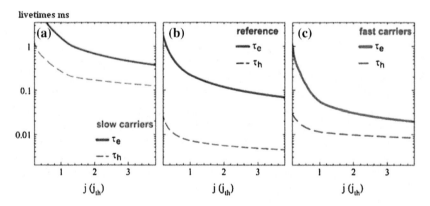

Fig. 4.39 The pump-dependent steady state values of three different QD structures: (**a**) slow carriers, (**b**) reference carriers, (**c**) fast carriers

Table 4.7 Numerical parameters of three different band structures of the QD–QW system are used in the simulation

Symbol	Value	Symbol	Value	Symbol	Value
W	0.7 ns^{-1}	A	4×10^{-5} cm^2	T	300 K
$\overline{W}$	0.11 μs^{-1}	N_a^{QD}	0.3×10^{10} cm^{-2}	L	1 mm
2κ	0.1 ps^{-1}	N^{QD}	1×10^{11} cm^{-2}	ε_{bg}	14.2
β	2.2×10^{-3}	B^s	540 ns^{-1} nm^2	τ_{in}	24 ps
a_L	15	j_{th}	$4.2 \times e_0 \times 10^{24}$ cm^{-2} s^{-1}	j	$3.5\, j_{th}$
Z_a^{QD}	1.8×10^6	m_c (m_h)	$0.043\, m_0$ $(0.45\, m_0)$	λ	1.3 μm
g	3.9×10^{11} s^{-1}	ΔE_e (ΔE_h)	210 meV (50 meV)	ν_L	230 THz

controlling the growth mode during epitaxial it is possible to create QDs with different sizes and compositions.

Figure 4.39b results from square-based pyramidal QDs with a base length of 18 nm and a ratio between the effective masses of holes and electrons of 10. Differences in the effective masses (e.g., obtained by changing the QD composition) will show up in a different ratio between electron and hole confinement energies. Figure 4.39a depicts the case of large confinement energies that are similar for electrons and holes, i.e. $\Delta E_e \approx \Delta E_h$, resulting in long Auger scattering lifetimes. Increasing the size of the dots leads to smaller confinement energies (shallow dot with smaller energetic distance between the GS and the QW transition) and thus to fast scattering rates (Fig. 4.39c). By comparing the injection properties of the QD laser for these three different cases, can show the crucial impact of the growth mode and thus the QD size on the laser dynamics. The calculation of the scattering rates has been described thoroughly and the values implemented for the simulations can be found. All other numerical parameters that were used in the simulation are shown in Table 4.7.

The reference lifetimes (as Fig. 4.39b) are implemented, the dynamics of the three different structures are compared.

4.4.1   One-Parameter Bifurcation Diagrams

In this paragraph, the dynamics of the optically injected QD laser modeled by Eqs. (4.41)–(4.46) is discussed. Outside a certain detuning range around $\Delta \nu_{inj} = 0$, the so-called locking tongue, the QD laser under optical injection shows complex oscillatory behavior with increasing injection strength K. To illustrate the dynamics simulated time series of the photon density for different input detuning frequencies. As an example of the complex dynamics is shown Fig. 4.39a depicts the results for constant injection strength $K = 0.39$, where sustained multipulse emission is found at $\Delta \nu_{inj} = -4.1$ GHz (solid line) and regular pulsations are seen at $\Delta \nu_{inj} = -7$ GHz

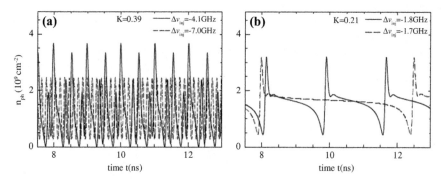

Fig. 4.40 Time series of the photon density n_{ph} for (**a**) $K = 0.39$ and detuning of $\Delta_{inj} = -4.1$ GHz (*solid*) and $\Delta_{inj} = -7$ GHz (*dashed*) and (**b**) $K = 0.21$ with detuning $\Delta_{inj} = -1.8$ GHz (*solid*) and $\Delta_{inj} = -1.7$ GHz (*dashed*)

(dashed line). On the other hand, Fig. 4.39b depicts the results for $K = 0.21$ for negative detuning $\Delta v_{inj} = -1.7$ GHz and $\Delta v_{inj} = -1.8$ GHz showing multipulses with different pulse shapes and a pulse repetition frequency that is sensitive to the detuning (compare the solid and dashed lines in Fig. 4.39b). The injection started at $t = 5$ ns. Parameters as in Table 4.1 with $\alpha = 0.9$ and reference scattering rates of Fig. 4.40b.

These SN bifurcation points are characterized by an abrupt change from cw operation to periodic pulsating light output. In fact, a saddle and a stable node collide on a cycle, disappear and leave a stable limit cycle (periodic pulsing). The period of the oscillations goes to infinity at the bifurcation point and follows an inverse square root law while the amplitude is constant. This global bifurcation is called SN. SN on an invariant cycle bifurcation and has already been studied in QW lasers under optical injection, as well as in QW lasers with saturable absorbers and in the framework of delay equations. The time series of the pulsating photon density observed behind the SN point are plotted in Fig. 4.39b. The small local maximum found in between the spikes is caused by oscillations around the reminiscent ghost of the formally stable fixed point that changes the shape of the limit cycle. In the bifurcation diagrams, e.g., Figure 4.40b, this ghost causes additional lines behind the SN point that suddenly disappear beyond a certain detuning. In this region close to the SN line the system is excitable and noise-induced dynamics, such as irregular spiking, can be observed.

By plotting the extremal points of the time series shown in Fig. 4.3 as a function of the input detuning Δv_{inj}, as was done in Fig. 4.40a–d for $\alpha = 0.9$, one-parameter bifurcation diagrams are formed. In these diagrams a single point indicates continuous-wave (cw) operation, while two points (maxima and minima) indicate periodic pulsing, marked by light shading. In Fig. 4.40a–d, bifurcation diagrams of the photon density are shown as a function of the detuning for different injection strengths. For all these figures, cw operation is found for zero detuning. In

Fig. 4.40a, b, saddle-node (SN) bifurcations can be found at the edges of the range of cw operation (locking range marked with dashed vertical lines in Fig. 4.40).

For higher injection strengths K the situation changes and the bifurcation diagrams plotted in Fig. 4.40c, d show supercritical Hopf bifurcations at the positive side of the locking range. This means that at these points a stable limit cycle is born. In contrast to the SN point the period of the oscillations is constant, while the amplitude is detuning dependent and goes to zero at the Hopf point. For $K = 0.39$ this Hopf bifurcation occurs at $\Delta v_{\text{inj}} = 2.94$ GHz (see Fig. 4.40c) and for $K = 0.52$ it occurs at $\Delta v_{\text{inj}} = 3.56$ GHz (as Fig. 4.40d).

SN indicates saddle-node (SN) bifurcations. The blue dashed lines indicate the input detuning region where cw operation is found, the light shading marks periodic pulsing. Parameters as in Table 4.7, $\alpha = 0.9$ and reference scattering rates of Fig. 4.39b. The $v_{\text{inj}} - v$ versus Δv_{inj} for increasing K, n_{ph} versus Δv_{inj} for increasing K and n_{ph} versus K for increasing Δv_{inj} respectively.

With increasing positive detuning the limit cycle that is born in the Hopf bifurcation increases in size until it touches the point of zero intensity where the locking behavior changes (discussed below). Behind this point the amplitude of the limit cycle starts decreasing.

A period doubling (PD) bifurcation that leads to period-2 oscillations can be observed for $K = 0.39$ at $\Delta v_{\text{inj}} = -5.98$ GHz and for $K = 0.52$ at $\Delta v_{\text{inj}} = -6.57$ GHz (while decreasing Δv_{inj}) as can be seen in Fig. 4.41c, d. Another PD bifurcation to a period-4 orbit can be seen for $K = 0.39$ at $\Delta v_{\text{inj}} = -4.38$ GHz. This leads to complex multipulses similar to those shown in Fig. 4.40a. Note also that with increasing injection strength (Fig. 4.41a–d) the amplitude of the oscillations increases as well. For a complete overview of the bifurcation diagrams as a function of K and $1m_{\text{inj}}$. Possibly chaotic emission behavior can be seen for $K = 0.52$ around $\Delta v_{\text{inj}} = -4$ GHz and $\Delta v_{\text{inj}} = -5$ GHz (Fig. 4.41d). Another important feature of an optically injected laser is the frequency locking, meaning that the output frequency v of the laser can be entrained to the frequency v_{inj} of the injected laser ($v - v_{\text{inj}} = 0$). Here $v_{\text{inj}} - \bar{v}$ is called output detuning. In order to extract the output detuning frequency, the momentary frequency $v(t) = \dot{\Phi}(t)\pi/2$ is defined, which is constant for the stable locking case. Outside the stable locking regime an average output frequency $\bar{v} = \langle \dot{\Phi} \rangle_t \pi/2$ is defined, which is determined by averaging over a long-time series. The point of interest here is the following: for which input detuning $\Delta v_{\text{inj}} = \Delta v_{\text{inj}} - v_L$ and which injection strengths K is the output frequency of the laser locked. Figure 4.41e–h answer this question by plotting the output detuning $\Delta v_{\text{inj}} - \bar{v}$ as a function of the input detuning Δv_{inj}. By comparing these figures with the bifurcation diagrams in Fig. 4.41a–d, it is obvious that the laser is frequency locked for the range where it shows cw operation, while it is unlocked behind the SN bifurcation.

This type of locking is usually referred to as Adler's locking. However, immediately beyond the Hopf bifurcation point the mean frequency of the emission on the limit cycle is still locked (see Fig. 4.41h for $\Delta v_{\text{inj}} \approx 4$ GHz) and the phase, although

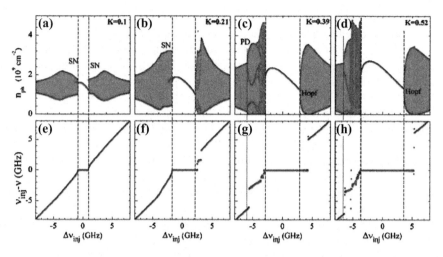

Fig. 4.41 **a–d** One-parameter bifurcation diagrams showing the maxima and minima of the photon density n_{ph} as a function of the input detuning Δv_{inj}. **e–h** Output detuning $\Delta v_{inj} - \bar{v}$ in terms of input detuning $\Delta v_{inj} = v_{inj} - v_L$. The injection strengths are (**a, e**) $K = 0.1$, (**b, f**) $K = 0.21$, (**c, g**) $K = 0.39$ and (**d, h**) $K = 0.52$

oscillating, is still bounded. Recently, this phenomenon was discussed for two coupled oscillators in. The type of locking transition, found at the Hopf bifurcation, is also called undamping of relaxation oscillations. The transition to an unbounded phase happens where the minima of the photon density reach zero while the output detuning performs a sudden jump (see Fig. 4.41c, g for $\Delta v_{inj} = 4.5$ GHz). An explanation can be given by looking at the complex E-field.

Before the minima of the photon density reach zero, the projection of the E-field to the complex plane is a cycle which does not include or touch the origin (thus the phase is bounded). When the minima of the photon density reach zero the projection of the E-field crosses the origin and with increasing positive input detuning surrounds the origin. This leads to a constantly increasing phase and thus to an emission at a different frequency. Details of that phenomenon have been discussed. Several bifurcation diagrams of local extrema of the photon density as a function of the input detuning as well as diagrams of output detuning over input detuning are concatenated to a film [6, 7].

4.4.2   Two-Parameter Bifurcation Diagrams and Path Continuation

The parameter dependence of the dynamics of the QD laser with optical injection can be visualized in two-parameter bifurcation diagrams. The results obtained so far by direct integration are summarized as contour plots in Fig. 4.42a, b, for $\alpha = 0.9$ and $\alpha = 3.2$, respectively.

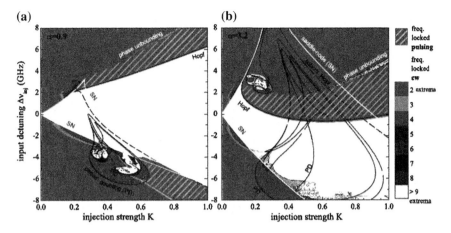

Fig. 4.42 Two-parameter bifurcation diagram for **a** $\alpha = 0.9$ and **b** $\alpha = 3.2$. Parameters as in Table 4.1 and reference scattering rates of Figs. 4.34 and 4.35b

The number of extrema found in the times series is used as a color code (for periodic waveforms, this corresponds to the number of local extrema per period). The region of frequency-locked cw operation appears yellow. For increasing injection strength K this locking range grows. The yellow–orange hatched area indicates the mean frequency locking where the photon density oscillates on a limit cycle with bounded phase. The transition to an unbounded phase (indicated by 'phase unbounding' in Fig. 4.42) found at the upper edge of the hatched region (where $n_{ph} \approx 0$) is not a bifurcation, but it appears as a numerical artifact due to division by zero. The orange region corresponds to periodic pulsations that show one maximum and one minimum, while more complex time series are found for parameters chosen within the dark-colored and the white regions in Fig. 4.42. In order to obtain a clear picture of the bifurcations, have to perform path continuation with the software tool matcont. The results are plotted as solid and dashed lines superimposed upon the data from direct numerical integration of the dynamical equations in Fig. 4.42. SN, PD and Hopf bifurcation lines are indicated by white, blue, and black lines, respectively. First, the dynamics for a small α-factor, i.e., for a typical QD laser, as depicted in Fig. 4.42a, is discussed. It can be seen that the cusp of the hatched area is a codimension two SN–Hopf-point where a Hopf and a SN line intersect. At this point the Hopf bifurcation undergoes a change from subcritical in the lower part (dashed line) to supercritical in the upperpart (solid line) and becomes thus detectable in direct numerical integrations. Furthermore, the path continuation gives proof that the elliptically shaped red region (four extrema per period) on the negative detuning side is bordered by a PD bifurcation. In this region, distinguish PD lines, where a formerly stable limit cycle undergoes a PD bifurcation (solid line), from those where an unstable saddle-cycle with one Floquet multiplier larger than unity undergoes a PD bifurcation (thin-dashed lines) as shown in Fig. 4.42. The color code marks the number of extremal values, i.e., the number of maxima and minima of the photon density n_{ph}. The light white area marks the average frequency locking with steady-state photon density, i.e., cw-lasing. The white–gray hatched area shows oscillating photon density (with one minimum and one maximum) but with a locked average frequency. Blue, black and white solid lines indicate the PD, Hopf and SN bifurcation lines, respectively. PD lines of saddle-cycles having two or more Eloquent multiplies larger than unity are not plotted. For positive detuning only PD lines involving stable limit cycles are plotted in Fig. 4.42a. That can find two large PD loops, one for positive detuning and one for negative detuning. For the lower PD loops also secondary PD bifurcations giving birth to period-4 limit cycles are plotted (solid and dashed blue lines). Stable period-4 limit cycles as detected by direct numerical integration are marked by dark violet areas (eight extrema perperiod). White regions inside the PD loops indicate more complex dynamics, i.e., more than eight extrema. Within the large lower PD loop, that find multistability between period-2 and different period-4 orbits. A description of the full bifurcation scenario is beyond the scope of this paragraph. Here, plot only PD bifurcations that enclose the main regions of more complex dynamics detected by direct numerical integrations.

The narrow sickle-shaped darker regions found for small K close to the SN lines for positive and negative detuning (regions with three (gray) and four (red) extrema) indicate the appearance of oscillations around the ghost of the formally stable fixed point. In agreement with the path continuation these are not PD bifurcations but period-1 oscillations. Only the shape of the limit cycle changes, resulting in additional extrema (see Fig. 4.40b for a possible time series). As noted before, the system is excitable near these regions.

Figure 4.42a depicts the situation for a small α-factor as commonly predicted for QD lasers. In Fig. 4.42b a larger α-factor is studied, which is more typical of QW lasers. As expected the locking range shrinks as the Hopf bifurcation line bends toward zero input detuning. Thus, pulsating behavior is found also for $\Delta v_{inj} = 0$ at high injection strength. Close to the codimension-2 SN–Hopf-point (cusp of the hatched area), the laser shows complex bifurcations leading to chaotic emission. However, the complex dynamics found for QW devices within the homoclinic teeth on the negative detuning side of the locking range is absent. Nevertheless, large regions of multistability between the simple cw solution and solutions with periodically modulated intensity, i.e., limit cycles, are found within the yellow area of frequency locking.

The data shown in Fig. 4.42 were obtained by up-sweeping the value of the injection strength K and thus show parts of the stable limit cycle operation inside the yellow area of cw operation. Down-sweep instead reveals only stable cw operation up to the SN line for negative detunings. In the experiment, up-sweeping and down-sweeping indicate the direction in which the injection strength is changed. For example, starting with zero injection and increasing it slowly can lead to different dynamic scenarios from starting with high injection strength and decreasing it slowly. In the simulations this is done by choosing the last value of the time series as initial conditions for the next simulation. Qualitatively, the results agree well with results for QD lasers that are modeled with equal and constant lifetimes for electrons and holes. However, in our model the lifetimes are not fitted parameters but nonlinear functions of the carrier density in the reservoir that are given by the band structure. Thus changes concerning the operation point of the device can be studied without re-adjusting parameters. The next section will explore the impact of different pump currents on the dynamics of the QD laser with injection [8].

4.5   Impact of the Pump Current on the Dynamics

The pump current affects many characteristics of the QD laser, e.g., with increasing pump current the photon density and the electron and hole densities in the QD and QW increase. This in turn influences the scattering rates and therefore the turn-on behavior of the QD laser. The analytic relation between the damping of the turn-on dynamics and the value of the scattering lifetimes is derived in. The section focuses on the impact of the pump current with the frequency locking of the optically injected QD laser.

Figure 4.43 shows the two-parameter bifurcation diagrams for current densities $j = 2.1\, j_{th}$ and $j = 4.9\, j_{th}$. As in Fig. 4.42, where $j = 3.5\, j_{th}$, the range of frequency locking at cw operation is marked by the yellow area, while the number of maxima observed in the time series is color coded. The figure shows that the codimension-2 SN–Hopf-point, at which the subcritical Hopf bifurcation changes to a supercritical one, shifts toward higher injection strengths K and higher input detunings Δ_{inj} with increasing pump current. For $j = 2.1\, j_{th}$ (as Fig. 4.43a) the Hopf line starts at $K = 0.14$ (for $j = 3.5\, j_{th}$ at $K = 0.22$) and for a pump current of $j = 4.9\, j_{th}$ the Hopf line starts at $K = 0.27$ (as Fig. 4.43b). Along with the increasing pump current, the

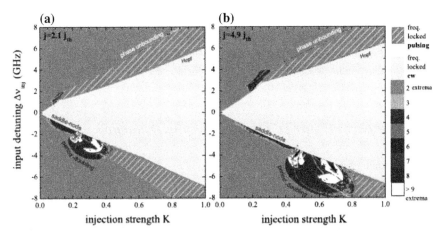

Fig. 4.43 Two-parameter bifurcation diagram for $\alpha = 0.9$ for injected current densities of (a) $j = 2.1\, j_{th}$ and (b) $j = 4.9\, j_{th}$

Table 4.8 Energy differences ΔE_e and ΔE_h between QW and QD GS for the three different scattering rates of the simulations

Data set	ΔE_e (meV)	ΔE_h (meV)	j_{th}	τ_e (ns)	τ_h (ns)
Slow	140	120	$0.28\, j_{th\text{-ref}}$	0.395	0.129
Reference	210	50	$1.00\, j_{th\text{-ref}}$	0.071	0.005
Fast	74	40	$2.14\, j_{th\text{-ref}}$	0.021	0.01

region where the frequency is locked at cw operation grows in the region of positive input detuning. For the negative detuning region, the different pump currents show no influence on the frequency locked region but the occurrence of the PD bifurcation shifts to higher K values. Furthermore, the regions where the oscillations around the ghost of the fixed point are found next to the SN bifurcation shrink in size.

As in Fig. 4.43, the white area indicates the locking region at cw operation and the white–gray hatched area indicates the region of frequency locking of a limit cycle. The color degree indicates the number of extrema per period of the photon density. The SN, PD, and Hopf bifurcation lines are indicated. Parameters as in Table 4.7 and reference scattering rates is in Fig. 4.39b.

If the lifetimes τ_e and τ_h are chosen to be constant, the locking range is invariant under changes of the pump current. Thus, the changes in the locking behavior with the pump current can be attributed to the nonlinearity of the carrier lifetimes, which makes them a crucial ingredient for the quantitative modeling of injected QD lasers. High current densities lead to high electron and hole densities in the QW, which lead to short carrier lifetimes of electrons and holes, τ_e and τ_h, respectively as in Table 4.8. The energy differences ΔE_e and ΔE_h between QW and QD GS for the three different scattering rates used for the simulations (second and third columns), threshold current density j_{th} compared to the threshold current of the reference rates ($j_{th\text{-ref}}$) (fourth column) and carrier lifetimes for electrons (τ_e) and holes (τ_h) at steady state without injection ($K = 0$) and with pump current $j = 3.5\, j_{th}$ (fifth and sixth columns). Predictions of certain laser dynamics of experimental operation parameters are possible only if the nonlinear dependence of the scattering rates on the QW carrier densities could be incorporated into the model.

4.5.1   Impact of Carrier Lifetimes on the Dynamics

Many characteristics of the QD laser dynamics can be traced back to the carrier lifetimes in the QD. In previous section, has already pointed out their importance for modeling current-dependent injection properties. No discuss different QD structures and their impact on the frequency locking behavior. Different carrier lifetimes can be implemented in the simulations by using scattering rates that have been calculated for different QD size sand thus different band structures. Figure 4.39 displays the dependence of the electron and hole scattering lifetimes on the pump current for three different structures. In Table 4.8, the confinement energies (as see Fig. 4.38b) of the corresponding QD–QW band structures are given as well as their corresponding carrier lifetimes in the QD at steady state for a pump current of $j = 3.5\, j_{th}$. The threshold current density also changes with the scattering rates. So far, only the reference rates (Fig. 4.44b) have been used for the simulations. The three panels in Fig. 4.44 show time series of the photon density for three different QD lasers that have been modeled by using the sets of carrier lifetimes given in Fig. 4.39. The upper panel (Fig. 4.44a–c), middle panel (Fig. 4.44d–f) and lower panel (Fig. 4.44g–i) show the dynamics for the slow rates, reference rates and fast rates, respectively. In the plots the optical injection starts at $t = 5$ ns after the laser has been turned on. The most important difference in the turn-on dynamics (without injection) of the three lasers is the damping of the relaxation oscillations, which is related to changes of the carrier lifetimes. For the slow rates the damping is small as for a class B laser, i.e., a typical QW semiconductor laser (Fig. 4.44a–c). For the reference rates the relaxation oscillations are strongly suppressed, which is typical of QD lasers, but one can still observe several oscillations (Fig. 4.44d–f). For the fast rates the relaxation oscillations are over damped as for a class A laser, i.e., a typical gas laser (as Fig. 4.44g–i). This dependence was analytically described; however, the different rates also alter the behavior of the laser with optical injection. This can be seen by comparing QD lasers with different sets of carrier lifetimes for various detuning frequencies. In Fig. 4.44, the dynamics has been compared for three different detuning frequencies $\Delta v_{inj} = 2.3$ GHz (left column in Fig. 4.44), $\Delta v_{inj} = -0.5$ GHz (see central column in Fig. 4.44) and $\Delta v_{inj} = -2.5$ GHz (see right column in Fig. 4.44).

The Fig. 4.44 shows that the different scales of the photon densities. From left to right, the detuning Δv_{inj} takes values of 2.3 GHz (left column) is 0.5 GHz (center column) and 2.5 GHz (right column). Parameters is in Table 4.7, $\alpha = 0.9$ and $K = 0.25$.

In agreement with the results regarding the modulation response of QW lasers, the relaxation oscillations of the laser become less damped under injection at positive detuning inside the locking range (as in Fig. 4.44d), while damping and frequency of the relaxation oscillations increase inside the locking range for negative detuning (see Fig. 4.44e). Furthermore, the different carrier lifetimes influence the size of the locking range in the $(\Delta v_{inj}, K)$-plane, as can be seen by comparing the yellow areas in Fig. 4.45a, b. For lower turn-on damping (larger carrier lifetime, slow rates) the codimension-2 SN–Hopf-point at the positive detuning side of the frequency locked region shifts to lower K and the size of the white–gray hatched area indicating mean frequency locking on a limit cycle increases.

A second crucial effect of the lifetimes on the bifurcation diagram can be found at the negative detuning side of the locking range. For large lifetimes and thus small damping, the range of sustained multipulse emission with complex pulse shape (e.g., shown in Fig. 4.44c) is large and the lower PD bifurcation line is already found for small K values (Fig. 4.45a). Further chaotic transients and complex bifurcations are found outside the locking range for negative detunings. Instead, the bifurcation diagram plotted in Fig. 4.45b for the strongly damped QD laser shows a

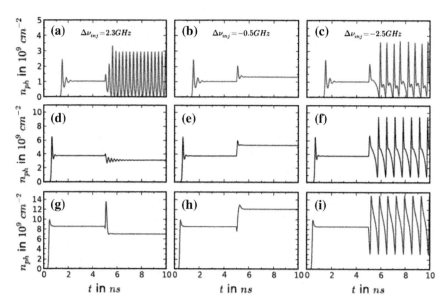

Fig. 4.44 Turn-on dynamics of the QD laser with injection starting at $t = 5$ ns for the slow scattering rates (**a–c**), the reference rates (**d–f**) and the fast rates (**g–i**) has been introduced in Fig. 4.39

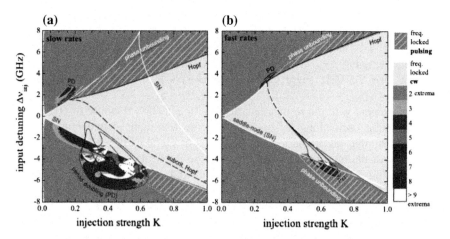

Fig. 4.45 Two-parameter bifurcation diagram for the slow scattering rates (**a**) and fast scattering rates (**b**) introduced in Fig. 4.34. Parameters is in Table 4.7 and $\alpha = 0.9$

large range of frequency locked cw operation. Outside this region only period-1 oscillations (see Fig. 4.44i for the time series of the photon density) are found up to $K = 0.35$. A very small area with a more complex pulse shape is found near the SN–Hopf-point and for $K > \psi 0$ (5 a small region surrounded by a PD bifurcation is found at the negative detuning side). It is interesting to note that the period-2 oscillations within the PD area have a bounded phase, and thus show mean frequency locking as indicated by hatching. The stronger damping of the turn-on dynamics leads to a reduction of the areas with complex dynamics and enlarges the range of stable frequency locked cw operation. This relation between turn-on damping and locking behavior should be interesting for experimentalists as the knowledge of the α-factor and the small-signal modulation response should be sufficient to predict and optimize the locking behavior of a laser.

If changing the current density as in Fig. 4.43, also leads to changes in the locking behavior that have a similar tendency as found for the lifetime variations. Long carrier lifetimes have a similar effect as low current density, and fast carrier lifetimes have a similar effect as high current density. Both can be traced back to the carrier lifetimes, v_e and v_h, respectively, and thereby to the damping of the turn-on dynamics. Increasing the damping shifts the beginning of the codimension-2 SN–Hopf-point to higher injection strength and leads to a shrinkage of the areas with complex dynamics found outside the locking range for negative detuning. The removal of the multipulse excitability was also found for class A lasers. These lasers can be approximately described by a single rate equation for the complex electric field and thus, as a two-variable system, cannot show chaos.

Using microscopic carrier–carrier scattering rates as the input for electron and hole lifetimes in the QD, have modeled an optically injected QD laser and described its nonlinear dynamics as a function of injection strength and input detuning. To have identified several changes in the locking behavior as well as in the bifurcation scenarios that are connected to variations in the carrier lifetimes of electrons and holes and thereby to the damping of the turn-on dynamics of the QD laser.

Most importantly the short carrier lifetimes are order of 10 ps lead to damped relaxation oscillations strongly, which in turn lead to large regions of stable cw operation. Sensitivity of the locking range to changes in the pump current is also found and attributed to changes in the pump current-dependent carrier lifetimes. Thus, it is crucial to incorporate the nonlinear dependence of the scattering rates on the carrier density into the surrounding reservoir in order to obtain quantitative results on the dynamics of an injected QD laser, i.e., the range of stable injection locked operation.

Scattering rates calculation of the carrier–carrier scattering rates implemented in the above simulations is described in detail in. The results of the microscopic simulations have been fitted with the following functions to enable their use with path continuation software and to allow the reader to follow the calculations. Using the following definition of normalized carrier densities in the reservoir:

$$W_e = w_e/(2N^{QD}) \text{ and } W_h = w_h/(2N^{QD}), \tag{4.50}$$

Table 4.9 Fit parameters for the reference rates from Table 4.8

Coefficient	Value	Coefficient	Value
a_e	-1.836×10^{-5}	a_h	3.326×10^{-5}
b_e	-7.89×10^{-6}	b_h	-8.064×10^{-4}
$c_{e,1}$	$-298\ 187.0$	$c_{h,1}$	-6886.56
$c_{e,2}$	38443.3	$c_{h,2}$	-7191.73
$c_{e,3}$	-3287.08	$c_{h,3}$	1117.15
$c_{e,4}$	112.303	$c_{h,4}$	-43.6502
$d_{e,1}$	53262.5	$d_{h,1}$	-17291.4
$d_{e,2}$	571.696	$d_{h,2}$	-13288.4
$d_{e,3}$	-72.5439	$d_{h,3}$	1000.69
$d_{e,4}$	0.683815	$d_{h,4}$	-52.8802

Table 4.10 Fit parameters for the fast rates from Table 4.8

Coefficient	Value	Coefficient	Value
a_e	-1.73923×10^{-5}	a_h	1.34424×10^{-5}
b_e	-1.30964×10^{-6}	b_h	-3.35376×10^{-4}
$c_{e,1}$	-707607.0	$c_{h,1}$	-626.353
$c_{e,2}$	106490.0	$c_{h,2}$	-3077.88
$c_{e,3}$	-8788.67	$c_{h,3}$	356.116
$c_{e,4}$	287.055	$c_{h,4}$	-17.8843
$d_{e,1}$	51376.6	$d_{h,1}$	-13087.5
$d_{e,2}$	7347.35	$d_{h,2}$	-15153.0
$d_{e,3}$	-663.515	$d_{h,3}$	1066.2
$d_{e,4}$	17.169	$d_{h,4}$	-55.0711

Can implement the following functions for the in-scattering rates as:

$$
\begin{aligned}
S_e^{in}(W_e, W_h) &= W \cdot (\tanh(a_e W_e + b_e)) \cdot \big(c_{e,1} W_e + c_{e,2} W_e^2 + c_{e,3} W_e^3 \\
&\quad + c_{e,4} W_e^4 + d_{e,1} W_h + d_{e,2} W_h^2 + d_{e,3} W_h^3 + d_{e,4} W_h^4\big), \\
S_h^{in}(W_e, W_h) &= W \cdot (\tanh(a_h W_e + b_h)) \cdot \big(c_{h,1} W_e + c_{h,2} W_e^2 + c_{h,3} W_e^3 \\
&\quad + c_{h,4} W_e^4 + d_{h,1} W_h + d_{h,2} W_h^2 + d_{h,3} W_h^3 + d_{h,4} W_h^4\big).
\end{aligned}
\tag{4.51}
$$

The parameter values of the coefficients are given in Table 4.9 for the reference rates, in Table 4.10 for the fast rates and in Table 4.11 for the slow rates. Due to the principle of detailed balance, the out-scattering rates can be calculated from the in-scattering rates with the help of Eqs. (4.48) and (4.49) for electrons and holes respectively.

Table 4.11 Parameters for the slow rates from Table 4.8

Coefficient	Value	Coefficient	Value
a_e	-2.6612×10^{-5}	a_h	1.94259×10^{-5}
b_e	-1.6475×10^{-6}	b_h	-4.74478×10^{-4}
$c_{e,1}$	-363381	$c_{h,1}$	-3601.34
$c_{e,2}$	50519.5	$c_{h,2}$	-15193.1
$c_{e,3}$	-4290.71	$c_{h,3}$	1441.14
$c_{e,4}$	146.18	$c_{h,4}$	-47.7236
$d_{e,1}$	69984.1	$d_{h,1}$	-19129.2
$d_{e,2}$	-74.2397	$d_{h,2}$	-5584.61
$d_{e,3}$	-86.6277	$d_{h,3}$	435.245
$d_{e,4}$	1.65736	$d_{h,4}$	-27.6885

4.6   Photon Rate Equation

The coherent interaction between a two-level system (e.g., QD) and a light mode can be described by semiconductor Bloch equations. Eliminating the fast microscopic polarization p of one QD (the probability amplitude for an optical transition) by assuming $\ddot{p} = 0$ leads to a quasi-static relation between p and the slowly varying complex amplitude E of the electric field. By further assuming equal energy for light mode and level spacing, this relation reads

$$p = -i\frac{\mu E T_2}{\hbar}(\rho_e + \rho_h - 1). \tag{4.52}$$

Here, μ is the associated dipole moment of the optical transition and T_2 is the lifetime of the microscopic polarization defining the homogeneous linewidth $\hbar/T_2$ of the levels. The term $(b_e + h_e - 1)$ describes the inversion of the two-level system with the electron and hole occupation probabilities in the QDs, ρ_e and ρ_h. For the derivation of the photon rate equation, start with the reduced field equation for the electric field without damping,

$$\dot{E} = \frac{i\omega_L \Gamma}{2\varepsilon_0 \varepsilon_{bg}} P, \tag{4.53}$$

where ε_0 is the vacuum permittivity, ε_{bg} is the background dielectric constant and ω_L is the transition frequency of the two-level system. Using the total macroscopic polarization inside one QW layer given by $P = 2N_a^{QD}\mu P/h^{QW}\mu p$ and the optical confinement factor 0 perpendicular to the direction of light propagation $\Gamma = a_L h^{QW}/h^w$ (height h^w of the waveguide, height h^{QW} of the QW layers that contain the self-organized QDs and the number of QW layers a_L), can arrive at the following field equation:

$$\dot{E} = \frac{2|\mu|^2 \omega_L T_2 a_L N_a^{QD}}{2\varepsilon_0 \varepsilon_{bg} \hbar h^w} (\rho_e + \rho_h - 1)E = Z_a^{QD} \bar{W}(\rho_e + \rho_h - 1)E \qquad (4.54)$$

with the Einstein coefficient of induced emission

$$\bar{W} = \frac{|\mu|^2 \omega_L T_2}{2\varepsilon_0 \varepsilon_{bg} \hbar V^w}, \qquad (4.55)$$

which measures the strength of the coherent interaction. $\bar{W}$ is depended on the volume of the optical waveguide $V^w = Ah^w$ and on the width of the optical transition $\hbar/T_2$. The number $Z_a^{QD} = a_L A N_a^{QD}$ is the number of active QDs inside the waveguide. Rewriting Eq. (4.50) for the photon density per unit area $n_{ph} = |E|^2 h\varepsilon_0\varepsilon_{bg}/2\hbar\omega$ gives to

$$\dot{n}_{ph} = 2Z_a^{QD} \bar{W}(\rho_e + \rho_h - 1)n_{ph}. \qquad (4.56)$$

In previous publications, a different value for $\bar{W}$ was implemented because the QD volume V^{act} was used instead of the optical waveguide volume V^w in Eq. (4.55). As a result the geometric optical confinement factor $\Gamma_g = V^{act}/V^w$ appeared in the equation for the photon density per unit area Eq. (4.56), which was somewhat misleading. Because the differing $\bar{W}$ was also used in the carrier equations, those simulations yield a rescaled photon density. Thus, the values for n_{ph} need to be multiplied by 6.6×10^3 to yield the real photon density in the cavity. Related to that, also the coefficient for spontaneous emission is scaled in those papers and the unscaled value should be $\beta = 2.2 \times 10^{-3}$. The Einstein coefficient for the spontaneous emission can be derived by calculating the incoherent interaction of the two-level system with all resonator modes in the framework of second quantization. It gives by:

$$W = \frac{|\mu|^{2^-} \sqrt{\varepsilon_{bg}}}{3\pi\varepsilon_0 \hbar} \left(\frac{\omega L}{c}\right)^3 \qquad (4.57)$$

4.6.1 High-Speed Quantum Dot Lasers

The modulation bandwidth of conventional 1.0–1.3 µm self-organized In(Ga)As quantum dot (QD) lasers is limited to $\sim$6–8 GHz due to hot carrier effects arising from the predominant occupation of wetting layer/barrier states by the electrons injected into the active region at room temperature. Thermal broadening of holes in the valence band of QDs also limits the performance of the lasers. Tunnel injection and p-doping have been proposed as solutions to these problems. In this paragraph, describe high-performance In(Ga)As undoped and p-doped tunnel injection self-organized QD lasers emitting at 1.1 and 1.3 μm. Undoped 1.1 μm tunnel injection lasers have ~ 22 GHz small-signal modulation bandwidth and a gain compression factor of 8.2×10^{-16} cm^3. Higher modulation bandwidth (~ 25 GHz) and differential gain (3×10^{-14} cm^2) are measured in 1.1 μm p-doped tunnel injection lasers with a characteristic temperature T_0 of 205 K in the temperature range 5–95 °C. Temperature invariant threshold current (infinite T_0) in the temperature range 5–75 °C and 11 GHz modulation bandwidth are observed in 1.3 μm p-doped tunnel injection QD lasers with a differential gain of 8×10^{-15} cm^2. The linewidth enhancement factor of the undoped 1.1 μm tunnel injection laser is ~ 0.73 at lasing peak and its dynamic chirp is <0.6 Å at various frequencies and ac biases. Both 1.1 and 1.3 μm p-doped tunnel injection QD lasers exhibit zero linewidth enhancement factor ($\alpha \sim 0$) and negligible chirp (<0.2 Å). These dynamic characteristics of QD lasers surpass those of equivalent quantum well lasers [9, 10].

Self-organized quantum dot (QD) lasers have been the subject of extensive study in the last decade and have demonstrated lower threshold current, linewidth enhancement factor, and dynamic chirp compared to quantum well (QW) lasers. Demonstration of high-speed QD lasers can, therefore, be envisioned as a breakthrough in terms of applications as coherent light sources for 1.0–1.3 μm short-haul local area network (LAN) and metropolitan are network (MAN) 10 Gbs^{-1} communication systems. However, achieving high modulation bandwidths with conventional SCH QD lasers has not been possible. There are unique problems that limit the modulation performance of conventional SCH QD lasers, compared to what is expected from an 'ideal' QD laser with a discrete density of states. First, the inhomogeneous linewidth broadening, associated with the stochastic size distribution of the dots, imposes a limit on the performance of QD lasers. More importantly, SCH QD lasers suffer from significant hot-carrier effects and associated gain compression due to the large density of states of the wetting layer and barrier states, compared with that in the QDs. As a result, the conventional devices cannot be modulated at bandwidths above 6–8 GHz. In addition, the hole distribution is thermally broadened into many available states with small energy spacing in QDs and a large injected hole density is required for a large gain in the ground state. This would also decrease the attainable gain and differential gain in conventional QD lasers. Two unique solutions have been proposed and implemented to overcome these problems in conventional SCH QD lasers: tunneling injection (TI) and acceptor (p) doping of the dots. In the tunnel injection scheme, 'cold carriers' are injected directly into the ground-state of the QDs by phonon assisted tunneling from an adjacent injector layer and are removed by stimulated emission at approximately the same rate. Therefore, the differential gain of the lasers can be optimized and hot carrier effects are minimized. With p-doping, extra holes are provided at the ground-state energy by either direct doping of the dots or by modulation doping in the GaAs barriers. These extra holes ensure population inversion with less injected holes from the contacts; consequently, the electron population in the dots and their leakage into barrier and waveguide layers is reduced as well.

In the paragraph first describe the intrinsic characteristics of QD lasers that determine the small-signal modulation bandwidth and the temperature dependence of the threshold current. This is followed by a description of tunnel injection and p-doping in QD lasers and a comparison of the two approaches. That although p-doping is helpful in improving the characteristic temperature, T_0 of QD lasers— especially at 1.3 μm, it does not help in realizing high modulation bandwidth lasers. On the other hand, tunnel injection not only decreases the temperature sensitivity of QD lasers, but also significantly enhances the high frequency response of the devices. Specifically, the properties of 1.1 and 1.3 μm QD lasers, in which tunnel injection and p-doping are incorporated, are described. The high-speed modulation characteristics of these devices are described and discussed. Finally, data on chirp and α-factor of these devices are presented. It will be evident that present high-speed QD lasers are promising candidates for applications in MAN and LAN systems.

4.6.2 Factors Limiting High-Speed Operation of QD Lasers

It is now recognized that the limitations to high-speed modulation of conventional SCH QD lasers is due to the electronic properties of the QDs arising from the nature of self-assembled growth. Due to the large lattice mismatch in the In(Ga)As/GaAs system (>1.7 %), the dots are formed in the Stranski–Krastanow growth mode, where zero-dimensional is lands (QDs) are formed on top of a wetting layer (two-dimensional electron–gas), as depicted in Fig. 4.46. The QDs and the wetting layer form a coupled electronic system, statistics cannot be described by quasi-Fermi equilibrium.

Due to the large number of the states in the two-dimensional electron-gas compared to the number of states in the dots, injected carriers predominantly reside

Fig. 4.46 Energy levels of 1.3 μm QDs showing the large energy spacing between the ground and first excited state in the conduction band and many (~ 10) levels with small spacing (8–10 meV) in the valence band

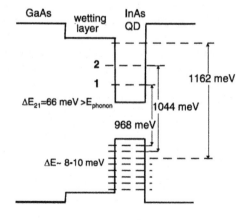

in the higher energy states in the wetting layer and QD lasers suffer from this undesired 'hot carrier' effect and associated gain saturation.

Differential transmission spectroscopy (DTS) measurements have indeed shown that electrons captured in the wetting layer/barrier states tend to remain there at temperatures above 180 K, i.e., they undergo very slow relaxation to the lasing energy state, that observed severe gain saturation in QDs at above 150 K and showed that it can be explained by incomplete population inversion in the ground-state of the QDs due to the occupancy of the wetting layer states. The gain saturation leads to low modulation bandwidth. Stated differently, the entropy change of carriers relaxing from the two-dimensional wetting layer states to the zero-dimensional QDs is responsible for the low modulation bandwidths measured in QD lasers.

An ideal QD laser should preferably have only one electron and one hole energy level. As shown in Fig. 4.46, the inevitable existence of multiple hole energy levels with small energy spacing (8–10 meV) results in thermal broadening of the hole population in energy. Consequently, the ground-state hole population is depleted, leading to a decrease in gain. A higher injection of holes, to compensate for this effect, necessitates increased injection of electrons due to requirements of charge neutrality. The excess carriers lead to leakage, nonradiative recombination outside the core, increased threshold current, and reduced differential gain.

4.6.3 Tunnel Injection and Acceptor Doping in QD Lasers

T_I and p-doping have been suggested and studied as two promising techniques to solve the hot-carrier related problems in QD lasers. Tunnel injection was originally proposed and demonstrated more than a decade ago to reduce hot carrier effects in QW lasers. In this scheme, 'cold carriers' are directly injected into the lasing energy state from an adjoining injector layer; thus hot carrier effects can be bypassed and the performance of the lasers would improve. High-performance GaAs- and InP-based QW lasers with high T_0, reduced chirp and improved modulation bandwidths have been reported for a long time. T_I, however, is more useful in enhancing the modulation bandwidth of QD lasers. As shown in Fig. 4.47a, cold electrons injected into the ground-state of the QDs by phonon-assisted tunneling can by-pass the hot carrier problems associated with the capture of electrons into the wetting layer/barrier energy states. Femtosecond DTS measurement of phonon-assisted tunneling confirms fast (~ 1.7 ps) temperature-independent tunneling times.

Tunneling also decreases carrier radiative recombination in the wetting layer/barrier regions, and based on Asryan and Luryi theoretically predicted a significant increase of T_0. Demonstrated a large increase in modulation bandwidth (~ 15 GHz) in our first tunnel injection QD lasers and as will be seen in the following, the characteristics of the lasers have steadily improved since then. p-Doping of QD lasers can be achieved by either director modulation doping of the

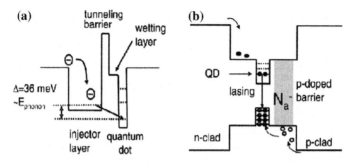

Fig. 4.47 a Injection of cold carriers into the ground state of the dot by tunneling from an adjoining injector layer. **b** Modulation p-doping of the QD barrier in order to increase the gain through the increase of hole ground state occupancy

dots. As shown in Fig. 4.47b for the modulation doping case, the holes of the p-doped barrier are transferred into the hole ground-state with lower energy in the adjacent QD layer; thus, fewer electron–hole pairs are required to be injected from the contacts to compensate for the thermal broadening of the hole distribution.

If increase of gain in p-doped QW lasers, that was followed by the prediction of enhancement of the relaxation oscillation frequency and reduction of linewidth enhancement factor in multiQW lasers. The p-doping is expected to be more beneficial in QD lasers due to the more pronounced thermal broadening of holes in the valence band with smaller energy spacing in QDs than in QWs. When the p-doping increase, that gain and threshold current will be reduction. Modeled the impact of p-doping on the modulation response and characteristic temperature, T_0 of QD lasers and have experimentally demonstrated T_0 as high as 213 K in p-doped InAs QD lasers. Measured temperature invariant operation ($T_0 = \infty$) in p-doped 1.3 μm QD SCH lasers and attributed this result to a significant role of recombination with its unique temperature dependence. However, contrary to theoretical predictions and observed only a slight improvement in modulation bandwidth, from 9 GHz in undoped lasers to 11 GHz in 1.1 μm p-doped lasers with otherwise identical heterostructures. The modulation response of the p-doped lasers is presented in Fig. 4.48, from which a differential gain of 6.9×10^{-15} cm^2 is obtained. The low bandwidth can be attributed to the inefficiency of p-doping due to the wetting layer states, inadequate enhancement of gain and differential gain, and the increased damping effect of Auger recombination in the modulation response [11, 12].

These observations confirm that although p-doping maybe beneficial in enhancing T_0, tunnel injection appears to be a better approach to achieve high-speed QD lasers. In the following, present the growth, fabrication, and characteristics of state of the art 1.1 and 1.3 μm tunnel injection QD lasers. As will be evident, in some of these devices, both techniques of tunnel injection and p-doping are simultaneously incorporated.

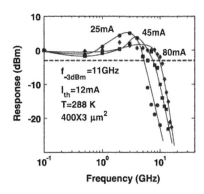

Fig. 4.48 Modulation response of single-mode 1.1 μm p-doped QDSCH laser at different biases with a maximum 3 dB bandwidth of ∼11 GHz

4.6.4 Tunnel Injection Lasers with p-Doping

Three types of tunnel injection lasers are discussed in this paragraph. They are undoped and p-doped 1.1 μm lasers and p-doped 1.3 μm lasers. All the laser hetero structures were grown by molecular beam epitaxy (MBE) on (001) GaAs substrates. The heterostructure of a 1.1 μm InGaAs tunnel injection QD lasers is schematically shown in Fig. 4.44a. The wavelength of the dot luminescence peak is controlled by adjusting the InGaAs dot charge during epitaxy. The active region consists of a 95 Å $In_{0.25}Ga_{0.75}As$ injector well, a 20 Å $Al_{0.55}Ga_{0.45}As$ tunnel barrier, and three coupled $In_{0.50}Ga_{0.50}As$ QD layers. The $In_{0.25}Ga_{0.75}As$ injector layer is grown at 450 °C and the QD layers are grown at 510 °C. The energy separation in the conduction band between the injector layer state and the QD ground state is ∼36 meV at room temperature. This energy separation ensures longitudinal optical (LO) phonon assisted tunneling from the injector layer to the dot ground states through the AlGaAs barrier. In the p-doped lasers, doping is provided by delta-doping (5×10^{11} cm^{-2}) the 500 Å barrier/waveguide region grown on top of the three layers of coupled QDs. 50 nm of the GaAs waveguide above the three coupled QD layers are p-type doped with beryllium ($Na = 5 \times 10^{17}$ cm^{-3}), averaging about 20 holes per dot. Figure 4.50a shows the band diagram in the active region of the 1.3 μm p-doped tunnel injection QD laser. All the depicted energy transitions are calculated and design values [13].

Long wavelength (1.3 μm) tunnel injection QD lasers are more difficult to realize due to the higher misfit-related strains involved in this system. In order to reduce the strain, the design of 1.3 μm QD lasers differs from the 1.1 μm lasers in that electrons tunnel into the first excited states of the dot, instead of the ground state. This eases the alloying requirements of the injected layer. Pump-probe DTS measurements show that the relaxation time from the dot first excited state to the ground state is very small, about 130 fs, if the excited states are filled with electrons, i.e., lasing conditions. One QD layer, which consists of 2.6 ML of InAs capped with 45 Å $In_{0.15}Ga_{0.85}As$, is grown on top of its neighboring 95 Å $In_{0.27}Ga_{0.73}As$ injection layer. In order to increase the modal gain, five periods of injector QDs/GaAs buffers are stacked. Compared to the 1.1 μm design, shorter tunnel barriers ($Al_{0.25}Ga_{0.75}As$) are employed to facilitate carrier injection across the SCH region.

It has to be noted that carrier lifetimes are long enough to ensure they reach all the five dot layers, as has been shown in 70-layer QD infrared photodetectors by simulations. The laser heterostructure exhibits strong photoluminescence (PL) with a narrow linewidth that room temperature that can be see in Fig. 4.45b later. This is an indication of efficient tunneling due to the selection process of the tunneling states in the dots. Mesa-shaped broad area (20–100 μm wide) and single mode ridge waveguide lasers (3–5 μm ridge width) were fabricated by standard lithography, wet and dry etching and metallization techniques. The 200–2000 μm long lasers were obtained by cleaving. Measurements were made on lasers with as-cleaved facets, as well as, facets with high reflectivity mirrors obtained with the deposition of dielectric DBR.

4.7 Static/Dynamic Characteristics of QD Lasers

4.7.1 Static Characteristics of QD Lasers

Light-current (L-I) measurements were made with the devices mounted on a Cu heat-sink, whose temperature was stabilized with a Pettier cooler. Pulsed biased (1 μs, 10 kHz) light-current measurements were performed on 1.1 μm TI-QD p-doped as-cleaved 200×3 μm^2 single-mode lasers and as the threshold current versus temperature plot in Fig. 4.49b indicates, $T_0 \sim 205$ K from 5 to 95 °C and slope efficiency of 0.465 W A^{-1} can be extracted for the devices. The inset shows the output spectrum of the device lasing at about 1090 nm. Similarly, the undoped 1.1 μm TI-QD lasers exhibit $T_0 \sim 363$ K for $5 < T < 60$ °C and a threshold current of 8 mA at 288 K for a 400×3 μm^2 device. From the L-I characteristics of undoped TI lasers of varying cavity length, that determine the value of internal quantum efficiency $\eta_i = 85$ % and cavity loss $\gamma = 8.2$ cm^{-1}, by plotting the inverse of differential efficiency η_d, against cavity length l. It is evident that the characteristic temperature of T_I lasers is much higher than typical values of $T_0 < 100$ K in conventional QD lasers, which is due to efficient (direct) injection of cold carriers into the ground state of QDs, minimal occupation of wetting layer/barrier states, and the consequent reduction in the radiative recombination component of threshold current from these higher energy states.

For the 1.3 μm p-doped TI QD lasers, L-I measurements were performed on devices with 95 % high reflectivity mirrors on one facet. The room temperature L-I characteristics of an 800×5 μm^2 single-mode laser, with its optical output spectrum, are shown in Fig. 4.51a. It is evident that the laser single-mode peak is from the ground state of the dots and is very close to the corresponding PL peak in Fig. 4.50b. The value of J_{th} is 180 A cm^{-2} for an 800 μm long cavity. Values of

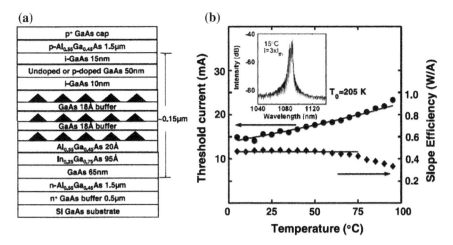

Fig. 4.49 a Heterostructure schematic of 1.1 μm undoped and p-doped tunnel injection QD lasers. **b** Variation of the threshold current and slope efficiency of $200 \times 3 \ \mu m^2$ single-mode 1.1 μm p-doped lasers with temperature. The *inset* shows the output spectrum of the laser at 3 times threshold

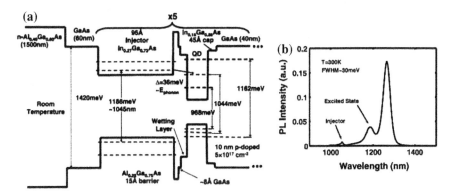

Fig. 4.50 a The energy band diagram in the active region of 1.3 μm p-doped tunnel injection QD lasers illustrating the phonon-assisted tunneling from the injector layer into the first excited state. **b** Room temperature PL spectrum of the heterostructure showing distinct peaks from the dot ground and excited states and the injector layer state

$\eta_i = 71 \ \%$ and cavity loss $\gamma = 6.3 \ cm^{-1}$ were determined for these devices under quasi-CW bias (10 % duty cycle) from the η_d versus l plot. A differential gain of $9.8 \times 10^{-15} \ cm^2$ is estimated in the devices from the same plot. The threshold current of a $400 \times 5 \ \mu m^2$ device versus temperature is shown in Fig. 4.51b. It is observed that $T_0 = \infty$ in the temperature range of 5–70 °C. This result is similar on temperature invariant operation of p-doped 1.3 μm QD lasers. As discussed in detail therein, although high T_0 has been predicted and previously reported in

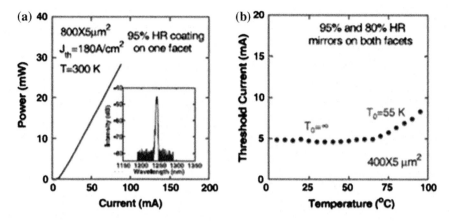

Fig. 4.51 a Pulsed light-current characteristics of 1.3 μm p-doped tunnel injection single mode QD lasers at room temperature. The *inset* shows the output spectrum. **b** Variation of threshold current of the 1.3 μm lasers with temperature

p-doped QD lasers, complete temperature independence of I_{th} may seem unlikely, since the inhomogeneous linewidth broadening of the gain, associated with the stochastic size distribution of self-organized QDs, should limit such ideal performances of QD lasers. The radiative recombination terms in the wetting layers and GaAs barrier/waveguide regions are functions of the Fermi–Dirac distribution, i.e., they contribute to finite T_0 as well.

Therefore, a recombination process whose rate decreases with temperature has to be considered in order to explain the experimentally observed $T_0 = \infty$. To have employed a self-consistent model to calculate the various radiative and nonradiative current components in p-doped and undoped lasers and concluded that the zero temperature dependence of the p-doped 1.3 μm QD SCH lasers is due to the significant role of Auger recombination in the devices, and its decrease with temperature according to the temperature dependence of the auger coefficient. The lower $T_0 = 205$ K in the 1.1 μm p-doped T_I lasers is a consequence of the following:

1. The Auger recombination is expected to bellower in higher band-gap materials, as confirmed by hydrostatic pressure dependence measurements. Therefore, auger recombination is less effective in playing the compensation g role to achieve high T_0;
2. The conduction and valence band off sets (ΔE_C and ΔE_V) are smaller in 1.1 μm QDs than in longer wavelength 1.3 μm dots. Consequently, the carrier leakage into the barrier/waveguide region is higher in 1.1 μm QDs, which also leads to a lower value of T_0.

4.7.2   Small Signal Modulation Response

The small signal modulation response of the lasers was measured with a high-speed photodetector, low-noise amplifier, as HP 8350B sweep oscillator and a HP 8562A electrical spectrum analyzer. The modulation response of the undoped 1.1 μm TI-QD laser at room temperature and at different injection currents is shown in Fig. 4.52. It is seen that the lasers have a maximum modulation bandwidth, $f_{-3\text{dB}}$ of ~ 22 GHz at a bias of 125 mA. By plotting the resonance frequency versus $(I - I_{th})^{1/2}$ a modulation efficiency of ~ 1.7 GHz mA$^{-1/2}$ is derived in these lasers. A differential gain, $dg/dn = 2.7 \times 10^{-14}$ cm^2 in these devices is derived from the measured modulation efficiency and a calculated optical confinement factor of $\Gamma = 2.5 \times 10^{-3}$. From the damping factor of the best fit to the modulation response, again compression factor, $\varepsilon = 8.2 \times 10^{-16}$ cm^3 is also obtained for these devices.

The modulation response for the 1.1 μm p-doped TI QD lasers is shown in Fig. 4.53a. The measurements were made under pulsed bias. The maximum $f_{-3\text{dB}}$ measured for an injection bias of $6.7 \times I_{th}$ is ~ 24.5 GHz, which is higher than the undoped sample and is indeed the highest modulation bandwidth reported to date in any QD laser. The resonance frequency of the devices is plotted versus $(I - I_{th})^{1/2}$ in Fig. 4.53b, from which a modulation efficiency and differential gain of 2 GHz mA$^{-1/2}$ and 3×10^{-14} cm^2, respectively are derived. The corresponding extracted gain compression factor of the lasers is $\varepsilon = 4 \times 10^{-16}$ cm^3. All these parameters show improvement compared to the undoped samples, which may be attributed to the slight impact of p-doping in the modulation characteristics of QD lasers. As that have also observed a few gigahertz enhancement of modulation bandwidth upon p-doping (from 9 to 11 GHz) in conventional 1.1 μm SCH lasers, which is due to the slight increase of gain and differential gain by the extra holes provided from the doped barriers.

Fig. 4.52 Modulation response of single-mode 1.1 μm undoped tunnel injection QD lasers at different biases at 15 °C

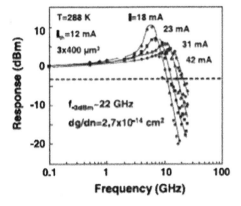

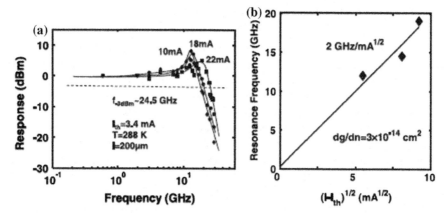

Fig. 4.53 **a** Modulation response of single-mode 1.1 μm p-doped QD tunnel injection lasers at different biases. **b** Resonance frequency of the lasers versus $\sqrt{I} - I_{th}$

Fig. 4.54 Modulation response of single-mode 1.3 μm p-doped QD tunnel injection lasers at different biases at 15 °C

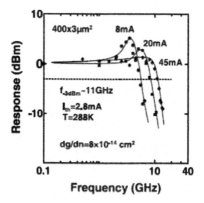

Figure 4.54 presents the modulation response characteristics of a single-mode $400 \times 3 \ \mu m^2$ p-doped 1.3 μm TI QD lasers. A maximum 3 dB modulation bandwidth of 11 GHz is measured in these devices at 45 mA. A modulation efficiency of 1.1 GHz/mA and a differential gain of $8 \times 10^{-15} \ cm^2$ are extracted, the latter of which is close to the estimation of the figure from the dc measurements presented before.

When speculate the following as a reason for the lower bandwidths observed in 1.3 μm TI QD lasers compared to that measured in 1.1 μm TI QD lasers (Figs. 4.52 and 4.53). As discussed before, very fast tunneling time constant and relaxation times from the dot excited states to the ground state have been measured by DTS. Therefore, the 1.3 μm tunnel injection lasers are most probably not limited by relaxation and tunneling times. The modal gain of a QD laser is proportional to dot density (fill factor) and density of states, of which the latter is inversely proportional to the average volume of a single dot. The average volume of the 1.3 μm dots is about a factor of 2 larger and the dot density is usually lower compared to 1.1 μm QDs. These geometrical differences translate to lower gain and differential gain in 1.3 μm lasers.

4.7.3 Dynamic Properties and α-Factor

The modulation characteristics of tunnel injection lasers presented in the previous section indicate that these devices have enormous potential for high-speed optical communication and super-speed data storage. Therefore, study on dynamic figures of merit for such applications in the present lasers, namely, linewidth enhancement factor α and dynamic chirp are important. The α-factor is a critical parameter in semiconductor lasers, since the laser linewidth is $(1 + \alpha^2)$ times larger than the Shawlow–Townes fundamental limit. The α is inversely proportional to the differential gain and it is evident that large differential gains are attainable in QD lasers. Therefore, low α-factors can be expected and have indeed been reported in conventional QD lasers. The linewidth enhancement factor of the tunnel injection lasers by the Hakki–Paoli method at threshold by using the formula:

$$\alpha = \frac{2}{\partial\lambda}\frac{\Delta\lambda_i}{\Delta\left\{\ln\left[\left(\sqrt{r_i}-1\right)\left(\sqrt{r_i}+1\right)^{-1}\right]\right\}} \qquad (4.58)$$

where $\partial\lambda$ is the mode spacing, r_i is the peak-to-averaged valleys ratio of the i_{th} competing mode in the optical spectrum and ΔN is the incremental carrier density for two differential bias values. It is worth noting that the subthreshold measurement results employed have shown excellent agreement with results from other measurements of α-factor above threshold, such as injection locking, in QW lasers. The results of the two techniques may be different in QD lasers if these devices exhibit multimode lasing from the excited states at high biases. However, spectral measurements of the lasers show a stable single-mode output spectrum from the ground state of QDs at all biases, and thus the subthreshold Hakki–Paoli technique should yield the true value of α. The subthreshold spectra were measured under pulsed bias (10 kHz, 1 %) at room temperature with a HP 70952B optical spectrum analyzer with a minimum resolution of 0.8 Å. The voltage increment, ΔV was kept below 0.1 V. The measured linewidth enhancement factors are shown against the peak wavelength of the subthreshold spectrum for the undoped TI 1.1 μm QD laser as in Fig. 4.55a.

The value of α at lasing peak ($\sim$ 1057 nm) is about 0.73, which is lower than what is measured in conventional SCH QD lasers. At other wavelengths, the value of α is ≤ 0.5. This is an indication of reduced hot carrier densities in the tunnel injection laser. The reduction of α in QD lasers, compared to typical values >2 in QW lasers, implies a very small refractive index change in the lasing core.

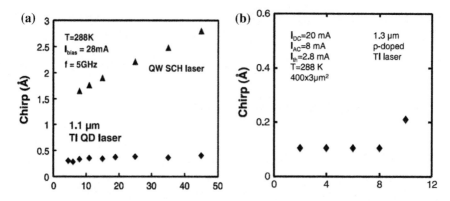

Fig. 4.55 Linewidth enhancement factor at the peak wavelengths of the subthreshold spectrum of (a) undoped and (b) p-doped 1.1 μm tunnel injection QD lasers. The *inset* in (**b**) shows two differentially close measured subthreshold spectra with bias voltage difference, $\Delta V \sim 0.1$ V in the p-doped lasers

Consequently, there is reduction of the self-focusing effect in these devices, which leads to absence of filamentation in their measured near-field pattern.

As can be seen in Fig. 4.55b, upon varying the voltage increment ΔV, from a differential value of 0.1 V to values as high as 0.5 V, no spectral differential shift of the longitudinal laser peaks, $\Delta \lambda_i$ was observed in p-doped 1.1 μm TI lasers. Therefore, α is virtually zero in these lasers (within the resolution of the spectrum analyzer). The α-factors of the 1.3 μm devices were measured by the same method and similarly, the values of the α-factors are essentially zero versus wavelength around the lasing peak. Finally, since chirp is directly proportional to α, tunnel injection QD lasers are expected to have ultra-low chirp, measured the chirp in both 1.1 and 1.3 μm TI QD lasers from the difference in the linewidth of single longitudinal modes with and without superimposition of an ac signal. The envelope of the dynamic shift in wavelength of the sinusoidal modulation signal was recorded with an optical spectrum analyzer. The evaluated chirp for the undoped 1.1 μm TI QD lasers versus peak-to-peak modulation current is shown in Fig. 4.56a at a modulation frequency of 5 GHz and dc bias of 28 mA. For comparison, the same figure presents our results from InGaAs 1.0 μm QW lasers, whose heterostructure design and device fabrication are similar to the TI QD lasers. The chirp of the QW lasers varies between 1.6 and 2.9 Å and is comparable to previously reported values, whereas the value is ~ 0.4Å in the TI QD lasers. Furthermore, upon changing the modulation frequency from 1 to 15 GHz and at a constant ac bias of 36 mA, the TI QD lasers show chirp of <0.6Å at all frequencies.

The measured chirp in 1.3 μm lasers as a function of modulating frequency, with a peak-to-peak modulation current of 8 mA and a dc bias of 20 mA, is shown in Fig. 4.56b. As can be seen, the dynamic chirp is negligible (<0.2 Å). Similar results were obtained for 1.1 μm tunnel injection QD lasers and are not discussed further herein. It can be clearly concluded that tunnel injection QD lasers have dynamic

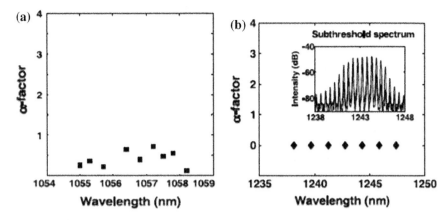

Fig. 4.56 a Measured chirp for undoped 1.1 μm tunnel injection QD laser and SCH QW laser at different peak-to-peak modulation currents. **b** Chirp versus modulating frequency in 1.3 μm p-doped tunnel injection QD lasers

properties that surpass those of QW lasers for 1.0–1.3 μm optical communication systems. The characteristics of very high performance 1.1 and 1.3 μm self-organized QD lasers are presented. The effects of tunnel injection and p-doping have been studied. While small signal modulation bandwidths up to 11 GHz can be measured with p-doping alone, bandwidths up to 25 GHz are measured in tunnel injection lasers. The p-doped lasers demonstrate $T_0 = \infty$, together with near-zero chirp and α-factors.

4.7.4   Quantum Dot Lasers Future Trends

The advantage of a discrete energy spectrum and efficient overlap of electron and hole wave functions in a quantum dot (QD) was recognized already early. The paragraph on the possibility of using QDs as active media of a semiconductor laser with strongly improved and temperature insensitive parameters appeared, many scientists and engineers started searching ways of fabrication of quantum dots and studying their properties. However, more than a decade passed until first lasers based on self-organized QDs have been fabricated and were proven to demonstrate the predicted properties.

Currently, the most promising way to fabricate QDs is based on the effect of spontaneous nanoislanding during heteroepitaxial growth. Flat (2D) nanoislands are usually formed by submonolayer deposition and the driving force relates to the surface stress discontinuity at the island edges. The elastic relaxation of the surface stress along the island boundary makes formation of uniform in size nanoislands energetically favorable. After overgrowth 2D islands represent ultrathin nanoscale pancakes inserted in a wide gap matrix. The localization energy of carriers and excitons in these islands is relatively small, except of materials with large electron and hole masses are used. In view of the small average thickness of the insertion, a possibility to stack strained 2D islands by keeping the average strain in the layer low exists. Arrays of 2D islands usually provide much narrower absorption or gain peaks. In the case of 3D islands the driving force relates to the elastic relaxation of the volume strain of the island formed on a lattice mismatched substrate. Possibility of stable with respect to ripening 3D islands appears if the total surface energy of the island is smaller than the surface energy of the corresponding area of the wetting layer occupied by it. The latter is possible if the strain-induced renormalization of the surface energy of the facets is taken into account.

Oscillator strength in a small QD is not a function of the QD volume. Thus, dense arrays of very small QDs (10^{12} cm^{-2}) provide much higher modal gain, as compared to a more dilute array of larger QDs (typically about 10^{10} to 10^{11} cm^{-2}). On the other hand larger QDs can provide much higher localization energy. This gives some flexibility in constructing of the device. In case, when one is interested to keep high maximum absorption or gain values, 2D islands are preferable. High temperature stability of the threshold current and a maximum long-wavelength shift of the emission (e.g., 1.3 or 1.5 μm range using GaAs substrates) are realized for 3D islands. Dense arrays of QDs can demonstrate lateral ordering due to their interaction via the strained substrate. Stacked 3D QD deposition demonstrate vertically—correlated growth. 2D islands demonstrate either correlated or anticorrelated growth depending on the relative thickness of the spacer layer. Several other promising ways to fabricate QDs using self-organization phenomena exist and references as:

1. spontaneous quasiperiodic faceting of crystal surfaces and heteroepitaxial growth of faceted surfaces;
2. spontaneous phase separation in semiconductor alloys during growth or slow cooling;
3. spontaneous alloy decomposition upon high-temperature annealing.

4.7.5 Edge Emitting and Vertical Cavity Quantum Dot Lasers

The structure of vertical cavity quantum dot lasers is shown in Fig. 4.57. Evident progress in using of QDs is achieved in the area of semiconductor lasers. Two basic device geometries have been applied also. In one case, the light propagates along the plane with QDs, and the resonator represents conventional Fabri–Perot cavity with natural cleavages as mirrors that can see Fig. 4.57b. In the other case, the light is emitted perpendicular to this plane as shown in Fig. 4.57a, while the cavity is confined in vertical direction by multilayer stacks of layers forming DBR.

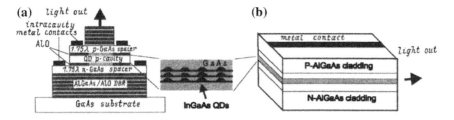

Fig. 4.57 QDs are used as active media of semiconductor heterostructure lasers in edge-emitting (**b**) and vertical cavity (**a**) geometry

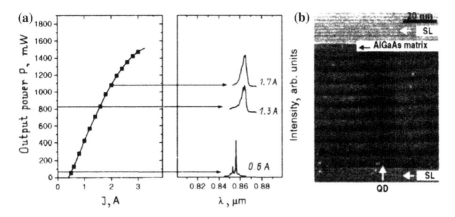

Fig. 4.58 **a** High-power operation of edge-emitting InAs/AlGaAs QD laser. **b** Transmission electron microscopy image of the active region of the high power QD laser. Stripe length 850 μm, width 100 μm and waveguide region of 0.3 μm

The first approach allows fabrication of high power lasers utilizing advantages of ultralow threshold current density due to QDs, possible preventing of dislocation growth and suppression of the laser mirror overheating by nonradiative surface recombination due to localization of carriers in QDs. In the second approach lasers with ultralow total currents can be fabricated, and, even more exciting, lasers based on single QD can be potentially realized.

The most important events in the QD laser field can be briefly listed that has been realized. First, QD injection laser has been fabricated in 1994 by a joint from Technical University of Berlin and Abraham Institute. Lasing via the QD ground state and the temperature insensitive threshold current have been demonstrated. Room temperature (RT) operation via quantum dots has been demonstrated. Ultrahigh material and differential gain in QD lasers have been manifested. RT lasing with 60 A/cm^2 has been realized. Continuous-wave RT high power operation of a QD laser (1500 mW) was realized as shown in Fig. 4.58.

Fig. 4.59 Threshold current
(J_{th}) of the QD VCSEL. The
emission spectrum at 1.3 J_{th} is
shown in the *insert*. Quantum
efficiency $\eta = 16$ % at 10 μm.
The basic parameters of edge
emitting and vertical-cavity
QD lasers are approached to
QW devices

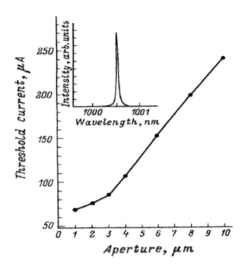

For copper heat sink and waveguide layer thickness of about 0.3 μm the
restructures show comparable results to the state-of-the-art quantum well
(QW) devices:

1. Low threshold InAs QD laser on InP substrate emitting at 1.84–1.9 μm has been
   fabricated.
2. Significant progress in the optical understanding of QD lasers with realistic
   parameters has been achieved.
3. QD lasers operating in the visible spectral range has been demonstrated.
4. Vertical-cavity surface emitting lasers (VCSELs) based on QDs with good
   properties have been demonstrated.
5. The QD vertical-cavity laser with parameters which fit to the best values for
   devices of similar geometry based on QWs (see Fig. 4.55) was demonstrated.
6. 1.31 μm lasing at room temperature with a threshold current density of
   240 A/cm^2 is demonstrated for the device based on InGaAs QDs in a GaAs
   matrix (Fig. 4.59).

4.7.6 Infrared Emission in Quantum Dot Lasers

In ultrathin layers or QWs, that exists a continuum of states at any energy above the
subband energy, as the in plane motion of charge carriers is not limited. If the
carrier is excited to the second subband, it relaxes to the first subband via emission
of a discrete quanta of energy an optical (LO) phonon. Due to the continuum nature
of electron states in a QW, there always exist states in the first subband to which
electron can scatter within 1 ps. Contrary, in QDs the relaxation time to the ground

sublevel takes typically 10–40 ps. The electron needs to emit a combination of different phonons to match the energy difference. This slowering increases the relative importance of the competing relaxation mechanism via emission of far-infrared (FIR) photons. The FIR emission was observed in QW and QD lasers. The intensity of the FIR spontaneous emission was about one order of magnitude higher in the QD case. Moreover, the FIR emission in QDs has a threshold character as it requires fast much higher intensity of the FIR emission in the QD case, hopefully, will be possible to create a new generation of FIR lasers [14].

4.7.7 *Extension of the Spectral Range of GaAs-Based Devices*

QDs allows a possibility to cover strategically important spectral ranges of 1.3 and 1.55 μm using GaAs substrates. This is particularly important for VCSELs where high quality monolithic AlAs. GaAs Bragg reflectors and developed oxide technology are available only on GaAs substrates. Recently it was discovered that associates of InAs QD is formed at low substrate temperatures emit light at wavelengths up to 1.8 μm at 300 K. The experimentally measured absorption coefficient for structures with stacked CdSe QDs in a ZnSe matrix in the direction perpendicular to the planes with nanoislands approaches $\alpha = 10^5$ cm^{-1}. High absorption coefficients and lack of exciton screening in dense arrays of QDs result in ultrahigh QD exciton (or, even higher exciton) gain values under generation of nonequilibrium carriers. Resonant waveguides are based on the effect of resonant enhancement of the refractive index (n) along the contour of the absorption (or gain) curve. To have a significant impact on the waveguiding properties of the media the absorption peak is to be strong enough ($\Delta n \sim 0.5$ for $\alpha \sim 10^5$ cm^{-1}). For resonant waveguiding it is not necessary to have external cladding of the active region with QDs by layers with significantly lower refractive indices. Practically, it means that lasers can be created in materials having no suitable lattice-matched heterocouple with lower refractive index.

In vertical-cavity surface-emitting (VCSEL) laser the effect of strong resonant modulation of the refractive index serves for self-adjustment of the cavity mode and lasing spectrum. As the material gain of a single QD reaches ultrahigh values due to δ-function-like density of states and negligible homogeneous broadening, even single quantum dots lasing may become possible. A highly reflective Bragg mirrors on both sides of the cavity are necessary for QW VCSELs, as relatively small maximum gain in these structures (about 10^3 cm^{-1}) require low external losses of the device. However, if the maximum gain can be made high enough, no necessity in highly reflective Bragg mirrors exists. For gain values exceeding 10^5 cm^{-1} and active layer thickness of 200 nm the facet (or mirror) reflectivity of the order of 30 % is enough to achieve vertical lasing. Due to the low finesse of the cavity and the self-adjustment effect no strict necessity in fitting of the cavity made and the gain spectrum exists. This effect was demonstrated for 20-times stacked CdSe submonolayer QD insertions in a ZnMgSe matrix grown on GaAs substrate.

4.7.8   Quantum Dot Composites

The gain of the array of QDs is not defined by a simple sum of gains of single QDs. Interaction of electromagnetic fields of anisotropic QDs or anisotropic QD lattices makes the splitting of the TE and TM modes for the same QD exciton transition in avoidable. This effect can result in splitting as large as several tens of meV, as was predicted theoretically and is proven experimentally. Maximum gain of the QD assemble is also a strong function of the relative arrangement of QDs, so-called quantum well lasers. Most of recent industrial QW lasers are based on thin layers of alloys used as active regions. It became clear now, that these layers, in most cases, exhibit quasiperiodic nanoscale compositional modulations creating in many cases dense arrays of quantum wire- or QD-like structures. By using the same average alloy composition the luminescence peak energy can be tuned by several hundreds meV by tuning the growth conditions. Careful evaluation of the impact of such effects on lasing characteristics of modern lasers is necessary to clear up the role of self-organized QDs or quantum wires in this case. QDs modified all the basic commandments of the double heterostructure (DHS) laser. DHS, DHS QW QD lattice matching undesirable, material gain orders of magnitude higher exciton screening is not important, homogeneous broadening at RT is small cladding with low n layers is not necessary VCSEL: Bragg reflectors and cavity are not necessary lasing in optical and near IR range and simultaneous FIR emission one family is not necessary (InAs/Si QDs) limited wavelength range on GaAs is extended to 1.8 μm. It appeared that the QD laser seems to be a completely new device with properties which can remarkably expand the possibilities in many application, rather than simply a laser with some parameters improved with respect to the DHS or DHS QW laser.

4.7.9   Advantages and Disadvantages of QD Lasers

Advantages

1. Adjustable wavelengths since energy levels rather than band-gap determine the wavelength;
2. Higher material gain for QD than for QW and quantum wires (QD gain is $10\times$ more than QW gain);
3. Material gain curve is narrower than for QW and quantum wires;
4. Small volume;
5. Low power needed;

6. High frequency operation possible;
7. Small linewidth of emission peak;
8. Low threshold current need compared to QW and quantum wires;
9. Superior temperature stability compared and suppressed diffusion of carriers compared to QW.

Disadvantages and problems:

1. Wavefunction is not zero at potential barriers and hence penetrate into it;
2. Modulation can be discontinuous, meaning that masses in wells and barriers differ;
3. Non-parabolicity of E-k which means that the mass changes with energy;
4. Multiple band in valance band (heavy and light holes);
5. Fabrication process can be complicated leading to nonhomogenous in size and shape leading to the broadening of the gain spectrum;
6. High material gain but low optical confinement factor leads to low modal gain;
7. Barriers are finite, not infinite, meaning that there is a carrier leakage out of the QD;
8. Strained wells might lead to shift in wavelength due to the existence of light and heavy holes in the valance band.

4.8   Femtosecon, UV, and Multiwavelength Lasers

4.8.1   Femtosecon Laser

The two photon optical storage need of ultrashort pulses laser which has to be under 100 fs or lower. In a ring fiber laser, a short length of core-pumped Er-doped fiber acts as the gain medium. A saturable absorber (or polarization-sensitive element exploiting the effects of nonlinear polarization rotation) embedded within the cavity favors mode-locked operation over continuous-wave laser activity. Pulse duration may be controlled by a variable dispersion control unit. The passively mode-locked ring oscillator can easily generate ultrashort pulses (≤ 150 fs) at a center wavelength of 1.55 μm with a repetition rate of 100 MHz. With the high gain available from Er-doped fibers, the optical power extracted from the oscillator may be used to seed one or more optical amplifiers. The pulse train typically exhibits an average power of more than 250 mW or 2.5 nJ/pulse. Since these ultrashort pulses are generated in single-mode fibers, their transverse profile has a perfect TEM00 shape, which can be focused to diffraction-limited spots. Several options further enhance the scope of femtosecond fiber lasers: Integrating a highly nonlinear fiber into the ring allows the generation of an octave-spanning super-continuum from 1050 to 2100 nm. Pulse compression of this broadband spectral distribution to sub-30 fs pulses and tunable frequency-doubling into the visible or near-infrared regime open new way for application. A frequency-doubling module connected directly to the amplifier output allows efficient conversion to 525 nm with output powers exceeding 60 mW or 0.6 nJ/pulse that was used to two-photon absorption multilayer storage experiment, can see Chap. 6 in this book in detail.

4.8.2 Supercontinuum Generation (SCG) Subpicosecond Laser

The short (typically subpicosecond) laser pulse can be converted to light with a very broad spectral bandwidth with photonic crystal fibers (PCFs). While temporal coherence is lost in the process, spatial coherence usually remains high. The spectral broadening is accomplished by propagating optical pulses through a highly nonlinear medium such as a PCFs, that unusual chromatic dispersion allows strong nonlinear interaction over a significant length of the fiber. Even with fairly moderate input powers, very broad spectra can be achieved. Although SCG can be observed in a drop of water given enough pumping power, PCFs are ideal media for SCG as the dispersion can be designed to facilitate continuum generation in a specific band. PCFs with a large nonlinear coefficient are available with a wide range of unique zero-dispersion wavelengths. For single-transverse-mode operation, the fibers are typically designed with relatively small air-holes. A potential application of supercontinuum pulsed lasers to optical data storage involves plasmatic nanostructures tuned to specific wavelengths within a broad range of optical frequencies. By providing simultaneous access to UV, visible, and near IR wavelengths, high-repetition-rate super continuum pulses may hold the key to substantial increases in storage density as well as data-transfer rates.

4.8.3 UV and Deep UV Lasers

Although some progress has been made in the development of UV laser sources, the major obstacle remains the extremely low external quantum efficiency (EQE) of the nitride emitters; there is a notable drop in efficiency as emission wavelengths approach deep UV. The EQE for LEDs is defined as the product of the internal quantum efficiency, carrier-injection efficiency, and light-extraction efficiency. The internal quantum efficiency is related to the crystalline quality of the epitaxial layers; fewer defects and dislocations lead to better performance.

If the device structures are grown on sapphire substrates, the lattice-mismatch—induced defects and dislocations (approaching 10^{10} cm^{-2} in some cases) cannot be reduced to a desired level on the order of 10^{4} cm^{-2} or lower. Two other issues that affect mainly the carrier-injection efficiency are a difficulty in achieving highly doped p-type AlGaN cladding and barrier layers, and reduced magnitude of the band discontinuities at key heterojunction interfaces. Also, UV light is absorbed at the p-side electrode.

To minimize the impact of lattice-mismatch-induced defects and dislocations, bulk AlN material has recently been proposed and used as a substrate for UV light emitters. The AlN substrate leads to lower lattice and thermal mismatches between it and subsequent AlGaN layers. To reduce dislocations when bulk AlN is used as the substrate, the AlGaN layers grown on top must be sufficiently thin if the Al composition in them is high to guarantee that the structure is pseudomorphic (that is, they maintain the crystalline structure of the substrate). This may represent an unacceptable device design compromise. Although improved performance characteristics in UV LEDs that use this approach has been reported, the EQE for deep UV LEDs (<350 nm emission wavelength) remains low (<2 %).

In addition to all the known problems that prevent the fabrication of highly efficient UV emitters, another limitation is the amount of residual strain in the device structures. Residual strain modifies the valence-band structure of AlGaN layers used in UV devices in a way that can negatively affect the emitted light polarization properties.

To complicate matters, UV light emitted from an active $Al_xG_{1-x}N/Al_yGa_{1-y}N$ multi-quantum-well structure can switch its polarization from transverse electric (TE) to transverse magnetic (TM), depending on the amount of residual strain present in the active layer—which further depends on the choice of template/substrate. When the emitted light is TM-polarized, it is difficult to extract along the c-axis (the direction along which the layers are grown). This leads to decreased light-extraction efficiency, and thus lower overall EQE. Thus, it is important to know in advance the aluminum composition of an $Al_xGa_{1-x}N/Al_yGa_{1-y}N$ quantum-well structure and its relationship to either the bulk AlN substrate used or the $Al_zGa_{1-z}N$ ($x < y < z$) template-on-sapphire substrate. These factors control the amount of residual strain in the active layer.

The ideal situation is where the Al content (and residual strain) does not lead to emitted light with a polarization that makes light extraction difficult. This basically implies that once the operating wavelength for a UV emitter has been selected, the active region and its associated layer structure should dictate the choice of template/substrate. This is not how UV emitters are conventionally designed. The usual approach is to accept the lattice and thermal constraints dictated by the template/substrate—which makes it difficult to decide how much residual strain can be managed.

4.8.4   Multiwavelength Lasers

The multiwavelength lasers are more important for multiwavelength optical storage. A model of three wavelength lasers is shown in Fig. 4.60, which are 405, 680, and 850 nm. The new multiwavelength lasers suited different multiwavelength lasers optical storage will be developed in the future with progress of materials science

Fig. 4.60 A model of three
wavelength laser

Fig. 4.61 Band gaps of UV
nitride ternary alloys (AlGaN
with a bowing parameter
$b = 0.89$ eV, InAlN-1 with
$b = 5$ eV) plotted against the
c-axis lattice constant

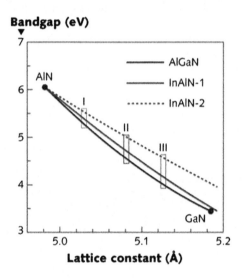

and IC process. But there are lot of technology obstructions, specially, the conventional materials of manufacturing semiconductor lasers cannot be adapt to make multiwavelength lasers yet (Fig. 4.61).

4.8.5 Application-Oriented Nitride

The one of most important methods to ultimate solution UV laser is the oriented nitride, that is to use a substrate whole lattice constant closely matches the lattice constant of thick barrier layers in the UV device structure. At best one should strive to design the device structure to be pseudomorphic—with minimal residual strain at the barrier/template interface. The rest of the layers can be chosen to have some

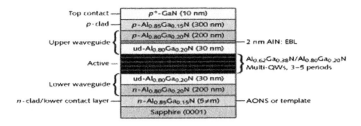

Fig. 4.62 A schematic shows the layer structure of a pseudomorphic LED designed to emit at about 240 nm. The device structure includes an AONS, multiple quantum wells (QWs), and an electron-blocking layer (EBL)

residual strain with the appropriate thicknesses necessary to ensure pseudomorphic growth. In the absence of bulk binary nitride substrates that could meet the lattice-matching requirements for photonic-device structures that span the potential spectral band enabled by nitrides and development of new application-oriented nitride substrates (AONS). In one example, a possible LED structure designed to emit at about 240 nm in the deep UV region is based on this concept as shown in Fig. 4.62. The dotted-red line (InAlN-2) is for a modified band gap of the InAlN alloy using a low value of $b = 2.5$ eV. Vertical bars denote the range of band gaps for a particular lattice constant accessible via AONS.

The need of AONS is best explained by considering the variation of the band gap and wavelength of $Al_xGa_{1-x}N$ as a function of lattice constant (can see Fig. 4.57). The bandgap is plotted as a function of the c-axis lattice constant C_a, one could just as well have plotted it as a function of the F_a lattice constant. For AlN with a bandgap of 6.13 eV ($x = 1$), c_o is 4.9816 Å; for GaN, with a bandgap of 3.43 eV ($x = 0$), it is 5.1815 Å. This range of bandgaps corresponds to UV wavelengths from about 202 nm to about 360 nm. For any other value of the Al mole fraction, x, the lattice constant lies between the two extremes. To minimize lattice-mismatch induced defects and dislocations, the $Al_xG_{1-x}N/Al_yGa_{1-y}N$ active structure should be grown on a substrate whose lattice constant closely matches that of the thickest structural layer. The substrates are necessarily bulk AlGaN ternary alloys; neither the bulk AlN binary alloy nor sapphire possesses the right lattice constant to match that of any ternary ($Al_xGa_{1-x}N$ or $In_xAl_{1-x}N$) or quaternary layers in any canonical nitride light-emitting device structure. As an alternative alloy for active layers, one could use InAlN as shown in Fig. 4.57 to cover the same UV spectral region and the other UV wavelengths between 359 and 400 nm.

Application-oriented nitride substrates are bulk $Al_xGa_{1-x}N$ ternary substrates with Al compositions chosen to lattice-match thick barrier layers. In some cases, an Al composition could be found where a fairly broad spectral region can be covered by a single AONS. The concept can be extended into the visible and to the IR region through use of $In_xGa_{1-x}N$ ternary substrates. No ternary nitride substrates currently exist. The challenges are balanced by the potential benefits for a range of electronic and photonic devices based on nitride compounds.

Development various wavelength lasers will be provided with following advantages for optical storage:

1. The various wavelength lasers, especially UV and deep UV, are very useful for multiwavelength and multilevel storage as mort photochromic materials are sensitivity to UV and deep UV;
2. Two-photon recording medium, information recording process from a single photon effect to the two-photon effect, to realize three-dimensional storage, and 3D optical solid state memory;
3. In favor of variety of optical parameters use to store information at the same time with multichannel to the increased read and write data rate;
4. Can establish a new codec system as multidimensional codes is great multi-degree tour length coding to achieve the sector guidance, correction improved the effective capacity and reduce redundancy;
5. Various wavelength lasers, photodetectors, and the digital processing circuit hybrid integration of different wavelengths that conduce to miniaturization of storage system and greatly reduce the system size, energy consumption, and improve storage system reliability.

The alloy semiconductor family, the aluminum gallium nitride AlGaN alloys are the most versatile for design and fabrication of emitters that span the UV spectrum from the UV-A (400–320 nm) through the UV-B (320–280 nm) to the UV-C (200–300 nm) as Fig. 4.57. Most devices have been grown on sapphire substrates and active region is grown. A typically deposits of AlN buffer layer on top of the sapphire substrate as a lattice constant bridge that mediates between the sapphire and subsequent AlGaN layers. This is followed silicon-doped $Al_yGa_{1-y}N$ cladding layer on top with active region, composed of $Al_xG_{1-x}N/Al_yGa_{1-y}N$ ($y < x \geq 0.6$) multiple quantum wells. A top cladding layer made of p-type $Al_yGa_{1-y}N$ is grown. Before the top contact layer is grown, many schemes are used to block the escape of electrons from the active region. The first incorporates a high-aluminum-content $Al_zGa_{1-z}N$ quantum barrier ($z \geq 0.9$). This barrier is followed by a heavily doped p-type GaN contact layer. In a second form of the electron-blocking scheme, researchers have used a multi-quantum-barrier structure in place of the single quantum barrier. In the latter case, the implementation employs several periods of an $Al_xGa_{1-x}N/Al_z Ga_{1-z}N$ structure that yield slightly better operating characteristics.

Optical media in the next generations will be more complex. Required high capacity, data rate, throughput, and quality will be harder to achieve, regardless of the future technology researchers. The cost and complexity of processes and equipment and the unit cost of media will increase, perhaps significantly in some cases. A major challenge to the industry is to prevent or minimize those difficulty. New or modified processes, manufacturing equipment, and quality control methods will be required for N-layer MLD and NFR media. More sophisticated and complex in-line and off-line test and measurement equipment will be required. More materials scientists, chemists, and physicists will cooperate to provide a natural evolutionary path and focus on future systems and key components, including various laser sources of course.

References

1. C. Liu, S.-H. Park, Numerical analysis of an annular-aperture solid immersion lens. Opt. Lett. **29**, 1742–1744 (2004)
2. Y. Zhang, X. Ye, Three-zone phase-only filter increasing the focal depth of optical storage systems with a solid immersion lens. Appl. Phys. B **86**, 97–103 (2007)
3. D. Xu, *Super-Density and High-Speed Optical Data Storage* (Liao-Ning Science and Technology Publishing House, Shen-Yang, 2009). ISBN: 978-7-5381-6248-6
4. S.Y. Park, J.I. Seo, J.Y. Jung, J.Y. Bae, B.S. Bae, Photoluminescence of mesoporous silica films impregnated with an erbium complex. J. Mater. Res. **18**, 1039 (2003)
5. G.W. Burr, Three-dimensional optical storage. in *SPIE Conference on Nano-and Micro-optics for Information Systems Presented at SPIE Optics and Photonics*, paper 5225-16, 2007
6. J. Tominaga, H. Fuji, A. Sato et al., The characteristics and the potential of super resolution near-field structure. Japan. J. Appl. Phys. **39**(1), 957 (2000) (Number 2B)
7. C.-H. Li, C.-M. Zheng, H.-P. Zeng, synthesis and photochromic property of 1-(spirobi [fluorene]-2-yl)-3, 4-bis(2,5-dimethylfuran-3-yl) -2,5-dihydro-1H-pyrrole. Chin. J. Org. Chem. **31**(05), 659–664 (2011)
8. Y. Chen, M.-L. Pang, K.-G. Cheng, Y. Wang, J. Han, J.-B. Meng. Synthesis and properties of brominated 6,6-dimethyl-[2,2-bi-1H-indene]-3,3-diethyl-3,3-dihydroxy-1,1 -diones, Chin. J. Org. Chem. **28**(7), 1240–1246 (2010)
9. M.-L. Pang, T.-T. Yang, J.-J. Li, S.-H. Yang, Z.-G. Lou, J. Han, J.-B. Meng, Synthesis and properties of novel photochromic spiropyran compounds with n-heterocyclic residue, Chin. J. Org. Chem. **68**(18), 1895–1902 (2010)
10. D. Zhang, M. Wang, Y.-L. Tan, Preparation of porous nano-barium-strontium titanate by sorghum straw template method and its adsorption capability for heavy metal ions. Chin. J. Org. Chem. **68**(16), 1641–1648 (2010)
11. Y. Kozawa, S. Sato, Sharper focal spot formed by higher-order radially polarized laser beams. J. Opt. Soc. Am. A **24**, 1793–1798 (2007)
12. J. Hamazaki, A. Kawamoto, R. Morita, T. Omatsu, Direct production of high-power radially polarized output from a side-pumped Nd:YVO4 bounce amplifier using a photonic crystal mirror. Opt. Express **16**, 10762–10768 (2008)
13. T. Moser, H. Glur, V. Romano, F. Pigeon, O. Parriaux, M.A. Ahmed, T. Graf, Polarization-selective grating mirrors used in the generation of radial polarization. Appl. Phys. B **80**, 707–713 (2005)
14. Y. Kozawa, S. Sato, Generation of a radially polarized laser beam by use of a conical Brewster prism. Opt. Lett. **30**, 3063–3065 (2005)

Chapter 5
Multi-wavelength and Multi-level (MW/ML) Storage Systems

The chapter presents the principles of multi-wavelength and multi-level (MW/ML) optical storage system design including: mathematical model, analysis and calculation of effective factors and configuration parameters, optical channel characteristics, synthesis evaluation, crosstalk, non-destructive readout, multi-dimension encoding, error correction, multi-level CD-ROM, Blu-ray drive, mastering system for MW/ML storage. In addition, the chapter also introduces principles of the lithography and duplication process of MW/ML optical storage based on conventional CD and IC technology. Meanwhile, same important experimental apparatus and experimental results are introduced also, as nonlinear absorption super-resolution storage experiment, phase shift super-resolution mask making, high-speed multi-dimension read out signal testing, large numerical aperture optical system design for multi-wavelength lasers and blue light super-resolution storage experimental system. A experiment based on near-field optical principles and correlative technology with changing the space phase and intensity distribution of the focused beam to improve the optical system transfer function to raise the resolution of the optical system, that has been used for MW/ML storage. The multi-level error code correction and multi-level run-length limited (RLL) code are great potential projects for multi-dimension, whether mathematical model method or processing circuitry have potential development.

The CD-ROM of MW/ML disc copy is a new kind of mastering disc replication processes, including multi-wavelength multi-level optical disc mastering system designed, optical disc replication and testing technology, which is an important extension for MW/ML optical storage. At last in this chapter introduced two kind of practical multi-level optical disc storage systems: multi-level Blu-ray Disc drive and multi-level rewritable storage with phase change medium, experimental results and prototypes.

© Tsinghua University Press and Springer Science+Business Media Singapore 2016 301
D. Xu, *Multi-dimensional Optical Storage*,
DOI 10.1007/978-981-10-0932-7_5

5.1   Configuration of MW/ML Storage Optical System

5.1.1   Multi-function MW/ML Storage Experimental System

The section presents a typical MW/ML optical storage system which is very important experimental equipment for the MW/ML technology research and development with multi-function. It is a most representative result of OMNERC at Tsinghua University.

The multi-function MW/ML optical storage experimental system is shown in Fig. 5.1a. Functions of the experimental system are shown in Table 5.1. A static read and write system is shown in Fig. 5.1b. There are two especial requires to the system: first is very high positioning accuracy of 3D and exposure energy control for multiwavelength lasers indeed. So it has a three-dimensional precision stage of positioning accuracy better than 20 nm. Second is the combination of many different wavelength laser sources and high accuracy modulation independently. The experimental system is not application only to study of multi-wavelength storage process, verify and validate the theoretical model of the storage mechanism. It can be used to test characteristic, parameters of the recording medium, and improve the recording performances of the recording film in further engineering application with light-induced photochromic materials. Moreover, the experimental system has a high accuracy air-bearing spinning table, which radial runout is within 25 nm, angle positioning accuracy is better than 0.1 s. So it can be mastering system for MW/ML ROM also.

Therefore, the experimental system can be used to experiment of MW/ML continuous data storage write and read out on the disc without pregroove or servo track. On these devices to carry out the experimental study of a series of read/write signal of the channel characteristics, selective absorption for different wavelength, mechanism of different light intensity nonlinear absorption and multi-wavelength optical system signal processing methods for MW/ML optical storage. Its optical

(a)   **(b)**

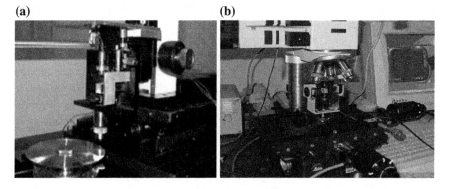

Fig. 5.1 The dynamic experimental system with air bearing high speed spinning table (**a**) and (**b**). Three-wavelength multi-level storage experimental system

Table 5.1 Main performances and functions of the MW/ML optical storage experimental system

Experimental performances	Features and functions
Read and write laser coupling and separation experiments	Modular of multi-channel beam combiner, the spectrophotometric device, the optical path adjustment mechanism
Data-parallel read and write flexible interface	Encoding and decoding of data, writing and read out control in parallel
Multilevel modulation and readout signal analysis evaluation	laser power, writing pulse width Control, and multilevel writing strategy experiment
Testing of absorptivity, reflectivity and sensitivity etc. properties of record materials	Reading/writing laser power calibration, photo-digital converter demarcation, cyclic fatigue experiments control and data processing software testing
Experiments of photo-electronic specification and parameters for system design	CCD detection and readout signal processing, testing analysis of the light intensity to read and write
Experiments of positioning precision, sensitivity, dependability and repeated stability of mechanical motion system	Four-dimensional closed-loop control precision work platforms: mobile mechanical positioning accuracy of ± 20 nm, angle positioning accuracy of ± 0.1 s
High-speed dynamic continuous reading and writing testing and results analysis experiments	Programmable laser pulse modulation coding control, high-speed reading/writing data processing, and repeatability of the experimental medium without servo track or pregroove and master
Overall system function and control software testing and analysis	PC interface, the D/A module, an appropriative control program and system analysis software

system is shown in Fig. 5.2, where: 1—semiconductor laser, 2—collimator, 3—calibration aperture, 4—polarization beam splitter, 5—1/4 wave plate, 6—assistant correction focus lens, 7—cylindrical lens, 8—narrowband filter, 9—detector, 10—beam combiner/splitter, 11—moving beam splitter, 12—lens, 13—detectors of CCD, 14—achromatic objective lens, 15—the objective lens of pickup, 16—disc or experimental samples. The experimental system employs two group lasers:

1. 405, 532 and 650 nm with an apochromatic objective lens;
2. 532, 650, 780 or 850 nm lasers with each assistant correction focus lens.

The first order Bessel function of the wide-band apochromatic objective lens is shown in Fig. 5.3. The block diagram of control system of the photochromic MW/ML storage experiment is shown in Fig. 5.4.

The photochromic MW/ML storage experimental system (as in Figs. 5.2 and 5.4) has 650, 532 and 405 nm wavelength semiconductor lasers (L_1, L_2, L_3) with modulation independently, that were conformed a coaxial beam by a beam coupling system to combine into a bunch of coaxial multi-wavelength collimation light for writing or reading. The coaxial multi-wavelength lasers beam into the achromatic

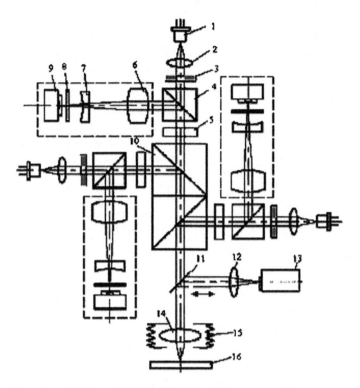

Fig. 5.2 The truth optical system for the MW/ML optical storage experiment

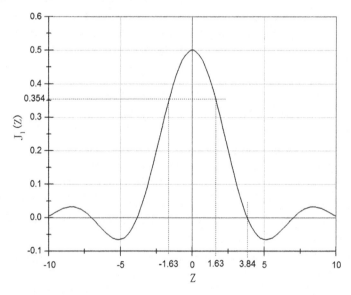

Fig. 5.3 The first order Bessel function (J_1) curves of the apochromatic objective lens for the MW/ML optical storage experimental system

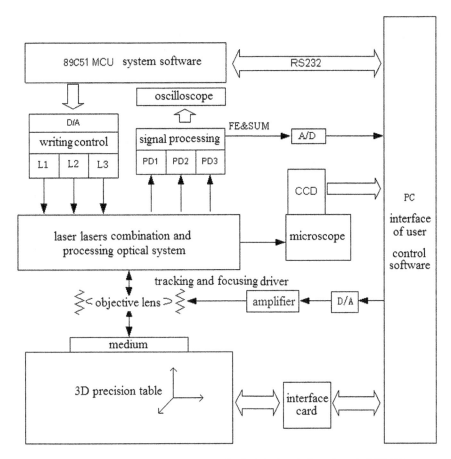

Fig. 5.4 The block diagram of control system for light-induced photochromic MW/ML storage experiment

objective lens focus on the experimental samples or disc for writing or reading when the beam splitter 11 moving out. If the beam splitter 11 moving in this system, it can be used to testing recording medium or master disc. Laser beams is reflected by sample reflective layer and back to the multi-wavelength beam coupling and splitter system, it was separated into three independent readout beam to photodetector PD_1, PD_2 and PD_3, then the read out signals are converted into electrical signals and complete readout signal extraction. Write and read lasers power and pulse width (exposure time) control are completed by the write control module. The module is consisted of the 89C51 microcontroller through the D/An interface. Microcontroller gets read and write commands from the RS232 interface of PC. Experimental samples on the plat of the three-dimensional precision working table, its movement was controlled by the PC via a card driver with accuracy of higher than 20 nm repeatability of positioning on the X, Y, and Z directions.

When the system startup, computer-controlled three-dimensional table and piezoelectric ceramic drive the optical head and objective lens to achieve the correct focus. Spot shape of read-write experiment can be analysis by the microscope with CCD camera. The microscopic images was observation by the CCD camera and analysis by the PC in real-time. Focus, displacement, read and write control by the SCM system software and PC control software, user interface software, unified and coordinated management. The drive mode and drive interface for each laser with highest modulation frequency that is need of different performances and the external circuit control to finalize the parameters and control program. The highly integrated semiconductor laser driver AD91660 and AD91661 were employed for laser read/write power control with a computer D/An interface. The AD91660 is suitable to the driver of the N-type semiconductor laser, but the AD91661 is available for the P-type laser also. The specific characteristics and working principles of the AD91660 driver chip was introduced as an example as following.

The parameters of the AD91660 chip are shown in Table 5.2, its internal structure and function is shown in block diagram Fig. 5.5. There are three different current output, write current (I_{WRITE}), bias current (I_{BIAS}) and the compensation current (I_{OFFSET}) of AD91660, by the logic control of their combination of output. The first two circuits are closed-loop control, the latter is open loop.

Laser constant power control is achieved by periodic calibration, the timing diagram of the calibration process is shown in Fig. 5.6. The power feedback signal

Table 5.2 The output parameters of the AD91660 chip	List	Parameters
	Up time	≥ 1.5 ns
	Down time	≥ 2.0 ns
	Output current	0–180 mA
	Bias current	0–90 mA
	Compensation current	0–30 mA
	Switching rate	0–200 MHz

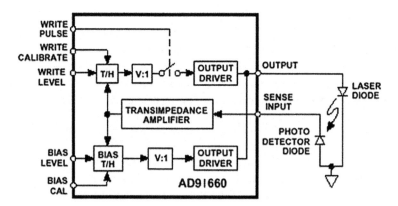

Fig. 5.5 The internal structure and function of the AD91660 in block

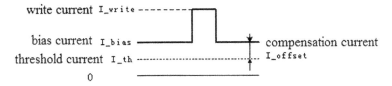

Fig. 5.6 The timing diagram of the calibration process for laser constant power control

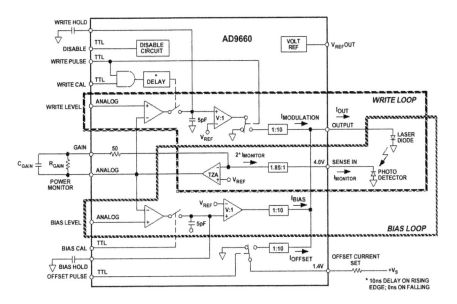

Fig. 5.7 Inner workings process and main function of the AD91660 chip

of AD91960 is from an external photodetector, and used to control the output current of write circuit, in order to achieve the most stable power output. Pulse-driven semiconductor laser with constant power drive are very different. Semiconductor lasers are damaged very easy by working electronic current. Its control current is shown in Fig. 5.7 that the drive current from 0 to instantly rise to the operating current may cause permanent damage to the semiconductor laser.

For continuous and shorter single-pulse frequency laser driving, requirement of slower start driving current, then rising to bias current I_{BIAS} that is greater than the threshold current I_{th} of the laser to set it into the lasing state as in Fig. 5.7. The semiconductor lasers can be very steep output power of less than 1 ns rise and falling edge based on the I_{BIAS}. It is appropriate bias current, write current and the compensation current that is the key to achieve high-quality laser pulse control. The working process and the main function of the AD91660 chip is described in Fig. 5.7, the bias circuit (bias loop) and writing loop to control the amplitude of the output current I_{out}, the detector (photodetector) from the SENSE IN pin output of the feedback current $I_{MONITOR}$, it is proportional to the input analog voltage of BIAS LEVEL and WRITE LEVEL pin.

On the other hand, the photodetector output current of the laser (LD) is proportional to the output power, therefore by changing the BIAS LEVEL, WRITE LEVEL input voltage can easily adjust the laser output power. The entire control system is periodic calibration with routing control logic to prevent the power down by the laser aging. Compensation current open loop output current I_{OFFSET}, I_{OFFSET}, it can be adjusted through an external resistor. The AD91660 can achieve continuous laser output, the output of the single pulse and continuous modulation of the pulse output function. Continuous laser output power can be adjusted by the BIAS LEVEL pin input voltage. The WRITE PULSE pin input, TTL or CMOS is compatible pulse signal, it can output a single pulse and continuous modulation of the pulse laser. The laser pulse width modulation mode was controlled by the microcontroller through IO port.

In order to ensure the stability of output power, the AD91660 in each start and continuous work has to be periodic calibration. The calibration process is shown in Fig. 5.8, when the DISABLE signal was low, in the first calibration bias current is closed, the emitted light was detected by detector PD, the current/voltage converter a feedback voltage is compared with reference voltage VREF. The calibration accuracy will reach the predetermined value by the control BIAS_HOLD.

Write current calibration and the bias current calibration are similar, when the modulation frequency is very higher, due to time lag, closed-loop control cannot work well, and so periodic calibration can improve the control accuracy. The AD9660's calibration is divided two steps, one initialized the calibration (the Initial Calibration), which is repeat calibration (recalibration) longer than the former need. Calibration and laser waveform is controlled by TTL level, therefore it is easy through the computer I/O to do. The writing power and bias power through the pin

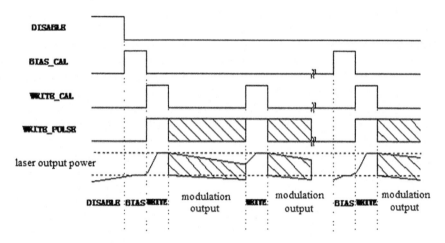

Fig. 5.8 Timing calibration process of AD91660 carried out periodic calibration

WRITE_LEVEL and BIAS_LEVEL are controlled by analog voltage compensation and adjusting the resistance on the OFFSET_CURRENT_SET pin.
Microcontroller through the I/O ports control D/A conversion device generated a
certain voltage. The voltage was sent to the AD91660 input to complete the writing
and bias power control. Compensation power control is adjusted by the poten-
tiometer. The AD91660 digital control signal is used to reading and writing control,
unify and coordinate by pulse width control circuit. Readout signal (from
four-quadrant photodiode) was amplification and generation readout signals and
focus error signal. The readout signal processing module is ensured the normal read
and focusing accurate.

The relative curves between sensitivity S and the incident wavelength λ of the
photodiodes is shown in Fig. 5.9a. Germanium photodiode spectral range is
between about 0.4–1.8 μm, the peak wavelength is about 1.4–1.5 μm. Silicon
photodiode spectral range is between 0.4 and 1.2 μm approximately, its peak was
wavelength of 0.8–0.9 μm. This experimental system used laser wavelength range
of 405–850 nm, so selected silicon photodiode as well. The property of
volta-current of the photodiode is shown in Fig. 5.9b. Its output characteristics are
very similar to transistor under the conditions of different conversion base current
I_0, but using illuminance or luminous flux to replace base current. In addition, when
the bias voltage is 0, the current is not normalized to 0, but is the corresponding
short-circuit current under photovoltaic effect by light. However, similar to the
transistor, the bias voltage, photodiode is working in the linear region always. The
four-quadrant photodetector D quadrant photoelectric signal amplifying circuit is
shown in Fig. 5.10 that the differencing combination of the four signals quadrants
(A, B, C, D) of photoelectric detector is available use to the auto-focusing signal,
i.e. UA + UB + UC + UD is read out signal SUM, but (UA + UC) − (UB + UD)
is focusing signal FE.

The focusing signal was utilized a differential amplifier, it can overcome adjust
gain to cause c changing of circuit symmetry. The system using the apochromatic
objective, its depth of focus is less than 1 μm. Auto-focusing control accuracy of

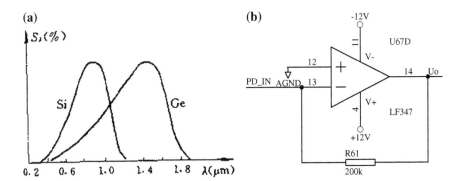

Fig. 5.9 **a** The curves of sensitivity S and the incident wavelength λ of silicon and germanium
photodiodes. **b** The property of volta-current of the photodiode

(a)   **(b)**

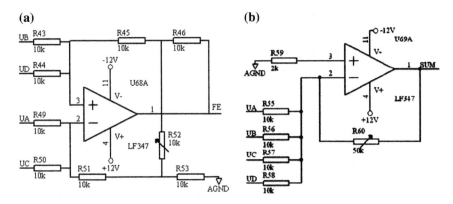

Fig. 5.10 The signal amplifying and process circuit of the four-quadrant photodetector D **a** for focusing control, **b** for read out

the system should be ± 0.5 μm. The 650, 532 and 405 nm laser beam are focused on the recording layer by the achromatic objective lens, so taken the 532 nm laser to be automatic focusing control signal. As the structure of objective lens of the system is more complex, that itself quality is up to 120 g, the existing moving coil actuator cannot drive it, so to adopt piezoelectric ceramic nano-displacement drive P720 with maximum moving displacement 100 μm, sensibility 10 nm and positioning accuracy of ± 20 nm. The drive voltage of P720 is between -20 and 120 V with the E503 amplifier module that can be matching with control voltage -2 to 12 V of computer, then through the computer and the DA converter drive piezoelectric ceramic.

The focusing of the system employs the Z-axis scanning focusing method, its focusing principles is shown in Fig. 5.11. Piezoelectric ceramics with apochromatic objective lens head is mounted on the Z axis of three-dimensional precision stage.

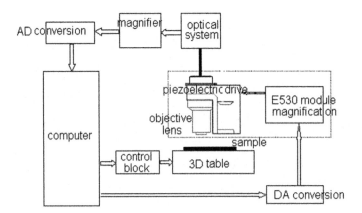

Fig. 5.11 The auto-focusing system on the Z-axis with piezoelectric ceramics device

The focusing at the beginning, by the computer control table along Z axis stepper movement with 20 nm resolution, computer acquisition AD signal from the four quadrant photodiodes and process to focusing. Computer through the DA converter determine the location of objective lens to control the piezoelectric ceramic drive stepping down within 50 nm to keep the move close to the focus on the linear region until the focus error signal to about 0. On the contrary, the objective lens to move in opposite directions to ensure that the objective has always been precisely focused on the recording layer of the experimental medium. Z direction of work-table moving control frequency is about 360 Hz, all the control process with computer, modify the program to adjust the focus accuracy and speed easily. This method is very high focusing accuracy, but the reaction rate and the working stroke are relatively lower. Computer can strictly control the moving distance after focusing, due to the glass substrate is high flatness with 1 μm/100 mm, so the system is relatively easy to implement and to ensure the focus accuracy when medium high speed moving.

5.1.2   Exposure Energy Control for MW/ML Recording

In this experimental system, the laser driver module is consisted of the 89C51A microcontroller and through IO port unified control. SCM can be achieved through a simple interface circuit and D/A converter links. Meanwhile, the processes of RS232 serial communication, microcontroller program, AD/DA data acquisition card programming, communication with microcontroller and DLL platform control are following functions:

1. The laser power control: three-channel D/A module, respectively, the bias of the three wavelengths of the read/write laser power setting. And completed the periodic calibration to the read/write laser power.
2. Read/write pulse control: according to the preset read/write signals of different frequency from programming and duty cycle pulse sequence for continuous reading and writing test.
3. RS232 serial communication: MCU and a PC computer through RS232 serial interface associated SCM various functional modules was realized by the control PC via RS232 interface. Microcontroller interface circuit and interface protocols ensured the correct communication.
4. Using modular assembly language microcontroller program achieved the function of each program from a PC via RS232 interface, scalable to add capabilities as need. The control computer using C++ programming and has the following features: Through the input interface of the laser writing power and pulse width, to control of the lasers source. Through the user interface set of data of the various functional modules of the system and send various control commands.

5. AD/DA data card programming focus error and readout signal acquisition and control the movement of the worktable. DA channel data card to control the PZT to drive the piezoelectric pottery drive optical head with objective lens to achieve precise focus.

6. The generated laser control parameters and control commands, and control of the user input data into control data for the microcontroller.

7. Serial communication with microcontroller, and pass the control data and commands; RS232 serial communication between PC and MCU interface protocol to send and receive data and instructions.

8. DLL platform control: programming and working platform for all kinds of sports mode, the platform motion mode is set through a unified user interface.

The experiment system was achieved precision: the X–Y worktable accuracy of 20 nm, the turntable resolution 0.1 arc-sec, the numerical aperture of apochromatic objective lens is 0.95, focusing accuracy ± 0.1 μm, CCD detector system optical magnification of $3000\times$ and the total magnification of $25,000\times$.

5.1.3 MW/ML Optical Storage Drive with Super-RENS Disc

Based on research above multi-function MW/ML optical storage experimental system, OMNERC of Tsinghua University developed a three wavelength and multi-level optical storage drive prototype and three wavelength recording disc with super-RENS layer The drive prototype is sameness with conventional CD drive except its pick up and compatible with DVD and Blu-disc, as the system employs 405, 650 and 780 nm wavelength lasers as in Fig. 5.12a. The miniaturization three wavelength optical pick up is shown in Fig. 5.12b. Its key part is hybrid integration and miniaturization module is shown in Fig. 1.7 of Chap. 1 in this book. Because 532 nm laser has not commercial product, so the triple-wavelength pick up is composed of 405, 650 and 780 nm lasers and corresponding photodetectores that are integrated on a polycarbonate substrate by precision injection.

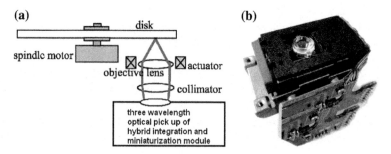

Fig. 5.12 The configuration of three wavelength and multi-level optical storage drive prototype (a) and miniaturization three wavelength optical pick up with 405, 650 and 780 nm lasers (b)

Due to the drive prototype was abrogated complex apochromatic objective lens, so its focus size has obvious chromatic aberration and have to adopt super-resolution technology to control and reform the focus of three wavelength lasers. The super-resolution technology has been introduced in Chap. 4 of this book. It's most important application is in multi-wavelength optical storage indeed. Except inorganic materials, some organic materials can be used to super-resolution structure (super-RENS) as the non-linear photochromic materials for example to utilize the super-RENS optical data storage in practice. The super-RENS mask layer has super-RENS effect to get smaller spots on the disc and increase the storage intensity. The photochemical super-resolution structure (P-super-RENS) mask was developed in OMNERC in 2000. Compared with inorganic materials super-RENS, P-super-RENS has many advantages, such as higher sensitivity, intensity and better manufacturability. Based on this technology, the $3T$ run-length is reduced from 0.84 to 0.48 μm and the spacing is reduced from 1.6 to 0.85 μm with good quality of the signal. So, P-super-RENS mask was adopted to multiwavelength and multilevel storage. The main reason is that can make the spots of indifferent wavelengths to be a smaller uniform size and increase the storage intensity of course.

The principles and structure of the disc with the P-super-RENS mask is shown in Fig. 5.13. Comparison with conventional discs, the P-super-RENS mask is added between substrate and the recording dye layer. When the laser beam scan the initialized P-super-RENS mask layer (colored), the irradiated zone to be bleached with a photochemical reaction only. Therefore other reading or writing lasers beam can penetrate mask layer to recording layer. Since the P-super-RENS mask layer has a threshold for bleached reaction, so the effective spot on P-super-RENS mask layer is smaller than the original diffraction spot and therefore achieved the super-resolution as shown in Fig. 5.13a. The reflectivity of bleach zone is increased rapidly as in Fig. 5.13b.

The schemes comprise a plenty of advanced technology, such as multi-layer coating, multi-color recording dyes, photochromic super-resolution mask materials and effective longitudinal coding. They can adopt both the similar focusing and the tracking way as conventional optical disc drivers, so it apply in industry more easily (Fig. 5.14).

Fig. 5.13 The structure of the disc with a P-super-RENS mask

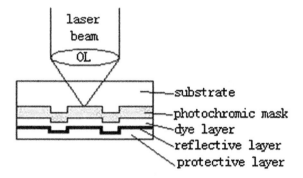

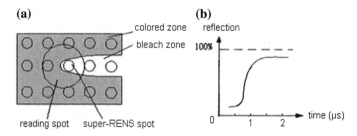

Fig. 5.14 The principium sketching of multi-wavelength disc with P-super-RENS mask layer

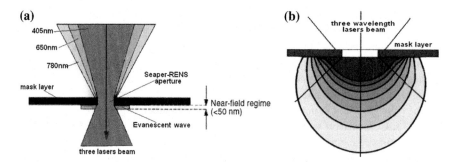

Fig. 5.15 **a** The principles of 405–780 nm wavelength lasers irradiate to mask layer, and come into being the near-field optics effect. **b** The light field distribution of three-beams penetrate through the super-resolution aperture on mask layer

The non-linearity of the photochromic media is caused by the macroscopic reaction speed nonuniformity. Increasing the difference between the total reaction speed and the local reaction speed can improve the super-resolution effect. Static testing experiment using two samples with reflection and transmission demonstrated that increasing the photochromic mask layer density can increase the media's non-linearity and the photochromic reaction macroscopic speed nonuniformity. The results also showed that decreasing the expose time can increase the media's non-linearity. The principles of super-resolution aperture come into being the near-field optics effect is shown in Fig. 5.15a, and the light field distribution after light emanating penetrate through the super-resolution aperture is shown in Fig. 5.15b with simulation calculation.

5.2 Optical Channel Characteristics

Photochromic medium grayscale reaction can be controlled by laser exposure energy and distribution light field energy of recording symbol at pupil plane. According to the characteristics of the linear region of the photochromic reaction,

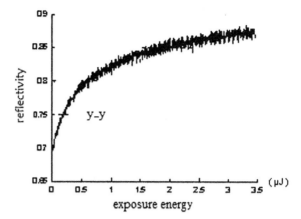

Fig. 5.16 The relationship between reflectivity and exposure energy of light-induced discoloration medium

using the superposition method and numerical calculation can obtain the optical channel signals of multi-level photochromic optical storage and theoretical value of its crosstalk between channels. The measuring experiment of reflectivity of the photochromic reaction process illustrated it has a linear and saturation regions as shown in Fig. 5.16. All experiments are accomplished with above Multi-function MW/ML experimental system in Sect. 5.1. In the photochromic reaction of initial region (reflectivity under 0.8), the reflectivity is linear with exposure energy which was used to multi-level photochromic storage that can be considered linear superposition of the amplitude of reflectivity in this region. Therefore the Fourier transform and convolution of linear superposition can be used to analysis for optical channel and crosstalk characteristics of photochromic multi-level optical storage.

According to linear superposition reflectivity of the various elements on light-induced discoloration disc can get:

$$U_\sigma(u', v') = U_F(u', v') + U_{M1}(u', v') + U_{M2}(u', v') \tag{5.1}$$

where U_σ is amplitude distribution function of reflectivity on disc surface, U_F is amplitude distribution function of reflectivity with pre-groove disc, U_{M1} is amplitude distribution function of reflectivity on recorded place, U_{M2} is amplitude distribution function of reflectivity on all adjacent tracks of the disc. After Fourier transform of Eq. (5.1) the amplitude reflectivity of the different elements on in the pupil plane reflected field of the disc as follows:

$$\tilde{U}_\sigma(x', y') = \tilde{U}_F(x', y') + \tilde{U}_{M1}(x', y') + \tilde{U}_{M2}(x', y') \tag{5.2}$$

where the x', y' is spherical pupil normalized coordinates. U'_{M1} and U'_{M2} represent the linear superposition of all photochromic recording spots go back to the pupil. According to read out spot center on the $U'V'$ plane, introduction a displacement factor for each record spot. The amplitude reflectivity of the blank disc surface with servo track of pre-groove can be expressed as:

$$U_F(u',v') = \begin{cases} r, & |v'| \leq \frac{\tau}{2} \\ re^{-j\alpha} & \frac{\tau}{2} \leq |v'| \leq \frac{q}{2} \end{cases} \tag{5.3}$$

where τ is the width of normalized servo pre-groove, α is differed phase of the shore and pre-groove. Consider the case $\Delta v'$ of the tracking error (TE), the amplitude distribution on the pupil as follows:

$$S_F(x',y') = \sum_n \tilde{U}_0\left(x',y' - \frac{n}{q}\right)\bar{U}_F(0,n)e^{-j\frac{2n\pi}{q}\Delta v'} \tag{5.4}$$

where $\bar{U}_F(m,n)$ is Fourier series of $U_F(u',v')$, $\tilde{U}_0(x',y')$ bit is amplitude distribution of readout light on pupil, and can be approximated as:

$$\tilde{U}_0(x',y') = \begin{cases} 0, & \sqrt{(x')^2 + (y')^2} \geq 1 \\ Ae^{-\frac{(x')^2 + (y')^2}{2\sigma}}, & \text{other} \end{cases} \tag{5.5}$$

If consider $n = 0$ and $n = \pm 1$ three level diffraction superimposed only. The amplitude reflectivity of single photochromic recording spot and ignoring initial reflectivity can be expressed as following:

$$U_i(u',v') = a\Delta t \cdot U_1(u',v')U_1^*(u',v') \tag{5.6}$$

where a is material sensitivity, Δt is exposure time, $U_1(u',v')$ is amplitude distribution on the spot of focusing surface that can be obtained by inverse Fourier transform of amplitude distribution on pupil $\tilde{U}_1(x',y')$. Equation (5.6) after Fourier transform becomes:

$$\tilde{U}_i(x',y') = a\Delta t \cdot \tilde{U}_1(x',y') \otimes \tilde{U}_1^*(-x',-y') \tag{5.7}$$

The amplitude distribution of return to the pupil $S_i(x',y')$ equals to the convolution integral between single of recording spots in the pupil plane amplitude reflectivity $\tilde{U}_i(x',y')$ and readout light intensity amplitude distribution $\tilde{U}_0(x',y')$. When the recorded point center has displacement $(\Delta u', \Delta v')$ to the optical center and consider the amplitude distribution of the circular symmetry, the amplitude distribution of return to the pupil can be written:

$$S_i(x',y') = a\Delta t \cdot \tilde{U}_0(x',y') \otimes \left\{ [\tilde{U}_1(x',y') \otimes \tilde{U}_1(x',y')]e^{-j2\pi(x'\Delta u' + y'\Delta v')} \right\} \tag{5.8}$$

According to the superposition principles, when the recorded spot 1 and closes recorded spot 2, intensity distribution of the readout signal on the pupil is:

$$I_\sigma \propto |S_\sigma|^2 = |S_F + S_1 + S_2|^2$$
$$= \left(|S_F|^2 + |S_1|^2 + 2|S_F S_1| \cos \Delta\phi_{F1} \right)$$
$$+ \left(|S_2|^2 + 2|S_F S_2| \cos \Delta\phi_{F2} + 2|S_1 S_2| \cos \Delta\phi_{12} \right) \qquad (5.9)$$

where S_1 and S_2 represent the complex amplitude of the recorded spot 1 and recorded spot 2. Expression in the first parenthesis is the contribution of the light intensity distribution on the record spot 1 on pupil, and expression in the second parenthesis is the crosstalk introduced by the record spot 2, $\Delta\phi_{F1}$, $\Delta\phi_{F2}$ and $\Delta\phi_{12}$ represent the phase differ between the complex amplitude. It can be seen that the composition of the crosstalk can be divided into the recorded spot 2 readout signal in a certain offset actually $\left(|S_2|^2 + 2|S_F S_2| \cos \Delta\phi_{F2} \right)$ and the crosstalk signal between recorded spots $(2|S_1 S_2| \cos \Delta\phi_{12})$.

Above theoretical analysis of reflected field of the recorded spot is suitable for the gray-scale modulation multi-level storage and photochromic multi-level optical disc to optimization design. As multi-gray modulation storage for example, the theoretical module of pupil function can be write:

$$\bar{T}(m,n) = \delta(m,n) + \frac{2\left(\sqrt{g} - 1\pi R^2\right)}{pq} E\left(2\pi R \sqrt{\left(\frac{m}{p}\right)^2 + \left(\frac{n}{q}\right)^2} \right) \qquad (5.10)$$

where $E(x) = J_1(x)/x$ and $\bar{T}(m,n)$ square are proportional to the intensity in the pupil, it can describe g modulation of the reflected light, i.e. the multi-level modulated signal by reflectivity. In optical disc storage system, only 0 and ± 1 order diffraction spot can be returned to the pupil, so m and n are possible combination for:

$$(M = 0, n = 0), (m = \pm 1, n = 0) \text{ and } (m = 0, n = \pm 1) \qquad (5.11)$$

The comparison of 532 nm and 650 nm photochromic materials with 8 level writing and read out experimental signal are shown in Figs. 5.17 and 5.18. The corresponding theoretical values of the pupil modulation function as: theoretical calculation result of readout signal amplitude distribution function $|S_\sigma|^2$, amplitude distribution function of reflectivity on recorded place S_F, readout signal single of recording spots 2 in the pupil plane amplitude reflectivity $|S_2|^2$, crosstalk of spot 2 readout signal $2|S_F S_2| \cos \Delta\phi_{F2}$, amplitude distribution of crosstalk of recorded spots 1 and 2 readout signal $2|S_1 S_2| \cos \Delta\phi_{12}$ are shown in Fig. 5.19. The comparison of theoretical calculation results of readout signal and its components of crosstalk are shown in Fig. 5.20, that are confirmed to correspond with experiments well.

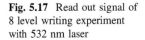

Fig. 5.17 Read out signal of 8 level writing experiment with 532 nm laser

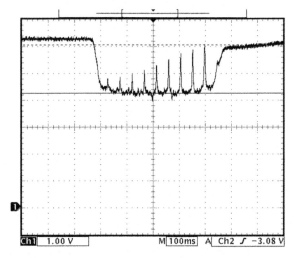

Fig. 5.18 Read out signal of 8 level writing experiment with 650 nm laser

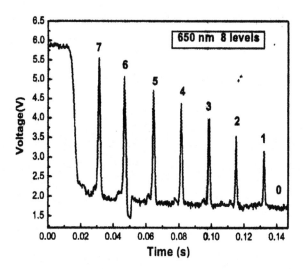

5.2.1   Equalization and Compensation of Optical Channel Signal

In the light induced photochromic multi-level optical storage, the optical channel signal compensation and equalization can be used to balance photochromic multi-level recording process and improve readout signal quality of multilevel storage, that the compensation and balanced approach is shown in Fig. 5.21.

As the impulse response function $h(t)$ of the optical channel signal from the optical pickup can be used to describe the readout signal characters and to be expressed as:

Fig. 5.19 The theoretical calculation results of readout signal of recording spots in the pupil plane: total readout light intensity amplitude distribution $|S_\sigma|^2$, amplitude distribution function of reflectivity on recording layer S_F, crosstalk of spot 2 readout signal $2|S_F S_2| \cos \Delta\phi_{F2}$ and crosstalk of recorded spots 1 and 2 signal $2|S_1 S_2| \cos \Delta\phi_{12}$

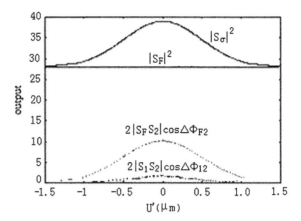

Fig. 5.20 The theoretical calculation results of total crosstalk: *I* total crosstalk, *II* crosstalk of reflectivity of recording layer, *III* the across crosstalk of reflectivity layer and recoding spot 2, *IV* across crosstalk of recoding spot 1 and 2

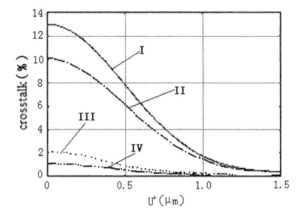

$$f(t) = a \sum_{i=-\infty}^{+\infty} D_i h(t - iT_b) \qquad (5.12)$$

If suppose to normalize exposure energy of recording spot 1 and spot 2 with gap T_b between P_1 and P_2, the read out signal could be described as:

$$f(t) = h_{P_1}(t) + h_{P_2}(t - T_b) + P_1 P_2 \cdot H_3(T_b) \qquad (5.13)$$

where $H_3(T_b)$ is all crosstalk components with the varies spacing of the two recording spots over the linear area. The equalized signal channel can restrain the nonlinear factors of the channels and control it within a consent range. So need of optimization of processing of the writing strategy, design and set up a non-linear compensation writing strategy, i.e. improve the channel characteristics of the input and output channel into linear area of writing, to ensure the quality of multilevel storage. It is different with conventional optical disc storage, that the quality of

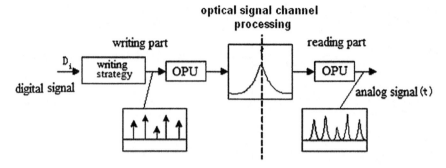

Fig. 5.21 The compensation and balanced processing to optical channel signals in photochromic multi-level storage

readout signal is related to the recorded spot and two adjacent recorded spots only, other crosstalk of signal can be ignored. But for the multilevel recording, the sequence of exposure energy of the channel nonlinear $\{P_i\}$ have to meet the following equation group (5.14), with the channel compensation as:

$$
\begin{cases}
\bar{f}(n) = \sum_{i=n-1}^{n+1} \hat{D}_i h(nT_b - iT_b) \\
f(n) = \sum_{i=n-1}^{n+1} P_i^2 H_1(nT_b - iT_b) + \sum_{i=n-1}^{n+1} P_i H_2(nT_b - iT_b) + (P_n P_{n-1} + P_n P_{n+1})H_3(T_b) \\
\sum_n |f(n) - \bar{f}(n)|^2 \quad \text{for minimum} \\
P_i \in [P_{\min}, P_{\max}]
\end{cases}
$$
$$(5.14)$$

The sequence $\{P_i\}$ satisfies the given conditions of the upper and lower bounds of $\bar{f}(n)$ and $f(n)$ for minimum variance. Therefore, using minimum variance search method of writing can achieve the optimized design to writing strategy. According to ensure the crosstalk of light-induced photochromic multi-level recording with equalizer can be reduced to zero, to design a zero crosstalk equalizer as shown in Fig. 5.22a and a partial response channel equalizer as shown in Fig. 5.22b.

5.3 Modulation and Code

The reflectivity curve of photochromic materials has a wide linear region that is more suitable for amplitude modulation for multi-level storage. The analytical calculations based on Fourier optics and optical channel diffraction theory, as well as the fast Fourier transform algorithm for traditional optical storage systems are suitable multi-level optical storage also. An overlay analysis method used to physical analysis and calculation of the process directly in the optical storage

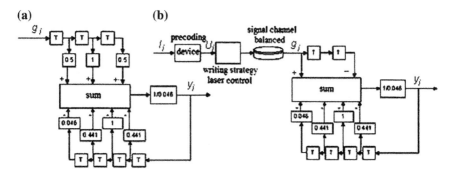

Fig. 5.22 a The zero crosstalk equalizer. **b** The partial response channel equalizer for photochromic multilevel optical recording

system for read out, that confirmed these analytical calculations to have important reference value for the modulation, coding and error correction of multi-level optical storage.

5.3.1 Modulation Characteristics

This section in order to analyze of optical channel response characteristics in light-induced photochromic materials for multi-level optical storage, based on the Fourier optics and overlay analysis method to establishment the coordinate system and application to optical readout signal model and analysis of the optical channel characteristics for photochromic multilevel recording.

The new coordinate system is shown in Fig. 5.23, initialization the refractive index of objective and image space are n and n', and the numerical aperture are (ξ, η) and (ξ', η') respectively, as well as the reference frame for object plane and image plane coordinates are (ξ, η) and (ξ', η'), then can establish Eq. (5.15) as following:

$$u = (n \sin \alpha / \lambda)\xi, v = (n \sin \alpha)/\eta$$
$$u' = (n' \sin \alpha' / \lambda)\xi', v' = (n' \sin \alpha')/\eta' \tag{5.15}$$

In the analysis, the entrance pupil and exit pupil calculation is using still the rectangular coordinates (X, Y) and (X', Y') of position, and the distance of P and P' (see Fig. 5.23) to optical axis are h and h', then normalized coordinates system can simplify and obtained:

$$x = X/h, \quad y = Y/h$$
$$x' = X'/h', \quad y' = Y'/h' \tag{5.16}$$

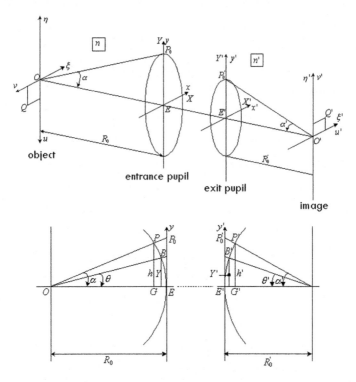

Fig. 5.23 The normalized coordinate system for analysis and calculation optical channel response characteristics in multilevel recording

In accordance above system of coordinate, the coordinates of any point of object plane $Q(u, v)$ has a correspondence coordinates $Q'(u' = u, v' = v)$ on image plane. If the optical system was satisfied the sine condition, the point on entrance pupil $B(x, y)$ will be conjugate to point $B'(x' = x, y' = y)$ on exit pupil. Therefore can omit the symbol "'" in mathematics, and the physical meaning is not difference between the entrance pupil and exit pupil.

As the laser beam is a Gaussian distribution, the complex amplitude of pupil in normalized coordinate system can be expressed as:

$$u_e(x', y') = Ae^{-\frac{(x')^2 + (y')^2}{2\sigma}}\text{Circ}\left(\sqrt{(x')^2 + (y')^2}\right)$$

$$\text{Circ}(r') = \begin{cases} 1 & r' \le 1 \\ 0 & r' > 1 \end{cases}$$

(5.17)

where x' and y' are the coordinates on pupil sphere, for the ideal objective lens, the light is a same phase spherical wave on exit pupils as in Fig. 5.23.

The normalized coordinate system in Fig. 5.23 can be used to superposition analysis conveniently also. The linear overlay analysis method based on Fourier

Fig. 5.24 Decomposition of the recording elements on disc surface

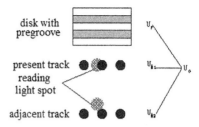

disk with pregroove

present track reading light spot

adjacent track

transform and convolution have a linear superposition features, but requires the record spot must not be coincidence a region. For the photochromic multi-level storage, recorded medium is in the linear region of the photochromic reaction, i.e. the light-induced color reaction amplitude reflectivity is a linear process with the exposure energy. Therefore, the recorded points overlap are conform the requirements of linear superposition, i.e. is not limiting the recorded point overlap for light-induced photochromic storage.

Photochromic optical disc can be decomposed into the current track and adjacent track or pre-grooves are shown in Fig. 5.24. According to the amplitude reflectivity is formation superposition on the recording surface of the disc can establish the equation $U_\sigma = U_F + U_{M1} + U_{M2}$. Where the U_σ is whole reflectivity distribution function on the surface of the disc, U_F is reflectivity distribution function on the disc surface with pre-groove, U_{M1} is amplitude reflectivity distribution function currently being reading recorded points in the all. Meanwhile U_{M1}, U_{M2} can be expressed the combination of reflectivity of series of photochromic recording point U_i which depends on the exposure energy U_i (integral of exposure energy in this disc surface).

The reflectivity of $U_\sigma = U_F + U_{M1} + U_{M2}$ via the Fourier transform can obtain the amplitude reflectivity function on pupil plane of all elements of the disc $\tilde{U}_\sigma = \tilde{U}_F + \tilde{U}_{M1} + \tilde{U}_{M2}$. But $\tilde{U}_{M1}$ and $\tilde{U}_{M2}$ can be expressed as series of amplitude linear superposition from light induced discoloration recorded points reflected returned into the pupil. According to the center of each recorded point relative to read out center is offset on plane of $u'-v'$, that have to introduce a weighting factor to superimpose analysis. Overlay superimpose analysis can be simplified to single point diffraction with some offset of recorded point and pre-groove. In write linear region, the light-induced discoloration recorded point amplitude reflectivity is decided to the amplitude distribution of written point, and therefore can be expressed as:

$$U_{i0}(x',y') = aI_0(x',y')\Delta t + r_L \tag{5.18}$$

where I_0 (x', y') is light intensity distribution function of recorded point, α is material sensitivity, r_L is amplitude reflectivity without writing state, Δt is the exposure time. If all the exposed time are same, can merge α and Δt to α_t. R_L is part of amplitude reflectivity U_F which is superimposed to recorded point on the disc, so subtracting the amplitude reflectivity, the reflectivity of single photochromic point should be:

$$U_i(x',y') = a_t U_0(x',y') U_0^*(x',y') \tag{5.19}$$

where $U_0(x',y')$ is amplitude distribution of writing point on image planet, and $U_0^*(x',y')$ as the conjugated form. $U_i(x',y')$ via Fourier transform can obtain:

$$\tilde{U}_i(x',y') = a_t \tilde{U}_0(x',y') \otimes \tilde{U}_0^*(-x',-y') \tag{5.20}$$

i.e. the light amplitude distribution of return to the pupil is amplitude reflectivity of recording point convolution integral of the of the light intensity distribution.

Meanwhile, the light intensity of entrance pupil surface is:

$$S_i(x,y) = b \int\limits_{-\infty}^{+\infty} \int\limits_{-\infty}^{+\infty} \tilde{U}_0(x-x',y-y')\tilde{U}_i(x',y')dx'dy' \tag{5.21}$$

where b is the ratio of the readout light amplitude to writing light amplitude, considering the $\tilde{U}_0(x',y')$ is circular symmetry, and therefore:

$$S_i(x,y) = C\tilde{U}_0(x,y) \otimes \tilde{U}_0(x,y) \otimes \tilde{U}_0(x,y) \tag{5.22}$$

where C is constant. The calculation of the readout signal of single recording point was transformed to the convolution integral of light amplitude distribution on three pupils. So the calculation of single recorded point read out signal can be simplified when use of fast Fourier transform.

Above analysis is assumed the read out center coincides with the recording point center, if the recorded point center has a displacement $(\Delta x', \Delta y')$ relative to the optical axis center, the calculation equation have to introduce a phase shift, namely:

$$S_i(x,y) = C\tilde{U}_0(x,y) \otimes \left\{ \left[\tilde{U}_0(x,y) \otimes \tilde{U}_0(x,y)\right] e^{-j2\pi(x\Delta x + y\Delta y)} \right\} \tag{5.23}$$

For disc with servo pre-groove, calculation of reflected field as shown in Fig. 5.24, that is a blank disc with a servo pre-groove, its surface amplitude reflectivity can be expressed:

$$U_F(u',v') = \begin{cases} r_L & |v'| \leq \frac{\tau}{2} \\ r_L e^{-j\alpha} & \frac{\tau}{2} \leq |v'| \leq \frac{q}{2} \end{cases} \tag{5.24}$$

where τ the normalized shore width, q is the normalized the space between pre-grooves, α is phase difference of shore and pre-grooves. Via Fourier transform can find its Fourier series as:

$$\overline{U}_F(m,n) = \delta(m)\frac{r_L\sin\left(\frac{n\pi\tau}{q}\right)}{n\pi}\left(1 - e^{-j\alpha}\right)n \neq 0 \tag{5.25}$$

When $m = n = 0$:

$$\overline{U}_F(0,0) = \frac{r_L\tau}{q}\left(1 - e^{-j\alpha}\right) + r_L e^{-j\alpha} \tag{5.26}$$

The light amplitude distribution on pupil is:

$$S_F(x,y) = b\sum_n U_0\left(x, y - \frac{n}{q}\right)\overline{U}_F(0,n) \tag{5.27}$$

where (x,y) is a circle with radius is (x,y) within center of $(0,0)$. $U_0(x,y)$ is the light amplitude distribution on disc recording plane, b is the ratio of light amplitude between read out and writing. According to the Hopkins analysis, the diffraction spots of order n are greater than $2q$ cannot be considered. Therefore, for blank optical disc, the amplitude on the pupil is the superposition of 0 and ±1 three diffraction spots. Such as conventional blank optical disc, it is 0.9231.

If TE exists, and has a displacement $\Delta y'$ in the vertical direction of track that the diffraction spots on the pupil will have corresponding phase shift, and the equation of light amplitude distribution on pupil will be became:

$$S_F(x,y) = \sum_n U_0\left(x, y - \frac{n}{q}\right)\overline{U}_F(0,n)e^{-j\frac{2n\pi}{q}\Delta y'} \tag{5.28}$$

According to above analysis, the readout signal along the track direction of the disc is a function of optical channel impulse response. Ideally, the light intensity distribution on the pupil of a blank light-induced discoloration disc can be expressed as:

$$\begin{aligned}I_\sigma &\propto |S_\sigma|^2 = |S_F + S_i|^2 \\ &= |S_F|^2 + |S_i|^2 + 2|S_F S_i|\cos\Delta\phi\end{aligned} \tag{5.29}$$

Read out signal can be decomposed into three components: $|S_F|^2$ is read out signal of blank disc with servo track, it is a constant always when without channel TE. $|S_i|^2$ is readout signal of a single light-induced discoloration recorded point that is proportional to the square of writing power. $2|S_F S_i|\cos\Delta\phi$ is the main components of read out single, where $\Delta\phi$ is their phase difference, that is superimposed of blank disc and recorded point, as well as proportional to writing power also. $|S_F|$ is larger than $|S_i|$ several times and almost over 10 times $|S_i|$, so it affects to readout signal slightly.

As based on a conventional DVD drive and substrate for multilevel recording experiment for example, its initial reflectivity is about 50 %, laser wavelength is

Fig. 5.25 The calculated
results of the total readout
signal and its components

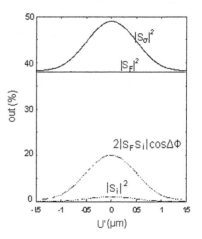

650 nm, numerical aperture of objective lens is 0.65, emitted light energy distribution coefficient on pupil σ is selected 1000, interval of track is 0.8 μm, shore width is 0.4 μm. The total readout signal and various components are calculated as shown in Fig. 5.25 that is normalized with the maximum value.

In addition, the shape of $|S_i|^2$ and $2|S_F S_i| \cos \Delta \phi$ is very similar as in Fig. 5.25, and similar to typical Gaussian distribution. Therefore, in analysis of the readout signal can consider that the exposure energy is approximated to read out. This calculation is static photochromic recording, but can also be equivalent to that is written with very short exposure time of dynamic recording, i.e. the writing pulse is an impulse function $\delta(t)$. In read out process, the disc rotation with constant line velocity v, read out signal $|S_\sigma|^2$ is changing to $h(t)$, so it is an optical channel impulse response function also.

A comparison experiment to the theoretical analysis is following: the medium is fulgide photochromic material, exposure laser is wavelength 650 nm and writing power is 12, 10 and 8 mW for four level recording experiment. The disc with constant linear velocity rotation, and read power is 0.5 mW. The read out signal to 4 level recording are shown in Fig. 5.26, as in Fig. 5.27a is written with the 12 mW, Fig. 5.27b for 8 mW write and Fig. 5.27c for 8 mW. Take three typical signals (as in the three virtual boxes of Fig. 5.26) compared with the corresponding theoretical calculations of the impulse response curve with its peak normalized (the direct-current component of the measured impulse response curve has been deducted) is shown in Fig. 5.27.

It can be seen from Fig. 5.27b, c the two curves of read out signal are good agreement between experimental and theoretical calculations results. But for the high-power recording as in Fig. 5.27a, the side lobe of the experimental curve is slightly higher than the theoretical value. The reason is that the larger exposure energy to record, the material in recording spot center is saturation, in accordance with the central area normalized to make side lobe relative enlarge.

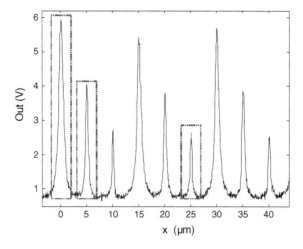

Fig. 5.26   The read out signal of 4 level recording

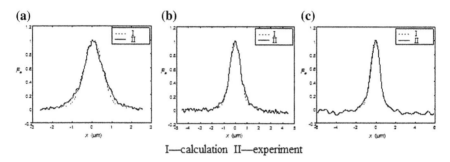

I—calculation II—experiment

Fig. 5.27   The comparison of experimental and theoretical calculation results of read out signal with 4 level storage

The modulation coding and channel detection is a key technology to MW/ML storage. Therefore establishment of a bit rate of 2/3 and 8 level (1, 2) RLL encoding method and base on run-length encoding and multi-level amplitude modulation for experiment of multilevel storage, that the process diagram is shown in Fig. 5.28, i.e. the user data was accepted and processed by multi-level encoder, then the output of the encoder of 2-bit. The encoding user data $a_2a_1a_0$ of 3 bits is converted to 2-bit channel data b_1b_0 of 3/2 bit rate with a FPGA logic control encoder and via data query and conversion of coding table mapping. User input data stream is cut into 3-bit data blocks. Channel encoded data b_1b_0 after waveform transformation (utilized non-antagonistic turn zeroing method) are used to writing laser power control with pulse intensity and width, to recording multi-level data on the photochromic disc, as Fig. 5.28b.

For MW/ML parallel storage, adoption the multi-channel and two-dimensional coding method, that can reduce coding redundancy and improve data rate and

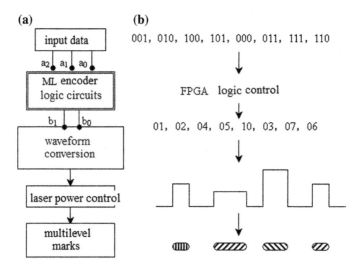

Fig. 5.28 The program to achieve the multi-level encoding and recording

capacity further. Employ MW/ML coding and finite-state encoder model, the single-channel one-dimensional user data can be converted to two-dimensional data array of many row and column with constraints run length. Each row of data in the two-dimensional data array corresponds to an optical storage data channel, the data of each data channel is used to control a specific wavelength of laser to write data on the disc. The two-dimensional array of data can be parallel written to disc by different wavelengths laser to achieve data-parallel recording for MW/ML optical storage. The medium of MW/ML optical storage is a mixture of several photochromic materials with different absorption spectra, which are match to different laser wavelength each other well respectively. Data of each channel was recorded in the same location of the recording medium at the same time that record multiple channels of information in a physical location, storage capacity and data transfer rate to improve synchronization. The design principles of the multi-level coding and channel coding method of combining high-density encoding scheme is shown in Fig. 5.29. The MW/ML parallel storage experimental results show that the

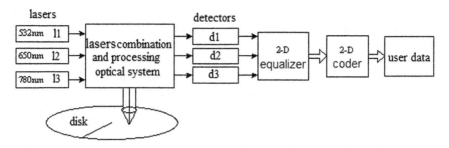

Fig. 5.29 Multi-wavelength and multi-level channel equalization and detection block diagram

application of the two-dimensional channel detection technology, can overcome to some crosstalk between different wavelength recording materials, and improve data capacity and recorded data reliability.

5.4 The Error Correction

For the computer system memory, actual error rate has to be down to 10^{-12}–10^{-13}. The original error rate of MW/ML is about 10^{-4}–10^{-5}. Must be studied and designed for photochromic MW/ML storage error correction coding scheme, error correction analysis and estimated systems. This section will discuss the principles and methods to reduce the error rate through the detection of defects and to establish a effective error correction code (ECC).

5.4.1 A Single-Wavelength Multi-level Error Correction Coding

According to the principles of the RS encoder, for the RS code without correction, its error adjoint polynomial as:

$$S(x) = S_1 + S_2 x + S_3 x^2 + \cdots + S_{n-k} x^{n-k-1} = \sum_{j=0}^{n-k-1} S_{j+1} x^j \tag{5.30}$$

where $S_j = \sum_{i \in \varepsilon} E_i \alpha^{ij} = R(\alpha^j) = E(\alpha^j)$

Constitute the error location polynomial:

$$\sigma(x) = \prod_{i \in \varepsilon} (x - \alpha^{-i}) \tag{5.31}$$

Error evaluation polynomial:

$$\omega(x) = \sum_{i \in \varepsilon} E_i \prod_{\substack{i \in \varepsilon \\ l \neq i}} (x - \alpha^{-l}) \tag{5.32}$$

As RS(n, k) is up to correct the $(n, k)/2$ errors only, so it can be set $|\varepsilon| = {}^{\circ}(\sigma) \leq (n - k)/2$, and ${}^{\circ}(\omega) \leq |\varepsilon| - 1$ must exist $\mu(x)$ and to meet:

$$\sigma(x)S(x) = \omega(x) + u(x)x^{n-k} \qquad (5.33)$$

Then can write in the following form:

$$\sigma(x)S(x) \equiv \omega(x)(\mathrm{mod}\ x^{n-k}) \qquad (5.34)$$

Errors are:

$$E_i = \frac{\omega(\alpha^{-i})}{\sigma^{l}(\alpha^{-i})} \qquad (5.35)$$

$$\sigma^{l}(x) = \sum_{i \in \varepsilon} \prod_{\substack{l \in \varepsilon \\ l \neq i}} (x - \alpha^{-l}) \qquad (5.36)$$

$$\sigma^{l}(\alpha^{-j}) = \prod_{\substack{l \in \varepsilon \\ l \neq j}} (\alpha^{-j} - \alpha^{-l}) \qquad (5.37)$$

If there are deleted error t, random error s with $s < (n, k)/2$, can use above method. The location of error is alike to an error and to be correction as set to 0. Meanwhile improvement S decoding, ε_1 to be random error domain, ε_2 to be deleted domain and ε is all error domain, i.e. $\varepsilon = \varepsilon_1 \cup \varepsilon_2$. And the random error location polynomial can write as:

$$\sigma_1(x) = \prod_{i \in \varepsilon_1} (x - \alpha^{-i}) \qquad (5.38)$$

Remove error location polynomial is:

$$\sigma_2(x) = \prod_{i \in \varepsilon_2} (x - \alpha^{-i}) \qquad (5.39)$$

All error location polynomial:

$$\sigma(x) = \prod_{i \in \varepsilon} (x - \alpha^{-i}) \qquad (5.40)$$

But, the error valuator polynomial as follows:

$$\omega(x) = \sum_{i \in \varepsilon} E_i \prod_{\substack{i \in \varepsilon \\ l \neq i}} (x - \alpha^{-l}) \qquad (5.41)$$

When $\mu(x)$ exists and to meet the:

$$\sigma(x)S(x) = -\omega(x) + u(x)x^{n-k} \tag{5.42}$$

As $\omega(x)$ is obtained, $\sigma_2(x)$ and $\omega(x)$ can be known soon.
Define a maximum number is $n - k - |\varepsilon_2|$, and the adjoint polynomial is:

$$\widehat{S}(x) = \sigma_2(x)S(x) = \left(\prod_{i \in \varepsilon_2} (x - \alpha^{-i}) \right) S(x) \tag{5.43}$$

The key Eq. (5.42) can be amended as follows:

$$\sigma_1(x)\widehat{S}(x) = -\omega(x) + u(x)x^{n-k} \tag{5.44}$$

Or write:

$$\sigma_1(x)\widehat{S}(x) = -\omega(x)(\mathrm{mod}x^{n-k}) \tag{5.45}$$

Using the Berlekamp-Massey (BM) iterative algorithm to solving $\sigma(x)$ and $\sigma_1(x)$, then access to Eq. (5.35) calculate the error as:

$$E_i = \frac{\omega(\alpha^{-i})}{\sigma^l(\alpha^{-i})} \tag{5.46}$$

For the RS(n, k) code, when the maximum distance of $n - k + 1$ code, can correct $(n, k)/2$ burst errors. However, if confined to correct the dele error only, it can correct $n - k$ error. To need of correct burst errors and dele errors at the same time, it can correct $2t + e \le d - 1$ error, where t is the burst error, e is delete errors. So using RS correct is more efficient with burst and delete error.

In the MW/ML storage, as each wavelength is independent coding, so it can be considered as a M layers disc and coding error correction will appear in corresponding layer. Therefore, as long as to know the wrong location of this layer, it can be deleted by other layer, to reduce the redundancy of other layer. For example, first layer using RS(n, k) coding can correct $(n - k)/2$ burst errors, but other layer of RS(n, $k + (n - k)/2$) coding can correct $(n - k)/2$ errors at least. When decoding, the wrong location of first layer can be gotten in the first, that the wrong location can be corrected by other layer decoding, simply remove the incorrect decoding can do to these locations. So this approach can improve the coding efficiency and decoding speed, and the M layer encoded process as shown in Table 5.3.

In this encoding process on each respective independent encode the first layer of the RS(n, k) code, $n - k$-byte checksum. The first i ($2 \le i \le M$) layer is made of RS(n, $k + [n - k]/2$) encoding, $n - k - [n - k]/2$-byte check sum. Than each layer uses the RS(n, k) code, to save the $M \times [n - k]/2$ bytes of space, can increase the effective data capacity. And the error correction performance on each of RS(n, k) encoding is the same as. This decoding process was used in this

Table 5.3 Encoding single for M wavelength

$B1, 1$	$B1, 2$	...	$B1, k$	$B1, k + 1$	...	$B1, k + [n - k]/2$	$B1, k + [n - k]/2 + 1$	...	$B1, n$
$B2, 1$	$B2, 2$	...	$B2, k$	$B2, k + 1$	...	$B2, k + [n - k]/2$	$B2, k + [n - k]/2 + 1$	...	$B2, n$
$\vdots$	$\vdots$	...	$\vdots$	$\vdots$	...	$\vdots$	$\vdots$	...	$\vdots$
$BM, 1$	$BM, 2$	...	BM, k	$BM, k + 1$	...	$BM, k + [n - k]/2$	$BM, k + [n - k]/2 + 1$	...	BM, n

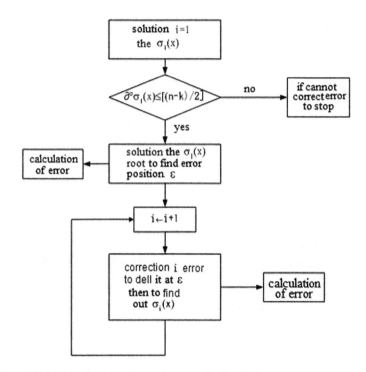

Fig. 5.30 Multi-layer storage decoding and error correction logic diagram

experimental system as in Fig. 5.30. The coding method logic is very original, and suitable to basic research for MW/ML coding only. For example, in the completion of the first layer decoding to get the wrong position, the layers can be used stand-alone decoder of the data at the same time to correct the error to delete the decoding, so that the decoding speed is greatly improved.

Actually the probability of wrong coding P_{wc} includes undetected error probability P_{ud}, decoding failure probability P_{df} and error probability of decoding error probability P_{de}. If definition decoding error probability P_{ef} is the sum of decoding failure probability P_{df} and decoding error probability P_{de}, i.e. $P_{ef} = P_{df} + P_{de}$. Undetected error probability P_{ud} refers to the error of transmission of code, i.e. the of cannot be found error by decoder. The P_{df} is that they may detect error, but it cannot correct probability when number of errors exceed the ability of correction. The P_{de} refers the probability of error by translated codes after decode. Therefore, the main consideration is P_{ef}.

The typical formula of correct decoding probability is:

$$P_{wc} = \sum_{i=0}^{t} \binom{n}{i} p_e^i (1 - p_e)^{n-i} \tag{5.47}$$

where p_e is the channel bit error rate (BER), so decoding error probability is:

$$P_{ef} = 1 - P_{wc} \tag{5.48}$$

For a given n, k, the relationship between probability P_{ef} of correct decoding and the original channel bit error p_e is shown in Fig. 5.31. The relationship between decoding error probability and $n - k$ is shown in Fig. 5.32. It can be seen from Fig. 5.32: when n fixed, as $n - k$ increases, the error correction performance will soon upgrade, when $n = 188$, $n - k = 16$, $P_{ef} = 6.54962 \times 10^{-22}$, $n - k = 10$, $P_{ef} = 5.56989 \times 10^{-14}$ and when $n = 200$, $n - k = 10$, $P_{ef} = 8.10498 \times 10^{-14}$. If need of the decoding error probability is limited to a certain value that can be adjusted by selection befitting n and k. For example, when $n = 200$, choose $n - k \geq 10$ can achieve bit error probability $\leq 10^{-12}$.

In MW/ML storage, for a number system based on units of q RS coding, the undetected error probability is $P_{ud} \leq q^{-(n-k)}$. For RS code in $GF(2^8)$ domain (as in

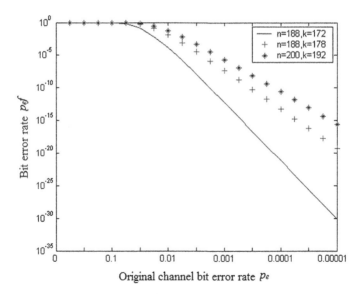

Fig. 5.31 The probability of correct decoding with the change of the channel bit error rate

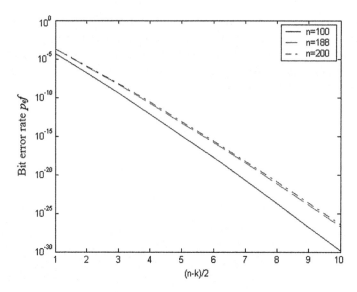

Fig. 5.32 The relationship between decoding error probability and $n - k$

three-wavelength and eight-level storage, which was used in this experimental system, the non-detection probability is indicated in Eq. (5.49), and the relationship between undetected error probability and $n - k$ is shown in Fig. 5.33.

$$P_{\mathrm{ud}} \leq 256^{-(n-k)} \tag{5.49}$$

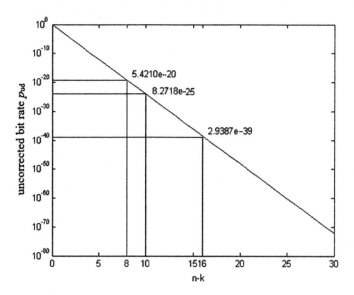

Fig. 5.33 The relationship between undetected error probability and $n - k$ in multi-level coding

It can be seen from Fig. 5.33, undetected error probability of rapid decline when $n - k$ increasing, i.e. when $n - k = 8$ detected error probability is 5.4210×10^{-20}, $n - k = 10$ undetected error probability is 8.2718×10^{-25}, $n - k = 16$ undetected error probability is 2.9387×10^{-39}. Undetected error probability is very small when $n - k$ is greater than a certain value.

For P_{de}, has:

$$P_{de} = \sum_{j=0}^{n} A_j P_{de}^j \tag{5.50}$$

where A_j is number of j code and is:

$$A_j = \binom{n}{j}(Q-1) \sum_{i=0}^{j-(n-k+1)} (-1)^i \binom{j-1}{i} Q^{j-(n-k+1)-i} \tag{5.51}$$

As $Q = 2^8 = 256$, for $j < n - k + 1$, $A_j = 0$, when the received codes are into the A_j domain, the error probability of decoder P_{de}^j is:

$$P_{de}^j = \sum_{v=0}^{t} \sum_{w=0}^{t-v} \binom{n-j}{v} \binom{j}{w}(Q-1)^{w-j} \left(1 - \frac{p_e}{Q-1}\right)^w (1-p_e)^{n-j-v} p_e^{j+v-w} \tag{5.52}$$

For $n = 28$, $k = 24$, and the BER $P_{ef} \approx P_{df}$, $P_{de} = P_{ef}$ as shown in Fig. 5.34, so just principal discussion P_{ef}.

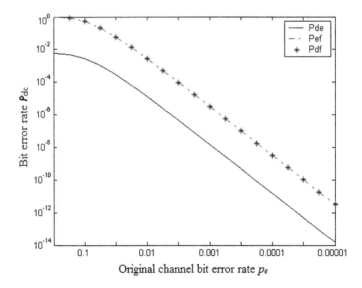

Fig. 5.34 The relationship between decoding error probability p_{dc} and the bit error rate p_e

5.4.2 Error Code Correction

For MW/ML storage with new light-induced photochromic medium that is need of introduction a concept of hard errors and soft errors. Hard errors include defects of materials of medium, coating process and substrate of the disc etc. that appear at the same position on the disc. Soft errors are random errors that will occur in each wavelength which is not effect to other wavelength (layer). But it has some new problems also. As the first layer (wavelength) use of the $RS(n - k)$ encoding, and the second layer use of $RS(n, k_1)$ encoding, which, $k_1 > k$ as shown in Table 5.4. For all layers, there are random errors that have two possibilities as the following:

1. The error location of first layer is hard wrong with no random error. But other layers have soft errors (random errors). It is relatively easy, that the first layer gets the wrong location, so the other layers can erasure it. Other layers need to correct t random errors only, which would require the rest layers RS coding to meet:

$$n - k_1 + 1 \geq (n - k)/2 + 2t \tag{5.53}$$

where t is determined by probability of random errors. Each layer is coding independently, first layer uses $RS(n, k)$ code and has $n - k$ byte correction codes. The i_{th} ($2 \leq i \leq M$) layer employ $RS(n, k_1)$ encoding and has $n - k_1$ byte correction code, that can save $M \times (k_1 - k)$ bytes space than each layer uses the $RS(n, k)$ encoding too. Decoding process is shown in Fig. 5.35.

2. If first layer has soft error also, it is more complex than previous case. Assumed the first layer has t_1 error total, including t_{1h} hard error and $t_s = t_1 - t_{1h}$ random error and to impact on each layer. When random error t_s existes, decoding process is shown in Fig. 5.36. First to decode the first layer, and get the wrong position, the second layer of error correction erasure decoding. The wrong location for the first layer as the second layer of deletion errors, while the second layer also need to correct t random errors, and requested k and k_1 to meet the $n - k_1 + 1 \geq [n - k]/2 + 2t$, where t according to the actual to choose. If $t_1 + 2t_2 \leq n - k_1$, (t_2 is the second layer number of random error), then the error of second layer can be corrected. However, if $t_1 + 2t_2 > n - k_1$, the second layer error will not be corrected. Because second layer correct error t_1 that is more t_s error. Therefore second layer has to be corrected without deleted, i.e. correct the $[n - k_1]/2$ errors, and get the wrong location of the second layer, then

Table 5.4 Multi-wavelength (multi-level) encoding error correction coding

$B1, 1$	$B1, 2$	...	$B1, k$	$B1, k + 1$	...	$B1, k_1$	$B1, k_1 + 1$	...	$B1, n$
$B2, 1$	$B2, 2$	...	$B2, k$	$B2, k + 1$	...	$B2, k_1$	$B2, k_1 + 1$	...	$B2, n$
:	:	...	:	:	...	:	:	...	:
$BM, 1$	$BM, 2$	...	BM, k	$BM, k + 1$	...	BM, k_1	$BM, k1 + 1$	...	BM, n

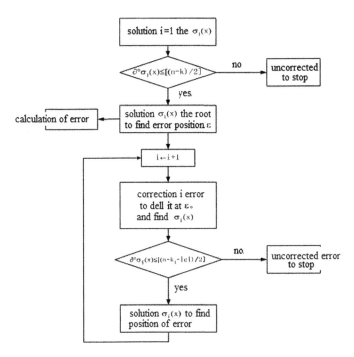

Fig. 5.35 The decoding process of first layer without random errors in multi-wavelength and multi-level storage

select the intersection of wrong location on first and second layers to be wrong location of third layer and correction until to completion of M_{th} layer error decoding correction. Actual, hard errors are much more than the soft error, therefore, according to the actual situation to choice t appropriately, this probability can be dropped to very small. From above analysis shows that this correction decoding speed is considerably limited. So may consider to correction of first layer error get its wrong location, then using deletion the location of error corrected the rest errors of M_{-1} layer. If the M_i layer error cannot be smooth correction, can make error correction without deleted, and take the intersection of its error location and the first error location to be the new location of the deleted, that can increase decoding speed appropriately. Finally the encoding method and capability of error correction are assessed as follows.

For soft and hard errors exist, each of the error will affect the correctness of the decoding. Assume that each layer of the error rate is the same as the first layer $P_{ud}^1 \leq 256^{-(n-k)}$. Then undetected error probability for the other layers is $P_{ud}^2 \leq 256^{-(n-k_1)} > P_{ud}^1$, undetected error probability for non-inspection of the entire coding error probability is $P_{ud} \leq 256^{-(n-k_1)}$, that should be shown in Fig. 5.37.

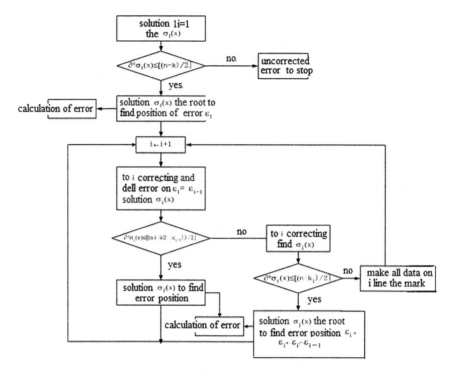

Fig. 5.36 The decoding process of first layer exist random errors in multi-wavelength and multi-level storage

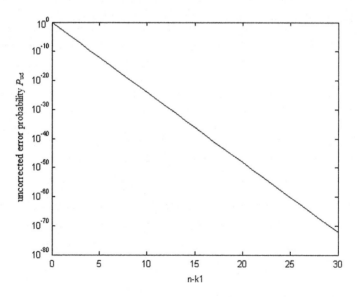

Fig. 5.37 The undetected error probability for error rate is the same of every layer

If the probability of correct decoding in the first layer P_{wc}^1 is follows:

$$P_{wc}^1 = \sum_{i=0}^{t_1} \binom{n}{i} p_e^i (1-p_e)^{n-i} \tag{5.54}$$

where p_e is the channel error rate, and $t_1 = [(n-k)/2]$. The error decoding probability P_{ef}^1 of first layer is $P_{ef}^1 = 1 - P_{wc}^1$.

For the second layer, if using the wrong location as the deleted location, in addition to correct t random errors, the correct decoding probability $P_{wc}^{2\,1}$ formula need to re-establish as the following:

$$P_{wc}^{2\,1} = \frac{\Pr(0)}{\Pr} \times P_{wc}^1 + \frac{\Pr(1)}{\Pr} \times P_{wc}^1 \times \sum_{i=0}^{1} \binom{n}{i} p_e^i (1-p_e)^{n-i} + \cdots \frac{\Pr(2t_2 - t_1)}{\Pr}$$

$$\times P_{wc}^1 \times \sum_{i=0}^{2t_2-t_1} \binom{n}{i} p_e^i (1-p_e)^{n-i}$$

$$P_{wc}^{2\,1} = \sum_{i=0}^{(2t_2-t_1)/2} \frac{\Pr(i)}{\Pr} P_{wc}^1 \sum_{j=0}^{i} \binom{n}{i} p_e^j (1-p_e)^{n-j} \tag{5.55}$$

where $\sum_{j=0}^{0} \binom{n}{i} p_e^j (1-p_e)^{n-j} = 1$, $\Pr$ is the probability of i soft error. $\Pr(i)$ is the total probability of soft errors and $\Pr = \sum_{i=0}^{n} \Pr(i)$. If the second layer using error location of first layer as the deleted position error correction is failure, the correct decoding probability is:

$$P_{wc}^{2\,2} = \sum_{i=0}^{t_2} \binom{n}{i} p_e^i (1-p_e)^{n-i} \tag{5.56}$$

Therefore, the error decoding probability of the second layer:

$$P_{ef}^2 = 1 - P_{wc}^{2\,1} - (1 - P_{wc}^{2\,1}) P_{wc}^{2\,2} \tag{5.57}$$

The third layer is more complex than the second layer, and can still be divided into the following two situations:

1. Third layer from second layer get the wrong location to delete errors and completion correction, the correct decoding $P_{wc}^{3\,1}$ should be:

$$P_{wc}^{3}{}^{1} = \begin{cases} \sum_{i=0}^{(2t_2 - t_2')/2} \frac{\Pr(i)}{\Pr} P_{wc}^{2}{}^{1} \sum_{j=0}^{i} \binom{n}{i} p_e^j (1 - p_e)^{n-j} & t_2' = t_1' \\ \sum_{i=0}^{(2t_2 - t_2')/2} \frac{\Pr(i)}{\Pr} P_{wc}^{2}{}^{2} \sum_{j=0}^{i} \binom{n}{i} p_e^j (1 - p_e)^{n-j} & t_2' \neq t_1' \end{cases} \quad (5.58)$$

where t_2' is the wrong location after error correction decoding of second layer, i.e. $t_1' = t_1$. If $t_2' = t_1'$, i.e. second layer using wrong location of first layer to be the deleted position decoding successful, its probability is $P_{wc}^{2}{}^{2}$. If $t_2' \neq t_1'$, second layer using wrong location of first layer to be the deleted position decoding is failure, but it is decoding successful independently, the probability is $P_{wc}^{2}{}^{2}$.

2. If the third layer using error location of above layer to error correction fails, but successful correct decoding independently also, its probability $P_{wc}^{2}{}^{2}$ is:

$$P_{wc}^{3}{}^{2} = \sum_{i=0}^{t_2} \binom{n}{i} p_e^i (1 - p_e)^{n-i} \quad (5.59)$$

So the third layer decoding error probability P_{ef}^{3} is:

$$P_{ef}^{3} = 1 - P_{wc}^{3}{}^{1} - (1 - P_{wc}^{3}{}^{1}) P_{wc}^{3}{}^{2} \quad (5.60)$$

For L layer is similar to third layer, can be divided into two cases:

1. Number L layer using error location of $L - 1$ layer as the deleted error correction decoding is successful, the probability of correct decoding $P_{wc}^{L}{}^{1}$ is:

$$P_{wc}^{L}{}^{1} = \begin{cases} \sum_{i=0}^{(2t_2 - t_{L-1}')/2} \frac{\Pr(i)}{\Pr} P_{wc}^{L-1}{}^{1} \sum_{j=0}^{i} \binom{n}{i} p_e^j (1 - p_e)^{n-j} & t_{L-1}' = t_{L-2}' \\ \sum_{i=0}^{(2t_2 - t_{L-1}')/2} \frac{\Pr(i)}{\Pr} P_{wc}^{L-2}{}^{2} \sum_{j=0}^{i} \binom{n}{i} p_e^j (1 - p_e)^{n-j} & t_{L-1}' \neq t_{L-2}' \end{cases} \quad (5.61)$$

where t_{L-1}' is the wrong location obtained from $L - 1$ layer decoded. If $t_{L-1}' = t_{L-2}'$ indicated $L - 1$ layer using wrong location of $L - 2$ layer as delete position decoding is successful, the probability is $P_{wc}^{L-1}{}^{1}$. If the $L - 1$ layer using wrong location of $L - 2$ layer to delete position decoding is failed, but successful correct decoding independently, its probability is $P_{wc}^{L-2}{}^{2}$.

2. If the L layer using wrong location of $L - 1$ layer to delete position decoding is failed, but self-correcting successfully. The correct decoding probability of the L layer $P_{wc}^{L}{}^{2}$ should be:

$$P_{wc}^{L\,2} = \sum_{i=0}^{t_2} \binom{n}{i} p_e^i (1 - p_e)^{n-i} \tag{5.62}$$

Therefore, L layer decoding error probability P_{ef}^L is:

$$P_{ef}^L = 1 - P_{wc}^{L\,1} - (1 - P_{wc}^{L\,1}) P_{wc}^{L\,2} \tag{5.63}$$

Decoding error probability with increasing number of layers is more complicated, because the changing of wrong location t'_{L-1} are more than delete on each layer often. So need of calculation in the decoding process, but can be estimated decoding error probability of upper and lower limits. If the first layer decoding success, other layers are used the wrong location as the deleted position decoding success also, that is best case and the probability of this time may be $P_{wc}^{L\,'}$. This method is very convenient, and decoding error probability is the lowest. If success of the first layer decoding, but other layers use the wrong location of last layer to be delete wrong location fail, and need to carry out their own error correction decoding, which is the worst-case and probability of decoding error P_{ef}^L is largest. When the correct decoding probability of first layer is follows:

$$P_{wc}^{1\,'} = \sum_{i=0}^{t_1} \binom{n}{i} p_e^i (1 - p_e)^{n-i} \tag{5.64}$$

If the second layer sues the wrong location decoding of the first layer success, the probability of correct decoding is:

$$P_{wc}^{2\,'} = \sum_{i=0}^{(2t_2-t_1)/2} \frac{\Pr(i)}{\Pr} P_{wc}^{1\,'} \sum_{j=0}^{i} \binom{n}{i} p_e^j (1 - p_e)^{n-j} \tag{5.65}$$

For L layer, all with the error location of before layer to decode success, the probability of correct decoding:

$$P_{wc}^{L\,'} = \sum_{i=0}^{(2t_2-t_1)/2} \frac{\Pr(i)}{\Pr} P_{wc}^{L\,'} \sum_{j=0}^{i} \binom{n}{i} p_e^j (1 - p_e)^{n-j} \tag{5.66}$$

As each layer using the correction position decoding of before layer to decode success, its error correction capability is equivalent to the first layer of error correction capability completely. The lower limit of the decoding error rate $P_{ef}^{L\,'}$ can be estimated by the equation:

$$P_{\text{ef}}^{L}{}' = 1 - \prod_{i=1}^{L} P_{\text{wc}}^{1}{}' \tag{5.67}$$

If each layer has to use own error correction decoding, which is the worst case, and the maximum limit $P_{\text{ef}}^{L}{}''$ is:

$$P_{\text{ef}}^{L}{}'' = 1 - \prod_{i=1}^{L} P_{\text{wc}}^{i}{}^{2} \tag{5.68}$$

where the $P_{\text{wc}}^{i}{}^{2}$ is correct probability of error correction decoding for the i layer, and which is:

$$P_{\text{wc}}^{1}{}^{2} = P_{\text{wc}}^{1} \tag{5.69}$$

In general, multi-wavelength (layer) storage channel original BER is $p_{\text{e}} \approx 10^{-4}$, if set $n = 188$, $k = 172$, $k_1 = 178$, then $t_1, = 8$. The probability of correct decoding is shown in Fig. 5.38. It can be seen, when $\text{Pr}(0) + \text{Pr}(1)$ is same, the probability $P_{\text{wc}}^{L}{}'$ of correct decoding is same almost. Therefore the probability of number of soft errors with "0" or "1" in n bytes is determined to the probability of correct decoding probability $P_{\text{wc}}^{L}{}'$, as shown in Fig. 5.38.

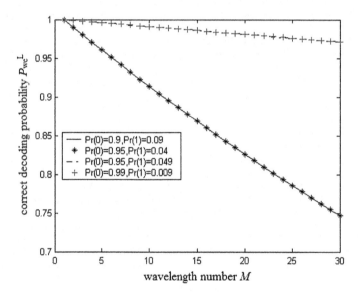

Fig. 5.38 The channel bit error rate of the multi-wavelength (layer) multi-level optical storage

Simplify treatment of $P_{wc}^{L}{}'$, with Eq. (5.66) is:

$$P_{wc}^{L}{}' = \sum_{i=0}^{(2t_2-t_1)/2} \frac{Pr(i)}{Pr} P_{wc}^{L-1}{}' \sum_{j=0}^{i} \binom{n}{i} p_e^j (1-p_e)^{n-j} \tag{5.70}$$

Normally, $2t_2 - t_1$ is very small, as $\sum_{j=0}^{i} \binom{n}{i} p_e^j (1-p_e)^{n-j} \approx 1$ so $P_{wc}^{L}{}'$ can be approximate:

$$P_{wc}^{L}{}' \approx \sum_{i=0}^{(2t_2-t_1)/2} \frac{Pr(i)}{Pr} P_{wc}^{L-1}{}' = \frac{\sum_{i=0}^{(2t_2-t_1)/2} Pr(i)}{Pr} \times P_{wc}^{L-1}{}' \tag{5.71}$$

The sum of $(2t_2 - t_1)/2$ of the probability of soft errors determined the probability of correct decoding. When the sum of the probability $(2t_2 - t_1)/2$ of soft errors is greater than 0.999, the probability of correct decoding is already close to 1, as shown in Fig. 5.39.

The number of level (layer) M is great influence the decoding error probability, it changes with the number of level (layer) M is shown in Fig. 5.40. And the curve of the maximum error probability and number of layers M is in Fig. 5.41. When $M = 18$, the minimum limit of the decoding error probability $P_{ef}^{L}{}''$ is $8.249603233561137 \times 10^{-15}$, and the maximum limit is $9.468807298581 \times 10^{-12}$ for example, as in Figs. 5.40 and 5.41.

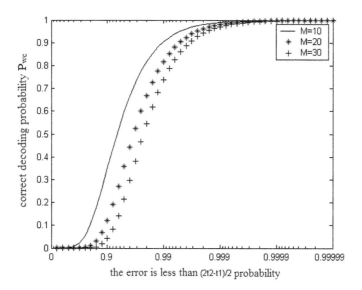

Fig. 5.39 The probability P_{wc} of correct decoding when sum of probability$(2_{t2-t1})/2$ of soft errors is within 1

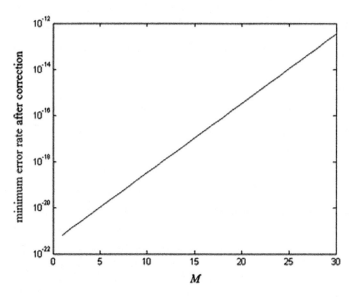

Fig. 5.40 The curve of relationship between decoding error probability and number of level (layer) M

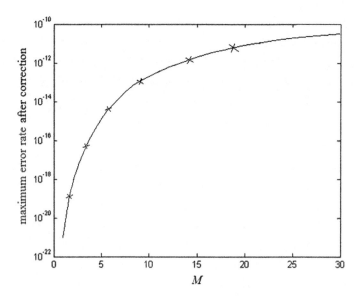

Fig. 5.41 The relationship of the maximum error probability with the number of layers M after correction

In above decoding method, calculation of P_{de} is more complex. But the $P_{de} = P_{ef}$, $P_{ef} \approx P_{df}$, so it is not need of analyze P_{de} in detail.

5.4.3   Reed-Solomon Error-Correcting Code

According to above calculation and analysis shows that this codec will not be able to read out any data when continued burst error at some time. In order to improve the performance of anti-burst error, introduce the Reed-Solomon error correcting code. It can correct a wide range of continuous burst error, that the principles are shown in Fig. 5.42. In accordance with the first data recording, data error detection code is arranged in a $(n_2 \times M) \times n_1$ ECC matrix. In order to preserve the

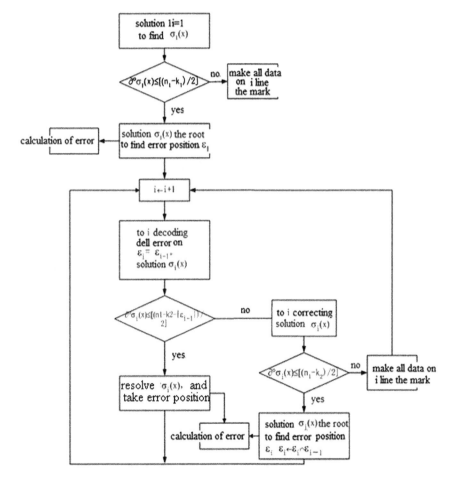

Fig. 5.42 Decoding algorithm and process of Reed-Solomon product codes

characteristics of the error between the layers, should be in the decoding first row decoder and then decoding the column. Therefore, the column is encoded as an internal code PI in the first, then to encode the row and to be outer-parity check code PO (Outer-parity Redd-Solomon code). Column encoding of M layer has higher probability of error, in order to avoid the error number exceeds the error correction capability in the columns of error correction, using a staggered column coding, i.e. for each j take the $j + M \times i$ $(0 \leq i \leq M; 0 \leq i \leq n_{2-1})$, total n_2 using $RS(n_2, k_3)$ to be encoded. The n_2 columns are corresponding data on the same layer in different regions. Therefore, in the column encoding, on the first layer of the k_1 column of the $RS(n_2, k_3)$ encoding k_2 columns of the RS, while the remaining layers (n_2, k_3) encoding. Column coding is done after the formation of a mixed array, as shown in Table 5.5. Finally, the line of $1 + M \times i$ $(0 \leq i \leq n_2$ is 1) $RS(n_1, k_1)$ encoding, for each j, $j + M \times i$ $(0 \leq j \leq M; 0 \leq i \leq n_2$ is $-1)$ $RS(n_1, k_2)$ coding gain is shown in Table 5.5 $(n_2 \times M)$ to $\times n_1$ coding matrix.

By this way to renew the original algorithm and process of encode and decode, and redesigned a new encoding and decoding algorithms and processing for MW/ML optical storage after experiments, it is shown in Fig. 5.43. According to computation, each row of M layer is to n^2 group decoding, and completed the outer code PO (outer-parity Reed-Solomon code) decoding process. If a row cannot be decoded correctly, then all the data of the row are tag. This tag in relation to error

Table 5.5 The $(n_2 \times M)$ to $\times n_1$ coding matrix

B1, 1, 1	B1, 1, 2	...	B1, 1, k_1			
B2, 1, 1	B2, 1, 2	...	B2, 1, k_1	B2, 1, k_1 + 1	...	B2, 1, k_2
:	:	...	:	:	...	:
BM, 1, 1	BM, 1, 2	...	BM, 1, k_1	BM, 1, k_1 + 1	...	BM, 1, k_2
B1, 2, 1	B1, 2, 2	...	B1, 2, k_1			
B2, 2, 1	B2, 2, 2	...	B2, 2, k_1	B2, 2, k_1 + 1	...	B2, 2, k_2
:	:	...	:	:	...	:
BM, 2, 1	BM, 2, 2	...	BM, 2, k_1	BM, 2, k_1 + 1	...	BM, 2, k_2
:	:	...	:	:	...	:
B1, k_3, 1	B1, k_3, 2	...	B1, k_3, k_1			
B2, k_3, 1	B2, k_3, 2	...	B2, k_3, k_1	B2, k_3, k_1 + 1	...	B2, k_3, k_2
:	:	...	:	:	...	:
BM, k_3, 1	BM, k_3, 2	...	BM, k_3, k_1	BM, k_3, k_1 + 1	...	BM, k_3, k_2
B1, k_3 + 1, 1	B1, k_3 + 1, 2	...	B1, k_3 + 1, k_1			
:	:	...	:	:	...	:
BM, k_3 + 1, 1	BM, k_3 + 1, 2	...	BM, k_3 + 1, k_1	BM, k_3 + 1, k_1 + 1	...	BM, k_3 + 1, k_2
:	:	...	:	:	...	:
B1, n_2, 1	B1, n_2, 2	...	B1, n_2, k_1			
BM, n_2, 0	BM, n_2, 1	...	BM, n_2, k_1	BM, n_2, k_1 + 1	...	BM, n_2, k_2

(continued)

Table 5.5 (continued)

$B1, 1, 1$	$B1, 1, 2$	...	$B1, 1, k_1$	$B1, 1, k_1+1$	...	$B1, 1, k_2$	$B1, 1, k_2+1$	...	$B1, 1, n_1$
$B2, 1, 1$	$B2, 1, 2$	...	$B2, 1, k_1$	$B2, 1, k_1+1$	...	$B_2, 1, k_2$	$B2, 1, k_2+1$	...	$B2, 1, n_1$
⋮	⋮	...	⋮	⋮	...	⋮	⋮	...	⋮
$BM, 1, 1$	$BM, 1, 2$	...	$BM, 1, k_1$	$BM, 1, k_1+1$	...	$BM, 1, k_2$	$BM, 1, k_2+1$	...	$BM, 1, n_1$
$B1, 2, 1$	$B1, 2, 2$	...	$B1, 2, k_1$	$B1, 2, k_1+1$	...	$B1, 2, k_2$	$B1, 2, k_2+1$	...	$B1, 2, n_1$
$B2, 2, 1$	$B2, 2, 2$	...	$B2, 2, k_1$	$B2, 2, k_1+1$	...	$B2, 2, k_2$	$B2, 2, k_2+1$	...	$B2, 2, n_1$
⋮	⋮	...	⋮	⋮	...	⋮	⋮	...	⋮
$BM, 2, 1$	$BM, 2, 2$	...	$BM, 2, k_1$	$BM, 2, k_1+1$	...	$BM, 2, k_2$	$BM, 2, k_2+1$	...	$BM, 2, n_1$
⋮	⋮	...	⋮	⋮	...	⋮	⋮	...	⋮
$B1, k_3, 1$	$B1, k_3, 2$	...	$B1, k_3, k_1$	$B1, k_3, k_1+1$	...	$B1, k_3, k_2$	$B1, k_3, k_2+1$	...	$B1, k_3, n_1$
$B2, k_3, 1$	$B2, k_3, 2$	...	$B2, k_3, k_1$	$B2, k_3, k_1+1$	...	$B2, k_3, k_2$	$B2, k_3, k_2+1$	...	$B2, k_3, n_1$
⋮	⋮	...	⋮	⋮	...	⋮	⋮	...	⋮
$BM, k_3, 1$	$BM, k_3, 2$	...	BM, k_3, k_1	BM, k_3, k_1+1	...	BM, k_3, k_2	BM, k_3, k_2+1	...	BM, k_3, n_1
$B1, k_3+1, 1$	$B1, k_3+1, 2$	...	$B1, k_3+1, k_1$	$B1, k_3+1, k_1+1$	...	$B1, k_3+1, k_2$	$B1, k_3+1, k_2+1$	...	$B1, k_3+1, n_1$
⋮	⋮	...	⋮	⋮	...	⋮	⋮	...	⋮
$BM, k_3+1, 1$	$BM, k_3+1, 2$	...	BM, k_3+1, k_1	BM, k_3+1, k_1+1	...	BM, k_3+1, k_2	BM, k_3+1, k_2+1	...	BM, k_3+1, n_1
⋮	⋮	...	⋮	⋮	...	⋮	⋮	...	⋮
$B1, n_2, 1$	$B1, n_2, 2$	...	$B1, n_2, k_1$	$B1, n_2, k_1+1$	...	$B1, n_2, k_2$	$B1, n_2, k_2+1$	...	$B1, n_2, n_1$
$BM, n_2, 0$	$BM, n_2, 1$	...	BM, n_2, k_1	BM, n_2, k_1+1	...	BM, n_2, k_2	BM, n_2, k_2+1	...	BM, n_2, n_1

code will be deemed to delete, when inner code PI (inner-parity Reed-Solomon code) is decoding. The decoding process of inner code PI is shown in Fig. 5.43, if request to improve error correction capability can be increased further algorithm, i.e. column correction cannot be corrected, that this column to be marked, and then run the error correction cycle again, and thus further reduce the uncorrectable error probability. While correcting coding scheme was estimated to error correction capability of soft and hard errors. After the introduction of Reed—Solomon code encoding method, on the column direction with the RS encoder at the same time, the corrected ability is improved greatly.

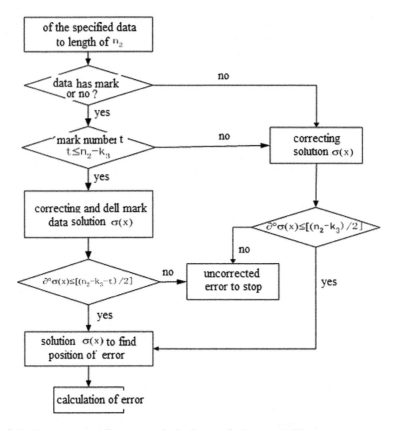

Fig. 5.43 The renew decoding process logic diagram for inner code PI

According to characteristics of the product codes, can calculate the length b of correct burst errors is:

$$b \leq \max(n_1t_3, n_1t_3, n_2t_1, n_2t_2) = \max(n_1t_3, n_2t_1) \qquad (5.72)$$

where n_1 is the number of columns of the product code, n_2 is the number of rows of product codes, t_1 is the RS(n_1, k_1) error correction coding and is $t_1 = (n_1 - k_1)/2$, t_2 is RS(n_1, k_2) burst length of error correction coding and is $t_2 = (n_1 - k_2)/2$, t_3 is RS (n_2, k_3) burst length of error correction coding and is $t_3 = (n_2 - k_3)/2$. In the first to decode the PO then decode the product code. In decoding decoding process, the soft and hard errors are decoded and corrected at the same time. Therefore, after the PO decoding is completed, the error probability of a row of data is:

$$P_{\text{ef}}^M \leq 1 - P_{\text{wc}}^{M\ 1} - (1 - P_{\text{wc}}^{M\ 1})P_{\text{wc}}^{M\ 2} \qquad (5.73)$$

The minimum is:

$$P_{\text{ef}}^{M\,\prime} \le 1 - \prod_{i=1}^{M} P_{\text{wc}}^{1\ \prime} \tag{5.74}$$

And the maximum is:

$$P_{\text{ef}}^{M\,\prime\prime} \le 1 - \prod_{i=1}^{M} P_{\text{wc}}^{i\ 2} \tag{5.75}$$

After PO decoding is completed, to decode PI. PI is encoding with the RS(n_2, k_3), and for consecutive decoding of M layers are divided in a different decoder, for avoiding the hard error and consecutive M error. In PO decoding, the RS(n_2, k_3) decoding the column is selected n_2 rows consisting of assumption by the L row of the M layer. The error correction probability of L row P_{wc}^{L} is described as follows:

$$P_{\text{wc}}^{L} = \begin{cases} P_{\text{wc}}^{L\ 1} + (1 - P_{\text{wc}}^{L\ 1})P_{\text{wc}}^{L\ 2} & t_L' = t_{L-1}' \\ P_{\text{wc}}^{L\ 2} & t_L' \ne t_{L-1}' \end{cases} \tag{5.76}$$

If the L row decoding fails, plus a label on each error on the row, as the column correction to delete, the probability P_{cd}^{L} is $P_{\text{cd}}^{L} = 1 - P_{\text{wc}}^{L\ 2}$.

It can be seen that if delete the wrong position is greater than $n_2 - k_3$ in the election n_2 rows, and cannot correct, the probability P_{cef}' is:

$$P_{\text{cef}}' \le \sum_{i=n_2-k_3-1}^{n} \binom{n}{i} (P_{\text{cd}}^{L})^i (P_{\text{wc}}^{L})^{n-i} = \sum_{i=n_2-k_3-1}^{n} \binom{n}{i} (1 - P_{\text{wc}}^{L\ 2})^i (P_{\text{wc}}^{L})^{n-i} \tag{5.77}$$

This is the column error correction can achieve maximum $P_{\text{cef}} < P_{\text{cef}}'$.
The Eq. (5.77) can be simplified, and rewritten as:

$$P_{\text{cef}}' = 1 - \sum_{i=0}^{n_2-k_3} \binom{n}{i} (1 - P_{\text{wc}}^{L\ 2})^i (P_{\text{wc}}^{L})^{n-i} \le 1 - \sum_{i=0}^{n_2-k_3} \binom{n}{i} (1 - P_{\text{wc}}^{L\ 2})^i (P_{\text{wc}}^{L\ 2})^{n-i} \tag{5.78}$$

$$P_{\text{cef}}' \le 1 - \sum_{i=0}^{n_2-k_3} \binom{n}{i} \left(1 - \sum_{j=0}^{t_2} \binom{n}{j} p_{\text{e}}^{j}(1 - p_{\text{e}})^{n-j}\right)^i \left(\sum_{j=0}^{t_2} \binom{n}{j} p_{\text{e}}^{j}(1 - p_{\text{e}})^{n-j}\right)^{n-i} \tag{5.79}$$

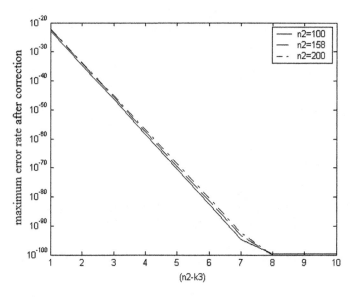

Fig. 5.44 The column correction can reach the maximum error correction probability

When given $n_1 = 188$, $k_1 = 172$, $k_2 = 178$, column error correction can achieve the maximum probability is shown in Fig. 5.44. It can be seen, after using the product code correction, the error probability decreased significantly. Therefore, choice parameters of the product code have to consider the length of correct burst errors. The correction probability can be guaranteed within length of burst error.

5.5   Multi-level CD-ROM

Above multiwavelength and multilevel storage systems is used the absorption of medium to realize multi-level easily. It is possible to make multiwavelength and multilevel CD-ROM for industrial mass production in the future. The mastering system of multiwavelength and multilevel CD-ROM is shown in Fig. 1.9 of Chap. 1, Figs. 5.1 and 5.2 in this Chapter. As the multilevel CD-ROM is not track so its mask manufacturing and replication process are rather simply. The contact exposure process of IC manufacture can be used to make multiwavelength and multilevel ROM replication. The replication system was developed in 2003 in OMNERC at Tsinghua University which is shown in Fig. 5.45, where 1—light source, 2—elliptical reflector, 3—plane mirror, 4—negative collimator lens, 5—honeycomb structures lens array, 6—collimator lens, 7—special fixtures with sealed vacuum press. The working spectrum of light source is based on absorption spectrum of medium that can choice 1.5–2 kW mercury or halogen lamp with spectrum range of 345–650 nm for 405, 532 and 650 nm multiwavelength and multilevel ROM or gray multi-level CD-ROM.

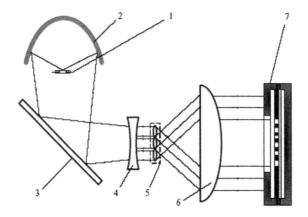

Fig. 5.45 Contact exposure replication system of multilevel CD-ROM

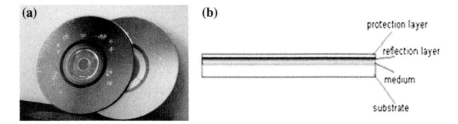

Fig. 5.46 The disc of gray multi-level CD-ROM (**a**) and its configuration (**b**)

The experimental products of gray multi-level CD-ROM with the contact exposure replication system is shown in Fig. 5.46.

The gray multilevel ROM disc is made by contact exposure replication system and its structure is shown in Fig. 5.47a that enlarged (6000×) view of surface structure of the gray multilevel ROM disc. Its readout signal is shown in Fig. 5.47b. It is not track so its mask manufacture and replication process are rather simply than traditional CD-ROM and its readout signal quality is better than traditional CD-ROM much more.

5.5.1 Injection Molding Multilevel CD-ROM

Above multiwavelength and multilevel storage disc is based on absorption of medium to light. Although it is very simpleness and effect. But the conventional CD injection process cannot make it. This section will discuss the injection multilevel CD-ROM.

(a)   **(b)**

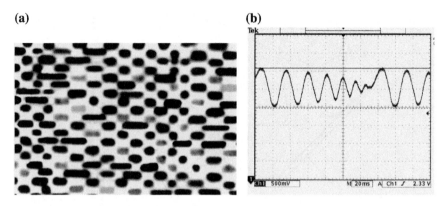

Fig. 5.47 The micrograph structure of the gray multilevel ROM disc (6000×) (**a**) and its readout signal (**b**)

5.5.1.1 Microstructures Design of Multilevel CD-ROM

The traditional CD-ROM is based on diffraction with pit depth and width modulation of recording symbol multilevel. Reflective lights from the pit and adjacent lands have a phase difference, which leads to a decrease of light intensity on the photon-electronic detector and a corresponding amplitude decrease in readout signal. The decrease amount is determined by phase difference and its distribution, that can calculate by Hopkin's classic optical disk diffractive model. So the microstructures of this kind of multilevel CD-ROM can be remodeled by this theory also. There are two multilevel technologies principally, i.e. pit depth and width modulation (PDWM) and signal waveform modulation (SWM). By this way, its most advantage is that the multilevel disc can produce with the traditional DVD industry manufacturing line. As SWM combined variation of size and position of inserted sub-pit/sub-land in an original land/pit to realize multilevel. So it has more scheme to realize multilevel modulation, more storage capacity, with lower BER (2×10^{-4}), better direct current (DC)-free control and servo performances. The writing strategy for SWM multilevel ROM can be explained by Fig. 5.48.

Figure 5.48 described the relationship between timing delay of laser beam recorder (LBR) and the micro-patterns of recording marks. Run-length Series

Fig. 5.48 Writing strategy of SWM system

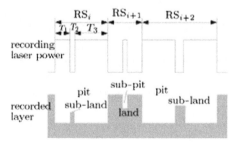

Table 5.6 Combined variation of size and position of sub-symbols with different levels

Level of a run-length	Level 0	Level 1	Level 2	Level 3	Level 4	Level 5	Level 6
size of sub-pit/sub-land	No	Large	Large	Large	Small	Small	Small
position of sub-pit/sub-land	No	Middle	Left	Right	Middle	Left	Right

(RS) and Level Series (LS) are two output data series after error correction and modulation coding. RSi and LSi represent the length of run-length and level of pit/land, respectively. After a sub-pit or sub-land is inserted, the original SWM recording mark is split into three parts, i.e. pits bland-pit or land-subpit-land, that define the vector TP = [T1, T2, T3], whose components are corresponding to three parts of a recording mark respectively. The LSi determines the numbers of T1, T2, T3 and their proportion, and RSi equals the summation of T1, T2 and T3. That means LSi determine the position and the size of inserted sub-land or sub-pit in an original recording mark, and RSi determines the length of the original recording mark. The adopted modulation code is a bimodulation code with a code rate of 7/8. There is no multilevel at 2–3T, and the level numbers of 4–11T are 2, 3, 4, 4, 5, 5, 7 and 7, respectively. Table 5.6 gives the relationship between the combined variation of size and position of inserted sub-pit/sub-land and level of a run-length. For example, if a large subland/sub-pit is inserted into the middle of the original 5T pit/land, the recording mark is called level 2 of 5T pit/land. Level 1 of 5T pit/land means inserting a small sub-pit/sub-land into the middle of the original 5T recording land/pit. Oppositely, if there is no inserted sub-mark, it is defined as level 0 of 5T pit/land.

5.5.1.2 Mastering Simulation

Richards, Boivin and Wolf gave the model of the focused intensity distribution of an incident laser beam coming through a focusing objective lens. Using cylindrical coordinate (ρ, φ, z) described the electric field near the focal point. The intensity profile of the beam spot is determined by the light electric field near the focal point. Kathleen and Thomas calculated the cylindrical-vector fields is $I = |e|^2$ and summarized as the following formula as:

$$e_\rho^s = A_s \int_0^\alpha \cos^{1/2}\theta \, \sin(2\theta) l_0(\theta) J_1(k\rho_s \sin\theta) \exp(\mathrm{i}\,kz \cos\theta) \mathrm{d}\theta, \qquad (5.80)$$

$$e_\phi^s = 2A_s \int_0^\alpha \cos^{1/2}\theta \, \sin\theta \cdot l_0(\theta) J_1(k\rho_s \sin\theta) \exp(\mathrm{i}\,kz \cos\theta) \mathrm{d}\theta, \qquad (5.81)$$

$$e_z^s = 2iA_s \int_0^\alpha \cos^{1/2}\theta \sin^2\theta \cdot l_0(\theta)J_0(k\rho_s \sin\theta) \exp(i\,kz\cos\theta)d\theta, \qquad (5.82)$$

where, e_ρ^s, e_ϕ^s and e_z^s are respectively the radial, the azimuthal and the longitudinal components of the electric field vector near focus in the cylindrical coordinates; function $J_n(k\rho_s \sin\theta)$ denotes a Bessel function of the first kind of order n; $k = 2\pi/\lambda$; $\alpha = \sin^{-1}(NA/n)$:

$$l_0(\theta) = \exp\left[-\beta_0^2\left(\frac{\sin\theta}{\sin\alpha}\right)^2\right]J_1\left(2\beta_0\frac{\sin\theta}{\sin\alpha}\right) \qquad (5.83)$$

with parameter β_0 is the ratio of pupil radius to beam waist. Dill and Pasman et al. researched the integrated circuit (IC)-lithography model, which was adopted to describe the exposure model in disc mastering. Considering the special characteristic of disc mastering, Song et al. and Yuan et al. predigested the model and described it as:

$$M(x, y, t) = M(x, y)e^{-CE(x,y)}, \qquad (5.84)$$

where M is the photo active compound concentration relative to the initial concentration, E is the exposure dose and can be calculated from:

$$E(x, y, t) = \int_0^t I(x, y, \tau)d\tau. \qquad (5.85)$$

Kim et al. Trefonas and Hirai et al. developed the model of IC-lithography separately. For the three models, Trefonas's model is the simplest, and the number of parameters needed to be demarcated is the smallest, which is expressed as:

$$R = R_0(1 - M)^q, \qquad (5.86)$$

where R_0 is the development rate of fully exposed photoresist and q is the photoresist contrast. The main parameters used in simulation and disc mastering are shown in Tables 5.7 and 5.8.

The simulation model is employed to generate the microstructure of each level of every run-length and optimize the writing strategy parameter TP. Based on simulated results, the optimum TP of each level of every run-length is designed and confirmed.

Figure 5.49 shows some parts of the simulation results in a recording track with optimum TP, and it illustrates the simulation profiles of level 0 of 3T pit, level 2 of 9T land, level 4 of 8T pit, and level 4 of 8T land (partly). The sizes of 2 sub-pits in this Fig. 5.50 are deferent from each other, which is used for explaining the levels

Table 5.7 Parameters used in simulation and master manufacturing

Parameters	Sign	Value
The maximum developing rate	R_o	12 nm/s
Developing contrast	q	2.3
Developing time	t_d	12 s
Photoresist constant	C	2.6 mm²/(mJ s)
Recording speed	v	3.52 m/s
Focus spot intensity	I_o	16.4 mW/μm²

Table 5.8 Optimum TP's of 5T, 8T and 10T

	T_P of 5T	T_P of 8T	T_P of 10T
Level 0	[22.5 0 22.5]/9	[36 0 36]/9	[45 0 45]/9
Level 1	[19 7 19]/9	[29.5 13 29.5]/9	[38.5 13 38.5]/9
Level 2	[16.5 12 16.5]/9	[24 13 35]/9	[31 13 46]/9
Level 3		[35 13 24]/9	[46 13 31]/9
Level 4		[32.5 7 32.5]/9	[41 8 41]/9
Level 5			[30 9 51]/9
Level 6			[51 9 30]/9

Fig. 5.49 Simulation of the recording symbols of a track

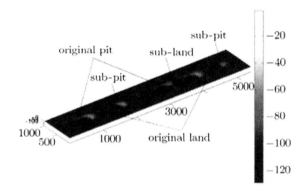

of run-lengths. gives the optimum TP of levels of 5T, 8T and 10T. Figure 5.49 shows the simulation results in a recording track with optimum TP, and it illustrates the simulation profiles of level 0 of 3T pit, level 2 of 9T land, level 4 of 8T pit, and level 4 of 8T land (partly). The sizes of 2 sub-pits in this figure is different from each other, which is used for explaining the levels of run-lengths.

According to Table 5.6 and the measured distances, the recording marks are in turn level 3 of 6T land, level 1 of 4T pit, level 0 of 2T land, and level 1 of 5T pit. According to Table 5.8, [36 0 36]/9, [29.5 13 29.5]/9, [24 13 35]/9, [35 1324]/9, [32.5 7 32.5]/9 are respectively set as the values of TP of levels 0–4 of 8T. The corresponding simulation profiles of 8T recording pits are shown in Fig. 5.50. The accuracy of simulation reaches up to 19 nm (T/9). Figure 5.50a–e show the simulated microstructures of 8T pits with levels 0–4. Figure 5.50f–j exhibit the contour

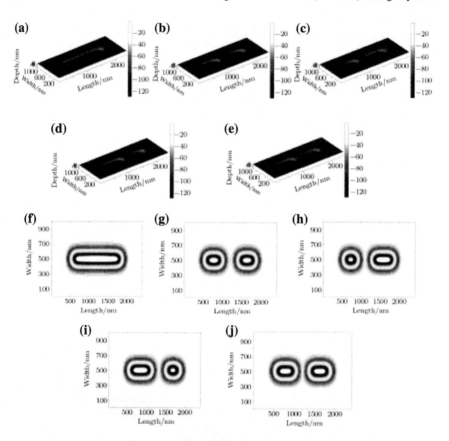

Fig. 5.50 Simulated microstructures (**a–e**) and contour maps (**f–j**) for 8T pits with levels 0–4

maps of 8T pits with levels 0–4. Comparing Fig. 5.50b with Fig. 5.63e show that the inserted sub-lands are both in the middle of 8T pits. However, the sizes of inserted sub-lands are changed, which is used for denoting levels 1 and 4 of 8T pits.

Figure 5.50b–d show the position variations in inserted large sub-land, which are used to differentiate levels 1, 2 and 3 of 8T pits. There is no inserted sub-land in 8T pits in Fig. 5.63a. Hence, Fig. 5.63a represents the microstructure of level 0 of 8T pit.

To verify the correctness of simulation results, the parameters was used to the SWM disc manufacturing. Figure 5.5 shows the actual profiles of 8T pits with levels 1–4. Figure 5.51a, d indicate the size variations of sub-lands, while Fig. 5.50a–c illuminate the position variations of inserted sub-lands.

Comparing Figs. 5.50 and 5.51 shows the simulated structures that are consistent with the actual ones. The simulation results reveal the actual profiles of different levels of 8T recording pits with an accuracy of 19 nm. Hence, the optimum writing strategy parameter TP based on simulation results is suited to form ideal microstructures of SWM recording marks. It is time-saving to manufacture the SWM disc designed with the simulation results. Moreover, changed the vector TP

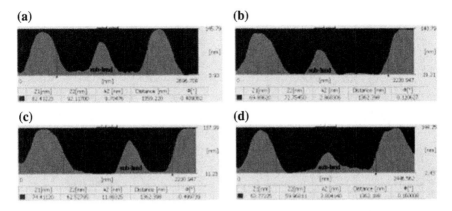

Fig. 5.51 Microstructures of 8*T* pits with levels 1 (**a**), 2 (**b**), 3 (**c**), 4 (**d**) of 8*T* pit

verify the realizable maximum level number of the run-length and provide a theoretical instructions for SWM disc mastering.

5.5.1.3   Write Strategy Optimization

The lithography model for optical disk mastering is need of calculation the pit profile on the disk in the first, that based on angle spectrum decomposition for simulation the readout signal. All the parameters are same to traditional DVD production. All the readout parameters are selected according to the DVD system. The 3*T* (*T* is the mark length corresponding to one channel bit), 4*T* and 5*T* remain the 2-level state, 6*T* and 7*T* realize 6 levels, 8*T* and 9*T* realize 10 levels, 10*T* and 11*T* realize 14 levels. Simulation results of typical run-lengths are shown in Figs. 5.52, 5.53 and 5.54 with 6*T* and 11*T*. For 6*T*, because it is short, the sub-mark

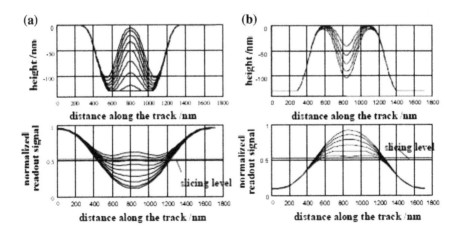

Fig. 5.52 Simulation results of writing vector trial for 6*T*: **a** is pit, **b** is land of recording mark

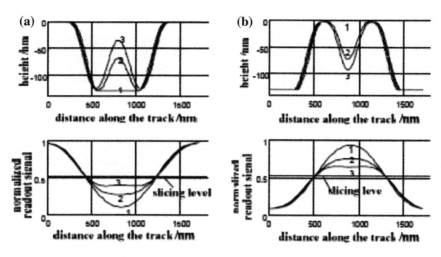

Fig. 5.53 The corresponding simulation results. Run-length is compensated with the sum of $T1$, $T2$ and $T3$ larger than 6

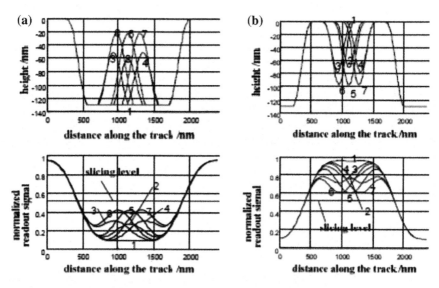

Fig. 5.54 Simulation results of $11T$ pit (**a**) and land (**b**) with selected writing vectors: *Above* is center cross-section along the pit track. *Below* is readout signal

is positioned in the middle, and only the length of the sub-mark is varied. Figure 5.52 shows the trial simulation results. For $6T$ pit, the calculated write vectors, from top to bottom, are [2 2 2], [2.125 1.75 2.125], [2.25 1.5 2.25], ..., [3 0 3] respectively. The unit is the duration of one channel clock. For $6T$ land, the calculated writing vectors, from top to bottom, are respectively [3 0 3], [2.5 1

Table 5.9 The selected writing vectors for 6T

Pit		Land	
Level no.	Writing vector (T)	Level no.	Writing vector (T)
1	[3 0 3]	1	[4.5 0 4.5]
2	[24.75 7 24.75]/9	2	[34 13 34]/9
3	[23.75 11 23.75]/9	3	[32 17 32]/9

2.5], [2.375 1.25 2.375], ..., [2 22]. The readout signal shows amplitude variation through the entire run-length, instead of waveform variation. Because 6T is shorter that reading light spot to the whole length. However, readout signal of different levels can be differentiated also. For the same T2 value in writing vector, the influence of a sub-land is more significant than that of a sub-pit. It means that different writing vectors should be selected for pits and lands to obtain symmetrical readout signals. The run-length is obviously shortened due to the inserting of sub-marks, so run-length compensation is needed (Table 5.9).

For other run-lengths, similar simulation is carried out, and then writing vectors are selected. Table 5.10 is the selected writing vectors for 11T, and Fig. 5.54 is the corresponding simulation results. 11T realize 14 levels. The sub-marks are positioned in the middle, middle left and middle-right. The length of sub-marks is also varied. The waveform of readout signal is obviously modulated with the level differentiation. More complex WS optimization is implemented for corresponding parameter selection and adjustment also.

The proposed SWM ML recording is implemented on the DVD platform. Commercially available mastering and injection molding equipment are used. All the processing parameters and conditions are the same as conventional 2-level DVD. Key mastering processing parameters are listed in Table 5.11. The only change is the formatter controlling the recording laser power. An ESP-7000 formatter by ECLIPSE corp. is employed to generate the required writing pulse.

Figure 5.55a shows the profile images of SAM ML and SWM ML disks scanned by an atom force microscope (AFM). Cross-section of sub-land/sub-pit is also shown in Fig. 5.55b, c. The sub-land is not as high as normal land. That is because

Table 5.10 Selected writing vectors for 11T

Pit		Land	
Level no.	Writing vector (T)	Level no.	Writing vector (T)
1	[5.5 0 5.5]	1	[5.5 0 5.5]
2	[45.5 8 45.5]/9	2	[43 13 43]/9
3	[31 9 59]/9	3	[30 14 55]/9
4	[59 9 31]/9	4	[30 14 30]/9
5	[43 13 43]/9	5	[55 17 41]/9
6	[32 13 54]/9	6	[30 17 52]/9
7	[54 13 32]/9	7	[30 17 30]/9

Table 5.11 Processing parameters for DVD mastering

Substrate	SiO$_2$ glass	
Photo-resist	Type	Microposit S1813
	Thickness	130 nm
Exposure	Laser wavelength	405 nm
	Objective lens numeric aperture	0.9
Development	Developer 3:1 D.I water microposit developer concentrate	
	Temperature	22 °C
	Time	12 s
Linear velocity	3.52 m/s (DVD 1×)	

Fig. 5.55 a Images of multilevel CD-ROM disc by atom force microscope (AFM), **b** cross-section of sub-land of 4-level and **c** cross-section of sub-pit with 4-level

the reading spot has a certain size, and the fringe part of spot will bleach the sub-land area. Also, the sub-pit is not as deep as normal pit. Because the exposure of photo-resist exhibits integral effect, and irradiation time of sub-pit area is not so long due to its short writing pulse.

A digital circuit based on field programmable gate array (FPGA) is developed to detect run-length and level data. A raw error BER of 10^{-4} is obtained. The feasibility of SWM ML recording on DVD is validated. The novel signal waveform modulation multi-level read-only recording is presented. Numerical simulation provides a helpful tool for the write strategy optimization, and the feasibility of this method is experimentally demonstrated on the DVD platform.

A commercial DVD pick-up and a commercial DVD servo circuit are used to readout the disk. The readout signal of random data is shown in Fig. 5.56. Actual signals of 6T and 11T are also shown in Fig. 5.57a, b. Comparing Fig. 5.57a, b with Figs. 5.53 and 5.54, it can be seen that the simulated and experimental signals agree well.

A 2-level-to-multi-level mapping modulation coding with 2 steps is employed in our method. Firstly, a 2-level RLL coding is carried out, where run-lengths smaller than 3T are allowed. Secondly, run-lengths smaller than 3T will be eliminated. These too short run-lengths will be combined with neighboring run-lengths to form a long run-length, and then mapped to a multi-level run-length. For example, 2T-2T-2T can be mapped to 6T of level 2, 2T-4T can be mapped to 6T of level 3,

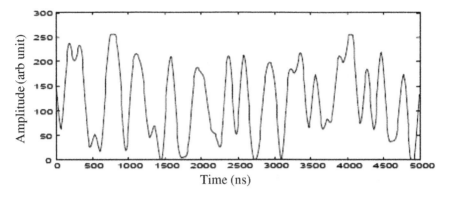

Fig. 5.56   Readout signal of random data on the SWM ML disk

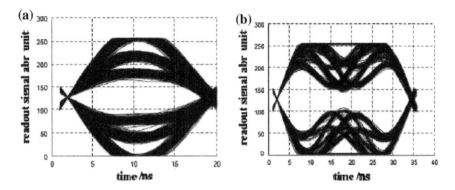

Fig. 5.57   Actual signals of 6*T* (**a**) and Actual signals of 11*T* (**b**)

3*T*-2*T*-3*T* can be mapped to 8*T* of level 2, 2*T*-2*T*-4*T* can be mapped to 8*T* of level 3. Without using all the levels of every run-length, to realize the same transfer rate and capacity as 2-level (1, 7) RLL coding, whose density ratio (DR) is 2 bits/(minimum symbol).

Experiments show that it is feasible for 5*T* to realize 4 levels, which means that (0, 6) RLL coding can be mapped to the ML coding and the DR will be 2.25 bits/ (minimum symbol). As the DR for 2-level DVD is 1.5 bits/(minimum symbol), an increase of 50 % in capacity can be expected if the min symbol length and track pitch are kept the same as DVD. As a comparison, the latest reported realizable DR of 4-level SAM ML is only 2 bits/(minimum symbol), whose corresponding increase in capacity is 33 %. Therefore, SWM ML shows superiority over SAM ML in capacity increase. This new SWM ML also has other merits:

(a)   In SWM ML, both land and pit can realize multi-level. It is naturally land-pit-spacing, which can facilitate the injection molding and RF signal DC-free control.

(b) Experiments show that the SWM ML has the same track following servo performance as 2-level DVD, while the SAM ML encounters the TE detection problem.

5.5.1.4 Other Schemes of SWM

There are many schemes of modulation except changing scheme of the mark for the multi-level CD-ROM as the accompanying sub-track construction and grating configuration etc. An accompanying sub-track construction multilevel CD-ROM and the improved adaptive partial response maximum likelihood (PRML) method combining modulation code for signal waveform modulation multi-level disc was used to multilevel CD-ROM. This improved adaptive PRML method employs partial response equalizer and adaptive viterbi detector combining modulation code. Compared with the traditional adaptive PRML detector, the improved PRML detector additionally employs illogical sequence detector and corrector. Logical sequence detector and corrector can avoid the appearance of illogical sequences effectively, which do not follow the law of modulation code for signal waveform modulation multi-level disc, and obtain the correct sequences. The improved PRML detector used a DSP and an FPGA chip. The experimental results show good performance that has higher efficient and lower complexity with the improved PRML method. Meanwhile, resource utilization of the improved PRML detector is not changed, but the BER is reduced by more than 20 %. This MD storage is realized to two formats as shown in Fig. 5.58. First format is based on a couple of tracks which consist of the spiral track and its accompanying sub-track. On a traditional optical disc the track is a spiral from inside to outside of the disc. In the first format an accompanying sub-track is inserted into the space between the inner tracks.

The 2D disc is based on a broad spiral, which consists of a number of parallel bit-rows that are aligned with each other in the radial direction in such a way that a 2D close-packed hexagonal lattice of bits results. Though this kind of 2D disc can utilize the area of the focused spot because of the hexagonal lattice of bits, it cannot

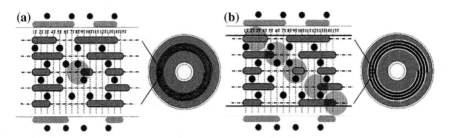

Fig. 5.58 Schematic format of multi-dimensional optical storage (**a**) schematic format for the first format and second format (**b**) of MD

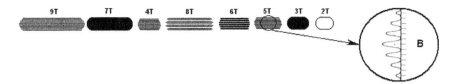

Fig. 5.59 The configuration of signal waveform modulation multi-level ROM disc with equidistant or unequidistant grating and the readout diffractive patterns B

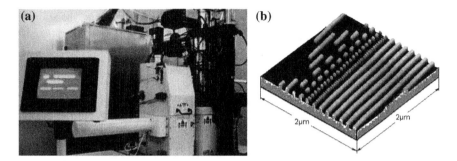

Fig. 5.60 The E-Beam mastering system (**a**) and the sample of experimental pattern (**b**) with the minimum size of the mark can be down to 20 nm

be compatible with the traditional one-single which is realized into kinds of the formats.

The grating configuration can be used to signal waveform modulation multi-level ROM disc as shown in Fig. 5.59. The readout signal is a diffractive pattern as in Fig. 5.60b. So the detector is an array with a difference processor that can take more exactitude multi-level modulation signal. But the difficulty of stomper manufacturing is great increase than traditional CO-ROM. Because the minimum size of the mark could be down to 50 nm, that has to use the E-Beam mastering system.

The E-Beam mastering system consists of a massive vacuum chamber with a special air bearing system for substrate rotation and translation. Mounted through the top of the chamber are a load-lock and the electron beam column as shown in Fig. 5.60a. The whole machine is mounted on active vibration isolators and is protected by a robust enclosure which provides the clean air environment as well as acting as an earthquake restraint that can set very tight specification pitch stability of with 10 nm and a repeatable local track pitch stability of 3.5 nm is achieved independent of the track pitch setting. The E-Beam mastering system is also ideally suited for production of the next generation optical disc or hard disc drives (HDD). The E-Beam mastering system is capable of writing the servo/gray code which can be embossed directly onto the platters, which will allow CD to go beyond the super-resolution size limit, with capacities beyond 1 tera bit per square inch.

A experiment pattern is shown in Fig. 5.60b note that the cutting direction is perpendicular to the grooves on the disc.

The E-Beam mastering system are almost identical to those required by conventional laser mastering. Standard process equipment is used for resist master preparation, developing, metallising and plating. The only change is the use of hot plate baking stations for soft bake and post exposure bake (PEB).The process work has concentrated on two resist types: positive e-beam resist (ER) and chemically amplified (CA). The ER resist process gives the best indication of electron beam spot size, however, it has low sensitivity and uses hazardous and environmentally unfriendly organic solvents for development. Most exposures have been made using the CA resist, with the ER resist used only as a reference to check spot size and column stability. The CA resist is up to 10 times as sensitive as the ER resist and uses aqueous developing solutions. The one complication with the CA resist is that it requires a PEB to fully form the features. The PEB controls the amplification process, to which the resulting feature size is very sensitive. The PEB hotplates to give temperature control of ± 0.1 °C and a temperature variation across the surface of the plate of ± 0.5 °C. This unique technologies make the system can easily produce very small pits required for multi-level CD-ROM master. The PEB technologies could reduce the stamper size physically and the number of steps dramatically in mastering process also. Based on the novel photochromic material multilevel optical storage that has good characteristics, so can achieve higher level RLL modulated as 8-level (1, 3) coding.

5.6 Multi-level Run-Length Limited Modulation Code

The multilevel modulation codes is key technology for multilevel optical storage, according to a new state segmentation method to design the 8 level RLL recording code, that have higher coefficient of coding rate and density. Its coding and decoding translation is concise that code logic can be used for high-density multi-level run-length modulated optical storage systems.

The current RLL modulation coding was used to multi-level optical storage. Since the variation of the amplitude of the optical disk read signal is not main random noise, so it is possible to employ length limited code in multilevel optical storage to increase the recording density. In the multilevel optical storage, if the length of recording symbol was changed, the multi-level run-length modulated will be achieved on a conventional recording medium that obtained a read signal amplitude characteristics of the multi-level optical storage. This method is suitable for most multi-level optical storage, such as using different pit depths multi-order ROM. Above multi-level RLL modulation amplitude modulating signal and combined, can read signal amplitude (multi-level) and record character length (run-length) two-dimension while multi-level length modulation information storage, and further increase the recording density of the data as shown in Fig. 5.61. As can be seen, the symbols of the multi-level RLL modulation codes have different

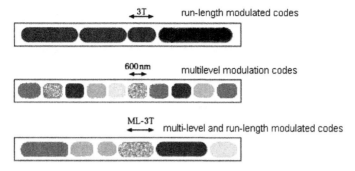

Fig. 5.61 The comparison of record marks of traditional RLL modulation, multi-level amplitude modulation and multi-level RLL modulation amplitude modulating

lengths and the "gray", which represents a number of different gray-physical states of the recording medium.

But the recording symbol of the multi-level RLL modulation replaces the information of corded signal amplitude and length at the same time. The signal to noise ratio of read-out will occurs serious loss that require more advanced coding and signal processing technology.

5.6.1   Multi-level Run Length Limited Coding

The multi-level run length limited encoding of an arbitrary data has to be conversion multi-level symbol sequences to meet specific channel restrictions, that is traditional RLL (d, k) code expansion indeed, also known as M-ray (d, k) code. In the M-array (d, k) length limited sequence, between any two non-zero characters, must have d '0' at least and a maximum of k '0', the character set is defined as a $\{0, 1, ..., M - 1\}$, where $2 < M < \infty$. On average, the encoder will transform m-bit of binary user data into the n channels symbols of multi-level. So the definition of coding rate is $R = m/n$ = maximum rate. The modulation coding capacity in theory C is calculated as follows:

$$C(M, d, k) = \ln \eta \qquad (5.87)$$

where, η is the largest real roots of the following characteristic equation:

$$zk + 2 - zk + 1 - (M - 1)zk - d + 1 + M - 1 = 0 \qquad (5.88)$$

Typically $\eta = RC$ indicates the coding efficiency coding bit rate to the coding capacity. The $(d + 1) \times C$ coefficient is defined as the density, in units of "bits minimum recording symbol" represents the M-array (d, k) coding recording density that can be obtained in Table 5.12.

Table 5.12 M-array (d, k) coded density coefficients

(d, k)	$M = 2$	$M = 4$	$M = 6$	$M = 8$	$M = 10$
(1, 2)	0.81	2.15	2.79	3.22	3.54
(1, 3)	1.10	2.31	2.91	3.31	3.62
(1, 7)	1.36	2.40	2.96	3.35	3.65
(2, 7)	1.55	2.67	3.23	3.62	3.91
(2, 10)	1.63	2.69	3.24	3.62	3.92

The density coefficients have several different parameters, including the $M = 2$, or traditional (d, k) code. In the same (d, k) parameters. The greater coefficients the higher density, but requirements for recording medium is higher. For the same order (level) of M, increasing the parameter d or k can also increase the density, but will increase the signal detection complexity. Therefore, in the design and selection of the coding parameter must balance various factors. As multi-level photochromic material experiment results for example analysis of 8-order (d, k) coding by Table 1: in $M = 8$ column, the density coefficient (1, 3) to (1, 2), the code density increase to about 9 %, while the (1, 7) code to (1, 3) code to improve 4 % only; for parameter $d = 2$, the density increased by about 10 %, but the complexity of the encoder will be increase that shows the 8-order (1, 3) code is better.

5.6.2   Construction of 8-Level (1, 3) Code

As shown in Fig. 5.62, the finite state transition diagram (FSTD) was described the order of 8 level (1, 3) codes, where in (7) represents a character in the {1, 2, ..., 7}. The marked circles represent different states of FSTD. The each connection line with arrow (path) indicates the status of the jump. The data, as (7), indicates the corresponding output character of jumps. The state '0' is output '0' character only, then into the state '1'. But state '1' and '2' can output '0' character or a multi-level character '1'–'7', then return to the state '0'. The state '3' can only output a multi-level character of '1'–'7' state, then into the state '0'. It is easy to verification that starting from any state in order to read the path identifier of FSTD can obtain all of 8 order (1, 3) limited character sequences.

Fig. 5.62 FSTD of the 8 level (1, 3) code

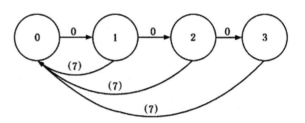

The relationship of FSTD connection can be expressed by a connection matrix $T = t_{ij}$ to $(k + 1) \times (k + 1)$ usually, while t_{ij} is the routes form state i jump to state j. Therefore the connection matrix of FSTD to 8-level (1, 3) code is:

$$T = \begin{bmatrix} 0 & 1 & 0 & 0 \\ 7 & 0 & 1 & 0 \\ 7 & 0 & 0 & 1 \\ 7 & 0 & 0 & 0 \end{bmatrix} \tag{5.89}$$

Theoretical analysis shows that the largest eigenvalues of the connection matrix T is the η of the Eq. (5.89). Thereby calculating capacity of 8 level (1, 3) code is $C = 1.66$. By this way, any $R < C$ based on FSTD finite state coder can be constructed. In order to obtain high efficiency and low complexity coding selected code rate is $R = m/n = 3/2 < C$. According to "coding theory" state segmentation algorithm, it is necessary a lot of states based 8 on the level (1, 3) code of $n = 2$ FSTD. In order to meet the requirements of the input data (total $2m = 2^3 = 8$ types of input data), the number of output routes of these states is over 8, that can be used to design the encoder compliance RLL limit. From each state in Fig. 5.63 FSTD jump two steps continuously, to obtain the corresponding two-FSTD codes, as in Fig. 5.63.

If the output number of routes in each FSTD is not less than 8, that can design 8 level (1, 3) the encoder directly of FSTD. By above analysis show that the output routes of state "3" are less than 8. So state "3" will be deleted directly, and get a new FSTD as shown in Fig. 5.64.

From analysis of state '1' and state '2' in the Fig. 5.64 can find out that there are two output routes with same label 0(7) and (7)0 to point to the same state '0' and '1'. According to the segmentation algorithms, that can be combined to simpler

Fig. 5.63 8 two-FSTD level (1, 3) codes

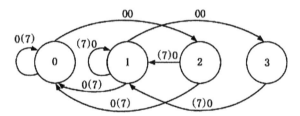

Fig. 5.64 A new FSTD was deleted state '3'

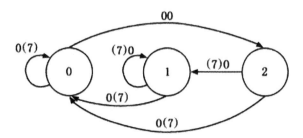

Fig. 5.65 The state "1" and "2" of FSTD are combined

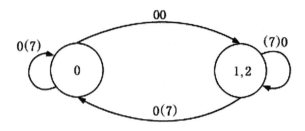

FSTD, as shown in Fig. 5.65, which the number of routes of two output states are over 8 and further simplification. Therefore, according to Fig. 5.65 design FSTD can obtain simplest encoder. Of course, there are many schemes of the output data corresponding to the source code in of the encoder. As a very convenient coding scheme: for the state '0', the eight kinds of input assigned to the appropriate output routes; for the state "1", from '000' assigned to the output 'X0', $X = \{1, 2, ..., 7\}$.

Above scheme determined the rule of conversion between source data and the output code word, that can use to constitute 8 level 32 (1, 3) encoder with code rate $R = 3.6$. Table 5.13 is the encoding table for the encoder, in which state "A" and "B" correspond to state "0" and "1, 2". When there are a lot of consecutive '000' in the source data, the encoder output code 'X0' in order to meet the maximum run length limited. Since a non-zero character X is any value from $\{1, 2, ..., 7\}$, this flexibility can be used to reduce the DC component of RLL recording waveform. As shown in Table 5.13, the encoder of 8-level (1, 3) has an packet decoding characteristics that can obtain input source data directly by corresponding to the current code, so it can control errors within one code effectively. A smart decoding rule is that removed the first code and other codes are converted to binary code to obtain the source data. The encoding and decoding rules are simple and implemented in hardware easily. So the 8-level (1, 3) encoder has better practicability.

The coding efficiency of the 8 level (1, 3) codes is up to $\varepsilon = RC = 91\ \%$, that a minimum recording symbol can be stored the $(1 + d)\ R = 3.0$ bits of user data, which is twice of conventional EFM + code. If the principles of the 8 level (1, 3) RLL modulation codes is implemented to DVD system, its capacity may be

Table 5.13 Two-state 8 level (1, 3) coding table with $R = 3/2$

Source data	State A		State B	
	Code	NS	Code	NS
000	00	B	X0	B
001	01	A	01	A
010	02	A	02	A
011	03	A	03	A
100	04	A	04	A
101	05	A	05	A
110	06	A	06	A
111	07	A	07	A

increased to 10 GB or over. If it was used to photochromic multi-level optical storage that the channel rate with 8 level 32 (1, 3) coding scheme, that the code density is up to 3.0 bits per record marks, and coding efficiency can reach to 91 %.

5.6.3 Rate 7/8 Run-Length and Level Modulation for Multilevel ROM

The Rate 7/8 run-length Multilevel recording technology can be used to conventional optical and mechanical units of CD. The modulation code scheme for signal waveform modulation multilevel (SWM) read-only optical disc has been implemented in OMNERC. The scheme is composed of RLL modulation and level modulation two-steps. RLL modulation is employed to meet the requirements of channel. To acquire higher code rate, the parameter d of RLL (d, k) is decreased to 0, which makes the presented scheme difference from other modulation codes of the optical storage systems. Increasing the number of k also contributes to the high code rate. Decreasing d and increasing k will respectively introduce more inter-symbol interference (ISI) and timing recovery error (TRE) to SWM optical system. Level modulation is used to resolve these problems. The decoding rule is simple and easy for implementation.

The signal waveform of SWM disc adopted the proposed code is also described. The information bits per 400 nm are 2.19, which is 46 % higher than that of DVD. With the requirement of high-definition television program (HDTV) increasing, the need for mass capacity disc is rising. Multilevel recording can increase the capacity of disc with no change the optical and mechanical units. The multilevel on all the pits ($3T$–$11T$) was realized in OMNERC with changing the pit-width and depth, which means the light intensity of the pits on the photo-detector will be differentiated according their depth and width. This multilevel recording disc is termed signal amplitude multilevel (SAM) disc. However, find the difficulty to implement the multilevel on the short run-lengths and the amplitudes of readback signal of pits with different levels cannot be clearly distinguished, when the number of level (M) is increasing. The M is limited, even if for long run-lengths also. Another way termed signal waveform multilevel (SWM) to implement multilevel recording on DVD discs. SWM recording is realized by inserting a sub-land/sub-pit into an original recording pit/land, which leads to the waveforms of readback signal of pits and lands differentiate according to the position and size of the sub-lands and sub-pits. In the SWM disc, there is no or less multilevel on the short run-lengths, which successively avoid the problem of SAM. The more multilevel is implemented, the longer the run-length is. Figure 5.66 is the atomic force microscopy (AFM) image of SWM disc. Figure 5.66b describes the profile shapes of the four recording symbols in Fig. 5.66a. The distances between three pairs lines are respectively 894, 1245, 910 nm, which means the three recording symbols are $5T$ (land), $7T$ (pit), $5T$ (pit). The variation of size and position of sub pit/sub-land, is the level information of run-length. Modulation code is one of the key technologies

(a)   **(b)**

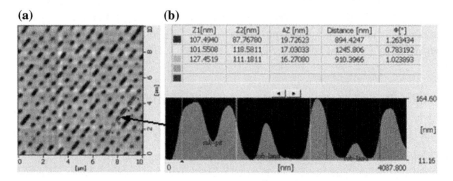

	Z1[nm]	Z2[nm]	ΔZ [nm]	Distance [nm]	Φ[°]
■	107.4940	87.76780	19.72623	894.4247	1.263434
	101.5508	118.5811	17.03033	1245.806	0.783192
	127.4519	111.1811	16.27080	910.3966	1.023893
■					
■					

Fig. 5.66 **a** AFM image of SWM of 7/8 run-length multilevel recording. **b** Cross sectional of sub-pit and sub-land

to improve the storage capacity of optical disc. The common modulation code used in currently optical recording disc is RLL code, which is described as RLL(d, k). The number of '0's between two neighboring '1's is at least d and at most k. The modulation code in compact disc (CD), digital versatile disc (DVD) and Blu-ray disc are respectively 8/17 RLL(2,10), 8/16 RLL(2,10) and 2/3RLL(1,7). There are some papers published for the modulation code of SAM disc. The modulation code for SAM is called M-array RLL(d, k). M is the level number of every run-length. As the difference between SWM and SAM, the M-array RLL(d, k) code cannot be directly used in SWM disc.

In this paragraph the modulation-coding scheme defined as run-length and level modulation (RLM) code is presented and the encoding and decoding rules are discussed detailedly. RLM is carried out through run-length modulating and level modulating two steps. Firstly, use run-length-limited (RLL) code to scheme out a RLL (0, 15) code with code rate $R = 7/8$. To avoid the appearance of two or more successive T, a substitution table is needed. Secondly, to alleviate the ISI and mitigate the TRE, level modulation is employed to eliminate the existence of T, continuous 2T and the long run-lengths (more than 10T).

5.6.4 Writing Channel of Run-Length Modulation Code

The writing channel for SWM optical disc is described in Fig. 5.67, which including ECC encoding, run-length modulation coding, level modulation coding and writing strategy. The presented modulation code scheme of this paper includes

Fig. 5.67 Writing channel for SWM disc

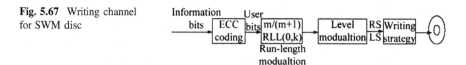

run-length modulation and level modulation process. The aim of run-length modulation process is to achieve high code rate $m/(m + 1)$ RLL(d, k) code. The level modulation code is used to resolve the problems of ISI and TRE, which are introduced in the run-length modulation process.

Run-length modulation process is higher capacity and higher code rate. K.A.S. Immink described the relationship between the capacity of (d, k) constrained channel $(C(d, k))$ and the parameters d and k. If d decrease or the k increase, the $C(d, k)$ will improve. To achieve high code rate $m/(m + 1)$, the parameter d of RLL (d, k) is selected as 0. If $d = 1$, the capacity of RLL(d, k) is limited to 0.6942 and the code rate $R = 7/8$ cannot be achieved. The number of kits set as large as possible. Table 5.14 gives the detailed encoding processing. The encoding table consists of

Table 5.14 Encoder of RLL (0, 15) main code table of concatenation rule

Main code table			Table of concatenation rule			
Basic code	Control code table		5 + 3 replacing table		111 replacing table 1	
	Input	Output	Input	Output	Input	Output
001	0001	10100	10101 11×	10011 01×	111 01010	010 00000
010	0010	10101	10101 111 1×	10011 100 1×	111 01001	011 00000
011	0011	10010	01001 11×	01011 00×	111 00100	111 00100
100	0100	10001	00101 11×	01011 01×	111 00101	111 00101
101	0101	01000	00101 111 1×	01011 100 1×	111 00101 11×	100 11100 10×
110	0110	01010	00001 11×	00011 00×	111 00010	011 10010
111	0111	01001	01101 11×	00011 01×	111 00001	100 00000
	1000	00100	01101 111 1×	00011 100 1×	111 01100	101 00000
	1001	00101	3 + 5 replacing table		111 00110	110 00000
	1010	00010	×11 10000	×00 11000	111 01101	001 11001
	1011	00001	×11 10100	×00 11010	111 01101 1×	010 11011 0×
	1100	01100	×11 10101	×00 11001	111 replacing Table 2	
	1101	00110	×11 10101 1×	×00 11011 0×	00110 111 00100	00001 110 01000
	1110	10110	×11 10010	×10 11000	10110 111 00100	10001 110 01000
	1111	01101	×11 10001	×10 11010	00110 111 00101	00001 110 01010
			×11 10110	×10 11001	10110 111 00101	10001 110 01010
			011 10110 11×	000 11100 10		
			111 10110 11×	101 11001 00		

main code table and table of concatenation rule. The main table is employed to convert information bits to satisfy the constraint of RLL(0, 15). Every 7 information bits is separated into two parts. The first 4 bits termed $I_{i,4}$ is called control code and the last 3 bits is called basic code ($I_{i,3}$). Basic code means that these bits are unchanged in the modulation processing. Control code $I_{i,4}$ is converted to 5 bits ($C_{i,5}$) according the control code table. Then, $C_{i,5}$ is concatenated with $I_{i,3}$ as the output series. When $C_{i,5}$ is concatenated with $I_i,3$ or I_i-1,3 concatenated with $C_{i,5}$, there will produce some specific bit-patterns, '111', '1111'and '11111' in the output sequences. If these bit-patterns are recorded onto the disc, successive pits with length of T will exist. Successive Ts will introduce serious ISI, which makes the signal unrecognized. To avoid the appearance of successive Ts, a concatenation rule table is needed. The 5 + 3 replacing table resolves the cases of successive Ts when $C_{i,5}$ is concatenated with $I_i,3$. The 3 + 5 replacing table is used for $I_{i-1,3}$ concatenating with $C_{i,5}$. Tables of 111 replacing are employed to eliminate the bit-patterns of '111'. Even if the bit-pattern of '111' is not completely eliminated, all the remaining '111' exist as the patterns of '111001', which is treated as a whole body termed '5n' in level modulation. After run-length modulating, the output data sequences will exist bit patterns '...0110...', '...111001...'. These bit patterns will introduce serious ISI to readout signals of SWM disc. With the number of k increasing, it will bring serious TRE to readout RF signal of SWM disc. To decrease TRE, the longest run-lengths, whose length is more than 10, should not exist in the run-length series.

If T and successive Ts exist in the actual system, the amplitude of the readback signal of T will be every lower and misidentify the slicer. Moreover, T with neighboring run-lengths will produce ISI and long run-lengths will bring TRE. To lighten the ISI and mitigate the TRE, level modulation is needed.

5.6.5 Level Modulation Process

The aim of level modulation is to combine T or $5n$ with the latter run-lengths to form longer run-lengths and separate the longest run-lengths to relatively short run-lengths. Run-length-level transition table is the processing of level modulation. The inputs of level modulation are outputs of run-length modulation and the outputs of level modulation are two vectors: run-length series (RS) and level series (LS). The two vectors are used to control the write strategy to record information data onto discs. The symbol 'm' in Table 5.14 represents the input series includes bit-pattern '11'. For instance, the series of '11001', '110100001', '110110001' are expressed as $4m$, $3m + 5T$, $3m + 5m$ respectively. Alphabets of A–F represent the levels of run-lengths after level modulation. Table 5.15 gives the detailed description of run-length-level transition process. The original run-lengths are the input ones, while the output sequences are the output run-lengths and levels. Run-lengths of $2T$–$10T$ remain unchangeable. Run-lengths of $11T$–$16T$ are cut into two parts to mitigate TRE. One is level 5 of $9T$, the other is level 0 of ($2T$–$7T$). $4m$–

Table 5.15   Run-length-level transition table

Original run-lengths					Output sequences				
2T					2T				
3T					3T				
4T	4m				4T	4A			
5T	5m	3m2			5T	5A	5B		
6T	6m	3m3	3m3m		6T	6A	6B	6C	
7T	7m	3m4	3m4m	5n2	7T	7A	7B	7C	7D
8T	8m	3m5	3m5m	5n3	8T	8A	8B	8C	8D
9T	9m	3m6	3m6m	5n4	9T	9A	9B	9C	9D
10T	10m	3m7	3m7m	5n5	10T	10A	10B	10C	10D
11T	*11m*	*3m8*	*3m8m*	*5n6*	*9E2*	*9F2*	*8E3*	*8F3*	*10F1T*
12T	*12m*	*3m9*	*3m9m*	*5n7*	*9E3*	*9F3*	*8E4*	*8F4*	*8E4A*
13T	*13m*	*3m10*	*3m10m*	*5n8*	*9E4*	*9F4*	*8E5*	*8F5*	*8E5A*
14T	*14m*	*3m11*	*3m11m*	*5n9*	*9E5*	*9F5*	*8E6*	*8F6*	*8E6A*
15T	*15m*	*3m12*	*3m12m*	*5n10*	*9E6*	*9F6*	*8E7*	*8F7*	*8E7A*
16T		*3m13*	*3m13m*	*5n11*	*9E7*	*9F7*	*8E8*	*8F8*	*8E8A*
		3m14	*3m14m*	*5n12*			*8E9*	*8F9*	*8E9A*
		3m15					*8E10*		
22222					*10E*				

10*m* are encoded as level 1 of (4*T*–10*T*). 11*m*–15*m* is separated into level 6 of 9*T* and level 0 of (2*T*–7*T*). To eliminate the patterns '1101' (3*m*), 3*m* is combined with next run-lengths to form multilevel of longer run-lengths. 3*m* and (2*T*–7*T*) are combined as level 2 of (5*T*–10*T*). 3*m* and (3*m*–7*m*) are united as level 3 of (6*T*–10*T*). To remove the bit-pattern '111001' (5*n*), it is also combined with neighboring run-lengths to constitute multilevel of longer run-lengths. 5*n* and (2*T*–5*T*) are incorporated to level 4 of (7*T*–10*T*). Assumed the output series of run-length modulation is '100011010001011100100010001101100010001…'. The input of level modulation: 4*T* 3*m* + 4*T* 2*T* 5*n* + 4*T* 4*T* 3*m* + 5*m* 4*T*. Where, '+' means the two run-lengths should be combined as a pit or land. The vectors RS and LS are RS = [4*T* 7*T* 2*T* 9*T* 4*T* 8*T* 4*T* …] and LS = [0 2 0 4 0 3 0 …]. The elements of RS are used to control the length of pits/lands and that of LS are employed to control the size and position of the sublands/sub-pits. Table 5.16 gives the realizable level number of each run-length after level modulation. The number of level of different length run-lengths is different.

Table 5.16   Realizable level number of run-lengths

Run-length	2*T*	3*T*	4*T*	5*T*	6*T*	7*T*	8*T*	9*T*	10*T*
Level number	1	1	2	3	4	5	7	7	7

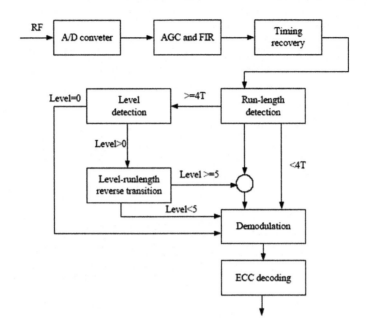

Fig. 5.68 The signal processing and decoding process for SWM

Figure 5.68 shows the signal processing and the complete decoding scheme for SWM system. When run-length is less than $4T$, the data is directly demodulated and ECC decoded. If the run-length is more than $4T$, the level of readback signal is detected with adaptive level detecting method. The data is also directly demodulated for level =0, while the level run-length reverse transition process is applied if level >0. If the level is 5 (E) or 6 (F), the reversed data must be combined with next input data before demodulated. The readout signal series is shown in Fig. 5.68, that the level 2 of $8T$ (P), level 1 of $10T$ (L), level 1 of $9T$ (P), level 0 of $5T$ (L), level 5 of $10T$ (P), level 0 of $4T$ (L) and level 3 of $8T$ (P) are represented severally. P means the recording pit and L island. Although the first pit and the last pit are both $8T$, the variation of waveforms is clearly and used to differentiate the level.

5.6.6 Performances Analysis of the Code

The DR of this code is calculated by DR = $(1 + d) \times R = 1.75$ bits/$T_{\min}$, where $T_{\min}$ is the minimum recording mark. DR of EFM plus code adopted in DVD is 1.5. They cannot be compared directly, because $T_{\min}$ in SWM and DVD is different. $T_{\min}$ is $3T$ and the physical length is 400 nm in DVD, while $T_{\min}$ is $2T$ and 320 nm in SWM disc. Hence, can define a new index termed information bits per 400 nm (IBP) to characterize the relationship between the code and the capacity of disc. The IBP of SWM is 2.19, which is 46 % higher than that of DVD. So the modulation code for SWM-ROM increased the capacity of 46 % also. The decoding system is shown in Fig. 5.68, and the hardware is simpler. Thus, the scheme of modulation could be used for SWM disc.

The preliminary study of the signal waveform (SW) ML (SWML) read-only disc which is based on the optical and mechanical units of the commercial digital versatile disc (DVD) is described above. The SWML read-only disc employs sub-lands/sub-pits, which are inserted on the conventional pits/lands, to achieve multi levels. By changing the length or the insertion position of sub-lands/sub-pits, readout signals with different waveforms can be generated, indicating different levels as shown in Fig. 5.69. Tracking servo which maintains the focused laser beam on the centre of the desired information track is indispensable to the optical disc system. In the high-density read-only disc, the differential phase detection (DPD) method especially DPD2 method is adopted as the standard TE detection method. The DPD method employs the phase (or time) difference between the high frequency components of readout signals from a four-quadrant photo detector to derive a TE signal. However, the application of ML technology always effects the characteristics of the TE signal. In the SWML read-only disc, readout signals of short run-length such as $3T$ and $4T$ (T is channel clock length) are more sensitive to the crosstalk of sub-lands/sub-pits, and these phase information which makes major contribution to the conventional DPD signal becomes inaccurate. Thus, he conventional DPD signal of the SWML read-only disc deteriorates in the uniformity and contains more noise.

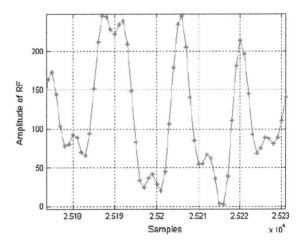

Fig. 5.69 Readback signal of multilevel pits and lands

5.6.7 Tracking Error Detection of Multi-level Discs

The TE detection method for SWML ROM discs is amplitude difference detection (ADD) method, it can improve the uniformity and signal-to-noise ratio (SNR) of the TE signal. Based on the diffraction theory and feasibility of the ADD method is analysed the implementation of the ADD method, to confirm the effectiveness of the ADD method in this section.

The readout model of optical discs can be described by the diffraction theory. Figure 5.70 shows optical disc where x-axis presents the tangential direction of the spiral track and y-axis presents the radial direction in which off-track occurs. The inner part of the zeros order is the four-quadrant detector denoted by A–D. The high frequency signal from each quadrant is the result of interference between these diffracted orders. For the reason of simplicity, the high frequency signals SA–SD from the four-quadrant detector can be expressed as:

$$SA(t, \phi_r) = \cos(2\pi\omega t + \psi) + \alpha\cos(2\pi\omega t + \phi_r + \psi) \tag{5.90}$$

$$SB(t, \phi_r) = \cos(2\pi\omega t - \psi) + \alpha\cos(2\pi\omega t - \phi_r - \psi) \tag{5.91}$$

$$SC(t, \phi_r) = \cos(2\pi\omega t - \psi) + \alpha\cos(2\pi\omega t + \phi_r - \psi) \tag{5.92}$$

$$SD(t, \phi_r) = \cos(2\pi\omega t + \psi) + \alpha\cos(2\pi\omega t - \phi_r + \psi) \tag{5.93}$$

where ω is the time frequency dependent on the scanning velocity of the spiral track, α is a small factor approximated to 0.1, ψ is the simplified phase between

Fig. 5.70 Diffraction pattern of the optical disc

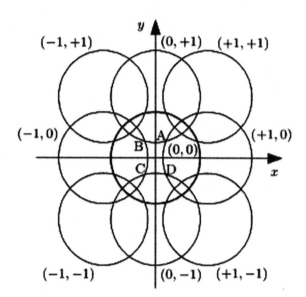

zeroth and first diffracted orders and ϕ_r is the phase derived from the TE. Here ϕ_r is proportional to the TE r and can be expressed as:

$$\phi_r = 2\pi r/q \tag{5.94}$$

where q is the track pitch.

The conventional DPD method measures the time difference between the diagonally summed signals SA + C and SB + D to generate the TE signal. And from Eqs. (5.90)–(5.93) can get SA + C and SB + D are given by:

$$SA + C(t, \phi_r) = 2[\cos(2\pi\omega t) + \alpha \cos(2\pi\omega t + \phi r)] \cos \psi \tag{5.95}$$

$$SB + D(t, \phi_r) = 2[\cos(2\pi\omega t) + \alpha\cos(2\pi\omega t - \phi_r)] \cos \psi \tag{5.96}$$

As shown in Fig. 5.71, the time difference Δt_i can be described as follows:

$$\Delta t_i(\phi_r) = t_2 - t_1. = SB + D(t_i, \phi_r)SB + D(t_i, \phi_r)$$
$$- SA + C(t_i, \phi_r)SA + C(t_i, \phi_r) \propto -\alpha \sin\phi_r\pi\omega(1 + \alpha \cos \phi_r). \tag{5.97}$$

According to Eq. (5.97), Δt_i is a bipolar function of the TE r, so it can present the off-track information.

Instead, the ADD method measures the amplitude difference of the radio frequency (RF) signal between time t_1 and t_2 in Fig. 5.71. The RF signal is the summation signal of the four-quadrant detector and can be expressed as:

$$SRF(t, \phi_r) = 4(1 + \alpha \cos \phi r) \cos \psi\cos(2\pi\omega t). \tag{5.98}$$

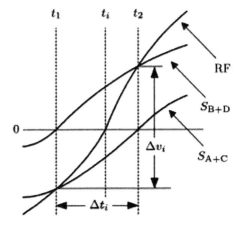

Fig. 5.71 Schematic diagram of time and amplitude difference between readout signals

So the amplitude difference Δv_i of the RF signal between time t_1 and t_2 can be written as:

$$\Delta v_i(\phi_r) = \text{SRF}(t_2, \phi_r) - \text{SRF}(t_1, \phi_r)$$
$$= \text{S RF}(t_1, \phi_r)\Delta t_i(\phi_r) \propto \alpha \cos \psi \sin \phi_r \tag{5.99}$$

From Eq. (5.99) can see that Δv_i is also a bipolar function of TE r. This means that the ADD method which measures amplitude difference Δv_i of the RF signal can also obtain the off-track information.

5.6.8   Implementation of the ADD Method

Figure 5.72 shows the block diagram of the ADD method for the SWML read-only disc. First, the high frequency signals from four quadrant detectors A–D are readout by the optical pickup unit. The digitized pulse signals AC and BD are produced respectively by slicers using zero-crossing detection according to the diagonally summed signals (A + C) and (B + D). The RF signal is generated by the sum of signals from the four-quadrant detector. Then these three signals AC, BD and RF are sent to the amplitude detector. The amplitude detector uses the rising and falling edge of signals AC and BD as trigger signals to sample and hold the amplitude of the RF signal, and then calculate the amplitude difference to output as signal E. Finally, the TE signal is produced by smoothing the signal E with a low-pass filter (LPF).

Figure 5.73 illustrates the signal processing of the amplitude detector. Take the rising edge for example, u_1 is the amplitude value of the RF signal when signal AC is rising; u_2 is the amplitude value of the RF signal when signal BD is rising. The amplitude detector outputs the amplitude difference $(u_2 - u_1)$ from time t_2 to t_5 which is the length of the current run-length. It can be noted that the time difference $(t_3 - t_1)$ is replaced by the amplitude difference $(u_2 - u_1)$ in the ADD method.

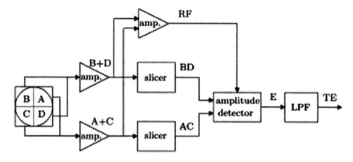

Fig. 5.72 Block diagram of the ADD method

Fig. 5.73 Signal processing
of the amplitude detector

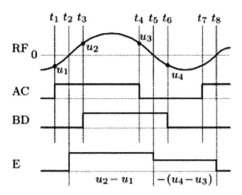

The output signal E with positive value means signal AC leads signal BD while that with negative value means signal AC lags signal BD. Because value u_2 is obtained after time t_2, the amplitude detector should output signal E with a certain time delay ΔT. As the longest recording mark in the SWML read-only disc is $14T$ (refer to the DVD-system, $1T$ represents 33 nm in length and 38 ns in time), the time delay ΔT can be selected as 500 ns which is sufficient and will not affect real-time of the final TE signal. After signal E is generated by the amplitude detector, the TE signal can be produced by the LPF.

In the SWML read-only disc, the ISI is more serious than the conventional optical disc, and the phase information from the long run-length is more accurate than that from short run-length. In the ADD method, the output of the amplitude difference is multiplied with the length of the run-length of the RF signal. In this way, the long run-length will contribute more phase information to the final TE signal. So the ADD method is expected to have an advantage for the SWML read-only disc and to improve on the uniformity and SNR of the TE signal.

To confirm the effectiveness of the ADD method, the experimental system based on a standard DVD driver is carried out. The signals, including A–D and RF, are readout by the standard DVD driver on the condition of focusing servo enabled and tracking servo disabled. Then the above signals are sampled by the analogue-to-digital converters and sent to a FPGA where the ADD method is implemented. The LPF is a first order filter and has the same cut-off frequency of (-3 dB) 30 kHz as that used in the conventional DPD method. For comparison, the conventional DPD signal, one of the output signals of the standard DVD driver, is also sampled.

Figure 5.74 shows the TE signals of the SWML read-only disc generated by the conventional DPD method and the ADD method. It can be found that the uniformity of the TE signal in the ADD method (called ADD signal for short) is much better than that of the conventional DPD signal. It was analysed statistically the two signals with about 500 crossing tracks. The distribution of the peak-to-peak (PP) values of signals is shown in Fig. 5.75 where the solid curve is the normal distribution curve with certain standard deviation by statistics. The normalized

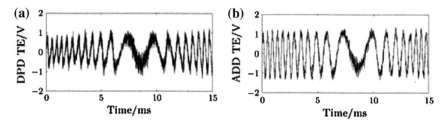

Fig. 5.74 The TE signals generated by conventional DPD method (**a**) and ADD method (**b**)

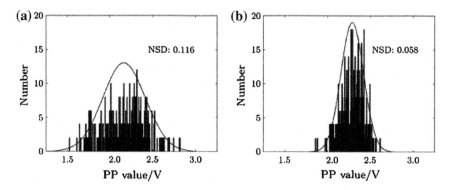

Fig. 5.75 The PP value distribution of **a** conventional DPD signal and **b** ADD signal

standard deviation of the PP value of the ADD signal is about 0.058 while that of the DPD signal is 0.116. The uniformity of the TE signal is improved by the ADD method in the SWML read-only disc.

Moreover, SNR of the ADD signal and DPD signal is also estimated. Because TE signals filtered by LPFs with cut-off frequency of 30 kHz and below 10 kHz are useful for tracking servo, signals within the frequency band from 10 to 30 kHz are considered as noise. The power of the ADD signal in this band is −18.5 dB with respect to that of the signal blow 10 kHz; while the power of the DPD signal in this band is −13.3 dB. The SNR of the ADD signal is increased about 5 dB than that of the DPD signal. This result shows that the ADD method can generate a less noisy TE signal than that of the conventional DPD method.

Based on the amplitude difference, a new TE detection method for the SWML read-only disc is proposed. Compared with the conventional DPD method, the ADD method raises the contribution of long run-length to the TE signal. The experimental results show that the ADD method improves the uniformity of the TE signal. In addition, SNR of the ADD signal is increased about 5 dB than that of the conventional DPD signal. The ADD method is more effective than the conventional DPD method in the SWML read-only disc.

5.7   Multi-level Blu-Ray Disc Drive

According to the requirements of the National Key Basic Research Development Program in China, the blue-ray multi-level storage systems are researched and developed to improve the data capacity and transfer rate, as an alternative techniques of the next generation optical disc. The experimental prototype is shown in Fig. 5.76. The objective lens of the optical head (pickup) is shown in Fig. 5.77. It was use of wavelength 405 nm blue semiconductor laser and the numerical aperture 0.80 objective lens, that the spot of focusing diameter can be reduced to 220 nm. The two technologies can increase storage density of 2.56× and 1.17×, respectively. Its main structure of the hardware was design to compatible with DVD drive. The system can use dye, phase change medium and new photochromic materials to achieve 4–8 level storage that can improve the capacity of 1.5× to 3× times. The drive adopt the ditch-shore recording, single-layer storage capacity is 20 GB, four-level single-layer storage capacity is 35 GB, single-sided double-layer storage capacity is about 65 GB with data transmission rate of 55.2 Mbps. This section focuses on the modulation and coding and physical formats, modulation code with

Fig. 5.76 Multi-level experimental prototype of the Blu-ray drive and write-once multi-level disc

(a)   **(b)**

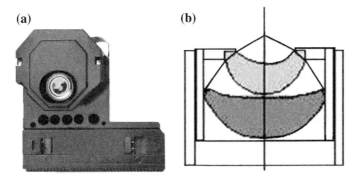

Fig. 5.77 The pickup **a** and layout of the objective lens, **b** of the Blu-ray and multi-level drive

Fig. 5.78 The structure of
the Blu-ray and multi-level
coding system RS linear
block codes

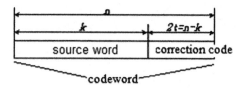

2/3(1, 7), error-correcting codes using Reed-Solomon product code [RS-PC (208, 192, 17) × (182, 172, 11)] and the corresponding new data format.

The system adopts Blu-ray and multi-level coding system and RS linear block codes. Therefore the format, codec algorithms and processing circuits of the disc have to be re-designed in the all. The new block codes is different from before and after data code convolutional code, the code word generated only with the current source data, that is shown in Fig. 5.78. Continuous data of block code bit stream is divided into fixed-length groups, each group further split the unit m-bit symbols. For example, take the 3-bit or 8-bit data to form a symbol, k symbols together to form the source word is encoded into a codeword of length n, referred to as m-bit symbols (n, k) block codes. Error-correcting codes using linear coding process are linear transformations, and through the matrix transformation. In this linear space of all possible m-bit source can transform coding, nothing to do with the m-bit data.

Source word and transform from the check sum code contains the source word is placed in the first half of part of the code, check sum attached in the second half. RS code in part by the parameters of m, n and k three 8-bit symbol RS(204, 188) code, i.e. DVB yards. The difference between the n and k (usually referred as $2t$) is length of code symbol. RS code can correct no more than (nk) /2 errors, that is up to can correct t errors. DVB code, source data is divided into 188 symbols of a group, after the transformation of the coding, the code word length of 204 symbols. Length of 16 symbol check character, to ensure the correct codeword up to eight errors. RS code is defined in a special limited domain, i.e. the Galois domain GF (2^m). The nature of the Galois domain with the integer domain, also a plus, subtract, multiply with the exception of matrix operations, polynomial operations, such as the operation was established in 2^m. Codeword on a GF (2^m) has a corresponding polynomial. The code in accordance with the order from high to low, the data bits as a polynomial corresponding coefficient. For example, in GF (2^4), i.e. code "1010" can be expressed in polynomial $x^3 + x$. Therefore, certain operations to the source word get the check sum and combined into a code word process, and available polynomial operator. RS code to select a suitable polynomial $g(x)$ (generator polynomial), and makes the codeword polynomial calculated for each source word are $g(x)$ times, the codeword polynomial divided by the generated polynomial from the remainder to 0. If the received codeword polynomial is divided by the generator polynomial remainder is not 0, that there is an error in the received codeword in need of correction. Further calculation shows that error to be corrected up to $t = (nk)$ / 2. RS code generator polynomial can be according to Eq. (5.100) to select:

$$g(x) = (x - \alpha)(x - \alpha^2) \cdots (x - \alpha^{2t}) = \prod_{i=1}^{2t} (x - \alpha^i) \qquad (5.100)$$

If $d(x)$ represents the source word polynomial can construct a codeword polynomial:

$$c(x)x^{n-k} \cdot d(x)/g(x) = h(x) \cdot g(x) + r(x). \qquad (5.101)$$

First calculate the quotient $h(x)$ and $r(x)$: the remainder $r(x)$ as a check word, so that the source word is placed in the first half of the codeword, the checksum $c(x) = x^{n-k} \cdot d(x) + r(x)$ is placed in the code word the second half, that can get:

$$\begin{aligned} c(x)/g(x) &= x^{n-k} \cdot d(x)/g(x) + r(x)/g(x) \\ &= h(x) \cdot g(x) + r(x) + r(x) = h(x) \cdot g(x) \end{aligned} \qquad (5.102)$$

where $r(x) + r(x)$ when modulo 2 is addition, the result is 0, so the codeword polynomial $c(x)$ must be generating polynomial $g(x)$ is divisible. The receiver detects the remainder is not 0, that can determine the received code word errors. In this system, each user sector includes 2048 bytes of user data, coupled with the former in the data sector header of 12 bytes (including 4-byte sectors marked code ID, 2-byte ID error detection code IED and 6 bytes of copy protection information CPR) and 4-byte sector tail (including 4-byte error detection code EDC), a total of 2064 bytes. These 2064 bytes form a 12 for 172 data blocks 2048 user data block and then through the scrambling process, the resulting block of data is the data sector, as shown in Fig. 5.79.

The system will be 16 consecutive data sector (total of $16 \times 2064 = 33{,}024$ bytes) combined to form a 192 OK 172 ECC blocks, the RS-PC code of the ECC block. By the above formula for each column 172 has a 16-byte external checksum PO. PO data will be attached to the end of the corresponding column, add 16 rows of data blocks. This 208-row (192 row of the original data plus PO data row 16) of each row can calculate the available length of 10 bytes of internal checksum PI.

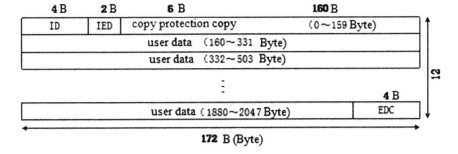

Fig. 5.79 The data sector structure ($172 \times 12 = 2064$ bytes)

Table 5.17 10 ECC blocks (a total of 182 × 208 = 37 856 bytes)

		172 Byt					P11o Byt	
$B_{0,0}$	$B_{0,1}$			$B_{0,170}$	$B_{0,171}$	$B_{0,172}$		$B_{0,181}$
$B_{1,0}$	$B_{1,1}$			$B_{1,170}$	$B_{1,171}$	$B_{1,172}$		$B_{1,181}$
$B_{2,0}$	$B_{2,1}$			$B_{2,170}$	$B_{2,171}$	$B_{2,172}$		$B_{2,181}$
$B_{190,0}$	$B_{190,1}$			$B_{190,170}$	$B_{190,171}$	$B_{190,172}$		$B_{190,181}$
$B_{191,0}$	$B_{191,1}$			$B_{191,170}$	$B_{191,171}$	$B_{191,172}$		$B_{191,181}$
$B_{192,0}$	$B_{192,1}$			$B_{192,170}$	$B_{192,171}$	$B_{192,172}$		$B_{192,181}$
$B_{207,0}$	$B_{207,1}$			$B_{207,170}$	$B_{207,171}$	$B_{207,172}$		$B_{207,181}$

192 row

P016 row Byt

PI data is attached at the end of the corresponding row of data blocks added 10, and finally can get a 208 OK 182 ECC block is shown in Table 5.17. The two generated polynomial factor for the aforementioned factors, despite the difference, but the final effect is still the same in GF (2^m) operations.

This ECC block 192 row per 12 row was splited into 16 block, for 16 row(PO data). In order to each row attach to the 16 blocks, that can get 16 × 13 × 182 = 2366 bytes data block to compose the actual recording sector finally. Record sector after the modulation and coding can be converted into the physical sector recorded on the disc. This system, the sector size of each record is 2366 bytes, 2048 bytes is the really useful data, in addition to the 10 bytes as a sector marked and copy the information protection, the other 308 bytes for the error detection and correction, including word of the 6-byte sector data error detection and 302 bytes of the RS code checksum (line 12 to the end of each row of 120 bytes, 172 at the end of each column of a total of 172 bytes and the last line and the last 10 cross-regional total of 10 bytes). Therefore, in order to achieve the required level of error correction, to arise about data redundancy.

In the process of receiving data, set $r(x)$ for receiving the code polynomial, $v(x)$ is the original code polynomial, $e(x)$ is polynomial for the error codes generated in the transmission process:

$$r(x) = v(x) + e(x) \tag{5.103}$$

where $v(x) = r(x) - e(x) = r(x) + e(x)$, if the transmission error $e(x)$ all coefficients are 0. The i to produce an error, then the coefficient is 0. Assumptions $c(0 \leq c \leq t)$, during transmission a mistake, and occurred in an unknown location, the error polynomial is:

$$e(x) = e_{i_0}x^{i_0} + e_{i_1}x^{i_1} + \cdots + e_{i_{c-1}}x^{i_{c-1}} \tag{5.104}$$

Otherwise accompanied by the polynomial:

$$
\begin{aligned}
S_i = r(\alpha^i) &= v(\alpha^i) + e(\alpha^i) = e(\alpha^i) \\
&= e_{i_0}\alpha^{i_0*i} + e_{i_1}\alpha^{i_1*i} + \cdots + e_{i_{c-1}}\alpha^{i_{c-1}*i}, \quad i = 0, 1, \cdots, 2t-1
\end{aligned} \tag{5.105}
$$

If the error number is less than t, solving the equations of the S_i composition may get the error location and error value, and to restore the original data. S_i solving is more complex, especially the introduction of an error location polynomial are as follows:

$$
\begin{aligned}
\Lambda(x) &= (1 - xX_1)(1 - xX_2) \cdots (1 - xX_c) \\
&= \Lambda_c x^c + \Lambda_{c-1}x^{c-1} + \cdots + \Lambda_1 x + 1
\end{aligned} \tag{5.106}
$$

Its roots X_1, X_2, ..., X_c is the inverse of the number of the error position, that the root of the error location polynomial can be obtained by Chien algorithm. So as get the wrong location number, the error value can be obtained by Forney's rule. The format parameters of Blu-ray multi-level read only disc and rewritable disc with conventional numerical aperture 0.65 objective lens are shown in Tables 5.18 and 5.19.

Table 5.18 The format parameters of Blu-ray multi-level read-only

Parameters	Single-side single-layer	Double-sided double-layer
User data capacity	20 GB/side	38 GB/side
Laser wavelength	405 nm	
Objective lens numerical aperture	0.65	
Data length	(a) 0.306 μm (b) 0.153 μm	
Channel length	(a) 0.204 μm (b) 0.102 μm	
Minimum record character length	(a) 0.408 μm (b) 0.204 μm	
Maximum record character length	(a) 2.652 μm (b) 1.326 μm	
Track spacing	(a) 0.680 μm (b) 0.400 μm	
Disc diameter	120 mm	
Disc thickness	1.20 (= 0.6 × 2) mm	
Disc center hole diameter disc	15.0 mm	
Inner diameter of information area	24.1 mm	
Outer diameter of information area	116.0 mm	
User data sector size	2048 bytes	
Error-correcting code	RS-PC (208,192,17) × (182,172,11)	
RS-to-PC correction region	32 physical sectors	
Modulation code	8/12 modulation, RLL(1,10)	
Burst error correction length	7.1 mm	
Standard speed	6.61 m/s	
Channel bit rate (standard rate)	(a) 32.40 Mbps (b) 64.80 Mbps	
User bit rate (at standard speed)	(a) 18.28 Mbps (b) 36.55 Mbps	

Table 5.19 Rewritable format parameters of Blu-ray multi-level disc

Parameters	Single-side single-layer
User data capacity	20 GB/side
Laser wavelength	405 nm
Objective lens numerical aperture	0.65
Data length	(a) 0.306 μm (b) 0.130–0.140 μm
Channel length	(a) 0.204 μm (b) 0.087–0.093 μm
Minimum length of the record symbol (2*T*)	(a) 0.408 μm (b) 0.173–0.187 μm
Maximum length of the record symbol (13*T*)	(a) 2.652 μm (b) 1.126–1.213 μm
Track spacing	(a) 0.680 μm (b) 0.340 μm
Physical address	WAP (Wobble Address in Periodic position)
Disc diameter	120 mm
Disc thickness	1.20 (=0.6 × 2) mm
Disc center hole diameter	15.0 mm
Disc information area inner diameter	24.1 mm
Disc information area outer diameter	115.78 mm
User data sector size	2048 bytes
Error-correcting code	RS-PC (208,192,17) * (182,172,11)
Correction region	32 physical sectors
Modulation code	8/12 modulation, RLL(1, 10)
Burst error correction length	(a) 7.1 mm (b) 6.0 mm
Standard speed	(a) 6.61 m/s (b) 5.64–6.03 m/s
Channel bit rate (standard rate)	(a) 32.40 Mbps (b) 64.80 Mbps
User bit rate (standard speed)	(a) 18.28 Mbps (b) 36.55 Mbps

5.7.1   Blu-Ray Super-Resolution Optical Head

The above-described Blu-ray drive used objective lens of numerical aperture 0.80, in order to further improve the Blu-ray multi-level optical disc storage capacity, as well as to be optical head of Blu-ray multi-level mastering system. In the optical system is supplemented the super-resolution technology to reduce the recording symbol size, to improve the optical disc storage density and can be used to Blu-ray multi-level mastering system further. The Blu-ray optical pickup with super-resolution and larger numerical aperture is shown in Fig. 5.80. The system is based on the characteristics of the halo ball to design a numerical aperture NA of 0.95, and after accurate calibration for the Blu-ray aberration. Using a special optical phase shift aperture and two collimating lenses are convergence and intercept the laser beam, that reduces the high-frequency diffraction light, and significant improvements in the energy utilization of the laser beam. The Blu-ray lager numerical aperture super-resolution optical system is shown in Fig. 5.81.

Fig. 5.80 The large
numerical aperture of Blu-ray
optical pickup

Fig. 5.81 Blu-ray high
numerical aperture
super-resolution storage
optical system

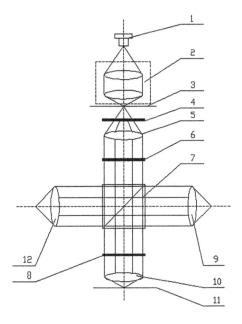

The optical system of Blu-ray and lager numerical aperture super-resolution storage is shown in Fig. 5.81: 1—405 nm semiconductor optical devices, 2—collimation and uniform lens group, 3—pinhole phase shift aperture, 4—uniform plane, 5—collimator, 6—light apodized, 7—prism 8—1/4 wave plate, 9—extract lens for readout signal, 10—objective lens, 11—medium, 12—to extract lens for the laser energy control signal. As introduces the pinhole phase shift aperture, uniform plane and phase shift of apodized, the space phase and intensity distribution of the focused laser beam are changed to improve the transfer function of the optical system that is shown in Fig. 5.82. Energy distribution of the focusing laser beam shows that the 80 % energy is focusing on range of radius of 110 nm.

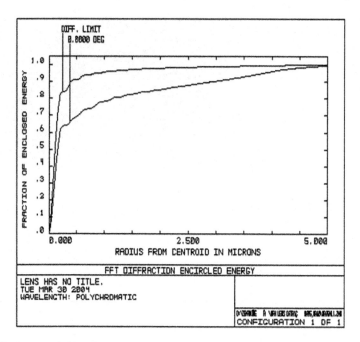

Fig. 5.82 The transfer function distribution of the super-resolution optical system

The traditional energy distribution of an optical system is described as:

$$I_i(x_i, y_i) = f\left[\lambda, \text{NA}, \sigma, t_0(x_o, y_o), H(x_s - f_x, y_s - f_y), I_{\text{eff}}(x_s, y_s)\right]$$
$$= \iint_\sigma I_{\text{eff}}(x_s, y_s) \left| \iint U(f_x, f_y) H(x_s - f_x, y_s - f_y) \exp\left[j2\pi(f_x x_i + f_y y_i)\right] df_x df_y \right|^2 dx_s dy_s$$

$$(5.107)$$

As the optical system adopted phase shift masks, optical spatial filtering and apodized technology, the light energy distribution of the optical system was changed to the form:

$$I_{\text{opt}}(x_i y_i) = f\left[\lambda, NA, o, t_{\text{opt}}(x_o, y_o), H(x_s - f_x, y_s - f_y), I_{\text{eff}}(x_s, y_s)\right] \approx I_{\text{ideal}}(x_i, y_i)$$

$$(5.108)$$

where $I_{\text{ideal}}(x_i, y_i)$ is light intensity distribution of the ideal image, $t_{\text{opt}}(x_o, y_o)$ is optimized complex amplitude transmittance function, $I_{\text{opt}}(x_o, y_o)$ is the optimized of the space light intensity distribution function. Since then, if reasonable to adjust the complex amplitude distribution of the light apodized to optimize the spatial frequency spectrum distribution with the phase shift mask, that can improve the light

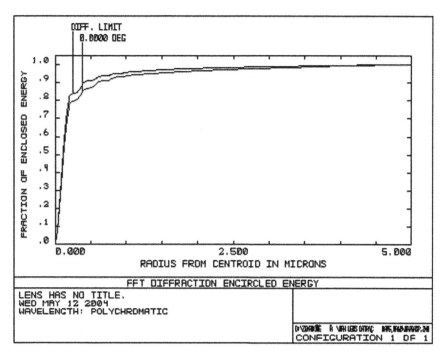

Fig. 5.83 The optical transfer function curve of Blu-ray objective lens

intensity distribution of the image plane, to improve the system resolution. Fully consider the features faint ball design, and use of super-resolution mask technology, to improve the transfer function of the optical system. Actual transfer function up to get the curve is shown in Fig. 5.83, and the modulation transfer function is shown in Fig. 5.84. From analysis of the diffraction energy distribution curve, 80 % of the energy distribution is close to the diffraction limit. That the distribution corresponds to 80 % energy on field of radius of less than 110 nm, the modulation transfer function curve of the cut-off frequency corresponding to a resolution of 210 nm.

In the development of Blu-ray objective lens, the focusing spot size is smaller than the resolution of optical microscopy in general. In order to precise determination of the actual optical properties of the super-resolution of the Blu-ray high numerical aperture optical system, using the knife-edge scanning method is specifically designed and developed by the diameter of the micro-spot detection system is shown in Fig. 5.85. In this system, the knife-edge scanning micro-displacement measuring device, which measurement accuracy reaches ± 10 nm sensitivity of 5 nm.

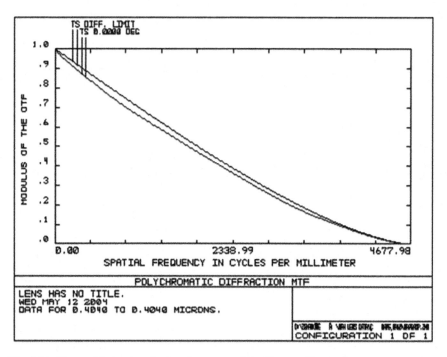

Fig. 5.84 The modulation transfer function curve of the Blu-ray objective lens

Fig. 5.85 The testing system
for diameter of the micro-spot
of Blu-ray super-resolution
optical pickup

The system measured the diameter of the super-resolution focusing spot is 193 nm actually. Above measurement results were compared to a standard DVD disc, that the results are shown in Fig. 5.86. It can increase single-layer storage capacity to 30, and 55 GB with four-level storage, double-layer storage capacity can be to 100 GB.

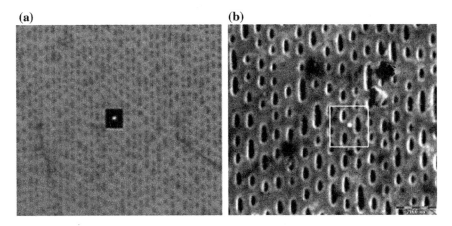

Fig. 5.86 The measurement and testing results of the Blu-super-resolution optical pickup comparison to standard DVD disc **a** magnification of 600×, **b** magnification of 1500×

5.8 Rewritable Multi-level Storage

Multi-level (ML) encoding has been demonstrated to significantly increase the linear densities achieved with standard methods of binary encoding in optical data storage systems. An overview of the channel is provided from write encoder and precompensation, to readout with a matched adaptive equalizer and Viterbi decoder. The multilevel channel has been implemented in silicon and designed to operate in parallel with the standard binary channel, thus maintaining full backward functionality of the underlying drive that developed an efficient prototyping and integration process to accomplish final system implementation with minimal impact to the host system. Here discuss servo and firmware functions that ease integration of the multilevel LSI into existing circuit-board architecture, firmware and software. Additionally, the same LSI can increase rewritable disc capacity of 9 GB (with 4-level storage) when combined with a 0.65NA DVD RW base drive with eight-level recording to 15 GB. If it combined with a blue laser, that capacity can be approximately 30 GB when combined multi-level recording and code system. A next-generation core will use the 12-level HD-ML encoding technique that, together with improved format efficiency, can increase capacity to 25 GB per layer. Meanwhile, based on DVD combination with a blue laser the capacity can be to 50 GB per layer.

The initial work on ML was with pit-depth modulation (PDM) on CD-ROM systems to achieve a multilevel reflectivity response by varying pit depth from zero to approximately a quarter of wavelength depth in the medium. That the "PDM" process was actually pit "volume" modulation, as later TEM pictures revealed. Proof of feasibility of this technology has been demonstrated on DVD-ROM as well, that focus to writable and rewritable forms of ML. The data encoding process of 8-level trellis coded, amplitude-modulation write channel are as following: The 8 levels of reflectivity are achieved via finely-controlled mark-size modulation. Writing is done in combination with ML write pre-compensation that minimizes system non-linearity. ML power-control compensates for slow power drifts during writing while servo error-signals are normalized so that the focus and tracking loops can be run continuously as in standard read mode. ML clock recovery is discussed in the context of the ML decoding process that is matched to an adaptive zero-forcing equalizer used for data readout. The generalized approach to system development, starting with signal processing algorithms is discussed for ML media research and development, then moving to FPGA prototyping and final system integration and testing of the ML Endec mixed-signal LSI.

Discussion of ML system implementation issues related to the LSI and drive circuit-board combination and the associated firmware and software. As multi-amplitude and multi-phase signaling in modem phone technology, ML optical technology transmits more information over a fixed bandwidth channel by using the available SNR more efficiently. The media manufacturers boosted the SNR of phase-change and dye-based media by fine tuning optical layer structure and composition for ML. By combining these two pathways and utilizing an efficient ECC, demonstrated the ability to manufacture a triple-density ML-R/RW disc and drive without changing optical pick-up or drive except to addition some chip. Based on conventional DVD technology, the physical specifications adopted ML coding in the system are shown in Table 5.20. The ML system data is encoded before

Table 5.20 ML storage system physical specifications compared to standard DVD

Parameter	Formula	DVD	DVD + ML
Cover thickness (mm)	t	0.6	0.6
Laser diode λ (nm)	λ	650	650
Objective lens	NA	0.65	0.65
Track pitch (μm)	P	0.4	0.35
Min mark length or ML data cell length (μm)	MML	0.31	0.21
Code rate	r = data bits ch bits	8/17	5/6
Channel bit length (μm)	$c = r/b \times$ MML	0.278	0.200
Density (μm^2/ch bit)	$d = p \times c$	0.310	0.221
Data bits per min. mark	b	1.41	2.50
Data bit length (μm)	MML/b	0.591	0.240
Linear velocity (m/s)	v	36.2	36.2
Channel bit rate (MHz)	$f = v/c$	52	95
User data rate (Mbps)	$f \times E$	9	39
Encoding efficiency	E = User bits ch bits	27 %	57 %
Total efficiency	E/r	57 %	69 %
Program area (mm^2)	A	8606	8653
User data capacity (GB)	$A/d \times E$	4.7	15

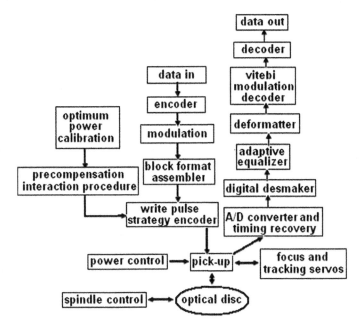

Fig. 5.87 Data flow of the multilevel encoding and decoding system

written on disc and is decoded after reading from the disc (see Fig. 5.87). Data is
ECC, modulation and format encoded before being translated into laser pulses that
write data on an ML disc using the laser of optical pick-up. For decoding ML
signals, the all-sum signal (RF signal) is processed by setting the gain, converting
from analog to digital, recovering the timing, adjusting for amplitude and offset,
equalizing and finally decoding for format, modulation and ECC. Spindle, tracking,
focusing and laser control systems are all similar to conventional optical disc
systems.

For the encoding, data is error correction encoded with the renew RS-PC (see
Figs. 5.88 and 5.43 renew decoding process logic diagram). The RS-PC adds 5
inner parity bytes per row and 16 outer parity bytes percolumn to the data.
Unlike DVD encoding, this ECC block is not interleaved with other blocks. Each
block is stored on the disc as an independent unit. The modulation encoder processes the ECC block bytes (see Fig. 5.89) and provides additional error correcting
capabilities that allows more effective to the SNR of the optical data storage system.
Each 5-bit group from the ECC encoder output, a parity bit is added an on volutional encoder to create a 6-bit group. These 6-bit groups are then mapped to two
8-level ML symbols.

After modulation encoding, the stream of ML symbols is placed into a complete
block structure that includes linking areas, a preamble, a data area and a postamble
is shown in Fig. 5.90. The ML data block is including the link-in and link-out areas
provide a buffer zone among each block, the preamble contains timing acquisition,

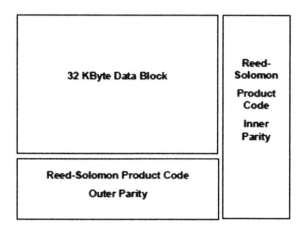

Fig. 5.88 The ECC block is adding error correction parity bytes to 64 kB data

Fig. 5.89 The diagram of modulation encoder

Link-In	Preamble	Data Area	Postamble	Link-Out

Fig. 5.90 Structure of ML data block

gain control, address, level calibration, equalizer adaptation information and data area contains timing and gain control information data and DC control information, the postamble provides for DC control clean-up and the link out contains a power control pattern. To each 5-bit group from the ECC encoder output, a parity bit is added by an on volutional encoder to create a 6-bit group. These 6-bit groups are then mapped to two 8-level ML symbols.

Each block is a separate data unit because it contains own timing, gain control, address, level calibration and equalizer adaptation information. Therefore this block can be written and read independently. Further, the gain control, level calibration and equalizer adaptation subsystems enable inter change between different drives by removing effects as mechanical and optical drive differences. Issues arising from disc defects are addressed by these subsystems also. There are subsystems to synchronize within data block and to control the DVD content of the ML signal.

5.8.1 Writing Calibration

The ML symbols are converted by the write strategy to laser pulses, it write the ML marks on the disc as Fig. 5.91. The write strategy is developed by an ML write calibration procedure that occurs when the first write command is issued.

This calibration system begins with ML optimum power control (OPC). ML OPC finds the optimum power(s) for writing pattern with a pre-selected range of different power levels, reading back this pattern, and measuring distinctive metrics of the pattern (similar to CD's asymmetry). The OPC pattern whose metrics are closest to the target value was written with the optimal powers.

Next, the calibration system develops a write strategy with the ML pre-compensation iteration procedure (PIP) as in Fig. 5.91. ML PIP also writes out a test pattern, reads the pattern back, and performs measurements of the writing distortion due to the neighboring marks as shown Fig. 5.92a. It is recovered reflectivity from random ML data: top is before PIP, bottom is after PIP as in Fig. 5.92a. Each of the 8 distribution contains 64 different traces resultant from every possible combination of the central level with all of nearest neighboring data cells. Figure 5.92b is equalized ML eye pattern showing another perspective of above level distributions after PIP process. PIP improves the ML writing strategy iteratively until the nearest-neighbor nonlinear writing distortion is below a threshold. The 8-level writing strategy has 512 different pulse definitions, one for each ML level with each combinations of neighboring marks. Because the writing

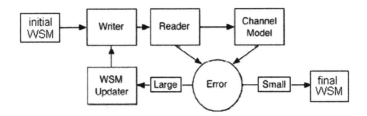

Fig. 5.91 ML PIP writes a pattern on the disc with the initial write strategy matrix (WSM)

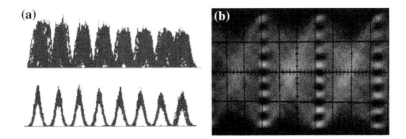

Fig. 5.92 **a** Recovered reflectivity levels from random ML data. **b** Equalized ML eye pattern

is tuned for a particular disc that has been inserted into drive, this adaptive write-strategy can correct for drive and disc variations.

When writing ML-R, laser power is constantly updated against a target power value: back-reflected light from the disc and laser light at the forward photodetector are measured during writing of a power-calibration pattern in the link-out of each block. The back-reflected signal is divided by a write-power control signal generated from the detector signal and then compared to adjusted again to target value. For ML-RW, constant power is maintained by sampling only the signal from the laser to photodetector (Fig. 5.93). These signals are sampled at specific times during the writing of a power control pattern in the link out of the ML data block. The sampled data is then processed to the CPU which in turn controls laser power through DACs. For both ML-R and ML-RW, sampling of the photodetector output is synchronized with the writing of the power calibration pattern in the link-out area of the block. The power-calibration pattern contains a repeated sequence of long periods with different constant laser powers. An example of writing results of the fine control laser pulse writing strategy is shown in Fig. 5.94. ML data cell length (mark edge to edge) is about 300 nm. Shape and extent of ML mark can be finely controlled with the write-pulse strategy. It is important to the laser writing power levels for the ML-R and ML-RW materials, that are similar to the conventional DVD-R and DVD-RW power levels. Details about the fundamental mechanisms and head-media interface will be discuss later.

In standard CD/DVD writing, the different lengths of the RLL modulation code are written on the disc using alternating marks and spaces of different lengths. Because this signaling is always a space following each mark and a mark following each space. Many CD/DVD focus-error and tracking-error signal-generators sample the PUH servo-signals during the spaces in the modulation code because during these times. There is a stable and constant power level–the reading power for

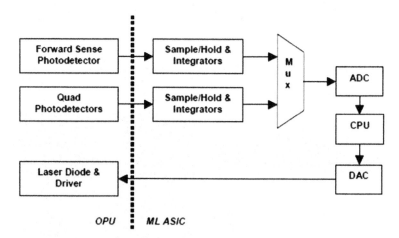

Fig. 5.93 ML power feedback control and monitors by forward sense and quadrant photodetectors

Fig. 5.94 The image of TEM of ML marks in phase change media on disc, which minimum data cell length (mark edge to edge) is about 300 nm

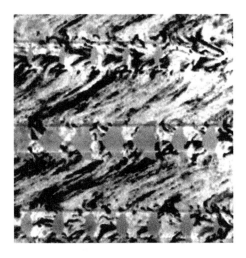

writing media and the erasing power for RW media. However, ML does not have these long spaces during writing, these types of focus- and tracking-error signal generation are not possible. Rather, the ML system separately normalizes the main quadrant photodetector signals and outrigger photodetector signals from the PUH (see Fig. 5.95). This is achieved by normalizing these signals before they go to the frontend chip. The outrigger photodetector signals are separately summed and used to adjust the gain-controlled amplifiers. Once adjusted, the main and outrigger signals are passed to the front-end chip, which uses these signals to calculate

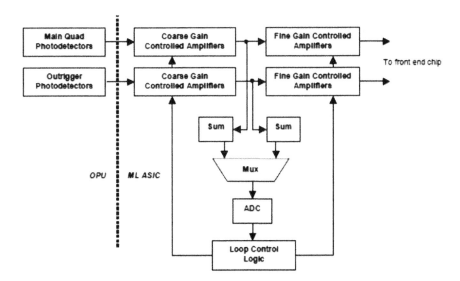

Fig. 5.95 During ML writing, the main quad and outrigger photodetector signals need to processed to remove the effects of the writing pulses

focusing and TE signals. These signals are then used to generate continuous focus-error and tracking-error signals. This pattern is read back and compared to a linear model of the channel. If the errors are large, the WSM is updated and the process repeated until the error falls below a threshold and achieves the final WSM.

For ML spindle speed control can be adjusted either from an error signal generated by the media's wobble groove or recovered data clock. That is similar to the DVD system, the wobble groove also contains addressing and other system information including target media positioning characters etc.

5.8.2 *Decoding*

On the decoding, all signal is read by a standard PUH. After this signal passes through again-controlled amplifier and digitized, timing is recovered with marks in the preamble (as in Fig. 5.96). This timing information is used to sample of all-sum signal to produce a digital stream at double ML symbol rate. Timing and gain/offset are maintained by a series of marks, that provide both a clean signal-edge for clock recovery as well as minimum and maximum signal levels for monitoring to AGC. Before the entire block is decoded, the block address is examined to ensure that the correct block is being read (see Fig. 5.97). This field contains an edge for timing recovery and maximum/minimum levels for gain/offset control. The block address is written in an easy read code for error correction. Once the block address is confirmed, the digital processing of the multilevel signal begins with a fine adjustment of the gain and offset with measurement of the envelope of the signals.

Fig. 5.96 PUH all-sum signal shows the transition from the link-into the timing acquisition section of the preamble

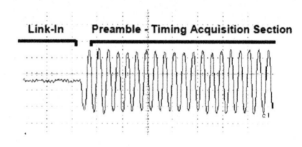

Fig. 5.97 Block address section of the preamble. As an AGC/timing field is on the left

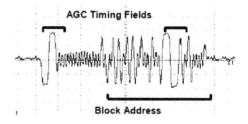

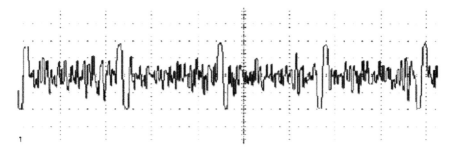

Fig. 5.98 Unequalized random ML data signal

An example of an unequalized data signal is shown in Fig. 5.98. It is an analog signal before the analog to digital converter (ADC) and the equalizer.

The ML data signal is equalized with a 11-tap fractionally-spaced equalizer. The taps are trained at the beginning of each block using an equalizer adaptation pattern. Equalization of the signal removes the intersymbol and interference by interaction of ML recording spot on disc. For this data, the standard deviation of each signal level of the dynamic range is about 3 %, if the hard decision level could give a symbol error rate of $\sim 10^{-2}$. However, the advance decoding system, this data has a $\sim 10^{-5}$ error rate after the Viterbi decoder. The low error rate is achieved not only as the convolutional coding provides a some error correction, but also as the level calibration system and the adaptive equalization system. The two systems help to compensate for interchange effects, that are from differences of discs and drives. After equalization, the deformatter removes the non-data marks and adjusts the signal according to the DC control system. The deformatted signal and level information are measured by the calibration pattern in the preamble (as in Fig. 5.99). The 8 levels of the ML signal was measured to counteract effects due to inter change, that is used the 256-state Viterbi decoder to recover the multilevel

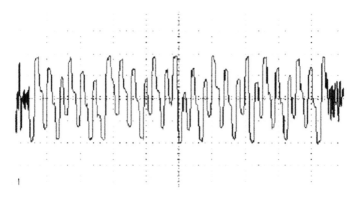

Fig. 5.99 Example of a level calibration section of the preamble

symbols. The data stream is then RS-PC decoded to produce the original 32 kB data block.

The system has written ML marks with commercially OPU successfully. But ML technology involves several key components that are researched, developed and implemented, including mixed-signal processing LSI, proprietary firmware and media, base-drive integration and existing software as: signal processing algorithms, media research and systems integration.

5.8.3   Special Processor and System Integration

Algorithm and media research form-factor of the ML system is need of large signal processing by building a custom media tester optimized for ML technology and using modular, swappable production opto-mechanic assemblies (OMA), that get an accurate estimation of the integration to various research results.

The signal processing algorithms are partitioned the analog and digital domain. Analog circuitry is designed and implemented rack-mount systems. The digital signal processing uses a special workstation with custom C++ software to ensure bit-exact and cycle-exact functionality for the eventual mixed-signal LSI. It operates precisely the LSI for system demonstration, development and testing. Meanwhile, it provides the LSI designers with a functional specification, and allows a powerful method to keeping test benches and vectors consistent throughout the entire design process.

Results from media research on system capacity, manufacturability, tests are carried out to evaluate various dye formulations, phase-change optical stack designs, and groove geometries, shows that this ML technology can increases capacity on commercial media, and achieved the original prototype as shown in Fig. 5.100. The form-factor of the system is great read-signal processing rate, and allows to be steadily operating long time. The FPGA emulation of the digital signal processing allows flexible validation of the LSI hardware descriptor language (HDL) coding for connecting to internal wires not out pins, and modified easily during debug to various fixes.

Based above original prototype system, the mixed signal processing is finalized with the optical system, and the functionality has been debugged with the FPGAs to lay out the LSI to replace the FPGA encoder and decoder for higher speed. Additionally, designed a general-purpose, low-level ATAPI package to test firmware. Then the media can be tested with the vendor's OMA, and made to fine-tune media-servo interactions to provide more margin performance. At all complex details with regards to hardware, firmware, and media can be determined and final systems integration to a second prototype system was shown in Fig. 5.101. An out-of-form-factor board is used to verify interfaces and functionality, while remaining flexible for rework and debug. In implementing basic functions to help with lower system debug. The same custom analog hardware is used, but now with an FPGA emulation of the digital signal processing in the LSI. The FPGA

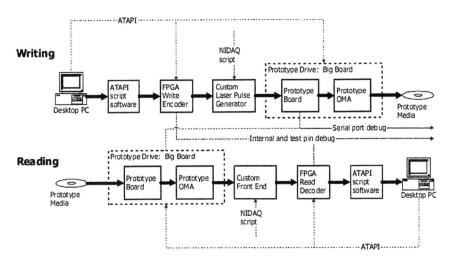

Fig. 5.100 The original prototype system, that the form-factor has been reduced, and the read-signal processing rate is increased to 12× DVD

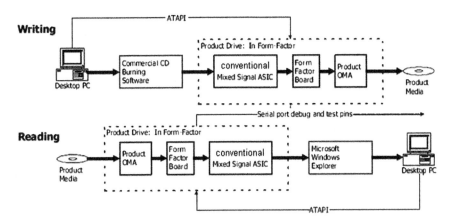

Fig. 5.101 The second prototype system, it is integration with conventional 120 mm disc drive, and the read-signal processing rate is 36× DVD by CLV technology

emulation allows flexible validation of the LSI HDL coding in two ways: The FPGA allows for connecting to internal wires that are not usually brought out to pins, and can be modified easily during debug to try various fixes. Meanwhile, the FPGA system allows for firmware design, coding and testing. Based on past experiment get a fully-functional system working at third LSI speed. When the mixed signal processing is finalized with the optical system, which replaces the FPGA encoder and decoder to operate at full speed. The drive is an ATAPI device, all drive functionality (e.g., servo, spindle and seek) can be scripted for extensive testing. To allow commercial software to provide ML functionality, the software

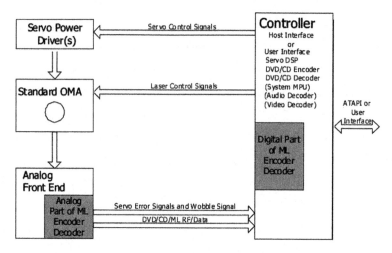

Fig. 5.102 The architecture of fine prototype system with current ML chip and ML-enabled drive

and firmware provide an ATAPI command-set to vendors to allow for easy integration.

Based on original prototype system and second prototype system with all design details set and tested, the final prototype system of ML integration system with conventional optical drive was developed and is shown in Fig. 5.102. Using an inform-factor drive base, the mixed-signal LSI is integrated on the circuit board. Firmware requires minor modifications from the Big Board base to the form-factor drive, but the serial port can still be used for debug. The ML CD-burning software is now finalized with the commercial CD software to ensure that their packages support ML writing and reading. The discs are tested with operating systems such as Microsoft Windows Explorer and Linux to ensure ISO and UDF support. Production media can be finalized and tested using special form-factor drives that allow for the same battery of speed, capacity, manufacturability, environmental and robustness tests that were carried out. The development process results in a design verification testing (DVT) procedure ready to aid in the drive production process.

A key point of ML technology is integrates with existing drive technologies. The ML LSI and firmware model are designed to simplify integration with a conventional optical drive. To ensure that the PIP procedure is fast, automatic, and compensates for a given drive-media environment. By its nature, PIP does not require the time-consuming characterization for every media type to be supported as in the conventional case. PIP, in combination with the adaptive equalization, timing recovery, and level recovery process described in paragraph, ensures a self-calibrating, adaptive, closed-loop system that minimizes the development needs for the system.

By keeping both the digital and analog signal processing within the ML LSI, the interface and signal integrity issues are minimized. In addition, the ML system has been designed to be programmable with many user options corresponding to different environments. This openness also extends to test ports and interrupts, to allow system diagnostic information be accessible for drive bring-up and media analysis. The ML Endec LSI makes the integration virtually a drop-in. Figure 5.102 shows a diagram of a typical ML-enabled drive architecture. That the standard drive components and signal paths of the underlying non-ML drive remain unchanged. The OMA, the analog front end, and the servo power drivers are identical to conventional drive. A minor ML interface is added to the controller chip that remains otherwise unchanged.

The controller's ML interface have three functions: (1) provides a local data input and output bus for the ML LSI; (2) provides a demodulated wobble signal, shown in Fig. 5.102 as the Bi-Clock signal and Bi-Data signal; and (3) accepts a spindle control output interface. The demodulated wobble signal and spindle control interface allow the ML LSI to provide addressing information and speed control when ML media is inserted. The shaded region of the drawing shows the necessary ML additions or connections required, while the remainder is a typical drive architecture. The system microprocessor, memory and associated connections are not included for clarity. At the board level, multiplexers are needed at two signal paths in the integration with the current ML chip. One of the signal paths is the laser control path to maintain backward compatibility with conventional EFM recording. The other signal path needed comes from the PUH photodetector outputs since ML recording uses servo-normalization during writing.

A future controller is integrated includes both DVD and encoder/decoders, of ML functions, it is designed to easily integrate with the analog front end chip and the controller chip is shown in Fig. 5.103. Adding firmware support for multilevel (ML) technology is accomplished: (1) develop drop-in ML-Library helper functions; (2) provide support for ML-media in disc initialization; (3) adjust the ATAPI

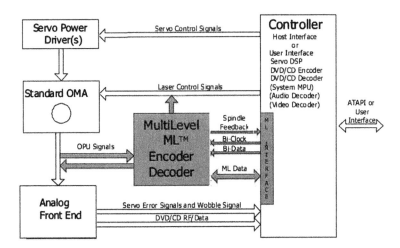

Fig. 5.103 An integrated ML drive architecture. ML chip is integrated into the controller and the analog front end. The chip number on board is same as conventional drive on the whole

command execution routines to account for ML-media; and (4) provide some simple "callback" routines. Designed the ML-code Library set (ML-Lib) to be easily added to pre-existing firmware sets or incorporated into the design of a new firmware set. ML-Lib is designed as a library of functions to be invoked by vendor firmware code when using the ML LSI. These functions have two purposes: to provide a clean "wrapper code" interface to the MLLSI and to provide calculations to aid ML functionality.

Rather, it is designed to extend the vendor firmware capability to handle the ML. ML-Lib code consists of C-language routines that are including functions to be directly by vendor firmware and interrupt routine. ML-Lib functions are code in the process of carrying out a large-scale task, such as reading blocks. ML-Lib provides interrupt service routine to be invoked by the interrupt from the ML LSI. When an ML disc is inserted, standard DVD firmware will not recognize the type of disc, so provided ML functions to read the MLAIP information in the wobble-groove. The library also supplies several routines to calibrate the analog portion of the ML LSI during startup.

In order to make ML-Lib as complete and logically consistent as possible, to define firmware functions that are anticipated from the vendor's code that will also be made available to ML-Lib functions. These are referred to as callback functions. Functions in this category are used to set laser power, tracking and focus offsets etc. Table 5.21 shows the general sequence for reading data from ML-media. Each of functions sets up a few registers in the ML LSI and returns. The overall operation of the read process and its timing are entirely under control of the host firmware. The write command is similar, containing one additional step of producing a write strategy matrix for the disc and drive combination if necessary. The three main

Table 5.21 General read command sequence

Read commands	
Host firmware	ML-Lib routines
Receive command	
Call → FindBlock()	Locate block
Seek to Block	
Call → MLPrepRead()	Prepare chip for reading
Arrive it Block	
Set up buffer pointers	
Call → MLInitRead()	Initialization for reading
Call → MLRead 0	Begin reading. Enable interrupts
Count block received, send to host	Continue reading
Count block received, send to host	Continue reading
Count block received, send to host	Continue reading
⋮	⋮
When all blocks rcvd,	Stop read process
Call → MLRdStop()	Disable interrupts

factors responsible for capacity improvement are highlighted: reduction of data cell length, increasing the number of bits per data cell, and increasing the encoding and total efficiency of the system.

The communication interface between the PC host software and the ML drive is across the IDE bus using the ATAPI protocol. The ML ATAPI interface is identical specification, that have modified the responses to some of the commands and created a mode page to handle responses that do not fit within existing structures. It is straight forward to adapt existing software interfaces to CD/DVD drives to accommodate ML. It is worked with ahead software to interface the current product, with the ML optical data storage system. These ISO formatted ML discs are recognizable from Windows, and the files can be opened and read by simply double-clicking. ML methodology has been demonstrated on CD/DVD-ROM, CD-R/RW, DVD-R/RW and blue laser, high-NA systems. From this variety of experience ensure that the ML technology can be applied to virtually any optical media system available [1, 2].

References

1. D. Xu, L. Ma, H.X. Li, *Multi-dimensional storage optical head* (Tsinghua University, Tsinghua University, 2001) (CN00124422.1)
2. T. Erneux, E.A. Viktorov, B. Kelleher, D. Goulding, S.P. Hegarty, G. Huyet, Optically injected quantumdot lasers. Opt. Lett. **35**, 070937 (2010)

Chapter 6
Tridimensional (3D) Optical Storage

The mechanism of 3D digital data storage, especially for multilayer optical disc, seeking method and principles of data parallel reading/writing, as well as method of multidimensional codes are important proportion in tridimensional optical storage, that will be described in this chapter. In fact, a various principles of reading/writing and materials can be used to 3D multilayer optical storage, as above introduced photochromism for example, in which photochromic compounds and polymers are provided with good photochromic and photorefractive properties for digital 3D-optical storage [1, 2].

6.1 Two-Photon Multilayer Storage

Two-photon absorption was first predicted by Maria Göppert-Mayer in 1931, and then was demonstrated experimentally by Kaiser and Garrett, using a Ruby laser. Parthenopoulos and Rentzepis proposed two-photon multilayer storage with organic or inorganic materials in 1999 in the first. The basic principles of the two-photon storage is shown in Fig. 6.1 and may be classified of two types: single wavelength (hv) two-photon multilayer storage (type 1) and two wavelength ($hv + hv'$) two-photon 3D storage (type 2). Two-photon sensitivity absorption is dependent on the square of the intensity of the incident light. The nonlinearity establishes the fact that the probability of two-photon excitation is significantly weaker than that of a single-photon excitation process. As a result an ultrashort pulsed laser such as a femtosecond laser is usually used for efficient nonlinear excitation. Organic dye molecules are generally two-photon excitable, however, the low two-photon sensitivity impedes their wide application in terms of devices. The two-photon absorption cross-section of majority photochromic and

© Tsinghua University Press and Springer Science+Business Media Singapore 2016   407
D. Xu, *Multi-dimensional Optical Storage*,
DOI 10.1007/978-981-10-0932-7_6

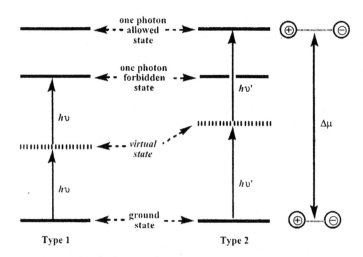

Fig. 6.1 The single wavelength (hv) two-photon multilayer storage (type 1) and two-wavelength ($hv + hv'$) two-photon storage (type 2) process

photoisomerization dyes generally is small in the range of 0.1–10 GM (1 GM = 10^{-50} cm^4 s photon^{-1}). Semiconductor nanoparticles such as quantum dots (QDs) and quantum rods (QRs) are promising candidates due to their broad spectral tenability over sizes and larger two-photon absorption cross-sections (reference Chap. 4 of this book). However, nanoparticles dispersing homogeneously into a recording medium itself are quite challenging.

But, most medium exist higher crosstalk that cannot be engineering practicability. The information recording process from a single-photon effect changes to the two-photon effects that can make the two-photon 3D space (multilayer) storage also. But for type 2 two-photon absorption storage, the virtual state can be established with two kinds of photon only that use to complete the two-photon multilayer storage rather easy than type 1 to control the access position of 3D storage.

In addition, several photochromic compounds have been investigated as two-photon photochromic data storage media also. The pioneering studies for 3D optical memory using photochromic compounds as the photochromic spiropyrans, etc., for example. The molecules of photochromic compounds exist in two isomeric forms of A and B as shown in Fig. 3.11 (see Chap. 3 of this book) which are a colorless cyclic form and a colored open form, respectively. Irradiation of light λ_1 to the colorless form A converts to the colored form B which exhibits fluorescence upon photoexcitation. Orthogonal two beams system was employed for reading and writing, where the molecule was excited at the intersection of the two beams simultaneously. For writing information, the excitation was performed by two-photon irradiation of λ_1 and λ_2 photon. Then isomer A at the intersection is photoisomerized to isomer B. The energy level diagram is changed along with the molecular structures. For reading data, only the λ_2 beam was used for irradiating the media and isomer B can be excited and emit fluorescence.

The two-photon absorption technology can create 3D optical data storage device with higher storage capacity that is how data is encoded on multiple layers of a fluorescent polymer. Data retrieval is done by a confocal microscope system, it filters out the interference of unwanted data from neighboring layers essentially. A multilayered fluorescent optical data storage device that increased the storage capacity of a typical DVD sized disc upto 500 GB or more. In the multilayer fluorescent discs (MFD) there are many layers on which data can be stored in a datatrack. The fluorescing and nonfluorescing bits instead lands and pits of conventional CD to interpret binary information. The nonfluorescing bits are normal fluorescent material that has been photobleached. Using the alternating coextrusion process of recording and buffer or insulating material can produce many hundred to thousand layers disc.

6.1.1   Single Wavelength (ħv) Two-Photon Multilayer Storage (Type 1)

Single wavelength multilayer storage is including two-photon (type 1) and two wavelength ($\hbar v + \hbar v'$) two-photon storage (type 2) that can be used for multilayer storage also. The photochromic reaction can be used to multilayer storage which performances same with two-photon (type 1) storage basically. So, it will be introduced with single wavelength two-photon multilayer storage (type 1). OMNERC of Tsinghua University developed a multifunction two-photon 3D storage experiment system which has Na 0.65 long focal length objective lens and adaptive aberration correction device to compensate spherical aberration as shown in Fig. 6.2a. The principles of compensate spherical aberration will introduce Sect. 6.5 of this chapter. The medium is a fulgide PMMA film (pyrryl substituted

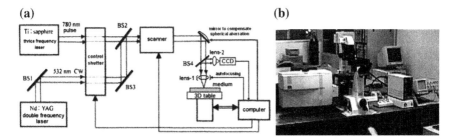

Fig. 6.2 **a** The multifunction two-photon multilayer storage experimental system with Na 0.65 long focal length objective lens, autofocusing, and mirror to compensate spherical aberration system. **b** The photo with three-dimensional (3D) precision positioning mechanism system

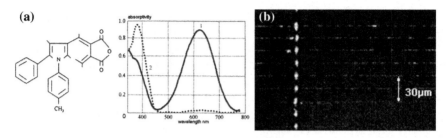

Fig. 6.3 The molecule structure of pyrryl-substituted fulgide and its absorption spectrum (**a**) and recording experimental result (**b**)

fulgide) in which the molecule structure is shown in Fig. 6.3a, and the absorption spectrum and recording experimental result is shown in Fig. 6.3b. The system includes single-beam two-photon and double-beam recording and readout system with 3D table that allow testing of materials (medium) and system components evaluation. As the medium is not truck generally that the main computer controls the tracking/layer addressing. Using a special designed, built Na 0.65 long focal length achromatic objective lens with aberration autocorrection for wavelength 780 and 532 nm (Nd:YAG CW 1064 nm lasers double frequency), and integrated into the autofocusing and objective lens movements by a linear magnetic driver control. The recording laser beam passes a beam expander to the objective lens that is focused inside the recording medium which can be precision moving with air bearing and magnetic drive. Since the laser beams into medium are changing with the depth, the recoding layers will cause additional aberration to objective lens. The main computer and processor can compensate with shifted binary optical lens 2 according refractive coefficient of medium and deepness into the medium by a precision positioning mechanism. A ultrafast Ti:Sapphire with wavelength 780 nm femtosecond laser for recording and 532 nm laser for reading out when the system is used for single wavelength two-photon multilevel storage experiment. The 780 nm laser beam focal point to generate two-photon effect (390 nm photon) to write down data on the focusing spot and then use 532 nm wavelength laser fluorescence effect to readout as shown in Fig. 6.3b. The medium employs pyrrole-spirobenzopyran with thickness of 500–1000 μm. Between the layers without other materials, the writing thickness was dependent on focus depth of the optical system and the distance between neighborhood data layers, that has to be greater than the depth of focus to avoid crosstalk.

After recording the 532 nm laser (operating at <0.5 mW) which is used to excite fluorescence from the recorded data bits. The fluorescence is then picked-up by the same objective lens and focused onto a detector. The distance between adjacent layers is about 15 μm in this experiment that can be used to record 30–60 layers at last. But readout signal is weakened obviously with writing depth increasing since recorded marks are interference with each other as shown in Fig. 6.3b.

6.1.2 Two Wavelength Two-Photon Absorption (Type 2) Multilayer Storage

The imperfection of type 1 two-photon multilayer storage is that need of great power and high frequency pulse lasers and is difficult to engineering application. In addition, the layer separation of type 1 two-photon multilayer storage is large that restricts its volume capacity. However, the type 2 two wavelength ($hv + hv'$) two-photon multilayer storage is better practicability for 3D memory. So type 2 two wavelength two-photon multilayer storage could be controlled by the layer separation with many methods as Figs. 6.4 and 6.5, for example. The medium at point P of spatial intersection of two different photons can only be recorded by altering the structure of the photochromic molecules as shown in Fig. 6.4a. Similarly, the medium will be write in data when λ_1 and λ_2 are working at one time, and readout by λ_1 only as shown in Fig. 6.4b.

Figure 6.5a shows a principium setup for type 2 two-photon optical storage system. The laser 2 has formed a cuneal plane light by cylindrical lens for

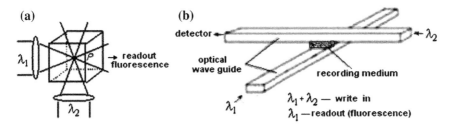

Fig. 6.4 The type 2 two-wavelength 3D storage. **a** The point P of spatial intersection of two different photons (λ_1 and λ_2) can be recording. **b** The recording medium will be control by optical waveguide with λ_1 and λ_2

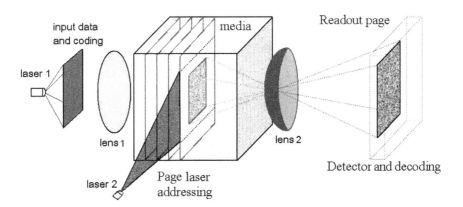

Fig. 6.5 The type 2 two-photon optical storage system, laser 2 for addressing page record with laser 1 for input data

addressing, writing, and reading. The laser 1 and objective lens 1 for writing and laser 2 for writing and reading with lens 2. The data is writing in by a spatial light modulator code and lens to project on the plane which is addressing page by laser 2 to record. In reading from the memory pages, the page addressing laser 2 beam is shaped to a sheet of light to this page and to readout the fluorescent image. The storage media is transparent to the laser light. The optical readout system projects the fluorescent image to the detector array and to the decoding system. Each pixel in the 2D page has different intensity which can also record multilevel data. If the system adopts an adjustable polarized light with different threshold medium to record, it can realize multidimension (4D, 5D or more) storage also [3].

An optical 3D storage system with variable-focus lenses for dynamic imaging system and with variable-focus lenses for 3D optical storage is shown in Fig. 6.6. It has enabled the automation of the 3D memories with a programmable controller, which gives commands for system operation and programs the data storage and retrieval processes. The optical system is based on the scanning line controlled by electrical signals. Actually, it was demonstrated as one-dimensional (1D) dynamic imaging system that can provide 1D spatial light modulator (SLM) and place cylindrical lens on the back focus plane. This 1D imaging is required for the linear-oriented information recording only. On the other hand, the principles of the system can be used to other 2D dynamic recording such as holographic recording 3D optical memory system, etc.

Beam 1 is the information input matrix SLM that is used to display the input image. The two Fourier lenses form a 4f system with a 2D Perovskite Lead Lanthanum Zirconate Titanate (PLZT) electro-optic lens at the Fourier plane. The input page information is imaged (following the arrows from left to right) by the three devices onto a plane inside the recording medium at the right-hand side of the figure. When a beam 1 configuration voltage is applied to the PLZT lens, the imaging plane moves back and forth along the optical axis of the system, thus it is permitting dynamic 3D recording. The applied voltage can be generated to form an address controller which determines input page to choose images. When the

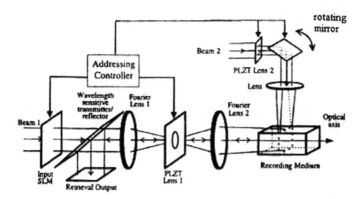

Fig. 6.6 3D optical recording and retrieval system with dynamic positioning imaging

imaging plane moves, the image size keeps unchanged. Beam 2 is used as the reference beam to produce a scanning line-of-light with the rotating mirror. The use of the PLZT electro-optic lens enables the line-of-light to move along the optical axis. The mirror is used to move the line perpendicular to the optical axis. Thus, the whole block of the recording medium could be covered. The rotating mirror can be replaced by an acousto-optic deflector controlled by electrical signals. When the recording is performed, a line information is displayed on the input SLM. The PLZT lens 1 adjusts its focal length to image that line information onto a certain place in the recording medium. The PLZT lens 2 and the acousto-optic modulator act accordingly to produce a line-of-light to coincide with the image of the line information. Therefore, a line of information is recorded. This system can also be used for the information retrieval purpose. The beam 1 is blocked, PLZT lens 2 and the acousto-optic lens conduct beam 2 to retrieve a line of information. The emitted fluorescent light from the recording medium travels back to the left (following the arrows from right to left as in Fig. 6.6). A corresponding voltage is applied onto the PLZT lens 1. Thus, the retrieved data is displayed at the retrieval output plane through the wavelength sensitive reflector. The wavelength sensitive transmitter/reflector is transparent to the wavelength of Beam 1 in the recording session, and it reflects the fluorescent light emitted from the recording media in the retrieval session since the emitting light of the recording media is different from the recording laser beam. Fast electrical operation ability also makes the system programmable and is easy to operate. The non-mechanical structure avoids movement of any element in the system, so that there will be no worn-out and thus the possibility of an error that occurs will be greatly reduced. Both the forward and backward dynamic imaging have been realized and demonstrated to the inspecting sponsor. The retrieval throughput is determined by the time response of the PLZT lens with reading speed of 100 Mbit/s [4, 5].

The capacity could be greatly increased by this way. But, there are many effects that can result in corrupted data as resolution of optical system and recording symbol interference between neighboring space to produce crosstalk. In addition to the readout head, laser or the media may not be aligned perfectly due to mechanical nature for addressing a page. Therefore, the image may be slightly shifted and pixels may lack some intensity or contribute some intensity to their neighbors. Thus, OMNERC of Tsinghua University focused on research of type 2, i.e., two wavelength ($h\nu + h\nu'$) two-photon multilayer storage. A new experimental system was completed in OMNERC at Tsinghua University in 2003. It is shown that in Fig. 6.7: where 1—laser I for writing and reading, 2—laser II for writing only, 3—lens, 4—light stop, 5—active fluorescence filter, 6—lens, 7—detector, 8—main synthesis prism, 9—splitting film, 10—reflective film, 11—objective lens, 12—binary optical components, 13—lens, 14—insulation layers of disc, 15—recording layers of disc, 16—protective layer, 17—reflective layer, and 18—disc substrate with pregrooves track. The thickness of the insulating layer (14) and recording layer (15) are 3.5 and 0.5 μm, respectively. The laser I and II can be focused on a same

Fig. 6.7 The type 2
two-wavelength ($hv + hv'$)
two-photon multilayer storage
system with higher resolution
on longitudinal direction and
pregrooves servo track

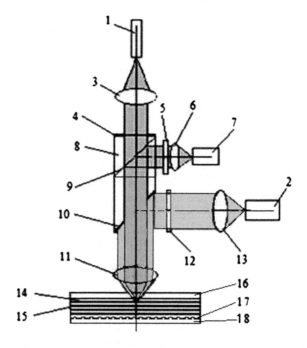

focal plane with a binary optical components (11) correction. When recording, the
focus of the laser beam (1) scans down to reflective layer (17) vertically for seeking
and tracking. When laser beam (1) arrives in the recording area, the laser (2) would
work just. The data was recorded when only the two photons at the same time
gathered at one position and that time two lasers beam bring two-photon effects on
this position in the medium. The recording materials are coated as layers between
transparent insulating layers, in which spacing is greater than the focus depth of the
laser I, therefore the crosstalk on different recording layers is avoided well as shown
in Fig. 6.8.

The absorption properties of photochromic medium will be changed when the
two laser beams gathered in the same focus through the lens center as shown in
Fig. 6.8a: The proportion of two wavelengths beams diameter is about 5:3 (laser 1
to laser 2) that the effective numerical aperture of laser 1 is 0.55. The energy
distribution of laser I and laser 2 beams on longitudinal direction is shown in
Fig. 6.8a.

Because, the two-photon absorption photochromical action is present at two
laser beams with certain intensity only. So, the recording symbol was depended on
focus depth of laser 1 beam and thickness of the medium. Accurate and effective
control of the laser 1 beam can gain smaller spot. Therefore, this scheme must be
higher than the stability of mechanical system and accurate actuator of pickup head
to realize focus and track.

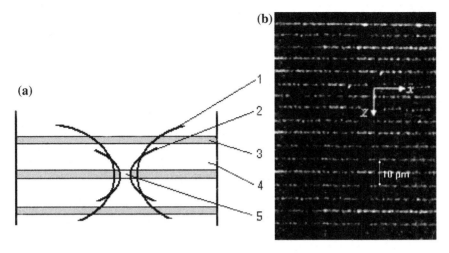

Fig. 6.8 The energy distribution near of focus plan on recording (**a**) where *1* laser 1, *2* laser 2, *3* recording layer, *4* transparent insulating layer, *5* recorded area. **b** is the recording experimental result, *z* is the depth direction, and *x* radius direction of the disc. The layer separation spacing is within 5 μm and the thickness of effective spot for recording is about 0.5 μm

The recording medium was made same shape as the general disc, that the thickness of medium is about 1 mm for 200 layers. The total capacity can be calculated $C_{total} \approx 1$ TB with Eqs. (6.3) and (6.4) later. The recording materials are coated as layer-by-layer with transparent isolation layer alternately, which spacing is greater than the focus depth, in order to avoid the crosstalk of different layers. The lights of the two wavelengths should focus at the same point with different binary optical components. When a disc rotates, the focus of the laser beam scans vertically on the bottom layers. The pregrooves are pressed on the substrate (bottom layers) by conventional CD process. In that case, the focusing, seeking, and tracking of a pickup head can be implemented. Because the two-photon absorption photochromical material would not react, unless two wavelength laser beams reach certain level, with the smaller one of the two spots, as area 5 is shown in Fig. 6.8a, in fact it determines the size of the effective spot. The layer separation spacing was reduced to 5 μm and the readout quality of signal was improved evidently. Accurate and effective control of the short wavelength laser beam can gain smaller spot. Though the scheme needs very stable worktable and more accurate actuator of pickup head to realize focus and track, the scheme is still more practical than the existing one of the 3D optical storage systems.

The written isomerous of the molecules was excited by 532 and 780 nm photons and then emission of fluorescence from the written isomerous by 532 nm laser beam. The intensity of the observed fluorescence was dependent on the excitation pulse energy. Thus, the readout is as fast as a few tens of picoseconds. As can be seen from expression, the intensity of the readout fluorescence signal is proportional to the content of photochromic molecules in the medium form in the area excited by reading radiation. In accordance with the experimental procedure, the system employed for the investigation of fluorescence from a photochromic sample with the optical multichannel analyzer detected via fluorescence film. In particular, detecting fluorescence is from certain area of the photochromic sample simultaneously irradiated by a pair of writing pulses of fundamental radiation and the second harmonic, to be able to visualize the accumulation of the colored form (by process of optical data writing) in the area where the writing beams intersected each other. The results of such experiments are presented in one-photon luminescence. Fluorescence reading in this case was carried out with the second-harmonic pulses, which were applied with a certain pairs of writing pulses and the numbers of writing pulses to specify the reading data. So, the readout signal can be processed with special digital amplifier with higher accuracy. The experimental systems achieved a total storage capacity of 220 GB per disc of 120 mm diameter, 1 mm thick with 50 layers. For demonstration purposes, these concepts have been successfully integrated in a portable demo unit.

The experiments show that the removable write-once-read-many times, random access digital storage system that is capable of very high data density, maybe TB per 120 mm disc removable media, fast access time (~ 100 ms), and fast data transfer rates. These figures will enable systems supporting high-speed data filtering, and content as well as index-based data searching algorithms. Recently, demonstrated the scalability of two-photon recordable photochromic doped polymeric WORM disc media to more than 200 layers with negligible interlayer crosstalk and excellent stability of the written bits. Raw bit-error-rates (BER) of 10^{-5} (similar to conventional CD) have been measured. Recently bit dimensions of $0.5 \times 0.5 \times 4.5$ µm. Eventually, data bit sizes could be down to the order of $0.3 \times 0.3 \times 2$ µm resulting in super high data densities. If this technology is combined with MW/ML, it may be stored in a given radial location resulting in data densities exceeding >1 Tb/in.^{2} or equivalently 1000 bits/µm. Meanwhile, a new faster manufacturing method and equipment for coating of recording and insulating layer alternately to make 100 discs of 100 layers at the same time.

For readout, a 532 nm CW laser diode operating at 0.5 mW is used. Readout of the recorded bits is carried out with very fast readout data rates and fast access times by optical excitation of the recorded bits and parallel detection of their fluorescence. Parallel readout is accomplished by using a class of optical systems known as depth transfer optical (DTO) systems. This architecture offers much higher data rates compared to serial readout or in-plane parallel readout, where the DTO reads data tracks across multiple layers in depth, as well as across a number of radial tracks. Theory, simulations, and experimental results of this 3D multilayer multiple-track parallel readout system have been characterized. A 64 parallel data channel readout system that reads 16 radial data tracks on 4 data layers in parallel from a two-photon recorded monolithic multilayer disc with a total data throughput of 128 Mbit/s with the disc rotating at 3600 rpm has been characterized and is operational. Crosstalk from adjacent tracks and adjacent layers has been shown to be 25–30 dB below the primary signal.

6.2 Vertical Resolution of 3D Optical Storage

The 3D volumetric optical storage approach used recording bits in a volume by two-photon (type 1) or photochromic action recording. A spot is written in the volume of polymer with photochromic molecules only at points of temporal and spatial intersection of two photons with sufficient photon energies to record by altering the structure of the photochromic molecules. Recording occurs only within a small volume around the focus of the laser beam due to two-photon response and follows the 3D point spread function (PSF) irradiance distribution of the focused beam. The recording response of the material follows the square of the optical system PSF resulting in a recorded bit size that is 30 % less than the Rayleigh criterion PSF. The recorded bits are read by fluorescence when excited by suitable optical radiation absorbed within the written spot volume [6].

Influence of various parameters in a 3D multilayer optical data storage system are analyzed to design advanced high capacity recording and playback optical systems that enable high data transfer speeds as a topic of active research and development. In the recording, the excited molecular distribution can be simply considered to be proportional to the square of the irradiance distribution of the recording laser beam. The recorded bit shape is modeled as:

$$P(x, y, z) = \alpha \times I^2(x, y, z) \tag{6.1}$$

where α is a constant, different from one type of molecule to another. $I(x, y, z)$ is the laser beam irradiance. At focus, the laser Gaussian-shaped is given by:

$$I(x, y, z) = \frac{I_0}{\omega_0^2 \left[1 + \left(\frac{\lambda_z}{n\pi\omega_0^2} \right)^2 \right]} \exp \left\{ \frac{-2(x^2 + y^2)}{\omega_0^2 \left[1 + \left(\frac{\lambda_z}{n\pi\omega_0^2} \right)^2 \right]} \right\} \tag{6.2}$$

where I_0 is the peak irradiance, $\omega_0 \approx 0.5\lambda/NA$ is the radius of beam waist, λ is the wavelength, and NA is the numerical aperture. The simulation result shows the irradiance squared I^2, distribution of a $\lambda = 532$ nm, NA = 0.65 system to have bit dimensions of $0.6 \times 0.6 \times 5$ μm^3 as shown in Fig. 6.9.

Bit size is determined at the $1/e^2$ value of the peak fluorescence. Figure 6.9b is the image of a real experimental recorded bit with $0.6 \times 0.6 \times 7$ μm^3. So the total areal (single-layer) raw capacity can be expressed as:

$$C_{areal} = \frac{\pi (r_{max}^2 - r_{min}^2)}{l_{bit} \times w_{pitch}} \tag{6.3}$$

where r_{max}, r_{min} are maximum and minimum recording radius of the disc, $l_{bit} \times w_{pitch}$ is the bit size in the X-Y plane and scales with λ/NA. The volume raw capacity of the 3D optical data storage disc is:

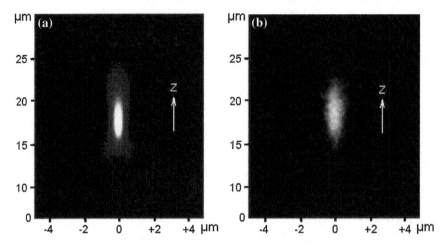

Fig. 6.9 Simulation and experiment result of multiple-layer recording bit size with 532 nm, 0.65NA and irradiance W/m^2

$$C_{total} = C_{areal} * N_{layer} \qquad (6.4)$$

where $N_{layer} = T_{disc}/S_{layer}$ is the number of layers. T_{disc} is the thickness of the disc; S_{layer} is the layer separation that decreases with increasing NA as shown in Fig. 6.10. Generally, T_{disc} should be smaller than the working distance of the objective lens. Figure 6.10 shows the versus volume capacity, NA for 405 and 632 nm wavelengths. Here, the T_{disc} is equal to working distance of the objective lens. Figure 6.10b shows that in a 3D multilayer optical data storage system, the total capacity of a system with long working distance lens (such as 0.65NA) is higher than a system with high NA but with short working distance [7].

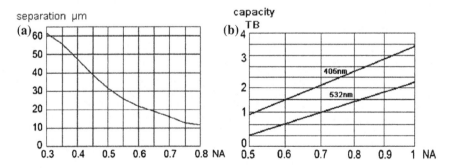

Fig. 6.10 Minimum layer separation of 3D multiple-layer storage (**a**) and volume capacity of multilayer optical data storage system with 120 mm diameter (**b**)

The numerical aperture influences not only the recorded bit size, but also the recording power. For two-photon recording, the required peak power P_{peak} of the laser can be expressed as:

$$P_{\text{peak}}^2 = \frac{\pi^2 2h\nu \cdot M_{\text{unit}} \cdot (0.5\lambda)^4}{D_M \sigma T \cdot t_p f_{\text{rep}}} \cdot \frac{1}{\text{NA}^4} \tag{6.5}$$

here $h\nu$ is photon energy, M_{unit} is recorded molecules per unit volume, λ is recording wavelength, DM is density of the original unrecorded molecules, σ is two-photon cross section, T is recording time, t_p is the pulse width of recording laser, f_{rep} is the repetition rate of recording laser. This shows that increasing the NA, the recording total recording power may be reduced by virtue of the increased irradiance in the focused spot. For example, a NA 1.0 objective lens is used only when 25 % of the recording peak power to the NA 0.5 recording system. The influence for aberration of objective lens in multilayer storage and its adaptive correction will be discussed in detail in Sect. 6.5 of this chapter.

6.3 Stereo-Multidimensional Storage Medium

Stereo-multidimensional medium is the base for research and development of the multidimensional storage with geometrical 3D stereo optical storage. Many foreseeable technology as multiwavelength and multilevel multilayer, volume holographic storage, EB mastering and super-REN, Blu-ray and UV laser, etc., that is in practice is based on various performances of excellence medium. The two-photon optical storage systems represent another volumetric optical storage technology with attractive attributes. But storage medium has to be preceded in development as it is needed of longer period from laboratory to commercial production. In these systems, the memory material consists of a photochromic dye embedded in a polymer host that changes its molecular state upon illumination with the appropriate light. When writing on the material requires the simultaneous excitation with two photons, and reading is accomplished by either illumination light with one or two photons which causes the written molecule to fluorescence. The requirement of two simultaneous photons allows addressing the individual volume elements in the medium by suitably directed optical beams. One beam can be configured to represent a data, a vector, or a waveguide such thatimpressed the second beam addresses the recording area by the location of its intersection with the data beam. These beams can intersect orthogonally or it can counter propagate. In the counter propagating case, the address is selected by appropriate timing of the intersection, that the short pulses are required to delineate the position of data sufficiently, and the optical pulse has to be shorter than 50 fs for a recorded plane of 10 μm depth [8].

6.3.1   Stereo 3D Storage Medium

Pty.Technology Ltd. developed a 3D optical storage materials technology based on identified by Daniel Day and Min Gu. Several large technology companies such as Fuji, Ricoh, and Matsushita have applied for patents on two-photon responsive materials for applications including 3D optical data storage, however, they have not given any indication that they are developing full data storage solutions. Mempile Inc. is developing a commercial system with the name TeraDisc. In 2010, they demonstrated the recording and readback of 100 layers of information on a 0.6 mm thick disc, as well as lower crosstalk, high sensitivity, and thermodynamic stability. They intend to release a red-laser 0.6–1.0 TB consumer product, and have a roadmap to a 5 TB blue-laser product. Constellation 3D developed the fluorescent multilayer disc, which was a ROM disc, manufactured layer by D-Data Inc. They provided various products and prototypes as shown in Fig. 6.11 to be a digital data multilayer storage. Storex Technologies has been setup to develop 3D media based on fluorescent photosensitive glasses and glass-ceramic materials. The technology derives from the patents of the Romanian scientist Eugen Pavel. At ODS-2010, conference presented results regarding readout by two non-fluorescence methods of a Petabyte Optical Disc. Landauer inc. develops a media based on resonant two-photon absorption in a sapphire single crystal substrate that showed the recording of 50 layers of data using 2 nJ of laser energy (405 nm) for each mark. The reading rate is limited to 10 Mbit/s because of the fluorescence lifetime. Difference 3D discs are developed with various materials as shown in Fig. 6.11, but no one was mass commercial production yet.

An important type of photochromic dyes as photochromic pyran for its fast photoresponse and photostability can be used in optical multi-D storage. It has been patent in USA that was the naphthopyran and photochromic arofused-pyran in which molecule structures are shown in Fig. 6.12a, b.

But its thermal-stability is rather poor, in order to increase optical density of colored forms and decrease the decay speed, the naphthopyrans have been redesigned and synthesized by Institute of Photographic Chemistry of Chinese Academy of Sciences, China in 2003. The molecule structures of compounds are shown in Fig. 6.13.

Fig. 6.11 Various 3D storage optical disc with materials are sensitive to difference wavelength

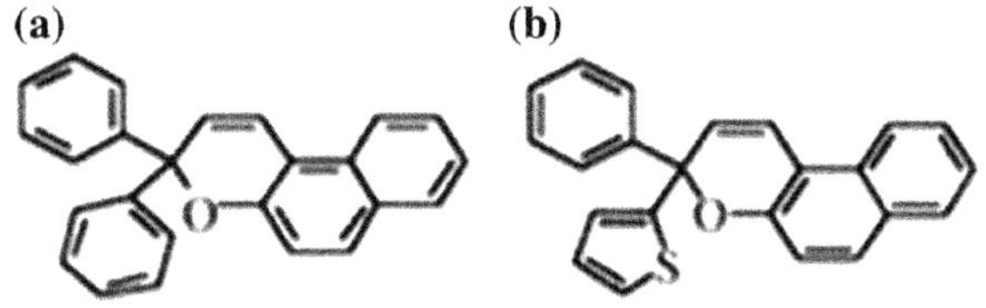

(a)   **(b)**

Fig. 6.12 The molecule structures of naphthopyran and photochromic arofused-pyran

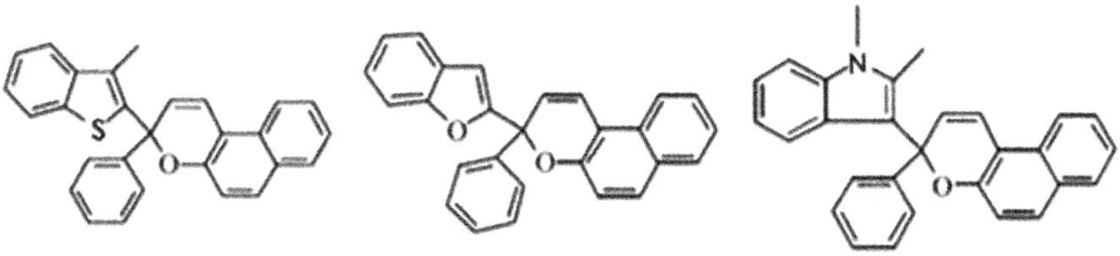

1   **2**   **3**

Fig. 6.13 The molecule structures of compounds of redesigned and synthesized photochromic arofused-pyran. *1* 3-phenyl-3-[3-methylbenzothiophene-2-yl]-3H-naphtho[2,1-b]pyran. *2* 3-phenyl-3-[benzothiophene-2-yl]-3H-naphtho[2,1-b]pyran. *3* 3-phenyl-3-[1,2-dimethylindol-3-yl]-3H-naphtho[2,1-b]pyran

Another requirement is that the organic photochromical medium to the absorption peak toward the short wavelength. So, the Institute of Photographic Chemistry of Chinese Academy of Sciences developed a new diarylethene and fulgide photochromic materials that the molecular structure is shown in Fig. 6.14.

Synthesized 6 different structures of perfluorinated resorcinarene crystal that the typical molecular structure and synthesis reaction are shown in Fig. 6.15. Its closed-loop state absorption spectrum of 400–700 nm wavelength and two-photon absorption cross section of 173 GM (1 GM = 10^{-50} cm^4 s $photon^{-1}$) and decomposition temperature is higher than 200 °C more than two monosubstituted compounds of 2–3 times. That has been made thickness of 100–500 μm PMMA film and used to multilayer recording experiments as shown that this material is greater potential.

Those materials have been made up to prototype 3D optical storage disc by OMNERC and Tongfang optical disc Inc. and applied in experiments of 3D storage, which is in Fig. 6.8.

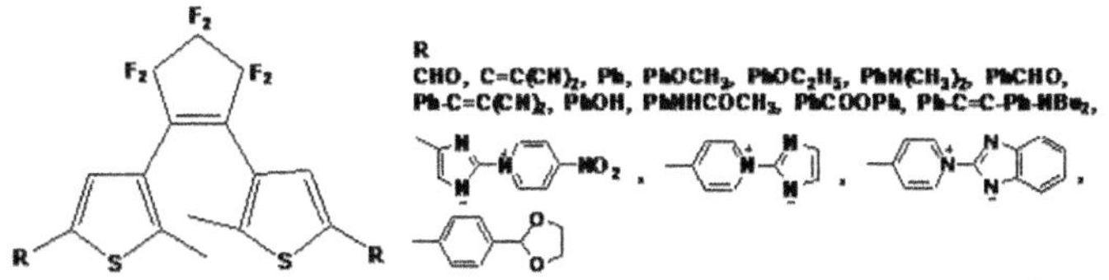

Fig. 6.14 The typical molecular structure of diarylethene and fulgide photochromic materials

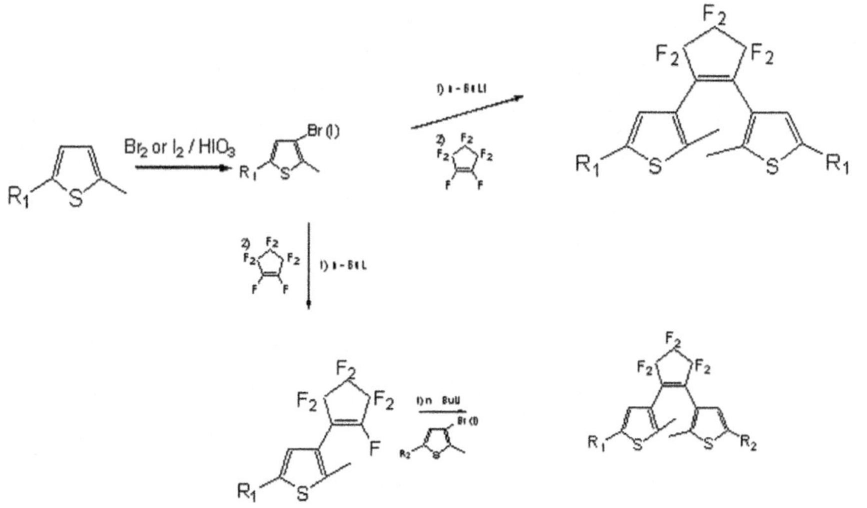

Fig. 6.15 The perfluorinated aryl vinyl photochromic compounds and schematic diagram of synthesis process

6.3.2 Multidimensional Holomem Materials

A new invention (International Application No. PCT/JP2006/307317) provides a hologram recording material that can efficiently modulate the refractive index with a visible light laser and is highly transparent even after the modulation. The hologram recording material is characterized by comprising a polymer of a monomer comprising, as an indispensable component, an acrylvinyl monomer represented by formula $CH_2=C(R_1)C(=O)O-R_2=CH_2$ where R_1 represents a hydrogen atom or a methyl group; and R_2 represents a saturated or unsaturated hydrocarbon group having 1–20 carbon atoms optionally containing a hetero atom or a halogen atom in its molecule, the polymer having a radical polymerizable side chain vinyl group in its molecule, and a photopolymerization initiator source which generates an active species upon exposure to visible light. In particular, the photopolymerization initiator source is a photopolymerization initiator, which absorbs visible light to generate active species, or a mixture of a visible light sensitizing coloring matter and a photopolymerization initiator.

The synthesis of the diarylethene material is shown in Fig. 6.16 is one of the Diarylethene materials synthesized by the Institute of Chemical Physics of Chinese Academy of Science. It has two forms. The open-ring form A is achromatous while the closed-ring form is blue. With the ultraviolet light λ_1 or the visible light ($\lambda_2 > 450$ nm), the molecules of the Diarylethene can be converted between the two forms.

Dissolve 20 g of the above materials in 5 liter solution of the PMMA-pimelin ketone (10 %, w/w), spin coat them on the substrate of diameter 120 mm, and of

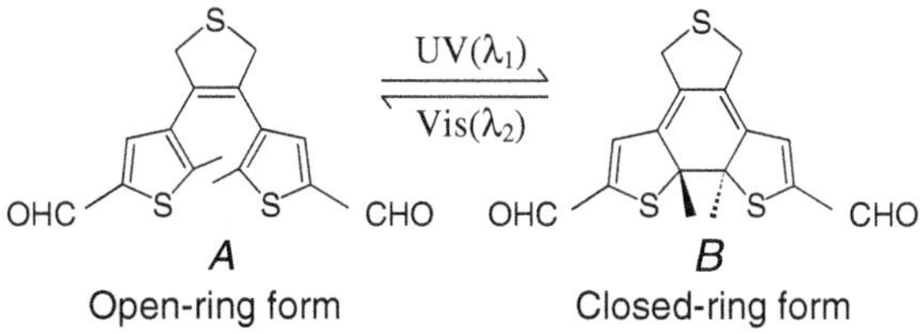

Fig. 6.16 The two forms of
the molecules of the
Diarylethene and their
photochromic process

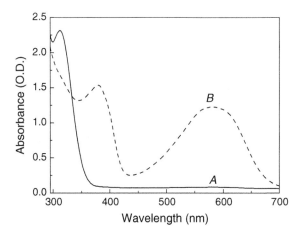

A
Open-ring form

B
Closed-ring form

thickness 0.8 mm disc through ultrasonic oscillations and the misce bene and dry
them. The thickness of the films attained to10 μm. The absorption lines of the two
forms of the films are illustrated in Fig. 6.17. The absorption peak of the
closed-ring form lies at 582 nm where the optical density of the films is 1.23, while
the absorption peak of the open-ring form lies at 320 nm.

The GE discs offer the same recording speed as Blue-ray and 20 times the
amount of storage space. GE first touted the technology in 2010. GE did not return
a request for comment on where that technology effort stands today. InPhase
Technologies took aim at the magnetic tape drive market with the industry's
800 GB holographic optical disc. InPhase, which was spun off from Bell
Laboratories, called its holographic product the Tapestry HDS-300R. InPhase had
also planned a second-generation 5 TB rewritable optical disc with data transfer
rates of about 80 MB/s in 2010. Some typical medium material which made in
China and performances are shown in Table 6.1.

For $LiNbO_3$:Fe Mg Mn, LN:Fe In, LN:Fe, Zn crystal phototropic materials
holographic storage add K_2O assist flux to get chemical computation class crystal
with parametric optimization of the crystal $LiNbO_3$ are shown in Fig. 6.18a. The
larger size (4–5 in.) with high uniformity is shown in Fig. 6.18b.

Any HODS is relation with storage material. So it has been put an important
position including Lithium Niobate crystals, including doped LN crystals, organic
and polymer materials, and testing technology and equipment. Doubly doped lithium
niobate crystals have been finished based on optimum doping impurity, ions doping

Fig. 6.17 The absorption
lines of the open-ring form
(*solid line*) and the
closed-ring form (*dashed*) of
the films of the Diarylethene

Table 6.1 Holographic materials comparison

Materials	Resolution	Scattering	Reliability	Sensitiviy (J/cm^2)	Stability	Max thickness (mm)
LNbO$_3$: Fe	+++	+++	+	0.02	+	10
LNbO$_3$ (Two-color)	++	++	+	0.02	++	10
Polaroid (Photopolymer)	+++	–	0	20	+	Q5
PQ/PMMA	+	–	+	0.2–0.5	++	2
Bayer photo-addressable polymer	+++	0	++	0.002– 0.02	++	Q1

(a)   **(b)**

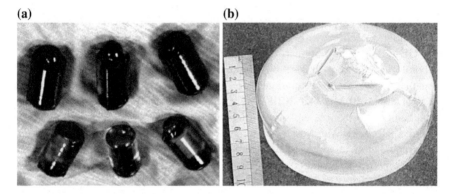

Fig. 6.18 a Chemical computation class crystals. **b** More than 4 in larger size LN crystal

concentration, and suitable oxidized treatment. The LiNbO$_3$ with 0.03 wt% Fe crystal is higher photorefractive sensitivity, dynamic range, and SNR. Another LiNbO$_3$:In:Fe crystal with 0.5 mol% In, 0.06 wt% Fe, and slightly oxidized state. It can be used to two-color nonvolatile holographic storage for restricted degressive exposure schedule. Doubly doped LN:Fe, Mg and LN:Fe, In and LN:Fe, Zn was used in experiment to prove its efficiencies which can be 60–80 %, refractive modulate degree of 10^{-4}, sensitivity of 10^{-5}–10^{-4}, writing/erasing time of 10^{-2} s, refractive gradient smaller than 3 × 10^{-5}, and larger dynamic grating of crystals by electrolytic. The near-stoichiometric LiNbO$_3$ crystals with 49.6 Li$_2$O by adding 11 mol% K$_2$O with 5 diameter and 300 mm length ±0.05 mol% have also been finished. Their main performances are shown in Figs. 6.19 and 6.20, respectively [9].

All experimental results based on photorefractive crystals of LN show that the LN crystals best advantage which has the high resolution. But its sensitivity, dynamic range, scattering noise, etc., are rather inferiority. It cannot be used to dynamic image record by appearances. Some photopolymeric and dye holographic

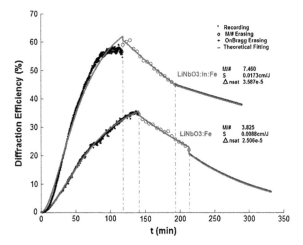

Fig. 6.19 Diffraction efficiency of near stoichiometric LiNbO$_3$ crystals with 49.6 Li$_2$O by adding 11 mol% K$_2$O

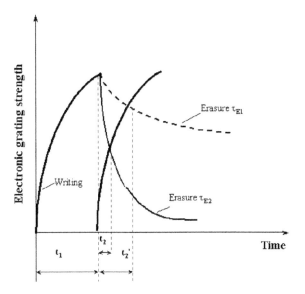

Fig. 6.20 Electronic grating strength of near-stoichiometric LiNbO$_3$ crystals with 49.6 Li$_2$O by adding 11 mol% K$_2$O

materials are emphasized on developing and experimenting at Chemical Department of Tsinghua University, since 2006. Most importance is the analysis and elimination of crosstalk noise of the materials, evaluation of SNR, that will be presented in this section. The dye sensitized matter system including dye/2 benzene-ketene peroxide matter serial and solicitation matter. Its configurations of molecule are shown in Fig. 6.21. The sensitized dyes for 532 nm laser and for 405 nm laser are shown in Fig. 6.22.

Other advanced organic stereo holography memory materials are photopolymers. It is great potential holography storage medium for industry application and

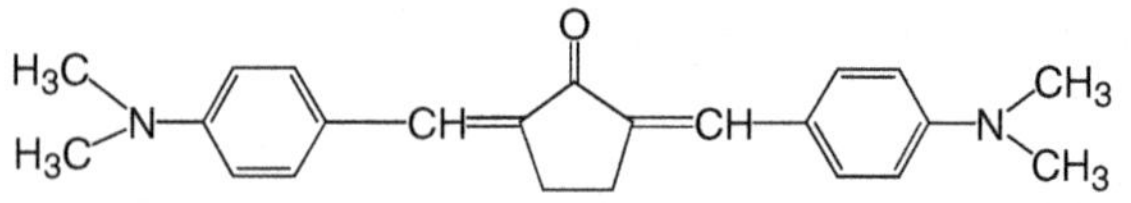

Fig. 6.21 The configurations of molecule of dye/2 benzene-ketene peroxide matter

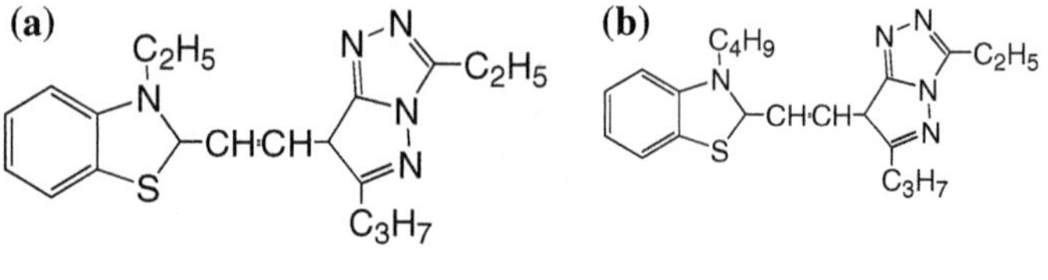

Fig. 6.22 The configurations of molecule of the sensitized dye by 532 nm laser (**a**) and sensitized by 405 nm laser (**b**)

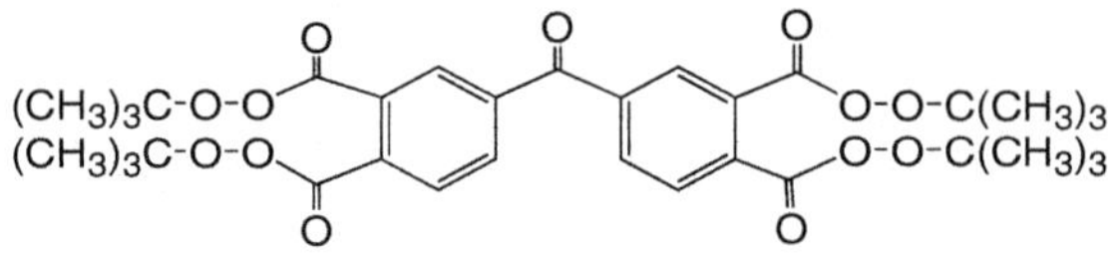

Fig. 6.23 The configurations of molecule of high sensitivity photo-polymers—Dye 6-aryl double imidazole is higher sensitivity photo-polymers and better stability

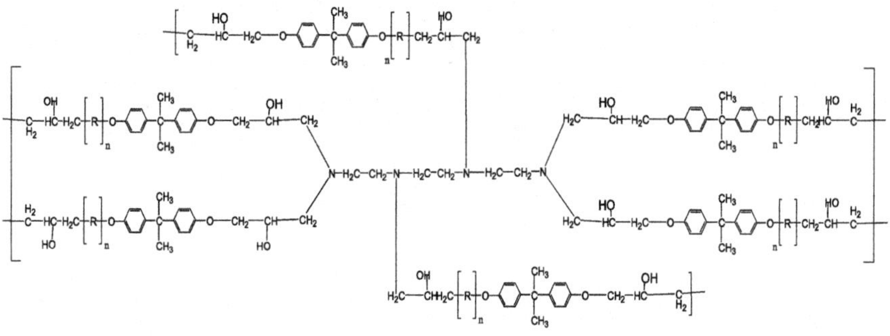

Fig. 6.24 The configurations of molecule of organic holography memory materials—colophony

has been developed at Tsinghua University also. The molecule of dye 6-aryl double imidazole is shown in Fig. 6.23. It is higher sensitivity photopolymers and has better stability. The advanced organic holography memory materials—colophony is shown in Fig. 6.24.

A molecule sediment film composed of phototype storage materials with PMo12/1-10-DAD system for 780 nm laser was brought. Its spectrum of absorption is shown in Fig. 6.25.

Fig. 6.25 The performances of molecule sediment film is composed of phototype material

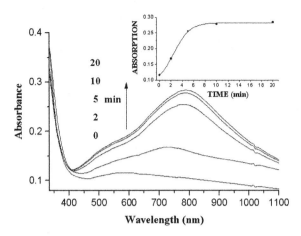

6.4   Multilayer Parallel Read/Write

The conventional approach to achieve high readout data rates in single-layer disc media data storage systems is to decrease the physical bit size and increase the velocity at the head/media interface. Further, increase in data throughput may be achieved by parallel readout architectures that read information across a number of data tracks simultaneously within a given data layer. A parallel readout architecture where data tracks are read in parallel across multiple layers in depth, as well as across a number of radial tracks by Call and Recall. Theory, simulations, and experimental results of this 3D multilayer multiple-track parallel readout system are presented later. The experimental result is from a 64 parallel data channel readout system that reads 16 radial data tracks on 4 data layers in parallel from a two-photon recorded monolithic multilayer disc.

In conventional CD and DVD optical data storage systems, the data channel is serial. Exploiting the potential of parallelism that exists in optical systems can increase data throughput. One way is to fan the readout beam into several beams by using a diffraction grating. This results in a linear $1 \times N$ array of focused illumination spots oriented radially, increasing the total data throughput by N times that of a serial channel device. An astigmatic line of illumination may also be used to readout $1 \times N$ multitracks. An alternative to the arrangement of $1 \times N$ focused spots in a radial line is to arrange them in the 2D spot array. This will not increase data throughput, but can reduce crosstalk and aid in matching the illuminated data tracks to a detector array by increasing the center-to-center pitch of the illumination spots.

Employing a large number of laser beams using a lens let array or a single high NA annular-field objective lens which can generate a 2D spot array. All of these architectures achieve parallel readout within a single layer. However, there is a limit as to how large N can be due to the limited object field of the objective lens. Figure 6.26 illustrates the multiple-track/multiple-layer parallel readout architecture within a multilayer 3-D disc geometry.

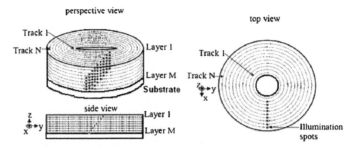

Fig. 6.26 The typical perspective graphics of multilayer 3D disc

The concept is to arrange the bit oriented 2D data page array on a tilted plane with respect to the optical axis so that multiple tracks ($1 \times N$) within multiple layers (M) are readout simultaneously. The experimental optical system is shown in Fig. 6.27, the integration of recording and readout illumination with the DTO layout indicating various system components. The data plane format is organized with $1 \times N$ multiple tracks being readout at a given layer inside the material with M more data layers being readout at the same radial location as shown in Fig. 6.28. Signal processing and ECC circuits were designed and fabricated. This architecture offers the potential for much higher data rates compared to serial readout or inplane parallel readout. For example, if each individual data channel within the 2D tilted data page is operating at 1 Mb/s a detector array format of 4×16 data channels readout simultaneously provides a total data rate of 64 Mb/s. Following this direction a 4×16 readout system is designed and built to simultaneously excite fluorescence of a tilted data plane containing 4 data layers and 16 data tracks on each layer. The readout is accomplished by using a class of optical systems of DTO systems. Depth-transfer imaging readout optics is used to collect the excited fluorescence of a tilted data plane and image the collected fluorescence to a tilted

Fig. 6.27 The integration of recording and readout illumination with depth transfer optical systems, various components, signal processing, and ECC circuits of 3D WORM storage

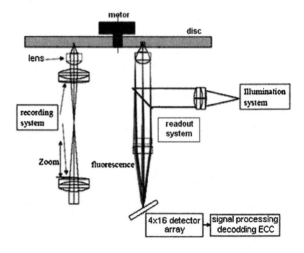

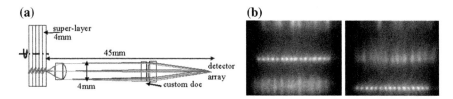

Fig. 6.28 a Readout system of 4 × 16 with 4 × 15 mW 635 nm laser diodes each split into 16 beams. **b** Fluorescent spots from layers 1 and 3 on detector

detector array. The Scheimpflug condition relates the lateral magnification to the tilt angles of the data plane/detector plane. Each detector element within the 2D array can be thought of as an individual serial data channel. The DTO systems are a focal relay telescope in a Keplerian configuration used as a 1 × imaging system. The focal configuration has the desirable first-order properties of constant lateral and longitudinal magnification.

6.4.1 WORM Multilayer Storage System

Figure 6.27 shows the architecture of WORM optical data storage system with separate recording and readout heads. The recording head consists of a 0.65NA objective lens with an adjustable conjugate changing spherical aberration compensation telescope. The bit size that this 532 nm laser system can record is 0.9 × 0.9 × 12 μm. This bit density corresponds to a raw capacity of ∼100 GB in a 120 mm diameter and 3 mm thick disc. The DTO systems design is indicated in by the 0.65NA of 4 mm focal length, 4 mm diameter objective lens that collimates the fluorescence, and then imaged by the lens (18 mm focal length, 5 mm diameter) to a detector array, tilted with respect to the optical axis, with a ∼4× inplane magnification. The 0.65NA lens is designed with commercially available optics and one custom element, which readout module is integrated with a collinear illumination module having 4(layer) × 16(track) parallelism.

This is part of a 64-channel experimental readout system including a parallel detector array. The second lens is a DTO systems used to achromatize the 40 nm of fluorescence bandwidth by providing the opposite sign of dispersion of glass. The objective lens fits in a standard CD voice-coil for focus/tracking actuation. The lens is designed for 1.2 mm of disc thickness, so above presents spherical aberration correction methods under consideration to employ in this system. The readout layers away from this design thickness a 4 × 16 collinear illumination module is used to excite the fluorescence from 16 tracks across 4 layers for a total of 64 channels as shown in Fig. 6.28. The fluorescence is imaged by the DTO systems to the detector plane. The 4(layer) × 16(tracks) are recorded at a depth of 1.2 mm having individual bit sizes of 1.2 × 1.2 × 20 μm, the tracks are recorded on a 3 μm track pitch, and the layers are recorded on a 30 μm layer pitch. The

Table 6.2 CMOS detector
array specifications

Array format	4 × 16
Pixel size	10 × 30 μm
Frame rate	1 MHz
Sensitivity	<2000 photons
Power supply	5 V
Power consumption	300 mW
Chip size	4 × 3 mm
Process technology	TSMC 0.35 μm
Input/output	CMOS compatible

illumination spots are arranged to match the recorded tracks and layers. The fluorescence spots at the detector (image) plane are observed with a video microscope as shown in Fig. 6.28b. Signals are detected with a PMT fitted with a 5 μm pinhole (comparable to a CMOS (active pixel sensor) APS detector array element) for sampling the fluorescent spots at the tilted image plane to observe adjacent track/layer crosstalk. Then a CMOS detector array is inserted in the image plane and the resulting signals observed. The properties of the CMOS detector array are shown in Table 6.2: array format 4 × 16 pixels of size 10 × 30 μm frame rate 1 MHz sensitivity <2000 photons [10].

Figure 6.28 shows the video microscope images of the 4 layers having 64 fluorescent spots, focused to adjacent recorded layers inside the disc generating fluorescence imaged by the DTO systems to detector. The fluorescent bits are very well-defined and the experimental size of ∼7 μm agrees very well with the simulation. The fluorescent spot pitch also agrees with the simulation at 12 μm. Note while all 4 layers are simultaneously illuminated in Fig. 6.28 that only one group of sixteen spots from a particular layer is in focus in a given image as the video microscope is imaging inplane, while the fluorescent data page exists on a tilted plane. Figure 6.29 shows the oscilloscope traces of data sampled in the fluorescent data plane of Fig. 6.28 by a PMT. The signals shown in Fig. 6.29 are characteristic of all 64 channels and may have amplitude variations of up to 20 %. The corresponding signals resulting from the Parallel Solutions CMOS detector are typical of all 64 channels as in Fig. 6.30. The specifications of the CMOS detector are shown

(a)   **(b)**   **(c)**   **(d)**

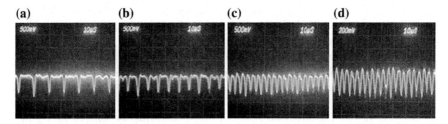

Fig. 6.29 Typical oscilloscope waveforms at the image plane in Fig. 6.27 of encoded data of a 6T data pattern, **b** 4T data pattern, **c** 3T data pattern, **d** 2T data pattern

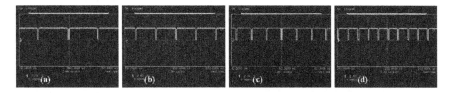

Fig. 6.30 Waveforms at the image plane in Fig. 6.18 encoded data of **a** 6*T* data pattern, **b** 4*T* data pattern, **c** 3*T* data pattern, **d** 2*T* data pattern are detected by an integrating detector array

in Table 6.2. The signal spectra from the signal of a 2T datatrack that is typical of the 4 layers are shown in Fig. 6.31. The 2T signal spectra show good uniformity across the 4 layers. Adjacent track and adjacent layer crosstalk is about 25–30 dB below the fundamental frequency indicating good performance for this track and layer pitch. If the layer and track pitch become closer it is expected that the crosstalk in the observed spectra will increase. Active servo was not used in this experiment as the disc was not removed after recording.

The performances of experimental two-photon addressed volumetric optical disc storage systems have been presented. The 3D parallel readout, simultaneous readout of multiple-tracks across multiple layers was realized experimentally with a DTO system to image a tilted object plane of 64 digital data channels in a 4(layer) by 16(tracks/layer) to a tilted image/detector plane. The fluorescent spots at the detector plane were observed using a video microscope and observed to be of good quality. A PMT was scanned in the fluorescent image plane to measure the signal quality where crosstalk from adjacent tracks and adjacent layers was shown to be 25–30 dB below the primary signal. A 64 channel CMOS detector array developed is integrated in the readout system providing a total readout data throughput of 64 Mb/s. Influence of the numerical aperture on a 3D multilayer optical data storage

Fig. 6.31 Typical signal spectra of encoded data of 2*T* data at the 4 layers showing the fundamental 2*T* frequency component and the adjacent track and layer components

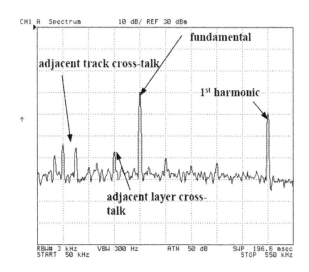

system is analyzed based on simulation and experiments. A high NA objective lens decreases bit size, layer separation, and increases the volumetric capacity. Increasing NA also decreases the required recording laser power for two-photon absorption recording. The extrapolation based on present experimental results indicates that 1 TB/disc capacities are well within the capabilities of two-photon recorded volumetric discs.

6.4.2 Multibeam Readout

Fluorescent Multilayer Disc (FMD) not only increases capacity, that data rate is increased by read in parallel across multiple layers in depth and across a number of radial tracks. An experimental system was developed by Call Recall that is from a 64 parallel data channel readout system that reads 16 radial data tracks on 4 data layers in parallel from a two-photon recorded monolithic multilayer disc that can increase data throughput as shown in Fig. 6.32a. One way is to fan the readout beam into several beams by using a diffraction grating. This results in a linear $1 \times N$ array of focused illumination spots oriented radially, increasing the total data throughput by N times that of a serial channel device.

An astigmatic line of illumination may also be used to readout $1 \times N$ multitracks. An alternative to the arrangement of $1 \times N$ focused spots in a radial line is to arrange them in a 2D spot array. This will not increase data throughput, but can reduce crosstalk and aid in matching the illuminated data tracks to a detector array by increasing the center-to-center pitch of the illumination spots. Employing a large number of laser beams using a lens let array or a single high NA annular-field objective lens can generate the 2D spot array. All of these architectures achieve parallel readout within a single layer. However, there is a limit as to how large N can be due to the limited object field of the objective lens. The multiple-track/multiple-layer parallel readout architecture is within a multilayer 3D disc. The

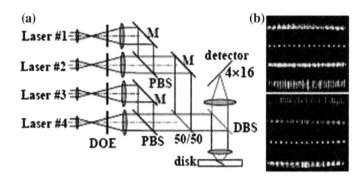

Fig. 6.32 a Optical system of 4×16 readout with 415 mW and 650 nm laser each split into 16 beams and focused to recording layers inside the disc generating fluorescence to detector. **b** Photograph of readout signals

concept is to arrange the bit-oriented 2D data page array on a tilted plane with respect to the optical axis so that multiple tracks ($1 \times N$) within multiple layers (M) are readout simultaneously. The data plane format is organized with $1 \times N$ multiple tracks being readout at a given layer inside the material with M more data layers being readout at the same radial location as shown in Fig. 6.32b. This architecture offers the potential for much higher data rates compared to serial readout or inplane parallel readout. If each individual data channel within the 2D tilted data page is operating at 1 Mb/s a detector array format of 4×16 data channels readout simultaneously provides a total data rate of 64 Mb/s. Following this direction, a 4×16 readout system is designed and built to simultaneously excite fluorescence of a tilted data plane containing 4 data layers and 16 data tracks on each layer. The readout is accomplished by using a class of optical systems known as DTO systems. Depth-transfer imaging readout optics was used to collect the excited fluorescence of a tilted data plane and image the collected fluorescence to a tilted detector array. Each detector element within the 2D array can be thought of as an individual serial data channel. The DTO systems are a focal relay telescope in a Keplerian configuration of a 4f one-to-one imaging system. The afocal configuration has the desirable first-order properties of constant lateral and longitudinal magnification.

6.5 Adaptive Aberration Correction

The quest for high density data storage devices has lead to the proposal of 3D optical storage as potential successors to the digital versatile disc and Blu-D technologies. Rather than writing data in a single plane, the data are written in a number of layers in a suitable recording substrate. Although different recording media have been suggested for example photorefractive, photochromic, or fluorescent media suffer from the same problem that affects both the recording and readout of these devices aberrations. So, an important problem for multilayer disc storage is the aberration correction especially when laser beam into medium deeply will bring out spherical aberration for objective lens. This section will present spherical aberration autocorrection and residual aberration measurement method.

6.5.1 MEMS Mirror to Compensate Spherical Aberration

Using large-stroke deformable membrane mirror at 45° incidence achieve a very compact optical system capable of fast multilayer focusing in multilayer optical disc. The MEMS mirror replaces a lens translation mechanism and liquid crystal compensator, resulting in a single optical element to control both focus depth and compensation of attendant focus-dependent spherical aberration. The membrane optical requirements in terms of stroke and aberration compensation are required for multilayer focusing. The adjustable range of at least 1.6 µm peak wavefront spherical aberration correction at a membrane displacement of 7 μm, which should be sufficient capability for quadruple layer Blu-D discs. Commercially, the Blu-D format has increased the capacity to 100 gigabytes (GB) of read/write storage on triple layer discs and 128 GB of write once storage on quadruple layer discs. Read only and rewritable 20-layer discs with 500 GB and 50-layer discs with 1 TB of storage have been developed. These multilayer discs not only need accurate and fast focus control, but the associated variation in optical path length through the glass disc medium comes with greater spherical aberration. The induced aberrations of the objective lens when not being used at its ideal infinite conjugate configuration may also prove significant at 0.85 NA. Current optical pickup heads utilize liquid crystals, magnification change of the objective lens, diffractive optical elements, diffractive refractive elements hologram optical elements as wavelength-selective filter, or deformable mirrors for spherical aberration compensation. The advantages for the deformable mirrors are that they are achromatic and typically have fast response times, where speed is important for reading and writing data quickly. Switching between layers also requires focus control. This may be achieved by translating lenses with motors. As a more compact alternative, some investigators have previously proposed deformable mirrors for focus control. A single-actuator elliptical boundary mirror at 45° incidence angle for compact focus control and spherical aberration correction was used for multi-layer Blu-D. This design alters the stress distribution of the device to control its shape as it deflects under piezo-electric actuation. They electrostatically actuated 3 mm × 4.24 mm elliptical deformable membrane mirror for both focus control and aberration correction at 45° incidence angle to simplify and reduce cost of the system by replacing two components with one compact element capable of fast response times in optical pickup unit (OPU) for multilayer disc. The MEMS mirror should also be able to compensate for any chromatic spherical aberrations resulting from a diffractive element in a multiwavelength unit to compensate for laser variations.

The 3 mm × 4.24 mm mirror is formed from a 2.5 μm thick membrane supported on a <100> silicon substrate using the process. A 6 nm chrome and 160 nm siller layer serves as the reflective coating and top electrodes, while a back side layer of 6 nm chrome and 200 nm gold provides good electrical contact with the silicon counter electrode as in Fig. 6.33. The 200 nm Au is coated onto the bottom side for the silicon counter electrode, 3 μm via incongruent melting quartz, and gold layers allow for dry-etching of the silicon to create an air gap [11].

Etching in XeF$_2$ through an array of small via the membrane surface yielded an air gap of approximately 25 μm between the membrane and the substrate. Although the reflective surface is gold, the fabrication process is amenable to aluminum, which is more reflective at 405 nm.

Three electrodes provide a means for shaping the mirror to compensate spherical aberration. Having an equal potential on all three electrodes naturally leads to a parabolic shape of the mirror that provides defocus for small deflections. Adjusting the potentials on each electrode independently allows for shaping the surface, and three electrodes offer adjustment of both primary (3rd order) and secondary (5th order) spherical aberration.

(a)   **(b)**

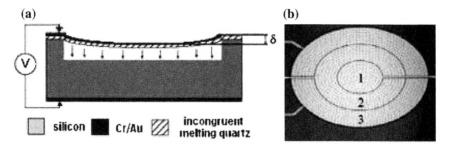

Fig. 6.33 The mirror to compensate spherical aberration **a** cross-section of the released mirror. **b** The 3 mm × 4.24 mm elliptical boundary mirror with three concentric electrodes

The elliptical boundary mirror used at 45° incidence in the back focal plane of a disc read objective lens as in Fig. 6.34. The MEMS mirror provides both focus control and spherical aberration correction to compensate change in the glass thickness for multilayer discs. Specifications for a quadruple layer BD-ML disc are shown in Fig. 6.34. The major and minor axes of the elliptical mirror and its electrodes are in proportion of, so that they project with circular symmetry along the optical axis. The variable focal length of the membrane lens is realized, where r is the radius of the small axis of the mirror. Assume the objective lens is well-corrected for spherical aberrations for infinite conjugate imaging with the focus occurring at the deepest layer in the disc (L_0 for BD-ML). To address other layers in the disc, the MEMS mirror is deflected, resulting in a shorter focal distance within the disc (L_1–L_3 for BDXL). For dual-layer disc, the full-range focus shift ΔZ between the two layers is 25 μm, while quadruple layer BDXL has four layers with aggregate change in the thickness ΔZ of 46.5 μm. Calculate the necessary change in MEMS focus and corresponding nominal deflection based on paraxial analysis. The results are tabulated in Table 6.3.

The spherical aberration due to refraction at the air-disc interface may be calculated assuming a point object is located at a depth Z in the disc medium, with its

Fig. 6.34 Disc read optical layout with quadruple layer Blu-D disc layer specifications shown. Deflection δ of the MEMS mirror shifts the focus of the beam from the deepest layer L_0 to layer L_3, which is 46.5 μm closer to the objective lens than L_0

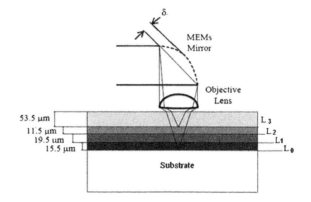

Table 6.3 Focus control and spherical aberration correction requirements for the multilayer Blu-D

	Maximum change in disc thickness (µm)	Objective lens focal length (mm)	MEMS focal length f_m (mm)	Deflection δ of MEMS required (µm)	Peak spherical aberration due to change in disc thickness
Dual layer DVD disc	25	2.33	340	2.4	0.16 µm, 0.26 λ
Quadruple layer BDXL™ disc	46.5	1.76	106	7.5	1.3 µm, 3.3 λ

Gaussian image point located in air at a distance Z/n from the air/disc interface. Assume the system is well-corrected for the deepest layer with $Z = Z_1$, and calculate the residual spherical aberration when the system is refocused at a different depth, where Z_2 is the distance between layers. This wavefront aberration W is given by,

$$W(\theta) = \Delta Z \left\{ n\left(\sqrt{\tan^2(\theta)+1} - 1 \right) - \frac{1}{n}\left(\sqrt{n^2 \tan^2(\theta)+1} - 1 \right) \right\} \qquad (6.6)$$

where θ is the ray angle in glass, and the maximum wavefront aberration corresponds to

$$\theta = \theta_0 = \sin^{-1}(NA/n). \qquad (6.7)$$

The peak spherical aberration due to the maximum change in disc thickness for DVD's and the quadruple layer BDXLTM disc are also shown in Table 6.5.

6.5.2 Residual Asymmetric Aberration Measurement

The MEMS focus and aberration compensation with a NA is 0.65 objective lens with a wavelength $\lambda = 630$ nm laser that is representative of focusing and aberration correction in a BD-ML disc also. Specifically, constructed the experiment so that the overall membrane peak deflection was 7.5 µm, corresponding to the range of focus control necessary for quadruple layer format, and adjusted the glass layer thickness so that the peak wavefront spherical aberration was 1.3 µm, corresponding to the peak spherical aberration incurred when changing focus from the deepest layer to the shallowest. Using the interferometer setup shown in Fig. 6.35, interferometric images of the objective lens aperture under different conditions were taken. The 630 nm light was spatially filtered using single-mode fiber, and expanded to achieve uniform illumination of the aperture. The two lenses with

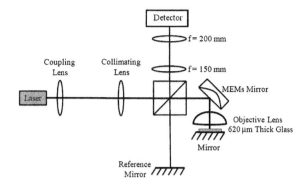

Fig. 6.35 Interferometer setup for testing the aberrations of the objective lens

$f = 150$ mm and $f = 200$ mm form an image of the objective lens aperture on the CCD camera, with a magnification of 1.33. Initially, placed a glass layer of 620 μm total thickness on top of a flat optical mirror behind the objective lens to represent the cover layer of an experiment disc for which this lens is well-compensated. The glass layer was composed of a stack of microscope cover-slips (BK7, $n = 1.52$), using index matching oil between each slide. Reference interferograms were recorded, verifying that the lens is indeed diffraction limited when focused through this thickness of glass.

The deformable mirror was deflected to approximately 7.5 μm and the thickness of the cover-slips was reduced to 370 μm, with the mirror/glass assembly repositioned to the location of best focus. This glass thickness change (250 μm differential) causes 1.6 μm of spherical aberration at NA = 0.65, similar to what is expected for NA = 0.85 when changing the focus from the deepest to most shallow layer (47 μm differential). That spherical aberration has a peak value of 1.3 μm, which is 3.3 waves of aberration at $\lambda = 405$ nm. Choose a value with slightly higher spherical aberration due to availability of cover-slips with only specific thicknesses. Interferograms were recorded for a variety of control voltage combinations. The control voltages were adjusted until the aberration was minimized [12].

The MEMS deformable mirror was characterized directly, using an 835 nm Michelson interferometer. The mirror deflection is calculated, and deflection curves are fit to a sixth-order polynomial with only even terms. Data were obtained for the membrane shape under the experiments with the objective lens. The relationship between the mirror displacement and the single pass optical path length in Fig. 6.35 for testing of the mirror with the objective lens is $\delta\sqrt{2}$. Figure 6.36 shows representative interferograms. The lens is well-corrected for the cover layer thickness of 620 μm of glass. Figure 6.36a, b verify that the initially flat MEMS mirror introduces no aberration into the system. Figure 6.36c shows the aberrated interferogram when the glass thickness is reduced by 250 μm to a total thickness of 370 μm. The peak spherical aberration introduced was measured to be 3.2 μm in double-pass reflection. The single-pass value is therefore in agreement with the 1.6 μm peak aberration predicted for this glass thickness using Eq. (6.6). With the MEMS mirror

(a)   **(b)**   **(c)**   **(d)**   **(e)**

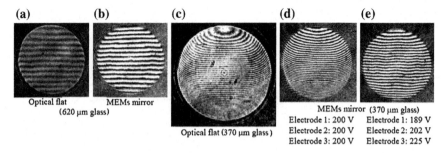

Optical flat   MEMs mirror
(620 μm glass)

MEMs mirror (370 μm glass)
Electrode 1: 200 V   Electrode 1: 189 V
Optical flat (370 μm glass)   Electrode 2: 200 V   Electrode 2: 202 V
Electrode 3: 200 V   Electrode 3: 225 V

Fig. 6.36 The representative interferograms for testing the aberrations of the objective lens

deflected to 6.7 μm at mirror center, with 200 V on all electrodes, that the peak round-trip spherical aberration is reduced by approximately one fringe as shown in Fig. 6.36d, showing that the shape of the mirror with uniform voltage on all three electrodes includes some spherical aberration that is partially compensating the system aberration. Then empirically adjust the voltages until the fringes were straight as in Fig. 6.36e. The required voltages were 189, 202, and 225 V on electrodes 1, 2, and 3, respectively. Based on fringe analysis, that finds the resultant residual single-pass spherical aberration of the system to be less than 130 nm peak to valley. The difference between the single-pass spherical aberration of the system with the optical flat is shown in Fig. 6.36c and with the MEMs mirror in Fig. 6.36e is approximately 1.6 μm. Figure 6.37 shows sixth-order curve fits for the membrane shape corresponding to the two different electrode voltage combinations used in the imaging demonstration. This data is taken along the short axis of the membrane, and plotted versus a normalized pupil radius. The quadratic coefficient of the polynomial curve fit describes a pure defocus of the system, while coefficients above second-order in the pupil coordinate indicate the amount of spherical aberration the mirrors introduce to the system. Using the polynomial coefficients determined in Fig. 6.37, the fourth- and sixth-order terms (corresponding to primary and secondary spherical aberration) are plotted in Fig. 6.38 for the two different membrane shapes. Figure 6.38a shows just the fourth- and sixth-order terms, while Fig. 6.38b shows the same data plotted with balancing defocus to better illustrate the deviation of the membrane shape from a purely parabolic profile. Using the MEMS mirror at 45° incidence shows that asymmetric aberration appears as the deflection increases. Figure 6.39 shows the system aberration with all tilt and defocus removed, with the best compensation of spherical aberration along the vertical aspect. Residual optical path difference along the horizontal aspect of the pupil is still present, which is a consequence of the loss of radial symmetry introduced by the off-axis deformable mirror. For the demonstration with approximately 7 μm membrane deflection, observe less than one fringe in double-pass reflection, can see that the single-pass aberration is less than 316 nm peak to valley.

The representative interferograms of testing the aberrations of the objective lens is shown in Fig. 6.36: (a) interferogram with an optical flat in place of the MEMS mirror and cover glass with thickness of 620 μm. (b) interferogram with

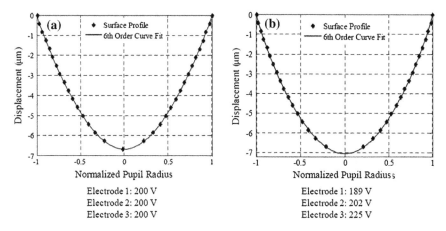

Electrode 1: 200 V
Electrode 2: 200 V
Electrode 3: 200 V

Electrode 1: 189 V
Electrode 2: 202 V
Electrode 3: 225 V

Fig. 6.37 Sixth-order even polynomial fits (—) to the measured displacements of the membrane: **a** with 200 V equipotential, a center displacement of 6.7 μm, and $a_2 = 7.72$, $a_4 = -1.85$, and $a_6 = 0.74$, **b** 189, 202, and 225 V on electrodes 1, 2, and 3, respectively, a center displacement of 7 μm, and $a_2 = 7.50$, $a_4 = -1.80$, and $a_6 = 1.23$

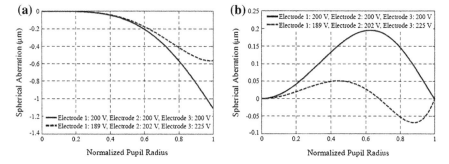

Fig. 6.38 Spherical aberration introduced by the surface shape of the membrane. **a** Plots of only, and **b** the same data with balancing defocus added incongruent melting quartz

Fig. 6.39 Asymmetric aberration evident when defocus and tilt are removed, and spherical aberration is balanced along the vertical aspect of the pupil (the short axis of the membrane)

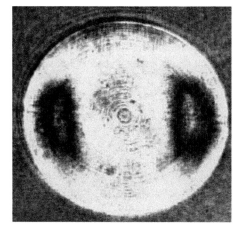

undeflected MEMS mirror and 620 μm cover glass, and both (a) and (b) show nearly flat fringes indicating the absence of spherical aberration. (c) interferogram with 370 μm thick cover glass shows approximately 3.2 μm of spherical aberration in double-pass reflection, corresponding to 1.6 μm single pass. (d) interferogram with 200 V on all three electrodes, peak mirror deflection is 6.7 μm and roundtrip aberration is reduced by approximately 1 fringe. (e) Three-zone control minimized single-pass spherical aberration to less than 130 nm peak-to-valley [13, 14].

The three-zone deformable mirror exhibits adjustable spherical aberration, introducing optical path variation with peak magnitude in excess of 1.6 μm superimposed on a defocus optical path difference of 10 μm (7 μm membrane deflection). This range of adjustment should be sufficient to correct for 1.3 μm spherical aberration introduced by the variable cover layer thickness of quadruple layer BDXL discs. Furthermore, the range of defocus, the membrane mirror can provide sufficient to address all four layers of the BDXL format. The complete design of a BDXL read/write head with integral MEMS deformable mirror would require attention to several other details. The deformable mirror causes the lens to be illuminated with a converging beam of light, while the lens was designed for a collimated illumination beam. This will introduce additional aberrations that must be considered. While, we did not have detailed information about the aspheric shape of the objective lens used in the experiments, that have simulated the aberrations of a similar 0.6 NA lens when illuminated with a converging beam, and observe the spherical aberration introduced by the lens partially that compensates the spherical aberration introduced by the change in cover layer thickness. In this case, the MEMS membrane mirror must balance the residual aberration of the system. This residual aberration in a typical BDXL system may be less than the 1.3 μm optical path difference calculated in Table 6.2, but may require compensation of higher order terms. At NA = 0.6, the best surface shape already contains significant coefficient values up to sixth order in the pupil coordinate as observed in Fig. 6.38b. It is possible that yet higher order correction may be necessary at NA = 0.85. A mirror to accomplish this balance may benefit from four or more electrode zones.

The residual asymmetric aberration is observed due to 45° off-axis illumination of the membrane lens which will represent an upper limit of the useful focus control range of this device. If larger focus shifts are necessary for future Blu-ray formats, then the membrane lens would require an electrode pattern that allows for correction of aberrations that lack axial symmetry (such as the residual aberration shown in Fig. 6.39). This could address more complex electrode structure. That showed spherical aberration correction with a 3 mm × 4.24 mm deformable membrane mirror, while having a large center displacement. Additionally, electrostatic actuation only requires connections at the electrodes for shaping of the membrane, thus allowing for easy implementation in small form factor optical systems. Therefore, conclude that the elliptical MEMS deformable membrane lens used at 45° incidence angle possesses the inherent characteristics necessary to realize an ultracompact multilayer read/write head with integral focus control and aberration correction.

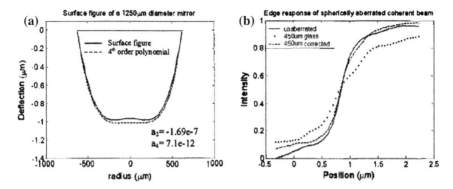

Fig. 6.40 a Surface profile of mirror measured with 170 V applied to the outer electrode (inner electrode grounded). Resulting peak deflection is 1.02 μm. **b** Measured response to an edge translated across the beam. The *solid line* corresponds to minimal spherical aberration. The *dotted line* is the edge measured with an uncompensated 450 μm thick glass layer in the beam. The *dashed line* shows an edge with the same glass layer but with compensation provided by the MEMS mirror

Fabrication of the deformable mirror has been described in detail elsewhere. The mirror shown in Fig. 6.33b consists of a circular silicon nitride membrane 1.5 μm thick and 1.25 mm in diameter, suspended over a silicon substrate with a 12 μm air gap. Two gold electrodes on the surface of the membrane, one in the center and another annular ring surrounding it, provide control over ρ^2 and ρ^4 curvature of the membrane. In these experiments it actuated only the outer annular electrode, maintaining the center electrode at ground potential, resulting in a nearly pure 1 surface profile. Optical profilometry was used to verify the surface shape, and a surface cross-section is plotted in Fig. 6.40a. With 170 volts applied to the outer electrode observed a surface profile $s(\rho) = 1.08\rho^2$ μm, leading to a wavefront modification $W(\rho) = -2.06\rho^4$ μm, or 3.2 waves of negative spherical aberration. In the first case, the beam passes through 280 μm of BK7 glass, near the design thickness of 250 μm for which this lens is optimized. No voltage is applied to the mirror. The 20–80 % edge width is 0.43 μm. In the second case, an additional 420 μm thick glass layer is inserted for a total thickness of 700, 450 μm more than what the lens is optimized for. The degraded edge response with no voltage applied to the mirror is illustrated in the second curve, with a 20–80 % edge width of 1.04 μm. In the third case, 170 volts is applied to the mirror, introducing compensating negative spherical aberration and resulting in the third curve, with a 20–80 % edge width of 0.56 μm, approximately twice as sharp as the uncorrected edge. The measurement results are shown in Fig. 6.40. Figure 6.40b plots the measured edge response under three distinct conditions. This result has been used in the multifunction two-photon multilayer storage experimental system as shown in Fig. 6.2a.

6.5.3   Correction Aberrations with Deformable Mirror of Liquid Crystal

The practical requirement that dry objective lenses must be used combined with the desire to use the highest aperture to minimize the size of the written data means that significant amounts of spherical aberration are introduced, a problem that is exacerbated as one focuses further into the recording medium. Moreover, a misalignment of the storage medium with respect to the optic axis results in the introduction of a combination of coma and astigmatism. All of these aberrations conspire to blur the focal spot, increasing the volume of the written bit, decreasing the resolution of the readout system, and effectively limiting the number of useable layers of data in the medium. Several recent developments in 3D optical data storage have used femtosecond pulsed lasers to induce multiphoton absorption effects. The nonlinear dependence of the multiphoton process on the light intensity means that the change of optical properties of the recording medium is confined to a small region. Bit data can therefore easily be written in closely spaced layers, permitting higher recording densities.

This is in contrast to single-photon phenomena, wherein the change in optical properties of the material occurs throughout the focusing cone and the resulting written bits are larger. Readout of the data is typically performed using an optical system similar to the confocal microscope. The confocal microscope is a point scanning microscope that employs a pinhole in front of the photodetector to obscure all light from the specimen except that from the focal spot. In this way, it only images a thin layer of the specimen and does not see the out-of-focus parts. As such, confocal optical systems are ideal for the readout of 3D optical memory devices. Both the recording and readout processes suffer from the effects of aberrations. It is important to note the way in which the induced aberration affects these two processes. Writing data involves only a single pass of the light, into the substrate. It has been shown that aberrations introduced here can be compensated by preshaping the light with an equal but opposite aberration, ensuring an aberration-free focal spot. The confocal microscope readout, on the other hand, involves first the illumination of the bit data, the beam passing into the substrate, and then the passage of light back out of the substrate. For readout, aberrations are introduced into both paths and therefore aberration correction is necessary in both paths. This section introduces the use of adaptive optics to overcome these problems. The techniques of adaptive optics were first developed for use in astronomical telescopes in order to compensate aberrations introduced by atmospheric turbulence. The techniques were later used in other applications such as ophthalmology and microscopy. In principle, an adaptive optics system consists of a method for measuring aberrations, an adaptive element for aberration correction and a control system. The aberration correction element would usually be a deformable mirror has been shown in Figs. 6.33 and 6.34 or as the liquid crystal device in Fig. 6.41.

This incorporated a titanium–sapphire laser, center wavelength 780 nm, pulse length 150 fs for writing data, and a helium–neon laser (633 nm) for readout. A 3D

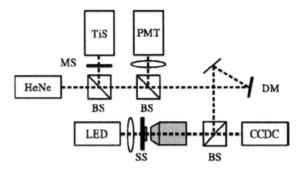

Fig. 6.41 Schematic of liquid crystal device aberration correction system: *HeNe* laser, *TiS* titanium–sapphire laser, *PMT* photomultiplier tube, *DM* deformable mirror of liquid crystal, *CCDC* CCD camera, *BS* beam splitter, *MS* mechanical shutter, *SS* scanning stage

piezostage was used for positioning and scanning of the recording medium. A green light-emitting diode was included to allow transmission images to be captured by a CCD camera. The readout signal was detected by a photomultiplier tube. Aberration correction was implemented using a membrane deformable mirror (DM). When focusing perpendicularly through a refractive index mismatch, as is the case with 3D optical memory, only spherical aberration is present. It can be represented by a rapidly convergent series of rotationally invariant Zernike polynomials. The aberration is dominated by the lowest order spherical aberration mode, and removal of this aberration mode is sufficient to improve performance of the system over significant depth. The control signals required to operate the DM in order to remove this spherical aberration mode were obtained similarly by using an interferometric method. In order to demonstrate the adaptive optics system which used multilayer recording media consisting of several photosensitive layers is separated by inert spacing layers. The 8 μm thick spacers were pressure sensitive adhesive layer, whereas the 1.5 μm thick recording layers consisted of polymethylmethacrylate doped with Trimethylindolino nitrobenzopyrylospiran that absorbs in the UV layer. The recording medium was mounted behind a 110 μm thick cover glass with a layer of glycerol for refractive index matching in order to reduce reflections from the upper surface. The unwritten recording layers were imaged in the adaptive confocal microscope using an oil immersion objective lens (NA 1.3, 40×). The amount of spherical aberration correction required for diffraction-limited imaging varies as the focusing depth is changed. The required correction can be obtained by optimizing the confocal microscope signal from a reflection of a surface in the recording medium. The presence of spherical aberration results in a reduced maximum signal. By adjusting the amount of correction, we could find the optimum setting for any particular layer. This was performed for the top and bottom layers of the recording medium. Since, the variation in spherical aberration is known to be linear in focusing depth (in homogeneous material), which could interpolate to get the optimum correction for any intermediate layer. Figure 6.42 shows images of the stack of recording layers for different aberration

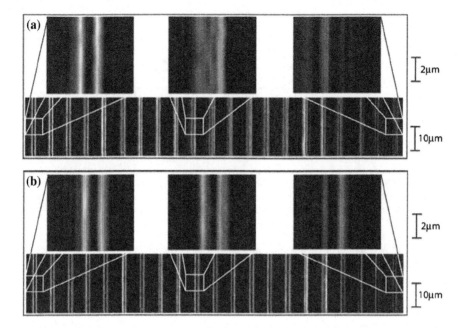

Fig. 6.42 Axial section confocal microscope scans of the whole multilayer recording medium with detail shown for the 1st, 10th, and 20th layers. The aberration correction was fixed at the level required for the top layer (**a**) and adapted for each individual layer (**b**). The optical axis is oriented down the page in these images

conditions. For each recording layer, two reflections should be visible: one from the top and one from the bottom surface of the 1.5 μm thick layer. The group of images on the left of Fig. 6.42 were obtained using the optimum aberration correction for the top layer, which is clearly imaged. The images of the other layers, situated deeper in the recording medium, are noticeably aberrated and the two reflections are no longer distinct. The three insets show enlarged detail of the 1st, 10th, and 20th layers, counting from the top of the specimen. The group of images on the right of Fig. 6.42 were obtained using varying aberration correction that was optimized for each individual layer. In this case, each layer is much more clearly imaged and for all layers the two separate reflections are clearly visible. Attribute the overall drop in reflected signal as one focuses deeper to a combination of reflective and absorptive losses and residual aberrations.

By taking advantage of nonlinear optical effects, it is possible to confine the written bit data to the focal spot of the objective lens. The data can therefore be written easily in a particular layer of the storage medium without affecting the adjacent layers. However, since aberrations reduce the focal spot intensity, the efficiency of a nonlinear process is considerably reduced. This may result in a negligible local change in optical properties and unreadable data. Although one could restore the intensity by increasing the laser power, the aberrations would cause an increase in focal spot size, particularly in the axial direction. In turn, this would lead to an increase in the required layer spacing and a corresponding reduction in storage density. The effect of aberrations by writing bitwise data is deep into the layered storage medium. The data, represented by an array of dots, were recorded in the form of voids created in the 19th layer, near the bottom of the recording medium. The femtosecond pulsed titanium–sapphire laser was used as the light source, the power at the objective lens pupil was approximately 20 mW, and the exposure time was 200 ms. The spacing between the recorded dots was 1.5 μm. When the aberration correction was turned off, set to the equivalent correction for the uppermost layer, no bits were written. These results show how depth-dependent aberrations in a multilayer optical data storage system can be corrected using adaptive optics. Benefits are obtained for both the recording and readout processes. In practical systems, dry objective lenses are required, the aberrations are larger and the optical effects would be more severe. In this case, predictive aberration correction would significantly extend the functional depth of both the recording and readout systems.

6.6   Stereo Optical Solid State Memory

Another important research project of stereo 3D optical storage devices is the optical solid state memory. The principles of two-wavelength ($\hbar v + \hbar v'$) two-photon absorption and multiwavelength and multilevel storage could be employed to dream up a new 3D storage device as 3D optical solid state memory. So OMNERC at Tsinghua University has studied 3D optical storage devices based on multiwavelength and multilevel storage in the early twenty-first century extensively.

6.6.1   Two-Wavelength ($\hbar v + \hbar v'$) Two-Photon (Type 2) Optical Solid State Memory

A configuration of two-wavelength ($\hbar v + \hbar v'$) two-photon absorption (type 2) memory cell is shown in Fig. 6.43. The laser 1 is used for writing and reading and laser 2 for writing only. The medium is two-photon (type 2) absorption material. The data was recorded when lasers $\lambda 1$ and $\lambda 2$ work at the same time, then readout by laser $\lambda 1$ with fluorescence detector. The memory array grid was connected each other with optical waveguide and binary optical switch control. Since, the optical waveguide and recording power of this memory cell are lower relatively, so there is no limit on the number of layers of the optical memory on principles. According to the current level of process technology calculate that the capacity per cubic centimeter can reach about 100 GB. But it has great development potential with progress of nanotechnology and photointegration manufacturing processing in the future.

Fig. 6.43 Stereo optical solid
state memory with
two-wavelength two-photon
absorption

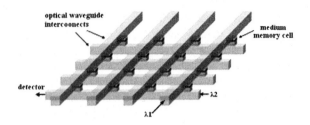

6.6.2   Photochromical Multiwavelength and Multilevel 3D
Optical Solid State Memory

Many kind of photochromic recording organic materials with wide-wavelength
range absorption spectra (400–1100 nm) have been developed and measured.
A photochromic recording organic optical solid state storage cells for absorption
bands due to the triplet excited state were observed at 400 and 1100 nm. Both
singlet and triplet excited state absorptions were also obtained. From this result, an
absorption coefficient of the singlet excited state was estimated. That also measured
absorption spectra in solution and its evaporated film. These functions of the
devices are controlled by the excited state properties of photochromic molecules in
the films. Therefore, these properties must be studied in detail to realize a high
performance. Although there are many studies to understand primary processes in
the medium, yet these problems have not been solved fully. Thus, many new
experimental techniques and equipment are required in the future. The measurement and analysis of the working parameters and relationship between medium and
light with respect to comparative magnitude, quantity, and writing in and readout
signals for a 5 wavelength recording medium was shown in Fig. 6.44. Among
them, materials of absorption peak on 405, 532, 650, and 780 nm have been
introduced in Chaps. 3 and 4. The molecular structure of one new organic photochromic fulgides for 930 nm is shown in Fig. 6.45.

Fig. 6.44 The absorption
spectrum curves of mixed
medium with sensitivity from
400–1000 nm wavelength

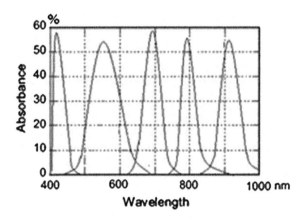

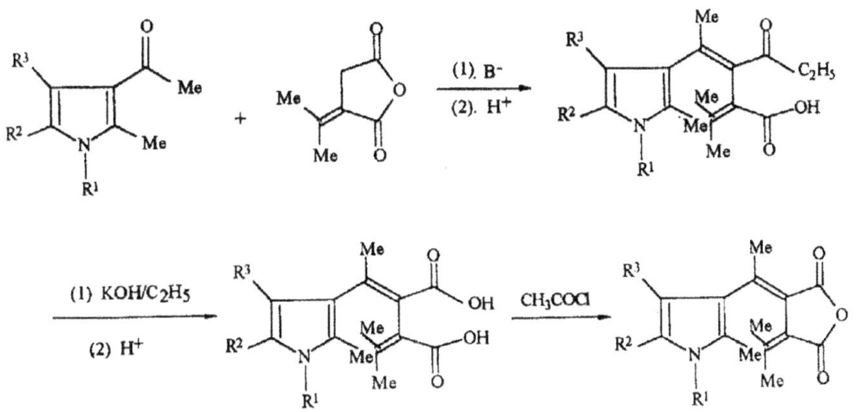

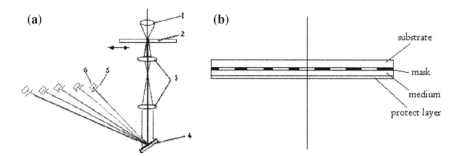

Fig. 6.45 A new organic photochromic fulgides, 3-[2-(N,N-dimethylaniline)-5-methyl-xazolemethylene]-4-isopropy -lidenetetrahydrofuran-2, 5-dione, and 3-(1,2-dimethyl-5-phenyl-pyrrolmethylene) 4-isopropylidenetetrahydrofuran-2,5-dione

Corresponding five wavelength lasers are 405 nm InGaN blue laser for violet laser Blue-D 405, 525 nm Green diodes, 650 nm AlGaInP red laser for DVD, 790 nm GaAlAs laser for compact disc drives, and 930 nm single-mode near-infrared laser diode in a diameter of 9 mm, (see Table 4.5 in Chap. 4) that can be employed for the light source of the memory. The multiwavelength memory cell experimental and testing system is shown in Fig. 6.46a. The five-wavelength laser probe light was focused on the sample which was composed of substrate, mask layer, medium, and protect layer. The mask made of chromium with thickness of 50 nm size are $1 \times 1, 2 \times 2 \ldots 5 \times 5 \ \mu m^2$ as shown in Fig. 6.46b. The storage cells on sample have exposure with ultraviolet light to ultraviolet color before experiment. The probe light of five-wavelength laser wasmodulated with higher power that introduced the medium with mixed photochromic recording organic

(a)   **(b)**

substrate

mask

medium

protect layer

Fig. 6.46 The diagram of an experiment system for testing optical storage cell of multiwavelength and multilevel (**a**): *1* multiwavelength lasers beam, *2* sample of storage cell, *3* optical system, *4* plane grating, *5* slot, *6* photodetectors array. **b** is the experiment sample of optical storage cell which was composed of substrate, mask layer, medium, and protect layer

materials to record. The exposure pulse duration was approximately 10–100 ns which is interrelated with concentration of medium in cell. That absorption spectrum curve of mixed medium for 405, 525, 650, 790, and 930 nm lasers were shown in Fig. 6.44. The materials added di(het)arylethenes and superoxochromium except that have been used in 3 wavelength storage system, but it is need of two new organic materials. The molecule structures of new organic photochromic materials: diarylethene fulgide are shown in Fig. 6.45. The mixed medium with susceptivity for 5 wavelength lasers that the crosstalk was 15–20 % in the mass, therefore it can be employed to 4 level storage at least.

In readout experiment, using lower power lasers beam irradiated cell on the sample, after passing through the cell, the light beam projected on a plan grating that was diffracted 5 spots and was detected by a photodetector array with the range of 400–950 nm. The readout single of multilevel recording experiment on the sensitivity wavelength for 405–930 nm material shows that the tiptop excited relative crosstalk of 20 %. It can be manufactured when the wavelength is in the range of 70–80 nm for different wavelength with confection in its configuration.

Thus, the absorption band observed in the near-IR wavelength range can be assigned to the absorption of the anion state and molecule. Experiment readout signals of sensitivity wavelength for 930, 790, 670, 532, and 405 nm photochromic materials as in Fig. 6.47.

Based on above experimental result to design a demonstration model of optical solid state storage device is shown in Fig. 6.48 that has brought out and demonstrated its basic functions in 2009 in OMNERC at Tsinghua University. Figure 6.48a composed of synthesizer of read/write lasers array and erasing laser (390 nm laser), plane waveguide, broadband photodetctor with high photoresponses for broad wavelengths from 400–1100 nm of light, and mixed photochromic organic recording materials layer to medium. The 6 laser beams (390, 405, 532, 650, 780 and 930 nm) was transmitted to a synthesizer with optical fibers. The 390 nm laser was used to original or erasing medium. Other 405, 532, 650, 780, and 930 nm lasers write in parallel with multilevel modulation when data is recorded to the cell. But readout signal has to be sequence by coordinated control from 400–1100 nm to take out as it is one photodetctor only. Because the readout pulse is very short (under 0.1 μs) one can hold higher data rate. According to this principle, it can be built a stereo 3D structure with optoelectronic integration and MEOMs technology. The waveguide has integral function to light with all reflective cavity. The position of recording cell can be controlled with two perpendicularity poles with each other by computer in the future.

If the multiwavelength with mico-aperture laser device will be developed its structure can be simplified as shown in Fig. 6.48b and manufactured expediently.

Advanced techniques of probing the primary process for photochromic organic materials. Many active species, such as singlet and triplet excitons and charge carriers are produced. These species are always able to absorb photons at specific wavelengths. Thus, it is possible to identify all of the active species and to follow all of the primary processes. Although absorption spectroscopy can be applied to the study on primary processes for photochromic organic materials, in reality only few

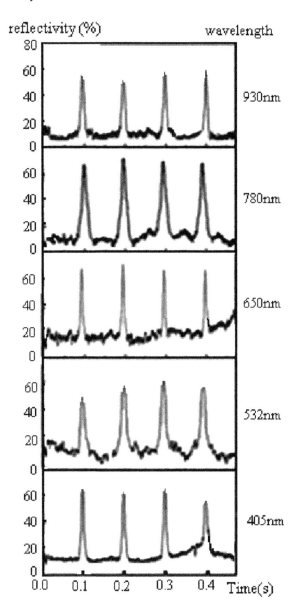

Fig. 6.47 The readout signals of 5 wavelength storage experiment with $3 \times 3 \ \mu m^2$ mask

studies have been carried out because of experimental limitations. There are mainly two serious problems, one is narrow wavelengths range and the other is high sensitivities. In a conventional absorption spectrometer, absorption spectrum in the visible wavelength range (400–700 nm) can be measured. In this wavelength range, there are many strong absorption bands due to excited molecules. However, in the near-infrared wavelength range, the absorption bands of the excited molecules

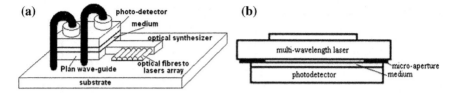

Fig. 6.48 The diagram of optical solid state storage device model

overlap with those of the ground-state molecules. Thus, it is difficult to divide an observed spectrum into the constituent absorption bands. Research discovered that the absorption is expected to appear a long wavelength range, due to charge carriers in photochromic organic materials. Meanwhile wide wavelength detection is important to test all active radiation in materials. It is very interesting that the nanodome photodetector provided remarkably high photoresponses between 300 and 1100 nm wavelength. The nanodome design is significantly effective to resolve the light-absorption limit of Si materials for short- and long-wavelength which need of multiwavelength storage [15].

For photochromic organic medium, their absorption coefficients are very high and therefore that are easily damaged by environment. Even below the threshold intensity of the damage, exciton interaction and charge carrier recombination occur very efficiently because of high-density excitation. Under this condition, it is difficult to evaluate the rate constants and yields of the primary processes in the medium and devices.

6.7 Multivalued Polarization-Sensitive Storage

Multivalued phase patterns were multiplex recorded by retardagraphy in order to improve recording density and data transfer rate. In the experiment, the phase pattern consists of four values which were recorded on a polarization-sensitive medium by focusing the recording beam, and three patterns were multiplex recorded by shifting the focal point. The recorded patterns could be independently reconstructed. Optical recording technique utilizing property of vector wave is expected for optical data storage with high density, large capacity, and high data transfer rate. The vector wave recording techniques such as retardagraphy and polarization holography have been investigated. In retardagraphy, optical information consisted of the phase retardation between two orthogonal polarization components is recorded on a polarization-sensitive medium using a single beam. The advantages of this technique are robustness for vibration and simplicity for optical system. Additionally, this technique can be regarded as a kind of in-line polarization holography, thus it is possible to the holographic multiplex recording

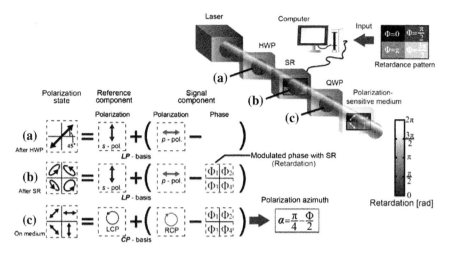

Fig. 6.49 The optical system and schematic of retardagraphy recording principles of the multivalued polarization-sensitive optical storage

case of Bragg selectivity in the case using a thick recording medium. Then, the optical information is recorded as a 3D birefringence pattern on the polarization-sensitive medium. Thus, it has good shift selectivity and can increase more crossover of an adjacent recording region. The improving method of recording density investigated only the multiplex recording by using binary pattern. The multivalued polarization-sensitive optical storage method adopted the multivalued phase pattern for multiplex recording to further improve recording density that the retardagraphy was investigated for multivalued phase data.

The optical system and recording principles of the multivalued polarization-sensitive optical storage is shown in Fig. 6.49. Where HWP is half-wave plate, SR is spatial retarder, and QWP is quarter-wave plate, respectively. The spatial phase retardation distribution Φ between p- and s-polarization components is given by using a spatial retarder. The phase of two orthogonal polarization components included in a recording beam can be independently modulated by using a spatial retarder. In retardagraphy, the phase retardation between two orthogonal polarization components as signal and reference components is modulated and recording by retardagraphy as in Fig. 6.49B. It is adequately converted into the polarization basis and recorded on polarization-sensitive medium. Most of polarization-sensitive media have sensitivity for polarization azimuth. Then, an optical anisotropy such as birefringence and dichroism are induced in the media. The principal axis of the optical anisotropy is dependent on the polarization azimuth. Thus, the phase retardation is correctly recordable by converting so that it may correspond to a polarization azimuth.

When 45° linear polarized light is illuminated onto the spatial retarder, the polarization distribution becomes elliptical and represented by:

$$U_1 = T_{SR} \begin{bmatrix} \exp(i\Phi/2) & 0 \\ 0 & \exp(-i\Phi/2) \end{bmatrix} \frac{1}{\sqrt{2}} \begin{bmatrix} 1 \\ 1 \end{bmatrix}$$

$$= \frac{T_{SR}}{\sqrt{2}} \begin{bmatrix} \cos(\pi/4) & -\sin(\pi/4) \\ \sin(\pi/4) & \cos(\pi/4) \end{bmatrix} \begin{bmatrix} \cos(-\Phi/2) \\ i\sin(-\Phi/2) \end{bmatrix} \qquad (6.8)$$

where TSR is the isotropic amplitude transmissivity of the spatial retarder. The polarization state is elliptical with the azimuth of 45° or −45°, and the ellipticity angle becomes −$\Phi/2$. Polarization-sensitive medium for polarization azimuth is sensitive to not only polarization azimuth but also to polarization ellipticity. However, the polarization-sensitive medium cannot recognize the handedness of the elliptical polarization. In order to precisely record the optical information, it is necessary to convert the polarization state corresponding to the properties of the recording medium. Using a quarter-wave plate with a fast axis of 45°, the polarization state becomes as follows:

$$U_2 = Q(\pi/4)U_1 = A_2 \begin{bmatrix} \cos(\Phi/2 - \pi/4) & \sin(\Phi/2 - \pi/4) \\ -\sin(\Phi/2 - \pi/4) & \cos(\Phi/2 - \pi/4) \end{bmatrix} \begin{bmatrix} 1 \\ 0 \end{bmatrix} \qquad (6.9)$$

where $Q(\pi/4)$ and A_2 are a Jones matrix of the quarter-wave plate with a fast axis of 45° and isotropic complex amplitude, respectively. Thus, the polarization state becomes linear polarization distribution with the azimuth of −$\Phi/2$ + $\pi/4$. When the polarization distribution is imaged onto the polarization-sensitive medium, birefringence distribution is induced and the optical information can be recorded as the principal axis of the birefringence. The polarization basis of Jones vector is linear. Here, the linear polarization basis is converted into circular polarization basis as follows:

$$U_2' = \frac{1}{\sqrt{2}} \begin{bmatrix} 1 & -i \\ i & 1 \end{bmatrix} U_2 = \frac{A_2'}{\sqrt{2}} \begin{bmatrix} \exp[i(\Phi - \pi/2)] \\ 1 \end{bmatrix} \qquad (6.10)$$

where upper and lower components are right- and left-circular components, respectively. In general, isotropic complex amplitude A_2' is spatially distributed. However, one can make A_2' homogeneous. In the special case, right- and left-circular polarization components can be regarded as a signal beam and a reference beam in polarization holography, respectively because the left-circular polarization component becomes homogeneous. In this case, it is not necessary to image the retardation pattern on the recording medium. For example, the information can be recorded by a focused beam. The isotropic amplitude is spatially distributed, whereas A_2' is homogeneous in the case of imaging on the medium. Then, the vector complex amplitude in circular basis can be written as

$$U_3 = \frac{A_3'}{\sqrt{2}} \begin{bmatrix} A' \exp(i\Phi') \\ 1 \end{bmatrix} \qquad (6.11)$$

where A_3', A', and Φ' are the distributed isotropic amplitude, the amplitude ratio to the reference component and the phase difference between the signal and the reference components, respectively. Then, the amplitude transmissivity tensor induced in the medium is expressed by

$$H = T_H R(-\Phi'/2) M R(\Phi'/2) \tag{6.12}$$

$$M = \begin{bmatrix} \exp(i\Delta\phi/2) & 0 \\ 0 & \exp(-i\Delta\phi/2) \end{bmatrix} \tag{6.13}$$

$$R(\varphi) = \begin{bmatrix} \cos\varphi & \sin\varphi \\ -\sin\varphi & \cos\varphi \end{bmatrix} \tag{6.14}$$

where T_H and $\Delta\Phi$ are isotropic amplitude transmissivity and retardance induced in the medium, respectively. The photoinduced retardance is dependent on the intensity and the ellipticity of the recording polarized beam, that is, isotropic amplitude and the amplitude ratio of right- and left-circular polarization components. In reconstruction, a left circularly polarized beam is illuminated onto the medium. Then, the complex amplitude vector is expressed by:

$$U_4 = H \frac{1}{\sqrt{2}} \begin{bmatrix} 1 \\ -i \end{bmatrix} = \frac{1}{\sqrt{2}} \left\{ \cos(\Delta\phi/2) \begin{bmatrix} 1 \\ -i \end{bmatrix} + i\sin(\Delta\phi/2)\exp(i\Phi') \begin{bmatrix} 1 \\ i \end{bmatrix} \right\} \tag{6.15}$$

The recorded retardation pattern is included in the second term of the equation. When $\sin(\Delta\Phi/2)$ is regarded to be proportional to the intensity of the recording signal component, the signal is completely reconstructed. The information in the signal component is extracted by imaging polarimetry. As shown in Fig. 6.50, when the recording pattern recorded on a polarization-sensitive medium by focusing the recording beam, the signal and reference components are changed Fourier pattern and spherical wave-like spherical wave shift multiplex in a holography, respectively. Then, the optical information is recorded as a 3D birefringence pattern on the polarization-sensitive medium. Similarly, the Bragg selectivity was formed on a recording medium. Thus, it has good shift selectivity and can increase more crossover of an adjacent recording region.

Figure 6.51 shows the experimental setup for Fourier-transform retardagraphy. In this experiment, a diode laser (406 nm) was used for recording and reconstruction. The polarization azimuth was adjusted using a half-wave plate so that the polarization state was elliptical with an azimuth of 45° or −45°. The polarizer after LC-SLM is used only in reconstruction.

The signal component included in is cording beam which was independently modulated by a parallel aligned liquid crystal spatial light modulator (LC-SLM) as spatial retarder. A recording beam was converted into linear polarization with the azimuth corresponding to the retardation by using quarter-wave plate (QWP). A phenanthrene quinone doped poly-methylmethacrylate (PQ-PMMA) film was

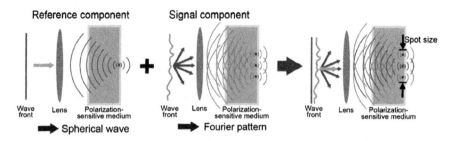

Fig. 6.50 Schematic of Fourier-transform retardagraphy: When the recording pattern recorded on a polarization-sensitive medium by focusing the recording beam, the signal, and reference components are changed Fourier pattern and spherical wave and recorded on medium

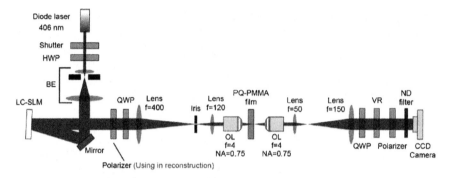

Fig. 6.51 Experimental setup for Fourier-transform retardagraphy: *BE* beam expender, *HWP* half-wave plate, *QWP* quarter-wave plate, *OL* objective lens, *LC-SLM* liquid crystal spatial light modulator, *VR* variable retarder, and *CCD* charge coupled device camera

used as a polarization-sensitive medium. The thickness was about 1 mm. The recording pattern was recorded on polarization-sensitive medium by focusing the recording beam. In reconstruction, a homogeneous image was displayed on the LC-SLM in order to obtain a homogeneous polarization pattern. The polarization state of the beam was adjusted to the polarization state of reference component in recording using a polarizer and a quarter-wave plate. The reconstructed beam transmitted through the PQ-PMMA film was analyzed by the 8-step phase-shifting method using an imaging polar metric system that consists of a QWP, a variable retarder (VR), a polarizer, and a charge-coupled device (CCD) camera. Then the recorded pattern was extracted on a computer as shown in Fig. 6.52, the images (a), (b), (c) are four-valued phase patterns on the LC-SLM. The images (d), (e), (f) extracted raw data using an imaging polarimetric system. The images (g), (h), (i) reconstructed four-valued phase retardation patterns. The images are a pattern in each recorded point. The shift value is 100 μm. The coded phase pattern consists of four phase values as the recording patterns inputted to the LC-SLM are shown in Fig. 6.52a–c. The pixel number of recording patterns was 320 × 320. The four values were every $\pi/2$ in the range of 0–2π rad, i.e., 0, $\pi/2$, $\pi/2$, and $3\pi/2$. The

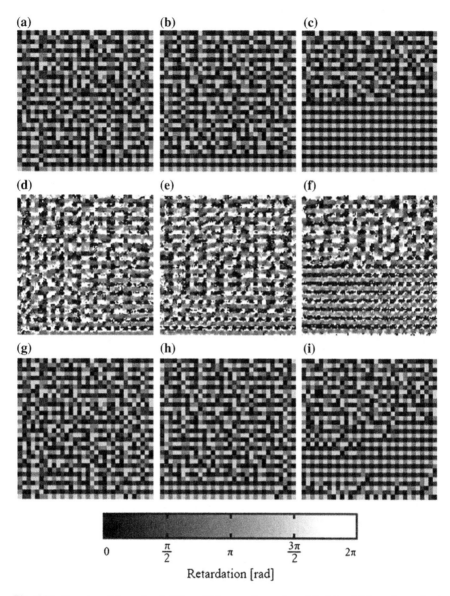

Fig. 6.52 Experimental results of shift multiple recording: where (**a**), (**b**) and (**c**) are four-valued phase patterns on the LC-SLM. (**d**), (**e**) and (**f**) are extracted raw data with a polarized system, (**g**), (**h**) and (**i**) are constructed four-valued phase retardation patterns

recording patterns were recorded on a polarization-sensitive medium by focusing the recording beam, and three patterns were multiplex recorded by shifting the focal point. The shift value and the diameter at the focal point were 100 and 200 μm, respectively. And the recording beam power was 2.05 W/cm², the exposure time was 33 ms. The reconstructed images are shown in Fig. 6.52d–i. The images (d), (e),

(f) extracted raw data using an imaging polarimetric system, and the images (g), (h), (i) reconstructed four-valued phase retardation patterns from the images (d), (e), (f), respectively. These images are a pattern in each recorded point, i.e., 0, 100, and, 200 μm, respectively.

In results, bit error rates of three patterns were obtained 12.3, 5.7, and, 13.8 %, respectively. It is predicted that these errors can be corrected by an error correction technique. As shown in Fig. 6.52, it was verified that three images consist of multivalued phase pattern which could be independently recorded and reconstructed.

Multilevel phase patterns consist of four values multiplex recorded on a polarization sensitive medium by Fourier-transform retardagraphy. The combined multiplex recording technique with the multilevel recording can effectively improve recording density and data transfer rate (DTR) [16].

6.7.1 Data Rate of Phase Multilevel Recording

Decoding performance of multilevel signals recorded using optical phase modulation was investigated by computer simulations. It was found that 4-ary phase modulated signals can be satisfactorily decoded by applying PR(1,2,1) ML provided that homodyne detector output signal-to-noise ratio (SNR) is equivalent to current optical drives. It was also found that 8-ary phase modulated signals require SNR to be 6.1–7.0 dB higher. Therefore, it can be concluded that the DTR can be more than doubled by using optical phase recording technology. The DTR in an optical disc system is determined by the disc rotation speed and the linear bit recording density. The former is physically limited and seemingly difficult to further improve significantly. The later is mainly limited by the optical conditions; however, it might be increased over a factor of two by introducing an optical phase multilevel recording technique. In this technique, multilevel symbols are encoded as phase symbols and recorded using microholograms, for example. Then their phases are readout using a homodyne detection technique. However, it is required to decode the phase modulated signals which suffer strong intersymbol interference (ISI) because the multilevel phase symbols have to be recorded with comparable symbol density with the current optical disc system.

It is readily imagined that the partial response most-likely (PRML) method will solve the ISI problem. However, negative side effects are also likely to occur due to increased numbers of the decoder inner states and branches. Therefore, read performance of an optical phase multilevel recording system has been investigated by computer simulation, and the feasibility of read DTR enhancement has been considered [17].

6.7.2 PRML for Phase Multilevel Signal Process

Figure 6.53 illustrates how ISI is reflected in phase modulated signal, while the optical spot moves from a region with phase 0 to that of phase of π. Here, phase

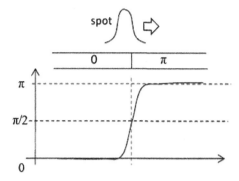

Fig. 6.53 ISI observed at phase transition

reference is the reference light used in the homodyne detector. The phase detected by using the phase-diversity homodyne detector is equivalent to the value obtained by weighted average of the phase within the optical spot. The range of the phase φ is determined as $0 \leq \varphi < 2\pi$. Naturally, the phase will pass through $\pi/2$ not $(3/2)\pi$. Figure 6.54 shows the phase plots (constellations) of 4-ary phase modulated signals (QPSK): Fig. 6.54a without ISI and Fig. 6.54b with ISI. The horizontal and the vertical axes, respectively represent amplitudes of the in-phase and quadrature-phase components of the homodyne detector outputs. Thus, the argument represents the phase. Four two-bit long symbols ('00', '01', '10', and '11') are allocated at phases of 0, $\pi/2$, π, and $(3/2)\pi$. If there is no ISI, the signal phase changes instantaneously, thus the phase plots appear as four distinct spots like in Fig. 6.54a. Contrarily, if the ISI exists, temporal phase transition becomes gradual, thus the corresponding constellation will appear like in Fig. 6.54b.

The PRML decoding technique can be extended for multilevel signal decoding in a straightforward manner. If the run-length is not limited, the number of inner states of a PRML decoder for M-ary signal with constraint length of L is increased to ML-1. The number branches exiting from each state increases to M, thus the total

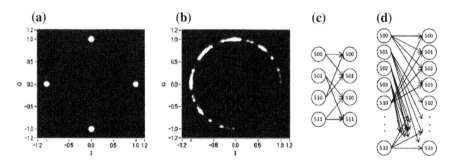

Fig. 6.54 Constellation examples for QPSK and trellis diagram: **a** without ISI, **b** with ISI, **c** trellis diagram for binary PRML decoder, and **d** trellis diagram for 4-ary PRLM decoder

number of the branches becomes ML. Shown in Fig. 6.54c, d are examples of trellis diagrams for systems with a constraint length of 3: Fig. 6.54c binary and Fig. 6.54d 4-ary signals.

Phase modulated signals used for the simulations were synthesized by convolving symbol data with an optical step response, which is obtained by an optical simulation. In this simulation, a wavelength of 405 nm and the objective numerical aperture of 0.65 were assumed. The procedure for synthesizing the signals is shown in Fig. 6.55a. First an optical step response was convolved with random 4(8)-ary symbol data, and then its sine and cosine were derived. Gaussian noises of equal power from different sources were superimposed over each of them, and they were regarded as I and Q output of a homodyne detector. The decoding procedure is illustrated in Fig. 6.55b. The phase signal is derived by calculating the instantaneous arguments from I and Q values. Then, it is equalized by using a 15-tap adaptive equalizer whose tap coefficients were obtained using the least squared error (LSE) algorithm. The target signal used for the adaptive equalization was synthesized from the same random symbol source used for the signal synthesize. The equalizer output is decoded by using the 4 or 8 leveled PRML decoder. The partial response classes used were either PR(1,2,1)ML or PR(1,2,2,1)ML [18, 19].

Simulation results and discussions in Fig. 6.55a shows the symbol error rate (SER) curves for a series of symbol lengths relative to the noise amplitude obtained for QPSK signal. The numerals in the legend represent the minimum symbol length in units of nanometers, and CBHD refers to the bit error curve obtained for HD DVD (15 GB/layer, channel bit length: 102 nm), which is meant for comparison with a binary recording system. It can be seen that the noise levels where curves cross SER of 10^{-5} are comparable to binary recording when symbol length is above 180 nm. Thus, the read DTR may be doubled when PR(1,2,1)ML is applied to a QPSK signal.

The cases for 8-PSK signal are shown in Fig. 6.55b. It is apparent that 8-PSK signals require SNR to be 6.1–7.0 dB higher to achieve SER of 10^{-5} due to the shrinkage of the Euclidian distance to the neighboring symbols. Thus, read DTR may be trebled if this noise requirement is acceptable for a homodyne detector in Fig. 6.56.

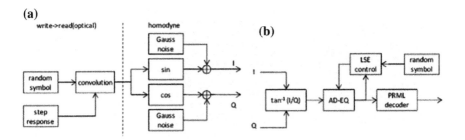

Fig. 6.55 Signal processes: **a** test signal synthesize and **b** decoding

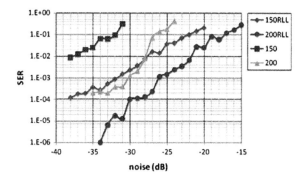

Fig. 6.56 SER curves for 8-PSK obtained by PR(1,2,2,1)ML

Figure 6.56 shows the SER curves for 8-PSK signals obtained using PR(1,2,2,1) ML, which shows higher performance against ISI when applied to binary recording. When the curves "150" or "200" are compared with corresponding results in Fig. 6.57b, extreme deterioration of the decoding performance is apparent. This is caused by the shrinkage of the Euclidian distance between the neighboring phase targets, which increases the possibility of path miss selection due to noise during the "add-compare-select" process [20, 21].

The curves denoted with "RLL" are the cases in which (1,7)RLL modulation for 8-ary symbols were used. With the (1,7) RLL modulation, the number of the states was reduced to 120 from 512, and the total number of the branches was reduced to 568 from 4096 (minimum symbol lengths were adjusted to 150 and 200 nm). The effects of these reductions are trivial, but significant SER improvements are valid at all noise amplitudes if compared with the cases without RLL. However, it is still difficult to state that PR(1,2,2,1)ML is superior to PR(1,2,1)ML, especially when considering the increased complexity of the system. Also, the physical nature of the medium should be considered in deciding whether to apply RLL modulation to the optical phase recording.

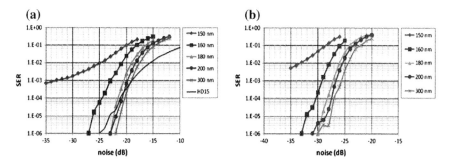

Fig. 6.57 SER curves obtained by PR(1,2,1)ML relative to noise amplitude: **a** QPSK, **b** 8-PSK

1. Read DTR may be increased by a factor of at least two compared to conventional system, if PR(1,2,1)ML decoding is applied to quadrature phase-shift keying (QPSK).
2. Read DTR may be further improved by a factor of three by 8-PSK recording and PR(1,2,1)ML if decoder input signal to noise ratio (SNR) is acceptable.
3. Longer constraint length does not necessarily lead to higher decoding performance due to an increased number of the decoder inner states [22, 23].

6.8 Multidimensional Codes

The multidimensional codes are used in several different areas to modern information coding with multidimensional modulation as in multidimensional optical storage. Trellis coding with multidimensional modulation uses multiple modulation symbols that map to a symbol of greater than two dimensions. Certain algebraic-geometry codes that are defined using projective algebraic curves over Galois fields are also referred to as multidimensional codes. Other types of codes that are multidimensional in nature are the 2D burst-identification codes. This section considers multidimensional codes in which the bits are encoded by a series of orthogonal parity checks that can be represented as coding in different dimensions of a multidimensional array. So here provide a brief introduction to product codes and some of their properties and discuss the properties of product codes constructed from single parity-check (SPC) codes, then discuss soft-decision decoding of these codes and present a class of multidimensional codes and provide performance results for two applications of these codes in multidimensional optical data storage [24–26].

6.8.1 Product Codes

Product codes were introduced by Elias in 1954 as a way to develop a code that could achieve vanishingly small error probability at a positive code rate. The scheme was used an iterative coding and decoding scheme in which each decoder improves the channel error probability for the next decoder. In order to ensure that any errors at the outputs of one decoder appear as independent error events in each codeword input to the next decoder. The encoding information bit using a series of orthogonal parity checks as an iterative coding. The coding scheme is now commonly referred to as a product code. In particular, the coding scheme is a systematic, multidimensional product code in which the number of dimensions can be chosen to achieve arbitrarily low bit error probability. Product codes are the most common form of multidimensional code. A product code of dimension p is generated in such a way that each information bit is encoded p times. Product codes are typically formed using linear block codes. Suppose that have p (not necessarily

Fig. 6.58 A two-dimensional product code

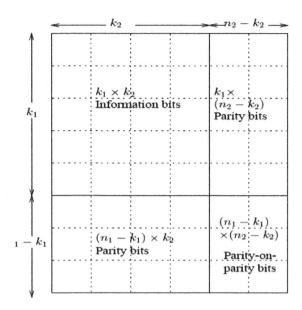

different) block codes $C_1, C_2, \ldots, C_p$ with block length $n_2, \ldots, n_p$ and information length $k_1, k_2, \ldots, k_p$. Then the p-dimensional product of these codes is a block code C with block length $n = n_1 n_2 \ldots n_p$ and information length $k_1 k_2 \ldots k_p$. The constituent codes $\{C_i\}$ are said to be subcodes of C. For a 2D product code, the code can be visualized as a rectangular array in which each column is a codeword in C_1 and each row is a codeword in C_2. Let n_i denote the block length of code C_i, and let k_i denote the number of information bits conveyed by each codeword of code C_i. That code C_i is a (n_i, k_i)-block code. Suppose that C_1 and C_2 are systematic codes. A diagram that illustrates the construction of Elias' systematic 2D product code is shown in Fig. 6.58. One possible way to encode the information is as follows:

1. Place the information bits in the $k_1 \times k_2$ submatrix.
2. For each of the first k_2 columns, calculate $n_1 - k_1$ parity bits using code C_1 and append those to that column.
3. For each of the first k_1 rows, calculate $n_2 - k_2$ parity bits using code C_2 and append those to that row.
4. For each of the last $(n_1 - k_1)$ rows, calculate $(n_2 - k_2)$ parity bits from code C_2 and append those to that row. This last set of parity bits uses the parity bits from code C_1 and encodes them with code C_2, and are thus known as parity-on-parity bits. In step 4, the parity-on-parity bits have the same values if they are instead constructed using code C_1 on the parity bits from code C_2. Product codes have many properties that are derived from their subcodes.

Some of the most commonly used block codes are the cyclic codes. If v is a codeword of a cyclic code C, then any cyclic shift of v is also a codeword of C. The following theorem for 2D product codes constructed from cyclic subcodes:

Fig. 6.59 Bit ordering to make 5 × 3 product code is a cyclic code. (a) original order, (b) order after right cyclic shift

(a)
Original order

0	10	5
6	1	11
12	7	2
3	13	8
9	4	14

(b)
Order after right cyclic shift

14	9	4
5	0	10
11	6	1
2	12	7
8	3	13

Suppose that C_1 and C_2 are cyclic codes with length n_1 and n_2, where n_1 and n_2 are relatively prime. Let $i_1 \equiv i \bmod n_1$, and let $i_2 \equiv i \bmod n_2$, for $i = 0, 1, 2 \dots$ $n_1 n_2 - 1$. Then the product of C_1 and C_2 is a cyclic code if the codeword $v = (v_0, v_1 \dots v n_1 n_2 - 1)$ is constructed such that the symbol v_i is the symbol in the (i_1, i_2)th position of the rectangular array representation. The mapping from i to (i_1, i_2) results in a cyclic enumeration of i that wraps around the edges of the $n_1 \times n_2$ rectangular array. For example, consider $n_1 = 5$ and $n_2 = 3$. Then the positions of i in the rectangular array are shown in Fig. 6.59a. Note that a right cyclic shift of the codeword v corresponds to a right cyclic shift and downward cyclic shift of the rectangular matrix, as illustrated in Fig. 6.59b. Thus, a cyclic shift of the codeword v results in another valid codeword [27, 28].

Products of more than two codes are also cyclic codes under similar constructions. Furthermore, the generator polynomial for the product code is shown to be a simple function of the generator polynomials for the subcodes. Let $g_i(X)$ be the generator polynomial for the ith subcode, and let $g(X)$ be the generator polynomial for the product code. The generator polynomial for the 2D product code is given by:

$$g(X) = \mathrm{GCD}\{g_1(X^{bn_2})g_2(X^{an_1}), X^{n_1 n_2} - 1\}, \tag{6.16}$$

where $\mathrm{GCD}(y, z)$ is the greatest common divisor of y and z, and a and b are integers satisfying $an_1 + bn_2 \equiv 1 \bmod n_1 n_2$. For example, for the 5 × 3 code of Fig. 6.49, $a = 2$ and $b = 2$ will satisfy: $(2)(5) + (2)(3) \equiv 1 \bmod 15$. These values come from the structure of the cyclic form of the product code. This is visible in Fig. 6.49, in which the separation (mod $n_1 n_2$) between neighboring positions is $an_1 = 10$ in any row and $bn_2 = 6$ in any column. Let $d_1, d_2, \dots d_p$ denote the minimum distances of subcodes $C_1, C_2 \dots C_p$, respectively. The minimum distance d for the product code C is the product of the minimum distances of the subcodes, $d = d_1 d_2 \dots d_p$. Thus, product codes offer a simple way to construct a code with a large minimum distance from a set of shorter codes with smaller minimum distances. Some results on the error-correction capability of product codes that are based on the structure of the codes. These results are not necessarily achievable with most hard-decision decoding algorithms. The random-error-correction capability, which is the maximum number of errors that a code is guaranteed to correct, is thus given by:

$$t = \left\lfloor \frac{d-1}{2} \right\rfloor = \left\lfloor \frac{\left(\Pi_{i=1}^{p} d_i\right) - 1}{2} \right\rfloor \tag{6.17}$$

Some error-control codes are able to correct more than t errors if the errors occur in bursts. A single error burst of length B_p occurs if all of the errors in the codeword are constrained to be consecutive symbols of the codeword. Cyclic product codes are particularly useful for burst error correction. Again, consider a 2D cyclic product code. Let t_1 and t_2 denote the random-error-correction capability of subcodes C_1 and C_2, respectively. Let B_1 and B_2 denote the maximum length of an error burst that is guaranteed to be corrected by subcodes C_1 and C_2, respectively. Then code C_{hi} can correct all errors that are constrained to B_i consecutive positions, regardless of the weight of the error event. The value B_i is said to be the burst-error-correction capability of subcode C_i. Let B_p denote the burst-error-correction capability of the product code. Then the B_p satisfies the following bounds:

$$B_p \geq n_1 t_2 + B_1 \tag{6.18}$$

and

$$B_p \geq n_2 t_1 + B_2. \tag{6.19}$$

Several researchers have investigated the burst-error-correction capability of product codes constructed from SPC codes. This research is discussed in the following section [29–31].

6.8.2   Products of Single Parity-Check Codes

A particular product code that has drawn considerable attention is the p-time product of SPC codes. These codes are product of SPC codes. SPC codes are $(k + 1, k)$ codes for which one parity bit is added for each k input bits. Typically, the parity bit is computed using *even parity*, in which case the sum of all of the bits in the codeword is an even number. The parity-check code is a cyclic code with generator polynomial $g(X) = X + 1$. Product codes that use SPC codes as subcodes have been called Hobbs' codes or Gilbert codes. Let C_i be a SPC code of length n_i for $1 \leq i \leq p$, and let $n = n_1 n_2 \ldots n_p$. Then the p-dimensional product of C_i, $1 \leq i \leq p$, is denoted by C and has block length n. Note that each information bit participates in exactly one parity-check equation in each dimension. The number of parity-check equations in which a bit participates is referred to as its density. Since each bit participates in exactly p checks, the parity-check matrix can be constructed to have exactly p ones in every column. Such a parity-check matrix is known as a regular low density parity-check matrix. Suppose that the 5×3 product code shown in Fig. 5.49 is constructed from even SPC codes. Let v_{ij} denote the (i, j)th code symbol in the rectangular array representation. Then, the code symbols must satisfy the row parity-check equations [32–34]

$$\sum_{j=0}^{2} v_{ij} = 0, \quad i = 0, 1, 2, 3, 4, \tag{6.20}$$

and the column parity-check equations as:

$$\sum_{i=0}^{4} v_{ij} = 0, \quad j = 0, 1, 2. \tag{6.21}$$

Thus, if the codeword v is constructed using the bit ordering described in Theorem 1, then a low-density representation for the parity-check matrix H is given by:

$$\begin{bmatrix}
1 & 0 & 0 & 0 & 0 & 1 & 0 & 0 & 0 & 0 & 1 & 0 & 0 & 0 & 0 \\
0 & 1 & 0 & 0 & 0 & 0 & 1 & 0 & 0 & 0 & 0 & 1 & 0 & 0 & 0 \\
0 & 0 & 1 & 0 & 0 & 0 & 0 & 1 & 0 & 0 & 0 & 0 & 1 & 0 & 0 \\
0 & 0 & 0 & 1 & 0 & 0 & 0 & 0 & 1 & 0 & 0 & 0 & 0 & 1 & 0 \\
0 & 0 & 0 & 0 & 1 & 0 & 0 & 0 & 0 & 1 & 0 & 0 & 0 & 0 & 1 \\
1 & 0 & 0 & 1 & 0 & 0 & 1 & 0 & 0 & 1 & 0 & 0 & 1 & 0 & 0 \\
0 & 1 & 0 & 0 & 1 & 0 & 0 & 1 & 0 & 0 & 1 & 0 & 0 & 1 & 0 \\
0 & 0 & 1 & 0 & 0 & 1 & 0 & 0 & 1 & 0 & 0 & 1 & 0 & 0 & 1
\end{bmatrix}. \tag{6.22}$$

This low-density parity-check matrix can be used in implementing a soft-decision decoder for the code. The SPC code is a cyclic code, so if $n_1, n_2, \ldots n_p$ are relatively prime, then the product code will be cyclic. For the 2D code with n_1 and n_2 relatively prime, the generator polynomial is given by:

$$g(X) = \text{LCM}\{X_1^n + 1, X_2^n + 1\} = \frac{(X^{n_1} + 1)(X^{n_2} + 1)}{X + 1}, \tag{6.23}$$

where LCM(y, z) denotes the least common multiple of y and z. For example, for the 5×3 product-SPC code, the generator polynomial is given by:

$$g(X) = \frac{(X^5 + 1)(X^3 + 1)}{X + 1} = X^7 + X^6 + X^5 + X^2 + X + 1. \tag{6.24}$$

The codes formed this way are not only cyclic, but are palindromes, in which the code is the same if each codeword is read backwards. Thus, for the 5×3 code, the *reciprocal* of $g(X)$ is:

$$X^7 g(X^{-1}) = 1 + X + X^2 + X^5 + X^6 + X^7 = g(X). \qquad (6.25)$$

A parity-check polynomial for a cyclic code can be found using the parity-check polynomial $h(X) = (X^n + 1)/g(X)$. However, the parity-check matrix formed using $h(X)$ is not usually a low-density matrix. The burst-error-correction capabilities of these codes were investigated. The burst-error-detection capabilities were investigated. A summary of some of the most important results were here. Neumann points out that some of the product-SPC codes are Fire codes, which are another class of block codes designed for burst-error-correction. Consider the maximum-length error burst that the code can correct. The first case of a 2D $n_1 \times n_2$ product code, where n_1 and n_2 are relatively prime. Let π_i denote the smallest prime divisor of n_i and define:

$$b_i = \left(\frac{\pi_i - 1}{\pi_i} \right) n_i. \qquad (6.26)$$

Then the code can correct all single error bursts up to length $B_p = \min\{b_1, b_2, b$ $(n_1 + n_2 + 2)/3\}$. Thus, for the 5×3 product-SPC code, the single-burst error-correction capability is:

$$B_p = \min \left\{ \frac{5-1}{5} \cdot 5, \frac{3-1}{3} \cdot 3, \left\lfloor \frac{3+5+2}{3} \right\rfloor \right\} = \min\{4, 2, 3\} = 2. \qquad (6.27)$$

A solid burst error is one in which every bit in the burst is received in error. Then the 2D product-SPC code can correct all solid burst errors of length min $\{n_1, n_2\}^{-1}$. If a 2D cyclic product-SPC code is used for single-burst error-detection, then its error-detection capability is easily derived. Any burst of length $n - k$ can be detected for an (n, k) cyclic code. Thus, any burst of length up to $n_1 + n_2 - 1$ can be detected for a $n_1 \times n_2$ cyclic product-SPC code. Neumann also shows that the code has the capability to simultaneously correct B_p errors while detecting burst errors of length almost equal to $\max\{n_1, n_2\}$ [35–37].

6.8.3   Decoding of Product-SPC Codes

Cyclic product codes can be decoded using a variety of hard-decision and soft-decision decoding algorithms. One of the advantages of product-SPC codes is that they are very simple. The some of the results contained errors that are corrected, and are efficient soft-decision decoding algorithms. However, some references to hard-decision decoding algorithms for these codes, particularly for application to burst-error correction. A decoding algorithm for single- and double-burst-error-correction, another threshold-decoding algorithm for product-SPC codes with multiple error bursts. A syndrome-based decoding algorithm provides better performance. Several soft-decision decoding algorithms exist for block codes. Product-SPC codes can be decoded in different ways that focus on an iterative decoding process that uses optimal maximum probability decoders on each subcode in each dimension. The resulting iterative decoding algorithm is typically not an optimal decoding algorithm but is usually significantly simpler than an optimal decoder. In order to understand the operation of this iterative decoding algorithm, first focus on the optimal symbol-by-symbol maximum likelihood decoding algorithm for binary linear block codes. The useful notion of a replica was first introduced in the context of soft-decision decoding. Consider a codeword $v = (v_0, v_1, \ldots, v_{n-1})$. A replica for bit v_i is information about bit v_i that can be derived from the code symbols other than v_i. Consider the $(7, 4)$ Hamming code. The parity-check matrix for this code can be written as:

$$\underline{H} = \begin{bmatrix} 1 & 0 & 1 & 1 & 1 & 0 & 0 \\ 0 & 1 & 0 & 1 & 1 & 1 & 0 \\ 0 & 0 & 1 & 0 & 1 & 1 & 1 \end{bmatrix}. \tag{6.28}$$

Let a codeword $v = (v_0, v_1, \ldots, v_{n-1})$. Then the parity-check equations that involve the first code symbol v_0 are:

$$\begin{aligned} v_0 + v_2 + v_3 + v_4 &= 0 \\ v_0 + v_1 + v_2 + v_5 &= 0 \\ v_0 + v_3 + v_5 + v_6 &= 0 \\ v_0 + v_1 + v_4 + v_6 &= 0 \end{aligned} \tag{6.29}$$

where all sums are modulo-2. These equations are linearly dependent, so not everyone conveys unique information. Suppose using the first three equations, which are linearly independent. Then the first symbol can be written in terms of the other six symbols in three ways as:

$$\begin{aligned} v_0 &= v_2 + v_3 + v_4, \\ v_0 &= v_1 + v_2 + v_5, \\ v_0 &= v_3 + v_5 + v_6. \end{aligned} \tag{6.30}$$

These three equations provide three algebraic replicas for v_0 in terms of the other symbols in the code. The parity-check matrix H is the generator matrix for the dual code. Thus, the replicas can be defined using code words of the dual code. Suppose that the codeword v is transmitted using binary phase-shift keying (BPSK). In the absence of noise, the symbols at the output of the demodulator can be represented by a vector $x = (x_0, x_1 \ldots x_{n-1})$, where x_i represents the binary symbol v_i. When noise is present, we denote the demodulator outputs by $y = (y_0, y_1 \ldots y_{n-1})$. Then, in terms of the demodulator outputs, there are three received replicas of v_0 in addition to y_0, which may be considered a trivial received replica. For example, for the $(7, 4)$ Hamming code described above y_2, y_3, and y_4 provide one received replica of v_0. With a random information sequence, let V_i be a random variable denoting the i_{th}

code symbol of a codeword $V = (V_0, V_1 \ldots V_{n-1})$. The codeword V is transmitted using BPSK over a memory less channel. Let X be vector that consists of the outputs of the demodulator in the absence of noise. For convenience, let $X_i = +1$ if $V_i = 0$, and let $X_i = -1$ if $V_i = 1$. The log-likelihood ratio (LLR) for X_i is defined by the:

$$L(X_i) = \log \frac{P(X_i = +1)}{P(X_i = -1)}, \tag{6.31}$$

where the natural (base e) logarithm is used. This term is called the a priori log-likelihood ratio. Considering noise, the received sequence consists of demodulator outputs $Y = (Y_0, Y_1 \ldots Y_{n-1})$. The conditional log-likelihood ratio for X_i to $Y_i = y_i$ is given by:

$$
\begin{aligned}
L(X_i|y_i) &= \log \frac{P(X_i = +1|Y_i = y_i)}{P(X_i = -1|Y_i = y_i)} \\
&= \log \frac{P(Y_i = y_i|X_i = +1)}{P(Y_i = y_i|X_i = -1)} + \log \frac{P(X_i = +1)}{P(X_i = -1)} \\
&= L(y_i|X_i) + L(X_i).
\end{aligned}
\tag{6.32}
$$

The LLRs are often referred to as "soft" values. The sign of the LLR corresponds to a hard-decision value, while the magnitude of the LLR corresponds to the reliability of the decision. A symbol-wise maximum a posteriori (MAP) probability decoder makes decision on X_i based on the larger of $P(X_i = +1|Y = y)$ and $P(X_i = -1|Y = y)$, or equivalently the sign of the:

$$L(X_i|Y = y) = \log \frac{P(X_i = +1|Y = y)}{P(X_i = -1|Y = y)} \tag{6.33}$$

In general, the log-likelihood ratio for X_i depends not only on Y_i but also other symbols in Y. This dependence corresponds to other received X_i. Hence, need to calculate the contribution of X_i to the above posteriori LLR. Define $\oplus$ as the addition operator over GF(2) with symbols $+1$ and -1, where $+1$ is the null element (since $+1$ corresponds to 0 in the original binary representation). Suppose that X_j and X_k form X_i via the relation as:

$$X_i = X_j \oplus X_k. \tag{6.34}$$

Want to find the log-likelihood ratio for the $X_j \oplus X_k$ by:

$$P(X_j = +1) = \frac{e^{L(X_j)}}{1 + e^{L(X_j)}} \tag{6.35}$$

with the relationship is:

$$P(X_j \oplus X_k = +1) = P(X_j = +1)P(X_k = +1) \\ + [1 - P(X_j = +1)][1 - P(X_k = +1)], \tag{6.36}$$

And can write as:

$$P(X_j \oplus X_k = +1) = \frac{1 + e^{L(X_j)}e^{L(X_k)}}{(1 + e^{L(X_j)})(1 + e^{L(X_k)})}. \tag{6.37}$$

Then, using $P(X_j \oplus X_k = -1) = 1 - P(X_j \oplus X_k = +1)$, the log-likelihood ratio for $X_j \oplus X_k$ can be written as:

$$L(X_j \oplus X_k) = \log \frac{1 + e^{L(X_j)}e^{L(X_k)}}{e^{L(X_j)} + e^{L(X_k)})}, \tag{6.38}$$

or equivalently:

$$L(X_j \oplus X_k) = \log \frac{[e^{L(X_j)} + 1][e^{L(X_k)} + 1] + [e^{L(X_j)} - 1][e^{L(X_k)} - 1]}{[e^{L(X_j)} + 1][e^{L(X_k)} + 1] - [e^{L(X_j)} - 1][e^{L(X_k)} - 1]}. \tag{6.39}$$

That using the relationship $\tanh(x/2) = (e^x - 1)/(e^x + 1)$, this expression can be simplified to following:

$$L(X_j \oplus X_k) = \log \frac{1 + \tanh(L(X_j)/2)\tanh(L(X_k)/2)}{1 - \tanh(L(X_j)/2)\tanh(L(X_k)/2)} \\ = 2a\tanh[\tanh(L(X_j)/2)\tanh(L(X_k)/2)] \tag{6.40}$$

IIn general, if X_i is given by $X_{j1} \oplus X_{j2} \ldots \oplus X_{jJ}$, then the log-likelihood ratio is given by:

$$L(X_j \oplus X_k) = 2a\tanh\left[\prod_{k=1}^{J} \tanh(L(X_{j_k})/2)\right] \tag{6.41}$$

For high signal-to-noise ratios, it can be approximated as:

$$L(X_j \oplus X_k) = \left[\prod_{k=1}^{j} \text{sgn}(L(X_{j_k}))\right] \cdot \min_{k=1,\ldots J} |L(X_{j_k})| \tag{6.42}$$

Thus, its reliability is determined by the smallest reliability of the symbols generally. More commonly, to determine the conditional log-likelihood ratio for a X_j given the received symbols $y_{j1}, y_{j2} \ldots y_{jJ}$. Then this conditional LLR can be written as:

$$2a \tanh \left[\prod_{k=1}^{J} \tanh(L(X_{j_k}|y_{j_k})/2) \right] \tag{6.43}$$

For the $(k + 1, k)$ even SPC code, the parity-check matrix is given by $H = [111 \ldots 1]$. Thus, there is one non-trivial (algebraic) replica for each code symbol. The algebraic replica for X_i is given by:

$$\sum_{j=0, j \neq i}^{n-1} \oplus X_j = X_0 \oplus X_1 \oplus \cdots \oplus X_{i-1} \oplus X_{i+1} \oplus \cdots \oplus X_{n-1}. \tag{6.44}$$

Thus, the conditional log-likelihood of this replica for X_i given Y is:

$$2a \tanh \left[\prod_{k=1, k \neq j}^{n-1} \tanh(L(X_k|y_k)/2) \right] \tag{6.45}$$

Considering this algebraic replica and the trivial replica, the a posteriori log-likelihood ratio $L(X_i|Y = y)$ for a code symbol X_i can be broken down into:

1. the a priori log-likelihood ratio $L(X_i)$,
2. the conditional log-likelihood ratio $L(y_i|X_i)$ of received symbol y_i given X_i, and
3. extrinsic information $L_e(X_i)$ that is information derived from the replicas of X_i.

For the SPC code, the a priori LLR for X_i is $L(X_i)$, and is set to zero initially. Consider transmission over an additive white Gaussian noise (AWGN) channel with code rate R_c and bit energy-to-noise density ratio E_b/N_0. Then the LLR for the received symbol y_i given X_i is then:

$$L(y_i|X_i) = \log \frac{\exp[-\sigma^{-2}(y_i - 1)^2]}{\exp[-\sigma^{-2}(y_i + 1)^2]} = L_c \cdot y_i. \tag{6.46}$$

where

$$L_c = \frac{2}{\sigma^2} = 4 \frac{R_c E_b}{N_0}. \tag{6.47}$$

The extrinsic information, i.e., the conditional LLR of the replica of X_i given y, is given by:

$$L_e(X_i) = 2a \tanh \left[\prod_{k=0, k \neq i}^{n-1} \tanh(L(X_k|y_k)/2) \right]$$

$$\approx \left[\prod_{k=0, k \neq i}^{n-1} \text{sgn}(L(X_k|y_k)) \right] \cdot \min_{k=0,\ldots n-1, k \neq i} |L(X_k|y_k)|. \tag{6.48}$$

Thus, the a posteriori log-likelihood ratio for X_i is given by:

$$L(X_i|\underline{Y}) = \underline{y}) = L(X_i) + L_c \cdot y_i + L_e(X_i). \tag{6.49}$$

The extrinsic information represents indirect information about the symbol X_i from other code symbols. In the context of product codes, due to the orthogonal parity-check construction, the extrinsic information about X_i that is derived from a particular component subcode does not involve any received symbols (other than X_i) of any other subcode that involves X_i. Thus, this extrinsic information represents information that is not directly available to the decoders of the other subcodes. The extrinsic information can be used by other subcodes by exchanging extrinsic information between the subcodes in an iterative fashion. Each decoder treats the extrinsic information generated by other decoders as if it were a priori information in its decoding [38, 39].

To illustrate this iterative decoding algorithm, consider a 2D product-SPC code that is formed from the product of two identical SPC codes. Conforming to the usual terminology in the iterative decoding literature, refer to the conditional LLR of X_i given y_i, $L(X_i|Y_i)$, as the soft input to a decoder of a SPC subcode. As previously discussed, this soft input is simply the sum of the a priori LLR of X_i, $L(X_i)$, and the conditional LLR of the received symbol y_i given X_i, L_{cyi}. The decoder generates the extrinsic information of X_i, $L_e(X_i)$, based on the soft inputs as described above. This extrinsic information is then used as the a priori LLR X_i for the decoding of the other component subcode. Extrinsic information may be calculated for only the systematic symbols or for all of the code symbols, which may offer some improvement in performance. For the results below, calculate extrinsic information for all of the code symbols. The decoding process alternates between the decoders for the two subcodes until a stopping criterion is met. Then the decoder outputs the a posteriori LLR of X_i, which is simply the sum of the soft input and the extrinsic information from each decoder for X_i. A hard decision on X_i is made based on the sign of this *soft output*. For example, consider the example illustrated in Fig. 6.60. The transmitted symbols are shown in Fig. 6.60a. The rightmost column and the bottom row contain the parity-check symbols and the other symbols carry information. The LLRs of the received symbols are given in Fig. 6.60b.

The decoder would make five errors (indicated by the shaded symbols) if decisions were made directly using these received values. Figure 6.60c shows the decoding results of the a posteriori decoding algorithm on the SPC defined along the rows. The number inside the upper triangle for each symbol denotes the soft input for that symbol, while the number inside the lower triangle is the extrinsic information generated by the decoder. Initially, we assume that the a priori LLRs of all the symbols are zero. For instance, the soft input for the symbol in the upper left-hand corner is given by $2 + 0 = 2$. To obtain the soft output for a symbol, simply need to add the values inside the upper and lower triangles corresponding to that symbol. For instance, the soft output of the symbol in the upper left-hand corner is $2.0 - 0.5 = 1.5$. We can see that three errors would result if the decoder were to make decisions based on the soft output after this decoder iteration. The decoding

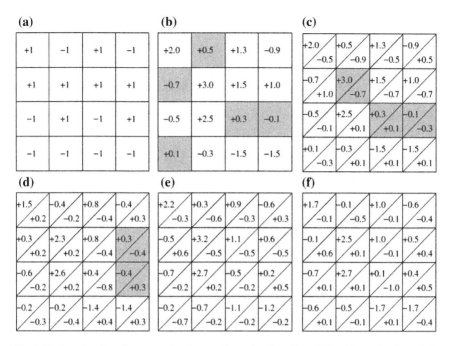

Fig. 6.60 Iterative decoding example of a two-dimensional product-SPC. **a** Transmitted symbols. **b** LLRs of received symbols. **c** 1st iteration horizontal SISO decoding. **d** 1st iteration vertical SISO decoding. **e** 2nd iteration horizontal SISO decoding. **f** 2nd iteration vertical SISO decoding

process continues for the SPC defined along the columns. The results are shown in Fig. 6.60d. The soft input of a symbol is now obtained by adding the received LLR of that symbol (from Fig. 6.60b) to the extrinsic information bit participating in parity check bit Legend information generated in the previous decoding process (see Fig. 6.60c). For instance, we obtain the soft input of the symbol in the upper left-hand corner as 2.0 − 0.5 = 1.5. Two errors would result if hard decisions were made at this time. The decoding process then returns to the decoder for the SPC along the rows (Fig. 6.50e) and then the SPC along the columns (Fig. 6.60e). All five errors that were initially present are corrected by this iterative decoding process. It is also easy to see that any further decoder iterations will not result in any changes in the hard decisions. So for this example, the decoding process converges. Interested readers are referred for additional discussion of the performance of iterative decoding with multidimensional product-SPC codes [40].

6.8.4 Multidimensional Parity-Check Codes

In this paragraph, define a class of multidimensional parity-check (MDPC) codes. These codes are punctured versions of the M-dimensional product-SPC codes

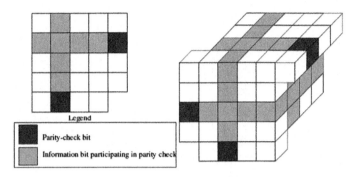

Fig. 6.61 Determination of parity-check bits for $M = 2$ and $M = 3$ MDPC codes

discussed above. In particular, the M-dimensional product-SPC codes have code rates that decrease as M is increased. By puncturing the majority of the parity-check bits for $M > 2$, the MDPC codes have code rates that increase as M increases. The parity bits for the MDPC code are determined by placing the information bits into a multidimensional array of size M, where $M > 1$. For a particular value of M, refer to the M-dimensional parity-check code as an M-DPC code.

For the special case of $M = 2$, this code is also referred to as a rectangular parity-check code (RPCC). In all that follows, we assume that the size of the array is the same in each dimension. This is not a requirement in general, but it does minimize the redundancy (and hence the rate penalty) for a given block size and number of dimensions. Suppose N is the block length (the number of input bits that are input to the code to create a codeword), where $N = DM$. Then D is the size of the M-dimensional array in each dimension. Each M-DPC code is a systematic code in which a codeword consists of DM information bits and MD parity bits. Then D parity bits are computed for each dimension, where the parity bits can be constructed as the even parity over each of the D hyper planes of size DM − 1 that are indexed by that dimension. This is illustrated in Fig. 6.61 for $M = 2$ and $M = 3$.

More formally, let $u_{i1,i2,\ldots,iM}$ be a block of data bits indexed by the set of M indices $i_1, i_2, \ldots, i_M$. The data bits are arranged in the lattice points of an M-dimensional hypercube of side D. Then the MD parity bits satisfy as:

$$p_{m,j} = \sum_{i_1} \cdots \sum_{i_{m-1}} \sum_{i_{m+1}} \cdots \sum_{i_M} u_{i_1,\ldots,i_{m-1},j,i_{m+1}\ldots,i_M} \tag{6.50}$$

for $m = 1, 2, \ldots, M$ and $j = 1, 2, \ldots, D$. Each sum above ranges over D elements, and modulo-2 addition is assumed. Since, the M-DPC code produces MD parity bits, the code rate is $D^M/(D^M + MD)$, or equivalently $(1 + MD^{1-M})^{-1}$. Clearly, as D is increased, the rate of the code becomes very high. For most values of N and M that are of interest, the rate of the M-DPC code increases as M is increased. The minimum code weight is $\min(M + 1, 4)$.

6.8.5 Burst-Error-Correction Capability of MDPCs with Iterative Soft-Decision Decoding

MDPC codes can achieve close-to-capacity performance with a simple iterative decoding algorithm in addition white Gaussian noise as well as bursty channels. The iterative decoding algorithm described can be employed with a slight modification. Since, the on-parity bits in the product-SPC code are punctured in the MDPC code, do not update the soft inputs corresponding to the parity bits in the iterative decoding process of the MDPC code. Using iterative decoding and with a minimal amount of added redundancy, MDPC codes are very effective for relatively benign channels with possibly occasional long bursts of errors. For example, a 3D parity-check code with a block size of 60,000 can achieve a bit error rate (BER) of 10^{-5} within 0.5 dB of the capacity limit in an AWGN channel while requiring only 0.2 % added redundancy. The same code can get to within 1.25 dB of the capacity limit in a bursty channel.

6.8.6 White Gaussian Noise Channel

Simulation results of a number of MDPC codes with block lengths of about 1000, 10,000, and 60,000 data bits over an AWGN channel with BPSK modulation are summarized in Table 6.4. The results are obtained after 10 iterations for all the codes. However, the decoding process essentially converges after 5 iterations for all of the MDPC codes that were considered. For instance, the convergence of the iterative decoding process for the 100^2 code is shown in Fig. 6.62. From Table 6.4, we conclude that with a block size of 1000 bits, the MDPC codes can achieve a BER of 10^{-5} within 2 dB of the capacity limit. When the block size increases to 10,000 bits, the performance of the MDPC codes is within 1 dB of the capacity.

Table 6.4 Performance of MDPC codes over AWGN channel

Code	Block size	Cade rate	E_b/N_0 at 10^{-5} BER (dB)	Coding gain at 10^{-5} BER (% of possible coding gain)	E_b/N_0 at capacity (dB)
32^2	1024	0.9412	6.3	3.3 dB (55.0 %)	3.7
10^3	1000	0.9709	6.75	2.85 dB(63.1 %)	4.75
4^5	1024	0.9808	7.25	2.35 dB (63.8 %)	5.3
100^2	10,000	0.9804	6.75	2.85 dB (70.8 %)	5.25
21^3	9261	0.9932	7.3	2.3 dB (80.4 %)	6.35
10^4	10,000	0.9960	7.75	1.85 dB(80.4 %)	6.8
245^2	60,025	0.9919	7.2	2.4 dB (79.4 %)	6.2
39^3	59,319	0.9980	7.9	1.7 dB (89.1 %)	7.4
9^5	59,049	0.9992	8.6	1.0 dB (88.1 %)	8.05

Fig. 6.62 Convergence of iterative decoding process for 100^2 code over AWGN channel

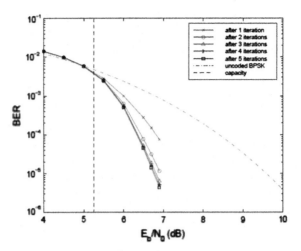

These results are comparable to the ones, in which codes based on pseudo-random bipartite graphs obtained from computer searches are employed. In comparison, the MDPC codes considered here have much more regular structures, faster convergence rates, and a simpler decoding algorithm. With a block size of approximately 60,000 bits, the 3DPC code 39^3 can achieve a BER of 10^{-5} at 7.9, 0.5 dB higher than the capacity limit. Moreover, significant coding gains over uncoded BPSK systems are achieved with very small percentages of added redundancy. For example, using the 21^3 code, a coding gain of 2.3 dB is obtained at 10^{-5} with less than 0.7 % redundancy. This accounts for 80.4 % of the maximum possible coding gain of 3.25 dB that is allowed by the capacity. It appears that the 3DPC codes are most efficient in terms of attaining the highest percentage of the maximum possible coding gain. Using the union bound technique, we can obtain an upper bound on the bit error probability of the MDPC codes with maximum likelihood (ML) decoding as follows:

$$p_b \leq \sum_{i=1}^{A^M} \frac{i}{A^M} \sum_{d=i}^{(M+1)i} W_{i,d} Q\left(\sqrt{\frac{2dE_b/N_0}{1+M/A^{M-1}}}\right), \tag{6.51}$$

where $W_{i,d}$ is the number of code words with information weight i and codeword weight d. Figure 6.63 shows the union bounds 3 obtained using Eq. (6.51) for the 32^2, 10^3, and 4^5 codes. Also shown in Fig. 6.63 are the bit error probabilities of these 3 codes obtained by the iterative decoder from simulations. From the figure that the BER performance obtained from simulations for the code 32^2 is very close to the corresponding union bound in the high E_b/N_0 region. Symmetric capacity restricted to BPSK is assumed here, the weight enumerator coefficients $W_{i,d}$ are obtained approximately by the Monte Carlo method. The first 30 terms in Eq. (6.51) are used to approximate the union bound [41].

Fig. 6.63 Union bounds and BER performance from simulations of 32^2, 10^3, and 4^5 codes over AWGN channel

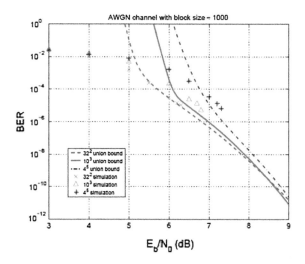

Table 6.5 Performance of MDPC codes over bursty channel

Code	Block size	Code rate	E_b/N_0 at 10^{-5} BER (dB)	Coding gain at 10^{-5} BER (% of possible coding gain)	E_b/N_0 at capacity (dB)
100^2	10,000	0.9804	13.7	4.15 dB (29.5 %)	8.4
21^3	9261	0.9932	14.8	3.05 dB (52.5 %)	12.0
10^4	10,000	0.9960	15.5	2.35 dB (57.5 %)	13.1
245^2	60,025	0.9919	14.1	3.75 dB (55.0 %)	11.5
39^3	59,319	0.9980	15.45	2.4 dB (75.0 %)	14.2
9^5	59,049	0.9992	16.6	1.25 dB (75.9 %)	15.4

This indicates that the performance of the iterative decoder is close to that of the ML decoder. For the 3- and 5-D codes 10^3 and 4^5, the BERs obtained from simulations are poorer than the ones predicted by the respective union bounds. This implies that the iterative decoder becomes less effective as the dimension of the MDPC code increases. Nevertheless, iterative decoding can still provide good coding gains, as shown in Table 6.5, for MDPC codes with more than two dimensions.

6.8.7 Bursty Channels

Although the MDPC codes have small minimum distances, they can correct a large number of error patterns of larger weights because of their geometric constructions. With suitable interleaving schemes, the MDPC codes are effective for channels with occasional noise bursts. To examine this claim, employ the simple two-state hidden Markov model, as shown in Fig. 6.64, to model bursty channels.

Fig. 6.64 Hidden Markov
model for bursty channel

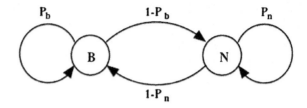

The system enters State **B** when the channel is having a noise burst. In State **N**, usual AWGN is the only noise. In State **B**, the burst noise is modeled as AWGN with a power spectral density that is B times higher than that of the AWGN in State **N**. It is easy to check that the stationary distribution of the hidden Markov model is $\pi_b = 1 - P_n/1 - P_b + 1 - P_n$ and $\pi_n = 1 - P_b/1 - Pb + 1 - Pn$. Assuming that the noise is independent from symbol to symbol after the deinterleaver at the receiver, the conditional LLR $L(y_i|X_i)$ is given by:

$$L(y_i|X_i) = 4R_c\frac{E_b}{N_0}\frac{x}{B} + \log\frac{\exp\left(-\frac{B-1}{B}R_c\frac{E_b}{N_0}(y_i - 1)^2\right) + \frac{\pi_b}{\pi_n}}{\exp\left(-\frac{B-1}{B}R_c\frac{E_b}{N_0}(y_i + 1)^2\right) + \frac{\pi_b}{\pi_n}}. \qquad (6.52)$$

Simulation results of a number of MDPC codes with block sizes of 10,000 and 60,000 bits are summarized in Table 6.4. In the simulation, $P_b = 0.99$, $P_n = 0.9995$, and $B = 10$ dB. This represents a case that long noise bursts occur occasionally. Random interleavers of sizes equal to the block size of the MDPC codes are employed. The results are obtained after 10 iterations for all the codes. The convergence of the iterative decoding process is similar to the AWGN case. From Table 6.4, with a block size of 10,000 bits, the 4DPC code 10^4 can achieve a BER of 10^{-5} within 2.4 dB of the capacity limit 4. When the block size is increased to 60,000 bits, the 5DPC code 95 can achieve a BER of 10^{-5} within 1.2 dB of the capacity limit. Although the MDPC codes are not as effective in bursty channels as in AWGN channels, they do provide very significant coding gains with very reasonable complexity as no channel state estimation is needed. The capacity is obtained by averaging the symmetric capacities under the normal and bursty states based on the stationary distribution of the hidden Markov model. This corresponds to the case that a perfect interleaver is employed so that for a given bit, the channel state is independent of the channel states of the other bits, and perfect channel state information is available at the receiver.

In fact, the complicated conditional likelihood-ratio calculation in Eq. 6.52 is not needed, since simulation results show that the degradation on the BER performance is very small if the second term on the right-hand side of Eq. 6.62 is neglected. Using the union bound and assuming a perfect interleaver is employed so that channel state changes independently from bit-to-bit, can obtain the following upper bound on the BER for an ML decoder with perfect channel state information:

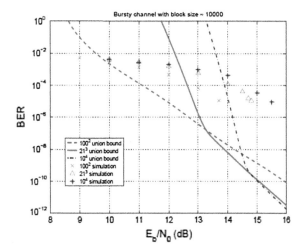

Fig. 6.65 Union bounds and BER performance from simulations of 100^2, 21^3, and 10^4 codes over bursty channel with $P_b = 0.99$, $P_n = 0.9995$, and $B = 10$ dB

$$P_b \leq \sum_{i=1}^{A^M} \frac{i}{A^M} \sum_{d=i}^{(M+1)i} W_{i,d} \sum_{k=0}^{d} \binom{d}{k} \pi_b^k \pi_n^{d-k} Q\left(\sqrt{\frac{2E_b/N_0}{1+M/A^{M-1}}} \cdot \frac{d-k+k/\sqrt{B}}{\sqrt{d}}\right)$$

(6.53)

Figure 6.65 shows the union bounds obtained by Eq. (6.53) for the 100^2, 21^3, and 10^4 codes. Also shown in Fig. 6.65 are the bit error probabilities of these 3 codes obtained by the iterative decoder from simulations. From the figure, the BERs obtained from simulations are poorer than those predicted by the union bounds. The reason is 3-fold. First, the iterative decoder only approximates the MAP decoder. The second, and perhaps the most important, reason is that random interleavers are of the same size as the codewords. These interleavers are not good approximations to the perfect interleaver assumed in the union bound, since such an interleaver would require interleaving across multiple codewords. Third, no channel state information is assumed at the receiver. Nevertheless, the simple iterative decoder and the imperfect interleaver can still give large coding gains as shown in Table 6.4.

6.8.8 Concatenated MDPC Codes and Turbo Codes

Turbo codes are parallel-concatenated convolutional codes that have been shown to provide performance near the capacity limit when very large interleavers (and thus codeword lengths) are used. These codes suffer from an error floor that limits their performance for shorter block lengths. The error floor is caused by error events that have very low information weight. Thus, these low-weight error events can be corrected by even a simple outer code. Several authors have investigated the use of an outer code to deal with these low-weight errors. BCH codes were considered in, and Reed-Solomon codes were considered. However, these codes are typically

Fig. 6.66 Bounds on the
performance of concatenated
outer multidimensional
parity-check codes with inner
rate 1/3 turbo codes

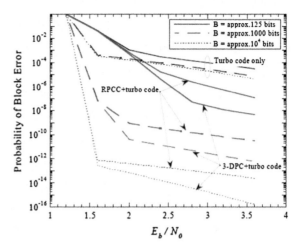

decoded with algebraic decoders because the complexity is too high for
soft-decision decoders for these codes. An alternative approach is to use the MDPC
codes that are discussed in the previous two sections as outer codes with a turbo
inner code. The MDPC codes have simple soft-decision decoders and typically
have very high rates that result in less degradation of the performance of the turbo
code than most of the other outer codes that have been used. Furthermore, the
MDPC codes are good at correcting bursts of errors, such as those that occur at the
output of a turbo decoder.

The results in Fig. 6.66 illustrate the potential of these codes with regard to
improving the error floor. The turbo code used by itself and in concatenation with
MDPC codes is the rate 1/3 turbo code with the constituent codes specified in the
standards for the CDMA and WCDMA third-generation cellular systems. The

Fig. 6.67 MDPC + turbo
codes and turbo codes with
block length of approximately
10^4 bits and rates
approximately equal to 1/3

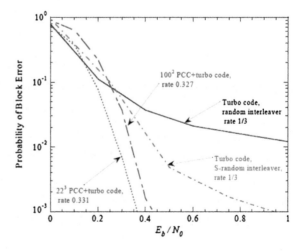

results indicate that the concatenated MDPC codes can reduce the error floor by many orders of magnitude in comparison to a turbo code by itself. These results have been verified by simulation, as illustrated by the results in Fig. 6.67. An alternate technique to improve the error floor is the use of S-random interleaving. The simulation results in Fig. 6.67 shows that the concatenated MDPC and turbo codes provide significantly better performance than turbo codes, even when S-random interleaving is used.

References

1. Q. Shen, D. Xu, Analysis of the effects of disk tilt on the differential-phase-detection signal in a high-density DVD read-only disk driver. Appl. Optic. **45**(17), 3998–4004 (2006)
2. J. Thévenin, M. Romanelli, M. Vallet, M. Brunel, T. Erneux, Resonance assisted synchronization of coupled oscillators: frequency locking without phase locking. Phys. Rev. Lett. **107**, 104101 (2011)
3. H. Hu, D. Xu, L. Pan, Modulation code and PRML detection for multi-level run-length-limited DVD channels. In *2006 Optical Data Storage Topical Meeting*, pp 112–114, 2006
4. E. Chow, S.Y. Lin, S.G. Johnson, P.R. Villeneuve, J.D. Joannopoulos et al., Three-dimensional control of light in a two-dimensional photonic crystal slab. Nature **407**, 983–986 (2000)
5. D. Xu, J. Liu, H. Wang et al., Multilevel read-only optical disk and method for producing the same, Patent No. US 7,680,024 B2, 2010
6. S. Aoki, M. Yamada, T. Yamagami, Development of deformable mirror for spherical aberration compensation. Proc. SPIE Optic. Data Storage 750513-6 (2009)
7. S. Aoki, M. Yamada, T. Yamagini, A novel deformable mirror for spherical aberration compensation. Jpn. J. Appl. Phys. **48**, 03A003 (2009)
8. S.J. Lukes, D.L. Dickensheets, SU-8 focus control mirrors released by XeF_2 dry etch. In *Proceedings of SPIEMOEMS and Miniaturized Systems X*, 793006-6, 2011
9. S. Kawata, Y. Kawata, Three-dimensional optical data storage using photochromic materials. Chem. Rev. **100**, 1777 (2000)
10. D. Yan, M. Wei, *Photofunctional Layered Materials*. Springer Copyright. ISSN 0081-5993 (Springer, Cham, 2015)
11. C.E. Olson, Three-dimensional fluorescent optical data storage in molecular glasses and highly crosslinked polymers, S.l. Boston College, 2003
12. S.J. Lukes, Surface micro-machined SU-8 2002 deformable membrane mirrors, M.S. Thesis, Department of Electrical Engineering, Montana State University, Bozeman, MT, 2011
13. Inphase Technologies, Inc. (Longmont, CO, US) and Nintendo Co., Ltd. (Kyoto, JP) Miniature Flexure Based Scanners For Angle Multiplexing Patent, 2008
14. Zhou et al., Waveguide self-coupling based reconfigurable resonance structure for optical filtering and delay. Opt. Exp. **19**, 8032 (2011)
15. D. Xu, N. Yi, L. Pan, K. Chen, J. Xiong, D. Lu, H. Wu, J. Ma, J. Pei, Multi-level read-only master disk, Tsinghua University, CN200610167683.5 (2007)
16. E. Togan, Y. Chu, A.S. Trifonov, L. Jiang, J. Maze, L. Childress, M.V.G. Dutt, A.S. Sørensen, P.R. Hemmer, A.S. Zibrov, M.D. Lukin, Nature(London) **466**, 730 (2010)
17. P. Tamarat, N.B. Manson, J.P. Harrison, R.L. McMurtrie, A. Nizovtsev, C. Santori, R.G. Beausoleil, P. Neumann, T. Gaebel, F. Jelezko, P. Hemmer, J. Wrachtrup, New J. Phys. **10**, 045004 (2008)
18. A.V. Gorshkov, A. André, M.D. Lukin, A.S. Sørensen, Phys. Rev. A **76**, 033804 (2007)
19. D.E. Pansatiankul, A.A. Sawchuk, Multidimensional modulation codes and error correction for page-oriented optical data storage. SPIE **4342**, 393 (2002)

20. J. Cai, W. Huang, Two-photon three-dimensional optical storage of a new pyrimidine photobleaching material. Optik-Int. J. Light Electron Optic. **126**(3), 343–346 (2015)
21. K. Heshami, A. Green, Y. Han, A. Rispe, E. Saglamyurek, N. Sinclair, W. Tittel, C. Simon, Phys. Rev. A **86**, 013813 (2012)
22. S. Felton, A.M. Edmonds, M.E. Newton, P.M. Martineau, D. Fisher, D.J. Twitchen, J.M. Baker, Phys. Rev. B **79**, 075203 (2009)
23. G. Heinze, A. Rudolf, F. Beil, T. Halfmann, Phys. Rev. A **81**, 011401(R) (2010)
24. Z. Li, H. Zhang, J. Shao, J. Appl. Optic. **30**(3), 442–447 (2009)
25. H. Yuan, D. Xu, Q. Zhang, J. Song, Dynamic model of mastering for multilevel run-length limited read-only disc. Optic. Express **15**(7), 4176–4181 (2007)
26. H. Yuan, D. Xu, H. Xu, J. Pei, Timing recovery method for multilevel run-length-limited read only disc. Jpn. J. Appl. Phys. **46**(9A), 5845–5848 (2007)
27. H. Yuan, H. Xu, L. Pan, D. Xu, Read channel for multilevel run-length-limited read-only disc. Jpn. J. Appl. Phys. **47**(7), 5859–5862 (2008)
28. H. Hu, H. Yuan, Y. Tang, L. Pan, New rate 6/9 run-length limited (2, 11) code with spaced pits/lands constraint for four-level read-only optical disc. Jpn. J .Appl. Phys. **47**(7), 5867–5869 (2008)
29. Y. Ni, W. Xiang, H. Yuan, L. Pan, C. Su, H. Wang, Improved mastering material for multilevel blue laser disc. Optic. Express **15**(20), 13244 (2007)
30. J. Pei, H. Hu, L. Pan, Q. Shen, H. Hu, D. Xu, Constrained code and partial-response maximum-likelihood detection for high density multi-level optical recording channels. Jpn. J. Appl. Phys. **46**(6B), 3771–3774 (2007)
31. E.P. Walker, J. Duparre, H. Zhang, W. Feng, Y. Zhang, A.S. Dvornikov, Spherical aberration correction for 2-photon recorded monolithic muiltilayer optical data storage, ODS 2001 Proc. SPIE. (2001)
32. H. Heng, X. Duanyi, 3-ary (2,10) run-length limited code for optical storage channels. Elec. Lett. **41**(17), 972–973 (2005)
33. E.D. Walker, W. Feng, Y. Zhang, H. Zhang, F. McCormick, S. Esener, 3-D parallel readout in a 3-D multilayer optical data storage system, ISOM/ODS meeting paper # TuB 4 (2002)
34. H. Hua, L. Pan, J. Xiong, Y. Ni, New efficient run-length limited code for multilevel read-only optical disc. Jpn. J. Appl. Phys. **46**(6B), 3782–3786 (2007)
35. J. Song, D.Y. Xu, G.S. Qi, H. Hu, Q.C. Zhang, J.P. Xiong, Multilevel read-only optical recording methods. Chin. Phys. **15**, 1788 (2006)
36. Y. Tang, J. Pei, L.F. Pan, Y. Ni, H. Hu, B.Q. Zhang, Experiments of multi-level read-only recording using readout signal wave-shape modulation. Chin. Phys. Lett. **5**(25), 1709 (2008)
37. W. Jia, Y. Luo, Yu. Jian, B. Liu, Effects of high-repetition-rate femtosecond laser micromachining on the physical and chemical properties of polylactide (PLA). Opt. Express **23**(21), 26932–26939 (2015)
38. Y. Tang, J. Pei, Y. Ni, L.F. Pan, H. Hu, B.Q. Zhang, Multi-level read-only recording using signal waveform modulation. Opt. Express **16**, 6156–6162 (2008)
39. Q.H. Shen, J. Pei, H.Z. Xu, L. Wang, D.Y. Xu, Multi-level read-only recording using signal waveform modulation. Jpn. J. Appl. Phys. **45**, 5764 (2006)
40. Q-H. Shen, J. Pei, D-Y. Xu, J-S. Ma, Analysis of the focus error characteristic in high density optical disk drive. Acta Physica Sinica **55**(8), 4132–4138 (2006)
41. G.A. Kaddoum, High data rate and energy efficient communication system. IEEE Commun. Lett. **19**(2), 175–178 (2015)

Chapter 7
Volume Holography and Dynamic Static Speckle Multiplexing

The hologram was invented by 1971 Physics Nobel laureate Dr. Dennis Gabor in 1948 at Imperial College (England) to solve an aberration problem in electron microscope images. The technology was practiced with the availability of the He–Ne gas laser and the invention of the off-axis (Fresnel) hologram by Emmett N. Leith and Uris Upatnieks at the University of Michigan in 1962. Finally, the concept of the 3D holomem (holography memory) was proposed by P.J. van Heerden, a research scientist at Polaroid in 1963. Van Heerden showed that a theoretical maximum volume storage density of $\rho_v \sim 1/\lambda^3$ was possible for binary data (for example $\lambda = 500$ nm, this calculates to 8 bits/μm^3, or 131 Tb/in^3). The corresponding theoretical areal density ρ_A is $\sim 8N$ bits/μm^2 ($\sim 5.2N$ Gb/in^2). N was interpreted as the number of holograms that can be independently stacked in a common volume of the storage medium. For $N = 1000$, the areal storage density is ~ 5.2 Tb/in^2. This storage density was so astounding a possibility in 1963 that industrial and government laboratories throughout the world effort to implement storage systems that exploited van Heerden's theory. Thus began the 38-year quest for a commercial 3-D holomem [1, 2].

From about 1963 through the mid-1970s, 2-D and 3-D holomems were viewed as universal storage solutions. Interest in holomems declined after this period because of the extremely difficult engineering challenges to systems implementation. The hologram storage medium was (and largely remains) the major item on the critical path. A few start-ups tried to develop either disk (notably, Tamarack Systems) or tape 3-D holomems in the late 1980s, but failed all. Finally, interest was rekindled in the early 1990s by significant improvements in key components. From about 1994 to the present, companies such as Holoplex, IBM, InPhase, and Polaroid-licensee Aprilis Optware, Lucent Bell Labs, and SIROS (Optitek) began serious attempts to commercialize. IBM is viewed by many as the 3-D holomem technology leader, but it does not appear to be ready to announce a product. Lucent Bell Labs has also done some very good research and development. Lucent and Imation private investors formed InPhase Technologies in 2001 located in Longmont, to develop a holographic disk storage system. Rockwell is addressing

© Tsinghua University Press and Springer Science+Business Media Singapore 2016
D. Xu, *Multi-dimensional Optical Storage*,
DOI 10.1007/978-981-10-0932-7_7

real-time correlators. Holoplex (funded by Hamamatsu) has actually shipped a few systems it is focused on human face and fingerprint identification. In 2008, Optware Company published a new 3D holographic volume discs that uses a recordable holographic driver of the WORM (Write Once Read Many) technology. Japanese companies, Hitachi, NHK, Pioneer, and Sony, for example, but none of these companies has shown anything at mass production and trade show yet [3].

7.1 Evolution of Volume Holographic Storage

One of important progress in 3-D holomem technology is the new medium, especially the macromolecule organic materials application in holomem. As the Diarylethene for example, it has two forms: the open-ring form A and the close-ring form B. With the incidence of the light with $\lambda 1$ wavelength (ultraviolet light) A is converted into B through the fading reaction while with the incidence of the light with $\lambda 2$ wavelength (visible light) B is converted into A through the chromic reaction. The reading and retrieving processes of the holographic gratings in the medium (Diarylethene) are illustrated in Fig. 7.1.

The molecules of the Diarylethene are converted into form B with the illumination of the ultraviolet light before recording the holograms. In recording, the intensity distribution of the sinusoidal wave field is formed by the interference between the object light and the reference light in the material. The molecules of the Diarylethene are converted into form A from form B with absorbing photos in the sinusoidal wave field. And the concentration distribution of the molecules with form A and form B changes linearly with the change of the sinusoidal wave field, as shown in Fig. 7.1. Because the absorption coefficients and refractive indexes of the two forms are different, the volume gratings formed in the materials are the hybrid gratings, including the amplitude gratings caused by the different absorption and the phase gratings caused by the different refractive indexes. In retrieving, regarding the original reference light as the retrieving light can erase the recorded gratings, because the retrieving light can accelerate the conversion of the molecules of the Diarylethene from the form B to the form A. Once the reading time is long enough,

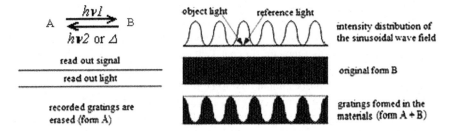

Fig. 7.1 The reading and retrieving processes of the holographic gratings in the Diarylethene

the molecules of the Diarylethene will be converted into form A completely so that the recorded gratings can be erased out. Now there are more than hundred macromolecule organic materials that can be used for 3-D holomems; to reference Chap. 2 in this book [4–6].

7.1.1 Common Volume 3-Dimension (3-D) Holomems

A major force behind the effort to implement real 3-D holomems in the early twenty-first century was the DARPA-sponsored Holographic Data Storage System (HDSS)/Photorefractive Information Storage Media (PRISM) initiative. The goals of this project included demonstrating a 1 Tb capacity and a 1 Gbps data rate. The program was largely successful (in fact, a 10 Gbps data rate demonstration was claimed), and the project was completed in early of this century. SIROS has exited the 3-D holomem part. A hologram is a true amplitude and phase representation of some object or data pattern. It is created with coherent laser light by interfering, the light transmitted or reflected from a data pattern (object or signal beam) with an unmodulated carrier (reference beam). The interference pattern created by the interaction of the two beams is captured by a storage medium, such as a pho-topolymer film or an electro-optical crystal and these are referred to as volume-phase storage media. For certain recording setups and storage medium thicknesses, a 3-D (volume) hologram is recorded, which has useful properties for data storage. The hologram is readout (reconstructed) by illuminating the hologram with only the reference beam. In principal, a perfect 3-D image of the original object is reproduced. Holograms have been used for things as practical as credit card and CD authenticators or interferometric profilometry to imposing full-color works of art [7, 8].

A 3-D holomem is a collection of 3-D holograms, which share a common storage medium volume (the hologram array), and the means to write and read these holograms as shown in Fig. 7.1. 3-D holograms are stored throughout a part of the volume of the storage medium defined by an area large enough to ensure adequate resolution of the input device pixels (see spatial light modulator (SLM) below). This permits multiple holograms to be recorded in a common volume and indi-vidually retrieved by, for example, (Bragg) angle or wavelength multiplexing. This collection of holograms is called a stack, even though all of the spatially varying information that represents the holograms is randomly intermingled and cannot normally be individually distinguished or erased. The areal density is defined as the number of stacks × data per page divided by the stack area and two or more stacks form a hologram array. Block (cube), disk, tape, and card formats have been evaluated, and each has its advantages and trade off. Add required hardware and software, and a 3-D holomem.

The concept of a 3-D holomem can understood by referring the schematic diagram of Fig. 7.2. Writing of the hologram in the storage medium delays the completion of imaging and output with photodetector array (PDA) detection, until hologram reconstruction. A collimated laser beam incident from left is shaped by a

Fig. 7.2 The typical
miniaturization holographic
optical tweezer array system
with common storage medium

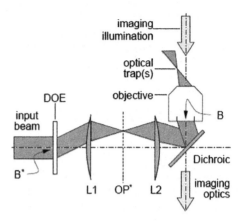

diffractive optical element (DOE), transferred to an objective lens' back aperture
(B) by lenses L1 and L2 and focused into a trapping array. OP denotes the plane
conjugate to the trapping plane. The point B is conjugate to B. The phase pattern on
the lower left (black regions shift the phase by π radians) produced the traps shown
in the lower right filled with 1 µm diameter silica spheres suspended in water.

The 3-D holograms each store a page of data. The page is created by an input
device called a SLM. The SLM typically is an array of N_p pixels (for example,
1024×1024), which can be thought of as on-off switches that convert electrons to
photons. Generally, the Fourier transform of the SLM is projected by a lens into the
volume of the storage medium. Fourier transform holograms maximize storage
density. A reference beam is added, and the resulting 3-D interference pattern is
captured (frozen) in the storage medium. This process is repeated for different
reference beam angles until a stack of N holograms are formed in a common storage
medium volume (a process called incoherent adding or superposition).

Using only the reference beam (the called reconstruction beam, which must be a
very accurate and precise replica of the original reference beam), the individual
holograms can be readout and imaged (inverse Fourier transformed) onto a PDA.
The amount of power in the reconstructed image of the nth of N holograms divided
by the power in the reconstruction beam incident on the hologram (PH) is called the
"diffraction efficiency" (η_n). Generally, "write scheduling" is used to make η_n about
the same for all N holograms in a stack. Write scheduling compensates for partial
erasure of prior-written holograms in real-time media such as lithium niobate and
for nonlinear characteristics in write-once media, such as photopolymers. The PDA
is the output device that converts photons back to electrons. Note that η_n PH divided
by the average number of pixels in the holographic image (1/2 N_p) is the average
power per reconstructed pixel (PD) incident on a photodiode of the PDA. The
corresponding energy per reconstructed pixel (ED) is PDτ_i, where τ_i is the PDA
integration time. ED, together with signal-dependent and random noise sources,
ultimately determines raw (also, soft or uncorrected) bit error rate (BER), read data
rate, and achievable user capacity [9, 10].

One last detail remains to be considered. What type of 3-D hologram page addressing method is to be employed:

1. A "moving parts" design, which is less expensive, but limits access speed significantly and may be less reliable.
2. A "no moving parts" (all-photonic) design, which is much more expensive, but is very fast and much more reliable.

The analogy to modern hybrid (E/O/E) versus all-optical network switching would not be too far afield. This is a very complex issue with many factors to be analyzed. The choice is generally application driven. It is but one illustration of the tradeoffs that make it difficult to obtain all the attractive features of a 3D holomem simultaneously. From an end-user/applications perspective, good reasons for a high level of interest in 3D holomems existed still do today. Some of the most important theoretical and idealized sense are: storage densities from 50 GB/in^2 up to 10 Tb/in^2 even more capacities from 200 GB disks up to 100 TB crystals, data rates from 50 MBps to many GBps, the results from the high degree of write/read parallelism inherent to 3D holomems, page access times from many ms down to <10 μs would be much less than magnetic tape and hard disc for most designs. Better reliability and stable long-term storage, assuming no moving parts and a completely photonic implementation, and probably true only for certain write-once storage media, with all of these exciting (potential) characteristics. But the characteristics, requirements, and specifications for 3D holomem storage medium are so complex that this component deserves a white paper in its own right. Lithium niobate, an electro-optical or "photorefractive" crystal is the best known, but has serious practical limitations. The best hope of the technology currently resides with certain types of WORM photopolymer. Meanwhile 3D holomem is a complex, analog, interferometric device, for example, a near-field recording (NFR) optical disk drive, which is well known to be extremely challenging. Very sophisticated servo systems will be required to compensate and control thermal variations (athermalization), vibration, and shock that affect everything from laser power to storage medium alignment. The latter is especially problematic, because the positioning accuracy and precision at the reconstruction beam-stack area interface are often measured in microradians and nanometers for ultra-high density 3D holomems. Creating a proforma bill of materials from a detailed system block diagram and applying reasonable component cost estimates will quickly convince the reader that a 1 TB holomem (block or disk format) will cost more than ten times a 1 TB magnetic disk drive. Of course, the cost/TB of the storage medium, which is removable, is likely to be very reasonable in the tradition of optical data storage. No photonic infrastructure or second sourcing exists for either the drive or storage medium. One must accept the fact that support for a large-volume, low-cost holomem storage product line does not exist today, despite the explosive growth of optical networking. The price/performance/reliability is very significant to induce market acceptance. A few subtle details about 3-D holomem performance and operation need also be shared: The capacity (C) of a 3-D holomem is upper bounded by the well-known equation: $C = N_H N_p = S N_L N_p$ where N_H is the total number of hologram, N_p is the number of pixels per page, and N_L is the number of hologram storage sites (each with a stack height $= S$). The value of C will be always less than the number given by the above equation, and often much less. User capacity will be determined by the acceptable level of corrected BER at the system level. As in optical disc drives, this implies the need for a relatively low raw BER as 10^{-4} for example. In turn, a minimum level of signal-to-noise ratio (SNR) is required. This helps explain why as stated earlier that ED is such an important variable. A 3D holomem is operationally a write-once (WORM) system, even if the storage medium itself is erasable. This is because individual holograms cannot be selectively erased some exceptions exist, but are still in the research stage or are impractical). In fact, for certain applications for example, medical records or movie film archiving that a true and stable write-once storage medium is preferable. Unless a very stable erasable storage medium with a fast erase/reset mechanism and very high speed writing sensitivity can be developed, write-once media is a compelling choice. A significant asymmetry exists between write and read data rates. This is mainly an issue of storage media write sensitivity. Electro-optical crystals require high power densities and relatively long exposure times to record holograms. Photopolymers, which have an inherent chemical gain mechanism, are orders of magnitude faster than electro-optical crystals. Recent advances in photopolymer storage media design have shown the potential to reduce the read/write data rate asymmetry significantly [11].

Current practice makes the diffraction efficiency of every hologram the same. Recently, the concept of making the raw BERs similar for each hologram evolved. Regardless, the implication is that "stack-at-once" (the term from CD-R) writing is required, or some method must be devised to control all the variables in real time during reconstruction of the holograms extremely complex. Hence, one should consider a 3D holomem as having a CD-ROM operational model without the replication as 3D hologram stacks cannot be copied with duplication directly. Dedicated writers located in well-controlled environment would fill hologram stacks in serial fashion (mastering), and dedicated readers would be used to read the data. This is a valid operational model for many applications, particularly archiving. Obviously, a 3D holomem using a disc format must anticipate a better operational model. Key performance parameters are coupled in nonobvious ways. For example, true page access time (TACC) is the sum of the page address time (a function of the page addressing mechanism) plus page read time. Page read time is inversely proportional to the (burst) read data rate. The read data rate (R) is proportional to diffraction efficiency (through the PDA integration time τ_i), which is inversely proportional to the square of the stack height N. Finally, capacity is directly proportional to N. For a block-oriented 3D holomem, $R = \text{constant}/C^2$, a nonobvious result. For those left more confused than enlightened by the preceding discussion, be advised that will be publishing a detailed report on 3D holomem technology (optical, X-ray, and electron beam) and products later. Optostor AG Company provided the first real trade show demonstration of a 3D holographic memory system prototype for an example of its hologram array as shown in Table 7.1.

Table 7.1 Optostor AG 3D holographic storage system specifications (I)

Specifications	Data
Capacity	5 TB
Write data rate	Not specified MBps
Page read time	10 ms
Read data rate	~ 12.5 MBps
Areal density +	~ 4.1 Tb/in^2 (theoretical max. = 12.6 Tb/in^2)
Areal density +	6640 (theoretical max. = 20,000, IBM ~ 400) bits/μm^2
Volume density +	~ 2.13 (theoretical max. = 6.64) bits/μm^3
System reliability (MTBF)	Over 10,000 h
Hologram storage life	Over 100 years
Storage medium	Doped lithium niobate (n ~ 2.3 assumed)
Fixing method	Not specified
Insertion loss	Not specified
Storage medium area	50×50 mm^2
Storage medium thickness	3 mm
Laser type	Frequency-doubled (SHG)YAG with $\lambda = 532$ nm (power, beam quality, and coherence not specified)
Spatial light modulator	1024×1024 pixel LCD
Photodetector array	1024×1024 element CCD
Hologram stacking method	Bragg-angle multiplexing
Hologram type	Fourier transform
Minimum hologram size + (surface area)	$\sim 2 \times 2$ mm^2 (double-Rayleigh resolution criterion for FT lens with $F = 2$)
Hologram size used +	$\sim 2 \times 2$ (based on a stated laser beam diameter of 2 mm) mm^2
Hologram thickness	3–5 mm

Although the system was operating in the read-only mode, the reading of bit pages was clearly demonstrated. The following background information will help in understanding Optostor's system: The storage medium (doped lithium niobate) was developed at the University of Cologne by Dr. Theo Woik, etc. The system is being developed by startup Optostor AG, a "32 %-subsidiary" of Eutelis Consulting GmbH. This has produced some 20 patents applications. The prototype uses Bragg-angle multiplexing to achieve hologram stacking that incoherent superposition of holograms, each separated by a minimum of one Bragg angle. Fourier transform holograms are recorded, which have a defined minimum area requirement and maximize areal storage density. The laser is a frequency-doubled YAG system emitting at 532 nm. Power output, beam quality, and coherence length are unknown. However, commercial devices are relatively mature, although expensive. Beam quality suitable for holographic recording with CW power in excess of 1 W can be obtained. The input device (SLM) is a Sony liquid crystal display (LCD) with 1024×1024 (1 K $\times$ 1 K) pixels (picture elements). The LCD uses switchable input cells with two reflective or transmissive states to create binary 0 and 1 s. Each pixel can be thought of as a mark or space, coding determines the number of bits each represents that similar to pulse position modulation used in early optical disk drives. The PDA is a charge-coupled device (CCD), type and frame rate were unspecified. Both CCD and CMOS PDAs are under consideration. The storage medium is doped-lithium niobate, a well-known type of real-time, volume-phase hologram recording material. It is either iron (Fe) doped or iron–silicon (Fe–Si) alloy doped. More importantly, this storage material can be coated in high-quality sheets at least 3 mm thick. This is an unusual format for data storage in lithium niobate, which is generally used in a block format. Lithium tantalate has also been mentioned as a potential storage medium. Since lithium niobate self erases during hologram readout, some type of fixing is needed to ensure long-term stability. The means for fixing the storage medium is unknown (heating has been mentioned as a means of stabilizing the holograms), that dimensional and environmental stability of its lithium niobate are very good [12, 13].

The hologram array has dimensions of $30 \times 30 \times 3$ mm^3. The 30×30 mm^2 area is subdivided into $100\times$ (3×3 mm^2) sub-areas (essentially contiguous cubes). This provides 100 hologram storage locations (stack sites). The total number of holograms per hologram array must therefore be $(8 \times 1$ TB)/$(1024)^2 = 7.63$ million. This means the stack height must be 76,294 holograms. The computed storage density is about 5.73 Tb/in^2. Basic technical specifications given by Optostor were: 5 TB system capacity; 10 ms page access time; and ~ 100 Mbps read data rate. A 100-year archival life (>500 years calculated) is claimed. The write data rate was not given, nor was any reliability data provided. Note that page size in bits divided by page access time, that is, $(1024)^2/0.01$ s is about 100 Mbps. This means that the 10 ms number is really the page read time, and 100 Mbps is the burst read data rate. Since no page addressing mechanism was identified, the time to access successive pages is unknown (1–5 ms is possible with a servo-controlled galvo mirror scanner), nor is average page positioning time as shown in Table 7.2.

CeBIT system capacity can be 5 TB and grow to 100 TB by design. Initial customers include banks, government health agencies, insurance companies, and video/film producers. The application is apparently archiving (European laws are much stricter about archiving medical, financial and government records than are those of the US or most Asian countries. A German law that requires that medical records be preserved for 30 years). The hologram array size is now $50 \times 50 \times 3$ mm^3 and the total number of holograms is 15.3 million. With the same 1 K $\times$ 1 K SLM as before, this does indeed calculate to 5 TB capacity. Assuming 2 mm 2 mm holograms (minimum size for double-Rayleigh resolved Fourier transform holograms with a F/2 lens), 625 hologram locations (stack sites) are required. To obtain 15.3 million holograms requires a stack height of 24,414 holograms, which is an improvement over 76,294, but is still a very large value for N, the storage density is 4.1 Tb/in^2 [14–16].

Another famous holography system IBM DEMON II 3D holomem system, the specifications are shown in Table 7.2 that test uses a 90° write/read geometry, which is the best case for Bragg angle multiplexing. This table provides a summary

Table 7.2 IBM DEMON II 3D Holomem

Specifications	Data
Max. volume density	400 bits/μm^3
Max. areal density	5.5×108 bits/mm^2
Max. areal density	0.36 Tb/in^2
Number of holograms (data pages)/stack, N	1350 (for 10^{-4} raw BER)
Hologram stacking method	Bragg angle multiplexing, 90° setup (best case)
Page access mechanism	mechanical (galvo scanner, ±15° range)
Coding	8 bit/12 pixel modulation
Raw bit error rate (BER)	10^{-4} (measured)
Corrected bit error rate	10^{-12} (Reed-Solomon EDAC)
Storage material	Fe-doped lithium niobate
Hologram type	Fourier transform
FT lens focal length	30 mm
Hologram area	1.6×1.6 (~ 28 % larger than minimum) mm^2
Storage medium thickness	5.5 mm
Laser type	Frequency-doubled YAG with $\lambda = 532$ nm
Spatial light modulator	1024×1024 pixel, reflective (IBM Yorktown, 12.8 μm/pixel)
Photodetector array	1024×1024 element CCD (12 μm photodiode pitch at 41 fps)
Photodetector threshold energy E_0	7.5×10^{-16} J (2000 photons for 10^{-4} raw BER; 1 photon yields 1 electron assumption)

of data reported by IBM for its DEMON II 3D holomem test system completely. This excellent result is probably the most complete reference available on the realities of 3D holomem design as AG 3D Holomem System (II) in Table 7.3. It functions as a pro forma 3D holomem spec sheet that have chosen a raw BER of 10^{-4} as the key analytical metric. This is a reasonable choice, given the anticipated applications. Some assumptions and design choices must be made on Optostor's behalf. The first has to do with write/read geometry. The 90° model cannot be used, because the hologram stacks are distributed over a 50×50 mm^2 hologram array. They chose reference and signal beam angles which ensure a minimum of 50 % beam overlap (it is 100 % for the 90° geometry) in the stack (a consideration forced by the 2×3 stack aspect ratio). The effective hologram thickness is 3 mm, but will ignore this. A 2D angular addressing scheme is assumed, because the Bragg angle for a 3-mm-thick hologram is too large to permit 1D angular addressing.

For 3D holomem, required system margins to ensure both short- and long-term reliability at customer locations are absent. The energy per pixel ED received at the photodiodes of the PDA is equivalent to 43 photons (below the noise floor of most PDAs). The SNR is less than 1. For archiving applications, the raw BER 10^{-4} should be an end of life specification, so the situation is actually worse than first appears. Scaling back N to be consistent with IBM's experimental data yields a

Table 7.3 AG 3D Holomem system (II)

Specifications	Data
Total number of holograms (data pages)	>25.3 million
Number of stacks (hologram storage locations)	625
Number of holograms (data pages)/stack, N	24,410 (maximum of 1350 for 10^{-4} raw BER)
Error detection and correction coding	Reed-Solomon type
External signal/reference beam angles	−46.7/+46.7 (internal angles of −18.4/+18.4; provides good beam overlap in the crystal) degrees
Minimum Bragg angle	1.27×10^{-4}/0.0072 radians/degrees
Hologram angular separation	2.54×10^{-4}/0.0144 (2 × min. Bragg angle to prevent crosstalk) radians/degrees
Min. angular access range $(1 - D, \theta$ only)	>2π/360 (not physically possible) radians/degrees
Min. angular access range (symmetrical $2 - D, \theta - \varphi$) +	>0.04/2.24 radians/degrees (157 θ × 157 φ holograms)
Read power, PH +	1 W(at hologram stack)
Max. diffraction efficiency ($N = 1$, optimum write)	20 % (post fixing, including insertion loss; 100 for ideal volume-phase change media)
1/N2 bias buildup loss (due to hologram stacking)	$1/(24{,}414)^2 = 1.68 \times 10^{-9}$ (signal strength loss compared to $N = 1$)
PDA frame read rate	100 fps (required for 100 Mbps read rate)
CCD integration time, τ_i	10 ms
Imputed IBM photodiode energy, E_0	$\sim 7.5 \times 10^{-16}$ (2000 photons for 10^{-4} raw BER per IBM; 1 photon yields 1 electron assumption) J
Power (per pixel) at photodetector, PD	$\sim 1.6 \times 10^{-15}$ W
Energy (per pixel) at photodetector, ED	$\sim 1.6 \times 10^{-17}$ J
ED/E_0 ratio	~ 0.02 (must be $\not\Rightarrow 1$)

capacity of 5 TB. If the end of life raw BER must be $<10^{-4}$, then achievable user capacity will again be decreased.

Going from laboratory research prototype to data center system will not be easy, even if the application is restricted to archiving. Based on the known, public domain facts, Optostor appears to have rediscovered a road well-traveled by other companies before it, all of which were turned back by very challenging engineering problems. Moreover, it seems unlikely that a research group at the University of Cologne and a 2-year startup have accomplished what the DARPA HDSS/PRISM project (comprised of some of the best US companies and universities in the field). But, given Germany's reputation for excellence is in chemistry. Also, it is unsettling that a company spawns the media frenzy before it appears to understand fully the basics of a complex data storage technology. An European company (Opticom, Oslo, Norway) claimed it could store 100 TB on a credit card-size storage medium comprising a polymer–protein matrix. After a flurry of publicity, Opticom delivered nothing. Optostor, most importantly, it must be determined whether or not Optostor's lithium niobate storage medium is exceptional and really capable of the writing and long-term preservation of archival records.

InPhase and Polaroid-licensee Aprilis have done excellent research in this area. Complutense University and the National Research Council of Canada announced a new photopolymer medium that has favorable properties for both 3D holomems and fiber optic gratings. The key requirements that PP storage media must have are high diffraction efficiency, relatively high write speed, no inherent or write/read-dependent noise, and no dimensional variations due to writing processes or temperature fluctuations. PP media approximately meet these basic criteria. Key component availability is limited. These include the laser (writing and reading device), SLM, PDA and page addressing mechanism or subsystem. These components do exist, but are very expensive often, designed and engineered mainly for laboratory use only, and lacking reliability characteristics suitable for commercial data storage devices, especially long-term archiving.

7.1.2 Two-Wavelength Holography

The two-wavelength holographic recording system is shown in Fig. 7.3. For two-color holographic recording, the reference and signal beam fixed to a particular wavelength (green, red or IR) and the sensitizing/gating beam is a separate, shorter wavelength (blue or UV). The sensitizing/gating beam is used to sensitize the material before and during the recording process, while the information is recorded in the crystal via the reference and signal beams. It is shone intermittently on the crystal during the recording process for measuring the diffracted beam intensity. Readout is achieved by illumination with the reference beam alone. Hence, the readout beam with a longer wavelength would not be able to excite the recombined

Fig. 7.3 The two-wavelength holographic recording system

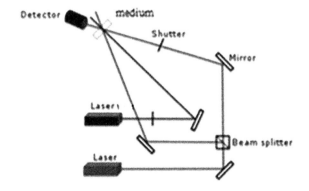

electrons from the deep trap centers during readout, as they need the sensitizing light with shorter wavelength to erase them.

Usually, for two-color holographic recording, two different dopants are required to promote trap centers, which belong to transition metal and rare earth elements and are sensitive to certain wavelengths. By using two dopants, more trap centers would be created in the lithium niobate crystal. Namely a shallow and a deep trap would be created. The concept now is to use the sensitizing light to excite electrons from the deep trap farther from the valence band to the conduction band and then to recombine at the shallow traps nearer to the conduction band. The reference and signal beam would then be used to excite the electrons from the shallow traps back to the deep traps. The information would hence be stored in the deep traps. Reading would be done with the reference beam since the electrons can no longer be excited out of the deep traps by the long wavelength beam. For doubly doped lithium niobate ($LiNbO_3$) crystals, there exists an optimum oxidation/reduction state for desired performance. This optimum depends on the doping levels of shallow and deep traps as well as the annealing conditions for the crystal samples. This optimum state generally occurs when 95–98 % of the deep traps are filled. In a strongly oxidized sample holograms cannot be easily recorded and the diffraction efficiency is very low. This is because the shallow trap is completely empty and the deep trap is also almost devoid of electrons. In a highly reduced sample on the other hand, the deep traps are completely filled and the shallow traps are also partially filled. This results in very good sensitivity (fast recording) and high diffraction efficiency due to the availability of electrons in the shallow traps. However, during readout, all the deep traps get filled quickly and the resulting holograms reside in the shallow traps where they are totally erased by further readout [17–19].

Hence, after extensive readout, the diffraction efficiency drops to zero and the hologram stored cannot be fixed. To solve the shrinkage of the material in volume holography established a more comprehensive evaluation system of polymer material properties based on light-induced discoloration of materials repeatedly rewritable nonvolatile readout. According to the reaction of the theoretical model, the dynamic process of grating formation in dual-channel recording conditions, S, P channels, such as light intensity, the same conditions as the respective modulation, and dual-channel records with the following characteristics: each channel to obtain the final refractive index modulation of the corresponding single-channel record 1/2 each channel to obtain the final diffraction efficiency is 1/4 of the corresponding single-channel records, each channel of the recording sensitivity similar. It can be seen from the characteristics of the first, then no influence on the material refractive index modulation, the dynamic range, and single-channel records of the materials are basically the same, so will not affect the limit of storage density, while the dynamic range sufficient conditions using the same complex structure as the single-channel recording, can get twice the storage density and capacity. In addition to channel recording sensitivity, similar characteristics can be seen that the orthogonal polarization dual-channel recording technology, ensure the same recording conditions, channel speed, improve memory parallel, and can effectively improve the rate of storage.

Cationic ring-opening polymer (CROP) holographic recording materials are a WORM-type holographic recording materials, but also superior performance of new materials. Holographic recording material shrinkage of studies CROP survey record in the case of deviation from symmetry incident $0°$, $10°$, $20°$, $30°$ saturated grating Bragg angle offset, experimental results show that the CROP holographic material Bragg angle offset some changes may be caused by the average material refractive index and measurement error. If selected, the appropriate reuse is expected to get higher multiplexing property. For rewritable material, use the open-loop light-induced discoloration of materials, the characteristics of the closed-loop state can be transformed into each other under the action of different wavelengths of light waves, designed based on the two resorcinarene materials reusable erasable, nonvolatile readout volume holographic deposit materials structures. Light-induced discoloration reaction and two-wave coupling theory and to study the holographic grating formation process in the two resorcinarene materials, the establishment of the dynamic model of the holographic gratings recorded, and theoretical and experimental curves of the change of refractive index modulation [20].

Storage aim to develop a 3D holographic optical storage technology based on photon-induced electric field poling using a far UV laser to obtain large improvements over current data capacity and transfer rates, but as yet they have not presented any experimental research or feasibility study. Microholas operates out of the University of Berlin, under the leadership of Prof Susanna Orlic, and has achieved the recording of up to 75 layers of microholographic data, separated by 4.5 µm, and suggesting a data density of 100 GB per layer.

7.1.3 Holographic Versatile Disc (HVD)

Holographic Versatile Disc (HVD) with capabilities of around 60,000 bits per pulse in an inverted/truncated cone shape has a 200 µm diameter at the bottom and a 500 µm diameter at the top. High densities are possible by moving these closer on the tracks: 100 GB at 18 µm separation, 200 GB at 13 µm, 500 GB at 8 µm, and most demonstrated of 5 TB for 3 µm on a 10 cm disc citation needed as shown in Fig. 7.4: where 1—writing/reading laser (532 nm), 2—positioning/addressing laser (650 nm), 3—hologram (data) (shown here as brown), 4—polycarbonate layer, 5—photopolymeric layer (data-containing layer), 6—distance layers, 7—dichroic layer (reflecting green light), 8—aluminum reflective layer (reflecting red light), and 9—transparent base, p—pit pattern.

The disc has to use a green laser, with an output power of 1 W which is high power for a consumer device laser. Possible solutions include improving the sensitivity of the polymer used, or developing and commoditizing a laser capable of higher power output while being suitable for a consumer unit by citation needed also. In the holography system, storage densities from 50 GB/in^2 up to many Tb/in^2, capacities from 200 GB disks up to many TB crystals, data rates from

Fig. 7.4 Holographic
versatile disc structure

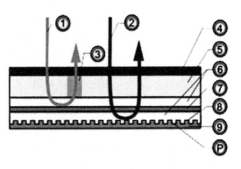

50 MB/s to many GB/s and page access times from many m secs down to <10 μs
are achieved. This results from the high degree of write/read parallelism inherent to
3D holomems. It is an optical disc technology developed that can store up to several
terabytes of data on an optical disc the same size as a CD, DVD, or Blu-ray disc
[21, 22].

Optware company Ultra-high density optical disc Capacity of HVD is 6 TB in
2008. InPhase Technologies, after several announcements and subsequent delays in
2006 and 2007, announced that it would soon be introducing a flagship product. In
2010, GE Global Research demonstrated their own holographic storage material
that could allow for discs that utilize similar read mechanisms as those found on
Blu-ray Disc players. There are very low-cost per TB (media cost/TB, for example,
would be much less than magnetic tape for most designs), very high reliability
(assuming no moving parts and a completely photonic implementation) and very
stable long-term storage (probably true only for certain write-once storage media).
With all of these exciting characteristics, one might reasonably ask why no commercial 3D holomem systems for data storage (particularly, archiving) have ever
come to market. Some of the basic reasons are listed below although their real
complexity cannot be examined or appreciated as absence of consummate storage
medium. No photonic analog to magnetic contains a substantial analysis of holographic memories. The motivation for this effort was a claim by a German startup
Optostor AG3 that it had essentially invented the primary read-only holography
data storage. The company provided a read-only demonstration of its technology
which came nowhere near matching its claims for storage density and capacity. The
bottom line is a detailed exposition of the basics of holomems, supported by
specific data provided by the excellent Holomem research of IBM. General Electric
Global Research Centers have created a new format of disc, dubbed the holographic
disc. It has the same physical dimensions as CDs, DVDs, or Blu-ray Discs, yet can
hold many times the data—up to 500 GB. General Electric has been working on
their HVD. As the technology is quite similar to CD, DVD and Blu-ray technologies, the players are also similar—GE states a HVD player will also be able to
play the other formats of discs. Despite its long and mainly unsuccessful development history, 3D holomems still generate excitement. A small company called
Optostor AG actually demonstrated a read-only system. It also made some claims
that may be too good to be true [23].

7.1.4 Advanced Structure and Performances

In the 3D holography storage system, the information is stored not only in the media surface but in the whole volume of the medium for holographic optical disc system (HODS) that provide a new direction for high-density storage technology. HODS is able to achieve higher storage density far beyond other technology because it can store many holographic figures in a volume. On the other hand, the requirements of large-capacity and super-fast performances are increased with the development of computer science and modern information process techniques. Holographic optical data storage has been under research and development for nearly half a century. Miniaturization 3D holographic optical data storage is shown in Fig. 7.5 with 405 nm laser and storage density of 3.6 Tb/in^2 by InPhase Technologies Inc. Many alternative methods of optical data storage have been proposed over the years, based either on new material systems or on novel read/write/erase schemes. The holographic storage is very high storage densities from 100 GB/in^2 up to many Tb/in^2 and higher capacities which is from 200 GB

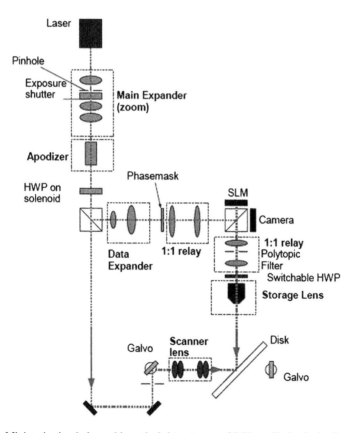

Fig. 7.5 Miniaturization holographic optical data storage of InPhase Technologies Inc.

disks up to 10 TB even more. But its data rates can be to many GB/s with the high degree of write/read parallelism inherent to 3-D holomems, and fast page access times has been down to <10 μs. So it is the important research progress still in optical storage area. The recordable holographic driver of the WORM volume holographic 3D discs has trusted that 6 TB data are stored in a disc of 12 cm diameter and 2 Gbit/s transfer rate are achieved that structure and record process as shown in Fig. 7.6a by Optware Company. The digital movie was stored in volume holographic 3D disc with the method of coaxial holography [24, 25].

Basic theories of the holography and new multiplexing principles were used for high-density digital holography storage. New-typed doped LN crystals, organic, inorganic composite and polymer materials, the system parameters influenced on performances of holography storage, diffractive effect, sensitivity, dynamic range, response time of medium and testing system were researched and developed in OMNERC. The chapter will present those results. Meanwhile, the encoding, decoding and super-high speed data transmission and processing will discuss. The new multiplexing principle—Dynamic Speckle Multiplexing (DSM) and its application in various holographic storage systems are introduced also.

The typical integration holography data storage system with TB capacity is the HVault's robotic library system which has from one to eight drives, as shown in Fig. 7.6b. A base model starts with 240 disc slots and can be expanded to as many as 540 slots. Up to three library cabinets can be daisy-chained together. A box can have petabytes of storage where anything in there is accessible in less than 10 s. The Vault's library systems will enable companies to archive vast collections of analog video that require digitization as well as content that has already been digitized. The product could be media for entertainment, medical imaging, and satellite imagery etc. The H-Vault's holography data storage robotic library system with 540 slots has petabytes capacity and storage where anything in less than 10 s as in Fig. 7.6b. H-Vault is also considering a consumer version of the holographic storage device

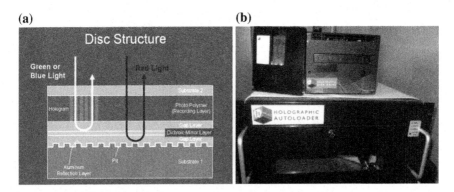

Fig. 7.6 **a** The structure of Holographic Versatile HVD disc and record process. **b** 3D holography data storage robotic library system

that would allow users to store more than a terabyte of data on a single platter, significantly more than Blu-ray Discs or DVDs. That amount of content would be onerous to transport on any kind of online distribution system [26, 27].

7.2 Performances and Evaluation of Medium

Theoretical model for evaluation and classification of holography storage materials, the algorithm model of the photorefractive effective time constant and electron field amplitude are flowing:

$$E^{SC} = m \frac{E_o \sqrt{E_{ph}^2 + E_d^2}}{E_q + E_d}$$

$$\tau_1(x) = \tau_{di}(x) \frac{1 + (E_d / E_u)}{1 + (E_d / E_q)} = \frac{\tau_0}{I_0 \exp(-\alpha x / \cos \theta)}$$

(7.1)

where E is the electron amplification and τ is the effective time constant. For evaluation of holography storage materials, based on above theoretical model, OMNERC developed a corresponding materials testing system; the optical system is shown in Fig. 7.7, where the space light modulator, scanning stage, laser source, and the detectors CCD1 and CCD2 are controlled by a computer, the testing process is roboticized as a standard holography storage system with DSSM also. The testing results for most usual materials in OMNERC as following: The maximum diffractive effect, photoscatter threshold (mw), and average writing time ($I \sim 1$ W/cm^2) of Fe,

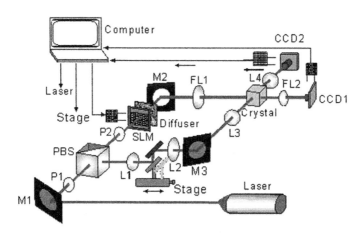

Fig. 7.7 Materials testing and evaluation system: *P1, P2, P3, L1, L2, L3, L4* lens, *FL1, FL2* Fourier lens, *SLM* space light modulator, *M1, M2, M3* mirror, *PBS* prism of beam, *SLM* spatial light modulator, *FL* Fourier transform lens, *CCD1, CCD2* detector array, *Crystal* tested sample. LN:Fe,Mg,In,Zn,Mn

Table 7.4 The recording performances with 532 nm laser testing results of Fe-, Mg-, In, Zn-, Mn-, Ce-doped LiNbO series photorefractive crystals

Crystals	Fe: LN	Fe,Mg: LN	Fe,In: LN	Fe,Zn: LN	Fe,Mn: LN	Ce,Fe: LN	Mn,Ce LN
Maximum diffractive effect	83	70	72	68	80	82	78
Photoscatter threshold (mw)	<0.1	>20	>30	>50	<0.1	<0.1	<0.1
Average writing time ($I \sim 1$ W/cm^2)	15	5	3	3	20	12	11

Ce,Fe, Tb,Fe, Fe,Sm, Mg, Fe series doped photorefractive LiNbO (LN) crystals are tested and the results are shown in Table 7.4. The LN:Fe,Mg series crystal are tested with 488 and 632 nm lasers that the results are shown in Table 7.5. Comparison of performances and experimental results for different materials are shown in Table 7.6. Meanwhile, the relationship between difference doped consistence (from 1.0 to 6.0 %) of LiNbO (LN) photorefractive crystal and the intensity of space electric field E_1 are shown in Fig. 7.8. The transverse resolution of a volume hologram, written with cw beams is limited by the numerical aperture of lens and resolution of the medium. However, when short-coherence-length light is used, the volume of the hologram is reduced if the coherence length of the light is much shorter than the beam diameters, as shown in Fig. 7.9. The relationship diffraction efficiency and angular deviation with different speckle size is shown in Fig. 7.10. The selectivity of different speckle size δ at proportion of selectivity in the criterion of optimization is shown in Fig. 7.11.

Above testing and evaluation results are very useful and those materials have been employed to the volume holography storage systems latter.

Table 7.5 The testing results of Fe- and Mg-doped LiNbO with different consistence series photorefractive crystals with 488 and 632 nm lasers recording

LiNbO:Fe,Mg (Fe 0.01 wt%)	Mg (0.0 mol %)	Mg (2.0 mol %)	Mg (4.0 mol%)	Mg (5.0 mol %)	Mg (6.0 mol %)
Anti photoscatter threshold 1 (mW)	<0.1 (488.0 nm) <0.1 (632.8 nm)	~ 8 (488.0 nm) ~ 20 (632.8 nm)	~ 20 (.488.0 nm) >30 (632.8 nm)	~ 80 (488.0 nm) >30 (632.8 nm)	>200 (488.0 nm) >30 (632.8 nm)
Photorefractive diffraction effect	~ 70 % ($I > 1$)	~ 70 % ($I < 1$)	~ 65 % ($I < 1$)	~ 55 % ($I < 1$)	~ 15 % ($I < 1$)
Photorefractive	~ 550 ($I > 1$)	~ 500 ($I < 1$)	~ 460 ($I < 1$)	~ 420 ($I < 1$)	~ 210 ($I < 1$)
Photorefractive respond time(s) (5 mW) (488.0 nm)	~ 160	~ 60	~ 35	~ 20	~ 15

Table 7.6 Comparison of testing experiment results of Fe, Ce, Tb, Sm, Mg-doped LiNbO (LN) series photorefractive crystals

Crystal	Diffractive effect (%)	Sensitivity (s)	Photoconduction (Ω)
Fe:LN	76.3	120	8.7×10^{-16}
Fe Sm:LN	68.7	27	2.1×10^{-14}
Ce:Fe:LN	76.0	100	8.0×10^{-16}
Tb:Fe:LN	70	25	1.1×10^{-14}
Mg:Fe:LN	56.4	34	0.9×10^{-14}
Zn:Fe:LN	58.5	30	6.1×10^{-15}
In:Fe:LN	13	3	1.6×10^{-15}

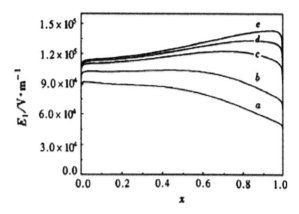

Fig. 7.8 The recording testing experimental results: relationship between different doped consistence (**a** 1.0 %, **b** 2.0 %, **c** 4.0 %, **d** 5.0 %, **e** 6.0 %) in LN and the intensity of space electric field E_1 [28]

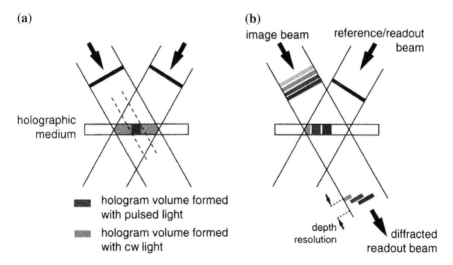

Fig. 7.9 **a** Narrowing of the volume hologram with short-coherence-length beams. **b** Decreased depth resolution as a result of angular separation of the writing beams

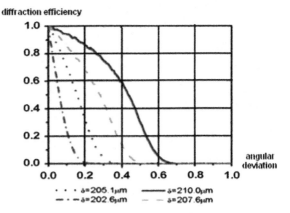

Fig. 7.10 The relationship between diffraction efficiency I and angular deviation θ with different speckle size δ of Fe: LN

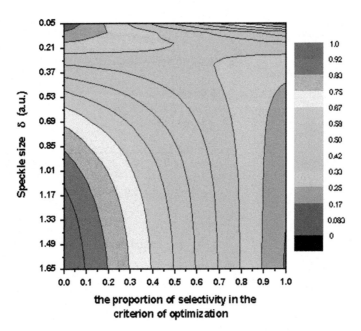

Fig. 7.11 The relationship between selectivity and speckle size δ and proportion of selectivity in the criterion of optimization

Experiments show that the oxygenation process can improve LN crystal sensitivity. The curves of backward scatter threshold effect to recording time of double-doped $LiNbO_3$ with different incident light intensity are shown in Fig. 7.12, where R is defined by Ir/Ii, in which Ir is the backward scattering light intensity, and Ii the incident light intensity [29].

The backward scatter threshold effect of double doped LN, erasure scatter noise and grating auto-expand effect in doubly doped lithium niobate crystals are shown

Fig. 7.12 The backward
scatter threshold effect R to
recording time of double
doped LiNbO$_3$: *curve 1*
corresponds to incident light
intensity of 10.3 W/cm^2,
curve 2 6.2 W/cm^2, *curve 3*
2.2 W/cm^2, *curve 4*
0.44 W/cm^2

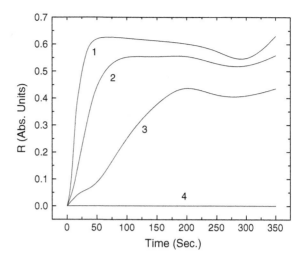

in Fig. 7.13: curve 1 corresponds to the results with illumination by another
incoherent laser with its intensity of 14.4 W/cm^2, curve 2 with 7.9 W/cm^2, curve 3
with 4.5 W/cm^2, and curve 4 without illumination.

Dependence of the diffraction efficiency of the space-charge-wave induced
grating on the incident position. The region represents the recorded area and only
outside the region the recording and reading beams separate completely.

The effective transmission rate as a function of the rotation angle, transmission
versus rotation angle ω, reading with difference wavelengths and intensity of
15 mW/mm^2, recording with wavelength $\lambda = 514$ nm exposure energy
120 Ws/mm^2 is Fig. 7.14. Transmission versus rotation angle ω for different
reading wavelength, intensity 15 mW/mm^2, recording with the wavelengths
$\lambda = 514$ nm, exposure energy 100 Ws/mm^2 is Fig. 7.15.

Fig. 7.13 The erasure scatter
noise and grating auto-expand
effect in doubly doped lithium
niobate crystals:
1 illumination by incoherent
laser intensity of 14.4 W/cm^2,
2 7.9 W/cm^2, *3* 4.5 W/cm^2
and *4* without illumination

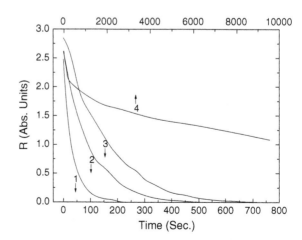

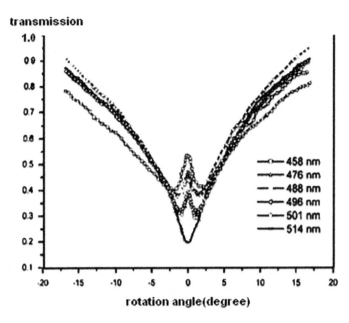

Fig. 7.14 The relationship curves of transmission rate to rotation angle ω, by difference wavelengths (from 485 to 514 nm) with intensity of 15 mW/mm^2 to read, but recording with wavelength $\lambda = 514$ nm and exposure energy 120 Ws/mm^2

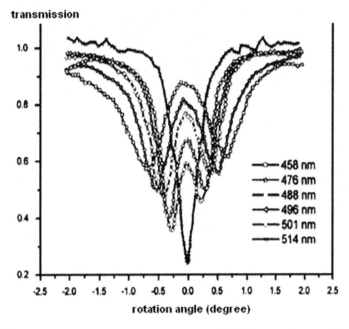

Fig. 7.15 Curves of transmission versus rotation angle ω for different reading wavelength (from 485 to 514 nm) with intensity 15 mW/mm^2, but recording with the wavelength $\lambda = 514$ nm and exposure energy 100 Ws/mm^2

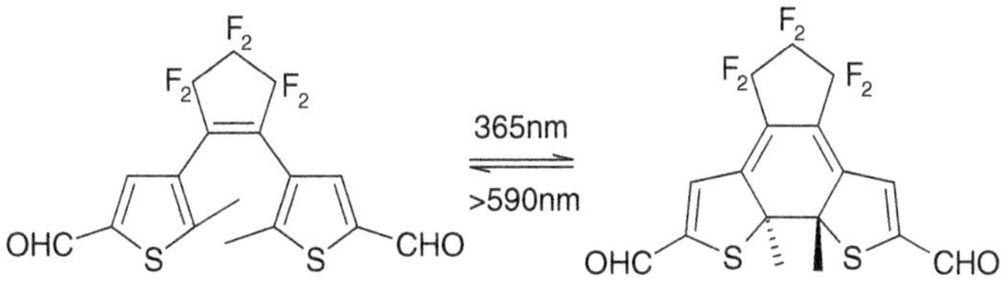

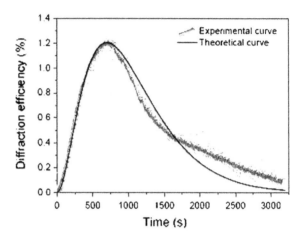

Fig. 7.16 Photochromic interconversion of the diarylethene

Fig. 7.17 The calculation and testing curves of dynamic diffraction efficiency to exposure energy (time) of the isomers Diarylethene

The testing experiment and evaluation for any new photopolymers materials have to be down from beginning. As Fig. 7.16 shows a reversible photo-induced transformation between the two isomers of diarylethene, it can be used for rewritable storage of 3D holomem. The calculation and experiment testing curves of diffraction efficiency of the diarylethene are shown in Fig. 7.17. The maximum diffraction efficiency is 1.2 % with the 10-μm-thick film [30–32].

7.3 Classification and Comparison of Multiplexing Schemes

7.3.1 Present Volume Holographic Storage System

The practical volume holographic storage system (VHS) can be implemented with the progress of the excellent media, such as photorefractive crystals and photopolymers, and some especial optical devices, such as high-resolution SLMs and charge-coupled detectors (CCD) etc. So VHS is a great potential information storage technology in the future. The typical high-speed rewritable holographic

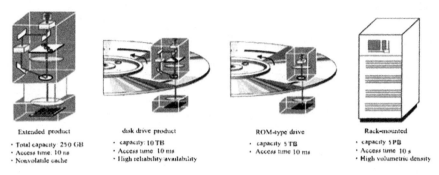

Extended product   disk drive product   ROM-type drive   Rack-mounted
· Total capacity 250 GB   · capacity: 10 TB   · capacity 5 TB   · capacity 5 PB
· Access time 10 ns   · Access time 10 ms   · Access time 10 ms   · Access time 10 s
· Nonvolatile cache   · High reliability/availability   · High volumetric density

Fig. 7.18 Development of volume holographic storage systems

volume is illustrated in Fig. 7.18 for example. That has lower than 1 ms of access time, much TB capacity, and some GB transfer rate can be attained in high-speed rewritable storage. It can be used to record huge file and high-resolution moving and 3D TV in the future.

A system produced by Lucent and Imation companies on the world employs the patent of correlation multiplexing. The reference light and the object light are set in one pickup so that the system structure is simplified and minimized, as shown in Fig. 7.19.

Longmont-InPhase Company put forward the prototype driver of HODS. The disc was made by Tapestry™ photopolymer WORM material with the thickness of recording material 2.5 mm. It is sandwiched between two 130–mm-diameter traditional plastic disk substrates. The prototype is the foundation for InPhase's family of Tapestry™ holographic drives, with data capacities of 4.6 TB on a side of the disc, and the life time is 50–100 years. Hitachi Maxell Ltd., designed and developed a new cartridge that provides maximum protection for the light-sensitive recording material, while maintaining the ease of integrating the cartridge into automated libraries. The prototype was completed that promote the development of key recording techniques and holographic disc that is illustrated in Fig. 7.20.

Fig. 7.19 A minimized HODS system and medium from Lucent and Imation Company

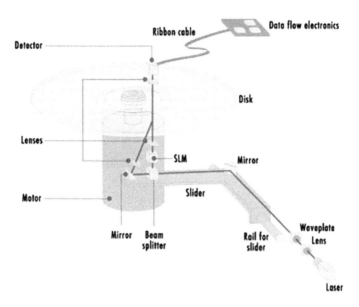

Fig. 7.20 Prototype driver of HVD by Longmont (CO)-InPhase

Volume holographic data storage, as a three-dimensional optical storage, records information based on the photorefractive effect of the crystal or the photochemical reaction of the photopolymer. In the course of magnetic storage and traditional optical storage, recording information bit depends on change of physical characteristics on the medium surface, such as turning of magnetic domain and pits formed by thermal effect. But HODS storage information with pages which can take very high-speed data rate than bit storage. As it allows data to be recorded and retrieved parallel. The process of reading and writing data in HODS also depends on the selectivity of holography gratings. The gratings are the interference fringes formed by two coherent beams. The interference fringes stored in the thick photosensitive material become the holograms. In the HDSS, a laser beam is divided into two beams. One is object light that loads data through the amplitude SLM after extension, the other is used as the reference light for recording and reading holograms.

In the process of recording the holograms, the reference light and the object light are interfered in the photosensitivity materials (such as the photopolymer or the inorganic doped $LiNbO_3$ crystal). The interference fringes are recorded in the photorefractive materials and the polymer with the dielectric absorption, the change of the refractivity or thickness in the result of a series of physical or chemical reactions caused by the interference in the media. Once the reference light with the proper incident angle or wavelength illuminates the holograms, the holograms will be reconstructed by diffraction light. And the optical signals will be converted into electric signals with the CCD detector array. One of the features of VHS is the capability of coherent addressing. Once all the images have been stored in a PR

Fig. 7.21 Principles of correlation addressing for volume holography

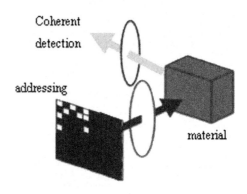

Fig. 7.21 Principles of correlation addressing for volume holography

crystal, then a certain reference beam can recover a corresponding image. While for a volume holographic correlator, a certain object image is input by an SLM, then it can retrieve a set of "reference beams" with different transmitting angles. The intensity of each beam stands for the correlation between the input image and the corresponding recorded image. In a word, not only data are recorded, but also a self-correlation and cross correlations of the images are generated in the volume holographic data storage as in Fig. 7.21.

For summing-up the HODS technology, it has the high data density, high capacity, high data rate of parallel transmission and correlation addressing. The key technologies in high-density volume holography storage are:

1. New principles of multidimensional holography with multiplexing.
2. Medium of new-typed doped crystals, organic, inorganic, and polymer materials.
3. The system parameters influenced on performances and minimization of holography storage systems.
4. Diffractive effect, sensitivity, dynamic range, response time of medium, and testing technology.
5. Encoding, decoding and super-high speed data transmission and processing.
6. The DSM is one of the important new multiplexing principles. So it will be described in detail as follow.

7.3.2 Comparison of Multiplexing Technology Schemes

The data recovery depends on the correlation between readout light and recorded reference light in HODS. So little deviation of the wavelength or the angle can cause the Bragg mismatch so that the diffraction efficiency of holograms can be decreased. While the readout light is completely same to the original reference light, the diffraction efficiency of holograms is the most high. Many holograms can be stored in a common volume due to the Bragg sensitivity to the angle or the wavelength so as to increase the recording density and capacity prominently. It is the multiplexing technology of VHS. The multiplexing technology is able to increase the recording density that decides the structure and complexity of the system. The angular multiplex was used in information storage. Common volume multiplexing, which is that many holograms are stored in common volume, includes angular multiplexing, wavelength multiplexing, phase-coded multiplexing, shift multiplexing, and speckle multiplexing so far. The comparison of the various methods and principles are shown in Table 7.7.

Since the change of the multiplexing parameters for recording and retrieving data pages in most of the multiplexing schemes depend on the mechanical structure, the transfer rate and the reliability of the HDSS are limited. However, the phase-coded multiplexing scheme can keep the static system structure, so it is superior to other multiplexing schemes. The random phase-coded multiplexing scheme is one of the simplest phase-coded multiplexing schemes. The scheme that the beams modulated randomly, called as speckle beams, are used to VHODS was proposed by T. Krile at first. A.E. Krasnov discussed the application of speckle reference light for recording holograms in thick media. But they did not consider the characteristics of the speckle beam in theory and its influences to retrieving data. The correlation effect of the speckle reference light in recording and retrieving holograms was illustrated by V. Markov in analysis for the shaping of the holographic laser beam. The group of V. Markov proposed the new multiplexing scheme of recording and retrieving holograms with the speckle reference light. Markov also proved the selectivity of the multiplexing scheme including speckle-shift selectivity and speckle-angular selectivity, and achieved high-density data recording and high-precision shift measure. The speckle multiplexing scheme proposed by Markov, which is a new and effective multiplexing scheme developed on basic of the conventional shift multiplexing, is called as the static speckle multiplexing scheme. It is not only used in the holographic volume storage, but also in the optical encryption that can prevent the users from attaining the unauthorized information. Otherwise, the application of random phase multiplexing in the holographic volume storage can increase the selectivity of shift, angular and wavelength multiplexing. So the random phase multiplexing becomes an appealing technology because it is potential to the storage capacity of system. Since then, the speckle multiplexing schemes developed rapidly and the achievements are followed as: Kamra K. implemented the speckle coding technique in beam fanning geometry. Markov V. proposed the relation between the speckle multiplexing and the laser transverse mode. Kim K. applied the fiber speckle patterns to HODS and enhanced the sensitivity of the multiplexing with the speckle-wavelength fixed multiplexing. Markov V. employed longitude selectivity of speckle field to achieve the multilayer speckle multiplexing. Sun C. implemented the all-optical fiber sensing system with the multimode fiber speckle. Zhang P. proposed the DSM scheme in 2003. Xiaosu Ma proposed the high-density optical storage based on the dynamic static speckle multiplexing scheme, and lustrated the insensitivity of speckle multiplexing in HDSS to the multilongitudinal mode of lasers in 2008 at Tsinghua University in China.

Table 7.7 Comparison of various multiplexing principles and methods for volume holographic storage

Multiplexing scheme	Implementation	Principles
Angular multiplexing	With the angular multiplex many holograms can be stored in a common volume. The difference of the holograms depends on the angle of incident light that can change in two–dimension	
Wavelength multiplexing	The wavelength multiplex that is based on the Bragg condition achieve recording many holograms in a common volume with the change of wavelength of reference light	

(continued)

Table 7.7 (continued)

Multiplexing scheme	Implementation	Principles
Shift multiplexing	The shift multiplexing that is based on the Bragg condition achieves recording many holograms in a common volume with the change of the position on the medium. The reference light of the spherical wave is employed and adjacent holograms can be superposed	
Phase-coded multiplexing	The phase-coded multiplexing that bases on the correlation between adjacent reference lights achieve recording many holograms in a common volume with phase-code of reference light depending on SLM. The wavelength and the angle of the reference light and object light are constant in the process of recording and retrieving	

(continued)

Table 7.7 (continued)

Multiplexing scheme	Implementation	Principles
Speckle multiplexing	The speckle multiplexing that base on the correlation between adjacent reference lights achieve recording many holograms in a common volume with speckle-code of reference light depending on random diffuser. The wavelength and the angle of the reference light and object light are constant in the process of recording and retrieving	
Hybrid multiplexing	The combination of the above multiplexing technology, including rotation-angular multiplexing, wavelength-angular multiplexing, space-angular multiplexing and angular-speckle multiplexing and so forth	

7.4 Multiwavelength Volume Holography Storage

The principles of multiwavelength optical storage can be used to volume holography storage to increase its capacity also.

7.4.1 Two-Wavelength Volume Holography Storage

The active part of 3D optical storage media is usually an organic polymer either doped or grafted with the photochemically active species. Alternatively, crystalline and solgel materials have been used. Sensitivities for different wavelength photopolymerizable materials mixed that can carry out multiwavelength volume holography storage so far as the crosstalk can be controlled availability. The two types of photopolymer holographic recording materials sensitive to blue and red light were presented. The photopolymerizable system comprises two monomers with a photoinitiation and an inactive component referred to as a binder. Its absorption spectrums are 480 and 630 nm that are suited wavelength 488 nm Ar+ laser and wavelength 632.8 nm He–Ne laser as shown in Fig. 7.22a. The experimental multiwavelength volume holography storage system is shown in Fig. 7.23. Main components of the medium are O–Cl–HABI and Diphenyliodonium hexafluorophosphate; the diffraction efficiencies are 30 % higher and sensitivities are within 25 J/cm^2 when exposed to red and blue lights, respectively. Spatial frequencies of red light ranged from 1600 to 2800 mm, and the best response spatial frequency is 2300 mm and the corresponding diffraction efficiency is 29 %. Spatial frequencies of blue light ranged from 1900 to 3000 mm, and the best response spatial frequency is about 2500 mm and the corresponding diffraction efficiency is 27 % with lower crosstalk among reconstruction images when

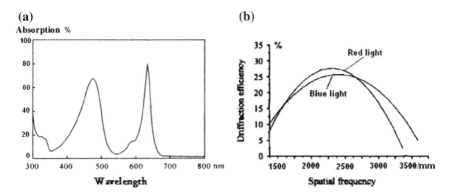

Fig. 7.22 Absorption spectrum (**a**) diffraction efficiency and Spatial frequencies (**b**) of two-wavelength volume holography storage medium

Fig. 7.23 Two-wavelength
volume holography storage
optical system: *SH* shutter,
M mirror, *L* lens, *SLM* space
light modulator, *BS* beam
splitter, *BE* beam epitome,
R recording media, *H* half
reflective mirror, *CCD*
detector

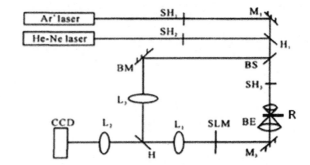

recording in the same point by red and blue lights as shown in Fig. 7.22b, so the
photopolymer is promising to two-wavelength multiplexing holographic storage.

The optical system of the two-wavelength volume microholography is shown in
Fig. 7.23, that focused beams of light are used to record submicrometre-sized
holograms in a photorefractive material, usually by the use of collinear beams. The
writing process may use the same kinds of media that are used in other types of
holographic data storage, and may use two-photon processes to form the holograms.

Data may also be created in the manufacturing of the media, as is the case with
most optical disc formats for commercial data distribution. In this case, the user
cannot write to the disc as it is a ROM format. Data may be written by a nonlinear
optical method, with very high power lasers to write speedy in. The fabrication of
discs containing data molded or printed into 3D structure has been demonstrated.
For example, a disc containing data in 3D may be constructed by sandwiching
together a large number of wafer-thin discs, each of which is molded or printed with
a single layer of information. The resulting holomem ROM disc can then be read
using a 3D reading method. Other techniques for writing data in three-dimensions
have also been examined, as persistent spectral hole burning, which also allows the
possibility of spectral multiplexing to increase data density. However, the media
currently requires extremely low temperatures to be maintained in order to avoid
data loss. Meanwhile, the void formation and microscopic bubbles are introduced
into a media by high intensity laser irradiation. Chromospheres poling, where the
laser-induced reorientation of chromospheres in the media structure leads to read-
able changes etc. can be used to ROM format holomem.

The reading of data from 3D holomem has been carried out in many different
ways. While some of these rely on the nonlinearity of the light-matter interaction to
obtain 3D resolution, others use methods that spatially filter the media's linear
response. Reading methods include: Two photon absorption such as resulting in
either absorption or fluorescence (Linear excitation of fluorescence with confocal
detection) etc. This method is two-photon microscopy and confocal laser scanning
microscopy essentially. It offers excitation with much lower laser power than does
two-photon absorbance, but has some potential problems because the addressing
light interacts with other data points in addition to one addressed. A method of
measurement of small differences in the refractive index between the two data states

is proposed. The method usually employs a phase contrast microscope or confocal reflection microscope. No absorption of light on the whole, so it is no risk to damaging data while reading, but requirement of refractive index mismatch to the disc. The thickness changing of the medium can bring accumulated random errors to destroy wavefront of optical system and focus quality. So some new optical storage technology has been demonstrated that as the method to write data on poled polymer matrix, optical coherence tomography parallel reading/writing etc.

7.4.2 3D Multiwavelength Medium

Medium for 3D multiwavelength medium holography storage could be several forms:

1. Disc—the disc medium offers a progression from CD/DVD/BD, and allows to carry out by the spinning process.
2. Card—the credit card form medium is portability and convenience, but maybe lower capacity than disc.
3. Crystal, Cube, or Sphere as a solid-state storage that can store big information; it could achieve MD holography storage in principles easily.

The manufacturing single layer disc is rather simple with the conventional molding process and may be possible for some conventional manufacturing equipment. The multilayer disc is more complex, that manufacturing this medium has to be constructed layer by layer. This is required if the data is to be physically created during manufacture. Another alternative method is analogous to a roll of adhesive tape to create the 3D multiwavelength medium.

7.4.3 3D Multiwavelength Holography Storage Drive

The drive for 3D multiwavelength medium holography storage was designed to reference common CD/DVD/BD drives technology to read and write to MD holography storage medium, particularly if the form and data structure is similar to conventional CD/DVD/BD. However, there are some notable differences that must account when designing such drive, including:

1. Laser sources and aberration correction: in the first, as lasers are high-powered and high-frequency accordingly that required especial bulky, cool, and safety concerns. Many solid-state lasers or pulsed lasers with 830, 780, 650, 532, and 405 nm, even more different wavelength are applied in a system, when 2-photon and difficult wavelength absorption are utilized for 3D holomem. Many existing optical drives utilize continuous wave diode lasers used to multiwavelength storage easily by these technologies, such as 532 nm laser. But these larger

lasers are difficult to integrate into the read/write optical systems of the 3D holomem. Variable spherical aberration correction is need to it, because the system must address different depths in the medium, and at different depths the different spherical aberration was induced in the wavefront. Many possible methods existed, including optical elements swap in and out of the optical system, moving elements and adaptive optics spherical aberration etc. Please referSect. 6.5 of Chap. 6 in this book.

2. Optical system: in multiwavelength 3D holomem systems, several wavelength (colors) of light are used together (e.g. reading laser, writing laser, sometimes even two lasers are required just for writing). Therefore, as well as coping with the high laser power and variable spherical aberration, the optical system needs to combine and separate these different wavelength light efficaciously with various MOEMs optical devices.

3. Detection: in conventional DVD or BD drives, the signal produced from the disc is a reflection by the addressing laser beam directly, it is therefore very intense. The multiwavelength 3D holomem storage however, that the signal must be generated within the tiny volume, so it is much weaker than the conventional CD signal evidently. In addition, it is radiated in all directions from the addressed point, so that must be used special optical collection devices, high-sensitivity photodetector and signal amplifier.

4. Data tracking: for 3D holomem systems, once it is moving along the z-axis layer by layer individual, that tracked and accessed cannot be similar to conventional CD, as it is not track on every layer. So possibility is using parallel or page-based addressing, it allows much faster data transfer rates, but requires the additional complexity of SLMs, imaging signal detectors and more powerful lasers with complex data handling. Despite the great attractive nature of 3D holomem storage systems, the development of commercial products has not started yet.

5. Destructive readout: Since both the reading and the writing of data are carried out with many laser beams, there is a potential for the reading process to cause a small amount of writing. In this case, the repeated reading of data may erase it eventually in medium, i.e., destructive readout. In this case, it has to be addressed by other approaches, such as the using of different absorption band for reading and writing, or using of a light which is not involved in the absorption spectrum to read.

6. Thermodynamic stability: Since most 3D holomem storage medium are based on photochemistrical reactions, therefore there is a risk that either the unwritten points will slowly become written points or revert to being unwritten at higher temperature even normal room temperature indeed. This issue is particularly serious for the spiropyrans, but extensive research was conducted to find more stable chromophores for 3D holomem memories which have very high thermodynamic stability as in Chap. 3 of this book.

7. Medium sensitivity: the medium sensitivity is a not a terminal research project. Although there are a lot of inorganic or organic photochemistry materials have been used to 3D holomem storage with good performances now, much higher

sensitivity of medium is need in order to increase storage density and capacity, decrease power of lasers, and to take semiconductor detectors to replace big Ti-sapphire lasers or Nd:YAG lasers to achieve excitation.

7.4.4 Exposure Process

In exposure process, holomem medium occurs photorefractive effect to recording information by light, so exposure is an important photochemical and photo-physical reaction in 3D holomem recording medium. The photopolymer is main 3D holo-mem medium, so choose it for example. The polymer medium after exposure (recording) with the optimized recording process, that generated a maximum gratings M direct result and equal refractive index modulation size at the same time ideally. Although researchers carried out some schedule to optimum scheduling experiments and compare various multiplexing techniques already, it is need for more repetitive examination and experimental procedures sequentially, due to that were not very satisfactorily, especially when changing the materials response characteristics and parameter of optical systems.

In the photochemical and photo-physical processes, during which occur pho-topolymerization, when attempting to improve the photopolymer material's application performances, and then carried out a modeling of the mechanisms which can denote the action of photopolymers in during- and postexposure. It has led to the development a mathematic tool, which can be used to predict the behavior of these materials under various complicated conditions. That is the Non-local Photopolymerization Driven Diffusion (NPDD) model. It is very useful to predict and analyze their implications on the improvement of photopolymer material performances. The NPDD introduced the research on photopolymer materials response characteristics to explain in a single model, as a most significant analysis method. In the NPDD, although shows that the diffusion is important in the grating formation process, but high-spatial frequency cut off is explained by assuming that chains grow away from their initial point. This material model is in the process of development, and its full range of validity, especially when combined with rigorous electromagnetic models. Meanwhile, the NPDD allowed the quantitative comparison of different materials and provided insights into the behavior of photopolymer materials.

The two-harmonic version of this model has a significant range of validity and can be used to approximate analytic solutions. It has been used in conjunction with Kogelnik's first-order coupled-wave mode to characterize photopolymer materials. Combined with rigorous electromagnetic methods, the NPDD has also been applied recently to examine commercial holographic materials for use in the fabrication of waveguide components. The two-harmonic NPDD is used to optimum exposure schedule for holographic storage medium; its main targets are as follows:

1. Production process of M sinusoidal gratings in a single material layer of equal refractive index modulation.

2. Optimization of the total polymer generated photorefractive effect, thus optimizing the modulation of each grating.

The NPDD with the Gaussian response function is valid for some photopolymer materials. This assumption is supported by many model's experimental results, carried out with different recording materials. Specifically, the approximate two-harmonic analytic was valid for low-exposure recordings. The monomer concentration distribution equalizes, through diffusion between exposures. It has been demonstrated that the relatively diffusion processes that take place in different materials:

1. Staggered exposures followed by repeated relaxation periods can be used to overcome constraints placed on photopolymer materials due to reciprocity failure.
2. Previously reported measurements of the relaxation of a strongly exposed acrylamide-based material containing both uncrosslinked polymer and monomer distributions.

That completed relaxation can require several minutes, when analysis presented place no constraint on the relaxation time necessary. Furthermore, no variation of the diffusion constant as a function of exposure, that the sufficient time is allowed for full monomer relaxation to take place and that any change in the diffusion constant is equivalent to change of the R parameter, see Eq. (7.5) later. In a commercial data storage system, allowing for long relaxation times between exposures may be impractical. It appears, however, that the analysis presented here might be modified, following the form of analysis to model a situation of partial/incomplete material relaxation between exposures. Although ultimately diffraction efficiency defines output signal strength, assume it can be controlled independent of grating formation by varying layer thickness. Therefore, an ideal uniform volume grating is formed and can ignore (1) grating nonuniformities, (2) layer shrinking/swelling, (3) induced surface relief gratings, (4) during each exposure, no previously recorded grating are replayed, and (5) that scatter by the material is negligible before and after recording. In this case, by varying the grating thickness (layer depth), the diffraction efficiency and angular selectivity can be controlled. Commercially available storage media have been reported that exhibit extremely good behavior.

The analysis presented is based on the storage of a highly idealized data pattern, i.e., exposure with a maximum fringe visibility, $V = 1$, sinusoidal interference pattern. The inclusion of a nonlinear material effect, however, effectively, a spectrum of gratings, simultaneously recorded is modeled. The predictions for several combinations of material and recording parameters are presented, where the parameters used are of the same order of magnitude as those determined experimentally. In this way, present a first-order study of the behavior of particular classes of photopolymer materials for use as holographic storage medium.

7.4.5 Materials Reaction Characteristics

Assume a sinusoidal interference pattern of amplitude I_0, period Λ and fringe visibility $V = 1$ is used for the ith exposure with duration t_i, and the holographic material with an initially uniform monomer concentration $u_0^{(i)}(0)$. Since the initial monomer distribution is uniform, all higher monomer harmonics are assumed negligible, i.e., $u_1^{(i)}(0) = 0$. In the two-harmonic NPDD approximation, the analytic solution found for the average, or zero-order harmonic, of monomer concentration is of the form:

$$u_0^{(i)}(\xi_i) = u_0^{(i)}(0) \times \bar{u}_0(\xi_i) \tag{7.2}$$

where $\xi_i = F_0 t_i = k I_0^\gamma t_i$, which is related to the ith exposure energy, $E_i = I_0 t_i$, k is a constant, and where previously derived the dimensionless factor as:

$$\bar{u}_0(\xi_i) = \exp\left[-(W + f_0)\frac{\xi_i}{2}\right]\left[\cosh\left(B\frac{\xi_i}{2}\right) + \left(\frac{W - f_0}{B}\right)\sinh\left(B\frac{\xi_i}{2}\right)\right] \tag{7.3}$$

For brevity, introduced parameters as:

$$W = S(f_0 + f_2/2) + R,$$
$$B = \left[(W - f_0)^2 + 2f_1^2 S\right]^{1/2} \tag{7.4}$$

And have:

$$S = \exp(-K^2\sigma/2)$$
$$R = DK^2/F_0 = R_0/(I_0^\gamma \Lambda^2) \tag{7.5}$$

where σ quantifies the effects of nonlocal polymerization due to chain growth (propagation) away from the point of initialization. $K = 2\pi/\Lambda$ is the grating vector magnitude. D is the diffusion constant of the monomer inside the material, which is assumed not to change during exposure, and therefore $\alpha = 0$. The rate F_0 of polymerization has been assumed to be either a linear function of the illumination intensity I_0 or a function of the illumination intensity I_0 raised to the power $\gamma = 1/2$. The parameters (f_0, f_1, f_2) are Fourier coefficients introduced to take account of γ and to assume F_0 does not vary as a function of time. The corresponding expression for the first-harmonic polymer concentration is:

$$N_1^{(i)}(\xi_i) = u_0^{(i)}(0) \times \bar{N}_1(\xi_i) \tag{7.6}$$

Once again emphasize the dependence on the initial concentration of monomer and define the dimensionless function, as:

$$\bar{N}_1(\xi_i) = \frac{4f_1S}{(W+f_0)^2 - B^2}\left\{R + \exp\left[\frac{(W+f_0)\xi_i}{2}\right] \times \left[\left(\frac{L}{B}\right)\sinh\left(\frac{B\xi_i}{2}\right) - R\cosh\left(\frac{B\xi_i}{2}\right)\right]\right\}$$

(7.7)

where:

$$L = (f_0 - R)R + [-f_1^2 + (f_0 + f_2/2)(2f_0 - R)]S$$ (7.8)

To assume the first-harmonic amplitude of the grating refractive index modulation is linearly proportional to this polymer concentration. In order to observe the general characteristics of materials governed by these equations, present four growth curves, as in Figs. 7.24a, b and 7.25a, b. These show the variation of N_1 as a function of j, for a representative range of material (D, s, a, g, k) and exposure parameters (I_0, t, L, V). In each figure, two curves are shown, for the cases $R = 10$ and $R = 0.1$. From the definition of R given in Eq. (7.5) that can see that each of these growth curves would be produced in several different physical situations. For example, keeping the spatial frequency (grating period) and exposing intensity constant, set $D = 10$, $F_0 \rightarrow R = 10$ and then set $D = F_0/10 \rightarrow R = 0.1$. However, that is having exactly the same R values occur for completely different recording situations. For example, keeping the material parameters D and k constant, and choosing an initial $I_0^\gamma \Lambda^2$ value for $R = 10$, then set $I_0^\gamma \Lambda^2$ to a different value more 100 times, and giving $R = 0.1$, that R can be maintained at a constant value if the exposing period varies, since this can be compensated for by a variation of the exposing intensity, while the corresponding j values can remain unchanged by varying the exposure time t. In general, a material with a high R value would be expected to produce a high-fidelity recording, while as R decreases, higher order grating harmonics will be generated, most noticeably during long exposures. From Figs. 7.24 and 7.25, we can see that, as R increases, both the maximum value attained by N_1 increases and the value at N_1 reached its maximum value ξ_M.

For $R = 10$, the maximum always occurs beyond the range of the figures, $\xi > 10$, while in all cases shown, when $R = 0.1$, a clear maximum always occurs for $\xi < 3$. This is significant when later discuss data storage scheduling. The effects of varying S and γ are also explored in these figures. In Figs. 7.24a and 7.25a, $S = 1$, and assume the materials exhibit local responses $s = 0$. In Figs. 7.24b and 7.25b, the materials are assumed to exhibit significant nonlocal behavior with $S = 0.265$ and thus $\sigma/\Lambda^2 = 1/8$. Examining the figures, as S decreases (σ/Λ^2 increasing), the polymer-concentration growth curves decrease in size and the $R = 10$ and $R = 0.1$ curves become more similar in size. Furthermore, the smaller the value of S, the more slowly the material reaches its final steady-state value. Finally, in Fig. 7.24, $\gamma = 1$, while in Fig. 7.25, $\gamma = 1/2$. Therefore, observe the effects of a nonlinear material response to the illuminating intensity. The smaller the value of γ, the lower the N_1 growth curve maxima. Using the analytic expression and this range of γ values, γ is observed to have relatively little effect on the results. The validity of the analytic expressions, derived retaining only two harmonics, has

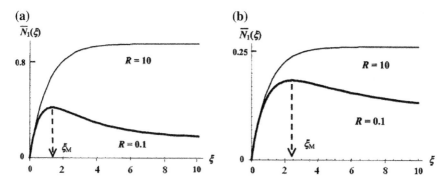

Fig. 7.24 **a** First harmonic of polymer-concentration growth curve, with $u_0^{(1)}(0) = 1$, $S = 1$ (local: $\sigma = 0$) and $\gamma = 1$ (linear) to see Table 7.11. **b** First harmonic of polymer-concentration growth curve, with $u_0^{(1)}(0) = 1$, $S = 0.265$ (nonlocal: $\sigma/\Lambda^2 = 1/8$) and $\gamma = 1$ (linear) as Table 7.12

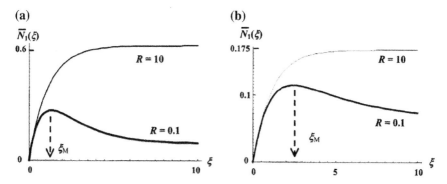

Fig. 7.25 **a** First harmonic of polymer-concentration growth curve, with $u_0^{(1)}(0) = 1$, $S = 1$ (local: $\sigma = 0$) and $\gamma = 1/2$ (nonlinear) to see Table 7.13. **b** First harmonic of polymer-concentration growth curve, with $u_0^{(1)}(0) = 1$, $S = 0.265$ (nonlocal: $\sigma/\Lambda^2 = 1/8$) and $\gamma = 1/2$ (nonlinear) to reference Table 7.14

previously been discussed in the case when it is assumed exposure continues until effectively all monomer present is polymerized, i.e., when ξ is large. An indication of the range of validity and accuracy can be provided using the percentage difference between the four- and two-harmonic NPDD prediction of N_1, as:

$$\text{error}(\%) = \frac{N_1(4 - \text{harmonic}) - N_1(2 - \text{harmonic})}{N_1(4 - \text{harmonic})} \times 100. \qquad (7.9)$$

It has previously been shown that, in all cases for which either $R > 1$ or $S < 0.25$, relatively good agreement, i.e., error % <10 %, occurs for long exposures; however, care must be taken to retain sufficient harmonics to ensure convergence for highly nonlinear and nonlocal materials. No examination of the performance of the approximate analytic solutions, for the case when ξ is small, has previously been undertaken. This case will be particularly important when many weak gratings are to be stored sequentially in a single medium. Therefore, compare the same material cases were examined in Figs. 7.24 and 7.25. Following the procedure used previously, plot the percentage error in N_1 (error %), over the ranges $-2 < \log_{10} R < +2$ and $0.1 < S < 1$, when $\xi < 0.1$ and 1. See Fig. 7.29a and b $\gamma = 1/2$ (nonlinear). See Fig. 7.29b. And 4, for the shorter exposures, the error % is at least an order of magnitude smaller than for the long-exposure case, $\xi > 10$, previously examined. In keeping with the previous results, the errors are always smallest when $R > 1$ and $S < 0.25$. Significantly, as γ decreases, the error % increases by a factor of ~ 10, for shorter exposures. The most significant result of this analysis is that the analytic expressions are reasonably accurate over useful ranges of ξ, S, R, and γ, and therefore can be used to estimate the holographic data storage schedule. The results presented reinforce and complement the detailed theoretical and experimental results presented elsewhere and can be best appreciated in that fuller context.

7.4.6   Recording Schedule Optimum

As assumed, sufficient time is available between exposures to allow the monomer distribution to completely equalize through diffusion. In this case, $u_0^{(i)}(\xi_i)$, the average monomer concentration at the end of the $(i + 1)$th exposure, is the starting monomer concentration for the $(i + 1)$th grating, and in general can write:

$$u_0^{(i)}(\xi_i) = u_0^{(i-1)}(\xi_{i-1}) \times \bar{u}_0(\xi_i)$$
$$= u_0^{(1)}(0) \times [\bar{u}_0(\xi_1) \times \bar{u}_0(\xi_2) \times \cdots \times \bar{u}_0(\xi_{i-1})] \times \bar{u}_0(\xi_i). \tag{7.10}$$

As it has been assumed that $u_0^{(1)}(0) = 100$, and have then studied the case of a single continuous recording, $\xi = \xi_1$. Similarly, for the ith exposure polymer concentration as:

$$N_1^{(i)}(\xi_i) = u_0^{(i-1)}(\xi_{i-1})\bar{N}_1(\xi_i) \tag{7.11}$$

Therefore, $N_1^{(i)}(\xi_i)$ depends on the monomer concentration left over after the previous recording. That all the gratings be equal, therefore:

$$N_1^{(i)}(\xi_i) = N_1^{(i+1)}(\xi_i) \Rightarrow u_0^{(i-1)}(\xi_{i-1})\bar{N}_1(\xi_i) = u_0^{(i)}(\xi_i)\bar{N}_1(\xi_{i+1}) \tag{7.12}$$

After M recordings, the material contains the following total amount of polymer is:

$$N_1^{\text{TOT}} = \sum_{i=1}^{M} N_1^{(i)}(\xi_i) = u_0^{(1)}(0) \sum_{i=1}^{M} \left[\prod_{m=1}^{m=i-1} \bar{u}_0(\xi_m) \right] \bar{N}_1(\xi_i) \tag{7.13}$$

Since as much of the dynamic range as possible should be used, N_1^{TOT} must be maximized. Combining Eqs. (7.10), (7.11), and (7.12), can get:

$$\bar{N}_1(\xi_{i+1}) = \bar{N}_1(\xi_i)/\bar{u}_0(\xi_i) \tag{7.14}$$

Given any suitable value of ξ_i, can iteratively find all the j values that guarantee uniformity. For example, if an optimum value for the last exposure ξ_M is known, and can immediately find ξ_{M-1} by numerically solving the equation. An optimum ξ_M value can be found. For the last exposure, $i = M$, the maximum possible grating that can be recorded for any initial monomer concentration will occur when:

$$\left. \frac{\partial \bar{N}_1(\xi)}{\partial \xi} \right|_{\xi=\xi_M} = 0 \tag{7.15}$$

ξ_M can be found using any standard root-finding procedure. The ξ_M values for $R = 0.1$ are shown in Figs. 7.24 and 7.25. Choosing this value, have found the exposure automatically, which maximizes the amplitude of the last grating. Substituting this value into Eq. (7.13), ξ_{M-1} can be found and so on. Once the complete set of ξ_i values is found for the M gratings, the associated polymer-concentration harmonic strength $N_1^{(i)}(\xi_i)$ can then be found by first substituting the ξ_{i-1} values back into the monomer formula, to give the initial monomer concentration for the ith exposure, $u_0(\xi_{i-1})$, and then calculating the polymer-concentration value. Regarding the accuracy of the schedule results, that ξ_M is both the first and the largest ξ value to be estimated. The analytic expressions will therefore, in general, be most inaccurate in estimating it. Since an iterative technique is being used, any error introduced in this first value will affect the entire schedule, so ξ_M should ideally be found to a high degree of accuracy. This may necessitate the use of higher order harmonic analysis or direct solution of the partial differential Eq. (7.9). In Table 7.8, the accuracy of the value of ξ_M calculated using the analytic function when $R = 0.1$ is tested. The ξ_M values obtained when retaining

Table 7.8 Exposures for maximum N_1, error% = $100[\xi_M(4)\xi_M(2)]/\xi_M(4)$

	R = 0.1; Linear, $\gamma = 1$			R = 0.1; Nonlinear, $\gamma = 1/2$		
	Four-Harmonic	Two-Harmonic	% Error	Four-Harmonic	Two-Harmonic	%
$S(\sigma)$	$\xi_M(4)$	Analytic $\xi_M(2)$		$\xi_M(4)$	Analytic $\xi_M(2)$	Error
Local: S (0) = 1	1.65	1.35	18 %	1.45	1.34	9.5 %
Nonlocal: S = 0.265	2.4	2.42	0.8 %	2.41	2.42	0.4 %

Table 7.9 Schedules and grating strengths

$R = 0.1, \gamma = 1, S = 1 \ (\sigma = 0)$				
$i(M)$	ξ_i	$N_1^{(i)}(\xi_i)$	$u_0^{(i)}(\xi_i)$	N_1^{TOT}
1	0.43	0.28	0.68	0.57
(2)	1.36		0.25	
1	0.29		0.76	
2	0.43	0.22	0.52	0.65
(3)	1.36		0.20	
1	0.22		0.81	
2	0.29	0.18	0.62	0.71
3	0.43		0.42	
(4)	1.36		*0.16*	
1	0.16		0.84	
2	0.22		0.69	
3	0.29	0.15	0.52	0.75
4	0.43		0.37	
(5)	1.36		0.14	

$*u_0(0) = 1, V = 1, \alpha = 0$ see Fig. 7.24(a)

four or two harmonics, $\xi_M(4)$ and $\xi_M(2)$, and the percentage error between these values are tabulated for four representative cases as in Table 7.9. As both S and γ decrease in size, corresponding to stronger nonlocal and nonlinear effects, the accuracy of the second-harmonic ξ_M will increase. These values are sufficiently accurate to provide a good first-order estimation of the optimum schedule.

The first apply the algorithm to the case when $R = 0.1$, $\gamma = 1$ (linear: $f_0 = 1$, $f_1 = 1$, $f_2 = 0$), and $S = 1$ (local: $\sigma = 0$). Start by examining the behavior of the medium for low numbers ($M < 8$) of recorded gratings. The optimum schedules when $M = 2, 4$, and 7 are illustrated in Fig. 7.26, and a detailed set of predictions is presented in Table 7.9, for $M = 2, 3, 4$, and 5. Table 7.9 is organized as follows:

1. In the first column, list the sequence of gratings, the bold numbers (in parentheses) indicating the total number of exposures.

Fig. 7.26 ξ_i plotted as a function of i for different numbers of exposures M. $R = 0.1$, $\gamma = 1$ (linear), $S = 1$ (local case), to see Tables 7.9 and 7.10

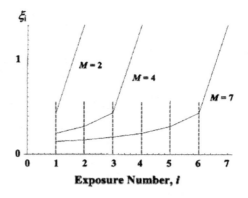

Table 7.10 Corresponding
to Fig. 7.24(a)

Local, $S = 1$; Linear, $\gamma = 1$ ($f_0 = 1, f_1 = 1, f_2 = 0$)				
	$R = 0.1$, $\xi_M = 1.3603$		$R = 10$, $\xi_M = 39.77$	
M	ξ_1	N_1	ξ_1	N_1
10	0.92545	0.0844	0.10608	0.0969
50	0.01952	0.0191	0.02012	0.01974
100	0.00986	0.0 098	0.01002	0.0099
500	0.00199	0.0020	0.00200	0.0020
1000	0.00090	0.0010	0.00100	0.0010

2. In the second column, give the values of ξ_i for each exposure. Examining the schedule, it is immediately clear that, for different values of M, the same ξ_i values reoccur.

3. In the third column are listed the first-harmonic polymer concentrations $N_1^{(q)}(\xi_q)$ in each grating (which are identical for uniform strength).

5. In the fourth column are listed the monomer concentrations remaining after each exposure. The values in italics in this column are the unused monomer concentrations left after the last exposure. This quantity decreases as M increases.

5. The values in the last column, on the right-hand side, are the total polymer concentrations formed, $N_1^{TOT} = M \times N_1^{(q)}(\xi_q)$. As can be seen, this value is much larger than the maximum value, which can be recorded in the single continuous exposure, shown for this case in Fig. 7.24a. Furthermore, this total amount increases as the number of gratings stored increases.

As M increases, the ratio ξ_M/ξ_1 increases dramatically, in practical storage systems, a large variation between the first and last exposures may be undesirable.

Practical implementations may therefore involve either leaving a significant amount of unpolymerized monomer within the material or the optimization of the recording schedule using criteria different than those proposed here. These results also hold true when even larger numbers of gratings are recorded. In Table 7.10, the behavior of this local ($S = 1$) linear ($\gamma = 1$) material, when $M = 10$, 50, 100, and 1000, is presented. For each value of R the common ξ_M, and the values of ξ_1 and N_1, are given. As R increases, ξ_M increases, while M increases, ξ_1 decreases and N_1 decreases, both are becoming almost identical for the two R values.

In Table 7.11, the analogous results for the case of $R = 0.1$ for a nonlocal ($S = 0.265$) linear material ($\gamma = 1$) are presented. The effect of introducing the nonlocality is to increase ξ_M and all the schedule values, while reducing the amplitude of each grating. To illustrate the general validity of the results, present results for this material when $R = 100$ and 0.01. For $R = 100$, $\xi_M = 37.1494$, and when $M = 1000$, $\xi_1 = 0.001000$ and $N_1 = 0.000264943$. When $R = 0.01$, then $\xi_M = 2.2440$, and for $M = 1000$, $\xi_1 = 0.0009997$ and $N_1 = 0.00026476$. Thus the general trends observed for $R = 0.1$ and $R = 10$ hold true.

When the nonlinear materials, $\gamma = 1/2$, governed by the behavior shown in Fig. 7.25a and b, are examined. To graph the schedules for $M = 10$, in Fig. 7.27a for

Table 7.11 Corresponding to Fig. 7.24(b)

$\gamma = 1$ $(f_0 = 1, f_1 = 1, f_2 = 0)$,
$S = 0{,}265, R = 0{,}1, \xi_M = 2.4546$

M	ξi	N_1
10	0.100610	0.025031
50	0.010980	0.005228
100	0.000990	0.002630
500	0.001999	0.000529
1000	0.000999	0.000265

$R = 10$, and in Fig. 7.27b for $R = 0.1$. In each figure, the schedules when $S = 1$ and $S = 0.265$ are presented. In Fig. 7.27a, only the first nine of the ten ξ_i values are plotted because both ξ_M values are so large (>40; see Tables 7.12 and 7.14). In Fig. 7.27b, ξ_i is plotted as a function of i for all $M = 10$ exposures. Examining the results, when $R = 10$, the schedules are almost identical for both S values; however, the corresponding ξ_i values are smaller than in the $R = 0.1$ case. The effect of the nonlinearity, γ has been to decrease N_1 and slightly increase the ξ_1 value. The most significant effect of decreasing S is to decrease the grating amplitude achievable by a factor of up to 4. To test the range of validity of these results, for the nonlocal ($S = 0.265$) nonlinear ($\xi = 1/2$) material presented in Table 7.13, examine the cases when $R = 10$ and $R = 0.1$. Setting $R = 100$, can get $\xi_M = 40.107288$, and in the case when $M = 1000$, $\xi_1 = 0.001111$ and $N_1 = 0.00017661$. Furthermore, if $R = 0.01$, then $\xi_M = 2.259032$, and for $M = 1000$, $\xi_1 = 0.001110$, and $N_1 = 5$ 0.00017646, that the trends observed for $R = 0.1$ and 10 hold true.

Using the nonlocal polymerization-driven diffusion model, a theoretical methodology allows an optimum recording schedule to be derived. The effects of varying M, R, γ, and S, on both the schedule and the resulting stored grating

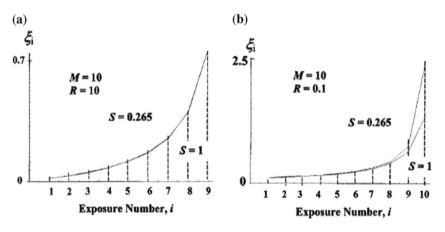

Fig. 7.27 **a** ξ_i plotted as a function of i for the first nine exposures. $M = 10$, $\gamma = 1/2$ (nonlinear). The almost identical $S = 0.265$ and $S = 1$ cases are shown. **b** ξ_i plotted as a function of i for all $M = 10$ exposures. $\gamma = 1/2$ (nonlinear). Both the $S = 0.265$ and $S = 1$ cases are shown [33]

Table 7.12 Corresponding to Fig. 7.25(a)

	Local, $S = 1$; Nonlinear, $\gamma = 1/2$			
	$R = 0.1$. $\xi_M = 1\,349$		$R = 10$, $\xi_M = 42.86$	
M	ξ_M	N_1	ξ_1	N_1
10	0.1004	0.05524	0.11616	0.06398
50	0.0215	0.01267	0.02223	0.01309
100	0.0109	0.00648	0,01110	0.00659
500	0.0022	0.00132	0.00220	0.00133
1000	0.0011	0.00066	0.00110	0.00067

Table 7.13 Corresponding to Fig. 7.25b

	Nonlocal, $S = 0.265$; Nonlinear $\gamma = 1/2$			
	$R = 0.1$ $\xi_M = 2.4388$		$R = 10$, $\xi_M = 40.211$	
M	ξ_1	N_1	ξ_1	N_1
10	0.11060	0.01653	0.11679	0.01747
50	0.02212	0.00348	0.02238	0.00351
100	0.01108	0.00180	0.01114	0.00176
500	0.00222	0.00035	0.00222	0,00035
1000	0.00111	0.00018	0.00111	0,00017

amplitudes' sizes for representative parameter values. The methodology is based on a clear set of physical assumptions and approximations. The results of the analysis include the following:

1. The individual ξ_i values in the exposure schedule are independent of the number of gratings to be recorded.
2. Implementing the optimum schedule may necessitate leaving unpolymerized monomer at the end of the recording process, u_0 $(\xi_M) > 0$.
3. The effects on the schedule values, ξ_i of the nonlocal response function becomes most significant as $i \rightarrow M$ and as R decreases in size. From the results, the largest percentage differences between schedules occur for $i = M$, $S = 1$, and $R = 0.1$.
4. The good data storage material is characterized by a low ratio between the first and last exposure ξ_M/ξ_1, indicating a narrow range of required exposure, so the material with a low value of R is best. That a low value of R implies $F_0 > D$, and for long exposures, this indicates a lower fidelity recording material (the production of significant higher grating harmonics).
5. If a good data storage material is primarily characterized by large uniform grating strength, then the more local the material, $S \rightarrow 1$, and the more linear the material $\gamma \rightarrow 1$, the better the material performance.
6. The value of R is of decreasing significance as the number of gratings recorded, M, increases. This is particularly true for the ξ_i values when $i < M$ and $M > 100$.
7. Having examined the accuracy of the approximate N_1 formulas for the case of short exposures, and also for use in estimating the value of ξ_M, that more exact numerical techniques will be required to more accurately identify the optimum schedule.

Several possible extensions of both the NPDD and the schedule optimization procedure method presented are currently being examined. It has been assumed that a linear relationship exists between monomer concentration and the rate of polymerization. This assumption motivated by variation of k as a function of ξ, which have observed in the experimental results. In the case of the schedule optimization algorithm, it is necessary that other optimization criteria and more general exposing intensity patterns be examined. In general, the pre-exposure encoding of data may be used to improve over all HDSS performance. Finally, the effects of imperfections in the recorded gratings, and thus on the readout diffraction efficiency, and the inclusion of relaxation time between exposures, must be explored as constraints on the optimization procedure.

7.4.7 Holographic Encryption

In optical storage and communication systems, optical encryption and data security are critical issues in preventing data storage and transmission from unauthorized access and attack. The optical image encryption method based on double random-phase encoding (DRPE) was proposed by Refregier et al. Since then, optical encryption technique is attracting increased research attention. Numerous proposals and improvements in optical encryption based on multiple-parameter fractional Fourier transform, Fresnel transform, joint transform correlator, digital holography, and phase-shift interference have also been made presented. In particular, the introduction of holography greatly broadens the flexibility, applicability, and variety of optical encryption. Holography can improve the security and can simplify the optical system by canceling the lens.

Aside from digital holography, increasing interest is given to computer-generated holography (CGH) because of its flexibility. Using computer can simulate the whole process of generating a hologram without optical exposure and development. Hence, many researchers implemented CGH into optical encryption, and proposed many useful methods. Among those methods, double random-phase encryption in Fresnel domain (DRPEiFD) is one of the most commonly used methods. For traditional DRPE, lenses make the optical system difficult to alignment of space. However, DR-PEiFD does not need any lens during encryption and decryption. This encryption method is completely lensless, which reduces the requirements in hardware. Therefore, DRPEiFD is more applicable and easier to implement. For these methods, noise is merely an incidental improvement along with new setups or algorithms. However, for the encryption and decryption technology, noise is always a vital problem. If great noise contaminates the decryption result from which people cannot recognize the original information, then the encryption method is not successful. More seriously, a lack of the original quality precisely may make potential users unwilling to use such encryption. Thus, it is in need of a new method to change the present circumstance in holographic encryption.

Quick-response (QR) code, which is a kind of two-dimensional (2D) barcodes, is widely used in security, cyber application, management system, etc. A QR code can conceal specific information in its special geometric pattern, which is composed of black and white dots in a square. QR code has become popular in our daily life because of its fast readability and great storage capacity. Since QR code has strong fault tolerance and error correction capability, even when nearly half of a QR code is broken, we can still recover the original information. Barrera et al. introduced a QR code in a standard optical encryption system. They used QR code to avoid speckle noise from polluting the outcomes of normal optical encryption and achieved attractive results. This new concept was reviewed as one of the research highlights in cryptography. For further improving the combination of QR code and optical encryption technology, introduce QR code into DRPEiFD to eliminate noise during encryption and decryption. First, transform the original information (such as texts) to a QR code with available software on the Internet, then encrypt this QR code with DR-PEiFD, and acquire a CGH. This QR code is a container in the temporary storage of information. In the decryption procedure, it is opposite to the encryption, obtain a contaminated QR code with noise. The distinctive error correction capability of this method provided the retrieval of clean and noise-free input information with appropriate applications in a smartphone. The numerical results demonstrate the obvious advantages because the final user is apparently satisfied with the retrieval information. DRPEiFD realizes lensless encryption with two Fresnel diffraction transforms and two white-noise phase plates within statistic independence. Figure 7.28a shows the optical setup of encryption. Supposing that λ

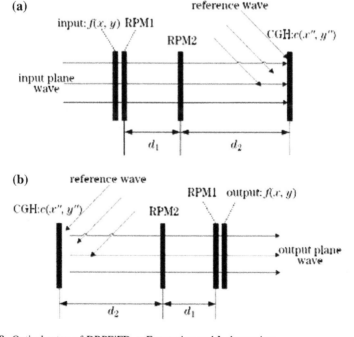

Fig. 7.28 Optical setup of DRPEiFD. **a** Encryption and **b** decryption

is the incident Fig. 7.39. Optical setup of DRPEiFD included encryption and decryption wavelength. The $f(x, y)$ is the input information to be encrypted. Two random-phase plates RPM1 and RPM2 are $\delta(x, y) = \exp[jn(x, y)]$ and $\varphi(x', y')$ $= \exp[jm(x', y')]$, respectively, wherein $n(x, y)$ and $m(x', y')$ are evenly distributed white noises between $[0, 2\pi]$ within statistic independence [34].

Two random-phase plates are placed parallel at planes where $f(x, y)$ is located; and where $f(x, y)$ is different distance d_1. The reference wave is $R(x'', y'')$, which has the same wavelength as the input plane wave. Then set a detector at the distance of $d_1 + d_2$ from the object plane to collect the encryption. The encryption $c(x'', y'')$ is:

$$c(x'', y'') = |FrT_{d_2}\{FrT_{d_1}[f(x, y) \cdot \delta(x, y)] \cdot \phi(x', y')\} + R(x'', y'')|^2 \qquad (7.16)$$

where FrT_{d_2} denotes the Fresnel transform with a distance of d, $c(x'', y'')$, which is actually a CGH, is the coherent superposition of input plane wave and reference wave. The decryption is shown in Fig. 7.28b, it is the opposite procedure of encryption. A reference wave, which is the conjugation of $R(x'', y'')$, it is needed to illuminate the CGH. The reconstructed beam propagates in Fresnel domain with a distance of d_2. After multiplying by the complex conjugation of $\varphi(x', y')$, transform the result in Fresnel domain with a distance of d_1 again. This process can be mathematically described as:

$$f(x, y) = FrT_{d_1}\{FrT_{d_2}[c(x'', y'') \cdot R^*(x'', y'')] \cdot \phi(x', y')\} \qquad (7.17)$$

Conduct numerical experiments of the encryption and decryption in computer, and the results are shown in Fig. 7.29. Figure 7.29a shows the input information "computer-generated hologram", and Fig. 7.29b is the encryption CGH by DRPEiFD. When one of the keys is incorrect, such as d_1 during decryption ($d_1 = 40$), the result is highly obscure to obtain any useful information, as shown in Fig. 7.29c. The correct decryption result is shown in Fig. 7.29d, which fades

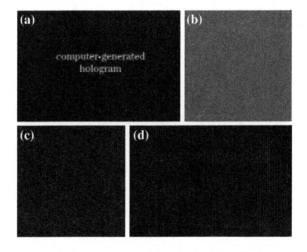

Fig. 7.29 **a** Input information, **b** corresponding encryption, **c** decryption with incorrect key d_1, and **d** decryption with correct keys, along with noise

(a)   (b)   (c)

Fig. 7.30 a QR code of text "COMPUTER-GENERATED HOLOGRAM", **b** retrieval when reading the QR code with a smartphone, and **c** reading result on the smartphone

greatly because of the noise. To recognize the original input from the decryption is difficult indeed. When the input information is more complicated, the decryption result definitely becomes more difficult to recognize. The QR code may be the most popular 2D barcodes for its fast readability and large storage capacity. With specific geometric patterns of black and white dots distributing in a square, a QR code can indicate the information of text, visiting card, email address, etc. Figure 7.30 shows a typical procedure of generating and retrieving a QR code. Black and white pixels in the matrix are similar to "0" and "1" in a computer. With appropriate applications in smartphones, which are massively used in daily life, and can retrieve the original input easily.

Figure 7.30a shows a QR code which represents a character string in capital form "COMPUTER-GENERATED HOLOGRAM". A QR code is designed in a 2D plane. Three bigger squares are located at three corners, and a smaller square is located near the fourth corner to normalize the position, orientation, and angle of viewing. Due to the error correction, information can be accurately retrieved even the QR code is partly damaged or contaminated. By appropriate applications in smartphones with a semiconductor image sensor, that can easily obtain the input information that a QR code represents.

Figure 7.30b shows an example. In this letter, we use a smartphone with some application on Android platform to scan the QR code in Fig. 7.30a. Afterwards, it can read the text "COMPUTER-GENERATED HOLO-GRAM" on the phone as shown in Fig. 7.30c. Subsequently, combined DRPEiFD and QR code together can eliminate noise in encryption and decryption. Free available software on the Internet can be used to generate the QR image with an H-level error correction. Therefore, the input can be retrieved correctly when 30 % of the QR image is occluded or contaminated.

Figure 7.31 shows the numerical results. As shown in Fig. 7.31a, the image of interest is a 256×256 pixel image of a character string "Tsinghua University", and Fig. 7.31b is its QR image. The QR code acts as a box to store the input

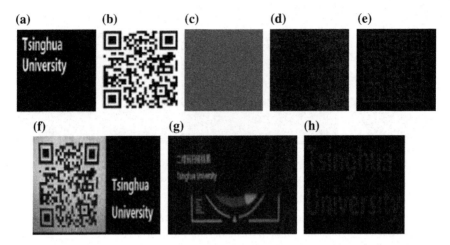

Fig. 7.31 **a** Original input, **b** corresponding QR image, **c** encryption hologram, **d** decryption with incorrect key d$_1$, **e** decryption with correct keys, along with noise, **f** retrieval when reading the QR code with a smartphone, **g** reading result on the smartphone, and **h** retrieval by previous method

temporarily. In our numerical experiments, d_1 and d_2 are 20 and 30 mm, respectively. Wavelengths λ of the input wave and reference wave are both 650 nm. All computational experiments are executed using a PC with a CPU of Intel T5750 (2 cores, both 2.0 GHz) processor and a memory of 2 GB. Using conventional DRPEiFD generated the CGH shown in Fig. 7.31c. Figure 7.31d, e are decryption results with incorrect and correct keys, respectively. When one of the keys (d_1) is incorrect ($d_1 = 40$), the decryption result shown in Fig. 7.31d is very vague to distinguish any useful information. While decrypting the CGH with correct keys, we can obtain a contaminated QR code with serious noise shown in Fig. 7.31e, which is for the next step of retrieval using a smartphone.

Scanning Fig. 7.31e with a smartphone directly reveals can result the pictures as shown in Fig. 7.31f, g by the error correction. This code is noise-free and completely similar to the original input. Figure 7.31h presents the noisy decryption without QR code. Serious noise decreases the image quality, and can hardly recognize the characters. In the proposal, the QR code acts as a box to bear and resist all the damage and noisy contamination. Using QR encoding, the decryption result is guaranteed clean and have potential in optical encryption systems. Combine DRPEiFD and QR code together to resist noise and damage in optical encryption and decryption. The DRPEiFD is commonly optical encryption system, which can improve the security and facility of holographic encryption. However, noise is always an inevitable problem. By encoding the input information to a QR code can make the QR code as a box to store the information temporarily. Numerical experiments show that although the QR code is contaminated by serious noise after decryption, but the retrieval outcome with a smartphone is absolutely noise-free.

7.5   Dynamic Speckle Static Multiplexing (DSSM)

The traditional multiplexing schemes are all based on the Bragg selectivity of volume holographic gratings. So the Bragg selectivity is the fundamental principles. The holograms are recorded based on the reference light with the distribution of single amplitude-phase including the plane wave, the spherical wave and the cylindrical wave, or more complex reference light with the wave front of the uniform distribution. The selectivity of the angle or the wavelength for plane wave can be analyzed by the coupled-wave theory and grating parameter. The diffraction efficiency η from a volume index grating with the direction of the readout beam tuned for the Bragg condition is given by:

$$\eta = \exp(-\alpha d / \cos \theta) \sin^2(\pi \Delta n d / \lambda \cos \theta) \qquad (7.18)$$

where η is the absorption coefficient, d is the thickness of the hologram, An is the amplitude of the index perturbation, and θ is the angle of incidence of the readout beam with respect to the normal to the surface of the hologram (assume an unslanted grating); θ and λ are assumed to be measured inside the medium. Typically, the amplitude is realized in most holographic materials it is only increasing part of the first period of η. As the grating amplitude and the thickness of the grating is increased beyond this regime, further coupling between the two waves results in a reversal of the energy-transfer direction to yield a drop in the diffraction efficiency was predicted by Eq. (7.18).

The η is the diffraction efficiency of the holograms and the bias factor of Bragg mismatch for the angle or the wavelength. The intensity of diffraction light depends on the angular mismatch between the reference light and the readout light of the holograms in angular multiplexing while the intensity of diffraction light depends on the wavelength mismatch between the reference light and the readout light of the holograms in wavelength multiplexing. Because of the limitation of the system geometry and the low multiplexing efficiency, the three-dimensional space and the dynamic range of the medium are not utilized fully. It is the bottleneck of achieving high-density storage. It is very different to apply the reference light with the random codes to the multiplexing scheme in the holographic volume storage of the high density and high capacity. On other hand, it can improve the quality of the Fourier hologram recorded because the randomization can increase the uniformity of the amplitude and the redundancy of the signals. At the same time, it can achieve the speckle multiplexing so can improve the sensitivity of multiplexing and the recorded density. The foundation of the speckle multiplexing is the randomicity and excellent self-correlation of the speckle field. In speckle field the half-width is very narrow and the peak value decreases rapidly, so that the many holograms can be recorded without or lower crosstalk as long as the correlation function between the adjacent speckle fields of the reference light becomes zero. The speckle multiplexing can also eliminate the side lobes by spherical wave.

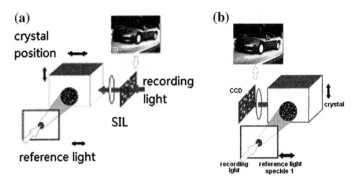

Fig. 7.32 The static speckle multiplexing storage. **a** Recording, **b** readout

The speckle multiplexing systems can be divided into the two kinds by its principles, as static speckle multiplexing is shown in Fig. 7.32 and the DSM is shown in Fig. 7.33. The former of static speckle multiplexing and the DSM are developed on base of the conventional shift multiplexing. The random diffuser is added to the reference light path in the static speckle multiplexing. Its shift selectivity is much more excellent than that of the shift multiplexing based on the reference light of the spherical wave. Although the structure is simple, the selectivity of the shift multiplexing is better than the space-angular multiplexing.

In Fig. 7.33, the speckle reference wave vector $R_W\left(\vec{r}\right)$ and the plane object wave vector $S_0(\vec{r}) = A\exp\left(i\vec{k}_{S_0} \cdot \vec{r}\right)$ cohere to form the phase holograms. The dielectric constant of the storage medium exposed $\varepsilon(\vec{r})$ can change locally $\varepsilon\left(\vec{r}\right) = \varepsilon_0 + \delta\varepsilon\left(\vec{r}\right)$, where $\delta\varepsilon\left(\vec{r}\right)$ is the increment of the dielectric constant and it is proportional to the intensity of the coherent field, as:

$$\delta\varepsilon\left(\vec{r}\right) \propto |E|^2 = \left|\vec{S}_0\left(\vec{r}\right)\vec{R}_W\left(\vec{r}\right)\right|^2 \propto S_0\left(\vec{r}\right)R_W^*\left(\vec{r}\right) \qquad (7.19)$$

where $R_w^*(\vec{r})$ is the conjugation of $R_w(\vec{r})$.

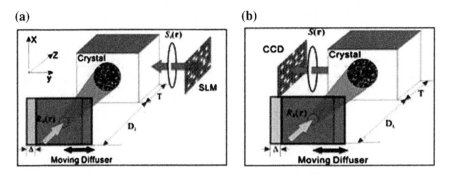

Fig. 7.33 The dynamic multiplexing storage. **a** Recording, **b** readout

If the readout wave vector modulated by the speckle $R_R(\vec{r})$ is employed to retrieve the holograms, the relation between $R_R(\vec{r})$ and the diffraction wave vector $s(\vec{r})$ will be described by the Maxwell equalizations. The Maxwell equalizations can be simplified to the scalar wave equations for the monochromatic field, the constant polarization state, and the isotropic materials. In the first-order Born approximation, with the perturbation theory the diffraction wave vector $s(\vec{r})$ is expressed as:

$$S(\vec{r}) = k_0^2 \int_{-\infty}^{\infty} \delta\varepsilon(\vec{r}') R_R(\vec{r}') \frac{\exp\left[ik_0\left|\vec{r} - \vec{r}'\right|\right]}{4\pi\left|\vec{r} - \vec{r}'\right|} dV' \qquad (7.20)$$

where V is the whole area of the holographic volume. In static speckle multiplexing system the distance that media moves each time (multiplexing distance) is defined as $\vec{\Delta}$. Suppose that the holograms move $\vec{\Delta}$ relative to the reference wave of writing $R_W(\vec{r})$ and the direction of propagation of the reconstructed object light is same to that of the reference light of writing. Then the difference between $R_R(\vec{r})$ and $R_W(\vec{r})$ is only caused by the movement of the holograms, can be expressed as $R_R(\vec{r}) = R_W(\vec{r} + \Delta)$, $\Delta = \Delta_\perp \hat{q} + \Delta_{//} \hat{Z}$ where $\Delta_{//}$ and $\Delta_\perp$ is the transverse component and the longitudinal component. With the three-dimensional correlation functions of the speckles $\left(C(r, r') = \langle R_w^*(r) \rangle\right)$ and the Fresnel–Kirchhoff diffraction integral, the normalized diffraction intensity $I_{DN}(\Delta) = I_D(\Delta)/I_D(\Delta = 0)$ can be expressed by:

$$
\begin{aligned}
I_{DN}(\Delta\perp) &= \frac{I_D(\Delta\perp)}{I_D(\Delta = 0)} \\
&= \frac{\left|\int_0^T \exp\left(\frac{ik_0 n_0 \Delta_\perp^2}{2(z + n_0 D_L)}\right) \int_{-\infty}^{+\infty} \int |K_D(q)|^2 \exp\left(-\frac{ik_0 n_0}{2(z + n_0 D_L)} q\Delta_\perp\right) d^2 q dz\right|^2}{T^2 \int_{-\infty}^{+\infty} \int |K_D(q)|^2 d^2 q}
\end{aligned}
$$

$$(7.21)$$

where $K_D(q)$, n_0 and D_L is the window function of the random diffuser, refractivity of the materials, and the distance between the random diffuser and the media is $k_0 = 2\pi/\lambda$, $q = q_x x + q_y y$. If it is simplified to one dimensional movement, i.e. $q = q_y y$, the displacement in the position of zero diffraction intensity is:

$$\Delta_{x/2} \propto \langle \sigma_\perp \rangle \approx 1.22 \frac{\lambda}{NA} \qquad (7.22)$$

The selectivity of the static speckle multiplexing is obviously more sensitive than the spherical wave multiplexing. And the static speckle multiplexing scheme can be compatible with the conventional disk system.

The random diffuser is moved instead of the medium so as to record all the holograms in the common volume and the number of multiplexing positions

Fig. 7.34 The point-source is
divided into many points by
the random diffuser. Each
point emits the spherical wave
with a special initial phase

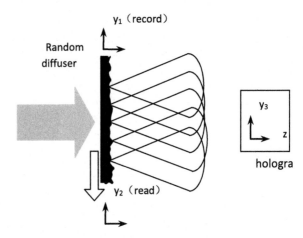

depends on the number of uncorrelated speckle patterns that can be caused by the
random diffuser with the DSM other than the static speckle multiplexing.

In the DSM scheme is shown in Fig. 7.33, that the reference light is the speckle
field formed relative to the reference light of writing after the random diffuser
moves $\vec{\Delta}$. The plane wave that permeates the random diffuser can be considered as a
group of the spherical waves emitted by the points with the random initial phase.
The complex amplitude distribution on the observation plane is different from the
initial distribution with change of the position of the random diffuser, as illustrated
in Fig. 7.34. So the complex amplitude distribution on the holograms plane could
be expressed as following:

$$P(y_3) = \int_{-d/2}^{d/2} \tilde{R}_1(y_1) \, \exp\{i\phi_1(y_1)\} dy_1 \qquad (7.23)$$

where d, $\phi_1(y_1)$, and $\tilde{R}_1(y_1)$ is the diameter of the illuminative area of the random
diffuser, the initial phase of each point relative to the roughness on the random
diffuser surface and the complex amplitude distribution of each spherical wave. In
the paraxial condition, $\tilde{R}_1(y_1)$ can be written:

$$\tilde{R}_1(y_1) = R_1 \, \exp\left\{ik(y_3 - y_1)^2/2z_0\right\} \qquad (7.24)$$

where R_1 and z_0 is the real amplitude and the distance between the random diffuser
and the hologram. $\tilde{R}_1(y_1)$ is a random function on the coded wave because the
readout light is modulated by the random diffuser. The diffraction result can be
expressed as:

$$D \propto \int\limits_{-l/2}^{l/2} \int\limits_{-d/2}^{d/2} \int\limits_{-d/2}^{d/2} R_1 R_2 S \exp(i\delta\phi_r) \times \exp(i\delta\phi_y) dy_1 dy_2 dy_3 \qquad (7.25)$$

where R_1, is the thickness of the hologram, R_2 is the real amplitude of the readout wave, and S is the real amplitude of the signal wave. $\delta\phi_r$ and $\delta\phi_y$ are both the phase difference between the reference wave and the readout wave:

$$\delta\phi_r = \phi_2(y_2) - \phi_1(y_1)$$
$$\delta\phi_y = \frac{k}{2z_0}\left\{y_2^2 - y_1^2 + 2y_3(y_1 - y_2)\right\} \qquad (7.26)$$

where $\phi_2(y_2)$ is the initial phase of each point on the random diffuser. Once the random diffuser of reading is different from writing, the diffraction will be destroyed, so it decreased the diffraction intensity. Therefore, the random diffusers of reading and writing are both uniform for the speckle multiplexing, i.e., $\delta\phi_r = 0$. While the illuminative position on the random diffuser by the readout light that is not same to the recording light, all the points on the random diffuser are moved laterally the distance $\Delta \neq 0$, as illustrated in Fig. 7.35.

So the phase difference between the modulated wave and the unmodulated wave in Eq. (7.29) can be rewritten as:

$$\delta\phi_y = \frac{k\Delta}{2z_0}(2y_1 - 2y_3 + \Delta) \qquad (7.27)$$

The integral of the random distribution function on the whole field of integration $\delta\phi_r$ will approximate zero if $y_1 \neq y_2 + \Delta$. The integral of the random distribution function can be expressed as:

Fig. 7.35 The point-source is moved Δ from the initial position laterally for reading

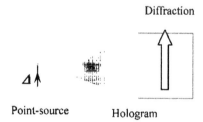

Diffraction

Point-source   Hologram

$$D_c = R_1 R_2 S \int_{-l/2}^{l/2} \int_{-d/2}^{d/2} \exp\left\{ i\frac{k\Delta}{2z_0}(2y_1 - 2y_3 + \Delta) \right\} dy_1 dy_3$$

$$= R_1 R_2 S \exp\left(i\frac{k\Delta^2}{2z_0} \right) \int_{-l/2}^{l/2} \exp\left(i\frac{k\Delta}{2z_0}y_1 \right) dy_1 \times \int_{-d/2}^{d/2} \exp\left(-i\frac{k\Delta}{2z_0}y_3 \right) dy_3$$

$$(7.28)$$

The diffraction result and the intensity relative to Δ are expressed as:

$$D_c = R_1 R_2 Sld \, \sin c \left(\frac{\pi d \Delta}{\lambda z_0} \right) \sin c \left(\frac{\pi l \Delta}{\lambda z_0} \right) \times \exp\left(\frac{i\pi\Delta^2}{\lambda z_0} \right) \qquad (7.29)$$

$$I \propto |D_c|^2 = (R_1 R_2 Sld)^2 \sin c^2 \left(\frac{\pi d \Delta}{\lambda z_0} \right) \sin c^2 \left(\frac{\pi l \Delta}{\lambda z_0} \right) \qquad (7.30)$$

The DSM scheme, which is simple and easy to implement, it is potential to the high-density storage. The DSM has the advantage over the static speckle multiplexing. Since many holograms can be stored in the common volume without moving the medium in the DSM, it is easy to be associated with the other multiplexing schemes for increasing the recorded density. It is also the foundation of the speckle-angular multiplexing and the DSM. So the DSM is fit to high-density storage for the next generation storage on photopolymer medium.

7.6 Hybrid Speckle Multiplexing with DSSM

All kinds of the multiplexing schemes stated as above, including angular multiplexing, wavelength multiplexing, phase-coded multiplexing, shift multiplexing, and the speckle multiplexing schemes, are all used to achieve the storage capacity of V/λ^3 in the volume V of the storage materials in theory. However, the practical storage capacity is limited not only by the dynamic range of media and the noise, but also by the requirements of the special multiplexing schemes. For example, the angular multiplexing is limited by the practical angular scanning range; the phase-coded multiplexing is limited by the resolution of the phase SLM; the multiwavelength multiplexing is limited by the adjusting range of the laser; the speckle multiplexing is limited by the incoherent random phase information supplied by the random diffuser. Single multiplexing often does not use the dynamic range of the medium sufficiently. Two or more kinds of the multiplexing schemes are combined to the hybrid multiplexing scheme in order to be the same with different recording materials and to exert the advantages of different multiplexing schemes in practical system. The present holographic 3D storage systems with high capacity are all implemented on base of the hybrid multiplexing schemes indeed. The hybrid speckle multiplexing scheme will be illustrated mainly next due to the excellent feasibility and performances of the speckle multiplexing.

7.6.1 Speckle-Angular Multiplexing Scheme

The speckle-angular multiplexing scheme is a random diffuser, which is added in the reference optical path of the conventional angular multiplexing scheme. It can increase prominently the storage density of angular multiplexing system, especially in the thin medium. On the other hand, speckle-angular multiplexing can eliminate the influence of the multilongitudinal mode of lasers as shown in Fig. 7.36.

In the speckle-angular multiplexing scheme, as shown in Fig. 7.36, the speckle reference wave vector $R_W(\vec{r})$ and the plane object wave vector $S_0(\vec{r}) = A \exp\left(i\vec{k}_{S_0} \cdot \vec{r}\right)$ cohere to form the phase holograms. The dielectric constant of the storage medium exposed $\varepsilon(\vec{r})$ can change locally $\varepsilon(\vec{r}) = \varepsilon_0 + \delta\varepsilon(\vec{r})$, where $\delta\varepsilon(\vec{r})$ is the increment of the dielectric constant, and it is proportional to the intensity of the coherent field, as followed:

$$\delta\varepsilon(\vec{r}) \propto |E|^2 = \left|\vec{S}_0(\vec{r}) + \vec{R}_W(\vec{r})\right|^2 \propto S_0(\vec{r})R_W^*(\vec{r}) \qquad (7.31)$$

If the readout wave vector modulated by the speckle $R_R(\vec{r})$ is employed to retrieve the holograms, the relation between $R_R(\vec{r})$ and the diffraction wave vector $S(\vec{r})$ can be described by the Maxwell equalizations. If the reference wave of reading $R_R(\vec{r})$ rotates $\delta\theta_A$ ($\delta\theta_A \ll 1$) relative to the reference wave of writing $R_W(\vec{r})$. The direction of propagation of the reconstructed object wave is same to the

Fig. 7.36 Scheme of the hologram recording with the speckle-angular multiplexing

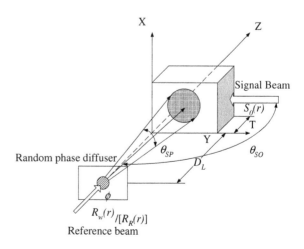

reference wave of writing. If the speckle is distributed normally and formed by the Gauss beam, with the Fresnel–Kirchhoff diffraction integral, the diffraction wave $S(\vec{r})$ can be expressed as:

$$S\left(\delta\theta_A, \vec{q}'\right) = \exp\left(\vec{k}_0 \sin\theta_S\right)t_0^2 \int\limits_0^T \frac{1}{z'\delta\theta_A} \exp\left\{\frac{-ik_0\delta\theta_A}{d_L}\left(z'\delta\theta_A + 2y'\right)\right\} J_1\left(\frac{k_0\phi_z\delta\theta_A}{2d_L}z'\right)\mathrm{d}z'$$

$$(7.32)$$

where $\vec{q}' = (x', y')$ or $\vec{r}' = (\vec{q}', z')$, J_1, d_L and d_L is type I Bessel function, the dimension of the spot on the random diffuser and the distance between the random diffuser and the medium. The relation between the normalized diffraction intensity $I_{DN}(\Delta) = I_D(\Delta)/I_D(\Delta = 0)$ and the angle change of the reference wave $\delta\theta_A$ is employed as selectivity of the DSM, where $I_D(\Delta = 0)$ is the diffraction intensity, when the displacement of the random diffuser is zero, i.e., as:

$$\frac{I_D(\delta\theta_A)}{I_{D\max}} = \frac{1}{\pi}\left(\frac{d_L}{k_0\phi_L T}\right)^2 \iint\limits_{0 \le q^2 \le D_H^2/4} \left|S\left(\delta\theta_A, \vec{q}'\right)\right|^2 d^2q' \qquad (7.33)$$

It means the mismatch between the angle and the space increases with the increment of $\delta\theta_A$.

The speckle-angular multiplexing scheme, indeed can be considered as the combination of the DSM and the angular multiplexing, and can be implemented with the conventional software and hardware of the angular multiplexing system. Meanwhile, the original time sequence and reference object ratio can also be used. The difference between the selectivity of the angular multiplexing and the speckle-angular multiplexing is only that oscillation does not appear in the intensity function $I_D(\delta\theta_A)$ for the speckle-angular multiplexing. The speckle-angular multiplexing scheme is very easy to realize and able to improve the selectivity of the system based on the multilongitudinal mode lasers.

7.6.2 Effect of the Dynamic and Static Speckle Multiplexing

The DSM scheme is the combination of the DSM scheme and the static speckle multiplexing scheme with the speckle-coded reference light. It uses the high density of the DSM and the high shift sensitivity of the static speckle multiplexing. As above, the high-density storage can be achieved in the common volume with the DSM, as long as the dynamic range of materials, the area of the random diffuser and the shift range of the precision shift system are all enough. Otherwise, the shift sensitivity of space multiplexing is also enhanced. With the capacity of the tiny shift of the static speckle multiplexing the hologram can be recorded in a new position by a tiny shift. The holograms in the new position are superposed over those in the

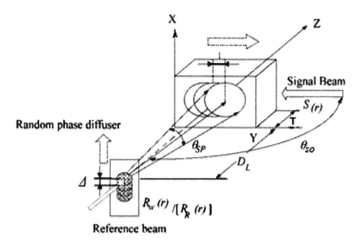

Fig. 7.37 The diagram of DSSM holographic storage with higher dynamic range and capacity, as the random phase diffuser can be moving

last position. But the DSSM scheme can obtain higher density of storage furthest application of dynamic range of the materials, which is illustrated in Fig. 7.37.

For the dynamic static speckle multiplexing (DSSM), in a common volume of the materials, the speckle pattern used to modulate the reference light. It is changed by moving the random diffuser and the holograms are recorded with the nonrelation of the adjacent speckles. After recording a group of the holograms in the volume, the medium is displaced slightly and the new group can be recorded in the new superposed position. The speckle pattern of the new group is either same to the last one, or not. The characteristics of the high density in a common volume for the DSM and the tiny displacement for the static speckle multiplexing are combined in the DSSM scheme. The DSSM is most propitious to 3-D volume holomem medium furthest, that can achieve the highest storage density in theory. The system implementation and optimization of the storage characteristics for DSSM are researched in order to achieve maximize holographic volume storage system.

The speckle multiplexing holographic storage employs the reference light modulated by the speckle instead of the homogeneous reference light in conventional holographic storage. At present there are many methods that can be used for the random phase modulation of the reference light, including the random diffuser method, the multimode fiber method and the pure-phase SLM method. There are a lot of methods that can be used to generate the speckle pattern with the random diffuser and the sector scattering effect. The simple and effect method of the random phase modulation of the reference light is setting the random diffuser in the reference light path. Through the random diffuser the reference light become the emanative random modulated reference light, which can be considered as the superposition of the spherical waves emitted from the point-sources with different initial phases. The selectivity of the multiplexing varies by changing the size of the

Fig. 7.38 The experimental optical system of the speckle multiplexing with the random diffuser

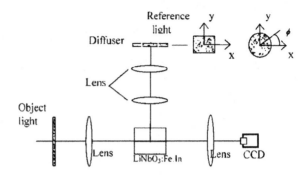

illuminative spot of the reference light on the random diffuser and the size of particle or the angle of divergence of the random diffuser. In the experiments of the speckle multiplexing, the random distribution of the reference wave should be changed, as generation different speckle patterns of the reference light, because the speckle multiplexing scheme is implemented in virtue of the cross correlation of the reference light after adjusting the random diffuser. The translation of the random diffuser in three dimension including transverse direction, vertical direction and the longitudinal direction is usually used to change the speckle pattern. Since the translation precision can be implemented by the precision moving stage controlled by the computer, the multiplex and the retrieve can be achieved perfectly. The random diffuser rotation multiplexing scheme can be rotated other than translation. The moving of the round random diffuser and the conventional one is not different essentially. The typical optical system with the random diffuser is shown in Fig. 7.38.

Generating the intensity distribution of speckles is by the multimode fibers. The multiplexing method for holographic storage based on the speckle generated from multimode fiber, that the various speckles are attained with the change on the fiber parameter and no electronics processing, so can be achieved all-optical communication. In this system reference light is coupled into the fibers. And the intensity distribution of speckle emitted from the fiber is changed when the parameters of multimode fiber, such as length and curvature are changed. The dynamic intensity distribution of speckles diffracted by multimode fiber can be used as the reference lights. The principles of the speckle multiplexing for holographic storage of multimode fiber can be employed in the all-optical sensing system because the speckle pattern depends on the distance of multimode fiber. Before the system starts to work, it is need to build a database in volume holograms in the first. Since the speckle pattern originates from the interference of light with random phase, can perform holographic multiplexing with random phase encoding. The stored holograms in the crystal can be regarded as a database used for interconnections between the incoming speckle and the corresponding output pattern. After the database is constructed, when a specific speckle from the sensing fiber is incident on the crystal, that the volume hologram automatically compares the phase of the incoming speckle with the stored ones. If the incoming speckle matches one in the

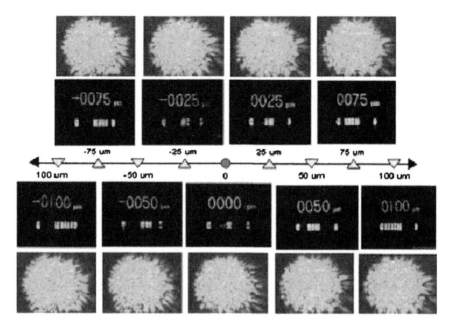

Fig. 7.39 The experimental speckle pattern, stage positions, corresponding speckle patterns and readout patterns in an all-optical sensing holograms processing system

database, or a pattern was diffracted through interconnection. Since there is not extra electronics processing, this sensing system is all optical. The diffraction images, the corresponding speckles and the corresponding positions of the stage in the system are shown in Fig. 7.39.

7.6.3 Elastomer Mask Phase Multiplexing

Prof. Raphael A. Guerrero proposed first that the elastomer mask made from the silicone is used in the phase multiplexing for holographic storage, illustrated in Fig. 7.40. The diffuser with the random phase distribution is pressed by the mould with inhomogeneous surface profile. Once the elastomer mask is biased from the equilibrium state, in the optical system, all the points on the wave front of the reference wave will change, so that the special phase addresses are formed relative to some strain. For the low cost of the silicone, the elastomer mask can replace the expensive SLM. In the system the holograms are recorded relative to the phase addresses that are formed by the strain of the elastomer mask. With difference from the conventional diffuser, the extension of the elastomer mask not only changes the position of the mask and the phase information, but also changes the intrinsic distribution function of phase information of the mask. It supplies the extra

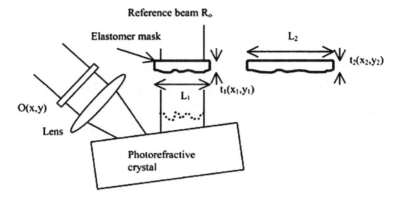

Fig. 7.40 Optical system of the multiplexing scheme with the elastic mask

parameter for the phase-coded of the holograms. The dynamic selectivity provided by the elastomer mask can bring the conversion from the holographic storage into the holographic cartoon immediately. The holograms can be retrieved in turn and made into the periodic video with the extension and contract of the elastomer mask. The image recorded can also be retrieved in real-time with the extension and contract of the elastomer mask. If the series of images are recorded in turn in the crystal and the speed of retrieve is enough, a holographic cartoon will be watched.

The researchers have found that the beam sector scattering effect of the photo refractive crystal. When laser beams go through the photo refractive crystal, the sector scattering effect happen, as in Fig. 7.41. The effect is the serious amplification of the scattering light caused by the rough surface, the inherent inhomogeneity and impurity of the refractive crystal due to the dual-beam couple of incident and scattering beams along the c-axis of the crystal. Because almost all the energy of the incident beams are scattered asymmetrically, the sector scattering speckle pattern is generated as in Fig. 7.42.

(a)   **(b)**

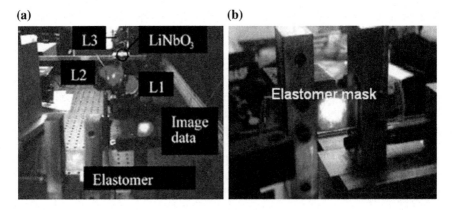

Fig. 7.41 Principles of the sector scattering speckle pattern generated by the elastomer mask

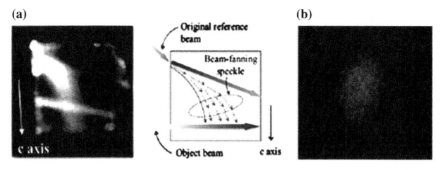

Fig. 7.42 Generating the sector scattering speckle in crystal (**a**) sector scattering phenomenon (**b**) sector scattering speckle observed outside of the photo refractive crystal

The sector scattering speckle pattern depends on the critical condition of incident light, such as intensity distribution, phase distribution, incident angle, and the incident position. At present, the sector scattering effect has been widely in all kinds of optical system, such as interferometer, new-type filtering, and so on.

Recording and retrieving the holograms with the sector scattering speckle multiplexing are simpler than those with other multiplexing. The reference light and the object light are not intersected in the whole volume of the photo refractive crystal while only the sector speckle reference light and the object light are intersected so as to inform the gratings. Since the external random diffuser and multimode fibers are not needed in the scheme, the optical system is compact. The scheme, which can also be used to read the holograms, ensures that the medium can move arbitrarily and be recorded or read in the different system, as shown in Fig. 7.43.

Fig. 7.43 Principles of the speckle multiplexing holographic storage with the photo refractive sector scattering effect

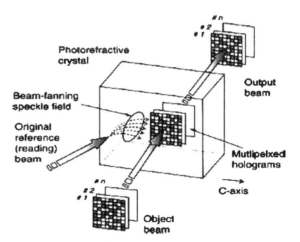

Obviously, the research further for DSSM should be concentrated on the characteristics of the speckle pattern, random diffuser requirements and the optimization of DSSM, as details are followed:

1. The characteristics, property, and implementation of the hybrid multiplexing schemes based on the speckle multiplexing in holographic storage system;
2. The relations among the speckle size, multiplexing selectivity, and the output of the multilongitudinal mode lasers;
3. The time sequence property, theoretical calculation and experimental research of the exposure in multitrack superposed dynamic static speckle hybrid multiplexing scheme;
4. The scattering noise of DSSM and the influence of the incident intensity and the access time to the scattering noise in the retrieved images;
5. Applications, such as the multitrack superposed and polarized volume holomem with DSSM multiplexing.

The improvement of multiplexing selectivity with DSSM Speckle multiplexing technology which makes the speckle coding for the reference beam with random phase diffuser in VHS has the advantages of high-density multiplexing and easy for use. By the former introduction, it is discovered that the randomness and self-correction of speckle field, that is narrow half width of self-correction and deep falling of the peak, are used in speckle multiplexing technology. So it gets good performances of multiplexing selectivity. Further, speckle multiplexing technology can be easily combined with the conventional multiplexing methods, such as angle multiplexing, wavelength multiplexing, and so on, which make it possible to improving the multiplexing sensitivity.

Static speckle multiplexing is the basic form of speckle multiplexing technologies. It is developed from the shifting multiplexing technology. Measuring the shifting selectivity with and without random phase diffuser and list the results are shown in Fig. 7.44, which vertical coordinate is the selectivity with random phase

Fig. 7.44 Comparison of selectivity of the speckle multiplexing: **a** the selectivity with speckle random phase diffuser, **b** without speckle random phase diffuser

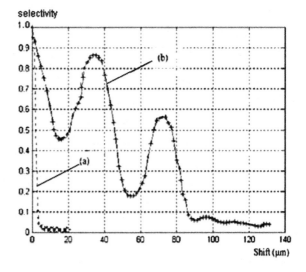

speckle multiplexing (a) and without random phase diffuser (b). It is obviously concluded that the shifting selectivity is well improved by adding the random phase diffuser. It is explained that the selectivity is following Bragg selectivity before adding the random phase diffuser while the selectivity is mainly determined by the correlation of speckle field of reference beam for reading out and that for writing in after adding the random phase diffuser. There is no side lobe for the selective curve which makes it possible to improve the selectivity evidently.

7.6.4 The Improvement of Angle Multiplexing Sensitivity

The angle selectivity is depressed by multilongitudinal mode of laser in the volume holographic system with semiconductor pumping laser diode. The combination of speckle multiplexing and angle multiplexing technologies can improve the angle selectivity well and reach the same multiplexing sensitivity as single-longitudinal mode laser. The experimental system with phase diffuser for speckle–angle multiplexing system is shown in Fig. 7.45. The selectivity curves are shown in Fig. 7.46, that the vertical coordinate is the selectivity, (a) is obtained with random phase diffuser, and (b) without random phase diffuser in the same system. It is seen that the angle selectivity is great improved and side lobe is suppressed by the speckle reference beam. The angle selectivity of the volume holographic system is improved by the speckle multiplexing technology. The multiplexing selectivity is increased 8 times to with angle multiplexing only.

7.6.5 The Improvement of Wavelength Multiplexing Sensitivity

Experiments substantiate that the speckle wavelength multiplexing technology can modify the performances of selectivity also. The experimental system is shown in Fig. 7.47, where temperature-controlled laser diode is used for laser source and

Fig. 7.45 Optical system with phase diffuser multiplexing: $\lambda/2$ half plate, *PBS* prism of beam, *M* mirror, *L* lens, *SLM* space light modulator, *RM* recording media

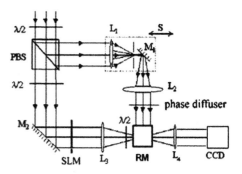

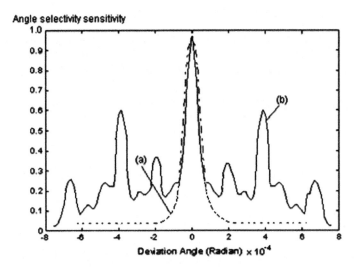

Fig. 7.46 Angle selectivity sensitivity curves to deviation angle: **a** it is obtained with random phase diffuser, **b** it is obtained without random phase diffuser

Fig. 7.47 The multimode fiber supplied the variable speckle field in holomem experiment system as wavelength speckle multiplexing

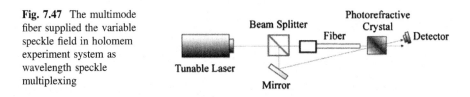

multimode fiber which is employed to supply the variable speckle field for reference beam.

As the wavelength change of laser diode, the speckle filed will be changed with a multimode fiber as shown in Fig. 7.48, which is the addressing picture in different holographic grating. The measurement result of multiplexing selectivity is shown in Fig. 7.49 with the wavelength speckle multiplexing. The wavelength multiplexing sensitivity can be improved by combining the speckle multiplexing in the volume holographic system observably which is very similar to speckle angle multiplexing.

7.6.6 Selectivity Sensitive of Medium Thickness

In conventional multiplexing technologies, the selectivity of thick recording medium is better than thin one often. So, massive crystal has advantages in storage density in this case, however, thin crystal has better dynamic range than massive ones which is partly limited by the deficient selectivity. The experimental measurement of the selectivity of angle multiplexing in the holography storage system,

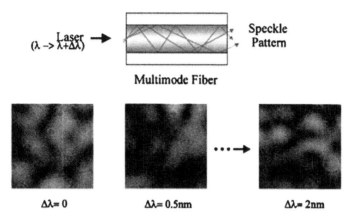

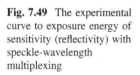

Fig. 7.48   The speckle field was varied with wavelength $\Delta\lambda$

Fig. 7.49   The experimental
curve to exposure energy of
sensitivity (reflectivity) with
speckle-wavelength
multiplexing

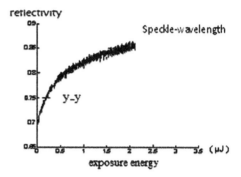

with thickness of 40 μm and thick, doped LiNbO$_3$ crystal of thickness 10 mm are
taken for without random phase diffuser are shown in Fig. 7.50. The comparing
results, the selectivity of crystal of thickness 10 mm is larger than thickness 40 μm
nearly to 100 times. But in volume holography storage with speckle multiplexing
technology, this difference of the sensitivity to thickness medium is less as shown in
Fig. 7.51 that was demonstrated by experiments also, the thin medium can overcome its shortage of bad selectivity to rely on random phase diffuser [35].

7.7   The Relationship of DSSM Selectivity and Speckle Size

Random phase diffuser is used to encode of the reference beam in volume holography storage system with speckle multiplexing technology. Speckle distribution is
imaging on the interface of storage medium by optical system, it is shown in
Fig. 7.52.

Fig. 7.50 The selectivity (diffraction efficiency) of different thickness medium (10, 40 μm of same material) in conventional multiplexing without random phase diffuser system

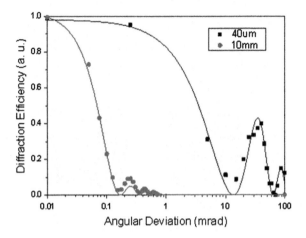

Fig. 7.51 Angle sensitivity (diffraction efficiency) of different thick medium in speckle multiplexing system

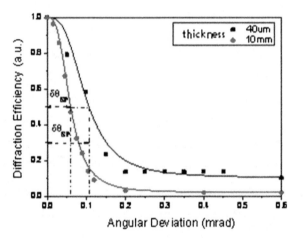

Fig. 7.52 Imaging system in holography storage

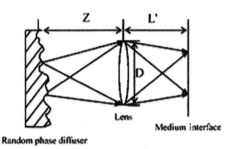

Light from the coarse surface of the random phase diffuser passed through the lens stop with different angles and fixed phase difference, it is exit window for adjacent light, so the stop could be considered as scattering surface. The distance

between the scattering surface and lens is not infinite, according to the imaging formula $L' = F(L'/Z + 1)$, so the speckle size could be calculated as:

$$d \cong 1.22(1 + L'/Z)\lambda F \qquad (7.34)$$

where Z is object distance, L' is the image distance, and F is the focal length.

The formula shows that speckle size is relied on the geometric configuration of optical system. So the speckle size will be changed if the lens is moved along the light axis direction. This section will discuss the influence of speckle size to selectivity, different effect (diffraction light intensity), and dynamic range in the holographic storage system.

The diffraction light intensity (different effect) varies with crystal shifting for different speckle size (from 14 to 34 μm) in static speckle multiplexing holomem system is shown in Fig. 7.53 by experiments. It is seen that the normalized diffraction light intensity is decreased with increasing the shifting amount, but the smaller speckle size is decreased rapidly. As smaller speckle size is narrower half width, so the shifting selectivity of system is better than the speckle size is larger one. Furthermore, the experiments discovered that the relationship between multiplexing selectivity and speckle size is a linear function almost, as shown in Fig. 7.54.

Fig. 7.53 Diffraction light intensity varies with crystal shifting for different speckle size in static speckle multiplexing system

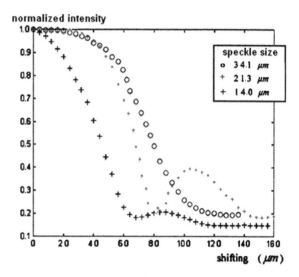

Fig. 7.54 The relationship curve between diffraction light intensity and speckle size

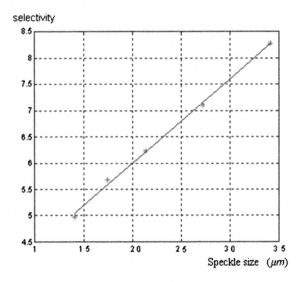

7.7.1 Speckle-Angle Multiplexing Selectivity and Speckle Size

The relationship between diffraction light intensity and crystal shifting with different speckle size in speckle-angular multiplexing system is shown in Fig. 7.55. The normalized diffraction light intensity is decreased with increasing the angular deviation $\delta\theta$, and it decreased more rapidly to smaller speckle size. So the angle selectivity of system will be better when the speckle size is smaller. The angular selectivity is a near-linear function of speckle size also, as shown in Fig. 7.56.

Fig. 7.55 Diffraction light intensity varies with angular deviation $\delta\theta$ for different speckle size in speckle-angle multiplexing system

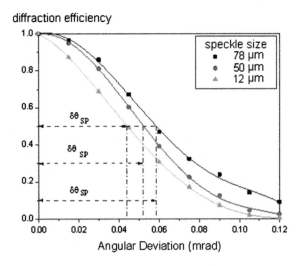

Fig. 7.56 The relationship curve between angular deviation and speckle size in speckle-angular multiplexing system

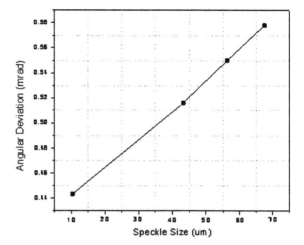

7.7.2 The Relationship Between Dynamic Speckle Multiplexing Selectivity and Speckle Size

In DSM mode, the diffraction light intensity varies with phase diffuser shift Δ_y to different speckle size is shown in Fig. 7.57. The speckle size δ for reference beam is modified with changing the distance between phase diffuser and storage medium or adjusting the aperture size as in Fig. 7.7. From Fig. 7.57 can conclude that the normalized diffraction light intensity is decreased with increasing the diffuser shift, and the smaller speckle size is decreased more rapidly. As smaller speckle size is narrower than the half width, the DSM selectivity will be better than the larger speckle size.

Fig. 7.57 Normalized diffraction light intensity varies with diffuser shift Δ_y for different speckle size δ in dynamic speckle multiplexing system

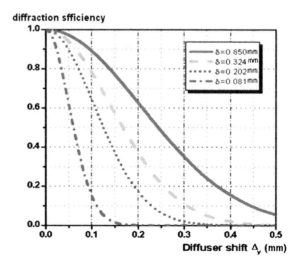

Except the relationship of dynamic range of speckle multiplexing and speckle size, another important index to evaluate the volume holography storage system is the dynamic range $M\omega$ of photo refractive crystal, which is expressed as:

$$M\omega = \frac{2\kappa}{\alpha/\cos\theta} e^{-\frac{\alpha L}{2\cos\theta}} \tanh\left(\frac{\alpha L}{2\cos\theta}\right) = m(M\omega) \tag{7.35}$$

where α is the intensity absorption coefficient, θ is half of the included angle of two recording beams and have:

$$\kappa = \frac{\pi}{2\lambda\varepsilon_0 \cos\theta\sqrt{n_1 n_2}} P_1 \cdot \bar{\varepsilon}_1 P_2 \tag{7.36}$$

It is a coupling constant of readout beam and diffraction light. P_1 and P_2 are the polarization unit vector of the readout beam and diffraction light, n_1 and n_2 are the refractive indexes for each light, and $\bar{\varepsilon}_1$ is the tensor.

While speckle reference beam is used in VHS, the light intensity modulation factor m which is caused by interference of recording beams during forming holographic grating could be modified as:

$$m_{\text{eff}} = \frac{2\sqrt{I_1 I_2}[1 - \exp(-I_1/I_2)]}{I_1 + I_2[1 - \exp(-I_1/I_2)]^2} \tag{7.37}$$

where $I_1 = |E_1|^2$ and $I_2 = \left\langle |E_2|^2 \right\rangle$. The average light intensity of the speckle field on the surface of crystal is:

$$I_2 = \left\langle |E_2|^2 \right\rangle = \left\langle E_2(\vec{r})E_2^*(\vec{r}) \right\rangle = \int_{-\infty}^{+\infty} |P(\vec{r}_0)|^2 |h(\vec{r}, \vec{r}_0)|^2 d\vec{r}_0 = \frac{\pi\omega_0^2}{\lambda^2 d_L^2}\left(1 - e^{-1}\right) \tag{7.38}$$

If $dL \gg L$, m_{eff} could be used to approximate the relationship of dynamic range of recording medium and reference speckle size.

$$\eta_{Mw}(\Delta) = \frac{Mw}{(Mw)_0} = \frac{2\omega_0\omega(z_0)\sqrt{\pi I_1(1 - e^{-1})}\Delta\left\{1 - \exp\left[-\frac{\pi\omega^2(z_0)I_1}{\omega_0^2(1-e^{-1})}\Delta^2\right]\right\}}{\pi\omega^2(z_0)I_1\Delta^2 + \omega_0^2(1 - e^{-1})\left\{1 - \exp\left[-\frac{\pi\omega^2(z_0)I_1}{\omega_0^2(1-e^{-1})}\Delta^2\right]\right\}} \tag{7.39}$$

where $\eta_{M\omega}$ is the normalized dynamic range of the recording medium, $\Delta = \frac{\lambda z}{\pi\omega(z_0)}$ is the speckle size in recording medium. The reference speckle beam intensity will be decreased by increasing the speckle size Δ. When the light intensity on the random

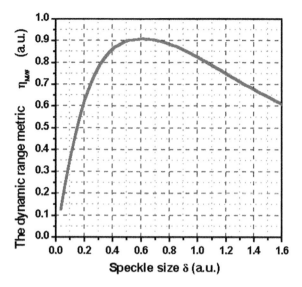

Fig. 7.58 Dynamic range varies with speckle size, the dynamic range is increased with increasing the speckle size to reach the maximum to decreasing

phase diffuser is certain, there is a limitation value for the dynamic range of the recording crystal varying with the speckle size. If the speckle size is very small, the dynamic range is increased with increasing the speckle size. However, oppositional trend occurs after reaching the maximum to decreasing, as is shown in Fig. 7.58.

Above all, the speckle size should be optimized according to different system parameters in practical VHS to guarantee the selectivity and dynamic range at the same time in speckle multiplexing mode. It depends on the actual requirements to the multiplexing selectivity and dynamic range of the system. By adjusting the weighting of selectivity and dynamic range, the optimized result could be suitable for system with different recording medium and multiplexing amplitude, by requirement of the system performance.

7.7.3   The Improvement Mode for Laser with Speckle Multiplexing Technology

Currently, most of the high light sources for VHS system are based on the single-longitudinal mode laser, such as Ar+ laser etc. However, the VHS system can hardly be suitable for commercial application, as it is restricted by the large-volume and complex water cooling system for this laser. Semiconductor diode pumping laser has been the first choice for the light source of the volume holographic system now, because its smaller volume and simple operations which is suited more to holographic storage. But the light output of semiconductor diode pumping laser is multilongitudinal mode laser always, which can decrease the angle selection of volume holographic and the storage density very much than

Fig. 7.59 Longitudinal mode of the Nd: YAG Nd:YAG solid-state laser double frequency laser

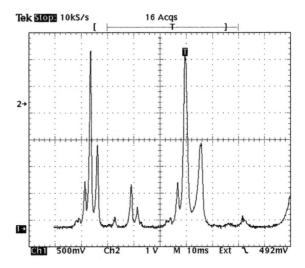

single-longitudinal mode laser. A double frequency Nd:YAG solid-state laser of the single-longitudinal mode was manufactured by Institute of Changchun Optics and Fine Mechanics technically, its output spectrum is shown in Fig. 7.59. This solid-state laser crystal has some advantages by testing. Its high gain, narrow linewidth, low threshold and physical properties make it a most versatile laser material for VHS applications. But it is found in the figure that several longitudinal mode output by the laser which will affect on VHS performances. In speckle multiplexing system, the reference beam is encoded by the speckle. The correlation of the speckle field with multiwavelength speckle is individual which is same as in single-wavelength speckle. So the selectivity of multi longitudinal mode semi-conductor pumping laser can be improved by speckle multiplexing technology which is similar with for single-longitudinal mode laser. Experiments are carried out that this method is validated very much as shown in Figs. 7.60 and 7.61. The solid line in Figs. 7.60 and 7.61 are curves of diffraction efficiency to angle deviation of Nd:YAG solid-state laser with and without phase diffuser separately. The wavelength of the Nd:YAG solid-state laser is 532 nm and the selectivity of angle deviation is within 3 % with diffuser. Same experiments for Ar+ laser with wavelength 514.5 nm, results are shown as the dash dot line in Figs. 7.60 and 7.61. It is seen obviously that the angle selectivity in VHS system with Nd:YAG solid-state laser for source has been greatly improved by speckle multiplexing technology. The angle selectivity for Nd:YAG solid-state laser is nearly same as the Ar+ laser. It has been used to the miniaturization VHS system well in OMNERC of Tsinghua University [36].

Speckle multiplexing technology is realized by adding a phase diffuser into the reference light path in the VHS system. The speckle-coded reference beam is used which make several advantages: good selectivity either for angle or shifting, monotone decreasing of the diffraction side lobe, multiplexing selectivity be linear

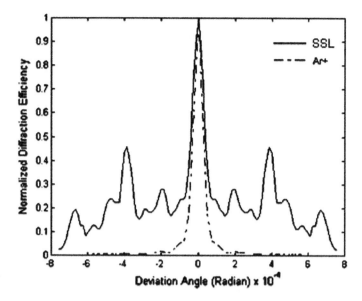

Fig. 7.60 The experimental diffraction curves of the angle selectivity of Nd:YAG solid-state laser (SSL) and Ar+ laser without diffuser

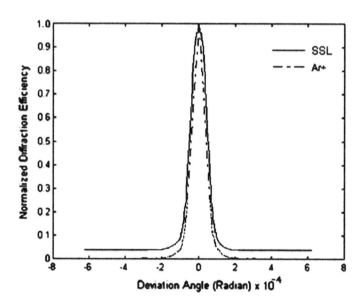

Fig. 7.61 The experimental diffraction curves of the angle selectivity of Nd:YAG solid-state laser (SSL) and Ar+ laser with diffuser

function of the speckle size, etc. Furthermore, the selectivity is less sensitive to thickness of the recording material in speckling multiplexing system. On the other hand, the research on how the multilongitudinal mode characteristic of DPL laser affects the angle selective in holographic system discovered that the angle selectivity will no longer be affected by multilongitudinal mode if speckle multiplexing technology is used.

7.8   Sequence Exposure Method in DSSM System

7.8.1   The Recording Timing Sequence of Multitrack Overlapping

The principles of static speckle multiplexing and DSM technology have been described above. The combination of static and DSM technology is also proposed. The section will discuss higher recording density and capacity with dynamic speckle and smaller multiplexing interval for static speckle or both. In order to efficiently use the three dimension space of the recording medium, multitrack overlapping dynamic static speckle multiplexing technology (DSSM) is proposed which is a significant way to more increase the recording density farther. Based on this technology developed the track overlapping recording sequence exposure method. If it was used to DSSM system, that the storage density, capacity and data transfer rate in volume holographic could be great increased, as shown in Fig. 7.62.

Comparing with the conventional holographic storage mode, the multitrack overlapping DSSM recording has great advantages in recording density and capacity. The stream of conventional holographic storage is serial concurrent page input or output. But the multitrack overlapping DSSM could realize higher capacity and data transfer rate in VHS. Meantime, the multitrack overlapping DSSM VHS is less sensitive to medium flaw than conventional VHS, and gets more rapid addressing speed, correlation addressing function. Besides the multitrack overlapping DSSM has more advantages than conventional holographic storage system, which recording marks should be separated from each other to avoid BER. But in multitrack overlapping DSSM system, the recording spot either transverse direction or longitudinal direction, either same track or different tracks are overlapped also.

Fig. 7.62 The multitrack overlapping recording sequence exposure method with DSSM: the recording spots are overlapped either transverse direction and longitudinal direction at the same time

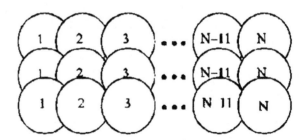

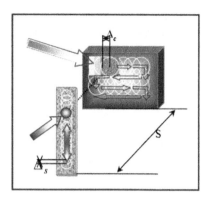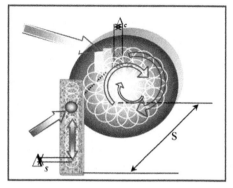

Fig. 7.63 Schematic diagram for card and disc media with multi track overlapping in DSSM recording system

Several volume holographic pictures could be recording at one point without interfering with each other, which equals several even several tens of layers of conventional holographic storage disc. The schematic diagram for card and disc medium for multitrack overlapping recording with DSSM is shown in Fig. 7.63.

In the overlapping VHS process, the subsequent holographic photo will erase the former ones partially due to the dynamic characteristic of recording crystal. So the diffraction efficiency of former photo will be lower than subsequent ones under same recording time and exposure intensity. The diffraction efficiency will increase gradually by recording holographic photo sequence, it is demonstrated by an experiment: where 20 holographic photos are recorded with the same exposure time of 15 s as shown in Fig. 7.64, where ordinate is gray degree and abscissa is recording number.

Fig. 7.64 The experiment of diffraction efficiency of serial holographic photo exposure continuously

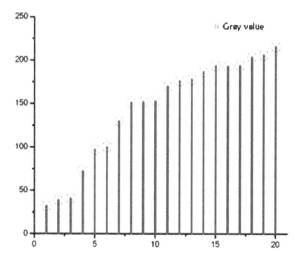

In multitrack overlapping DSSM system, the erasing process between different holographic photos is much more complicated. Especially in large-capacity storage system, former photo maybe totally erased and never be recovered again if improper exposure time sequence, that could make its advantage to forfeit completely. So, it is necessary to research and design the exposure time sequence for multitrack overlapping DSSM scheme. Two points referring to the diffraction efficiency should consider in the design of exposure time sequence: one is maximization of diffraction efficiency for all the holographic photos to certain numbers of photos, by which the diffraction efficiency of material could be used most efficiently, another point is that equalized diffraction efficiency for each recording and retrieving photo. A good sequence exposure method can widely apply in various VHS, which could uniformly diffract efficiency overall, reduce the requirement of reposition precision of the mechanical system, and get high diffraction efficiency at last.

However, from Fig. 7.64 can see that the sector effect and wave coupling effect have to be considered, i.e., the exposure time (energy) is required equilibrium for every recording photo. Even a bit of bug in the crystal will make it nonuniform for diffraction efficiency. The exposure time sequence was used in angle multiplexing VHS for example. Considering the volume holographic material has no bugs, the diffraction efficiency for angle holographic photo could be expressed with Kogelnik formula as:

$$\eta = \sin^2\left(\frac{\pi \Delta n d}{\lambda \cos \theta}\right) \qquad (7.40)$$

where η is for diffraction efficiency, Δn is for modulation ratio of material, d is for crystal thickness, λ is for wavelength of recording light, and θ is the Bragg angle. The result only comes to existence under noncoupling wave condition. Meantime, the relationship of diffraction efficiency and sequence exposure number is approximately. The diffraction of multiplexing holographic photos will reduce at η_0/N^2, where η_0 is the saturation diffraction efficiency of single photo and N is the number of exposure time. Maniloff et al., proposed an new way to evaluate the process of recording and erasing in holographic storage based on photo refractive crystal with coupling wave theory. The recording process for holographic photo with maximum uniform diffraction efficiency in angle multiplexing system was analyzed in theoretically. Asymmetry for recording and erasing constants was demonstrated by experiment also. Another classical volume holographic exposure time sequence was proposed by D. Brady, D. Psaltis et al., which is termed as incremental exposure time sequence. The recording process was divided in a serial of same cycle recording process in incremental exposure time sequence. Every holographic photo is recorded with same exposure time Δt by sequence during each cycle time. After that, next cycle is done until uniform diffraction efficiency is obtained. Incremental exposure method has several advantages. First, it is not necessary to consider the sector effect and wave coupling effect, because the exposure time is rather short during one exposure cycle. Second, the diffraction efficiency could be maximum also. However, there are a few difficulties in practice. As the uniform of diffraction efficiency is worth than the sequence exposure method, since the exposure time cannot be infinitely short. Meanwhile, the reference beam and recording beam has to be repositioned precisely during each exposure period. Any reposition error will make larger effect to uniform of diffraction efficiency, which is restrict to recording system. Lastly, if higher uniform is required, the cycle numbers will be great increased which makes it impractical. So, sequence exposure method is a practical choice for multitrack overlapping DSSM under current experimental condition.

Multitrack overlapping DSSM technology is a combination of DSM technology, static speckle multiplexing technology and multitrack overlapping technology indeed. In research, the time sequence is divided into several parts, the first is time sequence of DSM. Uniform reference beam is employed in volume holographic recording with conventional sequence exposure method. Considering the modulation ratio of refractive index in photo refractive crystal during recording and erasing process, it is an exponential function of time, and the recording process is:

$$\Delta n = \Delta n_s [1 - \exp(-t/\tau_r)] \tag{7.41}$$

and the erasing process is:

$$\Delta n' = \Delta n \exp(-t'/\tau_e) \tag{7.42}$$

where Δn is the amplitude of refractive index modulation variation at recording time t and Δn_s is the saturation refractive index modulation ratio, $\Delta n'$ is the amplitude of refractive index modulation ratio variation at erasing time t', τ_γ, and τ_e is recording and erasing time separately. t'_i is the exposure time for i holographic photo, and t'_i is the erase time for number i holographic photo also, which equals to the sum of exposure time for the photos of $i + 1$. If N holographic photos are recorded at one position of the photorefractive crystal in VHS, the refractive index modulation ratio of number 1 photo is deviated by:

$$\Delta n_1 = \Delta n_s \left[1 - \exp\left(-\frac{t_1}{\tau_r}\right)\right] \exp\left(-\frac{\sum_{j=2}^{N} t_j}{\tau_e}\right) \tag{7.43}$$

Equally, the deviations of refractive index modulation ratio for the holographic photos from number 1 to last could be present as:

$$\Delta n_i = \Delta n_s \left[1 - \exp\left(-\frac{t_1}{\tau_r}\right)\right] \exp\left(-\frac{\sum_{j=i+1}^{N} t_j}{\tau_e}\right) \tag{7.44}$$

$$\Delta n_N = \Delta n_s \left[1 - \exp\left(-\frac{t_N}{\tau_r}\right)\right] \tag{7.45}$$

It is considered that

$$\sum_{j=i+1}^{N} t_j = T_i, \tag{7.46}$$

then

$$\Delta n_i = \Delta n_s \left[1 - \exp\left(-\frac{t_i}{\tau_r} \right) \right] \exp\left(-\frac{T_i}{\tau_e} \right). \tag{7.47}$$

The saturation of refractive index modulation ratio Δn_s is shared by N holographic photos, so that is:

$$\Delta n_s = \sum_{i=1}^{N} \Delta n_i. \tag{7.48}$$

For the situation of nonuniform of the recording and reading constants, it is modified to:

$$\Delta n_s = \left(\frac{\tau_W}{\tau_E} \right) \sum_{i=1}^{N} \Delta n_i \Rightarrow \Delta n_i = \left(\frac{\tau_W}{\tau_E} \right) \frac{\Delta n_s}{N} \tag{7.49}$$

In order to obtain the uniform diffraction efficiency for each holographic photo, the refractive index modulation ratio should be same for all photos, i.e., as:

$$\Delta n_1 = \Delta n_2 = \Delta n_3 = \cdots = \Delta n_N = \frac{\alpha \Delta n_s}{N} \quad \text{where } \alpha = \frac{\tau_W}{\tau_E}. \tag{7.50}$$

It is derived from the:

$$
\begin{aligned}
\Delta n_N &= \Delta n_s \left[1 - \exp\left(-\frac{t_N}{\tau_r} \right) \right] \\
&\Rightarrow t_N = -\tau_r \ln\left(1 - \frac{\Delta n_N}{\Delta n_s} \right) = -\tau_r \ln\left(1 - \frac{\alpha}{N} \right)
\end{aligned} \tag{7.51}
$$

$$
\begin{aligned}
\Delta n_{N-1} &= \Delta n_s \left[1 - \exp\left(-\frac{t_{N-1}}{\tau_r} \right) \right] \exp\left(-\frac{t_N}{\tau_e} \right) \\
&\Rightarrow t_{N-1} = -\tau_r \ln\left(1 - \frac{\Delta n_{N-1}}{\Delta n_s} \exp\left(-\frac{t_N}{\tau_e} \right) \right) \\
&= -\tau_r \ln\left(1 - \frac{\alpha}{N} \exp\left(\frac{t_N}{\tau_e} \right) \right)
\end{aligned} \tag{7.52}
$$

$$t_{N-1} = -\tau_r \ln\left(1 - \frac{\alpha}{N}\exp\left(\frac{\sum_{j=0}^{i=1} t_{N-j}}{\tau_e}\right)\right)$$

$$\vdots \qquad\qquad (7.53)$$

$$t_1 = -\tau_r \ln\left(1 - \frac{\alpha}{N}\exp\left(\frac{\sum_{j=0}^{N-2} t_{N-j}}{\tau_e}\right)\right)$$

Considering the uniform reference beam, recording and erasing process meet the exponential rule. In speckle multiplexing system, no matter DSM, static speckle multiplexing, or their combination multiplexing, the reference beam is speckle beam, which is modulated by random phase diffuser. When reading out process, it is similar exponential rule with speckle reference beam always, whether it is suitable for sequence exposure method or not.

7.8.2 The Recording and Erasing

In order to research and validate the recording and erasing rule for DSM technology, one holographic photo is recorded by speckle coded reference beam, and its diffraction efficiency is measured with same interval time. The recording experimental curves are shown in Fig. 7.65a. Shift the diffuser to the position where the speckle reference beam is no longer correlating with former one. Turn on the reference beam to erase the photo and measure the diffraction efficiency with the same interval until, it is less than 1 %. The erasing curve is shown in Fig. 7.65b also.

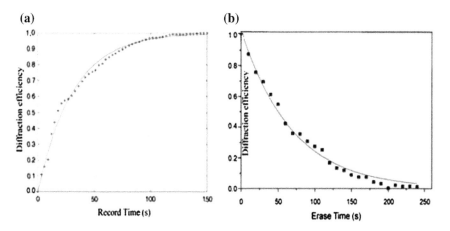

Fig. 7.65 a The diffraction efficiency curve to recording time by speckle coded reference beam. **b** The diffraction efficiency curve to erasing time without speckle reference beam

The results show that the either recording or erasing process is well with exponential rule that demonstrates it is similar with the uniform reference beam faultlessly. It shows that the sequence exposure method is suitable for speckle multiplexing system theoretically.

7.8.3   Exposure Time Sequence for Dynamic Speckle Multiplexing System

When sequence exposure method is used in DSM, the exposure curve could be calculated out and shown in Fig. 7.66.

200 holographic photos are recorded with DSM technology and the average gray value is measured every 10 photos. The deviation of the gray value is shown in Fig. 7.67, here ordinate is average gray degree and abscissa is sequence number of the holograms. It is seen easily from Fig. 7.67, that the diffraction efficiency is approximately uniform in DSM system. It is demonstrated that the same recording and erasing rule is obeyed by DSM system as a plain wave and the formula above can meet the requirement of larger capacity storage.

Hence the time sequence exposure for the combination of dynamic and static speckle multiplexing (DSSM) was determined consideration to practical sequence exposure experiment, that the storage sketch diagram is shown in Fig. 7.68. Some holographic photos (100 in the diagram) are recorded on the first point of recording medium and then the medium is moved a short distance which is far smaller than recording spot size, but larger than space selectivity in this direction. The recording mark is overlapped with each other and then moves it to second point to keep on recording photos and take turns.

Former theory and conclusion of time sequence and exposure time are all established on the basic of volume multiplexing, however in DSSM, not only the diffraction efficiency for the holographic photos at same point, but at the different points are uniform also. The storage process with DSSM is composed of two steps mainly: Step 1 is the DSM at one point, and step 2 is the shift process of the static

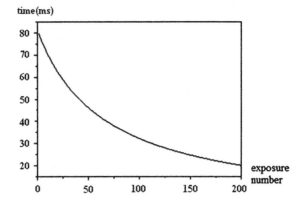

Fig. 7.66 Calculated sequence exposure curve of exposure time (ms) to exposure number

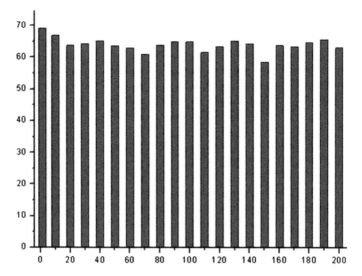

Fig. 7.67 Diffraction efficiency of 200 sequence exposure holographic photos in dynamic speckle multiplexing system

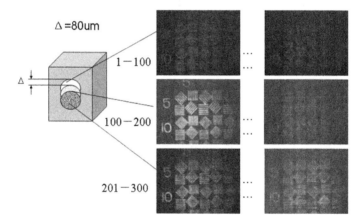

Fig. 7.68 Sketch diagram of practical sequence exposure experiment with DSSM

speckle multiplexing. The above time sequence formula was used to calculate the time sequence of dynamic multiplexing; however, for the holographic photos at different points it is much more complicated. The erasing effect on the first holographic photo by second one is weaker than before, which cannot be neglected since they have been partly overlapped.

The aim of uniform diffraction efficiency cannot realize only by conventional sequence exposure method in DSSM. The erasing factor for every recorded holographic photo in DSSM is divided into two parts: one part is the erasing effect by

Fig. 7.69 Sketch diagram of
overlapped area of two
exposure points

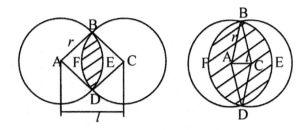

sequent holographic photo at the same point and the other part is the erasing effect
by all other holographic photos at different points. It is necessary to deduce the
exposure time sequence formula again. Two assumptions are taken for the
derivation according to above analysis. First, erasing rule of DSM is obeyed by
holographic photos at one point. Second, the erasing effect for the holographic
photo at different points is proportioned to the area overlapped. The reference
beam's shape is circular with the radius of r in DSSM and the center distance for
two reference beams is l ($l < 2r$), which is shown in Fig. 7.69.

From Fig. 7.69, the area overlapped can be calculated by:

$$S_{\text{shadow}} = S_{\text{ABED}} + S_{\text{CDFB}} - S_{\text{ABCD}} = 2S_{\text{ABED}} - S_{\text{ABCD}} \tag{7.54}$$

The ratio of overlapped area is:

$$\gamma = \frac{S_{\text{shadow}}}{S_{\text{circle}}} = \frac{2}{\pi} \arccos \frac{l}{2r} - \frac{1}{\pi r^2} \sqrt{r^2 - l^2/4} \tag{7.55}$$

According to the assumptions shown above, the erasing effect to the adjacent
holographic photo by current one could be modified as:

Recording: $\Delta n = \Delta n_s \left[1 - \exp(-t/\tau_\gamma) \right]$

Erasing:

$$\Delta n' = \Delta n \exp\left[-t' / \left(\frac{\tau_e}{\gamma_{uv}} \right) \right] = \Delta n \exp[-t'\gamma_{uv}/\tau_e] \tag{7.56}$$

where γ_{uv} is defined as the overlapped area ratio for number u holographic photo to
number v, which is termed as erasing factor.

The exposure time sequence for DSSM VHS is single track. The recording
model of DSSM VHS with single track is designed as shown in Fig. 7.70.

N points on one track are used for overlapping recording in DSSM VHS and
M holographic photos are recorded at every point. It is known that the recorded
holographic photos from number 2 to number $M \times N$ will erase the first one. Δn_g is
defined to be the deviation of the refraction index modulation of number g of
holographic photo, so it is:

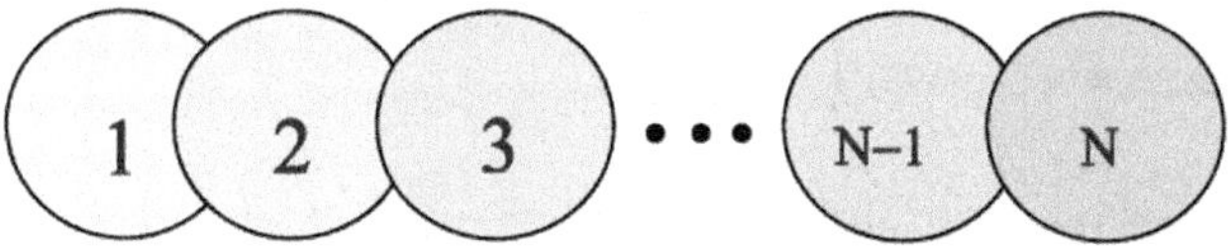

1	$M+1$	$2M+1$	$\cdots$	$(N-2)M+1$	$(N-1)M+1$
2	$M+2$	$2M+2$	$\cdots$	$(N-2)M+2$	$(N-1)M+1$
$\vdots$	$\vdots$	$\vdots$	$\ddots$	$\vdots$	$\vdots$
M	$2M$	$3M$	$\cdots$	$(N-1)M$	NM

Fig. 7.70 The overlapping recording model of DSSM volume holographic storage with single track

$$\Delta n_1 = \Delta n_s \left[1 - \exp\left(-\frac{t_i}{\tau_r} \right) \right] \exp\left(-\frac{\sum_{j=2}^{MN} \gamma_{j,1} t_j}{\tau_e} \right) \tag{7.57}$$

$$\Delta n_{M+1} = \Delta n_s \left[1 - \exp\left(-\frac{t_{M+1}}{\tau_r} \right) \right] \exp\left(-\frac{\sum_{j=M+2}^{MN} \gamma_{j,(M+1)} t_j}{\tau_e} \right) \tag{7.58}$$

$$\Delta n_{(i-1)M+i'} = \Delta n_s \left[1 - \exp\left(-\frac{t_{(i-1)M+i'}}{\tau_r} \right) \right] \exp\left(-\frac{\sum_{j=(i-1)M+i'+1}^{MN} \gamma_{j,[(i-1)M+i')]} t_j}{\tau_e} \right) \tag{7.59}$$

$$\Delta n_{NM} = \Delta n_s \left[1 - \exp\left(-\frac{t_{NM}}{\tau_r} \right) \right] \tag{7.60}$$

When the recording and reading time constants are different, it will get:

$$\Delta n_s = \left(\frac{\tau_r}{\tau_e} \right) \sum_{i=1}^{N} \sum_{j=1}^{M} \Delta n_{ij} \Rightarrow \Delta n_i \left(\frac{\tau_r}{\tau_e} \right) \frac{\Delta n_s}{MN} \tag{7.61}$$

It is required that the diffraction efficiency is uniform for each photo, i.e.,

$$\Delta n_{11} = \Delta n_{21} = \Delta n_{31} = \cdots = \Delta n_{N1} = \frac{\alpha \Delta n_s}{MN} \quad \text{where, } \alpha = \frac{\tau_r}{\tau_e} \tag{7.62}$$

The $M \times N$ is exposure time number for holographic photos as in Fig. 7.70 and t_i is the exposure time for number i photo. Then the 2D exposure matrix can be changed into 1D sequence. By this method, can get:

$$\Delta n_{NM} = \Delta n_s \left[1 - \exp\left(-\frac{t_{NM}}{\tau_r} \right) \right]$$

$$\Rightarrow t_{NM} = -\tau_r \ln\left(1 - \frac{\Delta n_{NM}}{\Delta n_s} \right) = -\tau_r \ln\left(1 - \frac{\alpha}{MN} \right) \tag{7.63}$$

It is exposure time for last photo. By the updated method, the exposure time for number $NM - 1$ photo is presented as:

$$t_{NM-1} = -\tau_r \ln\left(1 - \frac{\alpha}{MN} \exp\left(\frac{\gamma_{NM,(NM-1)} t_{NM}}{\tau_e} \right) \right) \tag{7.64}$$

The exposure time for the Mth holographic photo in column $N - 1$ can be expressed by:

$$t_{(N-1)M} = -\tau_r \ln\left(1 - \frac{\alpha}{MN} \exp\left(\frac{\sum_{j=(N-1)M+1}^{NM} \gamma_{j,(N-1)M} t_j}{\tau_e} \right) \right) \tag{7.65}$$

where the coefficient of erasing factor is added which present the erasing effect on the current photo by last M ones. The exposure time for the Mth holographic photo in column i is expressed by:

$$t_{iM} = -\tau_r \ln\left(1 - \frac{\alpha}{MN} \exp\left(\frac{\sum_{j=iM+1}^{NM} \gamma_{j,iM} t_j}{\tau_e} \right) \right) \tag{7.66}$$

And the exposure time for the i'_{th} holographic photo in column i is expressed by:

$$t_{(i-1)M+i'} = -\tau_r \ln\left(1 - \frac{\alpha}{MN} \exp\left(\frac{\sum_{j=(i-1)M+i'+1}^{NM} \gamma_{j,(i-1)M+1} t_j}{\tau_e} \right) \right) \tag{7.67}$$

The exposure time for the Mth holographic photo in the first column is then expressed by:

$$t_M = -\tau_r \ln\left(1 - \frac{\alpha}{MN} \exp\left(\frac{\sum_{j=M+1}^{NM} \gamma_{j,M} t_j}{\tau_e} \right) \right) \tag{7.68}$$

At last, the exposure time for the first holographic photo is by:

$$t_1 = -\tau_r \ln\left(1 - \frac{\alpha}{MN} \exp\left(\frac{\sum_{j=2}^{NM} \gamma_{j,1} t_j}{\tau_e} \right) \right) \tag{7.69}$$

The original derivation has been done for the sequence time exposure in DSSM with one track. These expressions for DSSM system are very much similar to other DSM system. The difference is the weighting coefficient of erasing factor only.

7.8.4   Simulation and Experiment of Single Track of Overlapping Sequence Time Exposure

The experiment of single track overlapping sequence time exposure is based on above conclusions. The experimental conditions are following. Light source is semiconductor laser with wavelength of 532 nm, radius of reference beam 3 mm, $M = 10$, $N = 10$, center distance of static speckles in neighborhood is 100 μm, recording medium is doped LiNbO$_3$:Fe crystal with 0.03 wt% Fe and $\tau_\gamma = 1050s$, $\tau_e = 1150s$, LiNbO$_3$:Fe:In crystal doped with 0.03 wt% Fe, 1.0 mol% In and $\tau_\gamma = 880s$, $\tau_e = 960s$.

The model of the storage experimental system is shown in Fig. 7.71. The laser beam is divided into two beams after collimating where one for recording and the other for speckle reference after modulated by random phase diffuser. Two beams entranced into the adjacent surface of the crystal with recording angle of 90°. Two

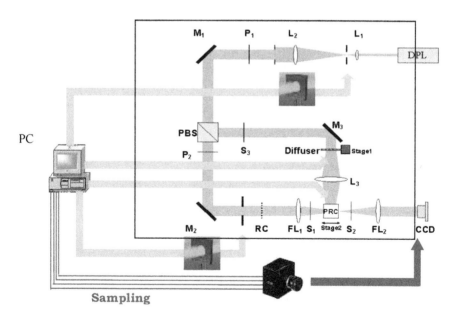

Fig. 7.71   The single track overlapping sequence time exposure experimental system with DSSM: M_1–M_3 reflecting mirror, P_1, P_2 Polaroid, L_1, L_2 lens, *PBS* polarization splitting prism, S_1–S_3 shutter, *RC* resolution chart, FL_1, FL_2 Fourier lens, *CCD* detector, *DPL* diode pump laser, *Diffuser* random phase board on the *Stage 1* and medium on *Stage 2*

Table 7.14 The sequences exposure time (unit: s) and coordinates in doped Fe:LiNbO$_3$ crystal in experimental system with DSSM

Poi.\Num.	1	2	3	4	5	6	7	8	9	10
1	42.79	41.26	39.83	38.49	37.24	36.07	34.97	33.94	32.96	32.04
2	32.04	31.17	30.34	29.56	28.82	28.12	27.44	26.80	26.19	25.61
3	25.61	25.05	24.52	24.01	23.51	23.04	22.59	22.15	21.74	21.33
4	21.32	20.93	20.56	20.20	19.85	19.51	19.19	18.87	18.57	18.27
5	18.24	17.96	17.68	17.42	17.16	16.90	16.66	16.42	16.19	15.96
6	15.92	15.71	15.49	15.29	15.09	14.89	14.70	14.52	14.33	14.16
7	14.10	13.93	13.77	13.60	13.44	13.29	13.14	12.99	12.84	12.70
8	12.64	12.50	12.37	12.24	12.11	11.98	11.86	11.74	11.62	11.50
9	11.43	11.32	11.21	11.10	10.99	10.89	10.79	10.69	10.59	10.49
10	10.42	10.32	10.23	10.14	10.05	9.96	9.88	9.80	9.71	9.63

shifting tables with high-precision position are used in the DSSM system. One is used to control the diffuser moving along the longitudinal direction and the other is for recording crystal to move at transverse direction.

The materials (doped LN:Fe:LiNbO$_3$, Fe:In:LiNbO$_3$, Zn:Fe:LiNbO$_3$ and Tb:Fe:LiNbO$_3$) are used to record experiment. The exposure time sequences and coordinates for both recording materials with DSSM system are calculated which for Fe:LiNbO$_3$ is listed in Table 7.14, and for Fe:In:LiNbO$_3$ crystal is listed in Table 7.15. The sequences exposure experimental results with Fe:LiNbO$_3$ crystal (Fe 0.03 wt %) in DSSM system is shown in Fig. 7.72. The recovered photo pictures of points with coordinates of 1-1, 3-3, 5-5 (as in Table 7.14) are shown in Fig. 7.73.

In order to reduce the scattering noise when recovering and speed up the recording, using the sensitivity higher doped crystal Fe:In:LiNbO$_3$ (Fe 0.03 wt%, In 1.0 mol%) for recording holographic recording experiment, that the results is

Table 7.15 The sequences exposure time (unit: s) and coordinates for doped Fe:In:LiNbO$_3$ crystal in experimental system with DSSM

Poi.\Num.	1	2	3	4	5	6	7	8	9	10
1	23.38	22.83	22.29	21.79	21.30	20.84	20.40	19.97	19.56	19.17
2	19.22	18.84	18.48	18.13	17.80	17.47	17.16	16.86	16.57	16.28
3	16.31	16.04	15.78	15.52	15.27	15.03	14.80	14.58	14.36	14.15
4	14.15	13.95	13.75	13.55	13.37	13.18	13.00	12.83	12.66	12.50
5	12.49	12.33	12.17	12.02	11.87	11.72	11.58	11.44	11.31	11.18
6	11.16	11.03	10.90	10.78	10.66	10.54	10.43	10.32	10.21	10.10
7	10.07	9.97	9.86	9.76	9.66	9.57	9.47	9.38	9.29	9.20
8	9.16	9.08	8.99	8.91	8.83	8.74	8.67	8.59	8.51	8.44
9	8.39	8.32	8.25	8.18	8.11	8.04	7.97	7.91	7.84	7.78
10	7.73	7.67	7.61	7.55	7.49	7.43	7.37	7.32	7.26	7.21

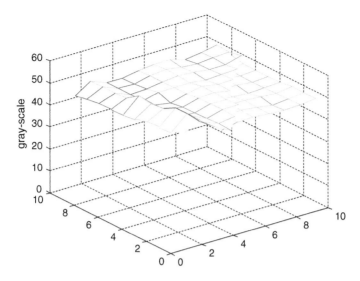

Fig. 7.72 The experiment results of single track superposed holographic with DSSM in doped Fe: LiNbO$_3$ crystal: diffraction efficiency to the exposure time (unit: s) and the coordinates to reference Table 7.14

shown in Fig. 7.74. Its deviation of diffraction efficiency is less than ±6.2 %. The experiment demonstrated that the exposure time sequence for DSSM with single track is acceptable.

The doped Fe:In:LiNbO$_3$ crystal is still used for experiments with 20 × 20 holographic photos are recorded further. The diffraction efficiency of each photo is measured and a deviation less than ±7.67 %, which is shown in Fig. 7.74. The recovered photos in LiNbO$_3$:Fe:In on coordinates of 1-1, 3-3, 5-5 (in Table 7.15). The theoretical formula to holographic photos with DSSM for single track overlapping sequence time exposure is demonstrated very well (Fig. 7.75).

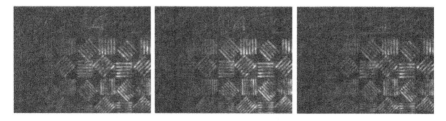

Fig. 7.73 The recovered photos pictures of points with coordinates of 1-1, 1-3, and 5-5 in doped Fe:LiNbO$_3$ crystal (coordinates to reference Table 7.14)

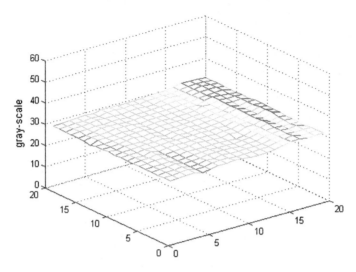

Fig. 7.74 The diffraction efficiency of doped Fe:In:LiNbO$_3$ crystal of holographic photos in single track superposed holographic experiment with DSSM system, the coordinates reference Table 7.15

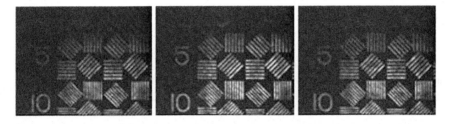

Fig. 7.75 The recovered photos of DSSM holographic photos (in LiNbO$_3$:Fe:In), its coordinates of 1-1, 3-3, and 5-5 reference to Table 7.15

References

1. GE Unveils 500-GB, Holographic Disc Storage Technology. CRN. April 27, 2009
2. G.L. Chen, C.Y. Lin, M.K. Kuo et al., Numerical suppression of zero-order image in digital holography. Opt. Express **15**(14), 8851–8856 (2007)
3. J.L. Zhao, H.Z. Jiang, J. Di, Recording and reconstruction of a color holographic image by using digital lensless Fourier transform holography. Opt. Express **16**(4), 2514–2519 (2008)
4. X.H. Hong, B. Yang, C. Zhang, Nonlinear volume holography for wave-front engineering. Phys. Rev. Lett. **13**, 16–17 (2014)
5. M. Kawana, J. Takahashi, S. Yasu, Characterization of volume holographic recording in photopolymerizable nanoparticle polymer at 404 nm. J. Appl. Phys. **117**, 053105 (2015)
6. E. Darakis, J.J. Soraghan, Reconstruction domain compression of phase-shifting digital holograms. Appl. Opt. **46**(3), 351–356 (2007)

7. W. Liu, *Holographic resolution and its application in memory and imaging* (California Institute of Technology, California, 2001)
8. H.H. Tang, P.K. Liu, Long-distance super-resolution imaging assisted by enhanced spatial Fourier transform. Opt. Express **23**(18), 23613–23623 (2015)
9. N.C. Pégard, J.W. Fleischer, Optimizing holographic data storage using a fractional Fourier transform. Opt. Lett. **36**, 2551–2553 (2011)
10. Z. Ren, P. Su, J. Ma, G. Jin, Secure and noise-free holographic encryption with a quick-response code. Chin. Opt. Lett. **12**(1), 010601 (2014)
11. M. Fan, J. Wei, A. Zhu, Y. Ming, Arylheterocycle substituted dibenzofuranpyran-type photochromic compound and preparing process and use thereof, CN1152109 C (2004)
12. M. Fan, C. Wang, X. Han, J. Xiao, Y. Ming, Colour threedimensional storage material for optical information storage and preparing method thereof, CN1204175 C (2005)
13. J. Men, H. Hoang, Y. Fu, T. Tan, P. Chen, 2-diaryl naphthopyran compounds and method for preparing same, CN1634916 A (2005)
14. J.T. Sheridan, F.T. O'Neill, J.V. Kelly, Holographic data storage: optimized scheduling using the nonlocal polymerization-driven diffusion model. J. Opt. Soc. Am. B **21**(8), 1443–1451 (2004)
15. G. Bianco, M.A. Ferrara, F. Borbone, A. Roviello, Volume holographic gratings fabrication and characterization, in *Proceedings of SPIE 9508, Holography: Advances and Modern Trends IV*, 950807 (2015)
16. D.Y. Xu, P. She, R. Liu, Z. Lei, G. Qi, *Multi-Level Phase Change Material Storage Optical Disc Reader*. (Tsinghua University, CN01139829.9, 2002)
17. J.M. Desse, P. Picart, F. Olchewsky, Quantitative phase imaging in flows with high resolution holographic diffraction grating. Opt. Express **23**(18), 23726–23737 (2015)
18. K. Feng, W. Streyer, Y. Zhong, A.J. Hoffman, Photonic materials, structures and devices for Reststrahlen optics. Opt. Express **23**(24), A1418–A1433 (2015)
19. H. Hu, L.F. Pan, D. Xu et al., Coding and signal processing for three level run-length limited optical recording channel. SPIE **6282**, 682271–682276 (2006)
20. H. Hu, D. Xu, L. Pan, Modulation code and PRML detection for multi-level run-length DVD chanels. SPIE **6282**, 682281–682286 (2006)
21. K. Otsuka, Self-mixing thin-slice solid-state laser Doppler velocimetry with much less than one feedback photon per Doppler cycle. Opt. Lett. **40**(20), 4603–4606 (2015)
22. S. Sweetnam et al., Characterization of the polymer energy landscape in polymer: Fullerene bulk heterojunctions with pure and mixed phases. J. Am. Chem. Soc. **136**, 14078–14088 (2014)
23. S. Gélinas, T.S. Van Der Poll, G.C. Bazan, R.H. Friend, Ultrafast long-range charge photovoltaic diodes. Science **343**, 512–517 (2014)
24. P.B. Deotare, W. Chang, E. Hontz, D.N. Congreve, Nanoscale transport of charge-transfer states in organic donor–acceptor blends. Nat. Mater. **14**, 1130–1134 (2015)
25. B. de Nijs, S. Dussi, F. Smallenburg, Entropy-driven formation of large icosahedral colloidal clusters by spherical confinement. Nat. Mater. **14**, 56–60 (2015)
26. Y. Peng, F. Wang, Z. Wang, Two-step nucleation mechanism in solid–solid phase transitions. Nat. Mater. **14**, 101–108 (2015)
27. M. Gu, X. Li, Y. Cao, Optical storage arrays: a perspective for future big data storage. Sci. Appl. (Published online) **3**, e177 (2014)
28. Z. Wang, Q. Meng, Z. Zhang, D. Fu, W. Zhang, Synthesis and photochromic properties of substituted naphthopyran compounds. Tetrahedron **67**(25), 2246–2250 (2011)
29. M. Shirakawa, N. Inoue, H. Furutani, K. Yamamoto, Advanced patterning approaches based on negative-tone development (NTD) process for further extension of 193 nm immersion lithography. *SPIE 9425. Advances in Patterning Materials and Processes*, vol. XXXII, p. 942509. doi:10.1117/12.2085744
30. J.D. Caldwell, I. Vurgaftman, Mid-infrared nanophotonics: probing hyperbolic polaritons. Nat. Photonics **9**, 638–640 (2015)

31. J.-Y. Zhang, J.-S. Ma, D.-Y. Xu, The servo system design of the optical disk storage based on the finite state machine. Opt. Tech. **33**(3), 453–455 (2007)
32. Z. Zhou, Q. Tan, G. Jin, Focusing of high polarization order axially-symmetrie polarized beams. Chin. Opt. Lett. **7**(10), 938–941 (2009)
33. A. Faraon, P.E. Barclay, C. Santori, K.M.C. Fu, R.G. Beausoleil, Resonant enhancement of the zero-phonon emission from a colour centre in a diamond cavity. Nat. Photonics **5**(5), 301 (2011)
34. W.J. Munro, A.M. Stephens, S.J. Devitt, K.A. Harrison, K. Nemoto, Nat. Photonics **6**, 777 (2012)
35. S. Crouch, B.M. Kaylor, Z.W. Barber, R.R. Reibel, Three dimensional digital holographic aperture synthesis. Opt. Express **23**(18), 23811–23816 (2015)
36. Y. Wang, B. Yao, N. Menke, Y. Chen, M. Fan, Optical image operation based on holographic polarization multiplexing of fulgide film. Chin. Opt. Lett. **09**(s1), s10302 (2011)

Chapter 8
Multitrack Superposed and Polarized Recording

8.1 Multitrack Superposed 3D Holomem

8.1.1 Exposure Time Sequence of Multitracks with DSSM

The volume space can be applied more efficiently, if it could use multitracks exposure with DSSM. Multi-photos storage at one point and 2D directions recording for card and disc medium have been realized with DSSM technology at OMNERC of Tsinghua University. Holographic photos are recorded at one point without crosstalk by changing the diffuser in reference beam with DSSM and multitrack overlapping recording. Then move the medium by N rows in transverse direction and M columns in longitudinal direction to exposure continuously, as shown in Fig. 8.1. As a result, $M \times N \times P$ holographic photos are recorded in a crystal, which equals to P information pages are recorded at $M \times N$ points. If the dynamic range of the material is good enough, DSSM with multitracks overlapped method is the easiest multiplexing method for high density volume holographic storage.

The same deducing theory is used to analyze compensation of the erasing effect with static speckle shift in DSSM in multitracks overlapped storage, where compensation factor γ is introduced. The recording and erasing process is related on the recording sequence of the multiplexing point, so the sequence is defined in Fig. 8.2.

In multitracks overlapped storage, the overlapped factor, i.e., compensation factors are replenished into the 2D matrix. To simplify the expression of the 2D matrix, it is changed to 1D sequence with $P_1, P_2, \ldots, P_{M \times N}$, as Eq. (8.1). The recording points are defined as number 1, 2, …, M, where P holographic photos are recorded for each. The matrix of overlapped factor is

© Tsinghua University Press and Springer Science+Business Media Singapore 2016   573
D. Xu, *Multi-dimensional Optical Storage*,
DOI 10.1007/978-981-10-0932-7_8

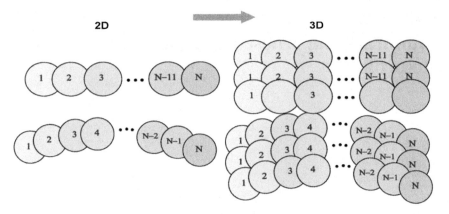

Fig. 8.1 The multitracks overlapped recording with DSSM

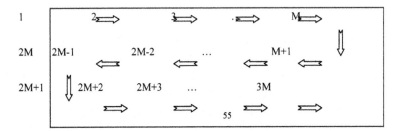

Fig. 8.2 Recording sequence in DSSM with multitracks overlapped

$$
\gamma =
\begin{bmatrix}
\gamma_{11} & & & & \\
\gamma_{21} & \gamma_{22} & & & \\
\gamma_{31} & \gamma_{32} & \ddots & & \\
\vdots & \vdots & & \ddots & \\
\gamma_{MN1} & \cdots & \cdots & & \gamma_{MN,MN}
\end{bmatrix}
=
\begin{bmatrix}
1 & 0 & 0 & 0 & 0 \\
\gamma_{21} & 1 & 0 & 0 & 0 \\
\gamma_{31} & \gamma_{32} & \ddots & 0 & 0 \\
\vdots & \vdots & & \ddots & 0 \\
\gamma_{MN1} & \cdots & \cdots & & 1
\end{bmatrix}
\qquad (8.1)
$$

where γ_{jk} presents the erasing factor to point k by point j. After P holographic photos are recorded at one point, another P photos could be recorded by shifting the crystal in transverse or longitudinal direction. DSSM volume holographic storage with N rows and M columns will be realized at last. However, the early one will be erased by following ones in some degree and the erasing effect is from both transverse and longitudinal directions with different extent. The deviation of refractivity index modulation and exposure time of yth holographic photo at point x are redefined as $\Delta n_{x,y}$ and $t_{x,y}$. Then

$$\Delta n_{1,1} = \Delta n_s \left[1 - \exp\left(-\frac{t_{11}}{\tau_r} \right) \right] \exp\left(-\frac{\left[\sum_{k=2}^{P} \gamma_{11} t_{1k} + \sum_{k=1}^{P} \sum_{j=2}^{MN} \left(\gamma_{j1} t_{jk} \right) \right]}{\tau_e} \right)$$

$$(8.2)$$

$$\Delta n_{(sM+l),q} = \Delta n_s \left[1 - \exp\left(-\frac{t_{(sM+l)q}}{\tau_r} \right) \right]$$

$$\exp\left(-\frac{\sum_{k=q+1}^{P} \gamma_{(sM+l)(sM+l)} t_{(sM+l)k} + \sum_{k=1}^{P} \sum_{j=sM+l+1}^{MN} \left(\gamma_{j(sM+l)} t_{jk} \right)}{\tau_e} \right)$$

$$\Delta n_{(NM),P} = \Delta n_s \left[1 - \exp\left(-\frac{t_{NMP}}{\tau_r} \right) \right] \quad \begin{array}{l} s = 0, 1, \ldots, N-1; \\ l = 1, 2, \ldots, M; \\ q = 1, 2, \ldots, P. \end{array}$$

$$(8.3)$$

where $\Delta n_{(sM+l)q}$ presents the refractivity modulation of number q holographic photo at row $s+1$ and column 1. In order to get uniform diffraction efficiency for each photo, make

$$\Delta n_{x,y} = \frac{\alpha \Delta n_s}{MNP} \quad \text{and} \quad \alpha = \frac{\tau_r}{\tau_e} \tag{8.4}$$

The exposure time for last photo is

$$t_{(NM),P} = -\tau_r \ln\left(1 - \frac{\Delta n_{(NM)P}}{\Delta n_s} \right) = -\tau_r \ln\left(1 - \frac{\alpha}{MNP} \right) \tag{8.5}$$

By reverse deducing, the exposure time for qth holographic photo at row $s+1$ and column 1 is

$$t_{(sM+l),q} = -\tau_r \ln\left(1 - \frac{\alpha}{MNP} \exp\left(\frac{\sum_{k=q+1}^{P} \gamma_{(sM+l)(sM+l)} t_{(sM+l)k} \sum_{k=1}^{P} \sum_{j=sM+l+1}^{MN} \left(\gamma_{j(sM+l)} t_{jk} \right)}{\tau_e} \right) \right)$$

Lastly, the exposure time for first photo is

$$t_{1,1} = -\tau_r \ln\left(1 - \frac{\alpha}{MNP} \exp\left(\frac{\sum_{k=2}^{P} \gamma_{11} t_{1k} + \sum_{k=1}^{P} \sum_{j=2}^{MN} \left(\gamma_{j1} t_{jk} \right)}{\tau_e} \right) \right) \tag{8.6}$$

Calculate the exposure time for DSSM with multitracks overlapped system according to the formula and apply it to the volume holographic system with doped LiNbO$_3$ crystal with different elements. The experimental condition is listed as follows: Light source is semiconductor with wavelength 532 nm, radius of reference beam is 3 mm, center distance for static speckles in neighborhood is 100 μm. The recording material specifications of seven kinds doped of LiNbO₃ crystal are listed in Table 8.1. The exposure time sequence for seven kinds of doped LiNbO₃ crystal is calculated out by above equations. Some modifications for optical system and control software are replenished in DSSM with multitracks overlapped volume holographic storage experiment system, in order to realize the movement of diffuser in 1D for dynamic speckle multiplexing (DSM) with the recording crystal shifting on 2D for static speckle multiplexing. The optical system and experimental set are shown in Fig. 8.3.

In order to get high repositioning precision and perfectly application of the calculated exposure sequence, some improvements are made in the software for two purposes. First, $2 \times 2 \times 10$ holographic photos are recorded in DSSM system with seven crystals, shown in Fig. 8.4.

Here the gray degree is proportion to diffraction efficiency. The experimental results of the seven doped LiNbO₃ crystal materials (see Table 8.1) are listed in Figs. 8.5, 8.6, 8.7, 8.8, 8.9, 8.10, and 8.11.

Summarizing the above experimental results for diffraction efficiency, the deviations of the diffraction efficiency are listed in Table 8.2. It is shown that the

Table 8.1 The specifications of various doped LiNbO₃ crystal material for DSSM with multitracks overlapped storage experiment

Crystal doped	Fe49	Fe54	Fe55	In:Fe1	In:Fe2	In:Fe3	In:Fe4
Component	Fe 0.05 wt %	Fe 0.03 wt %	Fe 0.03 wt %	In 0.5 mol % Fe 0.03 wt %	In 1 mol % Fe 0.03 wt %	In 2 mol % Fe 0.03 wt %	In 3 mol %, Fe 0.03 wt %
Selectivity for static speckle (μm)	9.75	6.5	9.75	7.15	5.2	4.55	6.5
Selectivity for dynamic speckle (μm)	18.4	33	16	23.2	24	22.4	25.6
Recording quality	★★	★	★	★★★	★★★	★★★	★★★

Fig. 8.3 The picture of the 2D static speckle multiplexing experimental equipment of volume holographic storage with DSSM multitracks overlapped

Fig. 8.4 $2 \times 2 \times 10$
recording array diagram

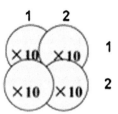

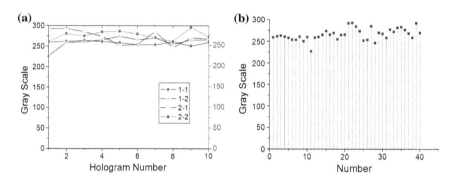

Fig. 8.5 Fe49 crystal (see Table 8.1). **a** Distribution curve for single point. **b** Comparing of the diffraction efficiency

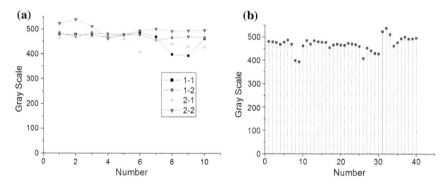

Fig. 8.6 Fe54 crystal. **a** Distribution curve for single point. **b** Comparing of the diffraction efficiency

diffraction efficiencies for the holographic photos in seven crystals are almost uniform which demonstrate the correctness of the deduced exposure time sequence for 3D DSSM storage system. Furthermore, the scattering noise could be strongly depressed if some elements like In, Mg, Zn, etc., are doped with the crystal of Fe: $LiNbO_3$, i.e., the Fe: $LiNbO_3$ doped with In is better than Fe: $LiNbO_3$.

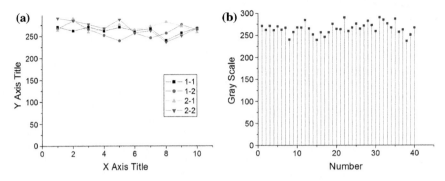

Fig. 8.7 Fe55 crystal. **a** Distribution curve for single point. **b** Comparing of the diffraction efficiency

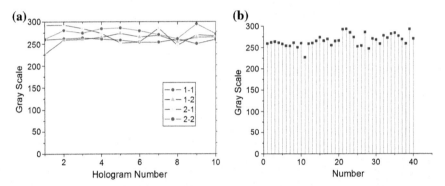

Fig. 8.8 In:Fe1 crystal. **a** Distribution curve for single point. **b** Comparing of the diffraction efficiency

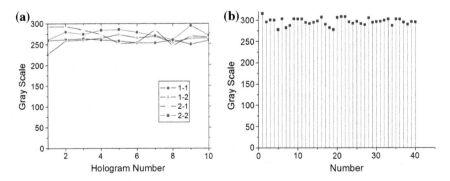

Fig. 8.9 In:Fe2 crystal. **a** Distribution curve for single point. **b** Comparing of the diffraction efficiency

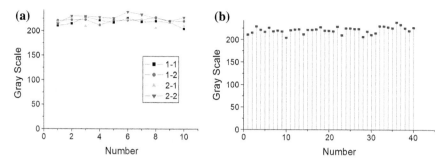

Fig. 8.10 In:Fe3 crystal. **a** Distribution curve for single point. **b** Comparing of the diffraction efficiency

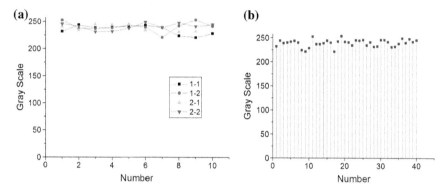

Fig. 8.11 In:Fe4 crystal. **a** Distribution curve for single point. **b** Comparison of the diffraction efficiency

Table 8.2 Deviations of diffraction efficiency for different crystals

Crystal	Fe49	Fe54	Fe55	In:Fe1	In:Fe2	In:Fe3	In:Fe4
Deviation (%)	±11.03	±13.89	±9.48	±9.00	±7.69	±8.33	±6.50

Crystal $LiNbO_3$:Fe:In (3#) doped with Fe 0.03 wt% and In 2.0 mol% is selected for further experiment with lower scattering noise of recovered photos. $4 \times 5 \times 5$ holographic photos are recorded in DSSM with multitracks overlapped system which is shown in Fig. 8.12.

The diffraction efficiency for retrieved 100 photos in DSSM with multitracks overlapped holographic storage system by conventional decreasing exposure method without introducing erasing factor is shown in Figs. 8.13 and 8.14. There is a grade of slope in the diffraction efficiency from the first photo to the last one which shows that the exposure time for sequent photos is too short to get uniform diffraction efficiency. The deviation is larger than ±18.4 %. For comparison, the calculated exposure time sequence is used for these 100 photos and the diffraction

Fig. 8.12 4 × 5 × 5 array
holographic exposure photos

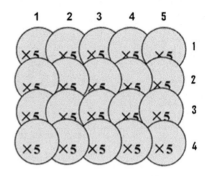

Fig. 8.13 Diffraction
efficiency for 100 retrieved
photos with conventional
decreasing exposure method

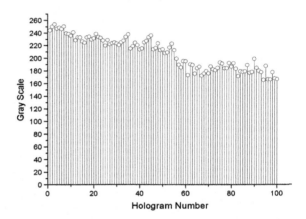

Fig. 8.14 The distribution of
gray scale of 100 retrieved
points pictures (5 × 20 array)

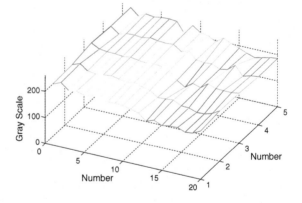

efficiency is shown in Fig. 8.15, which is almost uniform. The deviation is less than
±7.72 %. Some retrieved photos in high quality are shown in Fig. 8.16. These
results further demonstrate the correctness of the exposure sequence for DSSM with
multitracks overlapped system.

Fig. 8.15 Diffraction efficiency for 100 retrieved photos (5 × 20 array) with DSSM exposure method

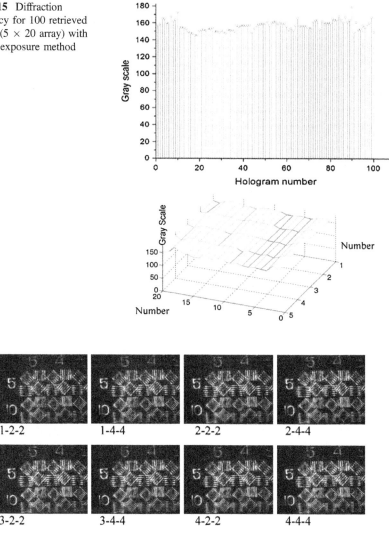

Fig. 8.16 Some retrieved photos with DSSM exposure method. The number of (*M-N-P*) denotes the position of retrieved image on the row *M*, column *N* and page *P*

In order to make efficient usage of recording material, DSSM with multitracks overlapped volume holographic storage method as well as its optimal exposure time sequence is proposed here. First, the exposure time sequence based on sequence exposure method is determined. Then the exposure method for DSSM with single track system is advanced by introducing the erasing factor γ. Lastly, exposure time sequence is confirmed for DSSM with multitracks overlapped system by further introducing the matrix of erasing factor γ. The numerical calculation and experiment result demonstrate its validity.

8.2   The Noise Analysis and Restriction

All kinds of noises in photorefractive crystal holographic optical storage system (HOSS), which deteriorate the quality of the images that are retrieved, decide the storage capacity of the holographic memory and whether it is practical or not. The SNR, definition, and fidelity can all be used to evaluate the image quality. So a great challenge of high-density holographic optical storage is how to reduce the noises and make the SNR and bit error rate meet the requirements in practical system.

The noise sources can be divided into two categories: system noise and hologram noise. Lens aberrations, SLM imperfections, detector noise, scattering and multiple reflections from lenses and other optical components, laser nonuniformity and fluctuations, and SLM-to-CCD pixel misalignment are examples of system noise. The rest of the noise arises from the hologram itself. Specifically, the hologram can introduce crosstalk between the recorded holograms, interpixel crosstalk, scattering from the recording material, multiple reflections in the medium, nonuniform diffraction efficiency in the recorded holograms, distortions that are due to surface imperfections, blurring that is due to limited spatial resolution of the material, and material shrinkage.

8.2.1   The Noise Generated by Multiple Reflections
from Optical Components

The laser with excellent spatial coherence and temporal coherence is usually employed as the light source in HHOS because the information is recorded on the materials with the coherent optics method. It is the spatial coherent beams that introduce the optical noise by interferences. Since the refractive indexes of the optical elements and the recording medium are different from the gas, each surface of discontinuity of the refractivity will reflect a part of incident light. The multiple reflections from the surfaces of all the optical elements form the interference wave, which is coherent to the output wave of the system, so that the complex interference fringes can be generated. The interference fringes are superimposed on the objective wave plane and recorded on the medium as information. In retrieving the optical elements in the retrieving system can also form the interference fringes on the image plane besides the interference fringes recorded on the medium. In order to reduce the adverse influence, the surfaces of the optical elements and the recorded medium in system are plated by the antireflecting film, which can decrease the modulation degree of the multiple reflection interference fringes.

8.2.2   Speckle Noise

The fine coherence of the laser beams causes that the waves scattered by the defect and impurity of the optical elements interfere and are superposed in the objective wave. It can introduce the speckle noise, the optical noise with Gauss distribution in the images, which can worsen the quality of the images. The speckle noise is the inherent noise in HHOS. The better the coherence of the beams is, the more serious the speckle noise is. Through the abundant research for the speckle noise, many methods of reducing the speckle noise in HOSS have been proposed with control of the temporal or spatial coherence of the beams.

8.2.3   Scattering Noise of the Recording Medium

The defect and impurity of the recording medium can scatter the incident wave and lead to the scattering noise of the image. When the photorefractive crystal is used, the defect and impurity of the crystal in nonuniform illumination can generate the photoinduced scattering light, i.e., a part of the incident light departs from the original direction of propagation and diffusers around. The part of the scattering light is recorded with user information. Retrieving a part of scattering light has the same direction of propagation as the retrieving beams and is superposed in the images. Although the energy of the light is little, the couple effect of the photorefractive crystal can amplify it and form stronger scattering noise. It can bring the loss of recording light energy, decrease of the intensity of the user information recording, and the retrieved images even influence the image quality and the diffraction efficiency.

It has been reported that blending two kinds of impurity with fit concentration, such as magnesium and iron or indium and iron, so as to improve the anti-refractivity capability of the crystal, can reduce the scattering noises.

8.2.4   Noise of the Detector and Circuit

The detector, electronic process elements of amplification and shaping, and A/D change circuit in HOSS can all generate all kinds of electronic noises so as to reduce the quality of images. The noises include photoelectronic detector noise, electronic element noise, the noise introduced by the quantization error in A/D change, and so on. Although the CCD detector is the low-noise element, the noises are superposed in ideal signal during injecting charge, transmitting charge, and detecting charge. The principle noise sources: the photon shot noise from photon or thermal excitation of the CCD, the fat noise, the capture noise by defects, the thermal noise in the resetting process of output circuit, the nonuniform dark current noise, and so on. The shot noise and the dark current noise are both the inherent noises of the CCD.

8.2.5 Crosstalk Noise

The crosstalk is the interference of the pixels in one page in the retrieving image. Because the numerical apertures of any practical optical imaging systems are limited, the spatial bandwidth of system is limited, too. After the objective wave with all the frequencies goes through the optical system, the components with the frequency over the spatial bandwidth of system are lost. Similarly, the high frequency information of the object is also lost in HOSS so that the crosstalk of the pixels in one page can be generated. The quality of the image is influenced by the crosstalk. The influence abides by the certain rules, but when there are random pixels in large scale in pages the influence among the pixels become random and regarded as noise.

8.2.6 The Noises of the Speckle Multiplexing System

The scattering noise is the main noise in the speckle multiplexing HOSS. Although the noises referred as the above are in the speckle multiplexing HOSS, the most important noise is the scattering noise. Plating antireflecting film can reduce the noise generated by multiple reflections from optical surfaces and decrease the modulation degree of the interference fringe formed by multiple reflections. The inherent laser speckle noise of the system and the noise of the CCD are much lower than the shot noise of the recording medium and the crosstalk noise.

The research on the experiment of reducing the crosstalk with many methods including the error corrected code, interleaving, equalization, and so on. The methods are effective for improving the crosstalk in HOSS. In particular, based on the analysis of the influence of diffraction limitation to crosstalk, employ the diffuser to modulate the objective light for reducing noise. It is proved that the reduction of crosstalk with the phase modulation of the objective function and imaging by the partially coherent light is effective. However, the most important factor is photoinduced scattering noise in DSM system. In practice, the scattering noise of the recording medium can deteriorate the quality of the holograms. The holograms retrieved from the shot noises in experiment are shown in Fig. 8.17.

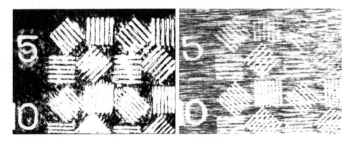

Fig. 8.17 The retrieved holograms in speckle multiplexing system with serious noise

8.2.7   The Generation of the Scattering Noise

The scattering noise of the crystal, which is the main noise source in the speckle multiplexing HOSS, is induced by the illuminating crystal with strong coherent light for long time. The noise is able to increase with prolonging illumination time. The scattering noise can reduce the effective dynamic range of the recording medium, the image quality of the optical imaging system, and the storage capacity. The background noise enhances with the extension of time so as to decrease the SNR of the system. At the same time, the energy loss of the pump light and signal light declines the signal gain. So the reduction of the scattering noise is one of the most important challenges in the implementation of the photorefractive crystal volume holographic memory. It is very significative to research the characteristics of the scattering noise and the influence of it to the quality of the holograms. In this chapter, we will discuss the effective method of depressing the scattering noise based on the research of the generation of the scattering noise.

The scattering of light often exists with different forms in the volume holographic storage. The nonlinear scattering is caused by the multiple wave couple process. Ashkin et al. gave the basic explanation as followed. When a beam incidents into the photorefractive crystal, the irregularity on the boundary between the air and the crystal, or the interior defect or nonuniformity of the crystal changes the direction of the propagation of a part of the incident wave so as to produce a very weak scattering. The weak scattering, like the seed, is amplified continuously in the recording process of the holograms, as illustrated in Fig. 8.18. It becomes the seed of scattering light. In the first step, the hologram of a scattering wave is recorded, that is, the interference fringes produced by the incident light and the scattering field are recorded in the crystal. And the holograms of the scattering field are read by the incident wave at the same time. So the reconstruction can be interfered with the scattering wave field. If the interference wave is amplified, the amplified wave field will be written into the hologram with the incident wave again. At that time the incident wave still has the correct phase and is amplified further. It produces a positive regeneration that enhances the scattering field continuously until some stable state.

Although the energy of the incident scattering light seed is little, the gain of the photorefractive material is high and the amplified noise can become the part of the incident light so that the further scattering is induced. Through the couple with the

Fig. 8.18 Multiple wave coupled process

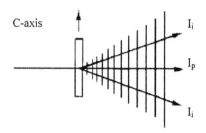

incident light, the intensity of the scattering light can be amplified strongly. As a result, a good deal of energy of the incident light is lost due to converting to the scattering light. The noise grating formed by the interference of the scattering wave and the incident wave is recorded in the crystal with the refractive grating of the signal light synchronously. Retrieving the noise grating meeting the Bragg condition is also read so as to influence the recovery of the holograms. Once the noise is serious, the signal will be submerged in the noise and cannot be read. The interference fringes recorded because of the scattering field are called as the noise grating.

On the other hand, the signal wave competes with the noise wave due to the same amplification principle in the course of the recording on lithium niobate. Once the signal wave can be amplified sufficiently, the noise intensity in whole will be decreased greatly. The competition depends on the photovoltaic field Eph in the crystal. In the model of the scattering noise multiple wave coupling, there are nonlinear relations between the sum of the scattering noise and the photovoltaic field in theory. In some incident angle the whole scattering intensity increases with the increment of the photovoltaic field and the photovoltaic field of the crystal enhances because of the increase of the incident light intensity. The intensity of the signal light declines with the enhancement of the photovoltaic field in some range, and it also drops when the photovoltaic field is lower than a special value. In Fig. 8.19, R and G indicate the ratio of the whole noise light intensity to the whole incident light intensity and the enlargement factor.

It means that the noise intensity will be reduced, the amplification of the signal light will be ensured, and the SNR of the holograms will be improved once the photovoltaic field of the lithium niobate crystal is decreased. So reducing the intensity of the incident light is a possible solution of decreasing the photovoltaic and reducing the noise.

In recent years, many researchers have proposed different methods to restrain the photoinduced scattering noise and attain prominent achievements. For example, in standard HOSS recording with the e wave or some signal wave that has a incident angle with the reference wave or c axis, can introduce a great deal of amplified

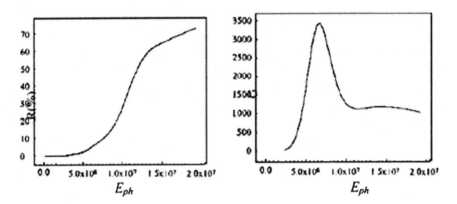

Fig. 8.19 Relations of the whole noise intensity, signal intensity, and the photovoltaic field

Fig. 8.20 The reference light spot modulated by the speckle in speckle multiplexing

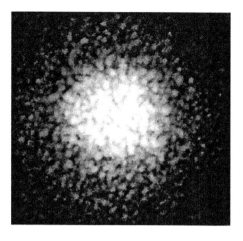

scattering noises into the retrieved images. Employing o wave can reduce some scattering noise. However, recording o wave can weaken the multiple wave couple efficiency greatly because the maximal electrooptic factor of the photorefractive crystal is not used for couple of the waves in that condition. Rajbenbach et al. proposed to use the method of the rotating crystal material slowly to weaken the scattering noise of the photorefractive crystal and obtain the amplified images with low noise, that Rabinovich and Feldman use the achromatic grating to control the light scattering. G. Zhang et al. achieved the reduction of the noise with the property that the time constant of the noise grating is very large while it is very small. The researcher for the photorefractive materials also found that doping $M(M = Mg^{2+}$, Zn^{2+}, In^{3+}, $Sc^{3+})$ with $LiNbO_3$:Fe can reduce the photovoltaic field, deaden the scattering noise, improve the quality of the photorefractive holographic storage, and obtain high diffraction efficiency, high definition, and fast responsible holograms.

Because the incident reference light is the scattering light through the diffuser in speckle multiplexing HOSS, as shown in Fig. 8.20, the nonuniform speckle coded reference light induces more serious scattering when it enters the crystal. It speeds the growth of the seed in scattering and the noise gratings are enhanced in further coupling amplification. In next parts, how to employ proper storage schemes to reduce the scattering noise and improve the equality of the images is discussed.

8.2.8   The Methods of Improving SNR and Reducing the Scattering Noise

According to the principles of the generation and amplification of the scattering in crystal, the scattering noise is the noise that the seed that generated by the incident speckle reference light is coupled and amplified. It becomes a stable state. So in order to cut down the scattering noise, two directions are followed.

Weaken the further amplification of the scattering noise: the scattering noise derived from the amplification of the seed. So the gain of the seed will be decreased if the recording time is reduced. Control the intensity of the scattering noise from the original process of the noise generation: if the intensity and the energy of the seed are weakened and incident scattering reference light is decreased, the photovoltaic field will be declined so that the gain of the noise light can be reduced.

In theory, both the methods can effectually control the scattering noise in crystal. However, further analyzing the practices of the two methods and combining the target of the experiment system, large capacity speckle multiplexing HOSS can draw the conclusion as follows.

When the requirement of the intensity of the grating is same, the more powerful incident light is necessary so as to generate very intensive photovoltaic field if the gain of the seed is reduced or the recording time is lessened. Then, in a short time, the intensive grating intensity can be attained, but the intensity of the noise grating can be superior to the intensity of the information grating as a result of the competition between the noise light and the signal light. So the noise is accumulated fast so that the signal grating is submerged in the noises in the retrieved image. In experiment the intensities of the incident reference light and signal light are 2.3 and 0.9 mW. For 6 s recording time, the recovered image is illustrated in Fig. 8.21. The holograms have been submerged by the scattering noise with the powerful incident light.

Although cutting the exposure time can decrease the noise in the image (recording time is shorter than 3 s in experiment). The proper exposure sequence is necessary for the large capacity HOSS. On the basis of the research on the exposure sequence of the system, the former the recording sequence is, the longer the recording time is. The exposure time of the first hologram often needs dozens of seconds, even longer. It will introduce large scattering noise into the hologram so that the bit error rate of the retrieved images is too high. So the method of shortening the gain time is not fit to the large capacity storage although it can reduce the scattering noise efficiently for single recorded hologram.

With the second method, decreasing the intensity of the incident light to some extent, which can control the scattering noise from the original step of generation,

Fig. 8.21 The signal grating is submerged by the noise grating as a result of recording with powerful incident light and short time

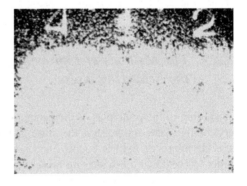

Fig. 8.22 The noise reduced obviously as result of recording with weak incident light and long time

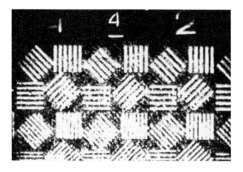

can reduce the photovoltaic field in the crystal. The method can not only weaken the energy of the scattering light noise, but also the sum intensity of the noise in the stable state. Different from the first method, the original intensity of the noise grating is lower than that of the signal grating in the competition between the noise grating and the signal grating. After recording for 60 s with the incident reference light 1.0 mW and the objective light 0.3 mW, the 2.0 mW light is used to read that the retrieved images are shown in Fig. 8.22. The scattering noise still exists, but the noise intensity is reduced relatively as shown in Fig. 8.22, even with long recording time.

8.2.9 The Experiment of Recording the Scattering Seed with Weak Energy

For the form mechanism of the scattering noise grating cannot be explained with simple two wave coupling, it is a complex multiple wave mixed process. The intensity of the scattering light depends on the material property, the direction of the crystal axis, the intensity of the incident light, and the spot size. So the next analysis is accorded to the experiment result.

In the system of the experiment, a continuous attenuating plate is set in the speckle multiplexing HOSS, as illustrated in Fig. 8.23. The intensity of recording light changes with the rotation of the attenuating plate.

The incident lights (including the objective light and the reference light) with different intensity and the resolution chart are used to record (the reference object ratio is optimized). In the recording, the diffraction effect and the modulation depth of the volume holographic grating must be constant so that the grating recorded with different intensity of the recording light is comparable. After recording the lights with same intensity are used to retrieve the holograms. The intensity and the quality of the retrieved images are compared in Table 8.3. The experiments adopt different doped $LiNbO_3$ including: Fe and $LiNbO_3$:Fe:In, $LiNbO_3$:Fe(54#): Fe 0.03 wt%, $LiNbO_3$:Fe:In (In 2#): Fe 0.03 wt% In 1.0 mol%, separately that the experimental results are shown in Table 8.3. Different intensities of the recording

Fig. 8.23 The optical system of scattering noise experiment with continuous attenuating plate: *RC* resolution chart, L_1, L_2 Lens; P_1, P_2 Polaroid; M_1–M_3 reflecting mirror; S_1–S_3 shutter; *PBS* Polarization splitting prism, FL_1, FL_2 Fourier Lens, Diffuser—random phase board on the Stage$_1$, Stage$_2$—precision tables for medium

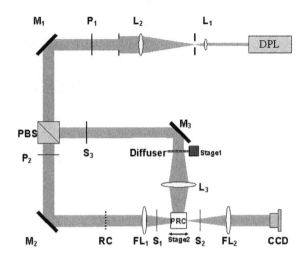

light (with constant reference object ratio) are used so as to obtain the uniform diffraction efficiency and modulation depth. At that time the lights with constant and enhanced intensity are employed to retrieve the holograms. The weaker recording light is the lower scattering noise of the retrieved images.

So the weak intensity of the incident light can lead to the low intensity of the seed, weak photovoltaic field, low intensity of the noise grating, and the retrieved images with weak scattering noise.

In Table 8.3, the scattering noise of $LiNbO_3$:Fe:In is weaker and its image quality is better because the photovoltaic field Eph decreases with the increment of the concentration of the antiphotorefractive impurity ions after codoped in the $LiNbO_3$ crystal. It agrees on the result of the experiment. However, it is recognized from the experiments that although the antiscattering crystal can reduce the scattering, the scattering noise of the retrieved images is not still ignored.

In the further analysis for the experiment results, calculating the intensity of the noise with the different intensity of the incident light can attain the curve between the intensity of the incident light and the intensity of the image noise, as in Fig. 8.24. It also is accorded with the theoretical curve.

The intensity of the noise cuts down rapidly with the decrease of the intensity of the incident light in constant modulation depth from the experiment. When the intensity of the reference light is 0.7 mW, the noise intensity can become the half of the original without change of the intensity of the signal. The lower the intensity of the reference light is, the weaker the noise intensity is. However, on the other hand, the very low intensity of the recording can cause the decrease of the signal intensity and the extension of the exposure time a hologram needs so that the requirements of the high capacity storage cannot be met. At last we chose the power of the reference light PR = 0.7 mW and the power of the object light PO = 0.25 mW.

The result of the experiment of the single page recording is shown in Fig. 8.25. Though the intensity of the image is weak (with constant diffraction efficiency), the

Table 8.3 The comparison of the retrieved images in the recording lights with different intensity

Recording light intensity		Retrieving light intensity (mW)	Retrieve images of LiNbO$_3$:Fe (54#)	Retrieve images of LiNbO$_3$:Fe:In (In 2#)
Reference (mW)	Object (mW)			
1.75	0.75	1.75		
0.9	0.55	1.75		
0.75	0.38	1.75		
0.5	0.22	1.75		

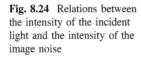

Fig. 8.24 Relations between the intensity of the incident light and the intensity of the image noise

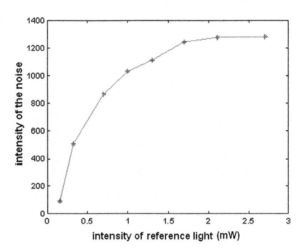

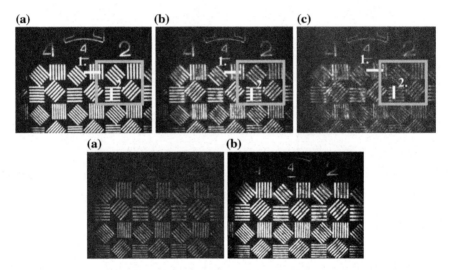

Fig. 8.25 Results of experiment of the recording with weak light and reading with powerful light in single page. **a** Holograms recorded with low intensity of the incident light. **b** Holograms read with high intensity of the incident light

retrieved images with low noise and high quality can be obtained by the high intensity of light.

It is attained from the above experiment that the recording mechanism of recording with weak light and reading with powerful light can weaken the intensity of the noise and reduce the speed of the noise setup so that the long exposure time can be used in storage. It is fit for the process of the large capacity.

8.2.10 The Generation of the Scattering Noise During Retrieving

In retrieving, the retrieving light beam is also the beam coded by the speckle as in recording. So the noise grating accorded with the matched condition can be read and can influence the retrieving the holograms correctly. At the same time, the retrieving light coded by the speckle is still scattered in the crystal. It will change the distribution of the electric field and the modulation degree in the crystal. So that the quality of the holograms recorded was influenced and new scattering noise grating can be introduced.

In order to discover the process of the scattering noise in retrieving, the experimental system shown in Fig. 8.26 was employed to read the holograms still.

The read out experimental results are shown in Fig. 8.31. After attaining a hologram with low-noise recording method just, the hologram image has good quality and SNR as Fig. 8.26a. However, when the retrieving light is opened, the quality of the holograms was decreased with the time. After 25 min the retrieved images are illustrated in Fig. 8.26b. Although the images are still resolvable, the fringes of the details begin to blur and the noise can be seen. When 50 min has passed, the retrieved images are illustrated in Fig. 8.26c. Serious crosstalk makes the images blurred and the noise becomes obvious.

The intensity distribution of the signal fringes with different erased time to cross fringes is shown in Fig. 8.27. The intensity distribution of the signal fringes with different erasion time to vertical fringes is shown in Fig. 8.28.

On the other hand, some nonsignal regions from the three above figures are chosen and used to measure the noise intensity. The results are illustrated in Fig. 8.29. With the increment of the erase time, the quality of the signal fringes reduces, the crosstalk increases, and the SNR cuts down. On the other hand, the noise in images appears so as to lead to the loss of the signal intensity. Finally, the whole noise of the images increases greatly relative to the original state and deteriorates the quality of the images.

In the whole retrieving the intensity of the signal and the intensity of the noise are measured separately and the SNR is calculated. The results are illustrated that

(a)   **(b)**   **(c)**

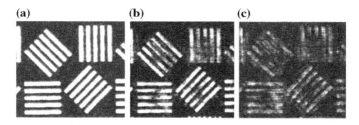

Fig. 8.26 The competition quality and SNR of image with different retrieved time. **a** $t = 0$ s. **b** $t = 25$ min. **c** $t = 50$ min

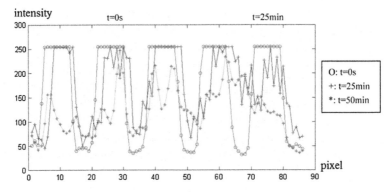

Fig. 8.27 The intensity distribution of the signal fringes with different erasion time to cross fringes pixel

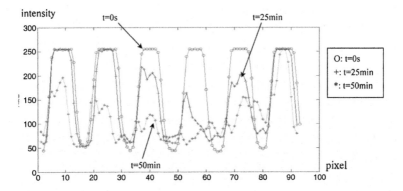

Fig. 8.28 The intensity distribution of the signal fringes with different erasion time to vertical fringes

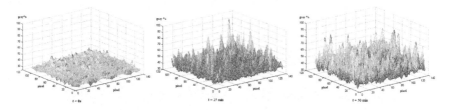

Fig. 8.29 The distribution of the noise intensity with different erasion time

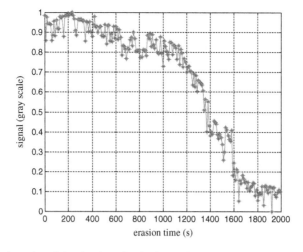

Fig. 8.30 The intensity of the signal grating in the erasion process

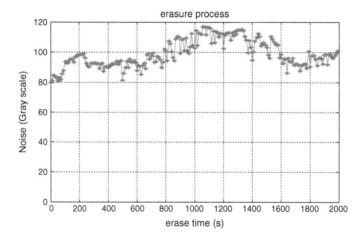

Fig. 8.31 The intensity of the noise grating in the erasion process

the intensity of the retrieving signal is in Fig. 8.30, and the intensity of the SNR is shown in Fig. 8.31 with different erasion time.

The curve in Fig. 8.30 describes the reducing process of the signal intensity. In fact, it is the erase curve of the hologram. The curve in Fig. 8.31 describes the change of the noise intensity. Based on the noise raises while the signal drops in erase process above in Figs. 8.30 and 8.31, can calculate the curve of SNR is shown in Fig. 8.32.

When increasing the reading time, the whole intensity of the retrieved images will reduce continuously and the SNR will cut down greatly so as to influence the quality of the images. There are some fluctuations in the intensity curve in images. But the fluctuations cannot influence the whole trend of the curve. So the effective

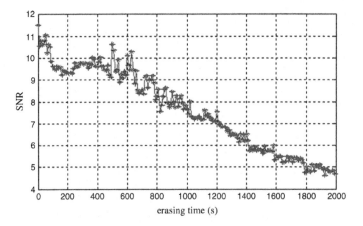

Fig. 8.32 The calculated curve of SNR in the erasing process

results are still obtained from the experiment. The fluctuations are caused by the fluctuations of the power of the laser after the measurement of the power of the laser.

By all results, in order to avoid introducing extra noise into the holograms, it is necessary to reduce the reading time of the grating in the crystal in retrieving. It can prevent the scattering noise from setting up in the retrieving process so that the retrieved images with low noise can be attained.

In a summary, the experiments of the recording and erasing proved that the recording with weak incident light can control the effect of the seed, reading with powerful light can ensure the recording intensity of the retrieved images, the reading time should be shortened as possible to prevent the scattering noise further. So in practice, it is considered that employing the powerful light of the pulse modulation to read holograms, it can obtain the retrieved images with high intensity and low noise as Fig. 8.33.

The storage mechanism of the recording with weak light and retrieving with powerful light in this partition can control efficiently the seed induced by the scattering of the speckle reference light of random coding in the crystal and

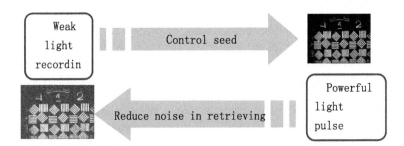

Fig. 8.33 Means to obtain the retrieved images with high intensity and low noise

scattering noise introduced by the seed. So it is hopeful to obtain the high-density and low-noise speckle multiplexing HOSS with the combination of the above recording and retrieving method and the optimized exposure sequence used in multitrack superposed DSSM storage. The method of recording with weak light and retrieving with powerful light and the optimized exposure sequence are used to achieve the experiment of the multitrack superposed DSSM volume holographic storage. The capacity of the storage is still $m \times n \times p = 4 \times 5 \times 5 = 100$ pages of holograms.

At first, in recording, the intensities of the reference light and the object light are set to 0.7 and 0.25 mW. (In the speckle multiplexing HOSS the spot of the object light is focused spot and the reference light is coded by the speckle so the reference object ratio 1:1 is not employed.) The exposure sequence of the multitrack superposed DSM is calculated based on the record and erase time constant remeasured. The light beam of speckle coding with the power 1.7 mW is used to retrieve light and the special software is used to control the switching time of the shutter in the optical path for simulating the pulse light. The diffraction efficiency is shown in Fig. 8.34.

It is known from Fig. 8.34 that uniform distribution of the diffraction efficiency is still attained in the multitrack superposed DSSM with the storage mechanism of the recording with weak light and retrieving with powerful light. The quality of the retrieved images is shown in Fig. 8.35.

On the contrary, the retrieved images with common laser power in retrieving without control of the read time are illustrated in Fig. 8.36.

The number of $(M\text{-}N\text{-}P)$ in Fig. 8.34 denotes the position of retrieved image on the row m, column n, and page P. On the other hand, through the numerical analysis of the above two group of holograms, the intensity distribution of the fringes is shown in Figs. 8.37 and 8.38.

Fig. 8.34 Distribution of the diffraction efficiency of the multitrack superposed DSSM

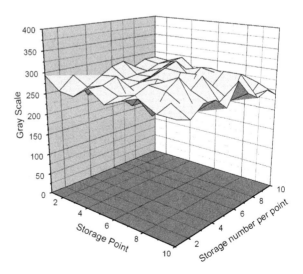

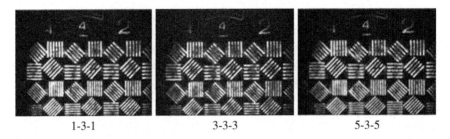

1-3-1   3-3-3   5-3-5

Fig. 8.35 The retrieved images of the recording with weak light and retrieving with powerful light

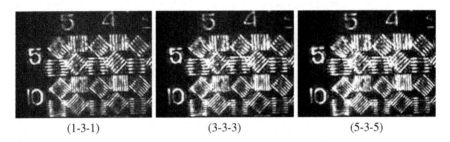

(1-3-1)   (3-3-3)   (5-3-5)

Fig. 8.36 The retrieved images of the multitrack superposed DSSM

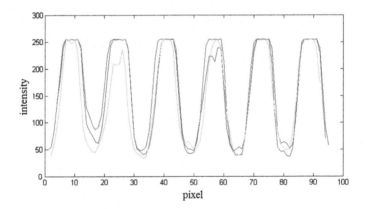

Fig. 8.37 The intensity distribution of the fringes relative to Fig. 8.35

It is recognized in Figs. 8.37 and 8.38 that the retrieved images with the method of recording with weak light and retrieving with powerful light have much better fringe resolution than those with the recording and retrieving lights of constant power.

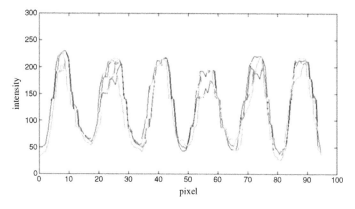

Fig. 8.38 The intensity distribution of the fringes relative to Fig. 8.36

In a summary, with the combination of the method of recording with weak light and retrieving with powerful light and the optimized exposure sequence, the quality of the holograms in the multitrack superposed DSM storage can be improved obviously. It can prevent multitrack superposed DSM from generating the powerful scattering noise in the recording and retrieving of the speckle reference light. With this method the retrieved images in multitrack superposed DSM system have not only large capacity but also the low noise and the high quality.

The prominent scattering noise is induced by the irregularity on the boundary between the air and the medium or the defect and the nonuniform in the medium. They bring the seed of the scattering in the crystal, which is amplified continuously until it becomes the serious scattering noise.

In order to analyze the influence of the intensity of the speckle reference light and the reading time to the scattering noise, great deals of experiments have been done. The method of recording with weak light and retrieving with powerful light, which is proposed based on the experiment, not only control the intensity of the seed and the gain of amplification, but also eliminate the deterioration for images caused by the long-time illumination on the grating so that the diffraction imaging with high quality can be attained.

8.2.11   Diarylethene and Application

Because in the holographic optical storage with the photorefractive crystal there are many types of noises that are difficult to control, it is required to change the materials in the experiments. The organic materials are the development direction of the media of the holographic optical storage. In this paper, the experiments are based on the Diarylethene. The Diarylethene was synthesized by Irie M., which is a kind of photochromic organic material. It has two different forms that can convert each other in light with special wavelength. In the conversion not only chemical characteristics but also physical characteristics change, such as refractive index, absorption coefficient, dielectric constant, and the ability of the oxidation–reduction. Otherwise, the two forms of the Diarylethene have no linear absorption effect in the infrared band. With the above characteristics, Diarylethene is used widely in the region of the bitwise memories and the elements of the optical waveguides. Through the research, the advantages of employing the Diarylethene in the rewritable HOSS include:

1. The difference of the refractive index of the two forms is large. It can reach 10^{-3} for most of Diarylethene, even 10^{-1} for some special ones.
2. Diarylethene is easy to perverse because of the excellent heat stability. Most of the materials can keep the chemical stability in 120 °C, even in 300 °C for some special ones.
3. The fatigue resistance of Diarylethene is excellent. After erasing over 10^4 times repeatedly, the performance of the materials is not reduced obviously.
4. The Diarylethene has a wide response range of spectral. The chemical reaction can happen in limit of the visible spectrum.
5. The Diarylethene has the high photosensitivity, rapid speed of the chemical reaction, and high quantum productive rate.
6. The resolution of the Diarylethene is higher. In the interference recording the reaction of the molecule magnitude happens. The size of a molecule of the Diarylethene is about 0.5 nm so as to achieve the information recording of the high resolution.
7. The reaction is easy to control for the Diarylethene without the dark reaction and the post-process. The recording process of the holographic storage is the process of the photon-type reaction.

As above properties of the photochromic, Diarylethene materials become prominent more and more in the research of the rewritable holographic storage. Next, the applications of the Diarylethene in the HOSS will be discussed. At first, the dynamic processes of the formation and reading of the holographic gratings for the photochromic Diarylethene are developed and the relative theoretical model is built and proved in the experiment. Second, the advantages of employing the Diarylethene to the rewritable holographic optical storage are discussed in the theory and the experiment.

8.3 The Dynamic Model of the Recording and Retrieving

Since Kermaisch proposed the influence of the thickness of the photochromic materials to the diffraction efficiency of the gratings, the researchers have been developing the mechanism of the holographic gratings. Tomlinson deduced the formulation of the form of the holographic gratings in photochromic materials and analyzed the results of the numerical simulation. Downie et al. described the physical model of the form of the holographic gratings in Bacteriorhodopsin film. The results of the experiments proved that the model is practical. Otherwise, they developed the influence of the recording light and retrieving light with different wavelengths to the diffraction efficiency of the holographic gratings. In 2002, Cattaneo et al. researched that the form of the Raman–Leitz gratings in the photochromic Diarylethene deduced the fitted expression and put forward the analysis and explanation for the form of the gratings. In this partition, the dynamic model of the recording and retrieving of the holographic gratings will be described, based on the theory of the coupled wave by Kogelnik and the research of photochemical reaction of the photochromic Diarylethene, and the relative evidences of the experiments will also be given.

Two coherent light beams with the common direction of the polarization (vertical to the incident plane) and the wavelength λ_2 enter symmetrically the materials with incident angle θ, and their intensities are I_1, I_2, as illustrated in Fig. 8.39. According to the Lambert–Beer equation in photochemical reaction the absorption light intensity of the Diarylethene $I_a(x, t)$ is

$$I_a(x, t) = I(x) \left\{ 1 - \exp\left[-2.303 \sum_{i=A,B} \varepsilon_i u_i(x, t) L \right] \right\} \qquad (8.7)$$

where ε_i is the molar extinction coefficient of the Diarylethene in the open-ring and the closed-ring forms, $u_i(x, t)$ concentration of the molecules of the relative form, L the thickness of the film, $I(x) = I_0 \left[1 + m \, \cos\left(\frac{2\pi x}{\Lambda} \right) \right]$ intensity distribution of the interference field, $I_0 = I_1 + I_2$ the average light intensity on the materials $m = \frac{2\sqrt{I_1 I_2}}{I_1 + I_2}$ modulation degree of the light intensity, and $\Lambda = \frac{\lambda_2}{2 \sin \theta}$ fringe separation of the interference field. For brevity, we can give two assumptions as

The molecules of the Diarylethene only have two forms, the open-ring form and the closed-ring form. The total number of the molecules of the Diarylethene is constant and the molecules distribute uniformly, that is, $\sum_{i=A,B} u(x, t) = u_0$, where u_0 is the total concentration of the molecules of the Diarylethene.

Fig. 8.39 Structure of the holographic gratings formed by the interference of the two waves in materials

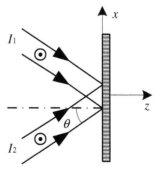

The molecules in form A do not absorb the photons with the wavelength λ_2, which only leads to the conversion of the molecules of the Diarylethene from form B to form A.

Usually, the films are thin (about 10 μm). The exponential term in Eq. (8.7) is expended for the binominal expression and only the constant term and the first-order term are reversed. The result is

$$I_a(x,t) = 2.303\varepsilon_B u_B(x,t)I(x)L \tag{8.8}$$

In sinusoidal wave field, the molecules of the Diarylethene are converted from form B to form A. The change of the concentration of the molecules in form A $u_A(x,t)$ is described as

$$\frac{\partial u_A(x,t)}{\partial t} = \frac{\phi_{B\to A}I_a(x,t)\lambda_2}{N_A hcL \times 10^{-3}} \tag{8.9}$$

where $\phi_{B\to A}$ is the quantum productive rate of the molecules from form B to form A, N_A Avogadro constant, h Planck constant, and c speed of light in vacuum.

It is presumed that the molecules all lie in form B before recording, i.e., $u_B(x,0) = u_0$ and for a long time all the molecules are converted into form A from form B, i.e., $u_B(x,\infty) = 0$. By Eqs. (8.2) and (8.3), the result is obtained as

$$u_B(x,t) = u_0 \exp\left\{-\frac{t}{\tau}\left[1 + m\cos\left(\frac{2\pi x}{\Lambda}\right)\right]\right\} \tag{8.10}$$

where τ is the temporal constant of grating recording as

$$\tau = \frac{N_A hc}{2303 I_0 \varepsilon_B \phi_{B\to A}\lambda_2} \tag{8.11}$$

It is recognized from Eq. (8.11) that τ is not relative to the thickness when the film is thin. $k = \varepsilon_B \phi_{B\to A}$ is relative to the properties of the materials, where k is generally called as sensitivity of the materials. The bigger the value of k is, the higher the sensitivity is and the shorter the recording time. I_0, λ_2 depend on the recording conditions. The stronger the light intensity is, the smaller the temporal constant of grating recording is. It is necessary to explain that reducing the wavelength of the recording light λ_2 do not always shorten the recording time because it also depends on the response sensitivity of the materials for the wavelength of the recording light.

The curve of the change of the normalized concentration of the molecules of the Diarylethene in form B with time in sinusoidal wave field with the modulation degree of the light intensity $m = 1$ is shown in Fig. 8.40. The materials are initialized at the beginning of recording (T_0) so that all the molecules can be in form B and the concentration of the molecules is constant in x direction. At T_1, T_2 the concentration that changes linearly with the wave field is distributed in sine-curve

Fig. 8.40 The simulation of the change of the normalized concentration of the molecules of the Diarylethene in form B distributed along *x* direction with recording time with the modulation degree of the light intensity *m* = 1. It is converted into the distribution of approximate rectangle in high exposure from the sinusoidal distribution in low exposure

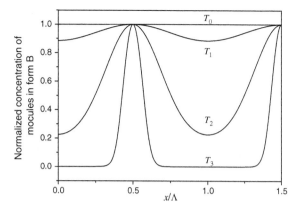

shape. At T_3 the sinusoidal distribution of the concentration of the molecules is broken due to the influence of the saturation effect of the gratings. Once the quality of the illumination is enough, the distribution of the concentration of the molecules is converted into the distribution of near δ function.

Since the refractive indexes and the absorption coefficient of the molecules of the Diarylethene in form A and in form B are different, the gratings recorded are hybrid gratings. The characteristics of the gratings can described with the coupled constant κ.

$$\kappa = \frac{\pi n_1}{\lambda_2} - j\frac{\alpha_1}{2} \tag{8.12}$$

where n_1 and α_1 are the modulation degree of the refractive index and the modulation degree of the absorption constant. According to the Kramers–Kronig equation, they have the linear relation, described as $\alpha_1 = \psi n_1$, where ψ is a constant that depends on the wavelength of the light.

If the retrieving light enters the recorded gratings in materials with the Bragg angle, the diffraction efficiency can be expressed as follows in response to the theory of the coupled wave:

$$\eta = \left[\sin^2\left(\frac{\pi n_1 L}{\lambda_2 \cos\theta_0}\right) + \text{sh}^2\left(\frac{\alpha_1 L}{2\cos\theta_0}\right)\right] \exp\left(\frac{-2\alpha L}{\cos\theta_0}\right) \tag{8.13}$$

where α is the average absorption constant of the materials and θ_0 the Bragg angle in the materials. $\theta_0 = \arcsin(\sin\theta/n_0)$, where θ is the incident angle and n_0 the average refractive index of the materials.

Referred as the above, since the films of the Diarylethene are very thin, its diffraction efficiency is usually less than 1 %. So the average absorption of the materials can be omitted and the expression of the diffraction efficiency can be simplified.

$$\eta = \left| \frac{\kappa L}{\cos \theta_0} \right|^2 = |\kappa|^2 \cdot \frac{L^2}{\cos^2 \theta_0} \tag{8.14}$$

It is proved in experiment that the change of the refractive index of the molecules of the Diarylethene is proportional to the concentration of the molecules. The refractive index of the molecules of the Diarylethene can be expressed as

$$\Delta n(x,t) = C u_A(x,t) = C[u_0 - u_B(x,t)]$$
$$= C u_0 \left[1 - \exp\left(-\frac{t}{\tau} \left[1 + m \cos\left(\frac{2\pi}{\Lambda} \right) \right] \right) \right] \tag{8.15}$$

where C is a constant relative to the property of the molecules of the Diarylethene. Equation (8.15) is expended for binomial formula with $\cos\left(\frac{2\pi}{\Lambda}\right)$ and the first-order term is reversed. The expression of the modulation degree of the refractive index can be obtained as

$$n_1(x,t) = C u_0 m \exp\left(-\frac{t}{\tau} \right) \frac{t}{\tau} \tag{8.16}$$

The module $|\kappa|$ of the coupled constant κ can be attained from Eqs. (8.12) and (8.16) as

$$|\kappa| = \sqrt{(\pi/\lambda_2)^2 + (\psi/2)^2} \cdot n_1$$
$$= C m u_0 \sqrt{(\pi/\lambda_2)^2 + (\psi/2)^2} \exp\left(-\frac{t}{\tau} \right) \frac{t}{\tau} \tag{8.17}$$

It is obtained from Eqs. (8.14) and (8.17)

$$\eta = (C m u_0)^2 \left[(\pi/\lambda_2)^2 + (\psi/2)^2 \right] \exp\left(-\frac{2t}{\tau} \right) \left(\frac{t}{\tau} \right)^2 \tag{8.18}$$

It is found that improving the concentration of the molecules of the Diarylethene can add the module of the coupled constant of the gratings and the diffraction efficiency of the gratings. But the precipitations will happen once the concentration of the molecules. So another practical method of improving the diffraction efficiency is making the Diarylethene grow to the film with the fixed shape. The derivative of the diffraction efficiency to time can be calculated from Eq. (8.18). The diffraction efficiency of the gratings can reach the peak at the time $t = \tau$.

In order to prove the validity of the above theory, the holographic gratings are recorded in real time and the diffraction efficiency of the gratings is also measured. The structure of the experimental facility is shown in Fig. 8.41. The linear polarized light beam from the He–Ne gas laser (wavelength is 632.8 nm) is divided into two beams. One is regarded as the object light, the other is regarded as the reference light. The direction of the polarization of both of them are vertical to the incident

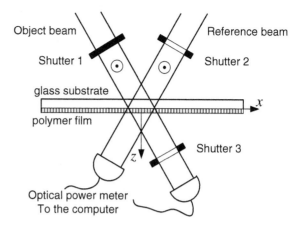

Fig. 8.41 The structure of the measurement of the diffraction efficiency of the gratings

plane. The diameters and the powers of the two beams are both 0.4 cm and 0.4 mW. The angle between the object light and the reference light is 60° in air. They enter the materials symmetrically and form the interference field. In recording the gratings, the frequency of the power collection and the sampling perdurability are controlled through the shutter. The sampling frequency is 10 Hz in the experiment. For reducing the erase effect of the recorded information of the retrieving light, the sampling perdurability is 10 ms. In the measurement of the diffraction efficiency we omit the influence of the average absorption. The diffraction efficiency is described as

$$\eta' = \frac{I_d}{I_d + I_t} \tag{8.19}$$

where I_d and I_t is separately the intensity of the diffraction light calculated based on the diffraction light power or the transmitted light power.

The measurement of the diffraction efficiency of the gratings is shown in Fig. 8.41, that the points in are obtained from the diffraction efficiency η' in the above formula (8.19). The time when the diffraction efficiency of the gratings reaches the peak (1.2 %) is 750 s. Since then, the diffraction efficiency will cut down with the extension of the recording time, which is caused by the saturation effect of the gratings of the Diarylethene. The measured and theoretical curves of the diffraction efficiency in recording gratings are shown in Fig. 8.42. The solid line in Fig. 8.42 is the theoretical curve from Eq. (8.18). The agreement of the two curves proves the validity of the above model. The little deviations are caused by the fluctuations of the output powers of the laser.

The holographic gratings are sinusoidal distribution before the gratings are saturated and the diffraction efficiency increases with the addition of the exposure while the sinusoidal distribution of the gratings are converted into the distribution of the rectangle and the diffraction efficiency cut down continuously. Once the recording time is enough long, the distribution of the gratings are changed into the

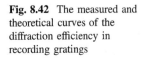

Fig. 8.42 The measured and theoretical curves of the diffraction efficiency in recording gratings

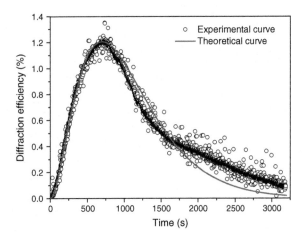

distribution of near δ function and the diffraction efficiency becomes zero. At that time, the recorded gratings have been erased.

With the conditions same as the above except that the light intensities of the reference light and the object light are regulated into 6.4 and 8.9 mW/cm^2, the peak of the diffraction efficiency of the gratings appears at 500 and 268 s. It proves that the saturation time of the gratings is in inverse proportion to the intensity of the recording light. The results are agreed with the above equations.

The recording process can be explained by the holographic storage of the images. In the experiment, the recording light field is regulated to the nonuniform distribution in which the intensity of light decreases continuously from right to left. In recording holograms, retrieve the recorded holograms each 10 s as shown in Fig. 8.43a–h. The image plane becomes distinct little by little from right to left with the extension of the exposure time. After 30 s, the definition of the right part of the image plane is on the top, as shown in Fig. 8.43c. After the time, the right part of the image plane is darkened little by little with the extension of the exposure time until it disappears completely. It means that the holographic gratings in the materials relative to the right part of the image plane are saturated at 30 s. If the recording continues, the gratings will be erased by oversaturation. In Fig. 8.43, the gratings on the left part of the image plane change obscurely instead of the right part. It may be caused by the unsuitable reference object ratio. The intensity of the light on the image plane is not uniform and only the reference object ratio is suitable on the right part of the image plane, as a result, the diffraction efficiency of the right part of the image plane is higher than that in other parts.

The change of the diffraction efficiency of the gratings in the sinusoidal light field has been illustrated. It will be discussed for the change of the diffraction efficiency of the recorded holographic gratings when the retrieving light illuminates the gratings continuously later.

As illustrated in Fig. 8.44, a retrieving light beam with the wavelength λ_2, the intensity I_3, the incident angle θ, and the direction of the polarization normal to the

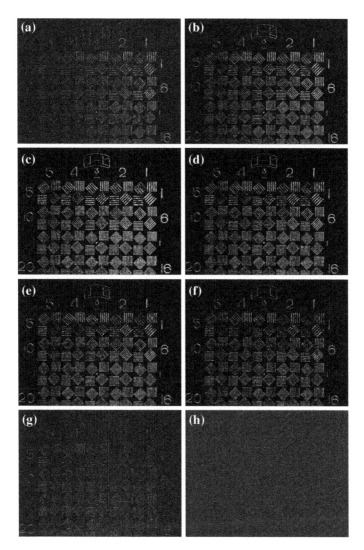

Fig. 8.43 The process of retrieving the gratings with nonuniform intensity of the light on the image plane, **a–h** show the holograms sampled each 10 s

Fig. 8.44 The general view of the retrieving light enters the gratings in the materials

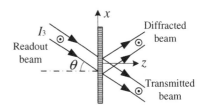

incident plane enters the gratings in the materials. The absorption intensity of the molecules of the Diarylethene $I_a^r(x,t)$ is

$$I_a^r(x,t) = 2.303\varepsilon_B u_B(x,t)I_3 L \tag{8.20}$$

In the retrieving wave field, the molecules of the Diarylethene in form B are converted into those in form A continuously. The change of the concentration of the molecules in form A is

$$\frac{\partial u_A(x,t)}{\partial t} = \frac{\phi_{B\to A}I_a^r(x,t)\lambda_2}{N_A hcL \times 10^{-3}} \tag{8.21}$$

where the implication of the parameters is the same as the above.

Next, the time at which the diffraction efficiency of the gratings reaches the peak in above is defined as the beginning time of retrieving the gratings. At that time, $u_B(x,0) = u_0 \exp\left[-1 - m \cos\left(\frac{2\pi x}{\Lambda}\right)\right]$. After the illumination of the retrieving light for a long time, all the molecules of the Diarylethene become form A, i.e., $u_B(x,\infty) = 0$. It is solved from Eqs. (8.14) and (8.15) that

$$u_B(x,t) = u_0 \exp\left[-1 - m \cos\left(\frac{2\pi x}{\Lambda}\right)\right] \exp\left(-\frac{t}{\tau_e}\right) \tag{8.22}$$

where τ_e is the erase temporal constant of the gratings as follows:

$$\tau_e = \frac{N_A hc}{2303 I_3 \varepsilon_B \phi_{B\to A}\lambda_2} \tag{8.23}$$

It is found from Eq. (8.23) that the erase temporal constant of the gratings is in inverse proportion to the intensity of the retrieving light. When the intensity of the retrieving light is doubled, the erase temporal constant of the gratings becomes the half. It can decrease the intensity of the retrieving light so as to the erase effect of the recorded holograms by the retrieving light and prolong the times of the data reading. Similarly, the erase time of the gratings does not depend on the thickness of the films of the Diarylethene when it is very thin.

The diffraction efficiency in retrieving is

$$\eta^r = \left(\frac{Cmu_0}{e}\right)^2 \left[(\pi/\lambda_2)^2 + (\psi/2)^2\right] \exp\left(-\frac{2t}{\tau_e}\right) \tag{8.24}$$

It is recognized that when the retrieving light illuminates continuously the recorded gratings, the diffraction efficiency decays with the exponential of the napierian base and the speed of decaying is proportional to the intensity of the retrieving light.

In order to prove the validity of the model, the gratings are recorded with keeping the above parameters of the experiment constant. After the gratings are saturated (1.2 % of the maximal diffraction efficiency), close the object light, control the shutter

Fig. 8.45 The experimental and theoretical results of the diffraction efficiency when the retrieving light with the intensity of 6.56 mW/cm² illuminates the holographic gratings

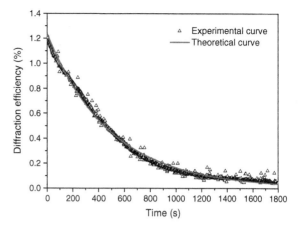

to open the reference light, and regulate the intensity of the retrieving light illuminating continuously the gratings in the Diarylethene to 6.56 mW/cm². The powers of the diffraction light and the transmitted light are sampled by the power meter with frequency of 1 Hz. Substituting the sampling results to Eq. (8.24) can obtain the curve or change of the diffraction efficiency, as illustrated in Fig. 8.45.

From Eqs. (8.21), (8.24) and the above experiment results, the following expression is attained:

$$\frac{N_A hc}{2303 \varepsilon_B \phi_{B \to A} \lambda_2} = 6400 \, \text{mJ/cm}^2, \quad \left(\frac{Cmu_0}{e}\right)^2 \left[(\pi/\lambda_2)^2 + (\psi/2)^2\right] = 1.2 \,\% \quad (8.25)$$

The theoretical curve of the change of the diffraction efficiency can be obtained from Eqs. (8.23) and (8.24) and above two numbers, as shown in Fig. 8.45. The erase temporal constant can also be calculated as $\tau_e = 976$ s. The agreement between the theoretical curve and the experimental curve proves the validity of the model. As a matter of fact, S. Cattaneo et al. also found that the diffraction efficiency of the gratings decays with the exponential of the napierian base with the extension of the retrieving time in the retrieving holographic gratings in the Diarylethene. But they only drew the conclusion through the data fitting of the experimental results and did not the deduction and the explanation.

8.4 Rewritable Performances

8.4.1 The Analysis for Single Hologram Storage on the Diarylethene

The experimental system of the recording and retrieving holograms is illustrated in Fig. 8.46. The linearly polarized light beam from the He–Ne laser is divided by the

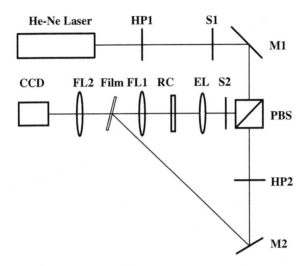

Fig. 8.46 The structure of the experimental system of the volume holographic storage. *HP1* and *HP2* are the half-wave plates; *S1* and *S2* are the shutters; *M1* and *M2* are the reflective mirrors; *PBS* is the polarized prism splitter; *EL* is the extending lens; *FL1* and *FL2* are the Fourier transform lens and the inverse Fourier transform lens; *RC* is the resolution chart; Film is the medium of the Diarylethene

PBS into two beams through filtering, shaping, and extending. One is the reference light beam; the other is the signal light beam. HP1 is the half-wave plate that used to regulate the intensity ratio between the reference light and the signal light. HP2 is the half-wave plate used to regulate the direction of the polarization of the reference light to be perpendicular to the incident plane. EL is the extending lens, which extends the signal light from the PBS and makes it corresponding to the image plane of the object (RC, resolution chart). FL1 is the doublet Fourier transform lens with the focus length 50 mm, which transmits the information on the image plane into the spectrum information in recording holograms. FL2 is the doublet Fourier transform lens same as FL1, which transmits the spectrum information on the films of the Diarylethene into image information on the CCD. The shutters S1 and S2 are employed to control the recording light and the retrieving light.

In experiment, the powers of the reference light and the signal light are separately 1.4 and 0.5 mW. The recording method is spectrum plane defocusing method that the medium does not lay on the rear focal plane of FL1, but the position at the distance of 8 mm. In order to test the pinch effect of recording holograms, the asymmetrical recording structure is used. The incident angles of the reference light and the signal light are separately −42° and 18°. Specially, the recording method of 90° is not used because the medium is the flaky material.

Figure 8.47a shows the image of the RC through the 4-f system. The minimal line width of the resolution is 13.3 μm. Since the resolution of CCD is limited (the size of a pixel is 9.8 μm×9.8 μm), the RC with thinner line width is not employed. Figure 8.47b shows the image of the RC through the 4-f system with the film of the

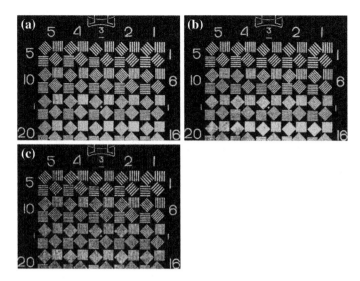

Fig. 8.47 **a** The images of the signal light through the 4-f system directly. **b** The images of the medium of the Diarylethene in the signal optical path. **c** The retrieved holograms with the original reference light as the retrieving light

Diarylethene in the signal optical path. Because of the absorption of the materials the intensity of the whole image plane darkens, but the resolution does not change. It proves that the quality of the film forming is good without scattering. Figure 8.47c shows the reconstruction of the retrieving light. The retrieving images have high resolution and high fidelity. The excellent fidelity also proves that the contraction of the materials is little and almost is omitted. According to the contraction theory of the polymer in the holographic storage, the contraction of the materials can lead to the distortion of the recorded images. That the contraction of the Diarylethene in recording is little is decided by the mechanism of the holographic storage. The information recording of the Diarylethene depends on the difference of the refractive indexes and the absorption of the two forms of the open-ring and the closed-ring. In recording, the reaction happens in original position and no migration happens for the molecules of the Diarylethene. However, the recording information on the photopolymers depends on the change of the refractive indexes induced by the polymeric migration of the monomer molecules in the modulation light field.

The storage life of the photochromic Diarylethene is an important parameter that evaluates the performance of the holographic storage. The stability of the Diarylethene in the closed-ring form is developed in experiment. Recording the holograms again after the medium of the Diarylethene initialized is reversed in room temperature in lucifugal condition for 10 months. Figure 8.48 shows the retrieved images in which the resolution is very good and no noise is seen. It means no change happens for the materials during the time. It is proved from the theory and the experiment that the stability of the Diarylethene in closed-ring form is decided by the type of the aromatic group. If it is the furan, the thiophene, or the thiazole ring, the

Fig. 8.48 The retrieved images of the holograms of the Diarylethene in closed-ring form after perversion in dark in room temperature for 10 months

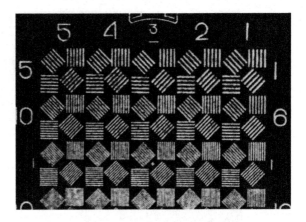

heat stability of the materials will be very good so that the conversion from the closed-ring form to the open-ring form cannot happen in the temperature of less than 80°. The aromatic group used in the experiment is the thiophene. Its stability is excellent, but we do not test it for longer than 10 months due to the limit of the experiment time. Similarly, after the materials on which the holograms have been recorded are reversed in dark for a long time, the quality of the retrieved images does not change. It is different from the crystal. The holographic gratings in the photorefractive crystal are formed as follows: The donors (the acceptors) are ionized and release the electrons (the holes) in the modulation light field. At the same time, the electrons (the holes) transit to the conduction band (the valance band) from the middle energy level with induced transition. In the conduction band (the valance band), the electrons (the holes) move due to the concentration gravity, the electric field, or the photovoltaic effect. The electrons (the holes) in transition can be captured again. Through excitation, transition, and capture repeat, the electrons (the holes) are captured by the traps in the dark light regions out of the illumination regions at last. It brings the change of the space charge distribution in crystal so that the space charge can be separated and the space charge field is relative to the distribution of the light field. Then, with the linear electro-optical effect, the phase gratings of the refractive index modulation are formed. The gratings formed by the transition of electron are easy to reverse. The gratings will be weakened, even disappear after they are put in air for some time. So it is necessary for reversing the gratings to convert the gratings of the electrons into the gratings of the ions with the heat fixed method.

8.4.2 The Rewritable Experiments of the Diarylethene

In order to develop the rewritable performance of the Diarylethene, the rewritable experiments proceed in the system is shown in Fig. 8.51. The ultraviolet light with the wavelength of 365 nm is employed to initialize the materials after each recording so that the Diarylethene in materials are all converted into the closed-ring

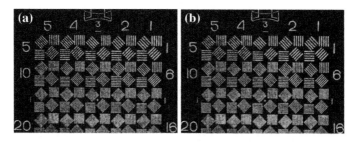

Fig. 8.49 a The retrieved images in the first recording; **b** The retrieved images in the recording after the hundredth erasion

form. Only 100 times of the rewritable experiments proceed due to the limit of the experimental condition. Figure 8.49a shows the retrieved images at the first recording. Figure 8.49b shows the retrieved images after erase for 100 times. It is found that the quality of the images does not decrease after erasing for 100 times.

The rewritable performance of the Diarylethene depends on the fatigue resistance. The better the fatigue resistance is, the better is the rewritable performance and vice versa. The photochromic reactions are always followed with the fracture and the recombination. In the recombination some secondary reactions may appear, which can influence the fatigue resistance of the Diarylethene. If the product of the secondary reaction is B', the photochromic reaction of the molecules of the Diarylethene is described as

$$B' \xleftarrow{\phi_S} A \quad \begin{matrix} \lambda_1, \phi_{A \to B} \\ \lambda_2, \phi_{B \to A} \end{matrix} \quad B \tag{8.26}$$

If the quantum productive rate of the secondary reaction ϕ_S is 0.001 (the molecules of the Diarylethene can be converted into the form A from the form B, i.e., $\phi_{B \to A} = 1$), 63 % of the molecules in form A will still be decomposed after 1000 times of photochromic reactions. So it is necessary to reduce the quantum productive rate of the secondary reaction in order to improve the fatigue resistance. The quantum productive rate of the secondary reaction depends on the structure of the molecules of the Diarylethene. After the thiophene is replaced by the benzothiophene, the fatigue resistance is improved prominently.

8.4.3   The Experiments of Multiplexing Storage of the Diarylethene

One of the advantages of volume holographic storage is multiplexing in a point. In order to prove the multiplexing performance of the Diarylethene, employ the experimental system as in Fig. 8.46 to record 10 pages of holograms. In experiment the recording method is the Fourier spectrum plane defocusing method and the defocusing amount is 8 mm. The powers of the reference light and the signal light are separately 1.4 and 0.5 mW. The angle between the reference light and the object light is 60°. The diameter of the recording region is 4 mm. A page of hologram is recorded on each 1.5° of rotation of the materials. The recording time of each figure is 4 s. The 10 pages of images are retrieved without crosstalk, as illustrated in Fig. 8.50a–j. Because the thickness of the medium is only 10 μm, it leads to the low diffraction efficiency and dark image plane in the imaging storage. In addition, the recording light field is not uniform. So the intensity of the image planes is also not uniform as shown in Fig. 8.50a–j.

Next, the relations between the sensitivity of the angle and the thickness of the materials in the multiplexing storage will be discussed. From the theory of the coupled wave, the sensitivity of angle in the multiplexing storage can be described as

$$\Delta\theta = \frac{\lambda}{2L\sin\theta} \tag{8.27}$$

where λ is the wavelength of the recording light, L the thickness of the materials, and θ Bragg angle. The sensitivity of angle in the multiplexing storage is in inverse proportion to the thickness of the medium. The thicker the medium is, the better the sensitivity of angle is. So it is necessary for improving the sensitivity of angle and the capacity of the multiplexing storage to increase the thickness of the medium.

The photochromic Diarylethene is employed to the rewritable volume holographic storage. The principles of recording and retrieving holographic gratings are developed. The theoretical model is set up and proved in experiment. The model is obtained from the combination of the chemical reaction of the photon absorption of the photochromic Diarylethene and the theory of the coupled wave. The conclusions are not only suitable to the volume holographic storage of the Diarylethene, but also those of all the organic photochromic materials.

Otherwise, the performance of the storage of the single page of hologram is developed in experiment. The multiplexing storage in single point is also achieved. They prove the advantages of high resolution, good fatigue resistance, and long life of the Diarylethene materials. It is potential to be employed in the rewritable HOSS.

8.4.4 Dithienylethenes Application in Polarization Holographic Storage

Volume holographic storage has been widely researched which is considered as a good candidate of next generation of optical storage. The usage of optical polarization for higher capacity will be discussed in next section. It has been demonstrated that optical polarization could be used for recording in holographic system which is termed as polarization holographic storage technology. It is popular in the areas of optical fake defending and polarization optical components. Polarization holographic storage with the anisotropic recording material could improve the

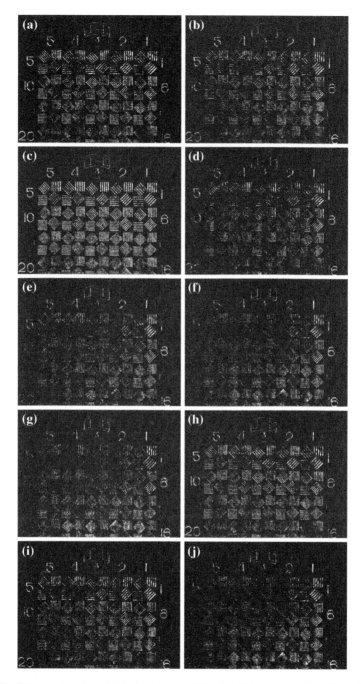

Fig. 8.50 Pictures of angle multiplexing storage of the Diarylethene materials with each 1.5° of rotation of the materials without crosstalk

signal-to-noise ratio of the retrieving photo by separating the polarizations of the
diffraction light and the noise. Photocromic dithienylethenes is a potential material
for rewritable holographic storage, near field optical storage, light modulator,
optical switch, and so on with the performance of high resolution, good hear
stability, and fatigue durability. Furthermore, it is a kind of photoisomerization
nonlinear optical material which is sensitivity to the polarization. It could be also
used for optical frequency doubling, polarization holographic storage, image processing, and so on.

The application of the dithienylethenes in polarization holographic storage was
also mainly researched. When activated by the polarized light, dithienylethenes
molecules will be anisotropic distributed. Photoanisotropy is composed of dichroism and birefringence. Dichroism expresses that the absorbing coefficient of
material will be changed with the polarization state of the light with the same
waveleghth and birefringence expresses that refractive index will be different for
different polarization states of the same light. Dithienylethenes is then suitable for
polarization multiplexing holographic storage, because not only the intensity
information of the light interference but also the polarization information could be
recorded. Two holographic photos could be recorded at one point and retrieved
without crosstalk based on the dithienylethenes by changing the polarization state
of recording light. Orthogonal polarized dual-channel optical storage system has
been advanced based on photochomic dithienylethenes material. The system could
reduce the recording and reading time when increasing the recording density.
Besides, it could be united with other multiplexing technologies, such as angle
multiplexing, phase multiplexing, circular multiplexing, shift multiplexing, speckle
multiplexing, and so on.

8.5 Polarized 3D Volume Holomem

8.5.1 Dithienylethienes Polarized Holomem Performances

The probability for the anisotropic molecules to absorb one photo relies on the
angle between the direction of absorbing principal axis of polarity molecules and
the polarization direction of the recording light. As a result, those molecules whose
absorb principal axis is parallel with light polarization direction will absorb more
photos than others to realize the photoisomerization. The absorbing coefficient for
anisotropic molecules is described as

$$
\begin{aligned}
\varepsilon = {}& \varepsilon_{\parallel} \left(\frac{a^2}{I_0} \cos^2 \theta + \frac{b^2}{I_0} \sin^2 \theta \cos^2 \varphi \right) \\
& + \varepsilon_{\perp} \left[\frac{a^2}{I_0} \sin^2 \theta + \frac{b^2}{I_0} \left(1 - \sin^2 \theta \cos^2 \varphi \right) \right]
\end{aligned}
\tag{8.28}
$$

Fig. 8.51 The directions of
the anisotropic molecules
relative to the polarization
direction

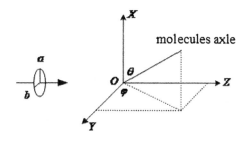

where $\varepsilon_{\parallel}$ and $\varepsilon_{\perp}$ are the absorbing coefficients whose absorbing principal axis direction is parallel and perpendicular to the light polarization direction separately. The molecules are considered to be single axle. θ and φ are the angels in polar coordinate which are expressed in Fig. 8.51. I_0 is the incidence light intensity, a and b are the polarization components along the OX-axis and OY-axis separately which accord with $a^2 + b^2 = I_0$. The angle between absorbing principal axis of dithienylethienes molecule and OX-axis is θ, the projection of which in YOZ plain is φ. So the projections of a and b along the molecule principal axis are $a \cos \theta$ and $b \sin \theta \cos \varphi$, separately, who will active the molecules by $\varepsilon_{\parallel}$. The component for a and b perpendicular to the absorbing principle axis are $a \sin \theta$ and $b\sqrt{1 - \sin^2 \theta \cos^2 \varphi}$ which will activate the molecules by $\varepsilon_{\perp}$. Commonly, $\varepsilon_{\perp}$ is far smaller than $\varepsilon_{\parallel}$ photoisomerization material.

As shown in Fig. 8.51 that the light polarization really affects the photoisomerization molecules. The photo birefringence and photochromic will be introduced in the material when the photoisomerization molecules redistributed according to the polarization even no light intensity is changed. In macroscopic, not only the light intensity but the polarization is also recorded. When the photoisomerization recording material is exposed by two orthogonally polarized beams, polarized holographic recording is then realized. If the molecules are isotropic, $\varepsilon_{\parallel} = \varepsilon_{\perp}$ is then obtained, that is the coefficient ε will be a constant which is not relative to the molecules directions and the light polarization. It is the reason why ordinary holographic material could not record the light polarization.

Photo induced anisotropic media could be divided into three types, amplitude type (absorbing type), phase type (refraction type), and hybrid type, which is similar with isotropic media. Amplitude type is caused by dichroism of the molecules while phase type is by photo birefringence. From former statements, for photo induced anisotropic media, the light axis induced by incidence light is always parallel to the dominated polarized direction. When linearly polarized light is used, the light axis induced is parallel to the vibration direction. When elliptically polarized light is used, the light axis is parallel to the major axis of the ellipse. When circularly polarized light is used, however anisotropy will not be induced.

The formation process of the polarized grating in anisotropic material is shown in Fig. 8.52. Two types of polarized modulation field distributions of the orthogonally polarized recording beams will be taken into consideration: (a) for orthogonally linearly polarized teams (b) for orthogonally circularly polarized beams. The molecules

Fig. 8.52 Formation process of the grating in anisotropic material with **a** orthogonally linearly polarized teams and **b** for orthogonally circularly polarized beams

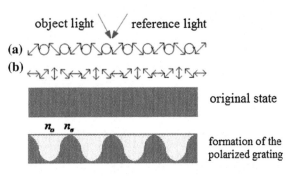

object light ＼／ reference light

(a)

(b)

original state

n_o　n_e

formation of the polarized grating

are originally isotropic before recording. During recording, molecules will then be induced to be anisotropic. Different absorption will made for different light polarization by each polarized molecules. Different transition velocity is then obtained for those molecules and two structure types of molecules are in each distribution state. For the part where the molecules have been transformed, the molecules have been polarized in the same direction. Different absorption and refraction will be obtained for different light polarizations. Polarized grating is then formed. The objective information could be read out by the light with same polarized direction with the reference beam.

8.5.2　Analysis for Polarized Holography Field

The reference beam R and objective beam O are entranced into the recording media surface XoY symmetrically with included angle 2θ who are orthogonally polarized, which is shown in Fig. 8.53. The entrance plain is parallel to XoZ. The phase of the reference beam in the direction parallel to oX φ_R, φ_O have the following relationship:

$$\varphi_O = -\varphi_R = \delta = \frac{2\pi}{\lambda_2} X \sin \theta \tag{8.29}$$

Fig. 8.53 Sketch of the orthogonally polarized beams

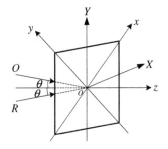

where λ_2 is the wavelength of recording beam and 2δ is the phase difference of objective and reference beams. The phase distribution along oY axis is constant.

The coordinate system for Jones vector is considered to be xoy, which is $45°$ separated from XoY. It is considered that the complex amplitudes of reference and objective beams are the same, that is $E_R E_R^* = E_O E_O^* = E_0 E_0^* = I_0$. For convenience, it is separated into two parts for the polarization of reference and objective beams in discussion.

The polarization direction of reference beam is perpendicular to the entrance plain, that is y-axis, while that of objective beam is parallel to x-axis. The expression of Jones vector is shown as

$$E_R = \begin{pmatrix} E_0 e^{-i\delta} \\ E_0 e^{-i\delta} \end{pmatrix}, \quad E_O = \begin{pmatrix} E_0 e^{i\delta} \\ -E_0 e^{i\delta} \end{pmatrix} \tag{8.30}$$

The composed complex amplitude E_a is

$$E_a = \begin{pmatrix} E_0 \cos \delta \\ iE_0 \cos \delta \end{pmatrix} \tag{8.31}$$

the reference beam is in right-hand circular polarization while the objective beam is in left-hand circular polarization. The expression of Jones vector is shown as

$$E_R = \frac{\sqrt{2}}{2} \begin{pmatrix} E_0 e^{-i\delta} \\ -iE_0 e^{-i\delta} \end{pmatrix}, \quad E_O = \frac{\sqrt{2}}{2} \begin{pmatrix} E_0 e^{-i\delta} \\ iE_0 e^{-i\delta} \end{pmatrix} \tag{8.32}$$

The composed complex amplitude E_b is by

$$E_b = \begin{pmatrix} E_0 \cos \delta \\ E_0 \sin \delta \end{pmatrix} \tag{8.33}$$

From Eqs. (8.29) to (8.32) we can see obviously that elliptical polarized light and polarized state is varied with phase difference 2δ, E_b is linearly polarized light and the vibration direction is varied with phase difference 2δ, as shown in Fig. 8.54. It is also got from the above expression that the period of polarization state modulation field is the same as intensity modulation field, it is $\Lambda = \frac{\lambda_2}{2 \sin \theta}$.

It is necessary to declare that media is isotropic before recording and it is in linear response to the polarized modulation field where the variation of the linear optical constant of the material induced by polarized light could be expressed by Hook's law.

The expression (8.31) is included that the intensity of the polarized modulation field will be constant of $I_a = E_a E_a^* = 2E_0 E_0^* = 2I_0$ for any phase difference value 2δ for orthogonally linear polarized beams. By the polarization performance of E_a, it is known that the principal axis direction of elliptical polarized light is always parallel

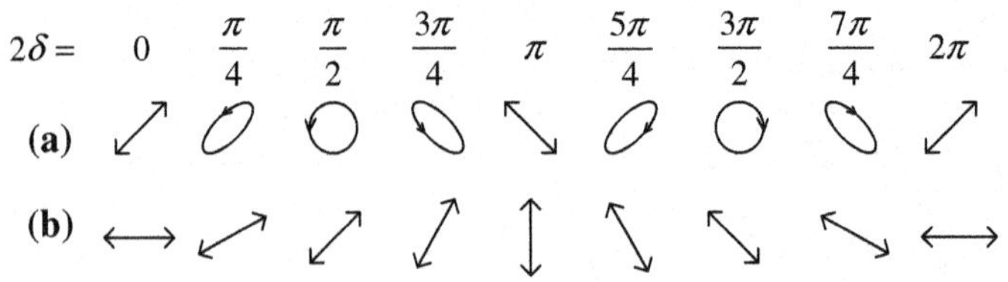

$$2\delta = \quad 0 \quad \frac{\pi}{4} \quad \frac{\pi}{2} \quad \frac{3\pi}{4} \quad \pi \quad \frac{5\pi}{4} \quad \frac{3\pi}{2} \quad \frac{7\pi}{4} \quad 2\pi$$

(a)

(b)

Fig. 8.54 Polarized modulation sketch of two orthogonally polarized beams along x–axis: **a** the polarization direction of reference beam is parallel to y–axis while, that the object beam is parallel to x-axis. **b** The reference beam is in right-circular polarization while objective beam is left-circular polarization

to x-axis and y-axis, the long axis for which is alternating however. Because of the anisotropy of molecules induced by elliptical polarized light, the light axis is always parallel to the long axis. For the circular polarized light could not induce the isotropy, maximum anisotropy of the material will be induced when E_a is linear polarized beam.

If E_a is considered to be the elliptical polarized light, the anisotropy induced is proportional to the difference of the light intensity along x-axis and y-axis, that is,

$$\Delta I = \cos^2 \delta - |i \sin \delta|^2 = \cos 2\delta \tag{8.34}$$

So, the transmission matrix of the amplitude type holographic photo could be expressed as

$$T_a^A = \begin{pmatrix} T_x & 0 \\ 0 & T_y \end{pmatrix} = \begin{pmatrix} T_0 + \Delta T & 0 \\ 0 & T_0 - \Delta T \end{pmatrix} \tag{8.35}$$

where up-corner mark A is the expression of amplitude type, $T_0 = (T_x + T_y)/2$, $\Delta T = (T_x - T_y)/2$. $(T_x - T_y)$ is proportion to ΔI with the coefficient of $(t_e - t_o)$, that is,

$$T_x - T_y = (t_e - t_o) \cos 2\delta \tag{8.36}$$

Substituting in Eq. (8.35), the transmission matrix of the holographic photo induced by orthogonal linear polarized beams is then

$$T_a^A = \begin{pmatrix} t_0 + \Delta t \cos 2\delta & 0 \\ 0 & t_0 + \Delta t \cos 2\delta \end{pmatrix} \tag{8.37}$$

where $t_0 = T_0 = (t_e + t_o)/2$ and $\Delta t = (t_e - t_o)/2$ is the expression of modulation degree of dichroism.

For pure phase type holographic photo, only the phase of the light is changed by the photo birefringence of the material. The transmission matrix could be obtained similarly

$$T_a^P = \begin{pmatrix} e^{i\psi_x} & 0 \\ 0 & e^{i\psi_y} \end{pmatrix} = e^{i\psi_0} \begin{pmatrix} e^{i\Delta\psi \cos 2\delta} & 0 \\ 0 & e^{-i\Delta\psi \cos 2\delta} \end{pmatrix} \tag{8.38}$$

where up-corner mark P is the expression of phase type, $\psi_0 = 2\pi n_0 L/\lambda_2$, $\Delta\psi = 2\pi \Delta n L/\lambda_2$, L is the thickness of the recording media, $n_0 = (n_e + n_o)/2$ is the average refraction index of e light and o light, $\Delta n = (n_e - n_o)/2$ is the modulation degree of birefringence.

The second case was discussed here. Equation (8.38) shows that the orthogonal circular polarized beams with the same amplitude are both linear polarized beams no matter what value it is for δ and the direction variation of vibration is δ, however. The optical constant will be the same everywhere in the photo dichroism recording media, while the light axis is varying with δ. It is considered that the coordinate system for the transmission matrix of holographic photo is still xoy. For amplitude type recording media, when the light axis turns δ degree from x-axis and the phase difference is 2δ, the transmission matrix for the holographic photo recorded will be

$$\begin{aligned} T_b^A &= R(-\delta) \begin{pmatrix} t_e & 0 \\ 0 & t_o \end{pmatrix} R(\delta) \\ &= \begin{pmatrix} t_e \cos^2 \delta + t_o \sin^2 \delta & (t_e - t_o) \cos \delta \sin \delta \\ (t_e - t_o) \cos \delta \sin \delta & t_e \sin^2 \delta + t_o \cos^2 \delta \end{pmatrix} \\ &= \begin{pmatrix} t_0 + \Delta t \cos 2\delta & \Delta t \sin 2\delta \\ \Delta t \sin 2\delta & t_0 - \Delta t \cos 2\delta \end{pmatrix} \end{aligned} \tag{8.39}$$

where $R(-\delta)$ and $R(\delta)$ are the spin matrixes for the coordinate system by clockwise and anticlockwise, separately, $t_0 = (t_e + t_o)/2$, $\Delta t = (t_e - t_o)/2$.

Similarly, for phase type recording media, the transmission matrix is

$$\begin{aligned} T_b^P &= R(-\delta) \begin{pmatrix} e^{i\psi_e} & 0 \\ 0 & e^{-i\psi_e} \end{pmatrix} R(\delta) \\ &= \begin{pmatrix} e^{i\psi_e} \cos^2 \delta + e^{i\psi_o} \sin^2 \delta & (e^{i\psi_e} - e^{i\psi_o}) \cos \delta \sin \delta \\ (e^{i\psi_e} - e^{i\psi_o}) \cos \delta \sin \delta & e^{i\psi_e} \sin^2 \delta + e^{i\psi_o} \cos^2 \delta \end{pmatrix} \\ &= e^{i\psi_0} \begin{pmatrix} \cos \Delta\psi + i \sin \Delta\psi \cos 2\delta & i \sin \Delta\psi \sin 2\delta \\ i \sin \Delta\psi \sin 2\delta & \cos \Delta\psi - i \sin \Delta\psi \cos 2\delta \end{pmatrix} \end{aligned} \tag{8.40}$$

where $\psi_0 = (\psi_e + \psi_o)/2$, $\Delta\psi = (\psi_e - \psi_o)/2$, ψ_0 and $\Delta\psi$ the definition of the same as expression (8.38).

The retrieving of the polarized holographic photo is the same as the ordinate holographic photo, where R is the readout light and S is the light after transmitting the holographic photo, that is,

$$S = TR \tag{8.41}$$

where T is the transmission matrix of polarized holographic photo.

When retrieving the holographic photo in the first case, same linear polarized beam is used for readout light with the same direction of the reference beam, that is,

$$R_a = \begin{pmatrix} E_0 e^{i\varphi_R} \\ E_0 e^{i\varphi_R} \end{pmatrix} = \begin{pmatrix} E_0 e^{-i\delta} \\ E_0 e^{-i\delta} \end{pmatrix} \tag{8.42}$$

From expressions (8.37) and (8.41) we can get

$$S_a^A = t_0 \begin{pmatrix} E_0 e^{-i\delta} \\ E_0 e^{-i\delta} \end{pmatrix} + \frac{\Delta t}{2} \begin{pmatrix} E_0 e^{i\delta} \\ - E_0 e^{i\delta} \end{pmatrix} + \frac{\Delta t}{2} \begin{pmatrix} E_0 e^{-i3\delta} \\ - E_0 e^{-i3\delta} \end{pmatrix} \tag{8.43}$$

where first item of the right side of the equation sign is for the direct light whose polarization state is the same as reference beam while the second and third items are for ± 1 level of diffraction light whose polarization state is the same as original objective beam. It is validated that the original polarization state could be recovered. This characteristic is different from ordinate isotropy recording media. The diffractive efficiency is t_0^2, $(\Delta t)^2/4$ and $(\Delta t)^2/4$, separately. In ideal situation, it is got $t_0 = 1/2$, $\Delta t = 1/2$ and the diffraction efficiency for ± 1 level of diffraction light is $\eta_{\pm 1} = 1/16 = 6.25\,\%$. It is the same as the amplitude holographic photo.

From expression (8.38) and (8.41) we can get

$$S_a^P = e^{i\psi_0} \begin{pmatrix} \sum_{n=-\infty}^{\infty} J_n(\Delta\psi) E_0 e^{i2n\delta} e^{i\delta} \\ \sum_{n=-\infty}^{\infty} J_n(\Delta\psi)(-1)^n E_0 e^{i2n\delta} e^{i\delta} \end{pmatrix} \tag{8.44}$$

where J_n is the Bessel function. For 0 and ± 1 level diffraction light it is

$$S_{a0}^P = e^{i\psi_0} J_0(\Delta\psi) \begin{pmatrix} E_0 e^{-i\delta} \\ E_0 e^{-i\delta} \end{pmatrix} \tag{8.45}$$

$$S_{a+1}^P = i e^{i\psi_0} J_1(\Delta\psi) \begin{pmatrix} E_0 e^{-i\delta} \\ - E_0 e^{i\delta} \end{pmatrix} \tag{8.46}$$

$$S_{a-1}^P = -i e^{i\psi_0} J_1(\Delta\psi) \begin{pmatrix} E_0 e^{-i3\delta} \\ - E_0 e^{-i3\delta} \end{pmatrix} \tag{8.47}$$

From expression (8.44) it can be found that ± 1 level diffraction efficiency is 33.3 % when $\Delta\psi$ is large enough which is the similar with isotropic recording media. However, it could recover the light polarization state which is different from the ordinary phase-type recording media.

It is concluded from the discussion above that the polarization state of the diffraction light is always perpendicular to that of the readout beam for the holographic photo in the first case whose polarization state will stay constant except the diffraction direction even the wavelength of the readout beam is changed. The holographic photo is equal to a half-wave plate. In the second case, when the readout beam is the circular polarized light, it is given as

$$R_b = \frac{\sqrt{2}}{2}\begin{pmatrix} E_0 e^{-i\delta} \\ -iE_0 e^{-i\delta} \end{pmatrix} \tag{8.48}$$

From expression (8.43) and (8.48), we can get

$$S_b^A = \frac{t_0\sqrt{2}}{2}\begin{pmatrix} E_0 e^{-i\delta} \\ -iE_0 e^{-i\delta} \end{pmatrix} + \frac{\Delta t\sqrt{2}}{2}\begin{pmatrix} E_0 e^{i\delta} \\ iE_0 e^{i\delta} \end{pmatrix} \tag{8.49}$$

From expression (8.44) and (8.48), we can get

$$S_b^P = \frac{\sqrt{2}}{2}e^{i\psi_0}\cos(\Delta\psi)\begin{pmatrix} E_0 e^{-i\delta} \\ -iE_0 e^{-i\delta} \end{pmatrix} + \frac{\sqrt{2}}{2}ie^{i\psi_0}\sin(\Delta\psi)\begin{pmatrix} E_0 e^{i\delta} \\ iE_0 e^{i\delta} \end{pmatrix} \tag{8.50}$$

The first item at the right hand of the equal mark is for the direct light whose polarization state is the same as the readout beam while the second item is for diffraction light whose polarization state is the same as original objective beam. The polarized grating recorded here is equivalent to polarized beam divider which could divided the right-hand and left-hand circular polarized light into two beam with different directions.

It is seen that only the original objective light could be recovered when original reference beam is used for readout light for amplitude-type or phase-type holographic photos recorded by orthogonal circular polarized beams. The diffraction efficiency of the amplitude type is $(\Delta t)^2$, the maximum value of which is 25 % in ideal situation while that of the phase type is $\sin^2(\Delta\psi)$ the maximum value of which is 100 %.

From the discussion for the holographic photos under two cases, it is found that larger diffraction efficiency is obtained for the grating which is recorded by the orthogonal circular polarized beams than by the linear polarized beams under the same recording and readout condition.

Okada Shudo et al. have researched the grating's characteristic formed by recording light with different polarized state and how the polarizations state and intensity of the diffraction light are affected by the polarization state of the readout beam. The recording media used in the experiment is the Bacteriorhodopsin film whose thickness is 30 μm. The wavelength of the recording beam is 515 nm and the diameter is 2 cm. The density for reference beam and objective beam is

Table 8.4 The polarization state of the diffraction light and the diffraction efficiency by the holographic grating

Reference beam	Objective beam	Readout beam	Polarization state	Diffraction efficiency
↕	↕	↕	↕	1.0
↕	↕	↔	↔	0.28
↕	↔	↕	↔	0.25
↕	↔	↔	↕	0.26
↔	↔	↔	↔	0.91
↔	↔	↕	↕	0.25
↔	↕	↔	↕	0.25
↔	↕	↕	↔	0.26
(R)	(R)	(R)	(R)	0.84
(R)	(R)	(L)	(L)	0.86
(R)	(L)	(R)	(L)	0.90
(R)	(L)	(L)	×	0
(L)	(L)	(L)	(L)	0.84
(L)	(L)	(R)	(R)	0.83
(L)	(R)	(L)	(R)	0.92
(L)	(R)	(R)	×	0

2.0 mW/cm^2 for each. They entrance to the media surface symmetrically with the included angle 8°. The wavelength of the readout beam is 633 nm and the intensity is 0.5 mW/cm^2. It entrances the grating of the media surface perpendicularly. The experimental results are listed in Table 8.4 where the diffractive efficiency for different polarization grating has been normalized by the numbers in the first row.

It is also shown from the experiment result that larger diffraction efficiency will be got if the circular polarized beams are used for recording rather than linear polarized beams when they are in orthogonal polarization state. It is corresponding with the theoretical result which gives us the basis of the orthogonal dual-channel storage system with circular polarized light.

8.6　Material Preparation for Polarized Storage

Material preparation for polarized holographic storage with dithienylethenes and the measurement for anisotropy. The molecular formula in the above part of Fig. 8.60 is one kind of dithienylethenes for polarized holographic storage which is supplied by Chemistry Department of Tsinghua University, the composed way for which could be refer to literature. It has two states. One is in open-loop state called state A with achromatic color, the other is in closed loop called state B with blue color. It could be transferred between two states if violet light λ_1 or visible light λ_2 is used, which is shown in Fig. 8.55.

Dissolve 15 mg dithienylethenes and 60 mg polymethyl methacrylate in 1 ml cyclohexanone liquor. Make it mix conformably. Spin it on the quartz plate with the size of 25 mm × 25 mm × 0.8 mm and airing it. The thickness of the thin is measured to be 10 μm. The absorption band for the open-loop and close-loop samples is shown in Fig. 8.56. The peak value in close loop is at 632 nm where the optical density is 0.4 while the absorption peak is at 305 nm for open loop.

The anisotropic characteristic of dithienylethenes is researched under polarized light. Its absorption performance is measured under linear polarized and circular

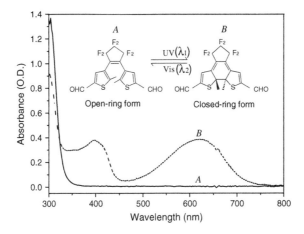

Fig. 8.55 Two states of dithienylethenes and their absorption band

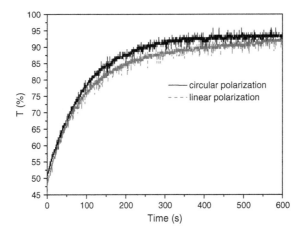

Fig. 8.56 Transmission curve for dithienylethenes under linear and circular polarized light

polarized beams. Semiconductor solid pumping laser with wavelength of 532 nm is used in experiment. The power for working light is 1.34 mW with diameter of 4 mm. The power of the transmitted light is sampled with power meter with the sampling rate of 10 Hz. The result is shown in Fig. 8.61 where black solid curve is for transmitted circular polarized light while red dashed line is for transmitted linear polarized curve. It is shown that larger absorption is made for the circular polarized light than linear polarized light which is the characteristic of the anisotropy. It is discovered by research that the characteristic of the photochromic material is corresponding with the structure of the molecule formula. The asymmetry of the structure will introduce the anisotropy of the material. Otherwise, little anisotropy will occur for symmetry structure of the molecule formula. Contrast experiments have been carried out, the result of which shows that the holographic photo could be recovered from the media of dithienylethenes whose molecule formula has bad symmetry.

8.6.1 Polarized Holographic Storage Experiment with Dithienylethenes

The polarized storage performance of dithienylethenes will be researched. Four holographic photos are recorded by recording light with different polarization states in the same dithienylethenes recording material and readout by corresponding light, which is shown in Fig. 8.57. Figure 8.57a, b is the recovered holographic photos which are recorded by orthogonal polarized light while Fig. 8.57c, d is recorded by the same polarization light. It is found that noise is serious for the latter two figures which are caused by the nonuniform of surface of the dithienylethenes recording media. It is much better for the former two figures for the stray light has different polarization state with the diffraction light which is filtered by the wave plate. It is why the holographic photos have higher signal-to-noise ratio when they are recorded by the orthogonal polarized beams. When the recording condition is the same expect the polarization state, the light exposure for 8.57a is two times larger than that for 8.57b which shows that the diffraction efficiency of holographic photos under orthogonal linear polarized beams is less than that of circular polarized beams.

8.6.2 The Polarization Multiplexing Characteristic of Dithienylethenes

Todorov et al. proposed polarization multiplexing storage by Methyl orange/PVA film. The first holographic photo is recorded by left-hand circular polarized reference beam and right-hand circular polarized objective beam; the second photo is by right-hand circular reference beam and left-hand circular polarized beam. The first photo, second photo, and their superposition are retrieved by left-hand circular polarized beam, right-hand circular polarized beam, and linear polarized beam,

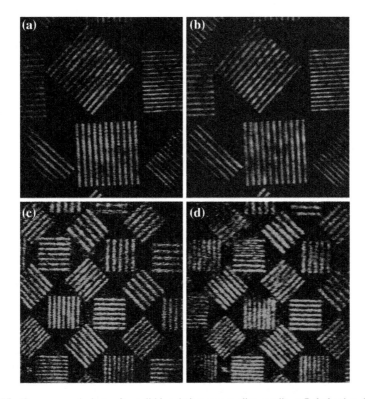

Fig. 8.57 The recovered photos from dithienylethenes recording media. **a** Polarization direction of the reference beam is parallel to incidence plane while that of objective light is perpendicular to it. **b** Reference beam is left-hand circular polarized beam while objective light is right-hand circular beam. **c** Polarization direction of reference and objective light are both perpendicular to incidence plain. **d** Both of reference and objective beams are left-hand circular polarization light

separately. In 2008, a new polarization multiplexing storage technology is realized on the basic of anisotropic Bacteriorhodopsin film by Koek et al. Two holographic photos are recovered at the same time. The first one is recorded with the left-hand circular polarized light for both reference and objective beams when the objective beam is changed to right-hand circular polarized for the second one. The same polarization light as reference light is used for retrieving both photos at the same time. They are separated by polarized beam split prism after passing though a quarter-wave plate. In this scheme, semiconductor solid frequency double laser with the wave length of 532 nm is used for recording and continuous laser with the same wavelength is for reading out.

In order to research the polarization multiplexing storage performance of dithienylethenes, two photos are recorded at the same point of c. The retrieved photos which are recorded by linear polarized multiplexing technology are shown in Fig. 8.58. Photo in Fig. 8.58a is the one recorded by intensity modulation where the polarization states of reference and objective beams are both parallel to the

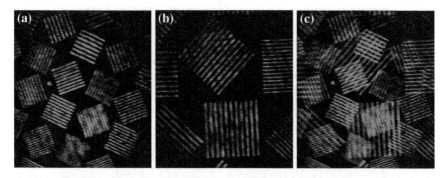

Fig. 8.58 Retrieved photos which are recorded by linear polarization multiplexing technology. **a** Polarization state are both parallel to incidence plane. **b** Polarization of reference beam is parallel to incidence plane while that of objective beam is perpendicular to incidence plane. **c** Both two photos retrieved at the same time

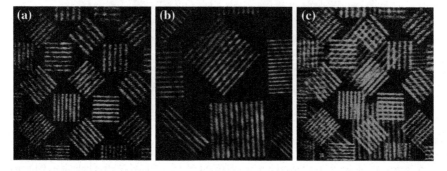

Fig. 8.59 Retrieved photos which are recorded by circular polarization multiplexing technology. **a** Left-hand circular polarized for both reference and objective beams. **b** Left-hand circular polarization for reference and right-hand circular polarization for objective beam. **c** Both two photos retrieved at the same time

incidence plain. Photo in Fig. 8.58b is the one recorded by polarization state modulation where the polarization state of the objective beam is changed to be perpendicular to the incidence plain. Both of two holographic photos could be retrieved at the same time if the reference beam is used as readout beam. They can be easily separated by a polarized beam split prism or a polarizing disk.

The retrieved photos which are recorded by circular polarized multiplexing technology are shown in Fig. 8.59. Photo in Fig. 8.59a is the one recorded by intensity modulation where both of the reference and objective beams are left-hand circular beams. Photo in Fig. 8.59b is the one recorded by left-hand circular reference beam and right-hand circular polarized objective beam where the polarization state is modulated. Both of two holographic photos could be retrieved at the same time if the reference beam is used as readout beam, shown as Fig. 8.59c. They can be separated by a polarized beam split prism or a polarizing disk after passing through a quarter-wave plate.

With the comparison of two polarization multiplexing scheme, it is found that the second one is little better. It is shown by the theoretical and experimental analysis that larger diffraction efficiency could be got if the orthogonal circular polarized beams are used for recording rather than orthogonal linear polarized beams. Higher light exposure is always needed in the holographic recording by orthogonal linear polarized beams which will consume the dynamic range of the material.

8.7   Orthogonal Polarized Dual-Channel Holomem

8.7.1   Based on Polarized Light Dual-Channel Storage System

OMNERC of Tsinghua University developed a orthogonal polarized dual-channel storage system based on series doped crystals and isotropic open-loop polymer of basic iron (CROP) material, as shown in Fig. 8.60. This scheme is an orthogonal polarized beams system, that the light does not interfere each other with isotropic

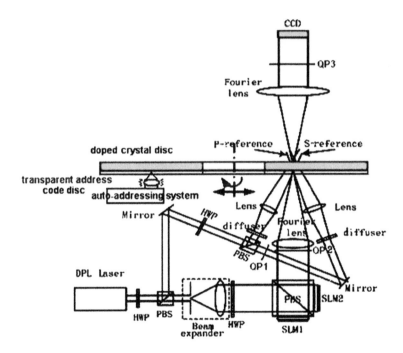

Fig. 8.60 Orthogonal polarized dual-channel storage system with transparent address code disc and auto-addressing system: *P* p-polarization light, *S* s-polarization light, *PBS* polarized beam splitter, *HWP* half-wave plates, *SLM1* and *SLM2* reflective type space light modulator, *QP1–QP3* quarter-wave plate, *DPL* diode pump solid laser

media. In holographic recording system, reference beam and objective beams with parallel polarization states and perpendicular polarization states could record independently. Only one of the reference beams need to be used for photo retrieving. If both of the reference beams are used at the same time, the overlap of two holographic photos will be obtained then. However, they could not be recovered independently at the same time because the recording media is not sensitive to polarization state of the readout beam. In fact, angle multiplexing technology is used in this dual-channel holographic storage system. A shortcoming of this system is that the light component of the background will be overlapped if parallel polarized beams are used for recording. A part of the dynamic range of the material will be consumed by which the recording density is reduced.

The polarized dual-channel storage system has a transparent address code disc and auto-addressing system, so it can sue more thickness anisotropic doped crystal material to increase capacity and addressing accuracy.

Unlike the scheme described above, a new orthogonal polarized dual-channel storage system based on the anisotropic dithienylethenes material is shown in Fig. 8.61. The characteristics of this scheme are listed below: Semiconductor diode pump solid laser with the wavelength of 532 nm is used for recording and reading out light source. Reflective type SLM1 and SLM2 are corresponding with horizontal and vertical polarized information light. Circular polarized light is used to read out, in order to achieve higher diffraction efficiency. Same information light is used to orthogonal polarized beams during recording. HP1 and HP2 are both half-wave plates where HP1 is used for adjusting the intensity of the whole information and reference beams with higher diffraction efficiency and better contrast, HP2 is for adjusting the light intensity of two orthogonal polarized beams

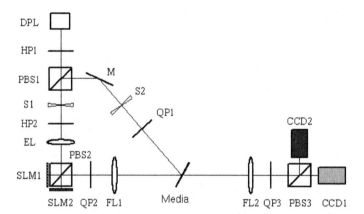

Fig. 8.61 Orthogonal polarized dual-channel holomem system: *HP1* and *HP2* are the half-wave plates; *S1* and *S2* are shutters; *M1* and *M2* are the polarized reflective mirrors; *PBS* is the polarized beam splitter prism, *EL* is the extending lens, *QP1*, *QP2*, and *QP3* are quarter-wave plate, *FL1* and *FL2* are the Fourier transform lens and the inverse Fourier transform lens, and the media is anisotropic dithienylethenes material

to achieve the balance of the reaction speed and diffraction efficiency. PBS1, PBS2, and PBS3 are polarized beam split prisms. PBS1 is used for separating the linear polarized beam from laser into reference and information light. PBS2 is for separating the linear polarized information beam into orthogonal polarized coaxial beams. PBS3 is for separating the two orthogonal polarized holographic photos and imaging to CCD1 and CCD2 separately. QP1, QP2, and QP3 are quarter-wave plate. QP1 is used for changing the reference beam from vertical polarization into right-circular polarization. QP2 turns the horizontal and vertical linear polarized beams from SLM1 and SLM2 to right-circular and left-circular polarized beams. QP3 changes the orthogonal circular polarized diffraction light to orthogonal linear polarized light.

In the experimental system, the powers of right-circular polarized is 1.45 mW for reference beam, left-circular polarized is 0.45 mW, and right-circular polarized is 0.05 mW for information beams separately. The diameter of the recording beam is about 4 mm. The reference and information beams are entranced in the recording media at the same point with the incidence angle of 14° and −23° separately. Defocusing method of Fourier spectra plane is used for holographic recording with the defocusing value of 10 mm. The recording time is 15 s, to make the photos of each channel by the reference beam and information beam, as shown in Fig. 8.62 where (a) for CCD1 from SLM1 and (b) for CCD2 from SLM2. No crosstalk is introduced for both photos expect for a little distortion which is caused by asymmetry of the recording field and the bug of the media.

The effect between two orthogonal polarized coaxial information beams will discuss the following. In the experiment, only two orthogonal polarized coaxial beams are used to expose the recording material without reference beam. After a short time, one of the information beams is closed. No corresponding holographic photo is found which expressed that there is little interfere between them. It could be explained theoretically. The distribution of first spectrum depends on the pixel number of imaging plain. The distributions of the first spectrums of them are not in superposition when the pixel number of the photos is different. Then there is surely not interfere during recording. For photos with the same pixel number, the optical difference is 0 from points on photos to the corresponding points on the material by passing through Fourier lens. Then there is still no interfere in the recording area [1].

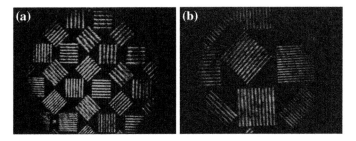

Fig. 8.62 Two holographic photos recovered independently by different polarized channels. **a** Photo of CCD1 from SLM1. **b** Photo of CCD2 from SLM2 (see Fig. 8.61)

The method of polarized holographic storage and the performances of above polarized holographic storage are founded the anisotropic dithienylethenes material, which is sensitive to the polarization state of the light. It is applied to the experiment of the polarized holographic storage very well.

The polarization state multiplexing holographic storage has been realized based on the photochromic dithienylethenes material and the orthogonal polarized dual-channel storage system was established. The recording and retrieving time is efficiently reduced in the system where the storage density of the material is improved at the same time. Moreover, it is well compatible with other multiplexing technologies, such as angular multiplexing, circular multiplexing, shift multiplexing, speckle multiplexing, and so on. It is demonstrated by experiment that not only dithienylethenes but also other anisotropic holographic material could be applied to this system.

8.7.2 Experiment for Retrieving of Nonvolatile Data

Because the chemical process of the photochromic response is reversible, the volatile data retrieving will occur for photochromic storage, that is the readout light will erase the information in the material when the writing light is used for readout light also. Long-time research has been made by scientists to overcome this shortage. The heat lock effect of dithienylethenes was discovered by Tatezono et al. Some molecules of dithienylethenes will not make chemical response until the temperature is beyond 100 °C in PMA. The electric lock characteristic of dithienylethenes is discovered by Kawai et al. Two molecule states could be obtained under the electric field for dithienylethenes liquor which light to photochromic chemical response is different. If the electric field is changed after the information recording is finished, nonvolatile data retrieving could be realized then. At present, the electric lock and heat lock technology for data fixing could only be done for some dithienylethenes in liquor state rather than in solid film.

Many resoling schemes for nonvolatile data retrieving are also proposed for photochromic material in solid thin, such as chiral readout, lock inner of molecule for nonvolatile data readout, readout by fluorescence, nonvolatile readout by light current detection, nonvolatile readout by multiphoton gate controller photochromic response, and so on. However, they are only suitable for bit storage system and not suitable for volume holographic storage system with big data page storage. Up till now, preferable scheme for nonvolatile data readout in rewritable holographic storage system based on photochromic material has not been proposed yet. Some new scheme for nonvolatile data readout will be explored for other photochromic materials in the future.

8.7.3 Two-Photon Recording and One-Photon Readout Experiment

Two-photon absorption (TPA) is the transition process from ground state to excitation state for one atom or molecule in dummy state by absorbing two photons directly, which is shown in Fig. 8.63. It is a photon response process of third nonlinear state absorbing for material in bright light condition. The TPA ability of the material is always expressed by TPA section δ. It is very important for the design and composition of material with large TPA coefficient to research the relationship between molecule structure and the value of δ. It is found by Marder et al. that the molecule with structure of D-π-D, D-π-A-π-D or A-π-D-π-A could improving the value of δ. Diphenyl derivative with high TPA section value up to 1250 GM(1 GM $= 1 \times 10^{-50}$ cm$^4 \cdot$ s $\cdot$ photon$^{-1} \cdot$ molecule^{-1}) has been composed by them. Kim et al. found that the value of δ could be greatly improved using chromophore [2, 3- b] thiophene and [2′, 3′- d] thiophene and thiophene (DTT) as the center of π electron system.

New two-photon fluorescence material with larger TPA section has been developed in recent years and the application research for two photons is widely developed, such as two-photon-pumped frequency upconverted blue laser, two-photon optical limiter, two-photon 3D optical storage, two-photon laser scanning fluorescence microscopy, and so on. For testing new medium development a recording testing experimental system, it is a typical original interferometer, as shown in Fig. 8.64 (up). The pulse laser is model CPA2001 femtosecond laser with wavelength 775 nm and pulse width 160 fs. Its peak intensity can be to 11 GM/cm^2. The space light modulator (MIL) is a grating with cycle of 155 μm and width of 13 μm. The included angle for two recording beams is 2° and the field is 3 mm. The medium was developed cooperation with Physical and Chemical Institute of CAS. The thickness of the recording material is 5 μm. The absorption section is $\delta = 1.03 \times 10^3$ GM. Four testing grating photos recorded on the photochromic material are shown in Fig. 8.64 (down), that the exposure time from left to right are 0.5, 1, 2, 3 min separately.

Fig. 8.63 The absorption process of two photons

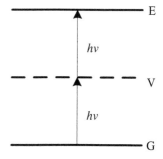

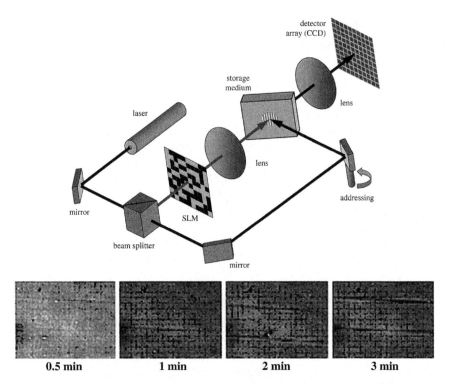

Fig. 8.64 The recording experimental system of two-photon holographic storage for medium (*up*) and the photo of grating with exposure time (*down*): 0.5, 1, 2, 3 min separately

The performance of photochromic dithienylethenes with two-photon recording and one-photon retrieving is easier to be realized in volume holographic system because high voltage field is not needed. It is based on the photorefractive polymer material with two-photon recording method by femtosecond laser with the same wavelength. However, high voltage field of 50 V/μm is needed during the recording process which will greatly improve the difficulty in practice. The photochromic dithienylethenes show good performances for TPA in infrared band however no linear absorption. The absorption spectrum is shown in Fig. 8.65, where curve 1 is the absorption spectrum in open loop which is in ultraviolet band and curve 2 is for closed loop which is in visible band. Curve 3 in dash line is for TPA spectrum in infrared band when the photochromoc dithienylethenes in closed loop will change to open loop after absorbing two photons with wavelength of λ_3 at the same time. It is seen from Fig. 8.65 that the absorption for dithienylethenes molecule in open–closed loop is 0 in infrared band, even so it is no linear absorption. So using wavelength of λ_3 to readout that may not be erase or destroy of the existing grating or holographic photo. A photo is recorded with wavelength of 632.8 nm, which is shown in Fig. 8.66a. The figure is exposed by light with wavelength of 780 nm and light intensity of 63.7 mW/cm^2 for 318 min. Then it is

Fig. 8.65 The absorption spectrum for dithienylethenes

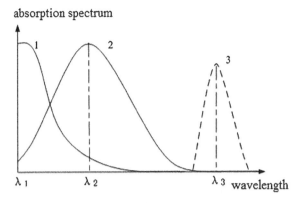

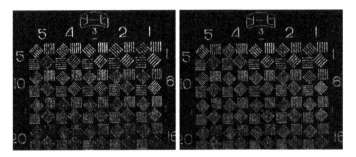

Fig. 8.66 The nonvolatile data readout experiment in infrared band

retrieved by the original readout light, which is shown in Fig. 8.66b. It is shown that the quality is almost the same for both photos which expresses that the material will not respond for the light of wavelength of 780 nm.

The volume holographic grating is pure phase-type grating which is recorded by the two-photon reaction in infrared band whose diffraction efficiency is higher than that of mixed type grating recorded by visible light. In order to analysis and validate the dithienylethenes molecule without linear absorption in infrared band, it is needed calculation relationship between dithienylethenes molecule and absorption coefficient.

During the TPA response of photochromic dithienylethenes, the TPA section has the relationship with absorption coefficient as

$$\alpha_{nl} = N_A u \delta I / h\nu \tag{8.51}$$

where u is the molar ratio, N_A is the Avogadro constants, I is recording light intensity, $h\nu$ is the energy of one photon. According to Lambert–Beer molecule absorption law, the absorbed light intensity I_a by material is

$$I_a = I(1 - \exp(-2.303\alpha_{nl}L)) \tag{8.52}$$

where L is the thickness of recording media. It could be simplified by binomial expansion as

$$I_a = 2.303\alpha_{nl}dI = 2.303N_A u\delta I^2/hv \qquad (8.53)$$

It is seen from the equation that the light intensity absorbed is direct proportional to recording light intensity. Because the response time is very short for two photons, the quick information recording could be achieved. If pico or even shorter than femtosecond laser with high power is used for recording light source, quick volume holographic recording could then be realized. The nonvolatile data retrieving could be obtained by the low power continuous laser with the same wavelength. Because the wavelength of recording light is the same as that of retrieving light, no photo distortion will be introduced by Bragg-mismatched readout.

8.7.4   Two-Wavelength Readout Experiment

Because no linear absorption is taken place in infrared band for photocharomic dithienylethenes material, nonvolatile data retrieving could be obtained by light infrared band which is recorded by visible light with the sensitive wavelength of the material. For holographic storage system in the future, recording and writing process could be separated. Two-wavelength readout technology could then be used for holographic read-only disk system. Because the requirement of the light source for readout is much lower than that for recording, cheap lasers with different wavelength in infrared band could be used in the holographic read-only system with nonvolatile data retrieving. The mechanism for two-wavelength readout technology is shown in Fig. 8.67.

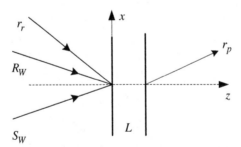

Fig. 8.67 The sketch of two-wavelength readout: S_W is information light, R_W is reference light, r_r is readout light, γ_p is the output diffraction field, and L is recording media thickness

From Fig. 8.67 we can see that the readout light r_r is exposed on the media that the diffraction grating is recorded on the material, and the diffraction field r_p could be described as

$$E_d(r_p) = \int_V r_r(r)\Delta\varepsilon(r)G(r, r_p)dr^3 \tag{8.54}$$

where $\Delta\varepsilon(r)$ is the phase of grating, $G(r, r_p)$ is the Green function in free space which is expressed as

$$G(r, r_p) = \frac{1}{j\lambda_r|r - r_p|}\left(jk_r|r - r_p|\right) \tag{8.55}$$

The expression of information and reference light grating is

$$S_w(r) = A_S \exp\left(j\vec{k}_w \cdot \vec{r}\right) \tag{8.56}$$

and

$$R_w(r) = A_R \exp\left(j\vec{k}_w \cdot \vec{r}\right) \tag{8.57}$$

The expression of readout light is

$$r_r(r) = A_r \exp\left(j\vec{k}_r \cdot \vec{r}\right) \tag{8.58}$$

and

$$\Delta\varepsilon(r) = R_w^*(r)S_w(r) \tag{8.59}$$

In equations from (8.56) to (8.59), A_S, A_R and A_r are the amplitudes of the information, reference, and readout light. $k_w = 2\pi/\lambda_w$, $k_r = 2\pi/\lambda_r$ where λ_w and λ_r are the wavelengths of the recording and readout light. Then Eq. (8.54) can be written as

$$E_d(r_p) = \int_V \exp\left(j\vec{k}_r \cdot \vec{r}\right)\exp\left(j\vec{k}_w \cdot \vec{r}\right)\exp\left(-j\vec{k}_w \cdot \vec{r}\right)G(r, r_p)dr^3 \tag{8.60}$$

where the constant factor is neglected. Substitute expression (8.55) into (8.60) and by Born approximation, it was integrated to get

$$E_d(r_p) = CL\exp\left(j\vec{k}_d \cdot \vec{r}_p\right)\sin c\left(\frac{L}{2\pi}\Delta k_z\right) \tag{8.61}$$

where C is the constant coefficient which is relative to the recording structure, L is the thickness of the volume grating, $\vec{k}_d$ is diffraction light vector which could be expressed as

$$
\begin{aligned}
\vec{k}_d = \left(\vec{k}_S - \vec{k}_R + \vec{k}_r\right) \cdot \hat{x}_p + \left(\vec{k}_S - \vec{k}_R + \vec{k}_r\right) \cdot \hat{y}_p \\
+ \sqrt{k_r^2 - \left(\vec{k}_S - \vec{k}_R + \vec{k}_r\right) \cdot \left(\hat{x}_p^2 + \hat{y}_p^2\right)} \cdot \hat{z}_p
\end{aligned}
\tag{8.62}
$$

where Δk_z is the Bragg-mismatched vector, that is

$$
\Delta k_z = \left(\vec{k}_S - \vec{k}_R + \vec{k}_r - \vec{k}_d\right) \cdot \hat{z}_p
\tag{8.63}
$$

It is presented in expression (8.63) that only the light which is satisfied the Bragg-mismatched angle could be diffracted by grating. When the wavelength and angle is of the readout light is the same as that of reference one, that is, $\lambda_w = \lambda_r$ and $\vec{k}_R = \vec{k}_r$, i.e., $\vec{k}_d = \vec{k}_S$, which presents the information of recording light could be completely reappeared by diffracted light. However, when the wavelength of readout light is different from that of reference one, such as $\lambda_w < \lambda_r$, then it is $\vec{k}_d \neq \vec{k}_S$. The direction of diffraction light will depart from the information light. From expression (8.63), it is seen that directions of diffraction light will be different for readout light with different wavelengths when the recording light is the same.

The diffraction efficiency of the holographic is obtained from expression (8.61) as

$$
I_d(r_p) = (CL)^2 \operatorname{sinc}^2\left(\frac{L}{2\pi}\Delta k_z\right)
\tag{8.64}
$$

It is seen that the diffraction efficiency will be greatly decreased when the mismatch value of Δk_z is increased. The diffraction efficiency is also corresponding with the thickness of material. The bad effect by mismatch will be weakened by decreasing the thickness.

It is considered that the incidence angles of the information light, reference light, and readout light are θ_S, θ_R, and ϕ_r, angle expanding value of the information light is $\Delta\theta_S$, the diffraction light angle for the readout light which satisfies Bragg angle is ϕ_s, which is shown in Fig. 8.68. It is necessary to point out that the angles ϕ_s are defined in the recording media without sign. The angle relationship for Bragg-mismatched could be obtained geometrically as

$$
\frac{\sin\left(\frac{1}{2}(\phi_r + \phi_s)\right)}{\lambda_r} = \frac{\sin\left(\frac{1}{2}(\theta_R + \theta_S)\right)}{\lambda_w}
\tag{8.65}
$$

$$
\phi_r - \phi_s = \theta_R - \theta_S
$$

Fig. 8.68 Wave vector for
Bragg-mismatched in
two-wavelength readout

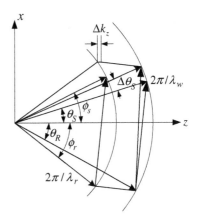

The solution of expression (8.65) can be obtained as

$$
\phi_r = \frac{1}{2}(\theta_R - \theta_S) + \sin^{-1}\left[\frac{\lambda_r}{\lambda_w}\sin\left(\frac{1}{2}(\theta_R + \theta_S)\right)\right]
$$
$$
\phi_s = -\frac{1}{2}(\theta_R - \theta_S) + \sin^{-1}\left[\frac{\lambda_r}{\lambda_w}\sin\left(\frac{1}{2}(\theta_R + \theta_S)\right)\right]
$$

(8.66)

From expression (8.66), it can be found that the Bragg-mismatched angle could be easily satisfied for single grating. However, in the storage of Fourier photos, the volume grating on the recording media which is located on the spectrum plain is formed by the interfere of the information light composed of a serial of plain wave with different angles and the reference plain wave with certain angle, so the Bragg angle could not be satisfied if only one readout light is used the wavelength of which is different from recording light. So the whole holographic photo could not be recovered by single plain wave in spectrum storage.

According to above relationship, expression of (8.63) could be presented as

$$
\Delta k_z = 2\pi\left(\frac{\cos\phi_r}{\lambda_r} + \frac{\cos(\theta_S + \Delta\theta_S) - \cos\theta_R}{\lambda_w} - \left\{\left(\frac{1}{\lambda_r}\right)^2\right.\right.
$$
$$
\left.\left. - \left[\frac{\sin\phi_r}{\lambda_r} - \frac{\sin\theta_R + \sin(\theta_S + \Delta\theta_S)}{\lambda_w}\right]^2\right\}^{1/2}\right)
$$

(8.67)

It is obviously seen that the maximum diffraction efficiency could be obtained if the angle of readout light is satisfied with expression (8.66) and $\Delta\theta_S = 0$. Keep the angle of readout light constant. The value of Δk_z will be increased if $\Delta\theta_S$ is increased. Then the diffraction efficiency is decreased according to expression (8.64). If $\Delta k_z = \frac{2\pi}{L}$, the diffraction efficiency is 0. If λ_γ/L is the maximum value of the information light, the angle expanding $\Delta\theta_S^M$ is

Fig. 8.69 The relationship
between amplitude width W
and its angle $\Delta\theta_S$ to spectrum
plain of holographic storage
photo, *SLM* is spatial light
modulator, *FL* is Fourier Lens

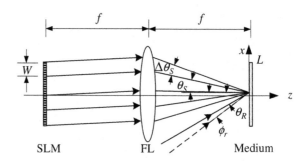

$$\Delta\theta_S^M = \sin^{-1}\left\{\frac{\lambda_w}{L}\left[\cot(\phi_r - \theta_R)\cos\theta_S - \sin\theta_S\right]\right\} \tag{8.68}$$

The relationship between amplitude width W of the readout holographic photo
by single plain wave and $\Delta\theta_S$ in the research of spectrum plain in volume holographic storage will be discussed, as shown in Fig. 8.69. Commonly, the size and
position of plus–minus first-order spectrum is mainly considered in the spectrum
storage. According to Fourier spectrum transition, it is known that the details of the
object are reflected by the high frequency part, the profile is by low frequency part,
and the zero-order spectrum is for the background without any information. If the
zero-order spectrum is blocked in the storage, then the contrast of the recovered
photo will be reversed without loss any basic information. In the experiment, zero
and plus–minus one-order spectrum will be often recorded. Only the volume
grating with plus one order of the Fourier spectrum is taken into consideration. The
relationship could be calculated geometrically as

$$W = \frac{2fn_0\sin\Delta\theta_S}{\sqrt{1 - n_0\sin^2\Delta\theta_S}} \tag{8.69}$$

where f is the focus length of the Fourier lens, n_0 is the average refractive index of
the recording media.

There are three methods that could be used to increase the retrieved photo area, if
single one plain wave is readout light in two-wavelength readout experiment. The
first one is reducing the thickness of recording media L, that is thin material is used
which will decrease the storage density of the material. The second one is to use
readout light with the close wavelength to recording light where sensitive wavelength response area is required for the material. The third one is increasing the
focus length of the Fourier Lens, which is not good for the miniaturization of the
system structure.

From above analysis found that the satisfied effect could be hardly achieved if
single plain wave is used for readout light only. In general, the modification of the
wave front of the readout light is tried to realize the nonvolatile retrieving of the
whole holographic photo. Spherical wave is used for readout light. The cylinder

lens is introduced in the readout system to retrieve the whole photo. The common point in these methods is to get the readout light with multi-angles. Because the plus–minus one-order spectrum of the object photo is composed of the light with different incidence angles and power factors, the uniformity of the light field and angle of the readout light are limited. It is not possible to get fully uniformity holographic photos. Therefore, a holographic diffuser is introduced to readout light with continuous angles and uniformity light field to realize the Bragg compensation in broadband by which uniform holographic photo could be retrieved more easily.

8.7.5 Other Dual-Channel Two-Wavelength Holomem

Another two-wavelength dual-channel holomem method is carried out to validate two-wavelength holographic storage scheme with Bragg compensation in broad-band. The optical system of the two-wavelength holographic storage system is shown in Fig. 8.70. Semiconductor pumping laser with wavelength of 532 nm is used for recording light source. The light changes to parallel light after passing the pinhole stop PH1, extender lens EL1, and optical stop D1. It is divided into objective light and reference light by polarized light splitting prism after passing

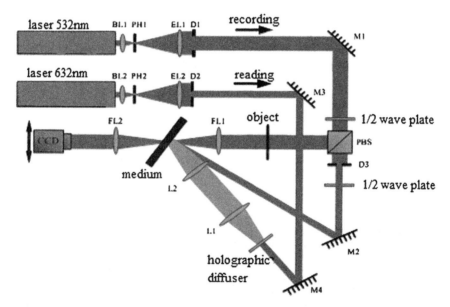

Fig. 8.70 The sketch of two-wavelength holographic storage system with Bragg compensation in broadband by holographic diffuser. Where: *BL1*, *BL2* condenser, *HP1*, *HP2* pinhole stop, *EL1*, *EL2* extender lens, *D1*, *D2*, *D3* stop, *M1*, *M2*, *M3*, *M4* mirror, *PBS* polarized beam splitter, *L1*, *L2* alignmet lens, *FL1* Fourier transform lens, and *FL2* the inverse Fourier transform lens, object resolution chart

reflecting mirror M1 and half-wave plate. The objective light finally enters into the recording material in frequency spectrum after passing resolution test pattern and Fourier Lens FL1 the focus length is 150 mm. The reference beam will get interfere with the objective light at certain angle after reflective mirror M2. He–Ne gas laser with wavelength of 632.8 nm is used for readout light source. Light changes into parallel light after passing the pinhole PH2, extender lens EL2, and optical stop D2. It then enters into material by certain angle after reflective mirror M3 and M4. Then the reference beam could be obtained after Lens L1 and L2 (focus lengths are both 100 mm and aperture sizes are 75 mm) and the holographic diffuser which is satisfied the Bragg compensation condition in broadband. The diffracted photos could be got by moving CCD meantime.

The recording material is doped $Fe:LiNbO_3$ in the experiment whose thickness is 3 mm and doped with 0.03 Wt%. The entrance angle is 27.56°. The reference light with original wavelength of 532 nm is for retrieving. The photo detected by CCD is shown in Fig. 8.71a. Then plain wave with wavelength of 632.8 nm is used to enter the media and detect the photo by moving the CCD. For it is difficult to change the angle of the reference light by changing the position and angle of mirror M4, so rotation of the crystal is easier for the angle compensation. The matched photo is shown in Fig. 8.71c. The different part of the original photo could be recovered separately by rotating the crystal. The angle for center part of the photo is chosen for the Bragg compensation angle in broadband. The incidence angle is 32.81°. Lens of L1 and L2 and holographic diffuser are then added between the crystal and mirror M4, the recovered photo is shown in Fig. 8.71b.

It is seen from Fig. 8.71 that only 1/3 of the whole photo could be recovered by plain wave for readout light and 4/5 of the photo could be recovered with the help of Bragg compensation technology in broadband by which the size of recovered photo could be greatly increased. Furthermore, the background noise is small and the signal-to-noise ratio of the recovered photo is high which demonstrates its feasibility. For only holographic diffuser with scatter angle of 0.2° could be obtained at present, it is still not possible to recover the whole photo. It is predicted that whole photo with good uniformity diffraction efficiency could be recovered if holographic diffuser with scatter angle of 0.3° is used.

For the diffraction angle and imaging position changed in the compensation readout system, the imaging lens and CCD should be moved correspondingly to

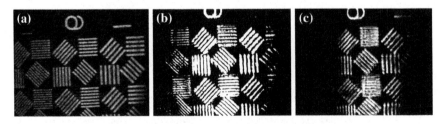

Fig. 8.71 The comparison of retrieved photos with different readout conditions

avoid the photo distortion where the optical collimation and image errors correction need further research. If the holographic diffuser is added in the reference optical path, parallel light could be used for retrieving the photos in the readout system which could simplify the readout system with nonvolatile readout.

8.7.6   Dual-Channel with Polarization Multiplexing Holomem

The polarization multiplexing of the holographic photos was considered in the two-wavelength holographic storage system with nonvolatile data retrieving. The technology could not be application if the nonvolatile data readout in the two-wavelength system is based on the loss of the multiplexing. According to the analysis in the above, it is found that page crosstalk and noise will be introduced in photo retrieving by light with another wavelength if the several photos are recorded at the same point of the media with multiplexing technologies, such as angular multiplexing, space multiplexing, circular multiplexing, and so on.

Two holographic photos are recorded at the same point of photochromic dithie-nylethenes media by polarization multiplexing technology. The retrieving of the photos is relative to light polarization state as shown in Fig. 8.72. It is a sketch of the two-wavelength holographic storage system with polarization multiplexing. The wavelength of the recording light and readout light are 532 and 632.8 nm separately. The polarization state of the reference light is perpendicular to the incidence plain in the recording. The polarization state of information light corresponding with space

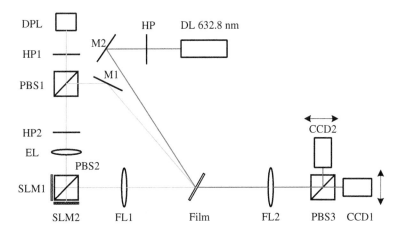

Fig. 8.72 The sketch of the two-wavelength holographic storage system with polarization multiplexing: *HP* half-wave plates, *M* mirror, *SML* space light modulator, *PBS* polarized beam splitter, *EL* extending lens, *FL1* and *FL2* Fourier transform lens and the inverse Fourier transform lens, Film is the medium of the Diarylethene

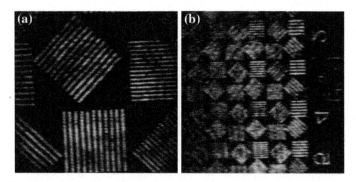

Fig. 8.73 The retrieved photos with two-wavelength readout with polarization multiplexing.
a From space light modulator SLM1. **b** From space light modulator SLM2

light modulator SLM1 is parallel to the incidence plain while that corresponding with space light modulator SLM2 is perpendicular to incidence plain. HP1 is the half-wave plain for adjusting the intensity ratio of the reference and information lights. HP2 is another half-wave plain for adjusting the intensity distribution of two information beams. HP is the half-wave plain for readout light which is used for getting the same polarization state of the readout light as reference light.

Because the recording media is very thin, only 10 μm and the band limitation of the light emitted from semiconductor laser, Bragg compensation is brought in some degree. The recovered photos are shown in Fig. 8.73, that most parts are retrieved. There is no crosstalk between two retrieved photos which demonstrated that the application of polarization multiplexing technology in two-wavelength holographic storage system is feasibility.

The nonvolatile data retrieving technology is based on photocharomic dithienylethenes material. Two-photon recording one-photon readout method is used to realize quick recording and nonvolatile data retrieving. According to the coupling-wave theory analyzed the Bragg matched in two-wavelength holographic storage system, that the diffuser is added in the reference optical path to make the reference light in certain special band. Readout light in infrared band and certain special frequency is used for nonvolatile data retrieving. The multiplexing technology in two-wavelength readout system is researched and the polarization multiplexing technology is adapted. Two holographic photos could be readout without crosstalk by the light with another wavelength. Because two states of photochromic dithienylethenes could be mutual transmitted by the exposure of ultraviolet light and infrared light, a beam of ultraviolet light with the same optical path of the readout light could depress the erasing effect of the readout light. The nonvolatile data retrieving was then obtained [2].

Based on the photochromic dithienylethenes, the rewritable and nonvolatile data retrieving technology for holographic storage system was developed. The main research contents and results are included the experiment of rewritable holographic storage system with the photocharomic dithienylethenes material, which refractive index and absorption coefficient are varied with the states of open and close loops. The recording and readout model of grating is established based on the wave coupling theory and the photon absorption law of photocharomic material which is experimentally validated. It is found in the experiment that some photocharomic dithienylethenes materials are anisotropic. A research of the polarized holographic storage is then carried out. The diffraction efficiency variation for the grating which is recorded by orthogonal polarized recording light with different polarization states is analyzed. The variation of polarization state of the diffraction state and diffraction efficiency is also analyzed for readout light with different polarization states. An orthogonal polarized dual-channel holographic storage system with polarization multiplexing technology is first proposed. Only one set of Fourier Lens, one reference beam and one orthogonal polarized coaxial information beam are used in the system. The recording and retrieving speed are improved as well as the storage density. Furthermore, it could be compatible to other multiplexing technologies, such as angular multiplexing, speckle shift multiplexing, circular multiplexing technologies which are good for application.

Nonvolatile data retrieving is always needed in the holographic storage system based on photochromic dithienylethenes material. One method is that pulse laser with high power in infrared band is used for quick holographic recording and continuous laser with same wavelength and low power is used for readout. Nonvolatile data retrieving could be obtained. The other method is that the laser with the sensitive wavelength of the photochromic dithienylethenes material is used for holographic recording and the readout light in infrared band is used for non-volatile data retrieving. The Bragg matching of the two-wavelength holographic storage is analysis according to wave coupling theory. The diffuser is put in the reference optical path to make the reference beam in certain spatial band. Nonvolatile date retrieving was obtained by the readout light in infrared band and certain spatial frequency.

The multiplexing technology in two-wavelength holographic storage system is researched and the polarization multiplexing technology is adapted. Two holographic photos could be readout without crosstalk by light with another wavelength. The recording and retrieving model of the holographic grating is established based on photochromic dithienylethenes which is experimentally demonstrated. The anisotropy of dithienylethenes is discovered through experiment. For the absorption coefficient and the refractive index are sensitive to polarization state, the polarized volume holographic storage is then researched. The orthogonal polarized dual-channel holographic storage system is proposed and experimentally demonstrated. For the photochromic dithienylethenes is without linear absorption and with strong TPA in infrared band, quick recording with two photons and nonvolatile data retrieving with one photon has been realized.

Nonvolatile photo retrieving has been realized with Bragg compensation in broadband for photochromic dithienylethenes. Two photos at the same point could be recovered without crosstalk from each other which are recorded by the polarization multiplexing technology.

Volume holographic storage is one of the most important subjects in the fields of super-high density optical digital data storage in next generation information storage technology. It has a lot of advantages such as fast data transfer rate, short access time, parallel processing, and the optical correlation. It may be the most potential storage techniques in the future. But the dynamic–static speckle multiplexing is an efficient multiplexing method to realize the high density of volume holographic system by increasing the storage capacity of unit volume. The characteristics, exposure parameters, reconstructive parameters, and scattering noise of dynamic–static speckle multiplexing have been analyzed and concluded. The influences to the multiplexing selectivity and thickness of the crystals are studied, as well as the relationship between the speckle sizes and selectivity, dynamic range, etc. The influence of DPL modes laser is also considered. In order to make the best use of the volume of the materials, DSM storage scheme with overlapped and multiple orbits is realized. Based on the sequential exposure, the exposure schedule of the DSM scheme is determined. By introducing an overlap factor to compensate for the erasure effect among adjacent storage positions during static multiplexing, a new exposure schedule for DSM scheme is proposed. And by further introducing an overlap factor matrix for compensating the erasure effect among different storage position, the exposure schedule for overlapped and multiple tracks DSM storage is also determined. Both theoretical calculations and the experimental results demonstrate the validity of the schedule. The serious scattering noises in the speckle multiplexing volume holographic storage system are studied. Based on the experimental analysis, an adaptable storage scheme is proposed to suppress the scattering light during storage process and avoid the deterioration of the reconstructive holograms during long-time readout process. The speckle multiplexing system could be improved based on the original system to realize the dynamic–static speckle multiplexing with multiple and overlapped orbits with higher storage density, low noisy in principles. An orthogonal polarization dual-channel volume holographic storage is presented to simultaneously record and reconstruct two orthogonally polarized holograms with negligible interchannel crosstalk independently, which improves the speed of input and output speed of data in volume holographic storage. The technique is combined with other multiplexing methods such as angle multiplexing, shift multiplexing, peristrophic multiplexing, shift multiplexing, etc. Several methods are proposed to realize the nonvolatile readout based on the properties of photochromic diarylethene. But there are following questions cannot be resolved yet or need study in the future. Correlation selectivity of the speckle encoding method with a diffuser in the reference beam or the object beam for volume holographic storage system should be experiment and analyzed continuing. The correlation selectivity has no sidelobes and does not depend on the thickness of the recording medium. What relation of the intrapage crosstalk, the thickness of the recording medium and sharpen the diffraction efficiency with the diffuser have to be studied. Research on manufacturing technology of diffusers can improve its performances for that the techniques can be effective and feasible in the volume holographic storage authentically.

The suitable recording material is the key to the development of the rewritable holographic storage. The research results from many countries are reported or developed. But there is not any commercial material indeed. Developing a photochromic diarylethene material could be application in rewritable holographic storage experiment. Some photochromic diarylethene are found to be sensitive to polarized light. Different polarized light will induce its anisotropy, which includes dichroism and birefringence. But its properties of fatigue resistance and thermal irreversibility are not good. The dependence of the performances of the rewritable volume holographic storage on the properties of photochromic diarylethene is investigating. A model to describe real-time grating formation in photochromic diarylethene-doped PMMA thin films is introduced based on photochromic chemicals reaction and the coupled-wave theory. But the model is needed to validate with more experiments. Crystal holography materials have better holography media with higher photorefractive efficiency. So it is better multiplexing performances than photopolymeric materials and can be used to storage more information at a point. The scattering noise in crystal materials is lower than photopolymers at the same storage density. But the sensitivity of the photorefractive crystal is rather poor and cannot be used to dynamic experiment. Another difficult of media is the process of the holography disk. The thickness of organic materials film is 10 μm only as the experimental condition limited. The media thickness is need increasing of course. But there is great difficulty in material process. The sampled disk is not protection layer. It cannot be against the high temperature and humidity without protection layer.

An auto-adapted electronic equalizer was employed in the digital signal processing system. Experiment shows that the part of crosstalk could be restrained in a certain extent. But the SNR is relation with materials intimately. The diffraction efficiency is related to laser power and exposure time. But it cannot test in detail in the experiment, that the experimental result is not exactitude very much and this technology is not matured very much.

The scattering noise in crystal materials is lower than photopolymers at the same storage density generally. But the experiment result of photopolymeric materials show that is better than photorefractive crystals holography storage. It could be brought as the thickness of organic materials film is thin very much. The thickness of organic materials film is 10 μm only as the experimental condition limited. Therefore the storage intensity is not increased distinctly than photorefractive crystals much more with different experiment system indeed.

The plain wave was used in a reference arm in the experiment system. When a speckle beam is used in a reference arm, the Bragg compensation condition will be difficult. The Bragg compensation technique could be suitable to DSSM system still or no that have to be analyzed and experiment much more. According to the past experiments, the selectivity of thick material is often better than thin one with lower crosstalk. At one time, the existent photopolymer thickness of disk sample is 10 μm. So a disc of diameter 120 mm and photopolymer of thickness 10 μm, a diffuser in reference beam and object beam with phase diffuser are used in this experiment. But the experimental results show that the intensity and capacity are poor than original crystal materials storage system. It could be redesigned and amended if would like to use the experiment system to engineering development indeed.

8.8   Focusing Properties with High Polarization Orders

The larger numerical aperture (LNA) focusing properties for axially symmetries polarized beams (ASPBs) especially with larger polarization orders has been used in orthogonal polarized dual-channel system as shown in Fig. 8.74. For an ASPB as shown in Fig. 8.74, the polarization orientation angle $\Phi(r, \varphi)$ of the electric field only depends on the azimuthally angle φ as $\Phi(r, \varphi) = P \times \varphi + \varphi_0$, where P is the polarization order number, φ is the azimuthally angle of the polar coordinate system, and φ_0 is the initial polarization orientation for $\varphi = 0$. The radially polarized beams ($P = 1$, $\varphi_0 = 0$) and azimuthally polarized beams ($P = 1$, $\varphi_0 = \pi/2$) are $P = 1$ ASPBs.

It has important influence for 3D volume holography indeed. Many researches on axially symmetric polarized beams (ASPBs) have drawn considerable attention, especially on radically polarized beams and azimuthally polarized beams for the especial applications in lithography, particle trapping, electron acceleration, scanning microscopy, and data storage for focusing. Many progresses have made for high numerical aperture focusing properties and practical applications. But the research on axially symmetric polarized beams with higher polarization order do not carried out and applications to such beams. Figure 8.75 illustrates the geometry of the problem, which adopted the similar notation. The incident field is an ASPB, which is assumed to have planar phase front. The f is the focal length of the objective lens, $S(r_s, \varphi_S, z_S)$ is an observation point near the focal plane, φ_s denotes the azimuthally angle with respect to the x_{axis} and θ represents the polar angle. A function to calculate the field distribution near focus of an aplanatic system for

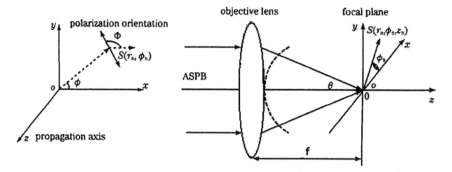

Fig. 8.74 Polarization orientation and focusing of the axially symmetries polarized beams (*ASPBs*)

Fig. 8.75 The relishing versus between the ratio I_z/I_r of the maximum intensities of the longitudinal and transverse field and the NA of the lens numerical aperture

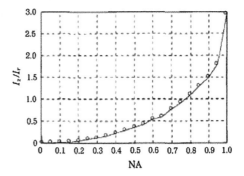

incident ASPBs with different polarization orders and initial azimuthally angles. The focal field of an ASPBe can be written with theory of Richards and Young-worth such as

$$E_r = \frac{-iA}{\pi} \int_{\theta_{\min}}^{\theta_{\max}} \int_0^{2\pi} l(\theta) \sin\theta \sqrt{\cos\theta} e^{ik[z_s \cos\theta + r_s \sin\theta \cos(\phi - \phi_s)]} \cos[(P-1)\phi + \phi_0]$$
$$\cos\theta \cos(\phi - \phi_s) d\theta d\phi,$$

$$E_\phi = \frac{-iA}{\pi} \int_{\theta_{\min}}^{\theta_{\max}} \int_0^{2\pi} l(\theta) \sin\theta \sqrt{\cos\theta} e^{ik[z_s \cos\theta + r_s \sin\theta \cos(\phi - \phi_s)]} \sin[(P-1)\phi + \phi_0]$$
$$\cos(\phi - \phi_s) d\theta d\phi,$$

$$E_z = \frac{-iA}{\pi} \int_{\theta_{\min}}^{\theta_{\max}} \int_0^{2\pi} l(\theta) \sin\theta \sqrt{\cos\theta} e^{ik[z_s \cos\theta + r_s \sin\theta \cos(\phi - \phi_s)]} \cos[(P-1)\phi + \phi_0]$$
$$\sin\theta d\theta d\phi, \tag{8.70}$$

where E_r, E_φ, and E_z are the amplitudes of the three orthogonal components $l(\theta)$ is the pupil apodization function, which denote the relative amplitude and phase of the field, k is the wavelength, $\theta_{\max}$ and $\theta_{\min}$ are the maximal polar angle and the minimal polar angle determined by the NA. It is based on the simulation results, can find some focusing properties of the beams, such as ever on axis energy, strong long initial field and intensity distribution at focus. That can use to 3D focused field distribution flexibly by diffractive optical elements (DOEs). Derive the mathematical expressions for the focusing fields, show the focusing field distributions near focus, and discuss some possible applications.

Based on Eq. (8.70), we can calculate the focusing field distribution of different ASPBs, but only present the results for $P = 4$ as typical $\theta_{\max}$, as the limited length requirement. The numerical $\theta_{\max}$ is a pupil apodization function, i.e.,

$$l(\theta) = \begin{cases} 1 & 0 \leq \theta \leq \theta_{max} \\ 0 & \text{otherwise} \end{cases} \qquad (8.71)$$

where $\theta_{max} = \sin^{-1}(NA/n)$, n is the index of reaction and θ_{min} is set to be 0 here. For all examples in the letter, n is set to be 1, and all length measurements are using units of wavelengths, therefore, $\lambda = 1$ is also assumed to be 1. The amplitude is assumed to be 1 also.

The ratio (I_z/I_r) of the maximum intensities of the longitudinal and transverse fields versus the NA of the lens is shown in Fig. 8.75. That revealed the longitudinal field is stronger than the transverse field (the sum of the azimuthally component and the radial component). This trend is shown in Fig. 8.75, which displays the plot for the ratio of the maximum intensities of the longitudinal and transverse fields versus the NA of the lens with $P = 4$. If NA is larger than 0.78, the longitudinal field I_z will be stronger than the transverse field I_r. When NA is close to 1, the ratio approaches 3. If adopting the appropriate pupil apodization function $l(\theta)$, that can increase the ratio further, which is the desirable application for orthogonal polarized dual-channel holography storage system design, electron acceleration, and near-field microscopy.

References

1. E.E. Narimanov, H. Li, Y.A. Barnakov, T.U. Tumkur, M.A. Noginov, Reduced reflection from roughened hyperbolic metamaterial. Opt. Express **21**, 14956–14961 (2013)
2. Eric R. Dufresne, Gabriel C. Spalding, Matthew T. Dearing, *Computer-generated holographic optical tweezer arrays* (Dept. of Physics, Illinois Wesleyan University, Bloomington, 2000)

Appendix A
Constants of Physics and Chemistry

Symbol	Name/meaning (definition)	Equation	Uncertainty (ppb)
c	Speed of light in vacuum	$299{,}792{,}458 \text{ m s}^{-1}$	Exact
h	Planck constant	$6.626070040(81) \times 10^{-34} \text{ J s}$	44
$\hbar$	Reduced Planck constant $\hbar \equiv h/2\pi$	$1.054571726(47) \times 10^{-34} \text{ J s}$ $= 6.58211928(15) \times 10^{-22} \text{ MeV s}$	44 22
k	Boltzmann's constant	$1.3807 \times 10^{-23} \text{ J/K}$	
e	Electron charge magnitude	$1.602176565(35) \times 10^{-19} \text{ C}$ $= 4.80320450(11) \times 10^{-10} \text{ esu}$	22 22
F	Faraday constant	$96485.3365 \text{ Å s/mole}$	21 ppm
$\hbar c$	Conversion constant	$197.3269718(44) \text{ MeV fm}$	22
$(\hbar c)^2$	Conversion constant	$0.389379338(17) \text{ GeV}^2 \text{ mbarn}$	44
N_A	Avogadro constant	$6.02214199 \times 10^{23} \text{ mol}^{-1}$	Exact
m_e	Electron mass	$0.510998928(11) \text{ MeV}/c^2$ $= 9.10938291(40) \times 10^{-31} \text{ kg}$	22 44
m_p	Proton mass	$938.272046(21) \text{ MeV}/c^2$ $= 1.672621777(74) \times 10^{-27} \text{ kg}$ $= 1836.15267245(75) \, m_e$	22 44 22
mu	Atomic mass constant	$1.660538921(73) \times 10^{-27} \text{ kg}$	44

(continued)

© Tsinghua University Press and Springer Science+Business Media Singapore 2016
D. Xu, *Multi-dimensional Optical Storage*,
DOI 10.1007/978-981-10-0932-7

(continued)

Symbol	Name/meaning (definition)	Equation	Uncertainty (ppb)
λ	Wavelength of light	Blue light $\lambda = 405$ nm	
λcp	Compton wavelength of proton	1.3214×10^{-15} m $= 1.3214 \times 10^{-6}$ nm	
λce	Compton wavelength of electron	2.4263×10^{-12} m $= 2.4263 \times 10^{-3}$ nm	
m_d	Deuteron mass	1875.612859(41) MeV/c^2	22
ϵ_0	Permittivity of free space	$1/\mu_0 c^2 = 8.854187817\ldots \times 10^{-12}$ F m^{-1}	Exact
μ_0	Permeability of free space	$4\pi \times 10^{-7}$ N A^{-2} $= 12.566370614\ldots \times 10^{-7}$ A^{-2}	Exact
α	Fine-structure constant	$e^2/4\pi\epsilon_0\hbar c = 7.2973525664(17) \times 10^{-3}$	0.32
re	Classical electron radius	$e^2/4\pi\epsilon_0 m_e c^2 = 2.8179403267$ (27) $\times 10^{-15}$ m	0.97
λe	$\lambda ce/2\pi$	$\hbar/m_e c = 3.8615926800(25) \times 10^{-13}$ m	0.65
a_∞	Bohr radius ($m_{\text{nucleus}} = 1$)	$4\pi\epsilon_0\hbar^2/m_e e^2 = r_e\alpha^{-2}0.52917721092$ (17) $\times 10^{-10}$ m	0.32
$hc/(1\text{ eV})$	Wavelength of 1 eV/c particle	$1.239841930(27) \times 10^{-6}$ m	22
$hcR\infty$	Rydberg energy	$m_e e^4/2$ $(4\pi\epsilon_0)^2\hbar^2 = m_e c^2\alpha^2 = 213.60569253$ (30) eV	22
σ_T	Thomson cross section	$8\pi r^2 e/3 = 0.6652458734(13)$ barn	1.9
μ_B	Bohr magneton	$e\hbar/2m_e = 5.7883818066$ (38) $\times 10^{-11}$ MeV T^{-1}	0.65
μ_N	Nuclear magneton	$e\hbar/2m_p = 3.1524512605$ (22) $\times 10^{-14}$ MeV T^{-1}	0.71
ω_{cycl}^e/B	Electron cyclotron freq./field	$e/m_e = 1.758820088$ (39) $\times 10^{11}$ rad s^{-1} T^{-1}	22
ω_{cycl}^p/B	Proton cyclotron freq./field	$e/m_p = 9.57883358(21) \times 10^7$ rad s^{-1} T^{-1}	22
G_N	Gravitational constant	$6.67384(80) \times 10^{-11}$ m^3 kg^{-1} s^{-2} $= 6.70837(80) \times 10^{-39}$ $\hbar c$ (GeV/c^2)$^{-2}$	1.2×10^5 1.2×10^5
g_N	Standard gravitational accel	9.80665 m s^{-2}	exact
k	Boltzmann constant	$1.3806488(13) \times 10^{-23}$ J K^{-1} $= 8.6173324(78) \times 10^{-5}$ eV K^{-1}	910 910
V_m	Molar volume (ideal gas at STP)	$22.413968(20) \times 10^{-3}$ m^3 mol^{-1}	910

(continued)

(continued)

Symbol	Name/meaning (definition)	Equation	Uncertainty (ppb)
b	Wien displacement law constant	$\lambda_{max}T = 2.8977721(26) \times 10^{-3}$ m K	910
σ	Stefan–Boltzmann constant	$\pi^2 k^4/60\hbar^3 c^2 = 5.670373$ $(21) \times 10^{-8}$ W m^{-2} K^{-4}	3600
$GF/(\hbar c)^3$	Fermi coupling constant	$1.1663787(6) \times 10^{-5}$ GeV^{-2}	500
$\sin^2\hat{\theta}$ (M_Z) $(\overline{M}S)$	Weak-mixing angle	$0.23116(12)$	5.2×10^5
GM (δ_{TPA})	Photoabsorption cross sections	1 GM $= 1 \times 10^{-50}$ cm^4 s molecules^{-1} photon^{-1}	
G	Gravitational constant	$G = 6.67408(31) \times 10^{-11}$ m^3 kg^{-1} s^{-2}	44
m_W	$W\pm$ boson mass	$80.385(15)$ GeV/c^2	1.9×10^5
m_Z	Z_0 boson mass	$91.1876(21)$ GeV/c^2	2.3×10^4
e	e constant/Euler's number	$e = 2.718281828459045235$	$e = \lim$ $(1 + 1/x)^x$
γ	Euler–Mascheroni constant	$\gamma = 0.57721566490153286$	22
$\alpha_s(m_Z)$	Strong coupling constant	$0.1184(7)$	5.9×10^6

Appendix B
Mathematical Symbols

Symbol	Name	Meaning/definition	Sample expression
+	Addition sign Logical OR symbol	Sum of a few values Logical disjunction	$3 + 5 = 8$ $\neg(A + B) = \neg A * \neg B$
*	Multiplication sign Logical AND symbol	Product of two values Logical conjunction	$3 \times 5 = 15$ $\neg(A * B) = \neg A + \neg B$
$\times (\cdot)$	Multiplication sign	Product of two values	$3 \times 5 = 3 \cdot 5 = 15$
$/(\div)$	Slash (division)	Division	$3/4 = 3 \div 4 = 0.75$
$\sum$	Sigma	Summation—sum of all values in range of series	$\sum x_i = x_1 + x_2 + \cdots + x_n$
$\sum\sum$	Sigma	Double summation	$\sum_{j=1}^{2} \sum_{i=1}^{8} x_{i,j} = \sum_{i=1}^{8} x_{i,1} + \sum_{i=1}^{8} x_{i,2}$
$\prod$	Product sign	Product of three up to infinitely many values	$\prod x_i = x_1 \cdot x_2 \cdot \cdots \cdot x_n, \ \prod_{n=1}^{\infty} \frac{1}{n}$
$f(x)$	Function of x	Maps values of x to $f(x)$	$f(x) = 3x + 5$
$\dfrac{dy}{dx}$	First derivative	Derivative—Lagrange's notation	$d(3x^3)/dx = 9x^2$
$\dfrac{d^2 y}{dx^2}$	Second derivative	Derivative of derivative	$d^2(3x^3)/dx^2 = 18x$
$\dfrac{d^n y}{dx^n}$	nth derivative	n times derivation	$f(x) = x \wedge n/(1 - x)$
$\dot{y}$	Time derivative	Derivative by time—Newton notation	dx/dt
$\ddot{y}$	Time second derivative	Derivative of derivative	d^2x/dt^2
$\dfrac{\partial f(x,y)}{\partial x}$	Partial derivative		$\partial(x^2 + y^2)/\partial x = 2x$
$\int$	Integral	Opposite to a derivation	$\int x^2 \ dx = x^3/3 + c$

(continued)

© Tsinghua University Press and Springer Science+Business Media Singapore 2016   655
D. Xu, *Multi-dimensional Optical Storage*,
DOI 10.1007/978-981-10-0932-7

(continued)

Symbol	Name	Meaning/definition	Sample expression
$\iint$	Double integral	Integration of function of 2 variables	$\iint f(x, y) \, dx \, dy$
$\iiint$	Triple integral	Integration of function of 3 variables	$\iiint f(x, y, z) \, dx \, dy \, dz$
$\oint$	Closed line integration	Closed contour/line integral	$\oint F(x) \, dx$
$\oiint$	Closed surface integral		$\oiint F(x, y) \, dx \, dy$
$\oiiint$	Closed volume integral		$\oiiint (x, y, z) \, dx \, dy \, dz$
$\otimes$	Tensor product	Tensor product of A and B	$A \otimes B$
$\langle x, y \rangle$	Inner product	A scalar function of two vectors, it is a generalization of the dot product	$\langle v, w \rangle = \langle w, v \rangle$
$\sqrt{a}$	Square root		$\sqrt{a} \cdot \sqrt{a} = a$, $\sqrt{9} = \pm 3$
$\sqrt[3]{a}$	Cube root		$\sqrt[3]{8} = 2$
$\sqrt[4]{a}$	Forth root		$\sqrt[4]{16} = \pm 2$
$\sqrt[n]{a}$	nth root (radical)		for $n = 3$, $\sqrt[n]{8} = 2$
^	Carat	Exponent	$2 \text{ ^ } 5 = 32$
!	Exclamation, factorial	$n! = 1 \cdot 2 \cdot 3 \cdots n$	$5! = 1 \cdot 2 \cdot 3 \cdot 4 \cdot 5 = 120$
$_nP_k$	Permutation	$_nP_k = \frac{n!}{(n-k)!}$	$_5P_3 = 5!/(5 - 3)! = 60$
$_nC_k$ $\binom{n}{k}$	Combination	$_nC_k = \binom{n}{k} = \frac{n!}{k!(n-k)!}$	$_5C_3 = 5!/[3!(5 - 3)!] = 10$
...	Continuation sign	Extension of sequence	$S = \{1, 2, 3, \ldots\}$
:	Colon, ratio sign	Division or ratio, symbol following logical quantifier or used in defining a set	$2{:}4 = 20{:}40$ $^\exists x{:} x > 4$ and $x < 5$ $^\forall x{:} x < 0$ or $x > -1$
|	Vertical line	Symbol following logical quantifier or used in defining a set	$^\exists x \mid x > 4$ and $x < 5$ $^\forall x \mid x < 0$ or $x > -1$ $S = \{x \mid x < 3\}$
::	Double colon	Arithmetic mean	$3{::}11 = 7$
∞	Lemniscate	Infinite summations, sequence limit	$^\forall x{:} x < \infty$
( )	Parentheses	Denotes a quantity, list, set of coordinates, or an open interval	$(a1, a2, a3, a4)$ $(3, 5)$
[ ]	Square brackets	Denotes a quantity or a closed interval	$w + [(x + y) + z]$, $[3, 5]$

(continued)

(continued)

Symbol	Name	Meaning/definition	Sample expression
( ]	Hybrid brackets	Denotes a half-open interval	$(a, b]$
[ )	Hybrid brackets	Denotes a half-open interval	$[a, b)$
{ }	Curly brackets	Denotes a quantity or a set	$E = \{2, 4, 6, 8, \ldots\}$
=	Equal	Indicates two values are the same	$-(-5) = 5$
~	Similarity	Proportional to	$\triangle ABC \sim DEF$
≈	Approximate equal	Two values are close to each other	$x \approx y$
≅	Congruent to	Equivalence of geometric shapes and size	$\triangle ABC \cong \triangle XYZ$
≠	Inequality	Indicates two values are different	$x \neq y$
<	Inequality	Left is smaller than value on right	$3 < 5$
≤	Inequality	Left is smaller than or equal to value on right	$x \leq y$
>	Inequality	Left is larger than value on right	$5 > 3$
| |	Absolute value sign	Distance of value from origin in number line, plane, or space	$\|-3\| = 3$
Δ	Increment sign, Triangle symbol	Indicates a small change, denotes vertices of triangle	$m = \Delta y / \Delta x$ $\triangle ABC = \triangle DEF$
⊥	Perpendicularity symbol	Geometry symbol	$L \perp M$
//	Parallel symbol	Geometry symbol	$L // M$
∠	Angle symbol	Geometry symbol	$\angle ABC = \angle DEF$
∃	Existential quantifier	Logical statements symbol	$\exists x : x > 4$ and $x < 5$
∀	Universal quantifier	Logical statements symbol	$\forall x : x < 0$ or $x > -1$
¬	Logical negation symbol	Logical statements symbol	$\neg(\neg A)\, A$
→	Logical implication symbol	Logical statements symbol	$A \rightarrow B$
↔	Logical equivalence symbol	Logical statements symbol	$A \leftrightarrow B$

(continued)

(continued)

Symbol	Name	Meaning/definition	Sample expression
$\in$	Element symbol	Set membership	$A = \{3, 9, 14\}, 3 \in A$
$\notin$	Not-element symbol	No set membership	$A = \{3, 9, 14\}, 1 \notin A$
$\subseteq$	Subset symbol	Subset has less elements	$A \subseteq B, \{9, 14, 28\} \subseteq \{9, 14, 28\}$
$\supseteq$	Superset	Set A has more elements or equal to the set B	$A \supseteq B \; \{9, 14, 28\} \supseteq \{9, 14, 28\}$
$\subset$	Proper subset symbol	Subset has less elements than the set	$A \subset B, \{9, 14\} \subset \{9, 14, 28\}$
$\supset$	Proper/strict superset	Set A has more elements than set B	$A \supset B, \{9, 14, 28\} \supset \{9, 14\}$
$\cup$	Union symbol	Objects that belong to set A or set B	$A \cup B = B \cup A$
$\cap$	Intersection symbol	Objects that belong to set A and set B	$A \cap B = B \cap A$
$A \not\subset B$	Not subset	Left set not a subset of right set	$\{9, 66\} \not\subset \{9, 14, 28\}$
$A \not\supset B$	Not superset	Set A is not a superset of set B	$\{9, 14, 28\} \not\supset \{9, 66\}$
$\varnothing$	Null symbol	Empty set $\varnothing = \{\ \}$	$C = \{\varnothing\}$

Appendix C
Calculation Symbols

Symbol	Symbol name	Meaning/definition	Example
$\lim_{x \to x_0} f(x)$	Limit	Limit value of a function	
ε	Epsilon	Represents a very small number, near zero	$\varepsilon \to 0$
y'	Derivative	Derivative—Leibniz's notation	$(3x^3)' = 9x^2$
y''	Second derivative	Derivative of derivative	$(3x^3)'' = 18x$
$y^{(n)}$	nth derivative	n times derivation	$(3x^3)^{(3)} = 18$
i	Imaginary unit	$i \equiv \sqrt{-1}$	$z = 3 + 2i$
z^*	Complex conjugate	$z = a + bi \to z^* = a - bi$	$z^* = 3 + 2i$
z	Complex conjugate	$z = a + bi \to z = a - bi$	$z = 3 + 2i$
∇	Nabla/del	Gradient/divergence operator	$\nabla f(x, y, z)$
$\mathcal{L}$	Laplace transform	$F(s) = \mathcal{L}\{f(t)\}$	
$\mathcal{F}$	Fourier transform	$X(\omega) = \mathcal{F}\{f(t)\}$	
δ	Delta function	It is a generalized function, can be defined as the limit of a class of δ sequences	
$\vec{x}$	Vector	A quantity having direction as magnitude, light and force etc	

(continued)

© Tsinghua University Press and Springer Science+Business Media Singapore 2016
D. Xu, *Multi-dimensional Optical Storage*,
DOI 10.1007/978-981-10-0932-7

(continued)

Symbol	Symbol name	Meaning/definition	Example
$\hat{x}$	Unit of vector	A unit vector in a normed vector space is a vector	$\hat{i} = \begin{bmatrix} 1 \\ 0 \\ 0 \end{bmatrix}, \hat{j} = \begin{bmatrix} 0 \\ 1 \\ 0 \end{bmatrix}, \hat{k} = \begin{bmatrix} 0 \\ 0 \\ 1 \end{bmatrix}$
AB	Line	Line from point A to point B	
$\overrightarrow{AB}$	Ray	Line that start from point A to point B	
$|x - y|$	Distance	Distance between points x and y	$|x - y| = 5$
a^b	Power	Exponent	$2^3 = 8$
$a \wedge b$	Caret	Exponent	$2 \wedge 3 = 8$
%	Percent	$1\% = 1/100$	$10\% \times 30 = 3$
‰	Per-mille	$1‰ = 1/1000 = 0.1\%$	$10‰ \times 30 = 0.3$
$\triangleq$ or :=	Equal by definition	Equal by definition	$A \triangleq B$
$\ll$	Much less than	Much less than	$1 \ll 1000000$
$\gg$	Much greater than	Much greater than	$1000000 \gg 1$
$\lfloor x \rfloor$	Floor brackets	Rounds number to lower integer	$\lfloor 4.3 \rfloor = 4$
$\lceil x \rceil$	Ceiling brackets	Rounds number to upper integer	$\lceil 4.3 \rceil = 5$
$(f \circ g)$	Function composition	$(f \circ g)(x) = f(g(x))$	$g(x) = x - 1 \Rightarrow (f \circ g)(x) = 3(x - 1)$
Δ	Delta	Change/difference	$\Delta t = t_1 - t_0$
Δ	Discriminant	It appears under the square root (radical) sign in the quadratic formula	$\Delta = b^2 - 4ac$
φ	Golden ratio	Golden ratio constant	
$|A|$	Determinant	Determinant of matrix A	
$\det(A)$	Determinant	Determinant of matrix A	
$\|x\|$	Double vertical bars	Norm	
A^T	Transpose	Matrix transpose	$(A^T)_{ij} = (A)_{ji}$
$A^\dagger$	Hermitian matrix	Matrix conjugate transpose	$(A^\dagger)_{ij} = (A)_{ji}$
A^*	Hermitian matrix	Matrix conjugate transpose	$(A^*)_{ij} = (A)_{ji}$
A^{-1}	Inverse matrix	$AA^{-1} = I$	
$\text{rank}(A)$	Matrix rank	Rank of matrix A	$\text{rank}(A) = 3$

(continued)

(continued)

Symbol	Symbol name	Meaning/definition	Example
dim(*U*)	Dimension	Dimension of matrix *A*	rank(*U*) = 3
$\mathcal{P}(A)$	Power set	All subsets of *A*	
A = *B*	Equality	Both sets have the same members	$A = \{3, 9, 14\}$, $B = \{3, 9, 14\}$, $A = B$
A^c	Complement	All the objects that do not belong to set *A*	
A*B*	Relative complement	Objects that belong to *A* and not to *B*	$A = \{3, 9, 14\}$, $B = \{1, 2, 3\}$, $A - B = \{9, 14\}$
A − *B*	Relative complement	Objects that belong to *A* and not to *B*	$A = \{3, 9, 14\}$, $B = \{1, 2, 3\}$, $A - B = \{9, 14\}$
A Δ *B*	Symmetric difference	Objects that belong to *A* or *B* but not to their intersection	$A = \{3, 9, 14\}$, $B = \{1, 2, 3\}$, $A \Delta B = \{1, 2, 9, 14\}$
A ⊖ *B*	Symmetric difference	Objects that belong to *A* or *B* but not to their intersection	$A = \{3, 9, 14\}$, $B = \{1, 2, 3\}$, $A \ominus B = \{1, 2, 9, 14\}$
A × *B*	Cartesian product	Set of all ordered pairs from *A* and *B*	
|*A*|	Cardinality	The number of elements of set *A*	$A = \{3, 9, 14\}$, $\|A\| = 3$
#*A*	Cardinality	The number of elements of set *A*	$A = \{3, 9, 14\}$, #*A* = 3
U	Universal set	Set of all possible values	
$\mathbb{N}_0$	Natural numbers set (with zero)	$\mathbb{N}_0 = \{0, 1, 2, 3, 4, \ldots\}$	$0 \in \mathbb{N}_0$
$\mathbb{N}_1$	Natural numbers set (without zero)	$\mathbb{N}_1 = \{1, 2, 3, 4, 5, \ldots\}$	$6 \in \mathbb{N}_1$
$\mathbb{Z}$	Integer numbers set	$\mathbb{Z} = \{\ldots -3, -2, -1, 0, 1, 2, 3, \ldots\}$	$-6 \in \mathbb{Z}$
$\mathbb{Q}$	Rational numbers set	$\mathbb{Q} = \{x \mid x = a/b, a, b \in \mathbb{N}\}$	$2/6 \in \mathbb{Q}$
$\mathbb{R}$	Real numbers set	$\mathbb{R} = \{x \mid -\infty < x < \infty\}$	$6.343434 \in \mathbb{R}$
$\mathbb{C}$	Complex numbers set	$\mathbb{C} = \{z \mid z = a + bi, -\infty < a < \infty, -\infty < b < \infty\}$	$6 + 2i \in \mathbb{C}$
Ω	Omega	Volume of an object, Ohms (resistance)	R2 = 330 Ω
ω	Omega	Transfinite ordinal, angular velocity, period	$\omega = 36{,}000$ rad/s, $\omega = 1/60$ s

(continued)

(continued)

Symbol	Symbol name	Meaning/definition	Example
$\mathbb{N}$,N	Enhanced or bold N	The set of natural numbers	$\mathbb{N} = \{0, 1, 2, 3, \ldots\}$
$\mathbb{Z}$,Z	Enhanced or bold Z	The set of integers	$\mathbb{Z} = \{0, 1, -1, 2, -2, 3, -3, \ldots\}$
$\mathbb{Q}$,Q	Enhanced or bold Q	The set of rational numbers	$\mathbb{Q} = \{a/b \mid a$ and b are in $\mathbb{Z}\}$
$\mathbb{R}$,R	Enhanced or bold R	The set of real numbers	It is the cardinality of $\mathbb{R}$

Appendix D
Symbols of Probability Analysis

Symbol	Symbol name	Meaning/definition	Example
$P(A)$	Probability function	Probability of event A	$P(A) = 0.5$
$P(A \cap B)$	Probability of events intersection	Probability that of events A and B	$P(A \cap B) = 0.5$
$P(A \cup B)$	Probability of events union	Probability that of events A or B	$P(A \cup B) = 0.5$
$P(A \mid B)$	Conditional probability function	Probability of event A given event B occured	$P(A \mid B) = 0.3$
μ	Population mean	Mean of population values	$\mu = 10$
$E(X)$	Expectation value	Expected value of random variable X	$E(X) = 10$
$E(X \mid Y)$	Conditional expectation	Expected value of random variable X given Y	$E(X \mid Y = 2) = 5$
$\mathrm{var}(X)$	Variance	Variance of random variable X	$\mathrm{var}(X) = 4$
σ^2	Variance	Variance of population values	$\sigma^2 = 4$
$\mathrm{std}(X)$	Standard deviation	Standard deviation of random variable X	$\mathrm{std}(X) = 2$
σ_X	Standard deviation	Standard deviation value of random variable X	$\sigma_X = 2$
$\widetilde{X}$	Median	Middle value of random variable x	$\widetilde{X} = 5$
$\mathrm{cov}(X,Y)$	Covariance	Covariance of random variables X and Y	$\mathrm{cov}(X, Y) = 4$
$\mathrm{corr}(X,Y)$	Correlation	Correlation of random variables X and Y	$\mathrm{corr}(X, Y) = 3$
$\rho_{X,Y}$	Correlation	Correlation of random variables X and Y	$\rho_{X,Y} = 3$
MR	Mid-range	$\mathrm{MR} = (x_{max} + x_{min})/2$	

(continued)

© Tsinghua University Press and Springer Science+Business Media Singapore 2016
D. Xu, *Multi-dimensional Optical Storage*,
DOI 10.1007/978-981-10-0932-7

(continued)

Symbol	Symbol name	Meaning/definition	Example
Md	Sample median	Half the population is below this value	
Q_1	Lower/first quartile	25 % of population are below this value	
Q_2	Median/second quartile	50 % of population are below this value = median of samples	
Q_3	Upper/third quartile	75 % of population are below this value	
x	Sample mean	Average/arithmetic mean	$x = (2 + 5 + 9)/$ $3 = 5.333$
s^2	Sample variance	Population samples variance estimator	$s^2 = 4$
s	Sample standard deviation	Population samples standard deviation estimator	$s = 2$
z_x	Standard score	It is a set of scores that have the same mean and standard deviation so they can be compared	$z_x = (x - x)/s_x$
$X \sim$	Distribution of X	Distribution of random variable X	$X \sim N(0,3)$
$N(\mu, \sigma^2)$	Normal distribution	Gaussian distribution	$X \sim N(0,3)$
$U(a, b)$	Uniform distribution	Equal probability in range a, b	$X \sim U(0,3)$
$\exp(\lambda)$	Exponential distribution	It is the probability distribution that describes the time generally	$f(x) = \lambda e^{-\lambda x}$, $x \geq 0$
gamma(c, λ)	Gamma distribution	*Gamma distribution* is a two-parameter family of continuous probability distributions	$f(x) = \lambda\ c\ x^{c-1}$ $e^{-\lambda x}/\Gamma(c)$, $x \geq 0$
$\chi^2(k)$	Chi-square distribution	It with k degrees of freedom is the *distribution* of a sum of the squares of k independent standard normal random variables	$f(x) = x^{k/2-1}e^{-x/2}$ $/(2^{k/2}\ \Gamma(k/2))$
$F(k_1, k_2)$	Distribution function	As to be used to define a particular probability distribution	F distribution
Bin(n, p)	Binomial distribution	Binomial distribution is basis for the popular binomial test of statistical significance	$f(k) = {}_nC_k$ $p^k(1 - p)^{n-k}$
Poisson(λ)	Poisson distribution	It is from a Poisson experiment	$f(k) = \lambda^k e^{-\lambda}/k!$
Geom(p)	Geometric distribution	It is a discrete analog of the exponential distribution	$f(k) = p(1 - p)^k$

Appendix E
Symbol of Measurement Unite

Symbol	Symbol name	Meaning/definition	Example
M(m)	Meter	1 m is the length of the path traveled by light during a time interval of 1/299,792,458 of a second	1M = c/299,792,458
$\text{Å}(cm^{-1})$	Angstrom	Long measure, 1 Å = 0.1 nm	$1\text{ Å} = M \times 10^{-10}$
s	Second	Unit of time measurement	1 h = 3600 s
′	Arcminute	Unit of angular measurement, $1° = 60′$	$\alpha = 60°59′$
″	Arcsecond	Unit of angular measurement, $1′ = 60″$	$\alpha = 60°59′59″$
m	Milli-	1×10^{-3}, as one millimeter = mm = $M \times 10^{-3}$	$1\text{ ms} = s \times 10^{-3}$
μ	Micro-	1×10^{-6}, as one micrometer = 1 μm = $M \times 10^{-6}$	$1\text{ μs} = s \times 10^{-6}$
n	Nano-	1×10^{-9}, as one nanometer = 1 nm = $M \times 10^{-9}$	$1\text{ ns} = s \times 10^{-9}$
p	Pico-	1×10^{-12}, as one picometer = 1 pm = $M \times 10^{-12}$	$1\text{ ps} = s \times 10^{-12}$
f	Femto-	1×10^{-15}, as one femtometer = 1 fm = $M \times 10^{-15}$	$1\text{ fs} = s \times 10^{-15}$
fa	Fatto-	1×10^{-18}, as one fattometer = 1 fam = $M \times 10^{-18}$	$1\text{ fas} = s \times 10^{-18}$
zm	Zepto-	1×10^{-21}, as one zeptometer = 1 zm = $M \times 10^{-21}$	$1\text{ zs} = s \times 10^{-21}$
ppm	Per-million	$1\text{ ppm} = 1/10^{-6}$	300 ppm = 0.0003
ppb	Per-billion	$1\text{ ppb} = 1/10^{-9}$	300 ppb = 0.0000003
ppt	Per-trillion	$1\text{ ppb} = 1/10^{12}$	$300\text{ ppb} = 3 \times 10^{-10}$
π	Pi	$\pi = 3.141592653589793238$, ratio of a circle to its diameter	Circular area = $r^2\pi$
rad	Radians	Radians angle unit	$360° = 2\pi$ rad

(continued)

© Tsinghua University Press and Springer Science+Business Media Singapore 2016
D. Xu, *Multi-dimensional Optical Storage*,
DOI 10.1007/978-981-10-0932-7

(continued)

Symbol	Symbol name	Meaning/definition	Example
grad	Grads	Grads angle unit	$360° = 400$ grad
h	Hour	A unit of measurement of time, one day = 24 h	1 h = 60 min
in	Inch	1 in = 0.0254 m	12 in = 31.2 cm
barn	Barn	1 barn = 10^{-28} m^2, unit of area, use to quantify absorption the cross-section of very small particles	
dyne	Dyne	1 dyne = 10^{-5} N (Newton), unit of force	1 kg = 9.8 N = 9.8 dyne 10^5
J	Joule	1 J = 1 kg m^2/s^2 = 2.78 kw h 10^{-7}	
erg	Erg	1 erg = 10^{-7} J, unit of energy	
V	Volt	Unit for electric potential, $1 \text{ V} = 1 \text{ kg m}^2 \text{ s}^{-3} \text{ A}^{-1}$	1 mV = 0.001 V
W	Watt	A unit of power, $1 \text{ W} = 1 \text{ kg m}^2 \text{ s}^{-3}$	1 mW = 0.001 W
W h	Watt–hour	A unit of energy, 1 W h = 3600 J	1 KW h = 1000 Wh
eV	Electron-volt	1 eV = 1.602176565 (35) $\times 10^{-19}$ J	
eV/c^2	Mass	1 eV/c^2 = 1.782661845 (39) $\times 10^{-36}$ kg (eV is converted to mass)	
0 °C	Celsius temperature	0 °C = 273.15 K (Kelvin temperature), water into ice at 0 °C	
Torr	Torr	760 Torr = 101,325 Pa (Pascal) = 1 Atm (atmosphere), unit of pressure	
b	Bit	Minimum unit of data	Data rate 1 Mb/s = 10^6 b/s
B	Byte	Unit of data: 1 B = 8 bit	
k	Kilo	10^3	1 kB = 10^3 B
M	Mega	10^6	1 MB = 10^6 B
G	Giga	10^9	1 GB = 10^9 B
T	Tera	10^{12}	1 TB = 10^{12} B
P	Peta	10^{15}	1 PB = 10^{15} B
E	Exa	10^{18}	1 EB = 10^{18} **B**
Z	Zetta	10^{21}	1 ZB = 10^{21} B
Y	Yotta	10^{24}	1 YB = 10^{24} B

Bibliography

1. J. Zhang, M. Gecevičius, M. Beresna, P.G. Kazansky, 5D data storage by ultrafast laser nanostructuring in glass, in *Proceedings of Conference on Lasers and Electro-Optics*, San Jose, CA, USA, 9–14 June 2013
2. J. Fischer, M. Wegener, Three-dimensional direct laser writing inspired by stimulated-emission-depletion microscopy. Opt. Mater. Express **1**, 614–624 (2011)
3. A. Jesacher, M.J. Booth, Parallel direct laser writing in three dimensions with spatially dependent aberration correction. Opt. Express **18**, 21090–21099 (2010)
4. G. Yuan, W.L. Tan, L.T. Ng et al., Multi-dimensional multi-level optical pickup head. J. Appl. Phys. **47**, 5933 (2008). doi:10.1143/JJAP.47.5933
5. L.H. Tingab, X.S. Miaoa, M.L. Leea et al., Optical and magneto-optical characterization for multi-dimensional multi-level optical recording material. Synth. React. Inorg. Met.-Org. Nano-Met. Chem. **38**(3), 284–287 (2008)
6. O.H. Park, S.Y. Seo, J.I. Jung, J.Y. Bae, B.S. Bae, Photoluminescence of mesoporous silica films impregnated with an erbium complex. J. Mater. Res. **18**, 1039 (2003)
7. J.Y. Bae, O.H. Park, J.I. Jung, K.T. Ranji, Photoionization of methylphenothiazine and photoluminescence of erbium 8-hydroxyquinolinate in transparent mesoporous silica films by spin-coating on silicon. Microporous. Mesoporous. Mater. **67**, 265 (2004)
8. K. Heshami, C. Healey, B. Khanaliloo, V. Acosta, C. Santori, Raman quantum memory based on an ensemble of nitrogen-vacancy centers coupled to a microcavity. Phys. Rev. A (2014)
9. M. Aspelmeyer, T.J. Kippenberg, F. Marquardt, Cavity optomechanics. Rev. Mod. Phys. **86**, 1391–1452 (2014)
10. M. Geiselmann, R. Marty, J.F. García de Abajo, R. Quidant, Fast optical modulation of the fluorescence from a single nitrogen-vacancy centre. Nat. Phys. **9**, 785–789 (2013)
11. J. Dovic, J. Wals, T. Ikkink, T. Tukker, M. Rieck, J.V.D. Eerenbeemd, A.L.V. Voorst, R. Rijs, Multi-track DVD-ROM. ODS 2001 **4342**, 112–114 (2001)
12. L. Chena, X. Pangb, G. Yua, In-situ coating of MWNTs with sol-gel TiO_2 nanoparticles. Adv. Mat. Lett. **1**(1), 75–78 (2010)
13. E.P. Walker, J Duparre, H. Zhang, W. Feng, Y. Zhang, Spherical aberration correction for 2-photon recorded monolithic multilayer optical data storage. ODS Proc. SPIE (2001)
14. R.S. Tucker, P.-C. Ku, C.J. Chang-Hasnain, Slow-light optical buffers: capabilities and fundamental limitations. J. Lightwave Technol. **23**(12) 4046 (2005)
15. V. Gorshkov, A. André, M. Fleischhauer, A.S. Sørensen, M.D. Lukin, Universal approach to optimal photon storage in atomic media. Phys. Rev. Lett. **98**, 123601 (2007)
16. I. Novikova, A.V. Gorshkov, D.F. Phillips, A.S. Sørensen, M.D. Lukin, R.L. Walsworth, Optimal control of light pulse storage and retrieval. Phys. Rev. Lett. **98**, 243602 (2007)
17. V. Gorshkov, A. André, M.D. Lukin, A.S. Sørensen, Photon storage in λ-type optically dense atomic media. i. cavity model. Phys. Rev. A **76**, 033804–033806 (2007)

© Tsinghua University Press and Springer Science+Business Media Singapore 2016
D. Xu, *Multi-dimensional Optical Storage*,
DOI 10.1007/978-981-10-0932-7

18. M.U. Staudt, S.R. Hastings-Simon, M. Afzelius, D. Jaccard, W. Tittel, N. Gisin, Investigations of optical coherence properties in an erbium-doped silicate fiber for quantum state storage. Opt. Commun. **266**, 720 (2006)

19. J. Simon, H. Tanji, J.K. Thompson, V. Vuletić, Interfacing collective atomic excitations and single photons. Phys. Rev. Lett. **98**, 183601 (2007)

20. Y. Omar, Y. Hida, H. Nakazato, Entanglement generation by a three-dimensional qubit scattering: concurrence vs. path (In) distinguish ability. Quant. Commun. Quant. Netw. Soc. Inf. Telecommun. Eng. **36**, 17–25 (2010)

21. S.A. Rice, A.R. Dinner, *Advances in Chemical Physics*, vol. 148 (2011)

22. G. De Ninno, E. Allaria, M. Coreno, F. Curbis, M.B. Danailov, Electron storage rings: a new bright light source for experiments. Phys. Rev. Lett. **101**, 053902 (2008)

23. S. Chaudhury, S. Merkel, T. Herr, A. Silberfarb, I.H. Deutsch, P.S. Jessen, Quantum control of the hyperfine spin of a Cs atom ensemble. Phys. Rev. Lett. **99**, 163002 (2007)

24. D. Gauthier, P.R. Ribič, G. De Ninno, Spectrotemporal shaping of seeded free-electron laser pulses. Phys. Rev. Lett. **115**, 114801 (2015)

25. A.R. Rossi, C. Vaccarezza, F. Villa, Large-bandwidth two-color free-electron laser driven by a comb-like electron beam. New J. Phys. **16**, 033018 (2014)

26. B. Kraus, W. Tittel, N. Gisin, S. Kröll, J.I. Cirac, M. Nilsson, Quantum memory for nonstationary light fields based on controlled reversible inhomogeneous broadening. Phys. Rev. A **73**, 020302(R) (2006)

27. S. Neergaard-Nielsen, B.M. Nielsen, H. Takahashi, A.I. Vistnes, E.S. Polzik, High purity bright single photon source. Opt. Express **15**, 7940 (2007)

28. J. Wrachtrup, F. Jelezko, Processing quantum information in diamond. J. Phys. Condens. Matter **18**, S807 (2006)

29. G. De Ninno, B. Mahieu, E. Allaria, L. Giannessi, S. Spampinati, Chirped seeded free-electron lasers: self standing light sources for two-color pump-probe experiments. Phys. Rev. Lett. **110**, 064801 (2013)

30. C. Spezzani, C. Svetina, M. Trovò, M. Zangrando, Two-colour pump-probe experiments with a twin-pulse-seed extreme ultraviolet free-electron laser. Nat. Comm. **4**, 2476 (2013)

31. A. Khan, K. Balakrishnan, Ultraviolet light-emitting diodes based on group three nitrides. Nat. Photon. **2**, 77–84 (2008)

32. M. Grajower, B. Desiatov, I. Goykhman, Direct observation of optical near field in nanophotonics devices at the nanoscale using Scanning Thermal Microscopy. Opt. Express **23**(21), 27763–27775 (2015)

33. W. Lee, X. Du, L. Li, W. Wang, An enhancing diffraction efficiency method and lens for China-definition Blu-ray drive, Hong Kong Polytechnic University, Guo Weigang, Shenzhen Suncheon, CN201120335419.4, 2012

34. S.W. McLaughlin, Y.-C. Lo, C. Pepin, D. Warland, Multilevel DVD: coding beyond 3 bits/data-cell, in *Joint International Symposium on Optical Memory and Optical Data Storage Technical Digest,* paper ThA.5, 2002

35. S. Spielman, B.V. Johnson, G.A. McDermott, M.P. O'Neill, C. Pietrzyk, T. Shafaat, D.K. Warland, T.L. Wong, M.P. O'Neill, T.L. Wong, Multilevel data storage system using phase-change optical discs, in *Optical Data Storage, Conference Digest*, 2000, pp. 170–172

36. M. O'Neill, K. Balasubramanian, J. Stinebaugh, Phase-change multilevel recording for 2GB CD-RW, in *Proceedings of the 13th Symposium on Phase Change Optical Information Storage*, 2001, pp. 43–50

37. H. Mikami, K. Osawa, K. Watanabe, Optical phase multi-level recording in microholo-gram, in *Technical Digest ODS'10, WD-01 [7730-68]*, 2010

38. J.C. Scott, L.D. Bolzano, Nonvolatile memory elements based on organic materials. Adv. Mater. **19**, 1452 (2007)

39. X. Li, J.W. Chon, R.A. Evans, M. Gu, Two-photon energy transfer enhanced three-dimensional optical memory in quantum-dot and azo-dye doped polymers. Appl. Phys. Lett. **92**, 063309 (2008)

40. H. Ishitobi, M. Tanabe, Z. Sekkat, S. Kawata, Nanomovement of azo polymers induced by metal tip enhanced near-field irradiation. Appl. Phys. Lett. **91**, 091911 (2007)
41. C. Hubert, R. Bachelot, J. Plain, S. Kostcheev, G. Lerondel, M. Juan, P. Royer, S. Zou, G.C. Schatz, G.P. Wiederrecht, S.K. Gray, Near-field polarization effects in molecular-motion-induced photochemical imaging. J. Phys. Chem. C **112**, 4111 (2008)
42. L. Novotny, B. Hecht, *Principles of Nano-Optics* (Cambridge University Press, 2007). ISBN:9780521832243.13
43. P. Biagioni, D. Polli, M. Labardi, A. Pucci, G. Ruggeri, G. Cerullo, M. Finazzi, L. Duò, Unexpected polarization behavior at the aperture of hollow-pyramid near-field probes. Appl. Phys. Lett. **87**, 223112 (2005)
44. I. Horcas, R. Fernandez, J.M. Gomez-Rodriguez, J. Colchero, J. Gomez-Herrero, A.M. Baro, WSXM: a software for scanning probe microscopy and a tool for nanotechnology. Rev. Sci. Instrum. **78**, 013705 (2007)
45. E.P. Walker, X. Zheng, F.B. McCormick, H. Zhang, N.H. Kim, J. Costa, A.S. Dvornikov, Servo error signal generation for 2-photon recorded monolithic multilayer optical data storage. ODS 2000 Proc. SPIE 4090, 179–184 (2000)
46. Y. Zhang, W.F. Ed Walker, H. Zhang, S. Esener, Numerical aperture influence on 3-D multi-layer optical data storage systems. ISOM/ODS meeting paper # TuP 28, 2002
47. D.E. Pansatiankul, A.A. Sawchuk, Multi-dimensional modulation codes and error correction for page-oriented optical data storage, in optical data storage 2001, in *Proceedings of SPIE*, vol. 4342, eds. by T. Hurst, S. Kobayashi, 2002, pp. 393–400
48. J.M. Sasian, M. Mansuripur, Design of lenslet array and high-numerical-aperture annular-field objective lens for optical data storage systems with parallel read-write-erase channels. Appl. Opt. **38**(7), 1163–1168 (1999)
49. I. Ichimura, K. Saito, T. Yamasaki, K. Osato, Proposal for a multilayer read-only-memory optical disk structure. Appl. Opt. **45**(8), 1794–1803 (2006)
50. A. Mitsumori, T. Higuchi, T. Yanagisawa, M. Ogasawara, S. Tanaka, T. Iida, Multilayer 500 Gbyte optical disk. Jpn. J. Appl. Phys. **48**(3), 03A055 (2009)
51. T.K. Kim, Y.M. Ahn, S.J. Kim, T.Y. Heor, C.S. Chung, I.S. Park, Blu-ray disc pickup head for dual layer. Jpn. J. Appl. Phys. **44**(5B), 3397–3401 (2005)
52. R. Katayama, Y. Komatsu, Blue/DVD/CD compatible optical head. Appl. Opt. **47**(22), 4045–4054 (2008)
53. Y. Komma, Y. Tanaka, S. Mizuno, Compatible objective lens for blu-ray disc and degital versatile disc using diffractive optical element and phase-step element which corrects both chromatic and spherical aberrations. Jpn. J. Appl. Phys. **43**(7B), 4768–4771 (2004)
54. F.B. McCormick, S. Esener, Massively-parallel writing and reading of information within the three-dimensional volume of an optical disk, particularly by the use of a doubly telecentric afocal imaging system, U.S. Patent Pending, Submission No. 09/225618, filed 5 Jan 1999
55. D. Barada, K. Tamura, T. Fukdua, M. Itoh, T. Yatagai, Retardagraphy: a technique for optical recording of the retardance pattern of an optical anisotropic object on a polarization-sensitive film using a single beam. Opt. Lett. **33**, 3007–3009 (2008)
56. G.J. Evans, P.A. Kirkby, Development and application of a ray-based model of light propagation through a spherical acousto-optic lens. Opt. Express **23**(18), 23493–23510 (2015)
57. D. Barada, K. Tamura, T. Fukdua, T. Yatagai, Optical information recording in films of photoinduced. Proc. SPIE **8281**, 828117-5; Jpn. J. Appl. Phys. **48**, 09LE02-1–09LE02-4 (2009)
58. D. Barada, Y. Kawagoe, K. Tamura, T. Fukdua, T. Yatagai, Self-imaging properties of fresnel retardagram recorded on azobenzene film. Jpn. J. Appl. Phys. **49**, 01AD02-1–01AD02-5 (2010)
59. B. Vial, Y. Hao, Topology optimized all-dielectric cloak: design, performances and modal picture of the invisibility effect. Opt. Express **23**(18), 23551–23560 (2015)

60. M. Engel, P.F. Damasceno, C.L. Phillips, Computational self-assembly of a one-component icosahedral quasicrystal. Nat. Mater. **14**, 109–116 (2015)

61. D. Barada, Y. Kawagoe, H. Sekiguchi, T. Fukuda, S. Kawata, T. Yatagai, Volume polarization holography for optical data storage. Proc. SPIE **7957**, 79570Q (2011)

62. T. Yatagai, D. Barada, Vector-wave holographic optical mass-storage. Proc. SPIE **8011**, 801106-1 (2011)

63. S. Wilson, G. Altshuler, A. Erofeev, M. Inochkin, L. Khloponin, V. Khramov, F. Feldchtein, Long pulse compact and high brightness near 1-kW QCW diode laser stack. Proc. SPIE **8241**, 8241-14 (2012)

64. H.F. Shih, W.C. Lu, J.Y. Chang, Design of single-path optical pickup head with three wavelengths using integrated optical unit. IEEE Trans. Magn. **45**(5), 2202–2205 (2009)

65. S. Reineke, F. Lindner, G. Schwartz, White organic light-emitting diodes with fluorescent tube efficiency. Nature **459**, 234–238 (2009)

66. L. Yaroslavsky, J. Astola, Introduction to digital holography, digital recording and numerical reconstruction of holograms, 37–58(22) (2009). ISBN:978-1-60805-079-6

67. D.P. Kelly, B.M. Hennelly, N. Pandey, In practical digital holographic systems. Opt. Eng. **48**(9), Article ID 095801-1-13 (2009)

68. E. Darakis, J.J. Soraghan, Reconstruction domain compression of phase-shifting digital holograms. Appl. Opt. **46**(3), 351–356 (2007)

69. P.A.M. Neto, H.M. Nussenzveig, Theory of optical tweezers. Europhys. Lett. **50**, 702 (2000)

70. C. Mio, T. Gong, A. Terray, D.W.M. Marr, Design of a scanning laser optical trap for multiparticle manipulation. Rev. Sci. Instr. **71**, 2196 (2000)

71. U. Schnars, W.P.O. Jüptner, Digital recording and numerical reconstruction of holograms. Meas. Sci. Technol. **13**, R85 (2002)

72. G.L. Chen, C.Y. Lin, M.K. Kuo et al., Numerical suppression of zero-order image in digital holography. Opt. Express **15**(14), 8851–8856 (2007)

73. S. Adrian, J. Bahram, Analysis obstruction from Fresnel fields. Zero-order and twin-image elimgraphy. Appl. Opt. **39** (2000)

74. X. Xiao, I.K. Puri, Digital recording and numerical reconstruction of hologram: an optical diagnostic for combustion. Appl. Opt. **41**(19), 3890–3899 (2002)

75. Miniature Flexure Based Scanners For Angle Multiplexing Patent, Inphase Technologies, Inc. (Longmont, Colorado, US) and Nintendo Co., Ltd. (Kyoto, Japan), 2008

76. S.-H. Lina, P.-L. Chenb, Y.-N. Hsiaob, Fabrication and characterization of poly(methyl methacrylate photopolymer doped with phenanthrenequinone (PQ) based derivatives for volume holographic data storage. Opt. Commun. **281**(4), 559–566 (2008)

77. R. Jallapurama, I. Naydenovaa, S. Martina, R. Howarda, V. Toala, S. Frohmannb, Acrylamide-based photopolymer for microholographic data storage. Opt. Mater. **28**(12), 1329–1333 (2006)

78. R. Henaoa, M. Tebaldib, R. Torrobab, Multiplexing encrypted data by using polarized light. Opt. Commun. **260**(1), 109–112 (2006)

79. C. Pégard, J.W. Fleischer, Optimizing holographic data storage using a fractional Fourier transform. Opt. Lett. **36**, 2551–2553 (2011)

80. M.R. Gleeson, J.T. Sheridan, A review of the modelling of free-radical photopolymerization in the formation of holographic gratings. J. Opt. A Pure Appl. Opt. **11** (2009)

81. K. Curtis, L. Dhar, A. Hill, W. Wilson, M. Ayres, *Holographic Data Storage: From Theory to Practical Systems* (Wiley, 2010)

82. GE Unveils 500-GB, *Holographic Disc Storage Technology*. CRN (2009)

83. S. Yang, M. Pang, J. Meng, Progress in research of bifunctional spirobenzopyran and spi-ronaphthooxazine photochromic compounds. Chin. J. Org. Chem. **31**(11), 1725–1735 (2011)

84. Y. Song, Y. Zhang, X. Ma, Z. Zhu, J. Xu, J. Liu, Photochromic properties of rare earth macrocyclic complexes of curcumin. Chin. J. Org. Chem. **69**(11), 1347–1353 (2011)

85. X. Deng, Z. Zhang, Y. Huang, F. Qing, Synthesis of photochromic 1,2-Dithienylethene derivative with a 2,2,5,5-Tetrafluoro-2,5-dihydrofuran bridge unit. Chin. J. Org. Chem. **30** (8), 1245–1249 (2010)

86. J. Liu, J. Han, J. Wang, M. Long, J. Meng, A new biindenylidenedione compound with two azobenzene units: synthesis and photochromic behavior both in solution and in the solid state. Chin. J. Org. Chem. **27**(9), 1839–1842 (2009)

87. J.B. Christensen, D.V. Reddy, C.J. McKinstrie, K. Rottwitt, M.G. Raymer, Temporal mode sorting using dual-stage quantum frequency conversion by asymmetric Bragg scattering. Opt. Express **23**(18), 23287–23301 (2015)

88. Z. Hua, Z. Li, C. Mei, Y. Shaning, Z. Qinglong, S. Yi, Synthesis and photochromic properties of formaldehyde induced WO3 powder. Chin. J. Org. Chem. **67**(2), 174–178 (2009)

89. D. Zhang, P. Hou, Preparation of nano-calcium titanate powder and its adsorption behavior for lead ion and cadmium ion in water. Chin. J. Org. Chem. **67**(12), 1336–1342 (2009)

90. Z. Liu, S. Zhang, G. Jia, Z. Liang, Y. Gao, Synthesis and photochromic property of three novel Calix arene-Schiff bases. Chin. J. Org. Chem. **29**(11), 1799–1803 (2009)

91. C. Niu, Y. Song, L. Yang, Synthesis of 5'-functionalized indolinospiropyrans with vinylene unit as linker. Chin. J. Org. Chem. **27**(10), 2001–2006 (2009)

92. Y. Chen, M. Pang, K. Cheng, Y. Wang, J. Han, J. Meng, Synthesis and properties of a new photo modulation magnetism molecular system. Chin. J. Org. Chem. **66**(9), 1091–1096 (2008)

93. W. Rong-Bao, Z. Da-Wei, L. Ya, L. Bo, Progress in photochromic spiro compounds containing O, N or S. Chin. J. Org. Chem. **28**(08), 1366–1378 (2008)

94. E. Pusztai, I.S. Toulokhonova, N. Temple, Synthesis and photophysical properties of asymmetric substituted silafluorenes. Organometallics **32**(9), 2529–2535 (2013)

95. I. Ichimura, S. Hayashi, G.S. Kino, High-density optical recording using a solid immersion lens. Appl. Opt. **36**, 4339–4348 (1997)

96. L.P. Ghislain, V.B. Elings, K.B. Crozier, S.R. Manalis, S.C. Minne, K. Wilder, G.S. Kino, C.F. Quate, Near-field photolithography with a solid immersion lens. Appl. Phys. Lett. **74**, 501–503 (1999)

97. M. Yoshita, K. Koyama, M. Baba, H. Akiyama, Fourier imaging study of efficient near-field optical coupling in solid immersion fluorescence microscopy. J. Appl. Phys. **92**, 862–865 (2002)

98. S.B. Ippolito, S.A. Thorne, M.G. Eraslan, B.B. Goldberg, M.S. Ünlü, Y. Leblebici, High spatial resolution subsurface thermal emission microscopy. Appl. Phys. Lett. **84**, 4529–4531 (2004)

99. J. Zhang, C.W. See, M.G. Somekh, Imaging performance of wide-field solid immersion lens microscopy. Appl. Opt. **46**, 4202–4208 (2007)

100. S.B. Ippolito, B.B. Goldberg, M.S. Ünlü, Theoretical analysis of numerical aperture increasing lens microscopy. J. Appl. Phys. **97**, 053105 (2005)

101. Z. Liu, B.B. Goldberg, S.B. Ippolito, A.N. Vamivakas, M.S. Ünlü, R. Mirin, High-resolution, high-collection efficiency in numerical aperture increasing lens microscopy of individual quantum dots. Appl. Phys. Lett. **87**, 071905 (2005)

102. P. Hu, Y. Niu, Y. Xiang, S. Gong, C. Liu, Carrier-envelope phase dependence of molecular harmonic spectral minima induced by mid-infrared laser pulses. Opt. Express **23**(18), 23834–23844 (2015)

103. G. Tessier, M. Bardoux, C. Boué, C. Filloy, D. Fournier, Back side thermal imaging of integrated circuits at high spatial resolution. Appl. Phys. Lett. **90**, 171112 (2007)

104. H. Hatano, T. Sakata, K. Ogura, T. Hoshino, H. Ueda, Plano-convex solid immersion mirror with a small aperture for near-field optical data storage. Opt. Rev. **9**, 66–69 (2002)

105. W.A. Challener, C. Mihalcea, C. Peng, K. Pelhos, Miniature planar solid immersion mirror with focused spot less than a quarter wavelength. Opt. Express **13**, 7189–7197 (2005)

106. Y. Zhang, Optical data storage system with a plano-ellipsoidal solid immersion mirror illuminated directly by a point light source. Appl. Opt. **45**, 8653–8658 (2006)

107. Y. Zhang, Optical intensity distribution of a plano-convex solid immersion mirror. J. Opt. Soc. Am. A **24**, 211–214 (2007)

108. Y. Zhang, H. Xiao, C. Zheng, Diffractive super-resolution elements applied to near-field optical data storage with solid immersion lens. New J. Phys. **6**, 75-14 (2004)

109. K.S. Youngworth, T.G. Brown, Focusing of high numerical aperture cylindrical-vector beams. Opt. Express **7**, 77–87 (2000)

110. S. Quabis, R. Dorn, M. Eberler, O. Glöckl, G. Leuchs, The focus of light-theoretical calculation and experimental tomographic reconstruction. Appl. Phys. B **72**, 109–113 (2001)

111. R. Dorn, S. Quabis, G. Leuchs, Sharper focus for a radially polarized light beam. Phys. Rev. Lett. **91**, 233901 (2003)

112. L.E. Helseth, Roles of polarization, phase and amplitude in solid immersion lens systems. Opt. Commun. **191**, 161–172 (2001)

113. Y. Kozawa, S. Sato, Focusing property of a double-ring-shaped radially polarized beam. Opt. Lett. **31**, 820–822 (2006)

114. K. Hirota, T.D. Milster, K. Shimura, Y. Zhang, J.S. Jo, Near-field phase change recording using a GaP hemispherical lens. Jpn. J. Appl. Phys. **39**, 968–972 (2000)

115. C. Liu, S.H. Park, Numerical analysis of an annular-aperture solid immersion lens. Opt. Lett. **29**, 1742–1744 (2004)

116. L. Shen, S. He, Studies of the imaging characteristics for a slab of a lossy left-handed material. Phys. Lett. A **309**, 298–305 (2003)

117. L. Chen, S. He, L. Shen, Finite-size effects of a left-handed material slab on the image quality. Phys. Rev. Lett. **92**, 107404 (2004)

118. D.R. Smith, D. Schurig, M. Rosenbluth, S. Schultz, S.A. Ramakrishna, J.B. Pendry, Limitations on subdiffraction imaging with a negative refractive index slab. Appl. Phys. Lett. **82**, 1506–1508 (2001)

119. S.T. Chui, L. Hu, Theoretical investigation on the possibility of preparing left- handed materials in metallic magnetic granular composites. Phys. Rev. B **65**, 144407 (2002)

120. J.Q. Shen, Z.C. Ruan, S. He, How to realize a negative refraction index material at an atomic level in the optical frequency range. J. Zhejiang Univ. Sci. **5**, 1322–1326 (2004)

121. S. Imanishi, T. Ishimoto, Y. Aki, T. Kando, K. Kishima, K. Yamamoto, M. Yamamoto, Near-field optical head for disc mastering process. Jpn. J. Appl. Phys. **39**, 800–805 (2000)

122. H. Hirayama et al., Marked enhancement in the efficiency of deep-ultraviolet AlGaN light-emitting diodes by using a multiquantum-barrier electron blocking layer. Appl. Phys. Exp. **3**, 031002 (2010)

123. T.K. Sharma, E. Towe, Impact of strain on deep ultraviolet nitride laser and light-emitting diodes. J. Appl. Phys. **109**, 086104 (2011)

124. H.J. Coufal, A.E. Craig, Z.U. Hasan, Advanced optical data storage—integrated optoelectronics devices, advanced optical data storage (proceedings volume), vol. 4988, in *Proceedings of SPIE*, 2003

125. Blu-ray Disc Association (PDF), *White Paper Blu-ray Disc Format-2.B Audio Visual Application Format Specifications for BD-ROM* (2009), p. 15

126. K. Kieu, M. Mansuripur, Femtosecond laser pulse generation with a fiber taper embedded in carbon nanotube/polymer composite. Opt. Lett. **32**, 2242–2244 (2007)

127. L. Gaeta, Nonlinear propagation and continuum generation in microstructure optical fibers. Opt. Lett. **27**, 924 (2002)

128. M. Mansuripur et al., Plasmonic nano-structures for optical data storage. Paper TD05-31, this conference, 2011

129. G. Skinner, K. Visscher, M. Mansuripur, Biocompatible writing of data into DNA. J. Bionanosci. **1**, 1–5 (2007)

130. H.P. Bazargani, J. Azaña, Optical pulse shaping based on discrete space-to-time mapping in cascaded co-directional couplers. Opt. Express **23**(18), 23450–23461 (2015)

131. P.K. Khulbe, M. Mansuripur, R. Gruener, DNA translocation through α-hemolysin nano-pores with potential application to macromolecular data storage. J. Appl. Phys. **7**, 104317-1:7 (2005)

132. Y. Yin, A.P. Alivisatos, Colloidal nanocrystal synthesis and the organic-inorganic interface. Nature **437**, 664–670 (2005)

133. Z.-Q. Zhou, W.-B. Lin, M. Yang, C.-F. Li, Realization of reliable solid-state quantum memory for photonic polarization qubit. Phys. Rev. Lett. **108**, 190505 (2012)

134. J.S. Lundeen et al., Direct measurement of the quantum wavefunction. Nature **474**, 188 (2011)

135. J.Z. Salvail, M. Agnew, A.S. Johnson, Full characterization of polarization states of light via direct measurement. Nat. Photon. **7**, 316–321 (2013)

136. L.N. Bairavasundaram, G. Soundararajan, V. Mathur, K. Voruganti, S. Kleiman, Italian for beginners: the next steps for SLO-based management, in *USENIX Conference on Hot Topics in Storage and File Systems (HotStorage)*, 2011

137. M. Mihailescu, G. Soundararajan, C. Amza, MixApart: decoupled analytics for shared storage systems, in *USENIX Conference on Hot Topics in Storage and File Systems (HotStorage)*, 2012

138. B. Niven-Jenkins, F. Le Faucheur, N. Bitar, *Content distribution network interconnection (CDNI) problem statement* (IETF, Internet-Draft, 2012)

139. D. Rayburn, Telcos and carriers forming new federated CDN group called OCX, in *StreamingMedia* (2011)

140. C. Osika, Cisco keynote: the future is video, in *CDN Summit* (2011)

141. A. Ben-Yehuda, M. Ben-Yehuda, A. Schuster, D. Tsafrir, The resource-as-a-service (RaaS) cloud, in *USENIX Workshop on Hot Topics in Cloud Computing (HotCloud)*, 2012

142. T. Zhu, A. Gandhi, M. Harchol-Balter, M. Kozuch, Saving cash by using less cache, in *USENIX Workshop on Hot Topics in Cloud Computing (HotCloud)*, 2012

143. A. Ben-Yehuda, M. Ben-Yehuda, A. Schuster, D. Tsafrir, Deconstructing Amazon EC2 spot instance pricing, in *Cloud Computing Technology and Science (CloudCom)*, 2011

144. G. Yadgar, M. Factor, K. Li, A. Schuster, Management of multilevel, multiclient cache hierarchies with application hints. ACM Trans. Comput. Syst. (TOCS) **29**, 5:1–5:51 (2011)

145. A. Lakshman, P. Malik, Cassandra: a decentralized structured storage system. ACM SIGOPS Operating Syst. Rev. **44**, 35–40 (2010)

146. B. Calder, J. Wang, A. Ogus, N. Nilakantan, A. Skjolsvold, S. McKelvie, Y. Xu, S. Srivastav, J. Wu, H. Simitci et al., Windows azure storage: a highly available cloud storage service with strong consistency, in *Proceedings of ACM SOSP*, Oct 2011

147. A.G. Dimakis, P. Godfrey, Y. Wu, M. Wainwright, K. Ramchandran, Network coding for distributed storage systems. IEEE Trans. Inf. Theory **56**(9), 4539–4551 (2010)

148. K. Esmaili, P. Lluis, A. Datta, The CORE storage primitive: cross-object redundancy for efficient data repair and access in erasure coded storage. arXiv, preprint arXiv: 1302.5192 (2013)

149. D. Ford, F. Labelle, F.I. Popovici, M. Stokel, V.-A. Truong, L. Barroso, C. Grimes, S. Quinlan, Availability in globally distributed storage systems, in *Proceedings of USENIX OSDI*, Oct 2010

150. Y. Hu, Y. Xu, X. Wang, C. Zhan, P. Li, Cooperative recovery of distributed storage systems from multiple losses with network coding. IEEE J. Sel. Areas Commun. (JSAC) **28** (2), 268–276 (2010)

151. K. Rashmi, N. Shah, P. Kumar, Optimal exact-regenerating codes for distributed storage at the MSR and MBR points via a product-matrix construction. IEEE Trans. Inf. Theory **57** (8), 5227–5239 (2011)

152. N. Shah, K. Rashmi, P. Kumar, K. Ramchandran, Interference alignment in regenerating codes for distributed storage: necessity and code constructions. IEEE Trans. Inf. Theory **58** (99), 2134–2158 (2012)

153. K. Shum, Cooperative regenerating codes for distributed storage systems, in *Proceedings of IEEE ICC*, 2011

154. K. Shum, Y. Hu, Exact minimum-repair-bandwidth cooperative regenerating codes for distributed storage systems, in *Proceedings of IEEE ISIT*, 2011

155. K. Shvachko, H. Kuang, S. Radia, R. Chansler, The hadoop distributed file system, in *Proceedings of IEEE MSST*, 2010

156. C. Suh, K. Ramchandran, Exact-repair MDS code construction using interference alignment. IEEE Trans. Inf. Theory **57**(3), 1425–1442 (2011)

157. S. Khadir, M. Chakaroun, A. Belkhir, A. Fischer, O. Lamrous, Localized surface plasmon enhanced emission of organic light emitting diode coupled to DBR-cathode microcavity by using silver nanoclusters. Opt. Express **23**(18), 23647–23659 (2015)

158. Z. Wang, A. Dimakis, J. Bruck, Rebuilding for array codes in distributed storage systems, in *IEEE GLOBECOM Workshops*, 2010

159. Y. Zhu, P. Lee, Y. Hu, L. Xiang, Y. Xu, On the speedup of single-disk failure recovery in XOR-coded storage systems: theory and practice, in *Proceedings of IEEE MSST*, 2012

160. D. Beaver, S. Kumar, H.C. Li, J. Sobel, P. Vajgel, Finding a needle in haystack: Facebook's photo storage, in *OSDI* (2010), pp. 47–60

161. Y. Hua, Y. Zhu, H. Jiang, D. Feng, L. Tian, Supporting scalable and adaptive metadata management in ultralarge-scale file systems. IEEE Trans. Parallel Distrib. Syst. **22**(4), 580–593 (2011)

162. Y. Zhang, A. Rajimwale, A.C. Arpaci-Dusseau, R.H. Arpaci-Dusseau, End-to-end data integrity for file systems: a ZFS case study, in *FAST'10*, San Jose, CA, 2010

163. A.A. Hwang, I.A. Stefanovici, B. Schroeder, Cosmic rays don't strike twice: understanding the nature of dram errors and the implications for system design, in *ASPLOS'12*, London, England, UK, 2012, pp. 111–122

164. Y. Oh, J. Choi, D. Lee, S. Noh, Caching less for better performance: balancing cache size and update cost of flash memory cache in hybrid storage systems, in *Proceedings of the 10th USENIX Conference on File and Storage Technologies (FAST'12)*, 2012, pp. 25–25

165. M. Canim, G. Mihaila, B. Bhattacharjee, K. Ross, C. Lang, SSD bufferpool extensions for database systems. Proc. VLDB Endowment **3**(1–2), 1435–1446 (2010)

166. J. Do, D. Zhang, J. Patel, D. DeWitt, J. Naughton, A. Halverson, Turbocharging DBMS buffer pool using SSDs, in *Proceedings of the 2011 International Conference on Management of Data*, ACM, 2011, pp. 1113–1124

167. M. Park, D. Geum, J. Kyhm, InGaP/GaAs heterojunction phototransistors transferred to a Si substrate by metal wafer bonding combined with epitaxial lift-off. Opt. Express **23**(21), 241271 (2015)

168. J. Guerra, H. Pucha, J. Glider, W. Belluomini, R. Rangaswami, Cost effective storage using extent based dynamic tiering, in *Proceedings of the 9th USENIX Conference on File and Storage Technologies*. USENIX Association 1, 2001, pp. 20–25

169. F. Chen, D.A. Koufaty, X. Zhang, Hystor: making the best use of solid state drives in high performance storage systems, in *Proceedings of the International Conference on Supercomputing, ser. ICS'11* (ACM, New York, NY, USA, 2011), pp. 22–32

170. R. Appuswamy, D. van Moolenbroek, A. Tanenbaum, Integrating flash-based SSDs into the storage stack, in *Mass Storage Systems and Technologies* (MSST), *IEEE 28th Symposium on. IEEE*, 2012, pp. 1–12

171. B. Debnath, S. Sengupta, J. Li, ChunkStash: speeding up inline storage deduplication using flash memory, in *Proceedings of University of Minnesota*, Twin Cities, USA, 2010

172. USENIX Conference on USENIX Annual Technical Conference, USENIX Association, 2010, p. 16

173. Q. Yang, J. Ren, I-CASH: intelligently coupled array of SSD and HDD, in high performance computer architecture (HPCA), in *2011 IEEE 17th International Symposium on. IEEE*, 2011, pp. 278–289

174. S. Im, D. Shin, ComboFTL: improving performance and lifespan of MLC flash memory using SLC flash buffer. J. Syst. Architect. **56**(12), 641–653 (2010)

175. J.-W. Park, S.-H. Park, C.C. Weems, S.-D. Kim, A hybrid flash translation layer design for SLC-MLC flash memory based multibank solid state disk. Microprocess. Microsyst. **35**(1), 48–59 (2011)

176. W. Song, D. Cheng, Y. Liu, Y. Wang, Free-form illumination of a refractive surface using multiple-faceted refractors. Appl. Opt. **54**(28), E1–E7 (2015)

177. Y. Oh, J. Choi, D. Lee, S.H. Noh, Caching less for better performance: balancing cache size and update cost of flash memory cache in hybrid storage systems, in *Proceedings of the 10th USENIX Conference on File and Storage Technologies (FAST)*, 2012

178. G. Wu, X. He, B. Eckart, An adaptive write buffer management scheme for flash-based SSDs. ACM Trans. Storage (TOS) **8**(1) (2012)

179. C. Min, K. Kim, H. Cho, S.W. Lee, Y.I. Eom, Random write considered harmful in solid state drives, in *Proceedings of the 10th USENIX Conference on File and Storage Technologies (FAST'2012)*, 2012

180. S. Park, K. Shen, Fios: a fair, efficient flash i/o scheduler, in *Proceedings of the 10th USENIX Conference on File and Storage Technologies (FAST'2012)*, 2012

181. F. Chen, R. Lee, X. Zhang, Essential roles of exploiting internal parallelism of flash memory based solid state drives in high-speed data processing, in *Proceedings of the 17th IEEE International Symposium on High Performance Computer Architecture (HPCA'2011)*, 2011

182. Y. Hu, H. Jiang, L. Tian, H. Luo, D. Feng, Performance impact and interplay of SSD parallelism through advanced commands, allocation strategy and data granularity, in *Proceedings of the 25th International Conference on Supercomputing (ICS'2011)*, 2011

183. G. Wu, X. He, Reducing SSD read latency via nand flash program and erase suspensions, in *Proceedings of the 10th USENIX Conference on File and Storage Technologies (FAST'2012)*, 2012

184. H. Kim, N. Agrawal, C. Ungureanu, Revisiting storage for smartphones, in *Proceedings of the 10th USENIX Conference on File and Storage Technologies*, 2012

185. H. Kim, M. Ryu, U. Ramachandran, What is a good buffer cache replacement scheme for mobile flash storage in *Proceedings of the 12th ACM SIGMETRICS/PERFORMANCE Joint International Conference on Measurement and Modeling of Computer Systems*, 2012

186. C.-C. Chang, C.-J. Lin, LIBSVM: a library for support vector machines. ACM Trans. Intell. Syst. Technol. **2**(3), 27:1–27:27 (2011)

187. S. Shepler, M. Eisler, D. Noveck (eds.), *Network File System (NFS) Version 4 Minor Version 1 External Data Representation (XDR) Description, RFC5662*, IETF (2010)

188. R. Tartler, A. Kurmus, A. Ruprecht, B. Heinloth, V. Rothberg, D. Dorneanu, R. Kapitza, W. Schröder-Preikschat, D. Lohmann, Automatic OS kernel TCB reduction by leveraging compile-time configurability, in *Proceedings of the Eighth Workshop on Hot Topics in System Dependability, ser. HotDep'12* (2012)

189. E.H. Nam et al., Ozone (O3): an out-of-order flash memory controller architecture. IEEE Trans. Comput. **60**(5), 653–666 (2011)

190. S. Lee et al., BlueSSD: an open platform for cross-layer experiments for NAND flash-based SSDs, in *Proceedings of International Workshop on Architectural Research Prototyping*, 2010

191. F. Mir, A.A. McEwan, A fast age distribution convergence mechanism in an SSD array for highly reliable flash-based storage systems, in *Proceedings of the 3rd International Conference on Communication Software and Networks (ICCSN'11)*, Xi'an, China, 2011

192. Y. Zhang, L.P. Arulraj, A.C. Arpaci-Dusseau, R.H. Arpaci-Dusseau, De-indirection for flash-based SSDs with nameless writes, in *Proceedings of the 10th USENIX Symposium on File and Storage Technologies (FAST'12)*, San Jose, California, 2012

193. K. Heshami, A. Green, Y. Han, A. Rispe, E. Saglamyurek, N. Sinclair, W. Tittel, C. Simon, Controllable-dipole quantum memory. Phys. Rev. A **86**, 013813 (2012)

194. J. Jin et al., Telecom-wavelength atomic quantum memory in optical fiber for heralded polarization qubits. Phys. Rev. Lett. **115**, 140501 (2015)

195. K. Heshami, N. Sangouard, J. Minar, H. de Riedmatten, C. Simon, Precision requirements for spin-echo-based quantum memories. Phys. Rev. A **83**, 032315 (2011)

196. J. Nunn, N.K. Langford, W.S. Kolthammer, T.F.M. Champion, M.R. Sprague, P.S. Michelberger, X.-M. Jin, D.G. England, I.A. Walmsley, Preprint arxiv:12081534 (2012)

197. N. Sangouard, C. de Simon, H. Riedmatten, N. Gisin, Quantum repeaters based on atomic ensembles and linear optics. Rev. Mod. Phys. **83**, 33 (2011)

198. J.J. Longdell, E. Fraval, M.J. Sellars, N.B. Manson, Stopped light with storage times greater than one second using electromagnetically induced transparency in a solid. Phys. Rev. Lett. **95**, 063601 (2005)

199. A.V. Gorshkov, A. André, M.D. Lukin, A.S. Sørensen, Photon storage in λ-type optically dense atomic media. i. cavity model. Phys. Rev. A **76**, 033805 (2007)

200. J. Nunn et al., Mapping broadband single-photon wave packets into an atomic memory. Phys. Rev. A **75**, 011401(R) (2007)

201. J. Nunn, K. Reim, K.C. Lee, V.O. Lorenz, B.J. Sussman, I.A. Walmsley, D. Jaksch, Multimode memories in atomic ensembles. Phys. Rev. Lett. **101**, 260502 (2008)

202. S.A. Moiseev, S. Kröll, Complete reconstruction of the quantum state of a single-photon wave packet absorbed by a Doppler-broadened transition. Phys. Rev. Lett. **87**, 173601 (2001)

203. N. Sangouard, C. Simon, M. Afzelius, N. Gisin, Analysis of a quantum memory for photons based on controlled reversible inhomogeneous broadening. Phys. Rev. A **75**, 032327 (2007)

204. P. Hedges, M.J. Sellars, Y.-M. Li, J.J. Longdell, Efficient quantum memory for light. Nature **465**, 1052 (2010)

205. M. Hosseini, G. Campbell, B.M. Sparkes, P.K. Lam, B.C. Buchler, Unconditional room-temperature quantum memory. Nat. Phys. **7**, 794 (2011)

206. V. Balzani, P. Ceroni, A. Juris, *Photochemistry and Photophysics* (Wiley, 2014)

207. M. Afzelius, C. Simon, H. de Riedmatten, N. Gisin, Multimode quantum memory based on atomic frequency combs. Phys. Rev. A **79**, 052329 (2009)

208. M. Afzelius et al., Demonstration of atomic frequency comb memory for light with spin-wave storage. Phys. Rev. Lett. **104**, 040503 (2010)

209. U. Staudt, S.R. Hastings-Simon, M. Nilsson, M. Afzelius, V. Scarani, R. Ricken, H. Suche, W. Sohler, W. Tittel, N. Gisin, Fidelity of an optical memory based on stimulated photon echoes. Phys. Rev. Lett. **98**, 113601 (2007)

210. W. Tittel, M. Afzelius, T. Chaneliére, R.L. Cone, S. Kröll, S.A. Moiseev, M. Sellars, Photon-echo quantum memory in solid state systems. Laser Photon. Rev. **4**, 244 (2010)

211. R.L. Ahlefeldt, A. Smith, M.J. Sellars, Ligand isotope structure of the optical $^7F0 \rightarrow {}^5D0$ transition in $EuCl_3 \cdot 6H_2O$. Phys. Rev. B **80**, 205106 (2009)

212. J.R. Maze, A. Gali, E. Togan, Y. Chu, A. Trifonov, E. Kaxiras, M.D. Lukin, Properties of nitrogen-vacancy centers in diamond: the group theoretic approach. New J. Phys. **13**, 025025 (2011)

213. H.J. Mamin, M. Kim, M.H. Sherwood, C.T. Rettner, K. Ohno, D.D. Awschalom, D. Rugar, Nanoscale nuclear magnetic resonance with a nitrogen-vacancy spin sensor. Science **339**, 557 (2013)

214. J.R. Maze, P.L. Stanwix, J.S. Hodges, S. Hong, J.M. Taylor, P. Cappellaro, L. Jiang, M.V. Gurudev Dutt, E. Togan, A.S. Zibrov, A. Yacoby, R.L. Walsworth, M.D. Lukin, Nanoscale magnetic sensing with an individual electronic spin in diamond. Nature (London) **455**, 644 (2008)

215. Y. Kubo, I. Diniz, A. Dewes, V. Jacques, A. Dréau, J.-F. Roch, A. Auffeves, D. Vion, D.Esteve, P. Bertet, Storage and retrieval of a microwave field in a spin ensemble. Phys. Rev. A **85**, 012333 (2012)

216. B. Julsgaard, C. Grezes, P. Bertet, K. Mølmer, Preprint arxiv:13011500 (2013)

217. M. Afzelius, N. Sangouard, G. Johansson, M.U. Staudt, C.M. Wilson, Preprint arxiv:13011858 (2013)

218. N. Bar-Gill, L.M. Pham, A. Jarmola, D. Budker, R.L. Walsworth, Preprint arxiv:12117094 (2012)

219. P.C. Maurer, G. Kucsko, C. Latta, L. Jiang, N.Y. Yao, S.D. Bennett, F. Pastawski, D. Hunger, N. Chisholm, M. Markham, D.J. Twitchen, J.I. Cirac, M.D. Lukin, Room-temperature quantum bit memory exceeding one second. Science **336**, 1283 (2012)

220. X.-H. Bao, A. Reingruber, P. Dietrich, J. Rui, A. Dück, T. Strassel, L. Li, N.-L. Liu, B. Zhao, J.-W. Pan, Efficient and long-lived quantum memory with cold atoms inside a ring cavity. Nat. Phys. **8**, 517 (2012)

221. H. P. Specht, C. Nölleke, A. Reiserer, M. Uphoff, E. Figueroa, S. Ritter, G. Rempe, A single-atom quantum memory. Nature (London) **473**, 190 (2011)

222. M. Bajcsy, S. Hofferberth, V. Balic, T. Peyronel, M. Hafezi, A.S. Zibrov, V. Vuletic, M.D. Lukin, Efficient all-optical switching using slow light within a hollow fiber. Phys. Rev. Lett. **102**, 203902 (2009)

223. M.R. Sprague, D.G. England, A. Abdolvand, J. Nunn, X.-M. Jin, W.S. Kolthammer, M. Barbieri, B. Rigal, P.S. Michelberger, T.F.M. Champion, P.St.J. Russell, I.A. Walmsley, Preprint arxiv:12120396 (2012)

224. A. Ferrier, C.W. Thiel, B. Tumino, M.O. Ramirez, L.E. Bausá, R.L. Cone, A. Ikesue, Ph. Goldner, Narrow inhomogeneous and homogeneous optical linewidths in a rare earth doped transparent ceramic. Phys. Rev. B **87**, 041102(R) (2013)

225. M. Sabooni, Q. Li, S. Kröll, L. Rippe, Preprint arXiv:1301.0636 (2013)

226. M. Afzelius, C. Simon, Impedance-matched cavity quantum memory. Phys. Rev. A **82**, 022310 (2010)

227. E. Saglamyurek et al., Broadband waveguide quantum memory for entangled photons. Nature (London) **469**, 512 (2011)

228. M. Bonarota, J.-L. Le Gouët, T. Chanelière, Highly multimode storage in a crystal. New J. Phys. **13**, 013013 (2011)

229. C. Clausen, I. Usmani, F. Bussières, N. Sangouard, M. Afzelius, H. de Riedmatten, N. Gisin. Nature (London) **469**, 508 (2011)

230. M. Hosseini, B.M. Sparkes, G. Campbell, P.K. Lam, B.C. Buchler, High efficiency coherent optical memory with warm rubidium vapour. Nat. Commun. **2**, 174 (2011)

231. S. Chen, Y.-A. Chen, T. Strassel, Z.-S. Yuan, B. Zhao, J. Schmiedmayer, J.-W. Pan, Deterministic and storable single-photon source based on a quantum memory. Phys. Rev. Lett. **97**, 173004 (2006)

232. K. Hammerer, A.S. Sørensen, E.S. Polzik, Quantum interface between light and atomic ensembles. Rev. Mod. Phys. **82**, 1041 (2010)

233. A.V. Gorshkov, A. André, M. Fleischhauer, A.S. Sørensen, M.D. Lukin, Universal approach to optimal photon storage in atomic media. Phys. Rev. Lett. **98**, 123601 (2007)

234. G. Hétet, J.J. Longdell, A.L. Alexander, P.K. Lam, M.J. Sellars, Electro-optic quantum memory for light using two-level atoms. Phys. Rev. Lett. **100**, 023601 (2008)

235. A. Louchet, J.S. Habib, V. Crozatier, I. Lorgeré, F. Goldfarb, F. Bretenaker, J.L. Le Gouët. Phys. Rev. B **75**, 035131 (2007)

236. A.V. Gorshkov, A. André, M.D. Lukin, A.S. Sørensen, Photon storage in λ-type optically dense atomic media. i. cavity model. Phys. Rev. A **76**, 033804 (2007)

237. J. Simon, H. Tanji, J.K. Thompson, V. Vuletic, Interfacing collective atomic excitations and single photons. Phys. Rev. Lett. **98**, 183601 (2007)

238. A. Kalachev, O. Kocharovskaya, Quantum storage via refractive-index control. Phys. Rev. A **83**, 053849 (2011)

239. H.P. Specht et al., Apart from saturation effects, the equations given below could also describe a single two-level system coupled to a cavity, as e.g. in H.P. Nature **473**, 190 (2011)

240. E. Saglamyurek, N. Sinclair, J. Jin, Conditional detection of pure quantum states of light after storage in a Tm-doped waveguide. Phys. Rev. Lett. **108**, 083602 (2012)

241. P.R. Hemmer, A.V. Turukhin, M.S. Shahriar, J.A. Musser, Raman-excited spin coherences in nitrogen-vacancy color centers in diamond. Opt. Lett. **26**, 361 (2001)

242. K.F. Reim, J. Nunn, V.O. Lorenz, B.J. Sussman, K.C. Lee, N.K. Langford, D. Jaksch, I.A. Walmsley, Towards high-speed optical quantum memories. Nat. Photon. **4**, 218 (2010)

243. S. Prawer, I. Aharonovich, *Quantum Information Processing with Diamond: Principles and Applications* (Elsevier, 2014)

244. P.E. Barclay, K.-M. Fu, C. Santori, R.G. Beausoleil, Hybrid photonic crystal cavity and waveguide for coupling to diamond NV-centers. Opt. Express **17**, 9588 (2009)

245. C. Santori, P.E. Barclay, K.-M.C. Fu, R.G. Beausoleil, S. Spillane, M. Fisch, Nanophotonics for quantum optics using nitrogen-vacancy centers in diamond. Nanotechnology **21**, 274008 (2010)

246. K. Heshami, N. Sangouard, J. Minár, H. de Riedmatten, C. Simon, Precision requirements for spin-echo-based quantum memories. Phys. Rev. A **83**, 032315 (2011)

247. A.I. Lvovsky, B. C. Sanders, W. Tittel, Nat. Photon. **3**, 706 (2009)

248. W. Tittel et al., Laser Photon. Rev. **4**, 244 (2010)

249. D. Giggenbach, R. Mata-Calvo, Sensitivity modeling of binary optical receivers. Appl. Opt. **54**(28), 8254–8259 (2015)

250. J. Nunn et al., Phys. Rev. A **75**, 011401 (2007)

251. K.F. Reim et al., Nat. Photon. **4**, 218 (2010)

252. R. Zhang, S.R. Garner, L.V. Hau, Creation of long-term coherent optical memory via controlled nonlinear interactions in Bose-Einstein condensates. Phys. Rev. Lett. **103**, 233602 (2009)

253. B. Zhao, Y.-A. Chen, X.-H. Bao, T. Strassel, C.-S. Chuu, X.-M. Jin, J. Schmiedmayer, Z.-S.Yuan, S. Chen, J.-W. Pan, A millisecond quantum memory for scalable quantum networks. Nat. Phys. **5**, 95 (2009)

254. R. Zhao, Y.O. Dudin, S.D. Jenkins, C.J. Campbell, D.N. Matsukevich, T.A.B. Kennedy, A. Kuzmich, Long-lived quantum memory. Nat. Phys. **5**, 100–104 (2009)

255. K. Wang, X. Zhang, Q. Chen, A naphthopyran type preparation of compounds, CN103087032 B (2015)

256. A.J. Musser, M. Liebel, Evidence for conical intersection dynamics mediating ultrafast singlet exciton fission. Nat. Phys. **11**, 352–357 (2015)

257. A. Darafsheh, C. Guardiola, A. Palovcak, J.C. Finlay, A. Cárabe, Optical super-resolution imaging by high-index microspheres embedded in elastomers. Opt. Lett. **40**(1), 5–8 (2015)

258. M. Gu, X. Li, Y. Cao, Optical storage arrays: a perspective for future big data storage. Sci. Appl. **3**, e177. doi:10.1038/lsa.2014.58,Published online 23 May 2014

259. L.P. Neukirch, E. von Haartman, J.M. Rosenholm, Multi-dimensional single-spin nano-optomechanics with a levitated nanodiamond. Nat. Photon. **9**, 653–657 (2015)

260. D. Xu, H. Hu, Q. Zhang, G. Qi, Photochromic multi-wavelength optical storage with multi-channel parallel writing method, CN200410009057.4 (2005)

261. D. Xu, L. Zhang, Principles of super-resolution mask in optical storage. J. Tsinghua Univ. **41**(4/5), 77–80 (2001)

262. J. Zhang, J. Ma, D. Xu, New measurement method of the traverse focus error detection system. J. Optoelectron. Laser **17**(2), 215–218 (2006)

263. H. Hua, X. Jian-Ping, X. Duan-Yi, Modulation principles and code design for multilevel optical data storage. Acta Phys. Sin. **56**(1), 208–212 (2007)

264. D. Xu, H. Hu, H. Hu, F. Zhang, Multi-level optical storage in photochromic diarylethene optical disc. Opt. Mater. **28**(8–9), 904–908 (2006)

265. Z. Zeng, X. Chen, H. Wang et al., Fast super-resolution imaging with ultra-high labeling density achieved by joint tagging super-resolution optical fluctuation imaging, Sci. Rep. **5**, Article number:8359 (2015)

266. D. Xu, P. Jiang, G. Qi, H. Li, Mastering and manufacturing method of the multi-wavelength and multi-level optical disc, CN03102682.6 (2003)

267. D. Xu, P. She, Z. Lei, R. Liu, Based on the phase detection method and apparatus for multi-levelr phase change material storage disc, CN01139828.0 (2002)

268. N. Singla, J.A.O. Sullivan, Influence of pit-shape variation on the decoding performance for two-dimensional optical storage (two DOS), Thesis, 2011

269. A.M. Kermarrec, N. Le Scouarnec, G. Straub, Repairing multiple failures with coordinated and adaptive regenerating codes, in *Proceedings of Net Cod*, 2011

270. A.D. O'Connell et al., Quantum ground state and single-phonon control of a mechanical resonator. Nature **464**, 697–703 (2010)

271. D.S. Papailiopoulos, J. Luo, A. Dimakis, C. Huang, J. Li, Simple regenerating codes: network coding for cloud storage, in *Proceedings of IEEE INFOCOM*, 2012

272. H. Heng, P. Long-Fa, Q. Guo-Sheng, H. Hua, X. Duan-Yi, Study of multi-level run-length limited photo-chromic storage. Acta Phys. Sin. **55**(4), 1759–1763 (2006)

273. M. Sathiamoorthy, M. Asteris, D. Papailiopoulos, A.G. Dimakis, R. Vadali, S. Chen, D. Borthakur, XORing Elephants: novel erasure codes for big data, in *Proceedings of the VLDB Endowment* (*to appear*), 2013

274. L. Xiang, Y. Xu, J. Lui, Q. Chang, Y. Pan, R. Li, A hybrid approach to failed disk recovery using RAID-6 codes: algorithms and performance evaluation. ACM Trans. Storage **7**(3), 11 (2011)

275. Y. Zhu, P. Lee, L. Xiang, Y. Xu, L. Gao, A cost-based heterogeneous recovery scheme for distributed storage systems with RAID-6 codes, in *Proceedings of IEEE DSN*, 2012

276. D. Xu, H. Hu, Q. Zhang, A photochromic multi-level optical storage with run length limited coded writing method, Tsinghua University, CN200310121702.7 (2004)

277. L. Pan, H. Hu, Y. Ni, J. Pei, H. Xu, D. Lu, D. Xu, H. Hu, Multi-level run-length data conversion method and apparatus, and blue multi-level optical storage device, Tsinghua University, CN200610169825.1 (2007)

278. Z. Qi-Cheng, N. Yi, X. Duan-Yi, H. Heng, Restriction of shot noise and material noise in a multilevel photochromic memory on signal-to-noise ratio. Chin. Phys. **15**(8), 1783–1787 (2006)

279. J. Poulin, C. Stern, R. Guilard, P.D. Harvey. Photochem. Photobiol. **82**, 171 (2006)

280. H. Zhang, A.S. Dvornikov, E.P. Walker, N.H. Kim, F.B. McCormick, Single-beam two-photon-recorded monolithic multi-layer optical disks. ODS 2000 Proc. SPIE **4090**, 174–178 (2000)

281. E.P. Walker, J. Duparre, H. Zhang, W Feng, Y. Zhang, A.S. Dvornikov, Spherical aberration correction for 2-photon recorded monolithic multilayer optical data storage. ODS 2001 Proc. SPIE (2001)

282. Ashley et al., Holographic data storage. IBM J. Res. Dev. **44**, 341–368 (2000)

CPSIA information can be obtained
at www.ICGtesting.com
Printed in the USA
LVOW02*1824090616

491948LV00001B/4/P